EXPLORATION OF THE DEEP CONTINENTAL CRUST

Edited by

H.-J. Behr, Göttingen
C.B. Raleigh, Hawaii

The German Continental Deep Drilling Program (KTB)

Site-selection Studies in the Oberpfalz and Schwarzwald

Edited by
R. Emmermann and J. Wohlenberg

With 259 Figures

Springer-Verlag Berlin Heidelberg New York
London Paris Tokyo Hong Kong

Prof. Dr. Rolf Emmermann
Institut für Geowissenschaften
und Lithosphärenforschung
Justus-Liebig-Universität Gießen
Senckenbergstraße 3
6300 Gießen, FRG

Prof. Dr. Jürgen Wohlenberg
Lehr- und Forschungsgebiet
für Angewandte Geophysik
RWTH Aachen
Lochnerstraße 4–20
5100 Aachen, FRG

ISBN-13:978-3-642-74590-4 e-ISBN-13:978-3-642-74588-1
DOI: 10.1007/978-3-642-74588-1

Library of Congress Cataloging-in-Publication Data. The German Continental Deep Drilling Program (KTB) / edited by R. Emmermann, J. Wohlenberg. p. cm. – (Exploration of the deep continental crust) 1. Boring – German (West) – Oberpfalz Region. 2. Earth – Crust. 3. Geology – Germany (West) – Oberpfalz Region. I. Emmermann, R. (Rolf), 1940– . II. Wohlenberg, Jürgen, 1933– . III. Series. QE511.G457 1989 551.1'3 – dc20 89-21998

2132/3145-543210 – Printed on acid-free paper

Preface

In October 1986 the German Minister for Research and Technology (Bundesminister für Forschung und Technologie), Dr. H. Riesenhuber, officially announced that the super-deep borehole of the Continental Deep Drilling Program of the Federal Republic of Germany (KTB) would be drilled in the Oberpfalz area of Northern Bavaria. The site selection was based on a recommendation from the Deutsche Forschungsgemeinschaft (DFG) made after an evaluation by the Project Management of the technical and financial risks involved.

This decision was preceded by a conference held from September 19 to 21, 1986 in Seeheim/Odenwald at which the results of the site studies in the Oberpfalz and the Schwarzwald were presented and thoroughly debated. The models and scientific targets resulting from these investigations formed the basis for a vote by the DFG Senate Commission for Geoscientific Interdisciplinary Research which was taken immediately after the conference. After evaluation of all scientific and technical aspects, the members of the commission voted almost unanimously for the Oberpfalz site. It was, however, strongly emphasized that both locations had a wealth of attractive research objectives and that despite clear-cut differences in some major aspects scientifically the two could be regarded as more or less equivalent. Both locations would be excellent sites for research drilling and would certainly contribute to the overall objectives of the program defined as "Fundamental research of the physical and chemical properties, conditions and processes in the deeper continental crust to achieve a better understanding of the structure, composition, dynamics and evolution of intracontinental crustal regions".

The decision to select the Oberpfalz target area was strongly influenced by the possible risks involved in drilling. Of particular significance were the results of the geothermal investigations which yielded considerably higher near-surface heatflow values and higher temperature gradients for the Schwarzwald than for the Oberpfalz. Thus, temperatures as high as 300°C might be reached at depths of 7-8 km in the Schwarzwald. Such temperatures would create many difficulties for the present drilling and logging techniques and would have a considerable impact on the budget provided for the project.

This book summarizes the results of the various research projects carried out in both target areas and presents the interpretations and models that provided the basis for the final site-selection. The contributions from the geoscientific groups involved in the pre-site investigations present a comprehensive picture of the present state of knowledge about the structure, evolution and geophysical characteristics of the basement in both areas.

The content of this book is complimented by the final two papers which were compiled during drilling of the KTB pilot hole. New ideas on the tectonic setting of the drill site area are put forward in the first of these articles and the second gives an up date on the state of the KTB project and reports first results of the pilot hole.

August 1989

R. Emmermann and J. Wohlenberg

Contents

KTB - How It All Began
E. ALTHAUS 1

The Crustal Structure at the KTB Drill Site, Oberpfalz
K. WEBER and A. VOLLBRECHT 5

The Geological Framework of the KTB Drill Site, Oberpfalz
W. FRANKE 37

Geochronology of the Moldanubian-Saxothuringian Transition Zone, Northeast Bavaria
B.T. HANSEN, S. TEUFEL and H. AHRENDT 55

Geochemical Characteristics of Metabasites in Different Tectonic Units of the Northeast Bavarian Crystalline Basement
M. OKRUSCH, E. SEIDEL, U. SCHÜSSLER and P. RICHTER 67

Metallogenic Evolution at the Saxothuringian-Moldanubian Plate Boundary in the Mid-European Variscides
H. DILL 81

Oberpfalz Deep Seismic Reflection Survey and Velocity Studies
J. SCHMOLL, R. BITTNER, H.-J. DÜRBAUM, T. HEINRICHS, R. MEISSNER, C. REICHERT, T. RÜHL and H. WIEDERHOLD 99

Crustal Structure in the Surroundings of the KTB Drill Site as Derived from Refraction and Wide-Angle Seismic Observations
H. GEBRANDE, M. BOPP, P. NEURIEDER and T. SCHMIDT 151

Processing of Reflection-Seismic Data in the DEKORP Processing Center, Clausthal
M. STILLER and R. THOMAS 177

Electrical Resistivity Studies in the Vicinity of the KTB Drill Site, Oberpfalz
V. HAAK 223

Petrography and Petrology of the KTB Location Schwarzwald
W. WIMMENAUER, H. KLEIN, H. MÜLLER and R. STENGER 243

Structural and Basement Evolution in the Central Schwarzwald Gneiss Complex
T. FLÖTTMANN and G. KLEINSCHMIDT 265

Ore and Mineral Deposits of the Schwarzwald
K. v. GEHLEN 277

Near-Vertical and Wide-Angle Seismic Surveys in the Schwarzwald
E. LÜSCHEN, F. WENZEL, K.-J. SANDMEIER, D. MENGES, TH. RÜHL, M. STILLER, W. JANOTH, F. KELLER, W. SÖLLNER, R. THOMAS, A. KROHE, R. STENGER, K. FUCHS, H. WILHELM and G. EISBACHER 297

Compressional and Shear-Wave Velocity Models of the Schwarzwald Derived from Seismic Refraction Data
D. GAJEWSKI 363

Electrical Resistivity Studies at the KTB Location Schwarzwald
A. BERKTOLD 385

Gravity and Magnetic Investigations at the KTB Locations Schwarzwald and Oberpfalz
H.C. SOFFEL, S. PLAUMANN, R. PUCHER, C. BÜCKER, H.-J. GÖTZE, M. WAGENER and V. HAAK 409

Geothermal Investigations at the KTB Locations Oberpfalz and Schwarzwald
H. BURKHARDT, U. HAACK, A. HAHN, H. HONARMAND, K. JÄGER, A. STIEFEL, P. WÄGERLE and H. WILHELM 433

Apatite Fission Track Dating of the Central European Basement. Postvariscan Thermo-Tectonic Evolution
G.A. WAGNER, I. MICHALSKI and P. ZAUN 481

A Reinterpretation of the Gravity Field in the Surroundings of the KTB Drill Site-Implications for Granite Plutonism and Terrane Tectonics in the Variscan
H.-J. BEHR, S. GROSSE, T. HEINRICHS and U. WOLF 501

The KTB Pilot Hole: Tectonic Setting, Technical Data and First Results
R. EMMERMANN 527

Contributors

Ahrendt, H. 55
Althaus, E. 1
Behr, H.-J. 501
Berktold, A. 385
Bittner, R. 99
Bopp, M. 151
Bücker, C. 409
Burkhardt, H. 433
Dill, H. 81
Dürbaum, H.-J. 99
Eisbacher, G. 297
Emmermann, R. 527
Flöttmann, T. 265
Franke, W. 37
Fuchs, K. 297
Gajewski, D. 363
Gebrande, H. 151
von Gehlen, K. 277
Götze, H.-J. 409
Grosse, S. 501
Haack, U. 433
Haak, V. 223, 409
Hahn, A. 433
Hansen, B.T. 55
Heinrichs, T. 99,501
Honarmand, H. 433
Jäger, K. 433
Janoth, W. 297
Keller, F. 297
Klein, H. 243
Kleinschmidt, G. 265
Krohe, A. 297
Lüschen, E. 297
Meissner, R. 99
Menges, D. 297
Michalski, I. 481
Müller, H. 243
Neurieder, P. 151
Okrusch, M. 67
Plaumann, S. 409
Pucher, R. 409
Reichert, C. 99
Richter, P. 67
Rühl, T. 99,297
Sandmeier, K.-J. 297
Schmidt, T. 151
Schmoll, J. 99
Schüssler, U. 67
Seidel, E. 67
Söllner, W. 297
Soffel, H.C. 409
Stenger, R. 243,297
Stiefel, A. 433
Stiller, M. 177,297
Teufel, S. 55
Thomas, R. 177,297
Vollbrecht, A. 5
Wagener, M. 409
Wägerle, P. 433
Wagner, G.A. 481
Weber, K. 5
Wenzel, F. 297
Wiederhold, H. 99
Wilhelm, H. 297,433
Wimmenauer, W. 243
Wolf, U. 501
Zaun, P. 481

KTB - How It All Began

E. Althaus

Chairman of the Senatskommission für Geowissenschaftliche Gemeinschaftsforschung der Deutschen Forschungsgemeinschaft.
Mineralogisches Institut der Universität (T.H.) Karlsruhe, Kaiserstr. 12, D-7500 Karlsruhe 1, FRG

The "Kontinentales Tiefbohr-Programm der Bundesrepublik Deutschland", well-known meanwhile under its abbreviation "KTB", is at present the most ambitious project in the geosciences in Germany. From the beginning, the "Senatskommission für Geowissenschaftliche Gemeinschaftsforschung der Deutschen Forschungsgemeinschaft", shortly the "Geokommission", was involved in this program; it was, as a matter of fact, the Geokommission from whom the idea originated and finally came to fruition.

It began at one of the meetings of the commission on May 5/6, 1977, when a paper was presented by Werner Schreyer - then member of the commission - with the title "Pionierbohrungen in Deutschland - warum?" In this presentation, the idea of a scientific deep drill hole was articulated for the first time - however with the idea of adding very modest wishes to drillings being performed with economic intentions. A major idea was, however, already the development of drilling techniques - particularly regarding the possibilities and abilities of the German drilling companies. It is no secret that the US Continental Drilling Program, which at that time was being debated intensively in the United States, had a pronounced influence on the conception of these ideas, especially regarding the scientific goals.

The Geokommission adopted the proposal and decided to follow this idea, and already in November of 1978 a more detailed study was discussed which mentioned a bunch of reasons why a deep drilling should be performed: both scientific and economic goals like energy and mineral resources, drilling techniques, borehole measurements, and acquisition of data on rocks of the deeper parts of the continental crust. The analogy to the Deep Sea Drilling Program was pointed out, as well as the fact that - due to the success of that program - our knowledge on the deeper part of the continental crust was beginning to lag behind that of the oceanic. The cost was - interesting enough - estimated at 30 to 50 million DM only. At that meeting a committee was formed chaired by Helmut Vidal, which had the task of preparing a study on the aims of such a drilling and on the chances of performing it. A member of this committee was, among others, also the late Henning Illies, from whose ideas the project profited to a great extent. At that time the present author, who had succeeded W. Schreyer as a commission member, also entered the project.

Already in spring of 1979, shortly after first contacts with the Federal Ministry of Research and Technology - BMFT - a first draft of a KTB study was presented to the commission, which encouraged the working group to continue its preparation, and in November of 1979 a meeting was proposed to be sponsored by the Deutsche Forschungsgemeinschaft (DFG), at which the ideas that had been worked out by the Commission were to be presented to a broader scientific public. At that time the study of the US National Academy of Sciences, "Continental Scientific Drilling Program", had appeared and had strengthened the view of the Geokommission that the time had come to propose a German Continental Deep Drilling to both the scientific community and the official institutions in charge of promoting scientific research in the Federal Republic of Germany, in first place the BMFT. Hence, a letter of Franz Goerlich, at that time responsible for geosciences at the DFG, made the first more realistic proposal to the ministry for sponsoring such a deep drilling. Nineteen localities were discussed, and the scientific goals remarkably resembled the final ones, containing matters like chemical and mineralogical composition of deep-seated rocks, their genesis, age and temperature, mineral reactions and mineralization, fluid budget, stress measurements, and the establishment of a deep seated laboratory. The depth was estimated at about 8 km, the time for drilling at 5 years, the investment necessary at 120 million DM and the running costs at 10 million DM per year - definitely more realistic than in the first tentative proposal. As a more exotic alternative the drilling of a deep (10 km!) shaft as a basis of a deep laboratory was also discussed.

From that time on, the project gained in momentum and the development was speeded up considerably. In the spring of 1980, both the Deutsche Geophysikalische Gesellschaft and the Deutsche Geologische Gesellschaft considered themselves to be committed to the program, as well as the Alfred-Wegener-Stiftung under its president Hans Closs, and they decided to sponsor a project study which was finally presented in the "Mitteilungen der Geokommission Nr. XI" in 1981. From the many proposals that had been discussed, four emerged finally which promised to attract the greatest number of geo-scientists, and which also seemed to be most promising with respect to the scientific output. These were: Hohenzollerngraben/Albstadt, Oberpfalz, Schwarzwald and Hohes Venn. In addition to the goals specified earlier, the development of new measuring devices and techniques was also accepted as a necessity.

Later in 1981, funds were granted by the BMFT for a 5-year pilot study to define more precisely aims and methods of research, find the best-suited locations, and develop a scheme for the project organization and management. It was stressed with emphasis that the scientific sponsorship and control by the DFG should be granted and that the organs of DFG - Senat, Geokommission, reviewers and referees - had to be responsible for the scientific quality of the program. For drilling techniques and performance, both scientific institutes and drilling companies were invited to integrate their knowledge in the project. Consequently, the KTB (this name had evolved meanwhile from different former titles) was published in the "Grauer Plan Nr. VII" describing the research program to be sponsored preferably by the DFG for the years to come. In 1982, an offer of President Hans Closs of the Alfred-Wegener-Stiftung was accepted that a coordinating office should be set up in Bonn with Franz Goerlich as the coordinator in charge of the development of the project. At the same time, Hans Jürgen Behr was appointed the leading geoscientist responsible for the scientific coordination.

After the unexpected death of Hans Closs in December of 1982, Peter Giese became president of the Alfred-Wegener-Stiftung and as such he took over the responsibilities of Closs. The program was presented at several meetings with international audiences, e.g., at the DGG Aachen and the EUG Strasbourg, and a first scientific meeting was held on April 18 to 19, 1983 at the DFG in Bonn during which the ideas on deep drilling that had developed so far were discussed controversively both among scientists and with representatives of the BMFT in order to define more precisely the intentions and goals of the program. At the same time, a study was completed by the "Kavernenbau- und Betriebsgesellschaft" (KBB) investigating the technical feasibility of the proposed project. The state of Niedersachsen, represented by its ministry for arts and science, signaled interest to take over responsibility for the organization for the first time and proposed the Bundesanstalt für Geowissenschaften und Rohstoffe (BGR) / Niedersächsisches Landesamt für Bodenforschung (NLfB) as the host institution.

From November 2 to 4, 1983, a colloquium was held at Neustadt/Weinstraße during which the status of the work related to the project was evaluated and the conditions for further proceedings were to be proposed. Both scientific reviewers from the DFG and a body of international scientists were present from whom recommendations were expected regarding both the scientific goals and the geographic location for the first deep drilling to be performed. All participants agreed that the subjects defined as objectives for the KTB deserved further support, and the scientific value of the project was regarded to be very high, even with very high, standards. The most important points that had been decided upon were: The drilling was to be performed in an area where crystalline basement rocks occur close to the surface, which would give a chance to reach rather deep-seated rocks, a depth of at least 8, in favorable conditions of even 12 to 15 km should be reached, and the temperature at final depth should be at least 200°C and, if possible even 300°C. The only two localities to meet these requirements were Oberpfalz and Schwarzwald, but it was not possible at that time to decide which one might be the best. These points were formulated more precisely in a memorandum by Behr, Goerlich, Maronde, Ziegler (then chairman of the Geokommission), and the present author, which was later published in Terra Cognita. By that time about 100 proposals had been submitted to the DFG for programs to be investigated both for the site selection and the development of methods and techniques. It was necessary to define a new scheme of organization for a project of that size, whose expenses were estimated now to be close to 500 million DM.

On the initiative of president Seybold of the DFG, a small working group consisting of the present author, Martin Kürsten of BGR and Dr. Kutschke of BMFT was established, which developed, as a compromise, the organization which was finally adopted for the program because it met the interests of all different groups involved in the project, i.e., scientists, engineers, and administrators. After considering different constructions - among others a company with several universities as members - it was finally agreed that (according to a proposal of president Bender of BGR/NLfB) the project management responsible for the technical performance should be installed at the NLfB, while the scientific aspects were to be concentrated in a "Schwerpunktprogramm" (preference research program) of the DFG. Much emphasis was placed both from BMFT and DFG on the fact that the program was to be opened for international cooperation. As coordinators for the scientific program, Rolf Emmermann, Peter Giese, and the present author were nominated, while H.-J. Behr was appointed the responsible geoscientist in the project management, which consists of two more members, P. Kehrer, and Heinrich Rischmüller as chairman.

On February 2, 1985, the project was presented to the public by minister Riesenhuber of BMFT, bringing the project into life officially. At that time 85 projects were being supported already which collected many data needed for the final decision about the project as a whole and the place where to perform it. It was presented to an international group of scientists at an Alfred-Wegener-Conference held in Seeheim/Bergstraße from October 4 to 10, 1985. Results of 105 working groups could already be discussed and the knowledge about the two target areas had increased considerably. The time began to ripen for a final decision between the two localities. The request to make a decisive proposal was put forward to the DFG from the BMFT in the spring of 1986, and the present author, who had succeeded W. Ziegler as the commission chairman, invited all working groups, the Geokommission, and scientists from Germany and from several other countries to a second conference at Seeheim from September 19 to 21, 1986 with the goal of recommending one of the two localities for drilling the borehole.

After 2 days of intensive, sometimes hot-tempered and controversial discussions, all arguments had been put up pointing out the specialities, advantages, and drawbacks of the two localities. In the end, neither of the two localities had gained strong preponderance over the other, and the Geokommission had a very difficult task deciding between the two of them. After long discussions, finally the decision was made unanimously in favor of the Oberpfalz, but the statement was added that the work that had been done in and the goals that had been defined for the Schwarzwald were by no means regarded as inferior to the Oberpfalz.

The recommendation to choose the Oberpfalz as the locality for the first KTB drillhole was conveyed immediately to the BMFT, and a few days later, on October 2, 1986 it was announced by minister Riesenhuber that the German Deep Drilling Project was to be started at that place. On September 18, 1987, Staatssekretär Probst of the BMFT pushed the red button which set the drilling rig to work for the pilot borehole; after 10 years of scientific and organizatory work, this started the active phase for the German Continental Deep Drilling Project, KTB.

The Crustal Structure at the KTB Drill Site, Oberpfalz

Klaus Weber and Axel Vollbrecht

Institut für Geologie und Dynamik der Lithosphäre der Universität Göttingen, Goldschmidtstr. 3, 3400 Göttingen, FRG

ABSTRACT

The Variscan basement at the Continental Deep Drilling Site (KTB) Oberpfalz on the western margin of the Bohemian Massif is composed of three polyphase-deformed structural units: the Saxothuringian, the Moldanubian and MP to HP metamorphic nappe complexes of the Zone of Erbendorf-Vohenstrauß (ZEV) and the Münchberg Massif (MM). The boundary between the Saxothuringian und Moldanubian, which is interpreted as a cryptic suture, is represented by the northwestern rim of the HT-mylonite belt of the Zone of Tirschreuth-Mähring (ZTM). This forms part of a formerly active continental margin, whereas the Saxothuringian terrane represents a formerly passive continental margin. Deformation of the Moldanubian active continental margin already began during the oceanic subduction stage and was therefore a longer-lasting, more penetrative deformational event than in the Saxothuringian, where the whole Variscan deformation is related only to the collisional stage.
The klippen of the ZEV and MM form part of a previously coherent nappe complex which was originally connected with the western part of the Bohemian (Zone of Tepla-Taus, ZTT) and now covers the Saxothuringian-Moldanubian suture. The Erbendorf Line near Erbendorf is a nappe boundary. Its interpretation as suture zone can no longer be upheld.
The Variscan crustal structure evolved during multiphase deformation in which two main events can be distinguished. The older events up to D_3 are characterized by NW-directed tectonic transport. This is revealed by the seismic crustal structure which is characterized by dominantly SE-dipping reflectors in all NW-SE-directed reflection seismic profiles. Based on reflection seismic data, D_2-backfolding and backthrusting of the Saxothuringian seems to be restricted to higher crustal levels and is interpreted as having resulted from crustal wedging and ramp tectonics. In the roof of a crustal wedge foreland-dipping backthrusts and antivergent folds develop, whereas foreland-directed overthrusting continues inside and below the wedge.
The younger event is characterized by NW-trending and SW-vergent D_4-folding and thrusting and merges into late-orogenic dextral wrench faulting and extensional tectonics. The ZEV nappe forms a D_4-syncline, in which the rocks have been preserved from erosion. The D_4-deformation, which increases in intensity to the south, documents a reorientation of the regional stress field at the southwestern rim of the Bohemian Massif. This is interpreted on the basis of geophysical and geological data as resulting from a late-collisional indentation process.

Variscan crustal convergence was probably much stronger than previously assumed. Initial oceanic subduction zones have been completely overridden by continental crust. Thereby, the complexly deformed cryptic sutures seen at the present outcrop level of the crystalline basement have lost their contact with the previous subduction zone. During collision, terranes previously separated by oceanic crust were thus thrust over one another, so that suture zones widely separated before collision are now closely neighboring or overridden by other terranes. The latter is presumably the case at the northwestern margin of the Bohemian Massif, where the Bohemian terrane overthrust the Saxothuringian-Moldanubian suture zone.

1 INTRODUCTION

The continental deep drilling site Oberpfalz is situated at the western margin of the Bohemian Massif, the largest exposed and continuous complex in the central European Variscan. This complex contains rocks of both the Saxothuringian and Moldanubian zones (cf. Kossmat 1927; for more details on the regional geology of the Bohemian Massif see Franke this Vol.). In the following, results of the presite studies on structural geology will be given and discussed together with DEKORP and KTB deep seismic reflection data.

The presite studies have yielded substantial new knowledge on the geology of the region and ultimately have also contributed to the selection of the deep drilling site. These studies are based on the research of the geological pioneers of this region, e.g., von Gümbel (1868, 1888, 1894), Suess (1912), Wurm (1925, 1926), Kossmat (1927) and von Gaertner (1941, 1951). Recently, the extensive geological mappings of Voll (1960), Schreyer (1965) and especially Stettner (1960, 1971, 1972, 1975, 1976, 1979, 1980, 1981) have provided a substantial basis for presite studies.

2 STRUCTURAL UNITS

In the area of the continental deep drilling three structural units can be identified, which have been intensively studied during the presite studies. The stratigraphy of these units is described by Franke (this vol.). The units are (Fig. 1):

The Saxothuringian of the Fichtelgebirge and the northern Oberpfälzer Wald,the Moldanubian to the south with the zones of Tirschenreuth-Mähring at its northwestern rim (ZTM), i. e., the boundary between Saxothuringian and Moldanubian, and the nappe complex of the zone of Erbendorf-Vohenstrauß (ZEV). In the west, the Franconian line (FL) separates the metamorphic basement from the Permo-Mesozoic cover of the South-German block.

2.1 Saxothuringian

On the geological map scale the Fichtelgebirge and the northern Oberpfälzer Wald is dominated by the NE-trending structures of the Fichtelgebirge anticline in the north and the Waldsassen-Hatzenreuth Synklinale in the south. Northwest of the Fichtelgebirge anticline the NE-striking nappe complex of the Münchberg Gneiss Mas-

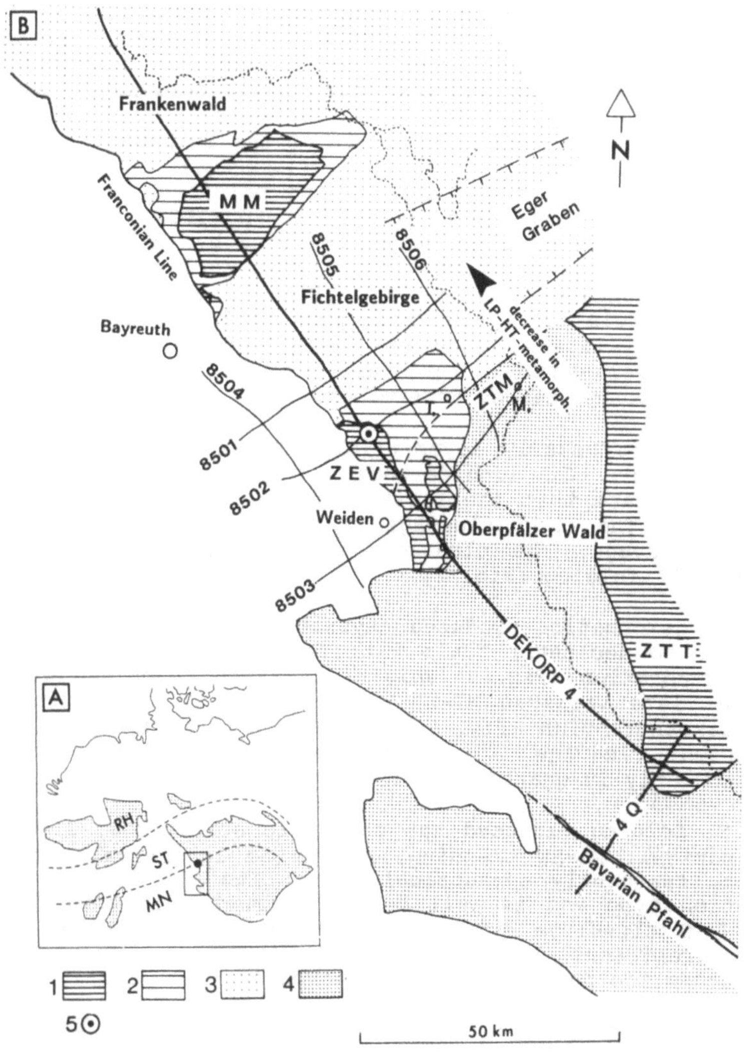

Fig. 1. Geological sketch map of the western rim of the Bohemian Massif in NE Bavaria with the DEKORP and KTB seismic network indicated. A Variscan basement outcrops in Middle Europe with zones after Kossmat (1927). RH Rhenohercynian; ST Saxothuringian; MN Moldanubian Region. Inset study area with deep drilling site (KTB). B Geological map with the tectonometamorphic units. 1 crystalline nappe complexes; 2 lower and less metamorphic parts of the nappe units; MM Münchberg nappe complex; ZEV nappe complex of the zone of Erbendorf-Vohenstrauß; ZTT Zone of Tepla-Taus, forming the western part Bohemian terrane (Bohemicum); 3 Saxothuringian; 4 Moldanubian; 5 KTB deep drilling site. Late to posttectonic granites are not shown. Reflection seismic network: heavy lines: DEKORP; thin lines: KTB; T Tirschenreuth; M Mähring; ZTM Zone of Tirschenreuth-Mähring

sif follows, preserved in the Vogtland Syncline (Franke this Vol.). According to Stettner (1980), the core of the Fichtelgebirge anticline consists of Precambrian metasediments, which supposedly are discordantly overlain by Cambrian metasediments. However, the presumed discordance has not yet been found in outcrops or drill holes. Von Gaertner (1941) correlated the Wunsiedel marble, which forms part of this core sequence, with Middle Cambrian limestones of other parts of the European Variscan.

The Saxothuringian rocks experienced polyphase deformation during the Variscan orogeny. According to Stein (1987) five phases of deformation can be distinguished, followed by late- to post-Variscan wrench and normal faulting. In contrast to the Moldanubian and ZEV unit, sedimentary bedding (s_0) (Fig. 2) and occasionally cross-bedding (Ludwig 1972) have been preserved, except in some strongly deformed phyllites. Based on these sedimentary fabrics and the generally well-known stratigraphy, vergence of folds and accompanying cleavage can be determined.

F_1 folds are usually asymmetrical, small-scale folds refolded by younger folds. The interference patterns of the folds overprinted by F_2 indicate that the first-generation folds are locally noncylindrical. The first cleavage (s_1) is generally a narrowly spaced slaty cleavage. The cleavage/bedding relationships indicate NW--facing F_1 folding (Fig. 2).

The D_2 deformation produced asymmetrical SE-vergent folds of centimeter to decameter scale with fold axes homoaxial to the first fold generation (Fig. 3).This deformation coincides with the peak of the late Variscan HT-LP metamorphism in the southern part of the Saxothuringian (Fig. 4). The second cleavage is a spaced crenulation cleavage with well-developed metamorphic layering, particularly in quartz-rich lithologies .

The F_3 folding produced the map-scale structures, i.e., generally cylindrical, NE-SW trending, medium to large-scale, open and upright, frequently symmetrical folds. Its cleavage is of the crenulation type and is well developed mostly in metapelitic rocks. The Fichtelgebirge anticline and the Waldsassen-Hatzenreuth syncline represent large-scale D_3 structures. The D_3 deformation postdates the peak of HT-LP metamorphism (Fig. 4).

The postmetamorphic D_4 deformation is documented by NW-SE to N-S striking, single or conjugate kink folds and kink bands with a dominantly W- to SW-facing reverse sense of shear (Fig. 2) Occasionally open, upright parallel folds have developed. The D_4 deformation indicates a reorientation of the regional stress field.

The D_5 deformation is presumably related to late-orogenic extension and collapse-tectonics and passes into D6 characterized by normal faulting and dextral shear. During the D_5 deformation horizontal crenulation and fracture cleavage as well as small-scale, cylindrical, open folds with horizontal axial planes formed in places where older metamorphic s-planes have been reoriented into a more or less vertical attitude. This fabric indicates horizontal extension or vertical compression under the force of gravity. Finally, the observed late-orogenic normal and wrench faulting might be related to transtensional tectonics and the formation of Permo-Carboniferous pull-apart basins.

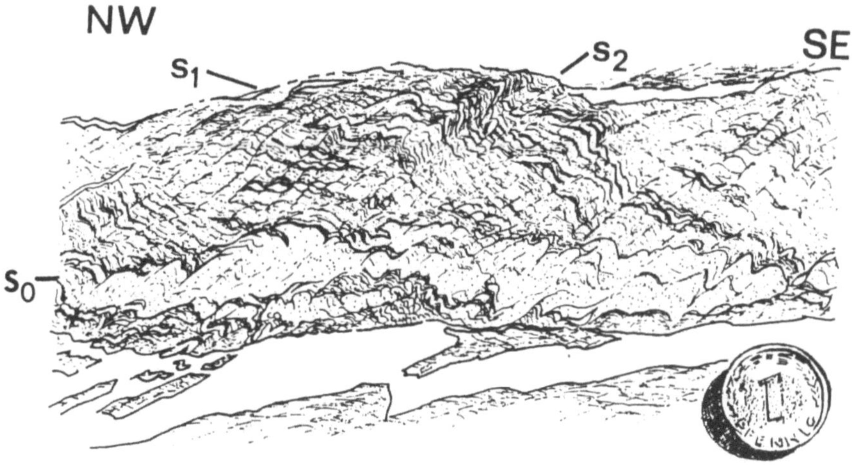

Fig. 2. Bedding-cleavage relationships in Saxothuringian mica-schists (Lerchenbühl near Neualbenreuth, north of Mähring); bedding (s_0): horizontal; s_1: NW-vergent and refolded by s_2 which is SE-vergent

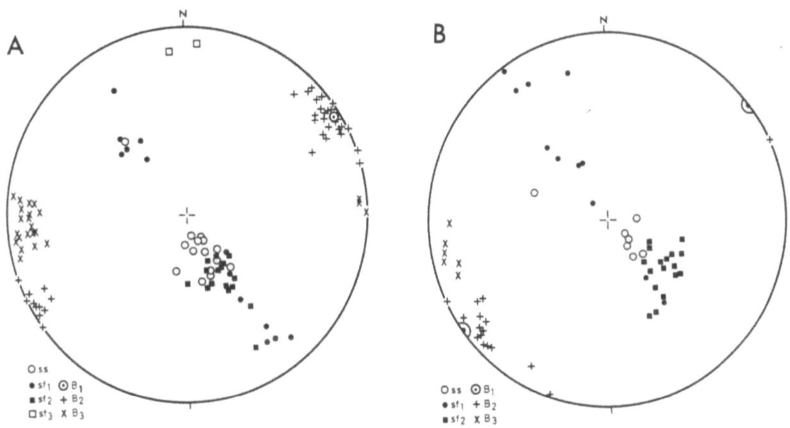

Fig. 3. D_1-D_3 fabrics (equal area projection, lower hemisphere) from Saxothuringian mica-schists near Neualbenreuth (A Lerchenbühl; B Froschfelsen). B_1 is reconstructed from s_0 and s_1 data and is oriented parallel to B_2. B_3 is more E-W-striking than B_1 and B_2.

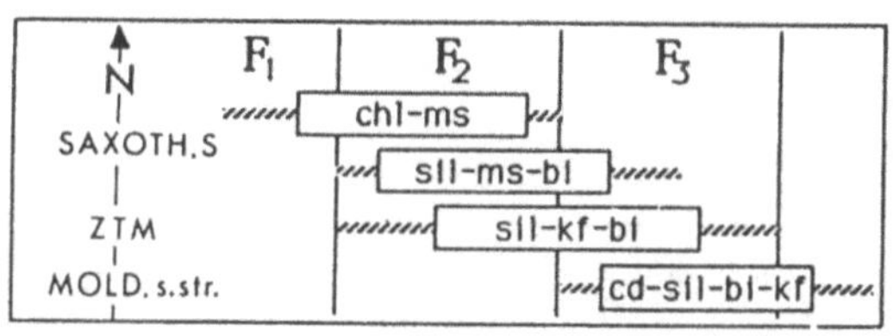

Fig. 4. Relationship between deformation and crystallization at the Saxothuringian-Moldanubian boundary zone. (After Stein and Wagener-Lohse, KTB 1986)

2.2 Moldanubian and ZTM

The Moldanubian is represented by the Monotonous Group predominantly composed of monotonous paragneises and micaschists with subordinate calcsilicates amphibolites and orthogneisses. The grade of the LP-HT metamorphism increases continuously from the upper greenschist facies of the Saxothuringian to the cordierite--kspar migmatites of the Moldanubian (Blümel this Vol.). Due to the higher grade of metamorphism and the monotonous lithology, the structural development of the Moldanubian is much more difficult to evaluate than the Saxothuringian. Sedimentary fabrics are generally obliterated and the compositional layering of sedimentary origin has been strongly overprinted by bedding parallel schistosity. Therefore, the determination of normal and overturned bedding is impossible.

In contrast to the Saxothuringian rocks,in which sedimentary bedding is frequently preserved, the oldest detectable fold generation in the Moldanubian rocks refolds a pre-existing schistosity which is parallel to the compositional layering. In the following, the formation of this schistosity is correlated with the earliest visible deformation phase D_1. The earliest detectable fold generation which refolds the D_1-schistosity is D_2 and corresponding folds are thus designated as F_2.

This earliest fold generation is represented by asymmetrical, tight internal folds in the centimeter to decimeter scale with fold axial planes parallel to s_1. They produce a new fold axial plane parallel schistosity in their fold hinges which develops parallel to s_1 outside the fold hinges. They frequently show the noncylindrical geometry of sheath folds. Due to these noncylindrical geometry the strike of the fold axis show a wide scatter between the NE and NW direction (Fig. 5). Because of the lack of any geopetal fabrics, the direction of facing is unknown. The dominance of NW-vergent asymmetrical folds, however, points to NW-directed tectonic transports. In some outcrops (Waidhaus, Zottbachtal) refolded F_2-folds can be observed (Stein 1987). There is, however, no reason to assume different folding events. It is more

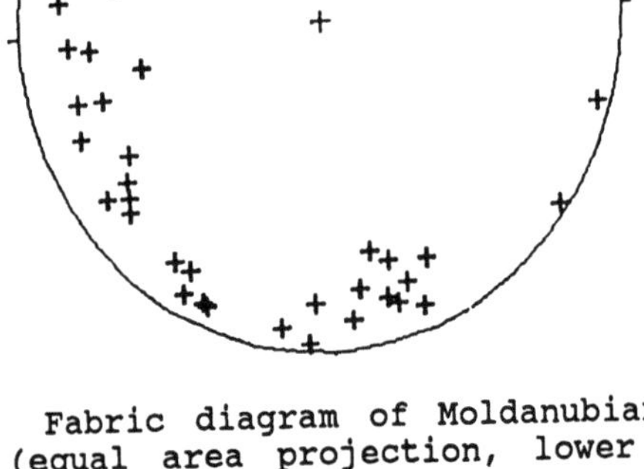

Fig. 5. Fabric diagram of Moldanubian noncylindrical F_2-folds (equal area projection, lower hemisphere); Zottbachtal (Fänger 1988)

probable that they result from a prograde, almost homoaxial deformation process, dominated by simple shear, during which earlier formed noncylindrical internal folds were refolded by younger ones.

F_3 folds are NE-trending, upright, and open folds of up to kilometer scale. They can be correlated to F_3-folds in the Saxothuringian (Stein 1987). Smaller second order folds in the decimeter to meter scale can be observed. In the cord-sill-kspar gneisses their formation coincides with the peak of HT-LP metamorphism (Fig. 4). They are refolded by large-scale, NW-SE- to N-S-trending, upright to slightly W- and SW-facing postmetamorphic F_4 folds. The nappe complex of the ZEV has been preserved in such a F_4 syncline (see below). The D_4 deformation is very important in the SE in the Bavarian zone (see below) and decreases in intensity northwestward, particularly in the Saxothuringian. The late-tectonic granites follow the strike of the dominant late-metamorphic structures. This is F_3 in the Fichtelgebirge (ENE) and F_4 in the Moldanubian and ZEV (NNW). Dextral shear observed in the ZTM (slickensides and kink bands, Stein 1987) compensates differences in the intensity of F_4-folding north and south of the Saxothuringian-Moldanubian boundary. The displacement vector is oblique upthrusting the Moldanubian to the NW to WSW.

Just as in the Saxothuringian, D_5 indicates extension. F_5 structures are developed dominantly in mica schists in which schistosity has a vertical or steeply dipping attitude. They display horizontal or gently dipping crenulation cleavage and F_5-fold axial planes. D_6 normal faulting and wrench faulting, predominantly with dextral shear, pass into post-Variscan faulting.

The structure of the boundary between Saxothuringian and Moldanubian, called the zone of Tirschenreuth-Mähring (ZTM) (Stettner 1979), differs from the Saxothuringian as well as from the Moldanubian. It contains somewhat higher amounts of calc-silicates, metacherts, and amphibolites than the Moldanubian further to the south. Therefore, Stettner (1979) has correlated the rocks of the ZTM with the Upper Proterozoic Variegated Group.

Structurally, the ZTM consists of largely mylonitic rocks of vertical attitude which formed in a wide temperature field from sillimanite-kspar (muscovite out) stability down to the lower greenschist facies. The peak of metamorphism coincides with the late D_2 and early D_3 deformation (Fig. 4), and the HT mylonitization presumably took place during these deformations. Folds which can be related to the HT mylonitization are very rare. Much more frequent are open folds in postmylonitic s-parallel quartz veins (see below). Kink bands can be related to horizontal extension as well as dextral shear and thus can be correlated with D_5 outside the ZTM shear zone. The mylonitization has developed discontinuously perpendicular to the general strike with relict lenses and layers of gneiss between the mylonite s-planes. These gneisses (Fig. 6) show the same succession of deformations as the gneisses outside the ZTM.

The ZTM has suffered two phases of diaphthoresis (Schreyer 1966). Diaphthoresis I took place at 2-3 kb and 500°-530°C (Kleemann 1986), in connection with hydraulic fracturing and the precipitation of large amounts of vein quartz along the schichtosity forming up to 30% of the rock volume (Weber 1985; de Wall 1987). The invasion of such hugh volumes of fluids is probably related to the underthrusting and dewatering of prograde metamorphic Saxothuringian sediments below the Moldanubian.

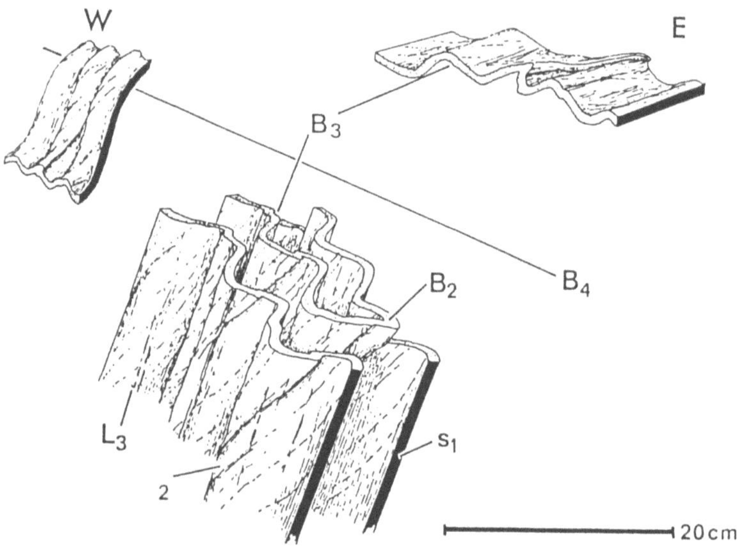

Fig. 6. D_1 to D_4 fabrics of Moldanubian gneisses, Mähring outcrop (Stein 1987)

Diaphthoresis II is related to the intrusion of numerous late tectonic sill granites (Weber 1985; de Wall 1987) at 1-2 kb and 380°-430°C (Kleemann 1986) and is connected with a second phase of hydraulic fracturing and quartz precipitation. The age of intrusion is 315 Ma and is only 5 Ma younger than the peak of the LP-HT metamorphism (U-Pb monazite, Teufel 1987; Hansen at al. this Vol.). This Variscan metamorphism of 320 Ma (Teufel 1987) coincides with the late F2 and early F_3 folding in the ZTM (Fig. 4). Since F_2 is the oldest clearly identifiable fold generation in the Moldanubian which is contemporaneous to the HT mylonitization, the ZTM shear zone must be of Herzynian (Upper Carboniferous) age.

Collectively, the results of structural, petrological (Blümel et al. this Vol.) and geochronological (Hansen et al. this Vol.) presite studies show that the rocks of the Saxothuringian and Moldanubian were welded together during the Middle Carboniferous (Sudetian) HT-LP collisional event. The structural boundary is very poorly exposed between Tirschenreuth and Mähring and also further to the east on Czechoslovakian territory. West of Tirschenreuth this boundary is obscured by posttectonic granites and further west is covered by the ZEV nappe complex (see below). Its internal structural development and sense of shearing during the different stages of mylonitization are widely unknown. A reverse sense of shear can be assumed during HT mylonitization, but has not yet been proven by means of detailed microstructural investigations. Rotation into its present vertical attitude could have taken place during the late D_2 and early D_3 deformations. Dextral shear is indicated by post-metamorphic kink bands and slickensides (Stein 1987) and began during D_4 deformation and was active probably also later. The possible bearings of the Tertiary Eger graben tectonics on the present state of this boundary in the Oberpfalz area are unknown. One major problem is the abrupt change of fold vergence when crossing the boundary from SE to NW, which will be discussed below.

2.3 Zone of Erbendorf-Vohenstrauß

In contrast to the Moldanubian, the MP unit (PT data see Blümel et al. this Vol.) of the ZEV consists of a more variegated sequence of dominantly paragneisses, amphibolites, calc-silicates, and minor amounts of orthogneisses. It is an inverted metamorphic nappe pile, similar to the Münchberg Massif (KTB 1986; Franke this Vol.) with mafic to ultramafic (serpentinites) embedded in greenschist-facies metavolcanics and metasediments at the base, forming the Erbendorf greenschist unit which may be correlated with the greenschist-facies metabasites and phyllonites at the southern margin of the ZEV. The poorly exposed eastern rim of the ZEV can be traced by several occurrences of mylonites (Heinicke and Vollbrecht in press) and lenses of serpentinites (Klinkhammer and Rost 1975), both regarded as parts of lower nappe units. The trace element pattern of the Erbendorf unit points to calc-alkaline island arc basalts, whereas the amphibolites of the ZEV overlying the Erbendorf greenschists reveal oceanic island basalt affinities. (KTB 1986, Okrusch et al. this Vol.). The character and age of metamorphism (390 Ma) and the cooling history are equivalent to the Münchberg Massif (Teufel 1987; Hansen et al. this Vol.). Lithology and the character of metamorphism are similar to the zone of Tepla-Taus (ZTT, Fig. 1) which forms the western part of the Bohemian Massif. V. Seidlitz (1921), Zoubek (1979), Stettner (1979), and Blümel (1983, 1985) have pointed out the lithofacies and metamorphic relationships between the MP metamorphic ZEV and MM and the likewise MP-metamorphic ZTT. Hence, a formerly coherent MP metamorphic nappe complex overlying the Moldanubian and the Saxothuringian of the Fichtelgebirge can be inferred. Further arguments can be derived from the structural point of view (see below).

The strike of lithologic boundaries and metamorphic s-planes of the ZEV (Stettner 1981) indicate a half-bowl-shaped, NW-SE-trending and SW-vergent syncline (Fig. 7). In the northwestern part the s-planes generally dip to the S and SW, whereas in the southeastern part gently NW-vergent folds developed. These NE-SW-striking folds overprint the older NW-trending mineral lineation. The angular relationship between schistosity and fracture cleavage indicates an overturned limb of a NW-vergent fold. After rotation of the schistosity into horizontal position (the assumed orientation prior to NW-vergent folding), the mineral lineation has the general NW-SE strike of the ZEV. This is interpreted as an indication of post-F_4 folding at the southern end of the ZEV.
This younger fold generation in the southern ZEV has been interpreted by Weber (1985) as having resulted from transpression during dextral shear along the Luhe line, which borders the ZEV to the S (Fig. 7). The attitude of S-dipping s-planes of the northernmost part of the ZEV might have resulted from the late-orogenic uplift of the Fichtelgebirge prior to the formation of the Erbendorf Rotliegend basin.

The oldest fabric perceptible in the paragneisses of the ZEV is a narrowly spaced metamorphic layering parallel to lithologic boundaries. Small-scale, strongly flattened internal folds have developed in some places. Garnet and particularly kyanite, which has also been transformed into fibrolite, are frequently cataclastically deformed. Quarz, feldspar, and biotite have equilibrated grain boundaries due to dynamic recrystallization. This texture can be interpreted as a HT-mylonitic fabric which formed during early stages of nappe transport and was subsequently folded around the NW-SE-striking fold axes of the ZEV.

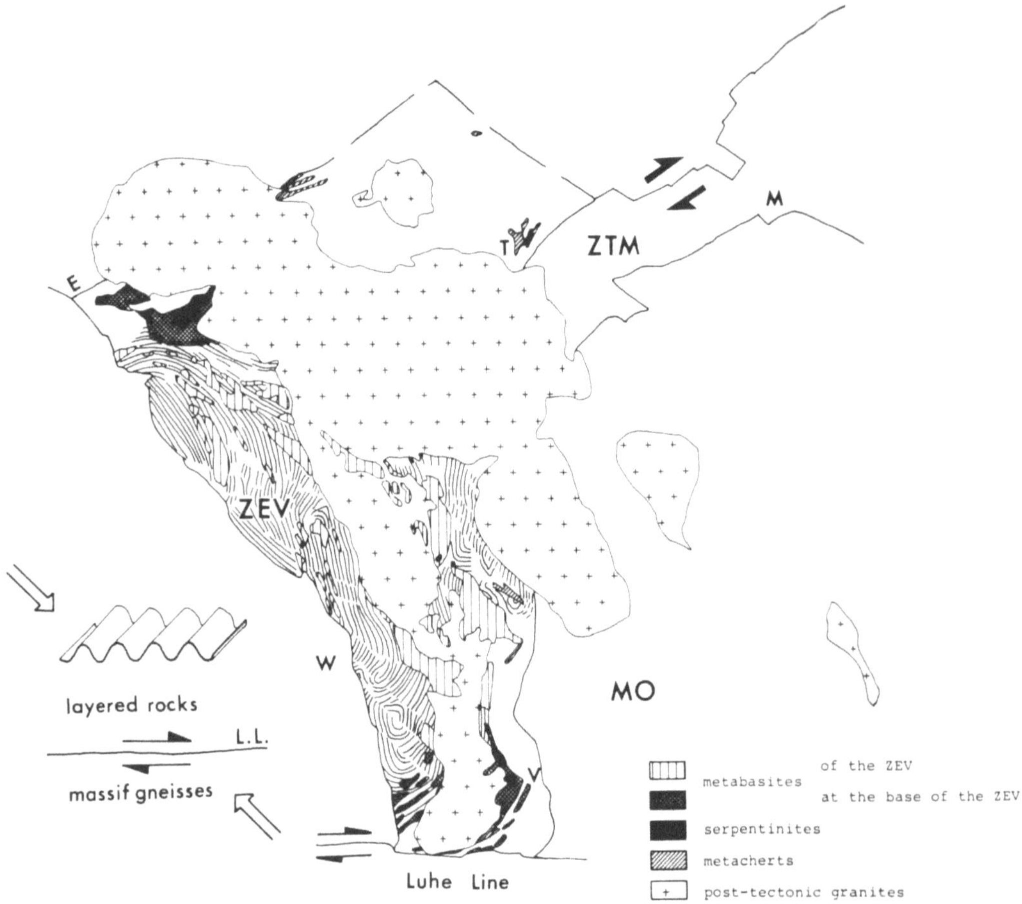

Fig. 7. Strike of schistosity and lithological boundaries of the ZEV. (After Stettner 1981). The ZEV forms a half-bowl-shaped, SW-vergent F_4-syncline. The general trend of the late-tectonic granites reflects the structural grain of the D_4-deformation. Dextral shear along the ZTM compensates differences in the intensity of F_4-folding in the Saxothuringian and Moldanubian. The NE-striking structures at the southern end of the ZEV result from transpression during dextral shear along the Luhe Line. Cataclastic normal faulting, observed along the Luhe Line is younger than dextral shear

Mylonitic fabrics with well-developed stretching lineation and recovered quartz fabrics which formed at lower temperatures are frequently observed along the eastern rim and in the lower part of the ZEV. The mylonites are regarded as part of the steeply NE dipping limb of the SW-vergent ZEV syncline, the stretching lineation strikes WSW when reoriented back to the horizontal position. Analysis of the corresponding sense of shear indicators (Heinicke 1987; Weber unpubl.) suggests WSW directed displacement vectors.

The fabric in the rocks of the ZEV has clearly developed in a different way than that of the Saxothuringian and Moldanubian rocks during all deformation phases older than D_4. The formation of NW--SE-trending ZEV syncline, in which the ZEV rocks have been preserved from erosion, can be correlated with the D_4 structures in the Saxothuringian and Moldanubian . The age of final nappe emplacement must therefore be younger than D_3 of the other units. This agrees with the fact that ZEV rocks were not influenced by the HT-LP metamorphism, which is why their higher metamorphic ages and cooling ages have been maintained. Still open to question are the younger Hbl cooling ages east and in the roof of the posttectonic granites bordering the eastern rim of the ZEV. It can be assumed that these granites, which intruded approximately along the boundary between the ZEV and the Moldanubian have changed the older Hbl cooling ages to younger ones. (Schüssler et al. 1986; Okrusch et al. this Vol.; Hansen et al. this Vol.).

The ZEV is bordered to the southwest by the Franconian Line, where it makes contact with the Permo-Mesozoic foreland. Near the surface, the Franconian line is indicative of a steeply NE dipping fault with an upthrusting sense of displacement, whereby the Permo-Mesozoic foreland was downfaulted. This interpretation is based on the Sanford model (Sanford 1959), according to which the stress distribution at the margin of an upthrusted block leads to reverse faulting on the downthrusted block. The simultaneously resulting extension produces secondary normal faults in the upthrusted block. Similar kinematics and geometry are also presumed for the Naab Valley fault zone.

2.4 Saxothuringian-Moldanubian Boundary and the Age of Nappe Transport

The character of the Saxothuringian-Moldanubian boundary changes over a distance of about 35 km from the Franconian Line (FL) in the west to the Czechoslovakian boundary in the east. East of the NS trending chain of posttectonic granites, Saxothuringian and Moldanubian rocks are welded together during LP/HT metamorphism which increases in intensity to the southeast (Wagener-Lohse and Blümel 1986; KTB 1986; Blümel et al. this Vol.). The small spacing between the isograds and their succession suggest steeply inclined isograd planes. The direction and minimum amount of inclination were obtained from the 900-m-deep Höhensteinweg drilling near Mähring which lies about 2 km NNW of the cord.-kfsp. isograd and remains in the sill.-kfsp. zone down to its final depth. The steeply inclined isograd plane may also be the cause of the continuous metamorphic transition in the area of the ZTM although considerable late- and postmetamorphic vertical displacements have taken place here. Relics of kyanite and garnet in feldspar and cordierite of the Moldanubian (Blümel in KTB 1986) might be related to an earlier, precollisional stage of pressure dominated metamorphism (see section 4).

The direct contact between Saxothuringian and Moldanubian rocks is not exposed. Hand and shallow drillings have shown that in the transition zone the dip of the s-planes decreases toward the NE (de Wall 1987). Since no oriented drill cores were able to be taken, the exact direction of dip of the s-planes is unknown. The southward deviation of the Poppenreuth borehole (Massalsky 1987) points to a N-dipping anisotropy in the rocks. Based on the studies of de Wall (1987) a gradual transition from vertical to steeply NNW dipping s-planes in the Saxothuringian over a distance of about 3 km can be assumed. The structural attitudes are similar to a tectonic ramp which formed and was rotated into a nearly vertical to NW-dipping position during D_2 to D_3 of the Saxothuringian.

There is no comparable metamorphic transtion between the ZEV and the Saxothuringian. The anchimetamorphic to lower greenschist facies rocks of the Saxothuringian Erbendorf Paleozoic are overlain here by the Erbendorf metabasites and MP-metamorphic gneisses of the ZEV with distinctly older metamorphic ages. This boundary is undoubtedly a nappe boundary. Its interpretation of a suture zone (Behr et al. 1982; Franke 1984) can no longer be upheld (Weber and Vollbrecht in KTB 1986; Vollbrecht et al. in press). An important argument for this is the fact that the present position of the ZEV nappe cannot simultaneously be its root zone. Rather, the ZEV nappe overlies the Saxothuringian-Moldanubian boundary. The northern portion of the ZEV overlies Saxothuringian rocks and the central and southern portions of the ZEV overlie Moldanubian rocks, when the ZTM is traced further to the SW below the Falkenberg granite and the ZEV. The assumed change in strike of the Saxothuringian/Moldanubian boundary below the ZEV from NE direction in the east to SSW direction in the west (Fig. 1) results from the D_4 deformation. F_4 folding, however, has not only affected the ZEV nappe but also the underlying geological units.

The southern ZEV appears more strongly uplifted and is thus apparently thinner than the northern portion. Evidence for this is given by reflection seismic data (see below) and the distribution by Bouguer anomalies with a stronger gravity maximum in the NW part of the ZEV and a weaker one in the SE which are related to metabasic rocks within the ZEV nappe (Plaumann 1986; Soffel et al. this Vol.; Weber 1985). The same picture is obtained for the distribution of magnetic anomalies. The apatite fission track data also point to an earlier uplift of the southern ZEV (Wagner et al. this Vol.).

The gravity map (Fig. 8; Plaumann 1986; KTB 1986) indicates that the ZEV continues toward the SW beneath the Permo-Mesozoic cover, where its SW and NW boundaries are suggested by contour lines. The NW-SE-trending contour lines in the Mesozoic foreland coincides with the Saxonian Freihungen-Kulmbach fault, which has the character of a steeply NE-dipping thrust. It may be interpreted as a reactivated lineament in the Variscan basement, corresponding to the southwestern boundary of the ZEV nappe.

If the tectono-metamorphic evolution of the various structural units (Saxothuringian, ZTM, Moldanubian, ZEV) are compared, a common structural evolution first begins with D_4 (Table 1). Since older structures which can be compared with those of other units have not yet been found in the ZEV, the ZEV nappe cannot have arrived here before D_4. Initial nappe transport is undoubtedly older than D_3 and must at least be connected with the HT mylonitization fabric described in section 2.3. Lower temperature mylonite fabrics in the basal part of the ZEV and the upper parts of the Moldanubian gneisses on the eastern margin of the ZEV presumably

Fig. 8. Gravity map of the Variscan basement of the Oberpfalz and the Permo-Mesozoic foreland. (After Plaumann 1986). T: Tirschenreuth; M Mähring; V Vohenstrauß; MM Münchberg Massif; ZEV Zone of Erbendorf-Vohenstrauß

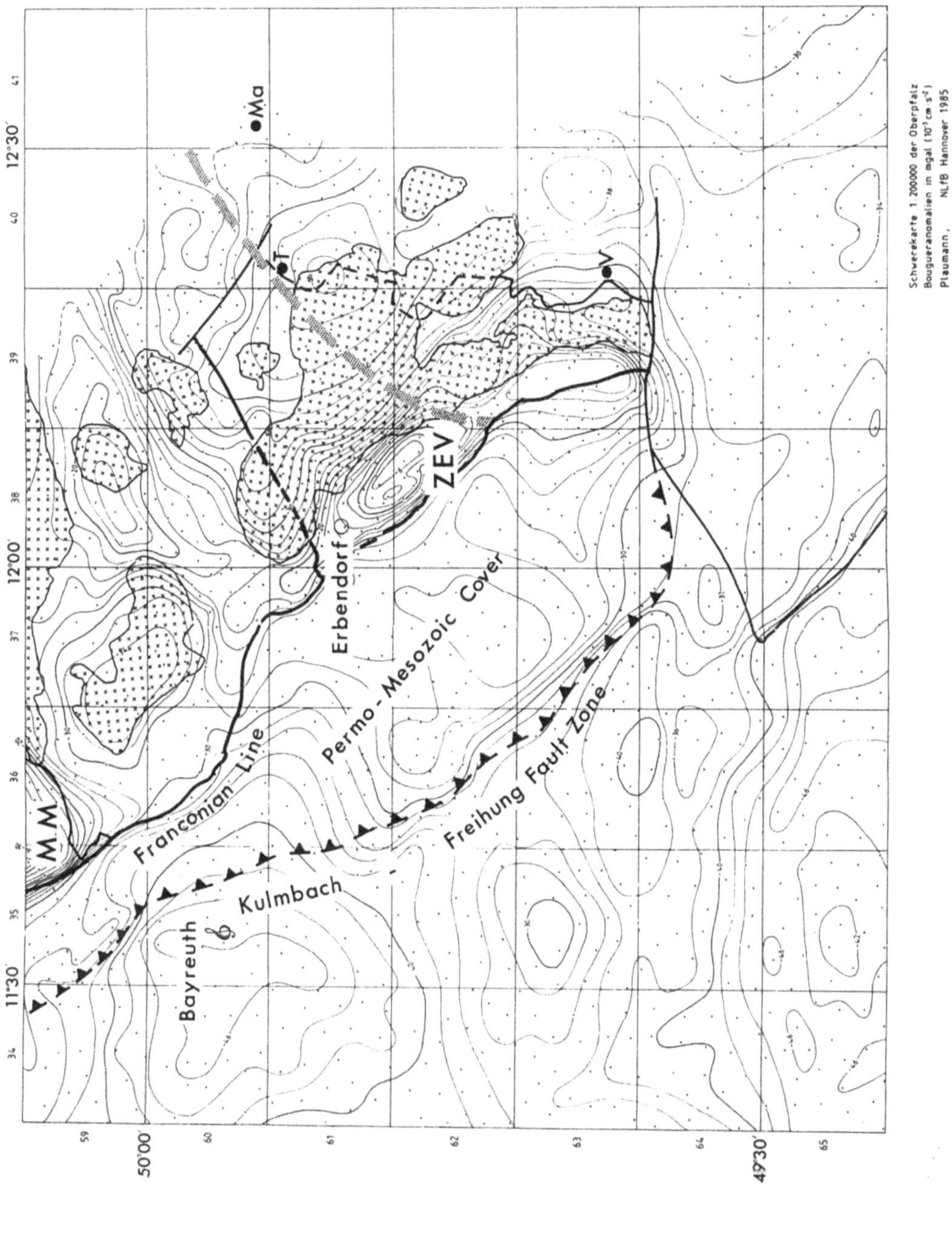
12°30'
12°00'
11°30'
50°00'
49°30'
Ma
T
V
M.M.
ZEV
Erbendorf
Franconian Line
Permo - Mesozoic Cover
Freihung Fault Zone
Kulmbach
Bayreuth
Schwerekarte 1:200000 der Oberpfalz
Bougueranomalien in mgal (10⁻³ cm·s⁻²)
Plaumann, NLfB Hannover 1985

developed during the late stage of transport. The final emplacement of the nappe and its folding in the present NW-SE-striking syncline probably occurred within a short time period at about 320 Ma after the peak of the LP-HT metamorphism and before the subsequent intrusion of post-tectonic granites. Reflection seismic data (see below) indicate that folding of the SW-vergent ZEV syncline before the intrusion of the posttectonic granites was connected with NE-dipping thrust faulting in the Moldanubian gneisses and possibly in the ZEV too. Tectonic thinning of the upper, weaker metamorphic Moldanubian gneisses and mica schists may have been connected with this SW-vergent folding and thrusting. Under these conditions the basal décollement of the ZEV may have been situated in a relatively high stockwork while being thrust over Moldanubian rocks, which may be the reason why the relatively weak LP metamorphism has not yet been substantiated in the ZEV. Possible evidence of such a situation is yielded from the occurrence of anchimetamorphic Saxothuringian sediments in the Erbendorf Paleozoic on the NW margin of the nappe pile. This question can only be answered after results from the KTB have been obtained.

There is also a similar problem regarding the inverted metamorphic nappe complex of the MM. Anchimetamorphic sediments (Bavarian Facies) form in the case as well as the basal unit of the nappe pile. These sediments must have been deposited in an area south of the Fichtelgebirge (Franke 1984). Nappe transport over the present Fichtelgebirge must have occurred later than the Lower Carboniferous. This must have occurred either before the LP metamorphism of the Fichtelgebirge (Franke this Vol.) or the nappe décollement zone was situated in a higher very weak metamorphic stockwork.

3 CORRELATION AND INTERPRETATION OF GEOLOGICAL AND REFLECTION SEISMIC DATA.

The network of the reflection seismic profiles of DEKORP and KTB covers the most important geological units and structural boundaries in the deep drilling area and its vicinity (Fig. 1). The key profile is DEKORP 4, which traverses the deep drilling site NW of Windischeschenbach. At first glance the line drawing in Fig. 9 (KTB 1986; DEKORP Research Group 1988) shows four structurally different domains:

1. The bowl-shaped structures N and S of the Fichtelgebirge can be correlated with the nappe complexes of the MM and ZEV. The base of the synformal MM is more distinct than that of the ZEV. This can be explained by different lithological contrasts between the nappes and their basement. The gneisses of the MM are underlain by anchimetamorphic sediments; whereas the ZEV rocks, at least in their southern parts, rest upon Moldanubian gneisses.

2. Along the entire profile a zone of prominent SE-dipping reflectors can be detected starting between 5 to 6 s TWT in the NW and descending down to the Moho from the northern ZEV southeastwards. These SE-dipping reflectors proceed into the Moldanubian zone below 5 s TWT to the SE end of the profile. Thus, it can be expected that SE-dipping reflectors will also be found in the lower crust even further to the SE over greater distances in prolongation of the DEKORP 4 profile.

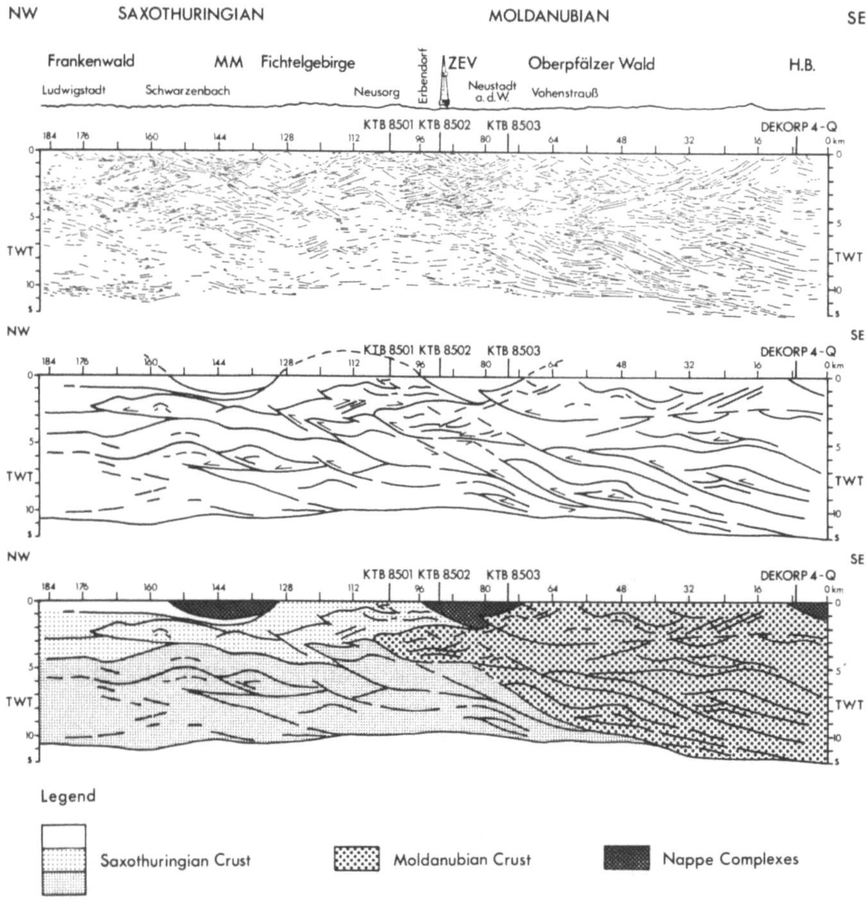

Fig. 9. DEKORP 4 profile; line drawing of the migrated and coherency-filtered section (A). B and C structural models based on the line drawing. MM Münchberg Massif; ZEV Zone of Erbendorf-Vohenstrauß; HB Hoher Bogen (part of the ZTT, Fig. 1). For explanation see text

3. At about 5 s TWT a distinct structural discontinuity separates the lower crust marked by SE-dipping reflectors from the upper crust of the Moldanubian in which NW-dipping reflectors are frequently encountered in addition to shallow, SE-dipping reflectors and broad undulations. At the NW end there is a highly reflective zone extending below the ZEV and into the Saxothuringian at 6-12 km depth, the so-called Erbendorf Body. Results of the wide-angle reflection survey (KTB 1986; DEKORP Research Group 1988) in the lower portion of this body at 12.5-13.5 km depth indicate vp-velocities of 7.0-7.5 km/s (DEKORP Research Group 1988; Schmoll et al. this Vol.).

4. At the NW end of the profile below 5-6 s TWT there is a relatively transparent zone wedging out to the SE with predominantly shallow reflectors. A comparison with ECORS, BELCORS and DEKORP 2 N suggests that this zone is pre-Variscan basement which was only slightly deformed during the Variscan.

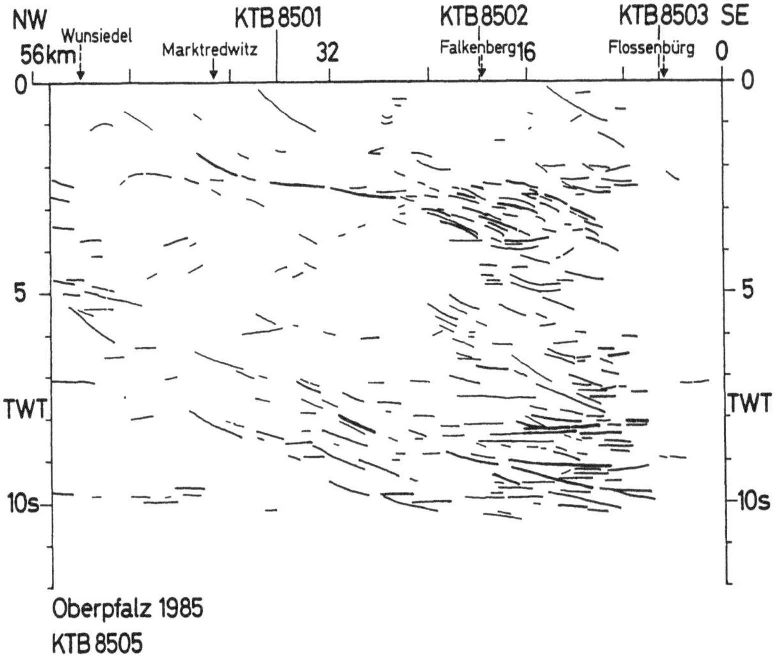

Fig: 10. KTB-profile 8505; line drawing of the migrated section

The reflection horizon NW of the MM at about 1 s TWT passes southeastward under the MM. It can presumably be correlated with Frasnian spilites, which are overlain by Late Devonian and Lower Carboniferous sediments. The large anticlinal structure under the northwestern part of the MM is the equivalent to the SW-plunging Hirschberg-Gefell Anticline at the surface. It represents a more or less autochthonous structure whose prolongation alongstrike can be traced to the brachyanticline of the Saxonian Granulite Mountains about 120 km to the NE.

The MM was preserved from erosion in the Vogtland syncline which is presumably a D_3 structure. The D_3 structure of the Fichtelgebirge anticline SE of the MM can be traced to a depth of about 3 s TWT. Its core at about 110 km can be correlated with the oldest exposed rocks (Stettner 1972). The relatively transparent domain between 112 and 128 km and down to 2 s TWT corresponds to the late- to posttectonic granites (320-290 Ma) forming the main part of the Fichtelgebirge.

The SE-vergent D_2 folds of the Saxothuringian can not be recognized in the seismic structures. The same is true for all other NW-SE profiles (Figs. 10 and 12). Only some of the NW-dipping reflectors above 3 s TWT below the Fichtelgebirge and NW of the MM may be interpreted as SE-vergent structural elements. Below 3 s TWT asymmetrical NW-vergent imbrication or fold structures prevail. Some of the structures resemble ramp structures, e.g., NW of the Erbendorf Body below about km 110 or reveal an internal duplex structure (Fig 11).

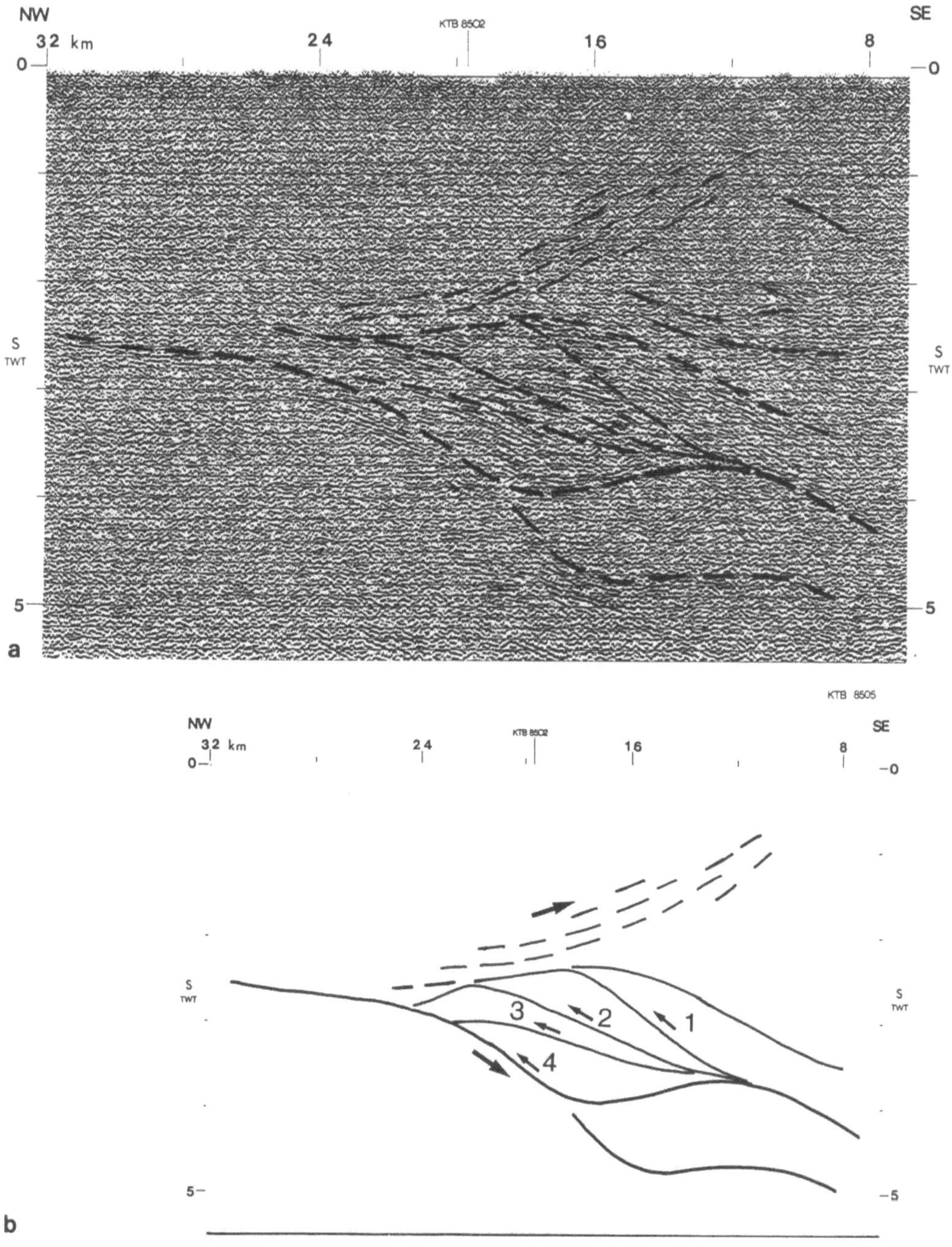

Fig. 11a,b. KTB-profile 8505; interpretation of the internal structure of the wedge-shaped body (Erbendorf Body) at about 3 s TWT, (migrated section). The internal structure of the wedge is that of a duplex-structure. The schuppen are numbered 1-4 according to their age. The duplex-wedge migrated to the NW. Faint, NW-dipping reflectors at the top of the wedge represent the duplex-roof, passively backthrusted and backfolded during the NW-directed migration of the wedge

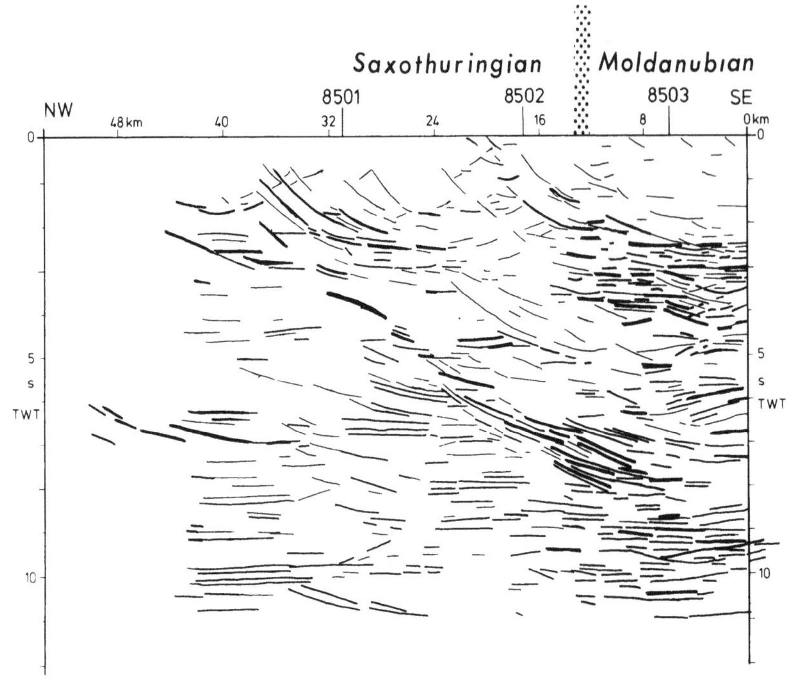

Fig. 12. KTB-profile 8506; line drawing of the migrated section. The Erbendorf Body between 2.5 s and 4 s TWT terminates just below the Saxothuringian-Moldanubian boundary

The seismic crustal structures revealed by the SE-dipping reflectors in all NW-SE reflection seismic profiles confirm the extensive overthrusting of Moldanubian crust onto Saxothuringian crust and the interpretation of this boundary as a suture zone. However, the boundary between Saxothuringian and Moldanubian rocks below the surface can not be recognized in the seismic patterns. The boundary drawn in Fig. 9c is based on the assumption that the Erbendorf Body forms part of the Moldanubian crust and that the high-velocity zone at its base can be traced SE-dipping down to the MOHO (Fig 13). Accordingly, Moldanubian rocks extend into the Saxothuringian Zone and must be overlain and underlain by Saxothuringian rocks. Based on this configuration the model of a Moldanubian crustal wedge penetrating Saxothuringian crust was developed by Weber (KTB 1986). Such wedge tectonics (or flake tectonics in the sense of Oxburgh 1972) is described from several other orogenic belts (Price 1986; Brown at al. 1986). This offers a reasonable explanation for D_2 backfolding and backthrusting observed in the Saxothuringian rocks at the surface. Below the wedge northwestward overthrusting and southeastward underplating continued during the wedging process. Hereby, the NW-vergent tectonic transport, which is indicated by SE-dipping seismic reflectors in the middle and lower crust in the DEKORP 4 profile can be considered to be dominant at least during D_1 to D_3. The SE-directed tectonic transport during D_2 can be explained as a secondary effect due to crustal wedging confined to higher crustal levels (Weber in KTB 1986; Vollbrecht et al. in press).

SE-vergent D_2 structures are observed in the Frankenwald (Schwan 1958; Franke 1984), at a long distance NW of the Saxothuringian/Moldanubian boundary. They are probably not related to the

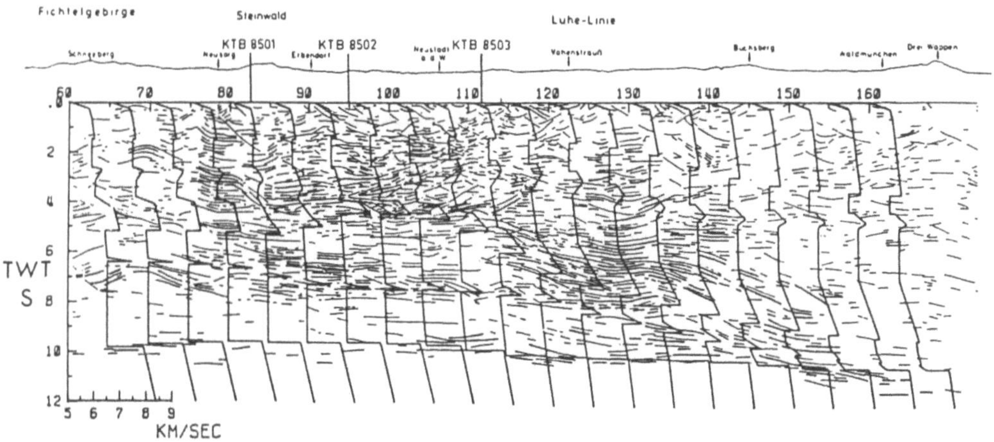

Fig. 13. Superposition of the velocity depth function (DEKORP research group 1988) and the line-drawing of Fig 11.

Moldanubian crustal wedge of the Erbendorf Body. However, wedge tectonics do not necessarily have to develop between colliding crustal segments. They can also develop within a crustal segment, for instance, through the successive formation of duplex structures. Movements of this type are similar to those of a passive roof duplex as described by Banks and Warburton (1986) from Pakistan. The leading edge of the wedge grows in the direction of tectonic transport, and the entire wedge becomes larger through the formation of new duplexes. In the roof of the duplex wedge foreland-dipping backthrust and antivergent folds develop, whereas foreland-directed overthrusting continues inside and below the wedge (Fig. 11a, b). If such wedge structures develop already during an early stage of deformation backfolding in their roofs may represent the first phase of folding. Several structures of the Saxothuringian crust reveal a wedgelike form, particularly the upper part of the large anticlinal structure under the northwestern part of the MM (Fig. 9). The structure, as a whole, is NW-vergent, and its upper part is thrusted subhorizontally toward the NW. The SE-vergent D_2 structures in this part of the Frankenwald might be related to this crustal internal structure (see also Vollbrecht et al. in press).

The internal structure of the Erbendorf wedge is difficult to discern. NW-dipping seismic reflectors are frequent here as well as further to the SE in the higher crustal level above 5 s TWT. They could represent SE-vergent D_2-structures, particularly inside the northwestern crustal wedge. Further to the SE another interpretation is more probable (see below). Otherwise, they could be explaned as foreland-dipping duplex structures or antiformal stacks. An equivalent wedge-shaped body seen in KTB profile 8505 has an internal NW-vergent duplex structure (Fig. 11). A SE-vergent passive roof complex may be indicated by weak NW-dipping reflectors on top of the wedge.

After the first period of polyphase homoaxial deformation and the formation of NE-SW-trending D_1 to D_3 structures, a second period of deformation can be defined by the NW-SE striking D_4 structures which indicate a reorientation of the regional stress field. The

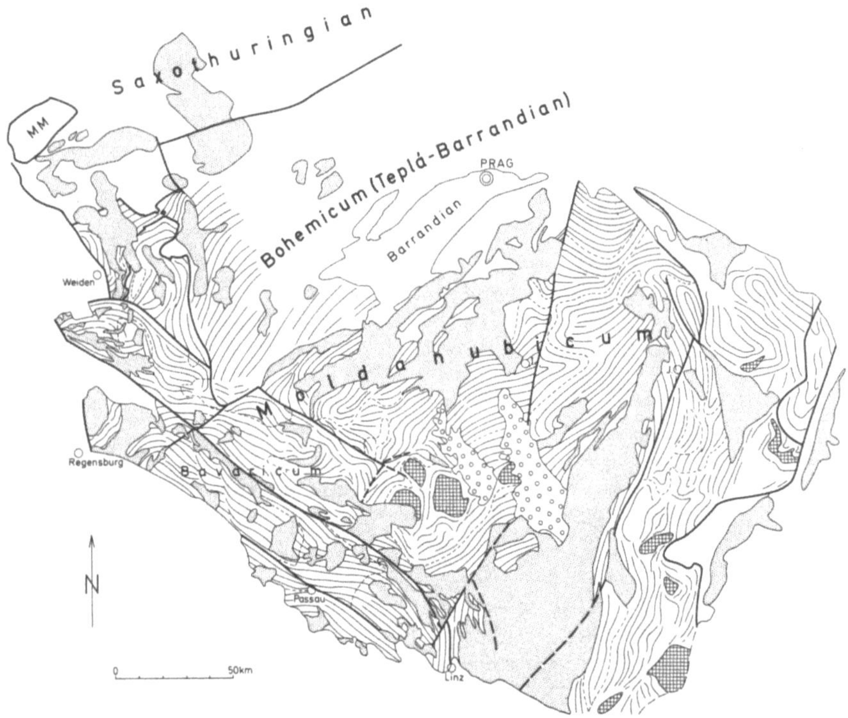

Fig. 14. Structural grain of the southwestern Bohemian Massif (Stettner 1986). NW-trending structures (D_4) are dominant along the southwestern margin of the Bohemian Massif. They decrease in intensity toward the NW

intensity of the D_4 deformation decreases from the SE to the NW. In the Bavarian SE of the Oberpfalz NW-SE striking structures are dominant (Fig. 14). They overprint older NE-SW-trending structures which are widely preserved in the Moldanubian of Czechoslovakia. NE-dipping metamorphic s-planes and mylonites as well as the SW--facing tectonic structures of the Bavarian indicate SW-directed tectonic transport.

The DEKORP profile 4-Q (Fig. 15) crosses the Bavarian SW of the Hoher Bogen (Fig. 1), forming the southwestern end of the allochthonous zone of Tepla-Taus (ZTT) of the western part of the Bohemian. In the southwest this profile intersects the important mylonite zone along the Bavarian Pfahl. At this intersection a steeply dipping reflection band (true dip angle 43°) is observed over about 10 km down to 2.5 s TWT (DEKORP Research Group 1988). Here it grades into a slightly NE dipping band of reflections at about 3-4 s TWT. This flat lying reflection zone can be correlated with the horizontal reflection band at the southern end of the DEKORP 4 profile (Fig. 11a, b). It represents the described structural discontinuity between upper and lower crust in the Moldanubian part of the DEKORP 4 profile. Hence, the structural discontinuity that formed during NW-directed overthrusting (D_1 to D_3) was also either active or had been reactivated as a décollement during the D_4 deformation at least in the Bavarian.

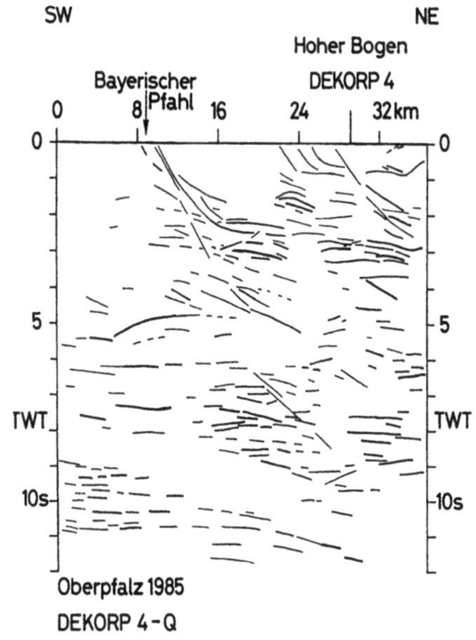

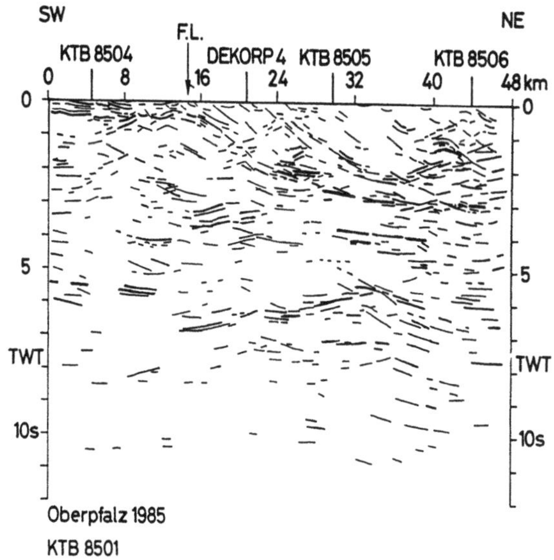

Fig. 15. DEKORP 4Q-profile; line drawing of the migrated section
Fig. 16. KTB-profile 8501; line drawing of the migrated section

NE-dipping seismic reflectors can also be observed in all SW-NE--oriented KTB profiles (Fig. 1) in the Oberpfalz area. This is true also for the Saxothuringian Zone (KTB 8501, Fig. 16) where the NW-trending D_4 structures are rather weakly developed at the surface. In KTB 8502 prominent, horizontal to slightly dipping reflectors are cut by and displaced along relatively transparent zones composed of relatively small-scale, steeply NE-dipping reflection elements (Fig. 17). The geometry and internal structure of these transparent zones is suggestive of reverse-fault zones in which the prominent subhorizontal reflectors are obliterated. These relatively young, NW-SE-trending structures are assumed to be equivalent to the NW-SE-striking strutures at the surface. They presumably represent the SW-vergent ductile or brittle reverse faults formed during D_4.

Coming back to the NW-dipping reflectors in the upper Moldanubian crust and their interpretation as SE-vergent structures, one has to realize that along the Moldanubian part of the DEKORP 4 line no SE vergent structures have yet been proven. The dominant structures at the surface, however, trend NW-SE. So, for the NW-dipping reflectors an interpretation other than backthrusting has to be considered. Figure 18 shows the course of strike of the metamorphic s-planes in the Moldanubian part of DEKORP 4. The ZEV is a bowl-shaped structure in seismic section like the NE-SW-striking syncline of the MM. However, in contrast to the MM, the ZEV is not a NE-SW striking structure, but a half bowl-shaped, i.e., a generally NW-SE-trending, and SW-facing syncline whose axes plunge to the SE in the northwest and to the NW in the southeast. Correspondingly, in the DEKORP 4 section the synform results from SSW-dipping reflectors in the northwest, horizontal reflectors in the middle and NW-dipping reflectors in the southeast.

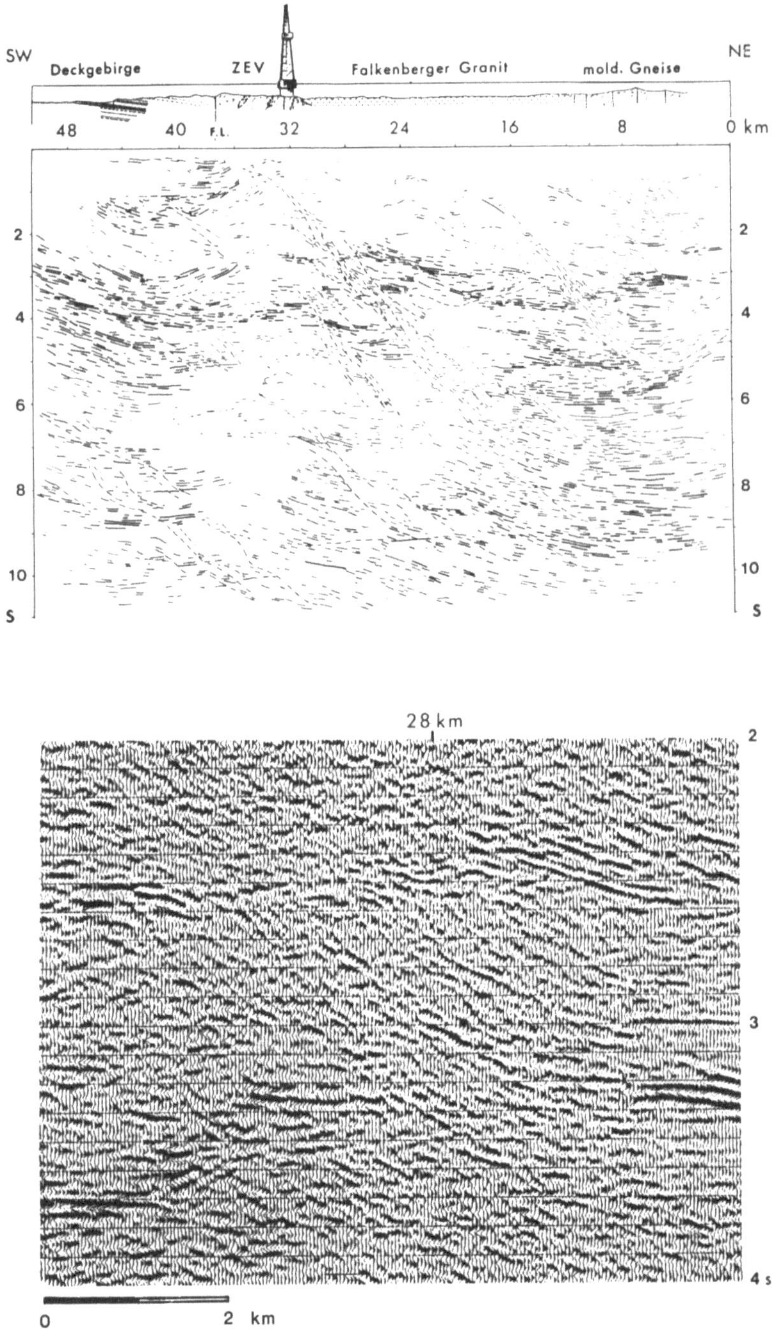

Fig. 17. KTB-profile 8502; line drawing of the migrated section, and (bottom) detail of the NE-dipping thrust-fault between 2 s and 4 s TWT below km 28

In the area of the Naabgebirge between the Luhe Line (LL) and Hoher Bogen (HB) (NGS in Fig. 18) there is an analogous situation. Here, the NW-SE striking, SW-vergent structures gains increase in intensity to the southeast toward the Bavarian Zone. Around the Naabgebirge the arc-shaped strike of the s-planes which generally dip to the NE indicates an anticlinal structure in the north, a synclinal structure in the middle and again an anticlinal structure in the south, just in front of the Hoher Bogen. This can be interpreted as resulting from a large-scale undulation of the axes of the generally NW-SE striking large-scale folds and thrusts. The described surface structures can easily be correlated with corresponding reflectors of the DEKORP 4 profile (Fig. 18b). It can be seen that NW-dipping reflectors result from an oblique traverse through the SW-vergent, NE-dipping structures and their respective reflectors.

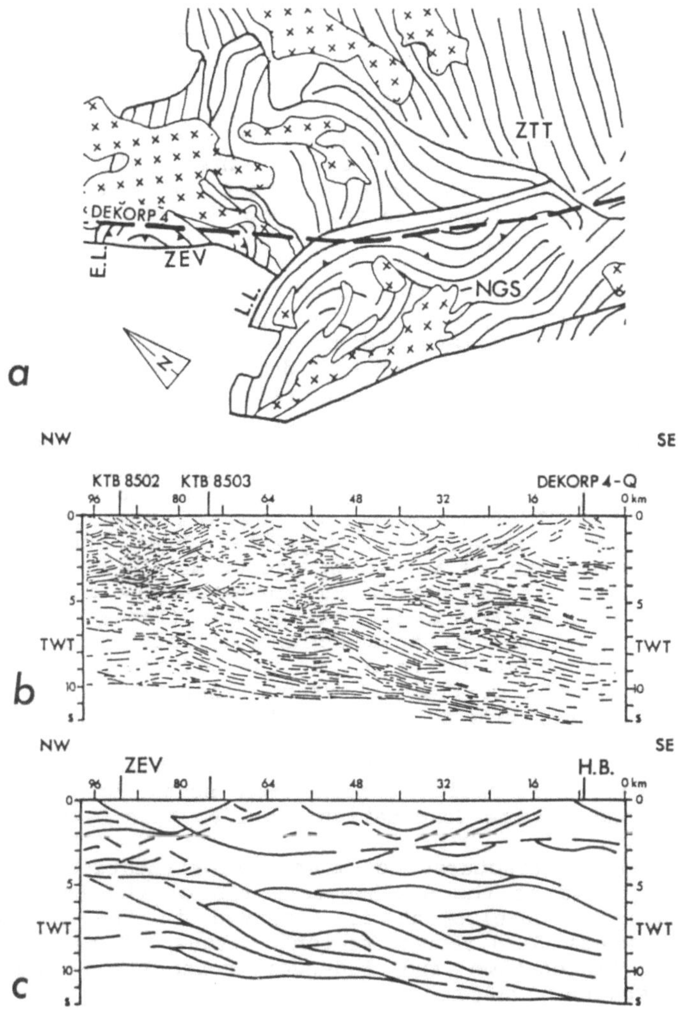

Fig. 18. Correlation between changes of dip and strike of the foliation in the Moldanubian part of DEKORP 4 (a), and the attitude of dipping reflectors above 4 s TWT (b). Additionally the structural model of the corresponding part of DEKORP 4 is shown (c)

4 THE STRUCTURAL DEVELOPMENT AT THE DEEP DRILLING AREA IN THE SETTING OF THE CENTRAL EUROPEAN VARISCAN BASEMENT

The Saxothuringian/Moldanubian boundary in the Oberpfalz reveals a complex and polyphase structural development. The dominant tectonic transport of Moldanubian thrusted onto Saxothuringian rocks and accompanied crustal wedging is NW directed and occurred during the collisional stage of the Variscan orogeny.

Crustal wedging connected with backfolding and backthrusting during collision led to tectonic overlapping of Saxothuringian onto Moldanubian rocks, as assumed for the Erbendorf wedge and the equivalent wedge shaped body of the KTB 8505 section (Figs. 9 and 11). A deeper crustal level is exposed east of Tirschenreuth than further to the west. Therefore, it can be assumed that Saxothuringian rocks formerly overlying the Moldanubian rocks of the ZTM and possibly further to the south were eroded. The former crustal wedge was rotated and transformed into the steeply dipping ZTM and upthrust to the present outcrop level. This interpretation is supported by the fact that the horizon of prominent reflectors between 2,5 s and 4.5 s TWT of the KTB 8506 section terminates below the northwestern rim of the ZTM (Fig. 12). Equivalent crustal wedges are seen further west in the KTB 8505 and DEKORP 4 (Erbendorf Body) profiles. Due to their F4-synclinal position and subsequent normal faulting they have not been exhumed as the ZTM. Therefore, their structural relationship to the Saxothuringian rocks has been preserved.

By the assumption that the Saxothuringian/Moldanubian boundary represents a plate boundary with a former passive continental margin at the Saxothuringian terrane and an active continental margin at the Moldanubian terrane, the Variscan deformation and metamorphism of the Saxothuringian can only be related to the collisional stage. In contrast, the D_1-D_2-deformation of the Moldanubian was probably a longer lasting event, which already began during the oceanic subduction stage. Relics of kyanite and garnet in feldspar and cordierite may be related to this pre-collisional metamorphic stage. The younger LP/HT metamorphism, however, must be attributed to the collisional stage, when augmented heat supply due to crustal thickening and delamination processes (Bird 1978; Weber 1981, 1984) as well as isothermal decompression due to rapid exhumation promotes LP/HT metamorphic overprint (Thomson and Ridlay 1987). With respect to the Moldanubian collision possibly began during D_2 and was active during D_3. In the Saxothuringian deformation began during the collisional stage. Therefore, D_1-D_3 of the Saxothuringian corresponds to D_2-D_3 of the Moldanubian (Table 1).

The correlation of Table 1 does not consider the assumed migration of the different deformation phases from south to north, which particularly concerns the deformations earlier than D_5/D_6.

The NW-trending D_4 structures document a reorientation of the regional stress field at the southwestern rim of the Bohemian Massif. An analogous development is observed in the southeastern Black Forest where SE-directed tectonic transport occurs in the area of the Zinken-Elme and Badenweiler-Lenzkirch shear zone (Krohe and Eisbacher 1988).

In the area between the Black Forrest and the Bohemian Massif the gravimetric and magnetic anomalies of the pre-Mesozoic basement reveal arclike contours to the N, outlining the structural fabric of the pre-Mesozoic basement. (Edel 1982; Weber 1986). Correspond-

Table 1. Structural development at the Saxothuringian/Moldanubian boundary in the Oberpfalz

		Moldanubian	Saxothuringian	ZEV
Subduction stage <450 and >330 Ma				
		D_1	No deformation (sedimentation)	Early stages of deformation not analysed; initial nappe transport
Collisional stage post lower carboniferous				
Crustal wedging	3	D_2 NE-striking	D_1 NW-vergent	D_2 not observed
	2	NW-vergent(?)		
and backfolding	0		D_2 SE-vergent	
uplift	M a	D_3 almost upright	D_3 almost upright	D_3 not observed
Reorientation of the regional stress field				**nappe emplacement**
Late collisional				
Indentation stage		D_4 SW-vergent	D_4 SW-vergent	D_4 SW-vergent
accompanied by		intrusion of late tectonic granites < 320 Ma		
wrench faulting		D_5	D_5	D_5
and extensional		intrusion of the Flossenbürg granite < 300 Ma		
tectonics		D_6	D_6	D_6

ing arc structures are also seen in the trend of the Permo-Carboniferous troughs of the South German Massif (Fig. 19). This configuration resembles an indentation structure similar to the Himalayas or the Ibero-Amorican arc (Taponnier and Molnar 1976; Matte 1986). The indenting body is represented by the crystalline basement of the South German Massif, which can presumably be traced into the Hercynian basement of the Alps. It is still open to question whether this basement represents a Variscan terrane (Vindelician Massif) as interpreted by Weber (1986b), or the indentation structure develops progressively inside the Moldanubian basement during the late collisional stage of the Variscan orogeny. The indentation process causes the observed reorientation of the regional stress field producing SE-vergent structures in the Black Forest and SW-directed tectonic transport at the southwestern rim of the Bohemian Massif. In both areas this process is more intensive and begins earlier in the south and fades out to the north as described in the Oberpfalz area.

In the central part of the described arc structure the coherent displacements should be directed to the SE in relation to the Black Forest and southwestern Bohemian Massif. The Tauber Valley structure at about km 110 in the DEKORP 2-South seismic profile (DEKORP Research Group 1985) might be interpreted as a corresponding NW-dipping structure. It represents a pronounced reflection band dipping 47° to the north (Behr and Heinrichs 1987). It roots at a depth of about 5 s TWT similar to equivalent structures in the Black Forest (KTB 1986) and at the western margin of the Bohemian Massif.

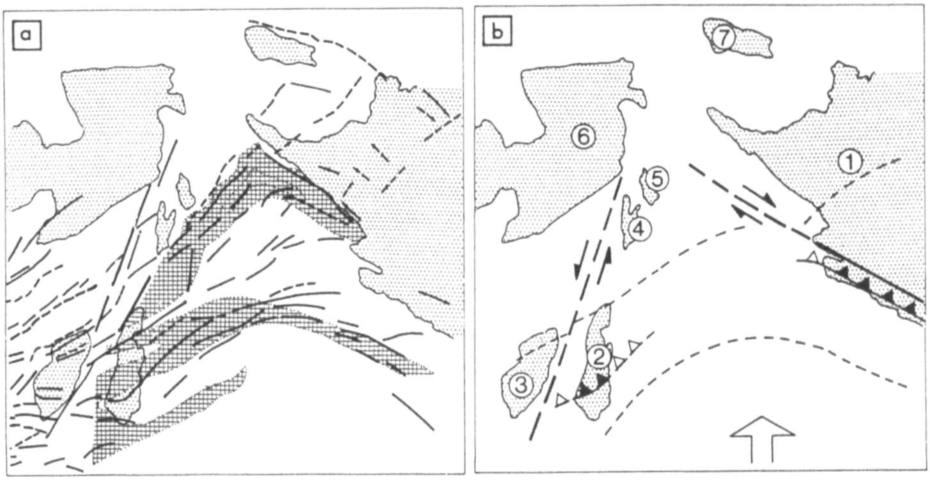

Fig. 19. a Axes of magnetic and gravimetric anomalies and corresponding arc-structures of the Permo-Carboniferous troughs (cross-hatched) between the Bohemian Massif and the Black Forest. b Sketch-map of the indenter and related thrusts and wrench faults. 1 Bohemian Massif; 2 Black Forest; 3 Vosges; 4 Odenwald; 5 Spessart; 6 Rheinisches Schiefergebirge; 7 Harz

Further manifestations of the indentation process are extensive dextral and sinistral shear zones. The dominance of dextral shear zones in the eastern part of the Central European Variscan basement, as shown by Arthaud and Matte (1977), resulted from the movement of terranes relative to the East European platform. More precisely, dextral movements in the mylonites of the Bavarian and along the Bavarian Pfahl and sinistral displacements between the Black Forest and the Vosges as well as in the northwestern Spessart (Weber in prep.) can be assigned to the indentation process. During the latest collisional indentation stage orthogonal collision zones can be transformed into oblique collision zones or wrench faults.

A further argument for the indentation model is the rotation of the magnetic declination observed on both sides of the arc, but in opposite directions. In the southern Vosges Edel et al. (1984) have shown that the mean direction ($D = 323^{o}$) deviates about 60° counterclockwise from the theoretical direction calculated with the Early Carboniferous European pole position. This deviation is interpreted as resulting from a counterclockwise rotation of the southern Vosges between late Visean and Westphalian. The same sense of rotation was also determined by Royer (1982) in the northern Vosges and by Bosum et al. (1973) in the crystalline Spessart.

In contrast, the rotation of magnetic declination in the region of the Bohemian Massif amounted to 20°-35° in clockwise sense during the Carboniferous and the Permian as shown by Krs (1978) and Kim and Soffel (1982). So the sense of rotation of the magnetic declination corresponds to the observed sense of shear both together supporting the proposed indentation model.

5 CONCLUSIONS

The presite studies on mineralogy, geochemistry, and geophysics in the Oberpfalz deep drilling area have yielded extensive new knowledge on the geological evolution of this portion of the Variscan orogen. These studies are based on comprehensive older studies whose results were partially confirmed and were also able to be revised in many cases. Fundamentally new data were obtained on the geophysical structure of the crust with which several of the structural models based on the surface geology were essentially confirmed. The most important results on the crustal structure are summarized in the following:

The Variscan crustal structure evolved during multiphase deformation in which two main events can be differentiated. The older, multiphase D_1-D_3 event was accompanied by extensive overthrusting of Moldanubian crust onto Saxothuringian crust. In the model presented here SE-vergent structures are a secondary fabric resulting from crustal wedging limited to upper crustal stockworks, particularly in the Saxothuringian. The Erbendorf line near Erbendorf does not represent the Saxothuringian-Moldanubian suture but the nappe boundary of the ZEV which overlies Saxothuringian rocks in the NW and the Saxothuringian-Moldanubian boundary further to the SE. The latter is exposed at the surface along the NW margin of the ZTM. There are still numerous open questions regarding the details of the development of this suture zone, particularly involving the development during the pre-collision stage, in the HT-LP metamorphism stage, and afterwards. The younger deformation events (D_4-D_6) are related to SW-directed tectonic transport and dextral shear. This deformation documents a reorientation of the regional stress field on the SW margin of the Bohemian massif whose cause is suggested to be an indentation process. D_5 and D_6 lead into late- and postorogenic extension tectonics which was modified in the southwestern prolongation by the Tertiary Eger Rift not treated here. The slight updoming of the mantle in the central portion of the DEKORP 4 profile may have formed in the course of this Tertiary extension during which the older SE-dipping overthrusts were possibly reactivated as low-angle normal faults.

There can be no doubt that, as in other orogens, deformation has wandered in the foreland direction. This means that the differing phases of deformation (particularly D_1-D_4) are older in the south than in the north. The relative age of nappe transport in this sequence of events is critically discussed (cf. Franke this Vol.). The NW-directed nappe transport can apparently be correlated with the stage of NW-vergent D_1 deformation. Yet, since based on the geophysical structure of the crust we conclude that the main tectonic transport was directed northwestward during the subsequent D_2 and D_3 deformations, and that the marked SE-vergent D_2 structures at the surface in the Saxothuringian can be interpreted as secondary effects of wedging and ramp tectonics, the supracrustal NW-directed nappe transport of the MM can also have occurred in a younger stage of deformation (older than D_4). In doing so, the relative age of the various phases of deformation in the different regions must be taken into account, e.g., D_1 in the northern Frankenwald may have occurred simultaneously with D_3 in the Moldanubian.

If the MM and ZEV nappes are interpreted as relicts of a nappe complex originally connected with the ZTT, the question arises as to the position and geometry of an additional suture zone, which can be regarded as the root zone for the MP metamorphic and eclogite-bearing nappes (MM). Variscan crustal convergence was probably much stronger than previously assumed. During this crustal convergence the initial oceanic subduction zones have been completely overridden by continental crust. Thereby, the complexly deformed cryptic sutures seen at the present outcrop level of the crystalline basement have lost their contact with the previous subduction zone. During collision, terranes previously separated by oceanic crust were thus thrust over one another (thick-skinned tectonics) so that suture zones widely separated before collision are now closely neighboring or overridden by other terranes. This is presumably the case at the northwestern margin of the Bohemian massif where the Saxothuringian-Moldanubian suture zone was overthrust by the Bohemian terrane. The present reflection seismic data obtained in European and North American basement rocks are strongly indicative of extensive thrusting of this type.

Acknowledgments. The funding of the KTB-presite studies by the Deutsche Forschungsgemeinschaft and the Bundesministerium für Forschung und Technologie, Bonn, is gratefully acknowledged. We are indebted to the intire Oberpfalz Working Group and to the DEKORP Group for many fruitful and encouraging interdisciplinary discussions. For continuous support by the local authorities, the Bayerische Geologische Landesamt and the public media, we extend our appreciation.

References

Arthaud, F. & Matte, Ph. (1977): Late Paleozoic strike slip faulting in Southern Europe and Northern Africa: result of a right lateral shear zone between the Appalachians and the Urals. - Geol. Soc. Amer. Bull., 88: 1305-1320.

Banks, C.J. & Warburton, J. (1986): "Passive roof" duplex geometry in the frontal structures of the Kirthar and Sulaiman Mountain Belts, Pakistan. J. Struct. Geol., 8: 229-237.

Behr, H.-J.; Engel, W. & Franke, W. (1982): Variscan Wildflysch and Nappe Tectonics in the Saxothuringian Zone (Northeast Bavaria, West Germany).- Amer. J. Sci., 282: 1438-1470.

Behr, H.J. & Heinrichs, T. (1987): Geological interpretation of DEKORP 2-S: A deep seismic reflection profile across the Saxothuringian and possible implications for the late Variscan structural evolution of Central Europe. Tectonophysics, 142: 173-202.

Bird, P. (1978): Initiation of intracontinental subduction in the Himalaya.- J. Geophys. Res.,83: 4975-4987.

Blümel, P. (1983): The western Margin of the Bohemian Massif in Bavaria.- Fortschr. Min., Bd. 61, Bh. 2: 171-175.

Blümel, P. (1985): Metamorphic History.-In Weber, K.: Excursion guide Oberpfalz.- 2nd Symp. on Observation of the cont. crust through drilling, Seeheim/Odenwald, 7.-9. Oct. 1985: 23-30, Göttingen.

Bosum,W.; Geipel, H. & Weinelt, W. (1973): Magnetische Untersuchungen im kristallinen Spessart und ihre geophysikalisch-geologische Deutung. Geol. Bavarica, 67: 11-34.

Brown, R.L., Journeay J.M.; Lane, L.S.; Murphy, C.D. & Ree, C.J. (1986): Obduction, backfolding and piggyback thrusting in the metamorphic hinterland of the southeastern Canadian Cordillera.-J. Struc. Geol., 8: 239-257.

DEKORP Research Group (1985): First results and priliminary interpretation of deep-reflection seismic recordings along profile DEKORP 2-South. - J. Geophys., 57: 137-163.

DEKORP Research Group (1988): Results of the DEKORP 4/KTB Oberpfalz deep seismic reflection investigations.-J. Geophys. 62: 69-101.

Edel, J.B. (1982): Le socle Varisque de l'Europe moyenne apports du magnétisme et de la gravimétrie.-Sci. Géol. Bull., 35,4: 207-224.

Edel. J.B.; Coulon, M. & Hernot, M.P. (1984): Mise en évidence par le paléomagnétisme d'une importante rotation antihoraire des Vosges méridionales entre le Viséen terminal et le Westphalien supérieur. Tectonophysics, 106: 239-257.

Fänger, R. (1988): Strukturgeologische Untersuchungen am Nordrand des Moldanubikums zwischen Pleystein und Neuenhammer (Oberpfalz, NE-Bayern). Dipl. Arbeit, Univ. Göttingen.

Franke, W. (1984): Variszischer Deckenbau im Raume der Münchberger Gneismasse, abgeleitetaus der Fazies, Deformation und Metamorphose im umgebenden Paläozoikum.- Geotekt. Forsch., 68: 253 pp.

Gaertner, H.R. von (1941): Die Schichtgliederung der Phyllitgebiete in Thüringen und Nordbayern und ihre Einordnung in das stratigraphische Schema.-Jb. Reichsamt Bodenforsch. 62: 54-80; Berlin 1944.

Gaertner, H.R. von (1951): Probleme des Saxothuringikums.- Geol. Jb., 65: 409-450.

Gümbel, C.W. von (1968): Geognostische Beschreibung des Ostbayerischen Grundgebirges.- 968 S.; Gotha.

Gümbel, C.W. von (1879): Geognostische Beschreibung des Fichtelgebirges mit dem Frankenwalde und dem westlichen Vorlande.- 698 pp; Gotha.

Gümbel, C.W. von (1988): Geologie von Bayern in zwei Theilen, Bd. 1 Grundzüge der Geologie.- 1144 pp; Cassel.

Gümbel, C.W. von (1894): Geologie von Bayern in zwei Theilen, Bd. 2 Geologische Beschreibung von Bayern.-1184 pp; Cassel.

Heinicke, F. (1987): Strukturgeologische Untersuchungen im Übergangsbereich zwischen der Zone von Erbendorf-Vohenstrauß nund dem Moldanubikum im Gebiet Vohenstrauß/Leuchtenberg, Oberpfalz - Oberflächengeoelektrische Kartierung und gefügekundliche Aufnahme.- Dipl. Arbeit, Univ. Göttingen, 84 pp.

Heinicke, F. & Vollbrecht, A. (in press): Gefügeuntersuchungen an Graphitquarziten im moldanubisch-saxothuringischen Übergangsbereich.-Erlanger Geol.Abh.

Kim, I.S. & Soffel, H.C. (1982): Palaeomagnetic results of Lower Paleozoic diabases and pillow lavas from the Frankenwald area (northwestern edge of the Bohemian Massif).- J. Geophys., 51: 24-28.

Kleemann, U. (1986): Petrographische Untersuchungen und thermobarometrische Analyse der metamorphen Überprägung in zwei tektonischen Bewegungszonen des nördlichen Oberpfälzer Waldes.- Dipl.Arbeit, Univ. Bochum, 67 pp.

Klinkhammer, B. & Rost, F. (1975): Die Serpentinite des Oberpfälzer Waldes.- Der Aufschluß, Sonderbd. 26: 39-64.

Kossmat, F. (1927): Zur Gliederung des varistischen Grundgebirgsbaues.- Abh. Sächs. Geol. L.-Anst., 1: 1-39

Krohe, A. & Eisbacher, G. H. (1988): Oblique crustal detachment in the Variscan Schwarzwald, southwestern Germany.- Geol. Rdsch. 77,1: 25-43.

Krs, M.. (1978): Palaeomagnetic evidence of tectonic deformation of blocks in the Bohemian Massif.- Geologie, 31: 141-150

KTB 1986 (1986): Kontinentales Tiefbohrprogramm der Bundesrepublik Deutschland, KTB, Ergebnisse der Vorerkundungsarbeiten Lokation Oberpfalz. In: Weber, K. & Vollbrecht, A. (eds), 2nd KTB Kolloquium Seeheim/Odenwald, 19-21 Sept. 1986.

Ludwig, V. (1972): Die Paragenese Chlorit, Muskowit, Paragonit und Margarit im "Griffelschiefer" des Ordoviziums in NE-Bayern.-N. Jb. Geol. Paläont.,Mh., 9: 546-560.

Massalsky, T. (1987): Bohrbeschreibung und Profil der Geothermiebohrung Poppenreuth.- KTB-Internal Report.

Matte, P. (1986): Tectonics and plate tectonics model for the Variscan Belt or Europe.- Tectonophysics 126: 329-374.

Oxburgh, E.R. (1972): Flake tectonics and continental collision.- Nature (Lond) 239: 202-207.

Plaumann, S. (1986): Die Schwerekarte der Oberpfalz und ihre Bezüge zu Strukturen der oberen Erdkruste.-Geol. Jb., E 33: 5-13.

Price, R. A. (1986): The southeastern Canadian Cordillera: thrust faulting, tectonic wedging, and delamination of the lithosphere.- J. Struc. Geol. 8, no. 3/4: 239-254.

Royer, J.Y. (1982): Etude paléomagnétique et géochronologique de Vosges cristallines du Nord.- Mem.DEA, Institut Physique du Globe, Strasbourg I.

Sanford, A.R. (1959): Analytical and experimental study of simple geologic structures.-Geol. Soc. Am. Bull. 70: 19-52.

Schreyer, W. (1966): Metamorpher Übergang Saxothuringikum-Moldanubikum östlich Tirschenreuth/Opf. nachgewiesen durch phasenpetrologische Analyse.- Geol. Rdsch., 27: 260-275.

Schüssler, U.; Oppermann, U.; Kreuzer, H.; Seidel, E.; Okrusch, M.; Lenz, K.-L. & Raschka, H. (1986): Zur Altersstellung des ostbayerischen Kristallins. Ergebnisse neuer K-Ar-Datierungen.- Geologica Bavarica, 89: 21-47.

Schwan, W. (1958): Untervorschiebungen und Aufbruchfalten.- N.Jb.Geol.Paläont., Mh.,1958: 356-377.

Seidlitz, W. von (1921): Tektonische Beziehungen der Münchberger Gneismasse zum Erzgebirge und nördlichen Böhmerwald.- Geol.Rdsch., 12: 270-292.

Stein, E. (1987): Die strukturgeologische Entwicklung im Übergangsbereich Saxothuringikum / Moldanubikum in NE-Bayern.- Diss. Univ.Göttingen, 125 pp.

Stettner, G. (1960): Über die geologische Entwicklung des Fichtelgebirges.- Aufschluß, Sonderband, 8: 9-22.

Stettner, G. (1971): Orogene Schollentektonik und granitoide Mobilisation im Westteil der Böhmischen Masse.- Geol.Rdsch., 60: 1465-1487.

Stettner, G. (1972): Zur geotektonischen Entwicklung im Westteil der Böhmischen Masse bei Berücksichtigung des Deformationsstils im orogenen Bewegungssystem.- Z. Dtsch. Geol. Ges., 123: 291-326.

Stettner, G. (1975): Zur geologisch-tektonischen Entwicklung des Oberpfälzer Grundgebirges.- Aufschluß, Sonderbd., 26: 11-38.

Stettner, G. (1976): Die Heterogenität des mitteleuropäischen Varistikums und das Problem der regionalen Gliederung - erläutert am Profil Saxothuringikum-Moldanubikum in Nordostbayern.- Nova Acta Leopoldina N.F. 224, 45: 59-66

Stettner, G. (1979): Der Grenzbereich Saxothuringikum-Moldanubische Region im Raum Tirschrenreuth-Mähring (Oberpfalz) und die Situation des Uran-führenden Präkambriums.- Z. Dtsch. Geol. Ges., 130: 561-574.

Stettner, G. (1980): Zum geologischen Aufbau des Fichtelgebirges.- Aufschluß, 31: 391-403.

Stettner, G. (1981): Geologische Karte 1:200000 CC 6334 Bayreuth.- Hannover (BGR).

Stettner, G. (1986): Structure and development of the Moldanubian Region in the Bohemian Massif.-Third EGT Workshop: The Central Segment. 14.-16. April 1986 Bad Honnef pp 141-148.

Suess, F.E. (1912): Vorläufige Mitteilung über die Münchberger Deckscholle. - Sitzungsber. Akad. Wiss. Wien, Math.-nat. Kl., 121 (Abt. IIa, H.10): 253 S.; Wien.

Suess, F.E. (1926): Intrusionstektonik und Wandertektonik im varistischen Gebirge.- 268 S.; Berlin.

Tapponier, P. & Molnar, P. (1976): Slip-line field theory and large-scale continental tectonics.-Nature (Lond) 264: 319-324.

Teufel, S. (1987): Vergleichende U-Pb und Rb-Sr-Altersbestimmungen an Gesteinen des Übergangsbereiches Saxothuringikum / Moldanubikum, NE-Bayern.- Diss. Univ.Göttingen 110 pp.

Thompson, A.B. & Ridley, J.R. (1987): Pressure-temperature-time (P-T-t) histories of orogenic belts. - Phil. Trans. R. Soc. Lond., A 321: 27-45.

Vollbrecht, A., Weber, K. & Schmoll, J. (in press): Structural model for the Saxothuringian-Moldanubian Suture in the Variscan Basement of the Oberpfalz (NE Bavaria, FGR) interpreted from geophysical data.-Tectonophysics.

Wagener-Lohse, C. & Blümel, P. (1986): Prograde Niederdruckmetamorphose und ältere Mitteldruckmetamorphose im nordostbayrischen Abschnitt der Grenzzone Saxothuringikum/ Moldanubikum. - Abstr. 76. Jahrestagung Geol. Vereinigung Gießen: 84-85.

Wall, de H. (1987): Mikrostrukturelle Untersuchungen und strukturgeologische Kartierung der Gesteine aus dem Übergangsbereich Moldanubikum/Saxothuringikum westlich Mähring, Oberpfalz.- Dipl. Arbeit, Univ. Göttingen 84pp

Weber, K. (1981): The structural development of the Rheinische Schiefergebirge.-Geol. en Mijnbouw, Vol. 60,1: 149-159.

Weber; K. (1984): Variscan events: early Palaeozoic continental rift metamorphism and late Palaeozoic crustal shortening. In: Hutton, D.H.W. & Sanderson, D.J. (eds) Variscan Tectonics of the North Atlantic Region. Geol. Soc. Spec. Publ. No. 14,3: 3-22.

Weber, K. (1985): Excursion-Guide Oberpfalz.- 2nd Int. Symp. on Observation of the Cont. Crust through Drilling.-FRG, 7.-9. Okt. 1985, Göttingen, 84 pp

Weber, K. (1986a): Metamorphism and crustal rheology - implications for the structural development of the continental crust during prograde metamorphism.- In: Dawson, J.B.; Carswell, D.A.; Hall, J. & Wedepohl, K.H. (eds). The nature of the lower continental crust.- Geol. Soc. Spec. Publ. No 24: 95-106.

Weber, K. (1986b): The mid-European Variscides in terms of allochthonous terranes.-Third EGT Workshop: The Central Segment. 14.-16. April 1986 Bad Honnef, pp 73-81.

Wurm, A. (1925): Geologie von Bayern. Nordbayern, Fichtelgebirge und Frankenwald 1. Teil.- 373 pp, Berlin

Voll, G. (1960): Stoff, Bau und Alter in der Grenzzone Saxothuringikum/Moldanubikum in Bayern unter besonderer Berücksichtigung gabbroider, amphibolitischer und kalksilikatischer Gesteine.- Beih. Geol. Jb., 42: 382 pp

Wurm, A. (1926): Über den Bauplan des variskischen Gebirges am Westrande der Böhmischen Masse.- Geol. Rdsch. , 17: 240-257.

Zoubek, V. (1979): Korrelation des präkambrischen Sockels der mittel- und westeuropäischen Varisziden.- Z. geol. Wiss., 7: 1057-1064.

The Geological Framework of the KTB Drill Site, Oberpfalz

Wolfgang Franke

Institut für Geowissenschaften und Lithosphärenforschung der Justus-Liebig-Universität, Senckenbergstr. 3, 6300 Giessen, FRG

1 Introduction

The crustal structure and the geotectonic development of the Variscan Belt of Europe have been controversial for al long time. Early in this century, F.E.Suess (1912), Kossmat (1927) and others had published thrust-tectonic concepts for the Variscan Belt, which are nothing less but anticipations of modern structural concepts stimulated by plate tectonics. The target area of the present KTB project served, in this early phase of research, as an outstanding example of horizontally directed tectonism. Since this time, the Variscan basement of NE Bavaria has occupied a key role in the ensuing geotectonic debate.

Later, the early mobilistic concepts were largely rejected, and replaced by models based upon vertical tectonic movements. This period lasted into the late 1960s, when H.Zwart (1967) defined differences between "alpinotype" and "hercynotype" orogens, the latter being characterized by the lack or scarcity of ophiolites, high-pressure metamorphic rocks and large-scale nappe thrusts, as well as by the abundance of granites instead of calc-alkaline volcanic rocks. Since that time, numerous studies have greatly diminished the apparent differences between the Alps and the Variscides. The discovery of high-pressure metamorphic rocks, ophiolite sequences and important nappe thrusts has justified the application of plate-tectonic concepts.

Though there is as yet no generally accepted plate-tectonic model, it is now possible to distinguish the main tectono-stratigraphic units, to propose a rough scheme for the opening and closure of the main Palaeozoic basins, and, above all, to define numerous open questions. For a survey of current ideas, the reader is referred to the works of Behr et al. (1984), Franke (in press), Matte (1986), Weber and Behr (1983), Weber (1984), Ziegler (1984,1986) and Zwart and Dornsiepen (1980).

Modern studies on the sedimentology, deformation, metamorphism and geochronology in the Palaeozoic rocks of NE Bavaria have been carried out since the late 1970s. These activities led to a revival of the classical nappe concepts (Behr et al. 1982, 1984, Franke 1984 a,b) and, eventually, revealed the area as a fascinating target for an ultra-deep drilling project. A pre-site study, carried out from 1982 to 1986, has produced a wealth of further information.

Though these more recent studies have greatly added to our knowledge about the Variscan basement, the KTB drilling project would have been impossible without the pioneer work of many colleagues, who have assembled, over many years or even decades, a broad factual base: a meticulous biostratigraphic survey of the unmetamorphosed Palaeozoic sequences (Geol. Institute Würzburg), basic information on the petrology of the magmatic rocks involved (Mineral. Institute Würzburg), the essentials of metamorphic petrology (Mineral. Institute Bochum) and, last but not least, a detailed geological mapping of the entire basement complex (Geol. Survey Bavaria, Munich).

The present work attempts to summarize the classical and the more recent studies on the surface geology of the target area, which may serve as a data base for the more specialized and/or interpretative contributions in this volume.

2 Position of the KTB Site Within the European Variscides

The Variscan Belt of Europe (Fig.1) is exceptionally wide. It ranges across approx. 1200 km, from the lowlands of northern France and Germany down into the Pyrenees or the Alps respectively. The width of the fold belt is due to the presence of a number of microplates which have been caught up in the collision of the major confining blocks, i.e. Laurussia and Gondwana. These units were separated from each other by a number of basins, which have occasioned, during the time of plate convergence, the formation of several subduction/collision zones. These are largely independent of each other, so that the Variscan Belt is best described as a collage of orogenic belts. The main microplates and basins are briefly discussed out below; for details and discussion of numerous open questions the reader is referred to the works cited above.

The northern border of the Variscan realm is marked by the <u>Laurussian megacontinent</u>, which was formed by the collision of Laurentia and Baltica during the Caledonian orogeny,i.e. in Ordovician and Silurian time.

Further south there followed an ocean of considerable width, whose trace is seen in the Caledonian suture of the British Isles and along the Tornquist Line, the southwestern margin of the East-European Platform (Fig.1). Closure of this <u>England/North Sea/Polish Ocean</u> was in progress at least from the Upper Ordovician onwards and was accomplished during the Silurian. The collision of Laurussia with the mid-European plate adjacent to the south created another "Caledonian" fold belt, which is almost completely concealed under younger cover in the English Midlands, the North Sea and the northern parts of continental Europe, and which is probably continuous with the belt of Caledonian deformation along the Tornquist Line.

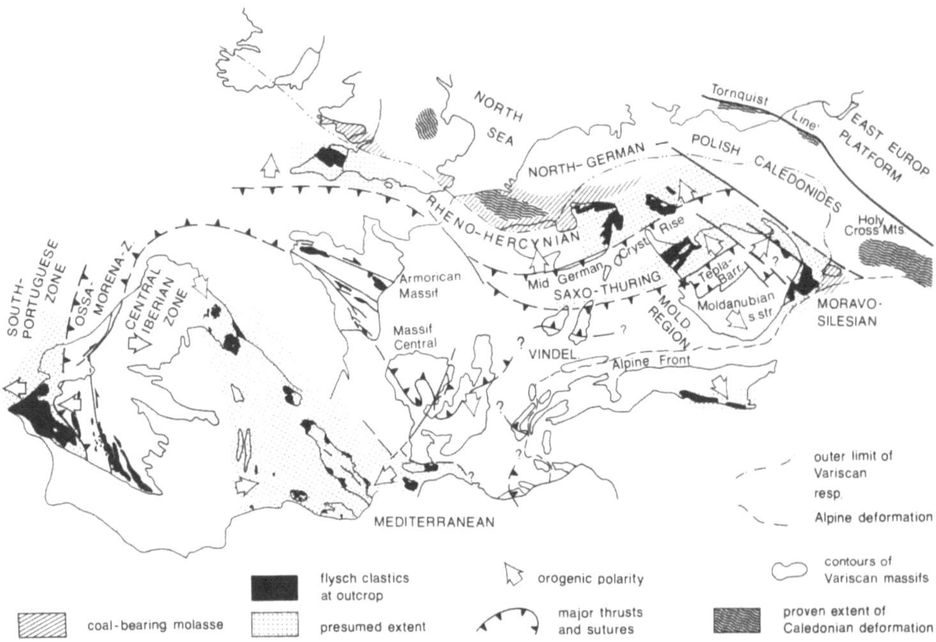

Fig.1: Main structural elements of the Variscan Belt of Europe, during the main period of synorogenic clastic sedimentation (Lower Carboniferous). The presumed extent of synorogenic clastics is marked only for foreland basins (stipple). The position of the KTB drilling site is denoted by an asterisk. After Franke and Engel 1986.

Subsequent to the closure of the northern ocean, the mid-European plate underwent crustal extension, which led to the formation of the Rhenohercynian Basin in Silurian(?), Devonian and Lower Carboniferous time. Spreading never went beyond the narrow ocean stage, since the newly formed Rhenohercynian crust was being consumed, at least from the Early Devonian onwards, by southward subduction.

The active margin which bordered the Rhenohercynian basin to the south, now deeply eroded, is known as the Mid-German Crystalline Rise. It shed Devonian and Carboniferous flysch clastics mainly toward the north, into the Rhenohercynian foredeep, and an almost negligeable amount of detritus into the basin adjacent to the south.

This southern basin is known as the Saxothuringian basin. Its origin goes back to an important Cambro-Ordovician phase of rifting, which is detectable almost everywhere in Europe. The Saxothuringian basin was bordered in the south by a pre-Variscan basement block, whose remains are detectable in the Bohemian Massif, in the Black Forest and Vosges, in the Armorican Massif of France and in Iberia. This Armorican/Bohemian microplate was probably subdivided, in itself, by minor rift structures (e.g. the Barrandian basin, see Fig. 5).
Like in the Rhenohercynian Belt, the Saxothuringian basin floor was being subducted, in Devonian and Carboniferous time, beneath the Bohemian microplate adjacent to the south. The KTB drilling site is located in the suture zone; the geology of the Saxothuringian basin and the Bohemian Massif will be discussed in greater detail below.

The Bohemian Massif is bounded to the SE by an important ductile shear zone, which gives evidence of dextral transpression (Matte et al. 1985,1987; Rajlich et al. 1986). The Moravo-Silesian Zone adjacent to the SE probably represents a lateral equivalent of the Rhenohercynian Belt in the north, with which it was originally linked by an arc structure.

The southern margin of the Armorican/Bohemian microplate is probably documented in the French Massif Central, where crystalline nappes carry remnants of a Massif Central or Proto-Mediterranean Basin. Possible equivalents exist in the western Alps. The pre-Variscan basement units situated further south (NE Iberia, France, southern Alps) probably represent the northern part of Gondwana.

The width of the Palaeozoic basins is constrained by palaeoclimatological arguments pertinent to the "Saharan" glaciation at the Ordovician/Silurian boundary, by radiometric ages on pre-Variscan basement rocks and by scant palaeomagnetic information. These data consistently indicate that the Variscan microplates up to the Caledonian sutures in England and along the Tornquist Line were not widely separated from each other, but considerably distant from Laurussia. This suggests that the only major or, at least, the main Palaeozoic ocean was the one marked by the Caledonides of the British Isles and Poland.

While the origin and age of the Palaeozoic basins differ from each other, their closure appears to have occurred during one major and more or less continuous period of plate convergence, in late Ordovician through Carboniferous time. Radiometric ages indicative of Late Ordovician/Silurian high- - to medium-pressure metamorphism have been recorded from the Massif Central and from the Saxothuringian Zone (see below); Devonian metamorphism is detectable at the Rhenohercynian, Saxothuringian and Proto-Mediterranean fronts, and Carboniferous low-pressure metamorphism and plutonism is ubiquitous. Radiometric evidence is backed up by the record of Devonian and Carboniferous flysch sedimentation, which likewise reflects orogenic activity in these zones (Engel and Franke 1983, Franke and Engel 1986,1988).

Tectonic vergence, the direction of sedimentary discharge, as well as the migration of deformation, metamorphism and foreland sedimentation consistently suggest a bilateral symmetry, with northward subduction at the Proto-Mediterranean front, and a tandem of southward directed subduction zones at the Rhenohercynian and Saxothuringian fronts (Fig.1).

3 Structure and development of the Saxothuringian Zone

3.1 Stratigraphy

Most of the Saxothuringian Zone (see the review in Franke 1984a) is made up of Precambrian(?) through Lower Carboniferous sedimentary and volcanic rocks in a conspicuous "Thuringian" facies, which is present in Thuringia, in the Frankenwald, in Saxony and, in a much more metamorphosed state, in the Fichtelgebirge and Erzgebirge Mts. (Figs.2,3).
The Cambrian and most of the Ordovician are made up of neritic clastic sediments and occasional carbonates, for which a continental basement must be inferred. By analogy with the neighbouring areas, a Cadomian age is probable, yet there is no evidence of the Precambrian/Cambrian boundary or any structural unconformity (see the discussion in Stein 1987).

The late Ordovician through Early Carboniferous (Tournaisian) consists of hemipelagic shales and carbonates. Few intercalations of quartzite turbidites among the Devonian shales are probably derived from the Mid-German Crystalline Rise to the NW.

A further source area existed, in late Givetian and Frasnian time, within the Saxothuringian basin. Coarse conglomerates and greywackes are arranged along a line which links areas to the NW of the Münchberg Klippe with the basement uplift of the Saxonian Granulite Mts., approx. 100 km farther to the NE (Granulitgebirge, see below and Fig.2). The clastics contain Ordovician through Middle Devonian rock fragments, including the Frasnian volcanics described below. All these lithologies clearly belong to the Thuringian facies.

Intra-basinal volcanism is most important in the Ordovician. It is bimodal in character, and dominated by felsic effusives and shallow-level granitoid sills. A second phase of vulcanicity occurred in the Frasnian, and has essentially produced mafic effusives of intra-plate character (Wirth 1978).
From the basal Viséan onwards, all of the Thuringian facies realm is flooded by flysch clastics, derived from the Bohemian Massif in the SE.

Thuringian facies rocks are also present in the immediate vicinity of the Saxothuringian/Moldanubian boundary, north of Erbendorf. They differ from the sequences north of the Fichtelgebirge, in that flysch sediments (greywacke turbidites) are already detectable in the early Famennian (Adam and Reuter 1981).

A distinctively different facies regime exists in areas which have turned out to represent tectonic klippen (see below). The most extensive outcrop of this "Bavarian" facies exists in the Münchberg Klippe NW of the Fichtelgebirge (Fig. 2).

Apart from a basal sequence of turbidites, the Cambrian and basal Ordovician of the Bavarian sequence consist of neritic clastic sediments. The main part of the Ordovician is made up of a volcano-sedimentary sequence, in which distal sandstone turbidites are intercalated among a bimodal, yet dominantly mafic suite of volcanic rocks (Wirth 1978).

The Silurian and Devonian are represented by radiolarian cherts with some greywacke turbidites.

The Tournaisian and Visean are present in the form of a wild-flysch sequence with all kinds of mass-flow deposits, which can be regarded as the proximal equivalent to the Thuringian flysch clastics.

3.2 Tectonic structures and metamorphism

The key to the structural development of the target area lies in a number of medium- to high-grade rock units, set within a major synform (Vogtländische Mulde), where they are in strange contrast with the surrounding sedimentary rocks. These are the metamorphic terrains of Münchberg, Wildenfels and Frankenberg (the latter two on GDR territory). The metamorphic blocks have alternatively been interpreted as diapirlike basement uplifts, or as tectonic klippen derived from areas south of the Saxothuringian Zone.

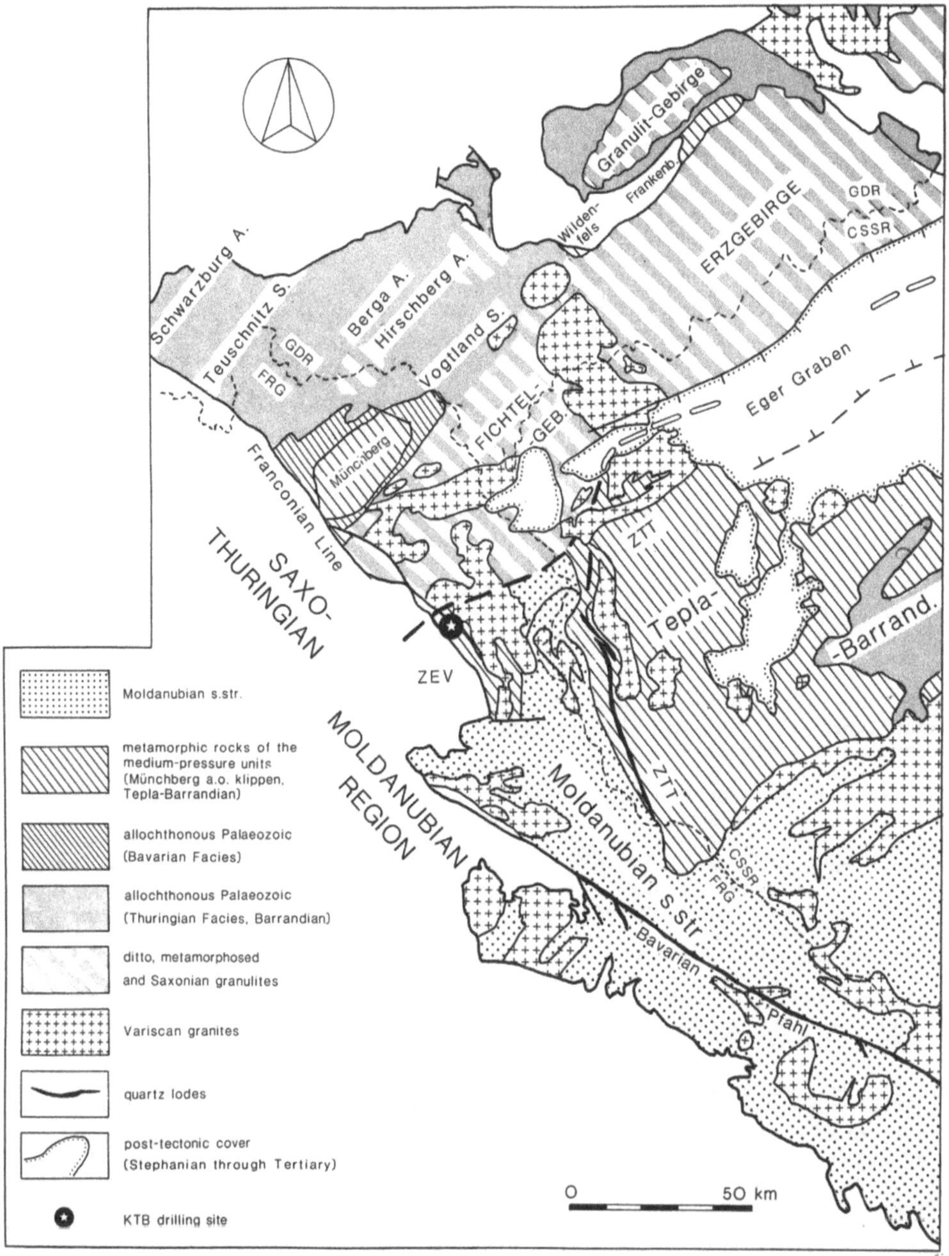

Fig.2: Geological map of the northwestern part of the Bohemian Massif, around the KTB drilling site.

The Münchberg Massif comprises the widest range of sedimentary and metamorphic facies. It represents a tectonic pile, in which both the metamorphic grade and the stratigraphic sequence appear in an inverted order (listed below in order from top to bottom).

1. "Hangend-Serie": a highly variable sequence of medium-pressure ortho- and paragneisses, with a predominance of mafic lithologies, probably developed from Proterozoic sedimentary and volcanic rocks.
   At the base of the sequence, there are intercalations of eclogites which reflect metamorphic pressures of 13 kb or more (Franz et al. 1986). The protoliths are gabbros and tholeiites (partly of oceanic affinity) of Cambrian or Ordovician age (Ducrot et al. 1983, Gebauer and Grünenfelder 1979, Stosch and Lugmair 1986,1987).

2. "Liegend-Serie": mainly medium-pressure paragneisses, probably developed from Proterozoic sediments, with a major proportion of ortho-augengneisses intruded in Early Ordovician time (approx. 480 Ma, Söllner et al. 1981).

3. "Rand-Amphibolit": a sheet of epidote-amphibolite of hitherto unknown derivation and age.

4. "Phyllit-Prasinit-Serie": a greenschist-grade sequence of metapelites and mafic volcanic rocks, which display calc-alkaline affinities (KTB 1986, Schüssler et al. 1986). The metapelites have produced a Proterozoic (Vendian) microflora (Reitz et al. 1988).

5. Very low-grade Palaeozoic sedimentary and volcanic rocks in three major thrust sheets, which are arranged in an inverse stratigraphic order, but show a predominantly normal sense of younging within the individual tectonic units.

This tectonic sequence has been emplaced on top of the Devonian and Carboniferous rocks of the Thuringian facies, which occur in a normal stratigraphic sequence.

The medium- to high-grade metamorphism of the Münchberg sequence reflects the earliest Variscan tectono-metamorphic activities in the target area. Ducrot et al. (1983) and Müller-Sohnius et al. (1987) suggest a Late Ordovician age (approx. 430 Ma) for the metamorphism of the eclogite and the Hangend-Serie respectively. Gebauer and Grünenfelder (1979) obtained 380 Ma on zircons from the eclogites (but see the comments in Ducrot et al. 1983!), and Stosch and Lugmair (1987) have reported Sm/Nd and Rb/Sr mineral isochrons of approx. 395 Ma.

A large number of cooling ages reflects the sequential upthrusting of the metamorphic units during the Early Devonian through Lower Carboniferous (Schüssler et al. 1986, Söllner et al. 1981).

As it was already envisaged by F.E.Suess (1912) and confirmed by modern studies (Behr et al. 1982, Franke 1984 a,b), the Münchberg complex can be interpreted as a nappe pile, which is derived from an area to the SE of the Thuringian facies sequences of the Fichtelgebirge. Since this interpretation is vital for the structural concept of the continental deep drilling site, the main arguments are briefly listed below:

1. The Ordovician through Devonian high- and medium-pressure metamorphism of the crystalline rocks must have taken place in another position than that of the underlying sedimentary rocks of the same age.

2. The Carboniferous flysch sediments are derived from the SE and contain numerous fragments conspicuous of the Bavarian facies, which suggests a southeasterly derivation at least for the Palaeozoic rocks within the Münchberg pile.

3. Shear criteria on originally subhorizontal thrust surfaces indicate a general transport toward NW.

The pre-site studies have brought further support:

1. Crystalline rocks of the same overall lithology, the same type of metamorphism (medium-pressure) and the same metamorphic age (approx. 390 Ma) exist at the NW margin of the Bohemian Massif, immediately to the SE of the Fichtelgebirge (see below).

2. The reflection seismic survey and conductivity experiments have revealed a bowl-shaped arrangement of reflectors in the Münchberg area, and the presence of highly conductive (i.e. sedimentary) rocks underneath the crystalline units (see contribution by Schmoll et al., and Haak and Blümecke in KTB 1986).

The deformation and metamorphism of the Münchberg nappes are largely imported from deeper structural levels in areas to the SE or, in the case of the Palaeozoic rocks riding low in the pile, were acquired during the emplacement of the nappes over the Thuringian parautochthon.

Within the latter unit, Franke (1984 a,b) and Stein (1987) have described a polyphase tectonic development. D1 is characterized by originally subhorizontal thrusts and tight folds with weakly inclined axial planes; this first phase corresponds to the emplacement of the nappes. While D1 dies out toward the NW of the Münchberg Klippe, D2 is present throughout the Saxothuringian Zone, from the SE margin of the Mid-German Crystalline Rise down to the Fichtelgebirge and even beyond. D2 has brought about a reversal of tectonic vergence (now toward the SE). Some of the first-order folds in the NW part of the Saxothuringian are probably D2 structures (Schwarzburg Anticline, Teuschnitz Syncline, Berga Anticline in Fig.2). D3 has upright pen folds; the Fichtelgebirge/Erzgebirge Antiform is a megascopic expression of this structural style.

Metamorphism in the parautochthon is of the low-pressure type (see contribution by Blümel). Higher grades (up to staurolite) are confined to the Fichtelgebirge/Erzgebirge Antiform, which were more deeply subsided under the tectonic cover of the Münchberg nappes. According to Stein (1987) and Wagener-Lohse and Blümel (1986), the peak of low-pressure metamorphism in the Fichtelgebirge coincides with the D2 phase of tectonic deformation. The delay of heating with respect to tectonic subsidence is readily explained by the notoriously poor thermal conductivity of rocks. The peak of metamorphism has been dated at 320 Ma (Teufel 1987, Teufel et al. 1986). The same phase of metamorphism has also affected the Moldanubian unit (s.str.) to the immediate south of the Fichtelgebirge (see below).

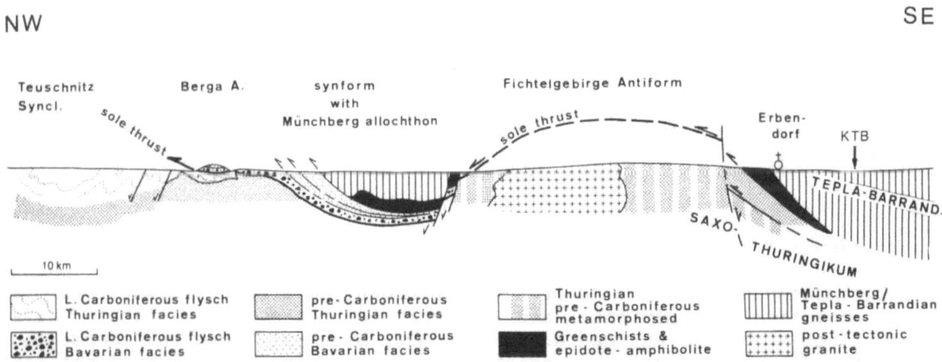

Fig.3: Diagrammatic geological cross-section through the Saxothuringian Zone and the northwestern part of the Tepla-Barrandian unit, with the location of the KTB drilling site. For deeper structural levels underlying the drilling site see Fig. 5. Approx. 2 times vertical exaggeration.

It is important to note that the D2 and D3 deformation phases, like the accompanying low-pressure metamorphism, are also detectable in the Moldanubian s.str. further south (Stein 1987, see also contribution by Blümel), which indicates a welding together of these units during D2. The Moldanubian has subsequently been affected by a D4 phase with roughly NS-trending folds, which, in the Saxothuringian Zone, are only represented by kink bands.

The principal relationships between the Saxothuringian Zone and the KTB target area further south may be summarized as follows (see Fig.3): Rocks akin to the Tepla-Barrandian unit of the Moldanubian Region were first affected by early Variscan subduction and metamorphism, and were subsequently upthrusted toward the NW. During this latter process, the reactivated Cadomian basement has accreted a deeper-water portion of the Saxothuringian basin (the Bavarian facies), and both units where subsequently emplaced on the epicontinental, external part of the Saxothuringian basin. Early folds and nappes were then refolded by younger deformation phases (D2, D3) and welded together with the Moldanubian further south during a phase of syn-D2 low-pressure metamorphism (see below and contribution by Blümel).

Palinspastic restoration of the Münchberg tectonic pile leads to a minimum displacement of the Münchberg Gneisses of about 200 km (Franke 1984 b). This figure does not include the crustal shortening required for the subduction of the Münchberg Gneisses to the depth where they acquired their high- and medium-pressure metamorphism.

The major features of the Saxothuringian Zone can be followed towards the NE into GDR territory (see Fig.2). The klippen of Wildenfels and Frankenberg are sisters to the Münchberg (see already Kossmat 1927), yet less well preserved. At Frankenberg, the thrust contacts are sealed by Latest-Lower Carboniferous intra-montane molasse. Since Upper Viséan flysch has been involved into the folding and thrusting, at least the D1 and D2 phases of tectonic deformation can be precisely dated as terminal Lower Carboniferous. These stratigraphic constraints are in reasonable accord with the radiometric age (see above), since "absolute" time scales place the Lower/Upper Carboniferous boundary at 320 (Odin 1982) or 333 Ma (Harland et al. 1982, see also Table 1).

The Saxothuringian Zone of the German Democratic Republic (GDR) encloses one peculiar feature, which is less prominent in West Germany: this is the brachyanticlinal structure of the Saxonian Granulite Mts. (Granulitgebirge, Fig.2). High-grade metamorphic rocks in the core of the structure show an extremely rapid transition into low- and very low-grade Palaeozoic rocks of the Thuringian facies. Facts and fiction pertaining to the Granulite Mts. may be taken from Weber and Behr (1983) and Weber (1984). For the present purpose, it is important to note that the Granulite Mts. correspond to the position of the source-area for the Frasnian clastic sediments (see above), which, in turn, is marked by a pronounced updoming of seismic reflections in the Dekorp 4 line (see contribution by Schmoll et al.). It is probable that the Frasnian clastics record part of the uplift history of the Granulite Mts. Since uplift and erosion were accompanied by basaltic volcanism, these features appear to relate to an extensional regime. They therefore invite comparisons with the metamorphic core complexes of the North American Basin and Range Province rather than with the compressional regime invoked, e.g. by Weber (1984).

4 Moldanubian Region

The Bohemian Massif to the SE of the Fichtelgebirge/Erzgebirge Antiform is mainly composed of plutonic and metamorphic rocks. The latter have conventionally been regarded as pre-Variscan basement, but an increasing number of palaeontological and radiometric datings have yielded evidence for a dominantly Variscan tectono-metamorphic imprint, and even the occurrence of Palaeozoic protoliths. Presumably, the geological development of the Moldanubian Region will have to be rewritten during the next years; the following lines, therefore, only summarize some well-established essentials and point out open questions.

As shown in Fig.2, the Moldanubian Region may be subdivided into the Tepla-Barrandian Unit (or Bohemicum), and the Moldanubian in a strict sense (referred to below as the Moldanubian s.str.). The western part of the Tepla-Barrandian is known as the zone of Tepla-Taus (ZTT). The KTB drilling site is situated in the zone of Erbendorf-Vohenstrauß (ZEV), which can be regarded as an outlier of the ZTT. As discussed below, the metamorphic rocks within the Münchberg and related Klippen of the Saxothuringian Zone likewise represent equivalents of the Tepla-Barrandian unit.

4.1 Stratigraphy

4.1.1 Tepla-Barrandian

The Tepla-Barrandian unit has a bowl-shaped structure, whose centre (the Barrandian Syncline) holds the least metamorphosed rocks. These are Cambrian through Givetian sediments and volcanic rocks of very low metamorphic grade. The unconformable, transgressive onlap of Cambrian conglomerates on a greenschist-grade, pre-Variscan basement is the main reason for the classical notion of a Cadomian age of the metamorphism in the Bohemian Massif.

Summaries of the Palaeozoic development are available in Chlupac (1968) and Havlicek (1981). The Cambrian and Ordovician are composed of fluviatile(?) to shallow marine clastic sediments. The Silurian through Middle Devonian are essentially composed of hemipelagic and neritic carbonate sediments including reef limestones. The sequence is topped by Upper Givetian sandstone turbidites, which herald Early Variscan tectonic activity. Volcanism was active in the Upper Cambrian (felsic), Lower Ordovician, Middle Silurian and Lower Devonian (the latter three mafic). The tectonic setting is clearly intra-plate.

Taken altogether, the Barrandian Palaeozoic sequence is another example of early Palaeozoic rifting. It is uncertain whether the Barrandian basin was continuous with the Saxothuringian basin or the Palaeozoic of the Moldanubian s.str., or whether it represents a separate rift structure.

The basement underlying the Palaeozoic sequence is composed of meta-sediments and volcanics of Proterozoic age, which have undergone a Cadomian deformation and metamorphism and, later, a varying degree of Variscan basement reactivation (see below). The Proterozoic age is inferred from the Lower Cambrian transgressive onlap and locally confirmed by palynological data (Konzalova 1981). The most characteristic feature is the abundance of metamorphosed mafic rocks. Geochemical studies have revealed a broad spectrum of tectonic affinities (suboceanic mantle, ocean-ridge, calc-alkaline and within-plate; see Machart 1984, Misar 1984, Kastl and Tonika 1984, Jakes and Waldhauserova 1987, Uhlig in KTB 1986, Schüssler et al. 1986).

Teufel (1987) has detected, within the ZEV, Early Palaeozoic mafic and felsic intrusives. The presence within the ZEV/ZTT units of Palaeozoic volcanic or sedimentary rocks appears unlikely, yet cannot be excluded at present.

4.1.2 Moldanubian s.str.

The Moldanubian s.str. is mainly composed of monotonous sequences of clastic sediments with occasional intercalations of volcanics, marbles and graphitic schists. Since the Moldanubian rocks have suffered metamorphism up to granulite grade, previous subdivisions were mainly based upon litho-stratigraphic correlations (e.g. Stettner 1975). It has been assumed by most authors that the Moldanubian rocks represent Precambrian basement. However, an increasing number of palaeontological findings (mainly palynological data) prove the presence of large amounts of Palaeozoic protoliths, ranging in age between the basal Cambrian and Silurian (Andrusov and Corna 1976, Gunia 1985, Pacltova 1981, Pflug and Reitz 1987, Reitz et al. 1988).

This is in accord with the occurrence of detrital zircons affected by thermal events at around 560 or 530 Ma, which are contained in Moldanubian clastic metasediments (Grauert et al. 1973, Teufel 1987).

Though the presence of a continental basement can be inferred from the shallow-water character of the Palaeozoic quartzites and carbonates, it must be stated that there is as yet no reliable evidence of pre-Palaeozoic rocks. The Palaeozoic rocks are so widely distributed over the area that they must be regarded at least as a major constituent of the Moldanubian s.str.

Though the thick sequences of Palaeozoic clastic rocks are in accord with the concept of Early Palaeozoic rifting, it is uncertain whether the Moldanubian Palaeozoic represents a separate structure, or whether it was connected with the Barrandian and/or Saxothuringian basins.

4.2 Main Tectonic Structures and Metamorphism

The main difference in the tectono-metamorphic records of the ZTT/ZEV and the Moldanubian s.str. lies in their metamorphic character (see contribution by Blümel). The ZEV/ZTT is characterized by medium-pressure assemblages with occasional high-pressure relics, whereas most of the Moldanubian is dominated by low-pressure metamorphic facies. Some relict kyanite inclusions may be taken to suggest that at least part of the Moldanubian has undergone a medium-pressure phase prior to the installment of the low-pressure facies.

Radiometric studies on the ZEV/ZTT rocks (Fischer et al. 1968, Teufel et al. 1986, Teufel 1987, Schüssler et al. 1986) have revealed ages around 390 Ma for the medium-pressure metamorphism, which is identical with that of the Münchberg gneisses.

The low-pressure metamorphism of the Moldanubian s.str. (and the Saxothuringian of the Fichtelgebirge) has been dated at about 320 Ma (Teufel 1987). A Rb/Sr-WR age on Moldanubian rocks east of the ZEV of approx. 385 Ma possibly reflects the earlier medium-pressure stage, which would then be time equivalent to that of the ZEV/ZTT rocks.

A still older event is reflected in U/Pb and Rb/Sr ages between 450 and 490 Ma, reported by Gebauer (1975), Grauert et al. (1974), Köhler (1981), Köhler and Müller-Sohnius (1976, 1985) and Gorochov et al. (1983). Since orogenic metamorphism at that time is not compatible with the continuous records of Cambro-Ordovician sedimentation in the neighbouring areas, these data might be taken to reflect rift-related metamorphism at deeper crustal levels (see the discussion in Weber 1984).

From the structural relationships around the drilling site (see below and the contribution by Weber et al.) it can be inferred that the ZEV/ZTT unit rests upon the Moldanubian s.str., i.e. Proterozoic rocks overlie a unit containing Palaeozoic sediments. Hence, the ZEV/ZTT must represent a nappe, as it had already been envisaged by Blümel (1983).

A ductile mylonite at the eastern boundary of the ZEV against the Moldanubian s.str., formed under amphibolite-facies conditions, has been dated at approx. 330 Ma (Teufel 1987), which closely pre-dates or coincides with the time of low-pressure metamorphism. It is probable, therefore, that the overthrusting of the ZEV/ZTT occurred immediately before or during the low-pressure metamorphism (320 Ma, see above), though a pervasive re-equilibration of the medium-pressure assemblages of the ZEV/ZTT is nowhere observable. A basal heating of the ZEV is indicated by a resetting of K/Ar hornblende ages at its eastern margin from around 390 to 320-330 Ma, which might reflect the low-pressure metamorphism , but could also be due to the intrusion of the nearby granites.

The subvertical shear zone has yielded evidence of sinistral movements (Heinicke 1987). If one assumes that the subvertical position of the mylonites is due to later refolding, and rotates the shear zone into its assumed original position (ZEV resting with a subhorizontal thrust contact on the Moldanubian s.str.), the tectonic fabrics indicate transport of the ZEV towards SW.

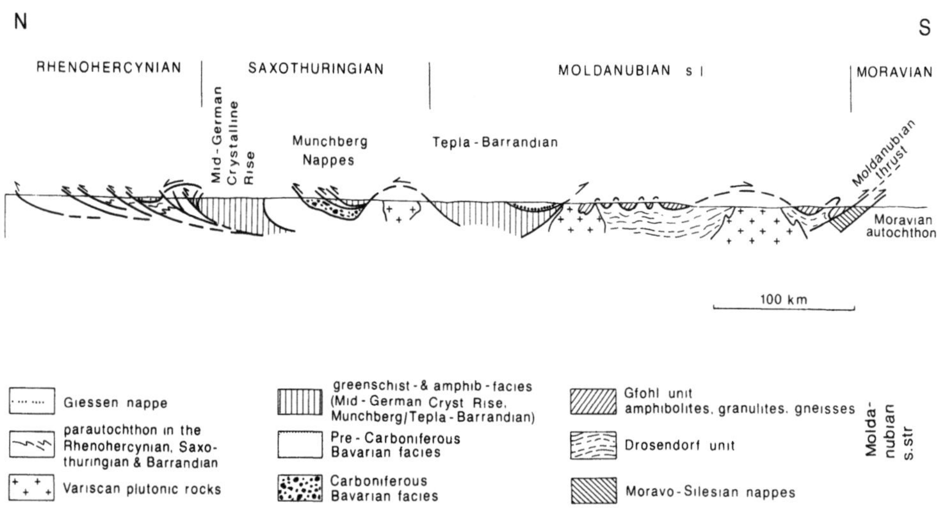

Fig.4: Diagrammatic cross-section through the Variscan Belt of Central Europe (after Behr et al., 1984).

The relationships between the ZEV/ZTT and the Moldanubian s.str. are, at present, insufficiently constrained and range high in the list of open questions.

Nappes are also present within the Moldanubian s.str.: granulite-facies rocks occur in the central and eastern parts of the area, where they rest structurally above rocks of lower metamorphic grades (Fig.4). The origin and the root zone(s) of the granulites are controversial (cf. Tollmann 1982 and Fuchs 1986). One plausible root zone lies at the SE margin of the Barrandian Syncline, where later the Central Bohemian Pluton was emplaced (see Fig.2, and Tollmann 1982). Southeastward directed tectonism is also reflected in the vergence of folds in the Barrandian Palaeozoic, and in fabric data from ductile mylonites related to the emplacement of crystalline nappes (Matte et al. 1985,1987; Rajlich et al. 1986).

NW-dipping reflectors (SE-verging thrusts?) are also prominent in the seismic line Dekorp 4 (see contribution by Schmoll et al.)

A number of radiometric data suggest that the SE-directed movements occurred between approx. 340 and 320 Ma (see Table 1), i.e. in Late Lower Carboniferous time. These movements correspond, in their direction as well as in their age, with the SE-verging D2 deformation characteristic of the Saxothuringian and the NW part of the Moldanubian s.str. (see above). If one accepts that the SE-derived Devonian and Carboniferous flysch clastics, and the metamorphism of the Münchberg and related units are early expressions of the NW-driving D1 deformation (caused by southeastward subduction), it appears that D1 is generally the older event, and the SE-facing D2 deformation has been caused by a general change of tectonic polarity.

The exact directions and the timing of thrusting events within the Moldanubian s.str. are another important item of future investigations.

Nevertheless, it can be stated that the Tepla-Barrandian/ZEV unit has undergone northwestward transport on its NW side, and appears to be involved in (grossly) S-directed movements on the SE (Figs.1,4). Hence, it occupies the central-Variscan zone of structural parting, which can be traced throughout the fold belt (see section 2).

Table 1: Timing of (grossly) S-facing folds and thrusts in the Bohemian Massif, as constrained by structural unconformities, post-tectonic plutons and radiometric ages. Positions of the Lower/Upper Carboniferous boundary after Harland et al. (1982) and Odin (1982, below). Tip of arrows marks minimum age of deformation indicated by post-tectonic granite intrusions. Horizontal extent of boxes gives margins of error.

LOWER- | UPPER CARBONIFEROUS

LOWER- | UPPER CARBONIFEROUS

350 — 300

Frankenwald, unconformity		≥ 333 resp. ≥ 320
Fichtelgebirge, deform. pre-granite		≥320
Saxoth.-Mold. boundary region, syn-D2 metam. U-Pb monazite (Teufel 1988)		ca. 320
ZEV-Mold. shear-zone U-Pb xenotim (Teufel 1988)		333 ± 3
Barrandian, S-ward thrust pre-granite (vanBreemen et al. 1982)		≥ 331 ± 4
Central-Bohem. granulites, thrust	U-Pb zir	341 ± 4
(vanBreemen et al. 1982)	U-Pb mon	338 ± 3
	Rb-Sr ms	331 ± 5
South-Bohem. granulites, thrust		
Matte et al. 1985	Ar-Ar bio	323 ± 7
metam.: Schenk & Todt 1985	U-Pb mon	340 ± 1

5. The Saxothuringian/Moldanubian Boundary Region in the Vicinity of the KTB Drilling Site

Unfortunately, the Saxothuringian/Moldanubian boundary is veiled along most of its extent by the Tertiary volcanics and sediments of the Eger Graben and by Late Variscan granites (see below and Fig.2). Further west, it is necessary to distinguish between two segments of the boundary region, which are separated by an important NW/SE-trending block fault throwing down to the SW. This fault can be traced from Bad Berneck (at the SW corner of the Münchberg Klippe) across the western part of the Fichtelgebirge Antiform (Fichtelnaab valley) into the Moldanubian Region, where it diverticulates and finally gets lost in the post-tectonic granites east of the ZEV.

West of the Fichtelnaab fault, the very low-grade rocks in the mantle of the Fichtelgebirge are overthrust by a greenschist unit with lenses of serpentinite. Its position and geochemical character (Schüssler et al. 1986) invite comparison with the "Phyllite-Prasinite Sequence" of the Münchberg Klippe. The greenschists have been overthrust by the ZEV amphibolites and gneisses. The contact between the greenschists and the Palaeozoic sediments is a brittle shear zone with formation of tectonic melange, which corresponds to similar tectonic surfaces in the Münchberg Klippe (Franke 1984 b). The situation west of the Fichtelnaab fault also pertains to the KTB drilling site (Fig.3 and contribution by Weber and Vollbrecht).

East of the Fichtelnaab fault, a deeper structural level is exposed. The ZEV/ZTT nappe has been eroded away, so that the Saxothuringian rocks in the north are in direct contact with the Moldanubian s.str.. Both of these units have been affected by the Lower Carboniferous low-pressure metamorphism and the D2, D3 and D4 phases of deformation. In fact, the outcrop of Moldanubian rocks which separates the ZEV and ZTT represents a vast NS-trending D4 antiform (Stein 1987).

The Saxothuringingian/Moldanubian boundary is contained, in this area, within a subvertical shear zone of several kilometres width, which was formed during the low-temperature metamorphism, and later underwent diaphtoritic overprinting (de Wall 1987). This shear belt is referred to as the zone of Tirschenreuth/Mähring (ZTM).

Seismic data evaluated in the contributions by Schmoll et al. and by Weber and Vollbrecht suggest that the primary contact between the Saxothuringian and the Moldanubian s.str., at greater depths, represents a SE-directed backthrust. The subvertical shear zone exposed at the surface is probably a later feature, which masks the original contact.

6 Late- and Post-Variscan Structures and Events

The central and southern parts of the target area are intruded by an important volume of post-tectonic granites. Their intrusion started immediately after the low-pressure metamorphism and the accompanying deformation, and range in age between approx. 320 and 280 Ma (Besang et al 1976, Köhler and Müller-Sohnius 1976, Lenz 1986, Richter and Stettner 1979). One larger group of granites occupies the core of the D3-Antiform of the Fichtelgebirge. Other granite bodies appear to follow NW/SE-directed fracture zones. In the Flossenbürg and Bürgerwald granites (both within the Moldanubian s.str.), the spatial array of trace-element zoning suggests platelike bodies, which possibly represent parts of laccolithic intrusions (Ackermann and Tavakkoli 1986).

The NW/SE alignment of some of the granites reflects a system of block faults, which was formed in Upper Carboniferous/Permian time, and reactivated repeatedly until Recent (Bachmann et al. 1987, Schröder 1987). One of these faults is the "Franconian Line" (Fig.2), along which the Permian through Tertiary rocks of the western foreland have been downfaulted for locally more than 1000 m (see also the evaluation of seismic data in the contributions by Schmoll et al. and Weber and Vollbrecht). Several Late Carboniferous/Permian depocentres appear to be aligned along that fault (e.g. Emmert 1981). It is possible that these basins represent pull-apart features associated with the period of dextral transpression, which has affected Variscan Europe in the terminal phase of Variscan plate collision (Arthaud and Matte 1977).

The main movement along these NW/SE faults, at least during the Mesozoic and later, is a more or less vertical downthrow to the SW. This also applies to the Fichtelnaab fault (see above), and the Kulmbach fault, which parallels the Franconian Line within the Mesozoic foreland. The Franconian Line, as well as the Fichtelnaab fault, dip steeply to the NE, as is observable in surface outcrops. Contrasting interpretations are presented in the contributions by Schmoll et al. and by Weber and Vollbrecht).

Another important NW/SE element is the Bavarian "Pfahl" (Fig.2), a dextral shear zone with ductile and subsequent cataclastic deformation and extensive quartz mineralization, which was still active in Triassic time (Horn et al. 1983). A similar quartz vein, the Bohemian "Pfahl", accentuates the SW margin of the ZTT in Czechoslovakia.

The youngest tectonic element in the target area is the Eger Graben (Fig.2), which roughly follows the Saxothuringian/Moldanubian boundary along approx. 200 km, and can be regarded as a late reactivation of the Variscan suture zone. The Eger Graben is characterized by Miocene sediments and a number of basaltic pipes and lava flows, which can be traced across the Franconian Line into the Mesozoic foreland.

7 Conclusions

The position of the KTB drilling site is representative of the Variscan Belt in Europe, in that it probes into an Early Palaeozoic rift basin, which was formed within the Cadomian basement of Gondwana, and then closed during the Late Ordovician through Carboniferous period of plate convergence.

The drilling site is situated (see Fig.5) in the suture zone between the Saxothuringian basin to the north and the Moldanubian (Tepla/Barrandian)

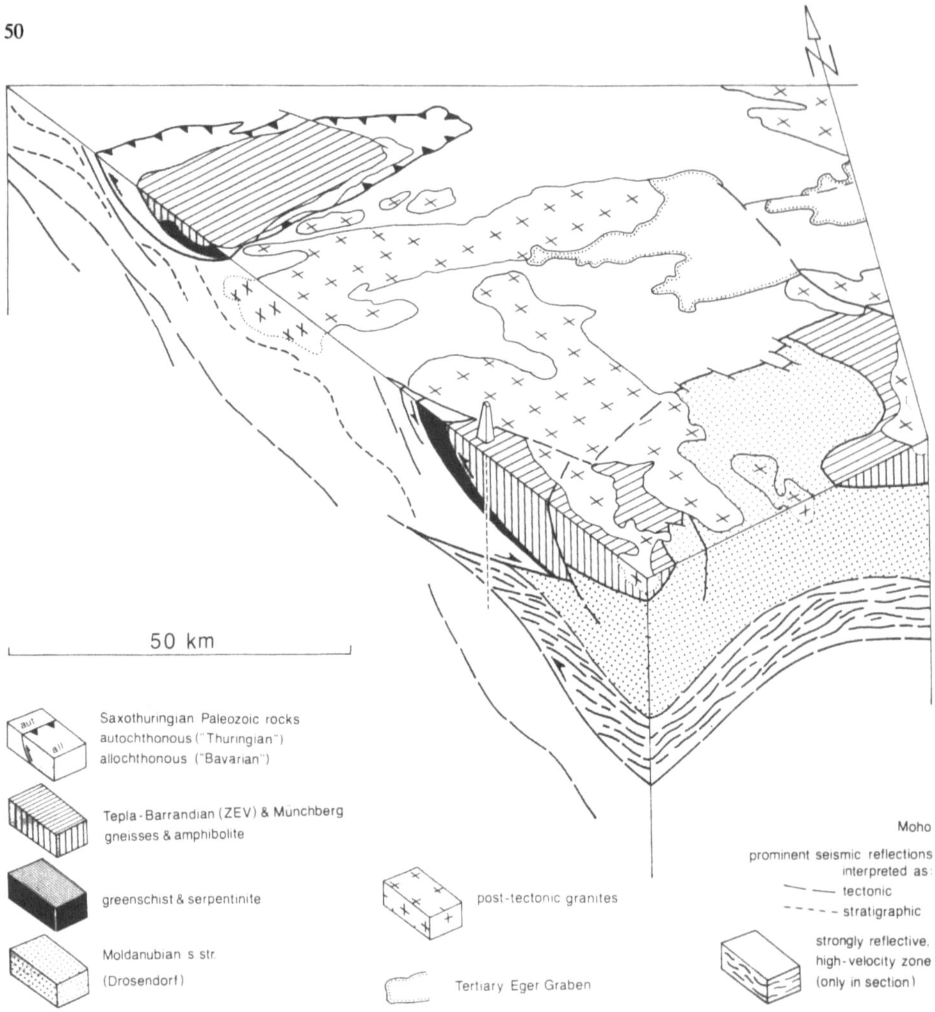

Fig.5: Diagrammatic 3-D representation of the crustal structure around the KTB drilling site. Depth to Moho is 11-12 sec TWT (approx. 33-36 km).

basement in the south, which itself has been reactivated by Variscan deformation and metamorphism.

After the emplacement of reactivated basement rocks (ZTT/ZEV, Münchberg gneisses) towards the NW over the infill of the Saxothuringian basin, the suture zone has been affected by low-pressure metamorphism, SE-directed backthrusting and multiple refolding.

The precise positioning of the drilling site is constrained by Late-Variscan granite intrusions and Post-Variscan block-faulting.
Seismic studies (contributions by Schmoll et al. and Weber and Vollbrecht) have largely confirmed the structural model deduced from the surface data, and added new important structural features. The most prominent new element is the "Erbendorf Body" with its pronounced reflectivity and high seismic velocities, which probably marks the Saxothuringian/Moldanubian suture at depth, below the level of backthrusting (see Fig.5).

The KTB site occupies a key position within this complex array of tectono--stratigraphic units, thrust surfaces and metamorphic facies, and will hopefully recover a representative cross-section through a Variscan suture zone.

REFERENCES

Adam,J.-F., and Reuter,A.,1981, Das Paläozoikum von Erbendorf (NE-Bayern). Ergebnisse einer Neukartierung. Unveröff. Dipl.-Arbeit Univ. Göttingen, 63 p.

Ackermann,H., and Tavakkoli,B., 1986, Granite - Form und räumliche Lage im metamorphen Grundgebirge (abs.): Jahrestagung Geologische Vereinigung, Giessen, p. 1.

Andrusov,D., and Corna,O., 1976, Über das Alter des Moldanubikums nach mikrofloristischen Forschungen: Geol.prace, spr. 65, p. 81-89.

Arthaud,F. and Matte,Ph., 1977, Late Palaeozoic strike-slip faulting in southern Europe and northern Africa: result of a right lateral shear-zone between the Appalachians and the Urals: Bulletin Geological Society of America, v.88, p.1305-1320.

Bachmann,G.H., Müller,M. and Weggen,K., 1987, Evolution of the Molasse Basin (Germany, Switzerland): Tectonophysics, 137(1987), p. 77-92.

Behr,H.-J., Engel,W. and Franke,W., 1982, Variscan Wildflysch and Nappe Tectonics in the Saxothuringian Zone (Northeast Bavaria, West Germany): American Journal of Science, v.282, p. 1438-1470.

Behr,H.-J., Engel,W., Franke,W., Giese,P. and Weber,K., 1984, The Variscan Belt in Central Europe: Main structures, geodynamic implications, open questions: Tectonophysics, v.109, p. 15-40.

Besang,C., Harre,W., Kreuzer,H., Lenz,H., Müller,P., and Wendt,I., 1976, Radiometrische Datierung, geochemische und petrographische Untersuchungen der Fichtelgebirgsgranite: Geologisches Jahrbuch, v. E 8, p. 3-71.

Blümel,P., 1983, The western margin of the Bohemian Massif in Bavaria: Fortschritte Mineralogie, 61, 2, p. 171-195.

Blümel,P., 1986, Metamorphic processes in the Variscan Crust of the Central Segment, in Freeman,R., Mueller,St.and Giese,P., eds.: Proceedings of the 3rd EGT workshop, Bad Honnef: European Science Foundation, p. 149-155.

Bogacz,W. and Krokowski,J., 1983, Rotation ot the Upper Silesian coal basin (abs.): 10. Congreso Internacional de Estratigrafia y Gelogica del Carbonifero, Madrid, p. 239.

Chlupac,I., 1968, Early Paleozoic of the Bohemian Massif. Guide to excursion 11 AC, Int.Geol.Congr. Prague 1968, 43 p.

de Wall,H., 1987, Mikrostrukturelle Untersuchungen und strukturgeologische Kartierung der Gesteine aus dem Übergangsbereich Moldanubikum/Saxothuringikum W' Mähring, Oberpfalz. Unveröff. Dipl.-Arbeit Univ. Göttingen, 76 p.

Ducrot,J., Lancelot,J.R., and Marchand,J., 1983, Datation U-Pb sur zircons de l'éclogite de la Borie (Haut-Allier, France) et consequences sur l'évolution anté-hércynienne de l'Europe occidentale: Earth Planetary Science Letters 622, p. 385-394.

Emmert,U., 1981, Perm nördlich der Alpen, in Erläuterungen zur Geologischen Karte von Bayern 1:500000, 3.Auflage, Bayerisches Geologisches Landesamt München, p. 34-40.

Engel,W. and Franke,W., 1983, Flysch-sedimentation: Its relations to tectonism in the European Variscides, in Martin,H. and Eder,F.W., eds., Intracontinental Fold Belts, Springer, p. 290-321.

Fischer,G., Schreyer,W., Troll,G., Voll,G., and Hart,S.R., 1968, Hornblendealter aus dem ostbayerischen Grundgebirge: Neues Jahrbuch Mineralogie Mitteilungshefte, v.11, p. 385-404.

Franke,W., 1984a, Variszischer Deckenbau im Raume der Münchberger Gneismasse - abgeleitet aus der Fazies, Deformation und Metamorphose im umgebenden Paläozoikum: Geotektonische Forschungen, v.68, p. 1-253.

Franke,W., 1984b, Late events in the tectonic history of the Saxothuringian Zone, in Hutton,D.W.H. and Sanderson,D.J., (eds.), Variscan tectonics of the North Atlantic region: Blackwell Scientific Publications, p. 33-45.

Franke,W., (in press), Variscan plate tectonics in Central Europe - current ideas and open questions. Tectonophysics.

Franke,W., and Engel,W., 1986, Synorogenic sedimentation in the Variscan Belt of Europe: Bulletin Société Géologique France, 1986,1, p. 25-33.

Franke,W. and Engel,W., 1988, Tectonic settings of synorogenic sedimentation in the Variscan Belt of Europe. In: Besley,B.M. and Kelling,G. (eds.), Sedimentation in a Synorogenic Basin Complex - The Upper Carboniferous of NW Europe. Blackie, Glasgow and London, p. 8-17.

Franz,G., Thomas,S., and Smith,D.C., 1986, High-pressure phengite decomposition in the Weissenstein eclogite, Münchberger Gneiss Massif, Germany: Contributions Mineralogy and Petrology, v.92, p.71-85.

Fuchs,G., 1986, Zur Diskussion um den Deckenbau der Böhmischen Masse. Jahrbuch Geologische Bundesanstalt, 129/1, p. 41-49.
Gebauer,D., 1975, Rb-Sr Gesamtgesteins- und Mineralsysteme, sowie U-Pb Systeme in Zirkonen während der progressiven Gesteinsmetamorphose: (Ph.D. Thesis), ETH Zürich, 1975.
Gebauer,D. and Grünenfelder,M., 1979, U-Pb zircon and Rb-Sr mineral dating of eclogites and their country rocks, Example: Münchberger Gneiss Massif, Northeast Bavaria: Earth and Planetary Science Letters, 42, p. 35-44.
Gorochov,I.M., Melnikov,N.N., Varsavskaja,E.S., and Kutjavin,E.P., 1983, Rb-Sr dating of magmatic and metamorphic events in the eastern part of the Bohemian Massif: Casopis pro mineralogii a geologii, 1983, 28, p. 349-361.
Grauert,B., Hänny,R. and Soptrajanova,G., 1973, Age and Origin of Detrital Zircons from the Pre-Pemian Basement of the Bohemian Massif and the Alps: Contributions Mineralogy and Petrology, v.40, p. 105-130.
Grauert,B., Hänny,R., and Soptrajanova,G., 1974, Geochronology of a Polymetamorphic and Anatectic Gneiss Region: The Moldanubicum of the area Lam-Deggendorf, Eastern Bavaria, Germany: Contributions Mineralogy and Petrology, v.45, p. 37-63.
Gunia,T., 1985, Biogenic Structures in Crystalline Limestones from Ledec on the Sasava River (Czechoslovakia-Moldanubicum). Preliminary Studies: Bulletin of the Polish Academy of Sciences, v.33, no.3-4, 1985, p. 102-106.
Harland,W.B., COX,A.V., Llewellyn,P.G., Pickton,C.A.G., Smith,A.G., and Walters,R., 1982, A geologic time scale: Cambridge University Press, Cambridge, London, New York, New Rochelle, Melbourne, Sidney, 131 p.
Havlicek,V., 1981, Development of a linear sedimentary depression exemplified by the Prague Basin (Ordovician-Middle Devonian; Barrandian area-central Bohemia): Journal of geological sciences (Journal geologickych ved), 1981, p. 7-42.
Heinicke,F., 1987, Strukturgeologische Untersuchungen im Übergangsbereich zwischen der Zone Erbendorf-Vohenstrauß und dem Moldanubikum im Gebiet Vohenstrauß/Leuchtenberg, Oberpfalz: Dipl.-Arbeit Univ. Göttingen, 84 p.
Horn,P., Köhler,H. and Müller-Sohnius,D., 1983, A Rb/Sr-WR-isochron ("fluid inclusion")-age of the Bayerische Pfahl, Eastern Bavaria: terra cognita, 3, (2-3), p. 199.
Jakes,P., and Waldhauserova,J., 1987, Orogenic sequences in the upper Proterozoic of the Bohemian Massif (abs.): Proterozoic geochemistry (IGCP 217), Lund 1987, p.14.
Kastl,E., and Tonika,J., 1984, The Marianske Lazne metaophiolite complex (West Bohemia): Krystalinikum, 17, 1984, p. 59-76.
Köhler,H.,1981, Rb-Sr Altersbestimmung und Sr-Isotopensystematik an Gesteinen des Regensburger Waldes (Moldanubikum NE Bayern) (Habilitation Thesis): Universität München, 1981.
Köhler,H. and Müller-Sohnius,D., 1976, Ergänzende Rb-Sr Altersbestimmung an Mineral- und Gesamtgesteinsproben des Leuchtenberger und des Flossenbürger Granits (NE-Bayern): Neues Jahrbuch Mineralogie, Mitteilungshefte, 1976, 8, p. 354-365.
Köhler,H., and Müller-Sohnius,D., 1985, Rb-Sr-Altersbestimmungen und Sr-Isotopensystematik an Gesteinen des Regensburger Waldes (Moldanubikum NE-Bayerns), Teil 1: Paragneisanatexite. Neues Jahrbuch Mineralogie Abhandlungen, 151,1: p. 1-28.
Konzalova,M., 1981, Some Late Precambrian microfossils from the Bohemian Massif and their correlation: Precambrian Research, 15, p. 43-62.
Kossmat,F., 1927, Gliederung des varistischen Gebirgsbaues: Abhandlungen Sächsisches Geologisches Landes-Amt, v.1, 39 p.
KTB 1986, Kontinentales Tiefbohrprogramm der Bundesrepublik Deutschland, Ergebnisse der Vorerkundungsarbeiten Lokation Oberpfalz, Weber,K. and Vollbrecht,A., (eds.), 186 p.
Lenz,H., 1986, Rb/Sr-Gesamtgesteins-Altersbestimmung am Weißenstadt-Marktleuthener Porphyrgranit des Fichtelgebirges: Geologisches Jahrbuch, v.E 34, p. 67-76.
Machart,J., 1984, Ultramafic rocks in the Bohemian part of the Moldanubicum and Central Bohemian Islet Zone: Krystalinikum, 17, 1984, p. 13-32.
Misar,Z., 1984, Ophiolites and related rocks of Czechoslovakia and their correlation: Krystalinikum, 17, 1984, p. 7-11.
Matte,P., 1986a, La chaine varisque parmi les chaines paléozoiques péri-atlantiques, modèle d'évolution et position des grands blocs continentaux au Permo-Carbonifère: Bulletin Société Géologique France, 1986, 1, p.9-24.

Matte,P., 1986b, Tectonics and plate tectonics model for the Variscan Belt of Europe: Tectonophysics, v.126, p. 329-374.
Matte,Ph., Maluski,H., and Echtler,H., 1985, Cisaillements ductiles varisques vers l'Est-Sud-Est dans les nappes du Waldviertel (Sud-Est du Massif de Bohème, Autriche). Données microtectoniques et radiométriques 39Ar/40Ar: Comptes Rendus Académie Sciences Paris, t.301, série II, no 10, p. 721-726.
Matte,Ph., Rajlich,P. and Lacassin,R., 1987, Variscan ductile thrusting and wrenching in the Bohemian Massif: strain pattern preliminary result: terra cognita, v. 7, no. 2-3, p. 173.
Müller-Sohnius,V., v.Drach,V., Horn,P., and Köhler,H., 1987, Altersbestimmungen an der Münchberger Gneismasse, NE Bayern, Neues Jahrbuch Mineralogie Abhandlungen 156, p. 175-206.
Odin,G.S., 1982, The Phanerozoic time scale revisited: Episodes, 1982, 3, p. 3-9.
Pflug,H.D., and Reitz,E., 1987, Palynology in metamorphic rocks: indication of early land-plants: Naturwissenschaften, 74, p. 386-387.
Rajlich,P., Synek,J., Sarbach,M. and Schulmann,K., 1986, Hercynian thrust-related shear zones and deformation of the Varied Group on the contact of granulites/Southern Moldanubikum, Bohemian Massif: Geologische Rundschau, v.75, no.3, p. 665-683.
Reitz,E., Pflug,H.-D. and Franke,W., 1988, Biostratigraphie im Kristallin. Kontinentales Tiefbohrprogramm der Bundesrepublik Deutschland, 1. KTB-Schwerpunkt-Kolloquium Giessen 1988, Zusammenfassung der Beiträge, p. 32.
Richter,P., and Stettner,G., 1979, Geochemische und petrographische Untersuchungen der Fichtelgebirgsgranite: Geologica Bavarica, v.78, 144 p.
Schröder,B., 1987, Inversion tectonics along the western margin of the Bohemian Massif: Tectonophysics, 137(1987), p. 93-100.
Schüssler,U., Oppermann,U., Kreuzer,H., Seidel,E., Okrusch,M., Lenz,K.-L., and Raschka,H., 1986, Zur Altersstellung des ostbayerischen Kristallins - Ergebnisse neuer K-Ar-Datierungen: Geologica Bavaria, 89, p. 21-47.
Söllner,F., Köhler,H. and Müller-Sohnius,D., 1981a, Rb-Sr-Altersbestimmungen an Gesteinen der Münchberger Gneismasse, NE-Bayern, Teil 1, Gesamtgesteinsdatierungen: Neues Jahrbuch Mineralogie Abhandlungen, 141, 1, p. 90-112.
Söllner,F., Köhler,H. and Müller-Sohnius,D., 1981b, Rb-Sr-Altersbestimmungen an Gesteinen der Münchberger Gneismasse, NE-Bayern, Teil 2, Mineraldatierungen: Neues Jahrbuch Mineralogie Abhandlungen, 142, 2, p. 178-198.
Stein,E., 1987, Die strukturgeologische Entwicklung im Übergangsbereich Saxothuringikum/Moldanubikum in NE Bayern. Diss. Univ. Göttingen, 125 pp.
Stettner,G., 1975, Zur geologisch-tektonischen Entwicklung des Oberpfälzer Grundgebirges: Der Aufschluß, Sonderband 26 (Oberpfalz): p. 11-38.
Stosch,H.-G., and Lugmair,G.W., 1986, Geochemistry and evolution of eclogites from the Münchberg Gneiss Massif/ W.Germany (abs.): terra cognita, 6, 2, p. 254.
Stosch,H.-G., and Lugmair,G.W., 1987, Geochronology and geochemistry of eclogites from the Münchberg Gneiss Massif, FRG: terra cognita v.7, no.2-3, p. 163.
Suess,F.E., 1912, Vorläufige Mitteilungen über die Münchberger Deckscholle: Sitzungsber. Akademie der Wissenschaften Wien, Mathematisch-Naturwissenschaftliche Klasse, v.121, no. 10, 253 p.
Teufel,S., 1987, Vergleichende U-Pb- und Rb-Sr-Alterbestimmungen an Gesteinen des Übergangsbereiches Saxothuringikum/ Moldanubikum, NE-Bayern. Diss. Univ. Göttingen, 110 pp.
Teufel,S., Ahrendt,H. and Hansen,B.T., 1986, Die Beziehung der Zone von Erbendorf-Vohenstrauss zum Moldanubikum aufgrund radiometrischer Altersbestimmungen (abs.): Jahrestagung Geologische Vereinigung, Giessen 1986, p. 75.
Tollmann,A., 1982, Großräumiger variszischer Deckenbau im Moldanubikum und neue Gedanken zum Variszikum Europas: Geotektonische Forschungen, 1982, 64, p.1-91.
Wagener-Lohse,C. and Blümel,P., 1984, prograde Metamorphose vom Niederdruck-Typ in der Grenzzone Saxothuringikum/Moldanubikum E' Tirschenreuth (NE-Bayern): Fortschritte Mineralogie, 62, Beiheft 1, p. 254-255.
Weber,K., 1984, Variscan events: early Palaeozoic continental rift metamorphism and late Palaeozoic crustal shortening, in Hutton,D.H.W., and Sanderson,D.J., eds., Variscan tectonics of the North Atlantic Region: Geological Society Special Publication, 14, p. 3-22.

Weber,K., and Behr,H.-J., 1983, Geodynamic interpretation of the Variscides, in Martin,H., and Eder,F.W., eds., Intracontinental Foldbelts: Springer, 1983, p. 427-469.
Wirth,R., 1978, Geochemie und Petrographie der paläozoischen Magmatite des Frankenwaldes. Diabase-Keratophyre-Pikrite. Diss. Univ. Würzburg, 130 p.
Ziegler,P.A., 1984, Caledonian and Hercynian crustal consolidation of Western and Central Europe - a working hypothesis: Geologie en Mijnbouw, 63, p. 93-108.
Ziegler,P.A., 1986, Geodynamic model for the Palaeozoic crustal consolidation of Western and Central Europe: Tectonophysics, 126, p. 303-328.
Zwart,H.-J., 1967, The duality of orogenic belts: Geologie en Mijnbouw, 46, p. 283-309.
Zwart., and Dornsiepen,U.F., 1980, The Variscan and pre-Variscan tectonic evolution of Central and Western Europe; a tentative model, in Cogné,J. and Slansky,M., eds., Geology of Europe, Colloque C6, 26th International Geological Congress, Paris 1980, p. 226-232.

Geochronology of the Moldanubian-Saxothuringian Transition Zone, Northeast Bavaria

B. T. Hansen[1], S. Teufel[2] and H. Ahrendt[2]

[1] Zentrallabor für Geochronologie, Corrensstr. 24, 4400 Münster, FRG
[2] Institut für Geologie und Dynamik der Lithosphäre, Goldschmidtstr. 3, 3400 Göttingen, FRG

1. Introduction

By the time the transition zone between the Saxothuringian and the Moldanubian units in northeastern Bavaria was selected as a possible location for the German continental deep-drilling project (KTB), great differences with respect to the interpretation of the geological evolution of the area were obvious, differences which had important consequences for the geochronological evolution of this area. Stettner (1975,1979) assumed that the high-grade Moldanubian gneisses were metamorphosed during Cadomian time, whereas the Saxothuringian sediments of Early Paleozoic age were supposed to have been subjected to an Early Hercynian reworking. In contrast, Schreyer (1966) proposed that the Saxothuringian and the Moldanubian units had suffered the same low-pressure metamorphism, which would mean that at least the last metamorphic overprinting of both units should be of Hercynian age. This was confirmed by structural studies, which showed a continuous structural development from the Moldanubian into the Saxothuringian (Stein 1987). The structural model of the transition zone between Moldanubian and Saxothuringian is also in agreement with the petrological studies of Wagener-Lohse and Blümel (1984) and Blümel (this Vol.).
However, the relationship between the nappe structure of the Münchberger Gneismasse (Vollbrecht 1981; Behr et al. 1982; Franke 1984) and the Zone of Erbendorf-Vohenstrauß (ZEV) was still an open question. Voll (1960) interpreted the ZEV as Saxothuringian, in contrast to Stettner (1975, 1979) who regarded it as the basement for the Bavarian Moldanubian.
An upper age limit for the metamorphic evolution is fixed by the age of the post-metamorphic granitic massifs, which were intruded in Upper Carboniferous times.
This chapter deals with data mainly obtained during the pre-site-studies, which so far are unpublished, or are only available as internal reports of the continental deep-drilling program and in the annual reports of the Zentrallabor für Geochronologie (ZLG) at Münster.

2. Previous Geochronological Investigations

2.1 Moldanubian and Saxothuringian Units

The first geochronological investigations in this area were made by Davis and Schreyer (1962). They obtained Rb-Sr ages in the range of 320-330 Ma on biotites from Moldanubian metasediments which they interpreted as the age of regional metamorphism during Hercynian times. K-Ar ages of hornblendes from several Moldanubian amphibolites dated by Fischer et al. (1968) also showed Hercynian ages.

However, the gabbro-amphibolite of the Neukirchener Massif in the Bavarian Forest showed a significantly higher age of 385 ± 12 Ma. Grauert et al. (1973) published U-Pb data obtained on detrital zircons from Cambro-Ordovician quartzites belonging to the Saxothuringian as well as on zircons from the Moldanubian Osser Quartzite. The data points of hypidiomorphic zircons plot close to the concordia at 600 Ma, whereas data obtained on rounded zircons plot over a wide range in the concordia diagram. Taken together, the data points define a discordia with intercepts at 2260 and 560 Ma. Both ages are regarded as inherited, and are thought to represent ages of orogenic processes within the source area for the sediments, as the growth of new zircons in quartzite can be excluded.
Ages interpreted as related to a Caledonian anatectic event were obtained by Rb-Sr whole rock analyses. Isochron ages in the range of 430 - 490 Ma were obtained on Moldanubian anatexites and diatexites (Grauert et al. 1973, 1974; Gebauer 1975; Köhler and Müller-Sohnius 1980, 1985). A strong Hercynian overprinting of the area is reflected in ages of about 320 Ma obtained on concordant zircons and monazites as well as from Rb-Sr isochrons on small slabs from Moldanubian paragneisses (Grauert et al. 1974, Tembusch and Grauert 1983, Schulz-Schmalschläger et al. 1984).

2.2 Nappe Units

The geochronological evolution of the "Münchberger Gneissmasse" (MM) is significantly different from that of the Moldanubian-Saxothuringian area. Rb-Sr determinations on augen gneisses of the "Liegendserie" yielded an age of 499 ± 20 Ma, which was interpreted as the age of the magmatic source rock (Söllner et al. 1981a). The sediments of the same series show a homogenization of the Sr-isotopes at about 490 ± 22 Ma ago which is, according to Söllner et al. (1981a), related to a Caledonian metamorphic event.
Gebauer and Grünenfelder (1979) interpreted U-Pb discordia ages of 525+40/-31 Ma and 380+14/-22 Ma obtained on zircons from the "Weißenstein" eclogite and an associated metagabbro as the time of formation and the high-pressure metamorphism respectively. The age of the high-pressure metamorphism was confirmed by Sm-Nd and Rb-Sr investigations carried out by Stosch and Lugmair (1986) on minerals formed during the high-pressure event, which yielded ages of 395 ± 4 Ma and 394 ± 14 Ma respectively.
These data are in agreement with previously published K-Ar and Rb-Sr mineral ages obtained on biotites, muscovites and phengites, which all fall in the range of 375-390 Ma (Söllner et al. 1981b; Gebauer and Grünenfelder 1979; Schüssler et al. 1986). Only a few determinations from the "Hangendserie" point to a later rejuvenation of the biotites (Söllner et al. 1981b).
A different interpretation of the geological evolution was given by Müller-Sohnius et al. (1987), who interpreted a Rb-Sr isochron age of 449 ± 37 Ma as belonging to the formation of the eclogites of Weißenstein and Martinsreuth. According to these authors, the time span between 435 and 370 Ma is only related to a cooling of these rocks.

2.3 Late to Post-Tectonic Granites

The late to post-tectonic granitic bodies in the investigated area have been subjected to numerous determinations, performed by the Rb-Sr method on whole rocks and on minerals, as well as by K-Ar dating on minerals (Köhler et al. 1974; Besang et al. 1976; Carl et al. 1985; Wendt et al. 1986; Lenz 1986; Kreuzer et al. 1988)

The Rb-Sr whole rock isochrons show that a suite of granites was intruded over a rather long period of time, starting at about 320 Ma (Weißenstadt-Marktleuthen and Leuchtenberg Granite) and ending 290 Ma ago (Zinngranite, Flossenbürg Granite). Age determinations on the micas, reflecting the closing temperatures, show a rather rapid cooling. Isotope data and geochemical investigations of the post-tectonic granitic bodies in the area point to a common source magma, which had a composition similar to that of the Leuchtenberg Granite (Köhler et al. 1974; Wendt et al. 1986).

3. Results of the Pre-Site Investigations

Based on the geochronological data obtained during the pre-site investigations for the German continental deep-drilling program a time evolution for the geological development was established, as well as a division into three distinct tectonometamorphic units, i. e. Moldanubian, Saxothuringian and the nappe units: Münchberger Gneismasse (MM) and the Zone of Erbendorf-Vohenstrauß (ZEV).

3.1 Moldanubian

Detrital zircons separated from several Moldanubian paragneisses show upper concordia intercepts in the range of 2400 - 2600 Ma (Fig. 1) (Teufel 1987). These ages are interpreted as the time of the primary crystallization of the zircons in the source region.

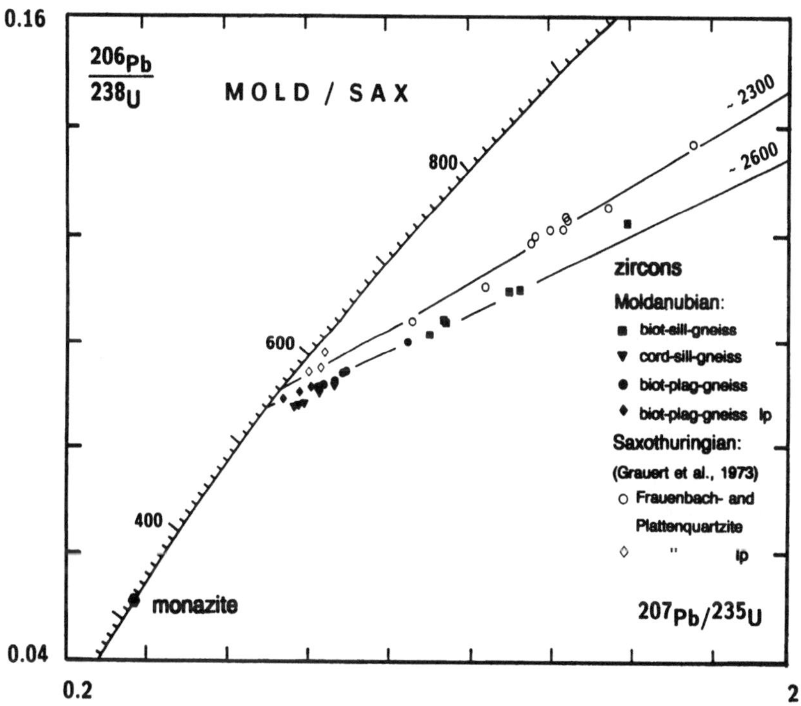

Fig. 1 Concordia diagram for zircons and monazites from paragneisses of the Moldanubian and the Saxothuringian (from: Teufel 1987)

The lower intercepts yield ages between 520 and 540 Ma. U-Pb analyses of the monazites of these samples give concordant ages at 320 Ma. As the zircon population consists mainly of detrital components and only a few grains show evidence of crystal growth, the lower intercept ages are also supposed to be inherited. Therefore the ages between 520 and 540 Ma are regarded as being a minimum age of a metamorphic event in the source area (Teufel 1987) (Fig. 1). In addition to the widespread monazite ages around 320 Ma (Grauert et al. 1974; Schulz-Schmalschläger et al. 1984; Teufel 1987), the analyses of monazites from one Moldanubian cordierite-sillimanite-gneiss show a concordant age of about 460 Ma. As there is as yet no evidence for this event in other gneiss samples the signifiance of this age for the metamorphic history of the area is still in question.
Rb-Sr data points of whole rock analyses on cordierte-sillimanite-gneiss samples scatter around a reference isochron of 390 Ma (Teufel 1987). At present, this data is best related to a medium pressure metamorphism, relics of which were found in the Moldanubian gneisses (Wagener-Lohse and Blümel 1984; Blümel 1984).
The last regional low-pressure metamorphic event is reflected in concordant monazite ages which all fall in the range of 320 - 325 Ma (Teufel 1987)

3.2 Saxothuringian

Based on U-Pb analyses on zircons Grauert et al. (1973) deduced an age of some 2000 - 2300 Ma for the source of the Saxothuringian sediments. The age is slightly lower than that claimed for the Moldanubian. However, this difference is not significant due to the long extrapolations and the influence of Hercynian low-pressure metamorphism on the Moldanubian zircons. Furthermore, Grauert et al. (1973) showed that the source region had been subjected to a metamorphic overprinting some 560 Ma ago.
In addition to the question of the metamorphic evolution within the area, the time for an early pre-metamorphic magmatic activity is also of importance. The widespreand "Epigneis" within the Fichtelgebirge represents just such an early type of igneous rock, which is intercalated with the Cambro-Ordovician sediments (Stettner 1981). The controversial origin of these rocks has been discussed by different authors (Gümbel 1879; Wurm 1925a, b 1961).
Zircons of a sample analyzed from this series yield U-Pb data which define a discordia with intercepts at 449 ± 4 Ma and 1960 +38/-40 Ma (Teufel 1987). As the epigneisses are interlayered with the Cambro-Ordovician sediments and as the zircon population exhibits a magmatic habit these rocks must be of volcanic origin. Therefore, the lower intercept age of 449 ± 4 is unlikely to represent the time of a metamorphic event. The Hercynian regional metamorphism which is dominant in this area has been dated by K-Ar analyses on synkinematic phengites, which yielded ages in the range of 290 - 310 Ma (Frentzel-Beyme 1985). The lower intercept age of the "Epigneis" zirkons can best be interpreted as the time of effusion which caused the formation of the volcanic rocks. The rather high proportion of inherited material must be related to assimilation of crustal material, a feature which so far has only been described for acid intrusive rocks (Gulson and Krogh 1973; Williams et al. 1983).
The U-Pb data from the monazites in this area show a concordant age of 320 Ma, which is consistent with the monazite ages from the Moldanubian. For the Saxothuringian monazites it can be shown that the age of 320 Ma is close to the peak of regional metamorphism, because monazites were found within the muscovite-sillimanite

zone but are absent in the andalusite zone. This means that monazite first appeared at temperatures in the range of 580 ± 50 °C, which is higher than the suggested closing temperature of 530 °C (Wagner et al. 1977).
Rb-Sr mica ages in the range of 310 to 290 Ma are interpreted as cooling ages for the low-pressure metamorphic event, or are related to the intrusion of the post-kinematic granites.

3.3. Frankenwald

The peak of regional metamorphism in the Franken wald area north of the Münchberger Gneismasse (MM) ist dated about 310 Ma, while the somewhat lower ages in the slightly higher metamorphic Frankenwälder Querzone represent cooling ages of the same event (Ahrendt et al. 1986). Samples dated southwest of the MM along the Franconian Lineament show a rejuvenation due to hydrothermal influence.

3.4 The Nappe Unit ZEV

The upper intercept ages obtained on detrital zircon from gneisses belonging to the ZEV unit also have a range in the order of 2400 Ma, which is consistent with the results on Moldanubian and Saxothuringian metadediments.
Like the Moldanubian unit, an event around 530 Ma is supported by Rb-Sr whole rock analyses on an amphibolite orthogneiss complex (Teufel 1987). However, because of the low initial ratio of 0.7045, this date can best be interpreted in terms of magmatic formation of these rocks rather than by metamorphism.
The lower intercept obtained on zircons from a paragneiss sample gives an age of around 480 Ma. It is obvious that the lead loss causes by this event was much less than that suffered by zircons in the corresponding Moldanubian gneisses. This is reflected in the much greater spreadof the data points, with values close to the upper intercept. Magmatic activity around 460 Ma is proven by idiomorphic zircons from an augen gneiss.

The age of the medium pressure metamorphic event is well documented within the ZEV. Zircons from a garnet-kyanite gneiss in the southern part display a lower intercept age in the order of 390 Ma, which is in agreement with the concordant monazite age of 380 Ma (Teufel 1987). This is confirmed by a Rb-Sr whole rock age of 380 Ma obtained on a late metamorphic orthogneiss intrusion.
Mineral ages on hornblende, muscovite and biotite fall in the range 380-370 Ma reflecting a rapid cooling of the area (Teufel et al. 1985; Schüssler et al. 1986).
The last thermal overprinting of the southeastern part of the ZEV is dated at 320 Ma (Schüssler et al. 1986). This age is consistent with the Rb-Sr mineral ages as well as with a Rb-Sr whole rock small slab determination on the garnet-kyanite-gneiss from the quarry at Michldorf (Köhler intern. report).
K-Ar ages obtained on amphiboles give ages around 380 Ma in the western part of the ZEV which are consistant with the above mentioned ages for the medium pressure metamorphic event. These amphibole ages, howewer, seem to be rejuvenated in the eastern part of the ZEV to values between 330 and 300 Ma (Kreuzer et al. 1988). Whether the ages in the range of 320 Ma are related to a thermal influence caused by the intrusion of the post-tectonic granites, or related to the low-pressure metamorphic event in the Moldanubian unit, which might have influenced the lower parts of the ZEV nappe during the overthrusting or shortly after, cannot be answered at present.
It is, however, interesting that xenotimes from a quartz mylonite in the border zone between ZEV and the Moldanubian near Vohenstrauß

give concordant ages of 330 Ma (Teufel 1987). The xenotime crystals are elongated, which could be related to the mylonitization and therfore might date the time of overthrusting of the ZEV onto the Moldanubian. The overthrusting must have taken place later than the Lower Carboniferous, as this unit is already overthrusted by the Münchberger Gneismasse. An upper time limit for the thrusting is given by the intrusion age of the earliest granites around 320 Ma. Therefore the field evidence supports the interpretation that the xenotime age of 330 Ma reflects the approximate time of the thrusting.

4. Conclusions and Implications for the Geochronological Evolution

The geochronological data obtained so far allow the following model for the chronological evolution of the investigated units (Table 1).

Moldanubian

1. First records of detrital material in Paragneisses on zircons of about 2600 - 2400 Ma.
2. A metamorphic event around 460 Ma.
3. Indistinct signs for a possible medium-pressure metamorphic event at about 380 Ma.
4. Final strong metamorphic overprinting under low-pressure, high temperature conditions at about 320 Ma which overprinted inherited metamorphic material and shows lower intercept ages rejuvenated to around 530 Ma.
5. At about this tim beginning of late to post-tectonic granitic intrusions, continuing until 290 Ma.
6. Regional cooling reflected in the mica ages of about 300 Ma.

Saxothuringian

1. The first records of detrital material in paragneisses on zircons of about 2300 Ma.
2. Minimum age of 560 Ma for a metamorphic event in the source area inherited in metasediments.
3. Volcanic activity around 450 Ma.
4. Last metamorphic overprinting in the same range, as found in the Moldanubian around 320 Ma.
5. Identical record of granitic intrusions between 320 and 290 Ma.
6. Regional cooling around 310 Ma in the southern part of the Saxothuringian, around 280 Ma in the north.

Nappe Units

1. ZEV
   a) First records of detrital material in paragneisses on zircons of about 2500 - 2400 Ma.
   b) Magmatic events between 530 and 460 Ma.
   c) Strongest metamorphic overprinting under medium-pressure conditions around 390 Ma.
   d) Subsequent rapid regional cooling until 360 Ma.
   e) Final emplacement of the ZEV as a nappe complex probably around 330 Ma.
   f) Emplacement of post-tectonic granites around 320 Ma, which probably influenced Rb-Sr and K-Ar systems in micas and amphiboles of the southeastern ZEV.
2. Münchberger Gneismasse
   a) Magmatic and metamorphic events between 530 and 470 Ma.
   b) High-pressure metamorphism at 390 Ma.
   c) Rapid cooling until 380 Ma.
   d) Low mineral ages of about 320 Ma, indicating a possible influence of the regional low-pressure metamorphism or of the post-tectonic granites.

Table 1. Geochronological evolution of the Moldanubian-Saxothuringian transition zone (Teufel 1987)

Moldanubian			Saxothuringian			Z E V		
Age (Ma)	Method[a]	Event	Age (Ma)	Method[a]	Event	Age (Ma)	Method[a]	Event
2600-2400	U-Pb- Zr	Approximate age of orogenic processes in the source area	2300-2000	U-Pb Zr	Approximate age of orogenic processes in the source area (Grauert et al. 1973)	2500-2400	U-Pb Zr	Approximate age of orogenic processes in the source area
540-520	U-Pb- Zr	Metamorphic event in the source area	560	U-Pb Zr	Metamorphic event in the source area (Grauert et al. 1973)	530	Rb-Sr Wr+ small slab	Magmatic formation of the amphibolite source rocks
						480	U-Pb Zr	Episodic lead loss, low-grade metamorphism?
455	U-Pb Mo	Metamorphic event	450	U-Pb Zr	Effusive formation of the epigneiss source rocks	460	U-Pb Zr	Intrusive formation of the augengneiss source rocks
385	Rb-Sr Wr	Medium-pressure metamorphism?				390	U-Pb Zr	Medium-pressure metamorphism
						380	U-Pb Mo	Medium-pressure metamorphism
						380	Rb-Sr Wr	Synkinematic granite intrusion
						370-365	Rb-Sr Mu Bi	Cooling
333	U-Pb Mo	Formation of mylonites in shear zones at the structural boundary Mold./ZEV				333	U-Pb Mo	Formation of mylonites in shear zones at the structural boundary Mold./ZEV
320	U-Pb Mo	Low-pressure metamorphism	320	U-Pb Mo	Low-pressure metamorphism in the transition zone to the Moldanubian	320	Rb-Sr Bi	Retrograde overprinting
315	U-Pb Mo	Intrusion of sillgranites	315	Rb-Sr Bi	Cooling in the southern Saxothuringian			
310-300	Rb-Sr Bi	Cooling	320-310	K-Ar Illite	Low-grade metamorphism in the Frankenwald area (Ahrendt et al. 1986)			

a) Abbreviations: Zr, zircon; Mo, monazite; Mu, muscovite; Bi, biotite; WR, whole rock.

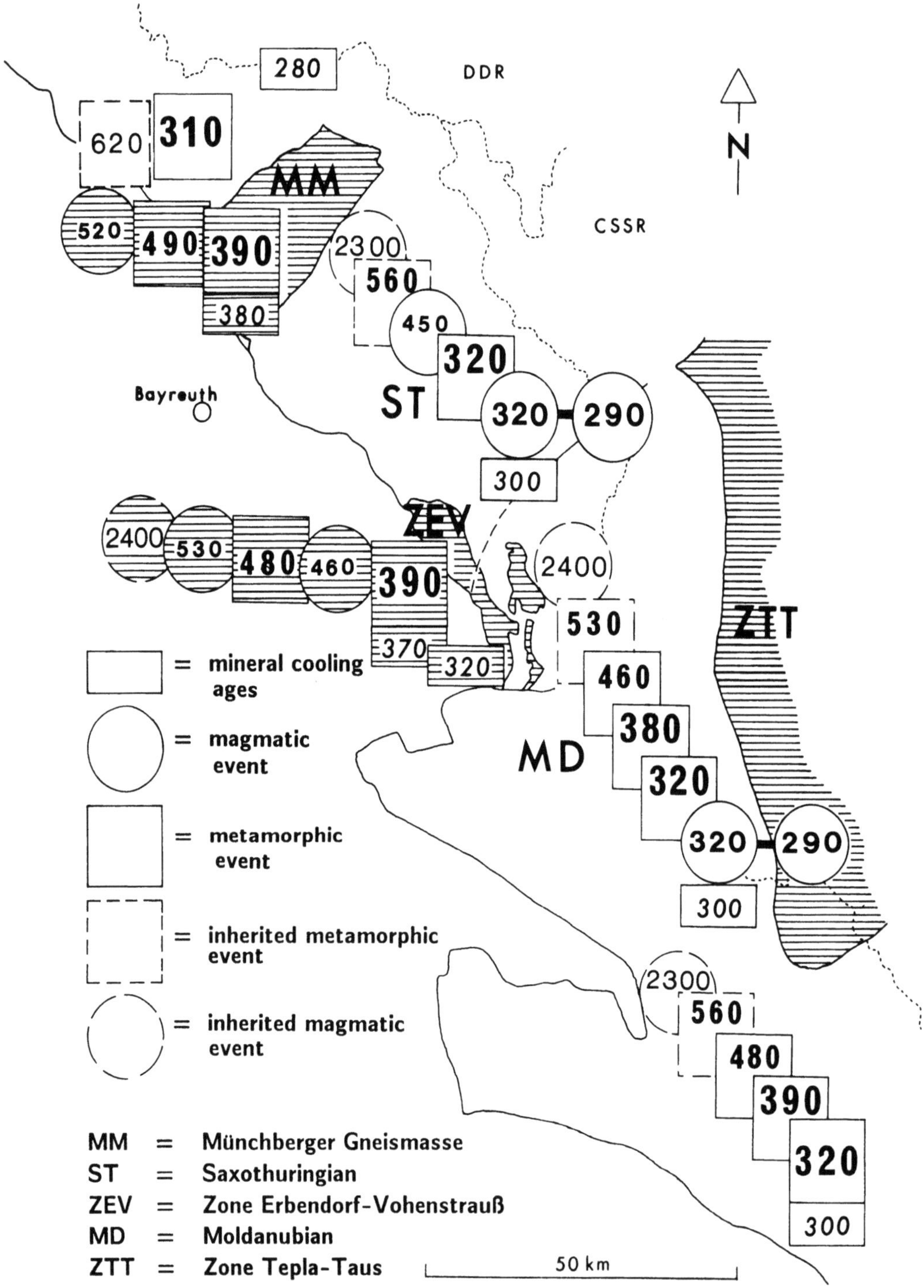

Fig. 2 Schematic compilation of the geochronological data of the northeast Bavarian basement

The age pattern of the MM coincides well with that of the ZEV (see above), especially in the range of 400 to 370 Ma. The pressure stressed metamorphism with its relatively rapid cooling is a dominant feature in both nappe units. On the basis of these data, the MM is well correlated with the ZEV as a nappe structure. According to the stratigraphic age of overthrust sediments, the final emplacement of the MM nappe complex is at least as young as the Lower Carboniferous. A schematic compilation of the geochronological evolution of the surroundings of the drill location is given in Fig. 2

REFERENCES

AHRENDT H, CLAUER N, FRANKE W, HANSEN B T, TEUFEL S (1986) Kontinentales Tiefbohrprogramm der Bundesrepublik Deutschland - Oberpfalz, Geochronologie. 2nd KTB-Colloq 19.9.-21.9.1986, Seeheim, Poster-Programm

BEHR H J, ENGEL W, FRANKE W (1982) Variscan wildflysch and nappe tectonics in the Saxothuringian zone (northeast Bavaria, West Germany). Am. J. Sci 282: 1438-1470

BESANG C, HARRE W, KREUZER H, LENZ H, MÜLLER P, WENDT I (1976) Radiometrische Datierung, geochemische und petrographische Untersuchungen der Fichtelgebirgsgranite. Geol. Jahrb. E8: 3-71

BLÜMEL P (1984) Mitteldruck- und Niederdruckmetamorphose in den außeralpinen Varisziden Mitteleuropas. Fortschr Mineral 62(1): 28-29

CARL C, DILL H, KREUZER H, WENDT I (1985) U/Pb- und K/Ar-Datierungen des Uranvorkommens Höhenstein/Oberpfalz. Geol Rundsch 74(3): 483-504

DAVIS G L, SCHREYER W (1962) Altersbestimmungen an Gesteinen des ostbayrischen Grundgebirges und ihre geologische Deutung. Geol Rundsch 52: 146-169

FISCHER G, SCHREYER W, TROLL G, VOLL G, HART S R (1968) Hornblendealter aus dem ostbayrischen Grundgebirge. Neues Jahrb Mineral Monats 11:385-404

FRANKE W (1984) Variszischer Deckenbau im Raume der Münchberger Gneismasse, abgeleitet aus der Fazies, Deformation und Metamorphose im umgebenden Paläozoikum. Geotekt Forsch 68: 253 p

FRENTZEL-BEYME K (1985) Fluidsysteme in Graniten, Metamorphiten und hydrothermalen Gangsystemen in der nördlichen Region der geplanten Tiefbohrung (Oberpfalz). Dipl-Arb, Univ Göttingen, 86 p (unpublished)

GEBAUER D (1975) Rb-Sr Gesamtgesteins- und Mineralsysteme sowie U-Pb Systeme in Zirkonen während der progressiven Gesteinsmetamorphose. Diss, ETH Zürich, pp 137

GEBAUER D, GRÜNENFELDER M (1979) U-Pb Zircon and Rb-Sr mineral dating of eclogites and their country rocks, example: Münchberg Gneiss Massif, northeast Bavaria. Earth Planet Sci Lett 42: 35-44

GRAUERT B, HÄNNY R, SOPTRAJANOVA G (1973) Age and origin of detrital zircons from the Pre-Permian basement of the Bohemian Massif and the Alps. Contrib Mineral Petrol 40: 105-130

GRAUERT B, HÄNNY R, SOPTRAJANOVA G (1974) Geochronology of a polymetamorphic and anatectic gneiss region: the Moldanubicum of the area Lam-Deggendorf, eastern Bavaria, Germany. Contrib Mineral Petrol 45: 37-63

GÜMBEL C W (1879) Geognostische Beschreibung des Fichtelgebirges mit dem Frankenwalde und dem westlichen Vorlande. Justus Perthes Verlag. 698 p

GULSON B L, KROGH T E (1973) Old lead components in the young Bergell massif, south-east Swiss Alps. Contrib Mineral Petrol 40: 239-252

KÖHLER H, MÜLLER-SOHNIUS D (1980) Rb-Sr systematics on paragneiss series from the Bavarian Moldanubicum, Germany. Contrib Mineral Petrol 71: 387-392

KÖHLER H, MÜLLER-SOHNIUS D (1985) Rb-Sr-Altersbestimmungen und Sr-Isotopensystematik an Gesteinen des Regensburger Waldes (Moldanubikum Ne Bayerns), part 1: Paragneisanatexite. Neues Jahrb Mineral Abh 151(1): 1-28

KÖHLER H, MÜLLER-SOHNIUS D, CAMMANN K (1974) Rb-Sr Altersbestimmungen an Mineral- und Gesamtgesteinsproben des Leuchtenberger und Flossenbürger Granits, NE Bayern. Neues Jahrb Mineral Abh 123(1): 63-85

KÖPPEL V (1974) Isotopic U-Pb ages of monazites and zircons from the crust-mantle transition and adjacent units of the Ivrea and Ceneri zones (southern Alps, Italy). Contrib Mineral Petrol. 43: 55-70

KREUZER H, SEIDEL E, SCHÜSSLER U, OKRUSCH M, LENZ K-L, RASCHKA H (1988) K-Ar geochronology of different tectonic units at the northwestern margin of the Bohemian Massif. Tectonophysics (in press)

LENZ H (1986) Rb/Sr-Gesamtgesteins-Altersbestimmung am Weißenstadt-Marktleuthener Porphyrgranit des Fichtelgebirges. Geol Jahrb E34: 67-76

MÜLLER-SOHNIUS D, DRACH V von, HORN P, KÖHLER H (1987) Altersbestimmungen an der Münchberger Gneismasse, Nordost-Bayern. Neues Jahrb Mineral Abh 156:175-206

SCHREYER W (1966) Metamorpher Übergang Saxothuringikum-Moldanubikum östlich Tirschenreuth/Opf., nachgewiesen durch phasenpetrologische Analyse. Geol Rundsch 55: 491-509

SCHULZ-SCHMALSCHLÄGER M, PROPACH G, BAUMANN A (1984) U/Pb-Untersuchungen an Zirkonen und Monaziten von Gesteinen des vorderen Bayerischen Waldes. Fortschr Mineral 62(1): 223-224

SCHÜSSLER U, OPPERMANN U, KREUZER H, SEIDEL E, OKRUSCH M, LENZ K-L, RASCHKA H (1986) Zur Alterstellung des ostbayrischen Kristallins, Ergebnisse neuer K-Ar-Datierungen. Geol Bavarica 89: 21-47

SÖLLNER F, KÖHLER H, MÜLLER-SOHNIUS D (1981a) Rb/Sr-Altersbestimmungen an Gesteinen der Münchberger Gneismasse (MM), NE-Bayern Teil 1, Gesamtgesteinsdatierungen. Neues Jahrb Mineral Abh 141 (1): 90-112

SÖLLNER F, KÖHLER H, MÜLLER-SOHNIUS D (1981b) Rb/Sr-Altersbestimmungen an Gesteinen der Münchberger Gneismasse (MM), Ne-Bayern Teil 2, Mineraldatierungen. Neues Jahrb. Mineral Abh. 142(2): 178-198

STEIN E (1987) Die strukturgeologische Entwicklung im Übergangsbereich Saxothuringikum/Moldanubikum in NE-Bayern. Diss, Univ Göttingen

STETTNER G (1975) Zur geologisch-tektonischen Entwicklung des Oberpfälzer Grundgebirges. Aufschluss 26: 11-38

STETTNER G (1979) Der Grenzbereich Saxothuringikum-Moldanubische Region im Raum Tirschenreuth-Mähring (Oberpfalz) und die Situation des Uran-führenden Präkambriums. Z Dtsch Geol Ges 130: 561-574

STETTNER G (1981) Grundgebirge. In: Geol Karte Bayern 1: 500 000. Erl. München, pp 7-29

STOSCH H-G, LUGMAIR G W (1986) Geochemistry and evolution of the eclogites from the Münchberg Gneiss Massif/W. Germany. Terra Cognita 6: 254

TEMBUSCH H, GRAUERT B (1983) Zur Diffusion von Strontium in heterogenen Metasedimenten. Fortschr Mineral 61(1): 255-256

TEUFEL S (1987) Vergleichende U-Pb- und Rb-Sr-Altersbestimmungen an Gesteinen des Übergangsbereiches Saxothuringikum/Moldanubikum, NE-Bayern. Diss, Univ Göttingen, 110 pp

TEUFEL S, AHRENDT H, HANSEN B T (1985) U-Pb und Rb-Sr-Altersbestimmungen im Übergangsbereich Moldanubikum/Saxothuringikum, NE-Bayern. Fortschr Mineral 63(1): 234

VOLL G (1960) Stoff, Bau und Alter in der Grenzzone Moldanubikum/ Saxothuringikum in Bayern unter besonderer Berücksichtigung gabbroider, amphibolitischer und kalksilikatführender Gesteine. Beih Geol Jahrb 42: 382 p

VOLLBRECHT A (1981) Tektogenetische Entwicklung der Münchberger Gneismasse. Goettinger Arb Geol Paläeontol, 122 p

WAGENER-LOHSE C, BLÜMEL P (1984) Prograde Regionalmetamorphose vom Niederdrucktyp in der Grenzzone Saxothuringikum/Moldanubikum E Tirschenreuth/NE Bayern. Fortschr Mineral 62(1): 254-255

WAGNER G A, REIMER G M, JÄGER E (1977) Cooling ages derived by apatite fission-track, mica Rb-Sr and K-Ar dating: the uplift and cooling history of the central Alps. Mem. Ist Geol Mineral Padova, 27 pp

WENDT I, KREUZER H, MÜLLER P, SCHMID H (1986) Gesamtgesteins- und Mineraldatierung des Falkenberger Granits. Geol Jahrb E34: 5-66

WILLIAMS I S, COMPSTON W, CHAPPELL B W (1983) Zircon and monazite U-Pb systems and the histories of I-type magmas, Berridale batholith, Australia. J Petrol 24: 76-97

WURM A (1925a) Geologie von Bayern, part I: Nordbayern, Fichtelgebirge und Frankenwald, 1st edu Bornträger, Berlin, 374 p

WURM A (1925b) Geologischer Führer durch Fichtelgebirge und Frankenwald. Bornträger, Berlin, 223 p

WURM A (1961) Geologie von Bayern, part I: Frankenwald, Münchberger Gneismasse, Fichtelgebirge, Nördlicher Oberpfälzer Wald, 2nd edu Bornträger, Berlin, 555 p

Geochemical Characteristics of Metabasites in Different Tectonic Units of the Northeast Bavarian Crystalline Basement

M. Okrusch[1], E. Seidel[2,3], U. Schüssler[1] and P. Richter[1]

[1] Mineralogisches Institut der Universität Würzburg, Am Hubland, D-8700 Würzburg, FRG
[2] Mineralogisches Institut der Technischen Universität Braunschweig, Gauß-Str. 28/29, D-3300 Braunschweig, FRG
[3] Present address: Mineralogisch-Petrographisches Institut der Universität zu Köln, Zülpicher Str. 49, D-5000 Köln, FRG

ABSTRACT

Comprehensive geochemical investigations of metabasites yielded constraints for a correlation of, or discrimination between the different tectonic units within the northeast Bavarian crystalline basement.

The Münchberg nappe pile consists of at least five large tectonic units which exhibit differences in lithology, in part also in metamorphic grade and in metamorphic history. The metabasites in each of these nappes show their own, significant geochemical characteristics. The lowermost tectonic unit, the Bavarian lithofacies, includes the anchimetamorphic Ordovician Randschieferserie which contains alkaline basalts. In their geochemistry, they are similar to the metabasites of the Fichtelgebirge crystalline complex in the autochthonous Saxo-thuringian. The next higher tectonic unit of the Münchberg nappe pile, the Prasinit-Phyllit-Serie contains metabasites which can be derived from subalkaline basalts with a clear calc-alkaline tendency. There is a striking geochemical resemblance to the metabasites of the Erbendorf Greenschist Zone (EGZ) underscoring the similar lithology of both allochthonous units which appear to be in a similar tectonic position. The Randamphibolit-Serie higher up in the Münchberg nappe pile consists of metabasites with tholeiitic characteristics and a pronounced differentiation trend. The next higher tectonic unit, the Liegendserie of the Münchberg gneiss complex s. str., contains metagabbros to metagabbronorites with a high-Al basaltic composition. The amphibolites and banded hornblende gneisses of the overlying Hangendserie are of subalkaline basaltic character with calc-alkaline affinity.

The Zone Erbendorf-Vohenstrauss (ZEV) is currently regarded as an allochthonous unit equivalent to the higher crystalline nappes of the Münchberg pile. However, the geochemical character of the metabasites do not encourage such a correlation. Neither the schistose and striped amphibolites nor the flaseramphibolites of the ZEV with their N-MORB and E-MORB character respectively, find convincing counterparts in the crystalline nappes of the Münchberg pile. However, an interestingly close resemblance exists between the schistose and striped amphibolites in the ZEV, on the one hand, and in the autochthonous Zone Tirschenreuth-Mähring (ZTM) and the adjacent Moldanubian sensu strictu, on the other.

Owing to the absence of age criteria, our results cannot be used, so far, to reconstruct the paleogeographical position of the individual tectonic units, based on the geochemical characteristics of their respective metabasites.

INTRODUCTION

During the pre-site studies for the German Continental Deep Drilling Program (KTB) our group investigated the geochemical characteristics of metabasites which form constituents of different tectonic units within the crystalline basement in northeast Bavaria. The aim of these studies was:

1. To evaluate constraints for a correlation and/or discrimination of these tectonic units.
2. To provide a data base for a sound recognition of metabasites and, consequently, their respective units in the future drill cores of KTB.

The geochemical characteristics of the various metabasites may serve as indications for the geotectonic positions of their respective protoliths, provided that the discriminating chemical elements and inter-element relationships were not affected seriously by post-magmatic alteration processes.

GEOLOGICAL POSITION AND PETROGRAPHICAL CHARACTERISTICS

The new geotectonic concept developed during the KTB pre-site studies distinguishes two major nappe complexes in the northwestern part of the Bohemian massif:

1. The Münchberg nappe pile
2. The Zone of Erbendorf-Vohenstrauss (ZEV) with the underlying Erbendorf Greenschist Zone (EGZ).

In contrast, the Saxothuringian of the Frankenwald and the Fichtelgebirge, the Moldanubian and the transitional Zone Tirschenreuth-Mähring (ZTM) are regarded as autochthonous (Weber and Vollbrecht, 1986).

The Münchberg Nappe Pile

The autochthonous Paleozoic of the *Thuringian lithofacies* in the southeastern Frankenwald is unconformably overlain by a stack of allochthonous units, the Münchberg nappe pile (Franke 1984).

The lowermost nappe unit consists of anchimetamorphic Paleozoic sediments whith intercalations of alkaline basalts to trachyandesites (Wirth 1978) and minor keratophyres, designated as *Bavarian lithofacies* by Wurm (1961).

The next higher unit, the *Prasinit-Phyllit-Serie* (Kraus 1954) consists predominantly of prasinites and quartz phyllites in variable proportions. Locally, serpentinites are intercalated.
The prasinites show the assemblage
actinolite + epidote + chlorite + albite + quartz + sphene + opaques
typical of the greenschist facies.

Higher up in the nappe pile a tectonic unit of considerably higher metamorphic grade occurs, the ***Randamphibolit-Serie***, which consists almost exclusively of amphibolites (briefly designated as "rand amphibolites") with the assemblage

hornblende + andesine + sphene ± grossular-rich almandine ± epidote + opaques.

The randamphibolites exhibit a strong cataclastic deformation. They underwent a retrograde overprint documented by the formation of actinolitic hornblende, albite, chlorite and epidote (Oppermann 1985).

The next higher crystalline nappe unit is the so-called *Liegendserie* of the Münchberg Gneiss Complex s. str. Within a sequence of para- and orthogneisses subordinate metagabbros to metagabbronoites (Matthes and Seidel 1977) are intercalated which exhibit distinct structural relics of the plutonic protolith as well as relics of the initial igneous An-rich plagioclase, hypersthene, clinopyroxene, brown hornblende, ilmenite and pyrrhotine. The regional metamorphic overprint led to the assemblage

oligoclase + hornblende ± garnet ± zoisite ± white mica + quartz + rutile/sphene.

The uppermost unit of the Münchberg nappe pile, the ***Hangendserie***, is characterized by a variegated lithology consisting predominantly of amphibolites and banded hornblende gneisses, and subordinate paragneisses, marbles and calc-silicate rocks. The metabasites contain the mineral assemblage

hornblende + oligoclase/andesine + quartz + rutile (partly replaced by sphene) ± almandine ± zoisite + opaques.

The banding is due to a variation in the hornblende vs feldspar + quartz ratio.

The famous *eclogites* of the Münchberg Gneiss Complex are concentrated in the border zone between the Hangend- and the Liegendserie. They can be interpreted either as constituents of the Hangendserie (Matthes et al. 1974) or as dislodged slices, tectonically intercalated between the two units (Stettner 1960). The eclogites testify to a high-pressure event which, according to Sm-Nd dating of Stosch (pers. comm., 1988), presumably took place 466 Ma ago. The main metamorphic event in the Hangend-, Liegend- and Randamphibolit-Serie is a medium-pressure metamorphism (Matthes et al. 1974; Blümel, in Weber and Vollbrecht 1986) which partly transformed the eclogites into eclogite-amphibolites. Recent K-Ar dating on hornblendes and micas from these units yielded ages around 380 Ma for the end of the amphibolite facies metamorphism, whereas the greenschist facies metamorphism of the underlying Prasinit-Phyllit-Serie seems to be slightly younger, i.e. about 365 Ma (Kreuzer et al. 1988).

The Erbendorf-Greenschist Zone (EGZ)

Situated between the autochthonous Fichtelgebirge in the north and the allochtonous Zone Erbendorf-Vohenstrauss in the south, the Erbendorf Greenschist Zone seems to be in a similar tectonic position as the Prasinit-Phyllit-Serie in the Münchberg nappe pile. The EGZ consists mainly of various metabasites, including metagabbros, and of serpentinites, but contains only minor metasediments. The main phase of regional metamorphism took place under conditions of the lowest amphibolite facies leading to the assemblage

oligoclase + plagioclase + epidote + opaques.

During a later retrogressive overprint, the mineral assemblages of this first metamorphic stage were partly replaced by albite, epidote, chlorite, carbonate and quartz. In the contact aureoles of the Steinwald and Falkenberg granites, the metamorphic sequences of the EGZ underwent a strong thermal overprint (Matthes 1951; Matthes and Olesch 1986).

The Zone Erbendorf-Vohenstrauss (ZEV)

The ZEV consists predominantly of pelitic metasediments which contain intercalations of various metabasites, ultramafics, calc-silicate rocks, graphite quartzites, graphite schists and orthogneisses. The metabasites can be divided into three main types:

1. ***Schistose and striped amphibolites*** prevail in the southern part of the ZEV, but are also recorded in a few outcrops in the northern ZEV. The mineral assemblage is
   hornblende + andesine/labradore + sphene + ilmenite + sulphides.
   The banding is caused by a variation in the hornblende vs plagioclase ratio or by calc-silicate intercalations, mainly consisting of
   salite + plagioclase ± garnet.

2. ***Flaseramphibolites*** are predominant in the northern and central parts of the ZEV. The flaser structure is caused by ill-defined lenses or wavy bands rich in plagioclase or, rarely, salite. The mineral assemblage is
   hornblende + oligoclase/andesine ± garnet ± salite + sphene + ilmenite + sulphides.
   The garnets are frequently replaced by fine-grained aggregates of plagioclase and hornblendes. At the western margin of the Falkenberg granite, the flaseramphibolites underwent a thermal overprint.

3. ***Metagabbros*** are concentrated in the southwestern corner of the ZEV. They are characterized by a coarse-grained, massive structure testifying to a plutonic protolith (Voll 1960). The rocks consist of platy, heavily altered plagioclase and hornblende which is partly replaced by felty aggregates of secondary amphiboles (Schüssler 1987). Some metagabbros contain considerable amounts of biotite.

U-Pb dating of zircons and monazites (Teufel 1988), and K-Ar dating of hornblendes and micas (Kreuzer et al. 1988) from various rock types of the ZEV indicate that the medium-pressure, amphibolite facies metamorphism in the ZEV (Blümel, in Weber and Vollbrecht 1986) took place in the Early Devonian, about 380 Ma ago. Younger K-Ar ages around 325 Ma which prevail in the eastern ZEV are presumably a result of contact metamorphism.

The Fichtelgebirge Crystalline Complex

Part of this zone is characterized by a variegated lithology with numerous intercalations of marbles, calc-silicate rocks and minor metabasites ("Bunte Gruppe" of Stettner 1975, 1980). There are, however, transitions to a more monotonous lithology with a predominance of pelitic to psammitic metasediments. The sequence underwent a low-pressure metamorphism ranging from the greenschist to the amphibolite facies (Mielke et al. 1979) dated at about 330 Ma by the K-Ar method (Kreuzer et al. 1988). Mineral relics of an older medium-pressure event were recently recognized by Lenz et al. (in prep.). The regional metamorphic assemblages were partly overprinted by the intrusion of the Fichtelgebirge granites.

Most of the investigated metabasites are fine-grained, schistose amphibolites with the assemblage

hornblende + plagioclase ± biotite ± quartz + sphene + ilmenite + sulphides.

Two of the investigated metabasites exhibit relics of an ophitic texture testifying to a doleritic basalt as a protolith. Between the platy plagioclase crystals, there are relics of augite and aggregates of serpentine after primary olivine.

The Zone Tirschenreuth-Mähring (ZTM)

In its lithology the ZTM is less variegated than the Fichtelgebirge crystalline complex, but less monotonous than the Moldanubian gneisses (Richter and Stettner 1983). The rare amphibolites of the ZTM and the adjacent Moldanubian s. str. can be hardly distinguished from the schistose and striped amphibolites of the ZEV. The mineral assemblage is

hornblende + andesine ± salite + sphene + opaques.

One amphibolite lense contains garnet which is widely replaced by aggregates of hornblende and plagioclase.

The metamorphism in the ZTM which was dated at about 320 Ma (Teufel 1988; Kreuzer et al. 1988) took place under low-pressure conditions ranging from the greenschist to the high amphibolite facies (Schreyer 1966; Blümel and Wagener-Lohse, in Weber and Vollbrecht 1986).

GEOCHEMISTRY

Analytical Methods

The major elements Si, Ti, Al, Fe^{tot}, Mn, Ca, K and the trace elements S, Sc, V, Cr, Co, Ni, Cu, Zn, As, Rb, Sr, Y, Zr, Nb, Ba, La, Ce, Nd were analyzed by standard XRF analysis using lithium tetraborate fusion discs and powder pellets respectively. Mg, Na and Li were determined by standard AAS methods after decomposing the samples in HF-H_2SO_4 or HF-$HClO_4$. P was determined by XRF or spectrophotometrically, Fe (II) by the vanadate method (Peters 1968) or spectrophotometrically; CO_2 was measured volumetrically, H_2O^+ by the Penfield method.
In part of the samples CO_2 and S were determined by IR spectrometric methods. W and Mo were analyzed by liquid-liquid extraction with Zn-dithiol (Richter 1984). Rare earth elements were measured by ICP-AES after decomposing the samples in HF-$HClO_4$ and separating the REE in chromatographic columns (Erzinger et al. 1984). Selected bulk rock analyses of metabasites from the Münchberg nappe pile, the Erbendorf Greenschist Zone, the Zone Erbendorf-Vohenstrauss, the Fichtelgebirge Crystalline Complex and the Zone Tirschenreuth-Mähring are presented in Tables 1 and 2. The REE data for the metabasites of the ZEV, EGZ and ZTM are given in Schüssler et al. (1988). A list of localities and the complete set of geochemical analyses can be obtained from the authors on request.

Table 1. Selected analyses of metabasites from different tectonic units of the Münchberg nappe pile

	Prasinit-Phyllit-Serie common samples				Hohenknoden		Randamphibolit-Serie					Hangendserie				
	P11	P22	P26	P35	P38	P40	RA16	RA17	RA42	RA44	RA23	S72/99	S72/100	S52	S72/108	H3
(wt.%)																
SiO2	48.2	47.4	48.0	50.1	47.2	46.2	48.5	47.7	47.4	47.4	50.0	50.8	48.4	52.0	50.6	51.9
TiO2	0.67	0.68	1.17	0.66	2.28	1.89	1.10	1.91	1.71	1.06	0.53	0.69	0.67	0.64	0.70	0.56
Al2O3	17.7	14.3	18.3	17.3	15.9	15.6	14.0	12.9	14.8	14.7	15.3	15.9	16.5	12.5	16.5	15.6
Fe2O3	4.04	1.68	5.77	4.07	3.80	2.77	2.00	1.90	2.00	2.34	1.71	2.09	1.29	3.5	3.7	2.40
FeO	3.11	5.54	4.11	3.38	5.83	7.06	9.11	13.4	9.15	6.16	6.25	6.23	7.33	7.85	7.96	8.28
MnO	0.11	0.12	0.12	0.12	0.16	0.18	0.21	0.34	0.21	0.17	0.16	0.15	0.17	0.23	0.24	0.18
MgO	6.13	13.9	5.27	6.52	6.99	8.74	8.48	6.69	7.28	10.4	8.70	7.92	9.03	9.91	6.11	6.79
CaO	13.7	9.13	9.60	10.1	8.98	9.06	11.8	10.8	12.4	12.9	12.7	10.8	12.0	9.26	8.68	7.92
Na2O	1.51	1.36	2.82	3.97	3.31	2.83	2.70	1.61	2.24	2.37	2.86	2.88	2.17	2.11	3.49	3.77
K2O	0.49	1.44	0.85	0.22	0.96	0.55	0.34	0.16	0.29	0.29	0.24	1.00	0.28	0.21	0.24	0.24
P2O5	0.14	0.28	0.22	0.14	0.36	0.33	0.10	0.15	0.18	0.12	0.05	0.05	0.05	0.10	0.13	0.08
H2O+	3.0	3.5	3.2	2.8	3.4	3.7	1.9	1.8	1.9	2.1	1.6	1.8	1.7	1.4	1.2	1.6
CO2	0.81	0.2	0.4	1.1	0.8	0.7	<0.05	<0.05	0.05	0.06	0.05	0.09	<0.05	<0.05	0.05	0.06
S	<0.02	<0.02	0.02	0.17	<0.02	0.02	<0.02	<0.02	0.06	<0.02	<0.02	<0.02	0.11	<0.02	0.22	0.11
Total	99.61	99.53	99.85	100.65	99.97	99.63	100.24	99.36	99.67	100.07	100.15	100.40	99.70	99.71	99.82	99.49
(ppm)																
Cr	161	592	475	205	184	318	376	158	153	811	469	255	300	482	49	83
Ni	66	319	137	89	87	152	95	45	71	253	89	148	212	83	30	30
Cu	28	44	56	74	42	83	75	28	115	48	83	19	69	81	290	64
Zn	64	54	99	64	80	90	88	129	93	73	62	95	73	92	118	83
Rb	10	20	19	5	19	12	8	4	4	9	6	23	16	8	10	9
Sr	241	412	383	220	351	323	160	58	261	250	168	125	130	93	130	170
Y	20	21	31	23	40	36	25	53	28	18	17	-	-	20	22	15
Zr	86	137	111	80	218	239	76	120	122	86	51	55	16	11	38	22
Nb	3	4	3	3	7	8	5	5	13	8	2	-	-	<3	<3	4
Ba	120	492	130	9	210	138	95	79	87	27	64	176	58	85	126	142
Pb	2	4	5	<1	3	5	<1	<1	3	<1	1	4	4	5	4	<5

Table 2. Selected analyses of metabasites from different tectonic units of the KTB target area

	Zone Erbendorf-Vohenstrauss flaseramphibolites		schistose & striped amphibolites				metagabbros			Erbendorf Greenschist Zone			Fichtelgebirge		Zone Tirschenreuth-Mähring	
	1-4	2-3	OP-84 -9	OP-84 -36	OP-84 -126	OP-84 -164	OP-84 -72	OP-84 -205	3-7	RBU-78 -66	RBU-78 -308	RBU-78 -342	Fi10a	Wa-1a	675	37
(wt.%)																
SiO2	46.2	47.2	47.9	46.6	46.8	48.8	50.8	51.6	48.4	51.2	48.6	46.5	51.5	51.0	47.9	47.4
TiO2	1.94	2.12	2.40	1.01	1.29	1.04	1.77	1.85	0.72	0.71	0.58	0.85	3.26	3.92	1.18	1.36
Al2O3	14.5	13.3	15.6	16.2	15.5	15.3	16.2	15.8	14.5	16.4	16.8	14.7	12.4	11.9	14.1	14.9
Fe2O3	2.86	2.00	2.86	1.22	1.34	1.66	1.42	1.66	1.17	2.60	3.57	5.28	1.78	3.33	0.78	1.13
FeO	8.82	9.70	8.41	8.62	7.88	7.60	8.35	8.22	6.60	5.21	2.43	1.55	10.3	10.9	9.73	8.78
MnO	0.17	0.17	0.15	0.15	0.13	0.15	0.16	0.16	0.13	0.13	0.11	0.15	0.15	0.16	0.17	0.18
MgO	7.80	6.70	4.98	9.65	8.37	8.54	7.13	6.38	12.9	3.65	9.90	9.80	6.90	7.80	7.94	7.8
CaO	12.3	13.1	10.9	11.0	12.9	13.0	8.82	9.00	12.6	14.6	15.1	8.60	10.0	8.65	14.2	12.5
Na2O	2.38	2.45	3.40	3.05	2.61	1.93	2.88	2.81	1.58	4.12	0.89	3.45	1.37	0.45	2.18	1.98
K2O	0.48	0.58	0.98	0.20	0.15	0.13	0.65	0.72	0.22	0.27	0.23	0.38	1.02	0.10	0.16	0.13
P2O5	0.42	0.46	0.29	0.06	0.08	0.07	0.17	0.18	0.07	0.23	0.19	0.14	0.48	0.50	0.15	0.19
H2O+	1.6	1.6	1.3	1.5	1.5	1.2	1.5	2.0	1.2	1.2	2.1	3.9	0.8	2.1	1.0	1.9
CO2	<0.1	<0.1	<0.1	<0.1	<0.1	<0.1	<0.1	<0.1	0.1	0.24	<0.1	5.1	<0.1	<0.1	<0.1	<0.1
S	0.04	<0.01	-	-	-	-	-	-	-	<0.01	0.02	<0.01	<0.01	0.13	0.27	0.66
Total	99.51	99.38	99.17	99.26	98.55	99.42	99.85	100.38	100.19	100.56	100.52	100.40	99.96	100.94	99.76	98.91
(ppm)																
Li	<10	<10	-	-	-	-	-	-	<10	<10	<10	33	50	<10	14	15
Sc	38	38	-	-	-	-	-	-	41	30	38	37	33	32	46	47
V	292	324	256	218	222	228	240	304	180	150	138	222	350	380	264	276
Cr	410	180	152	329	378	333	158	156	1350	380	425	200	100	105	370	380
Co	57	46	50	58	58	50	44	41	34	26	24	34	50	66	47	51
Ni	205	78	117	246	220	97	75	37	250	211	236	192	78	84	135	180
Cu	36	32	-	-	-	-	-	-	60	19	94	16	55	105	57	125
Zn	109	101	-	-	-	-	-	-	56	74	45	62	129	141	116	115
As	<2	8	<2	<2	3	6	-	-	-	<2	<2	<2	<2	4	<2	<2
Rb	5	9	8	<4	<4	<4	37	16	5	7	7	11	53	5	<4	6
Sr	385	385	366	231	266	227	301	282	150	412	345	283	790	470	154	144
Y	28	28	28	25	26	22	32	24	18	20	16	24	29	27	22	28
Zr	188	218	187	82	101	77	177	106	64	106	64	90	270	264	70	62
Nb	20	19	16	<3	<3	4	8	6	<3	6	<3	5	32	32	5	8
Mo	0.58	1.18	-	-	-	-	-	-	-	0.31	<0.10	<0.10	0.36	0.91	<0.10	0.25
Ba	532	382	265	91	55	38	223	240	74	<5	<5	12	228	216	80	65
La	17	17	-	-	-	-	-	-	<3	18	9	11	27	26	-	8
Ce	28	35	21	<10	<10	<10	21	17	<10	36	26	18	55	54	<10	15
Nd	21	15	-	-	-	-	-	-	9	18	12	12	23	25	19	12
W	0.36	0.31	-	-	-	-	-	-	-	0.65	0.34	<0.10	0.66	1.32	0.28	<0.10

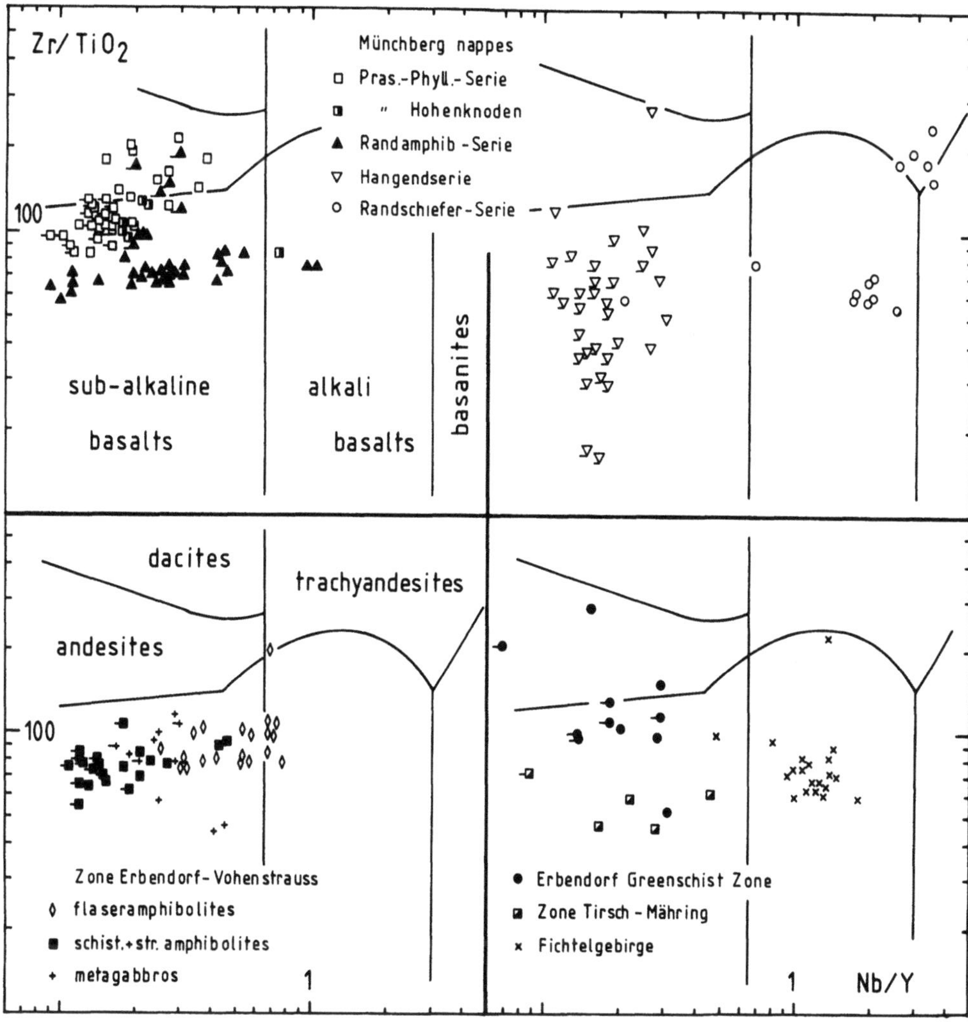

Fig. 1. Discrimination between subalkaline and alkaline series by Nb/Y vs Zr/TiO_2 (Floyd and Winchester 1978). Samples with Nb-contents below the detection limit of 3 ppm are indicated by bars. TiO_2 in wt.%, Nb, Zr, Y in ppm

The Münchberg Nappe Pile

Because of significant chemical differences, the metabasites in the individual tectonic units of the Münchberg nappe pile cannot be derived from one common type of protolith. This finding clearly contradicts former assumptions according to which the metabasites in the Prasinit-Phyllit-Serie, in the Randamphibolit-Serie and in the Hangend-Serie are metamorphosed equivalents of the Ordovician alkaline basalts in the anchimetamorphic Randschiefer-Serie of the Bavarian lithofacies. According to their chemical composition (Wirth 1978), these Ordovian basalts form two different groups of alkaline basaltic and trachyandesitic character respectively (Figs. 1, 3).

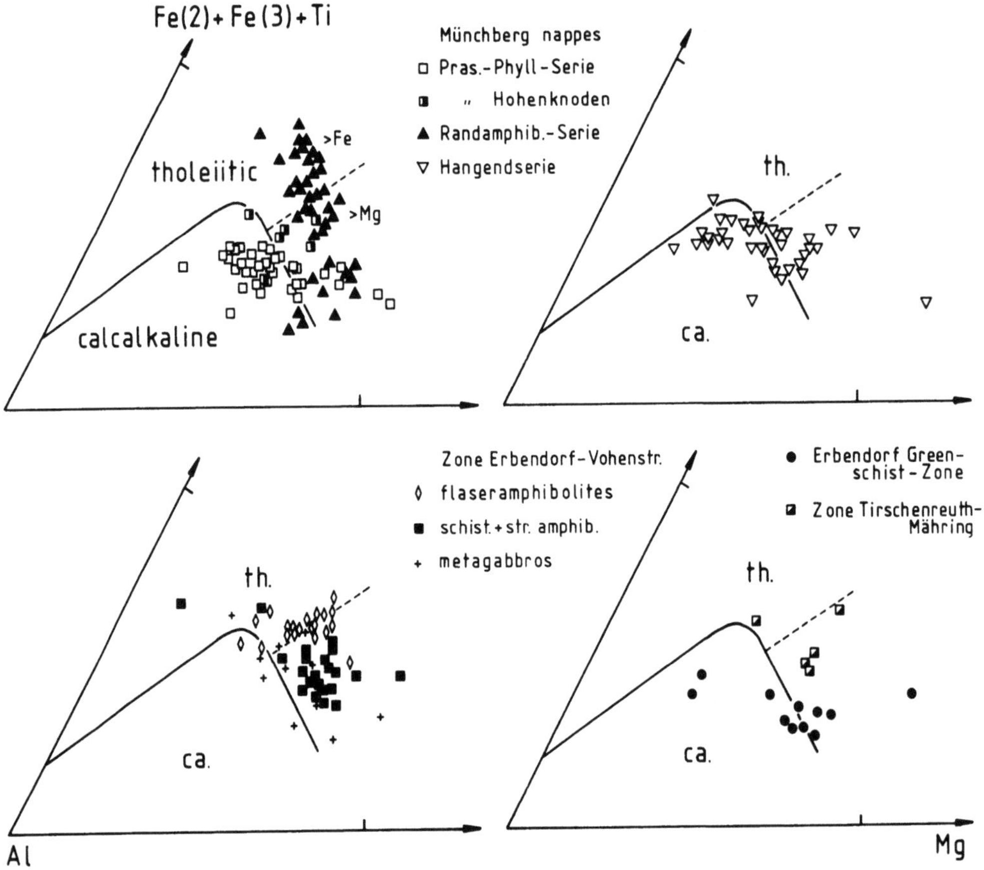

Fig. 2. Cationplot of Jensen (1976) used to subdivide subalkaline basalts into tholeiitic and calc-alkaline series respectively

Judging from their bulk rock chemistry as well as from structural and mineral relics, the metabasites of the *Prasinit-Phyllit-Serie* are clearly derived from a basaltic protolith including pyroclastics (Oppermann 1985). They are characterized by constantly low TiO_2 contents (mean value 0.9 ± 0.2 wt.%), but variable P_2O_5 and Zr values (Fig. 3). The prasinites are clearly subalkaline in their geochemical character (Fig. 1) and show a calc-alkaline tendency (Figs. 2, 3).

Prasinites of the abandoned quarry north of *Hohenknoden*, situated at the tectonic contact between the Prasinit-Phyllit-Serie and the Randamphibolit-Serie have distinctly higher values of TiO_2 and Zr than the bulk of the prasinites (Fig. 3). However, the Zr/TiO_2 and Nb/Y ratios in the Hohenknoden prasinites are the same as in the other prasinites (Fig. 1). Mineralogically the Hohenknoden prasinites are distinguished from the other, sphene-bearing prasinites by the presence of rutile which is more or less replaced by sphene. From the metabasites of the overlying Randamphibolit-Serie the Hohenknoden prasinites are also distinguished chemically. As a consequence we assume that an additional tectonic slice is exposed in the Hohenknoden quarry.

The amphibolites of the *Randamphibolit-Serie* conform to subalkaline basalts in their chemical composition, with tholeiitic characteristics (Figs. 1, 2). In contrast to the fairly homogeneous group of the prasinites (disregarding the Hohenknoden samples), the randamphibolites show a pronounced differentiation trend from Mg-rich to Fe-rich compositions (Fig. 2). This trend is matched by a parallel enrichment of TiO_2, Zr (Fig. 3), P_2O_5 and Y and by a depletion of Ni. The geochemical variation is expressed in the mineralogical composition, too: amphibolites with higher Mg/Fe ratios are free of garnet, whereas Fe-enriched amphibolites usually are garnet-bearing (Oppermann 1985).

In their bulk rock chemistry, the metagabbros to metagabbronorites of the *Liegendserie* conform to a high-Al basaltic composition, similar to that of the light eclogite type in the Hangendserie. However, judging from differences in e.g. the Zr and Ti contents, Matthes and Seidel (1977) deny a genetic relationship between these two rock types.

The amphibolites and banded hornblende gneisses of the *Hangendserie* are derived from a basaltic protolith (Pommerenke 1985) with consistently low TiO_2 and Zr contents (Fig. 3). They belong to the subalkaline group (Fig. 1) with a calc-alkaline trend (Fig. 2). In the relevant correlation diagrams the metabasites of the Hangendserie show an affinity to those of the Prasinit-Phyllit-Serie, but have usually lower Zr contents. The chemical differences to the tholeiitic Randamphibolites and the alkaline basalts of the anchimetamorphic Ordovician Randschieferserie (Bavarian lithofacies) are obvious (Figs. 1, 2, 3).

The *eclogites* of the Münchberg Gneis Complex can be divided into two principal groups: The dark eclogites have an N-MORB like composition, whereas the light, kyanite-bearing eclogites are high-Al basaltic in character (Matthes et al. 1974; Puchelt et al. 1978) and may be derived from plagioclase-rich, gabbroic cumulates (Stosch and Lugmair 1987). Despite some geochemical similarities, the light eclogites cannot be derived from the metagabbros and metagabbronorites of the Liegendserie (Matthes and Seidel 1977).

The Erbendorf Greenschist Zone (EGZ)

The predominant striped amphibolites and the subordinate metagabbros of the EGZ cannot be distinguished by their chemical compositions which conform to Mg-rich tholeiites to calc-alkaline basalts (Figs. 1, 2). In diagrams related to the tectonic setting of modern basalts the EGZ metabasites plot along the borderline between the midocean ridge and the island arc basalts or into the field of the island arc basalts (Fig. 3). The major and trace element contents of the metabasites in the EGZ and in the Prasinit-Phyllit-Serie of the Münchberg nappe pile agree strikingly well. This fact underscores the similar lithological assemblages of both tectonic units.

The Zone Erbendorf-Vohenstrauss (ZEV)

The *schistose and striped amphibolites* of the ZEV conform in their chemical composition to basalts of the subalkaline series (Fig. 1). Their Mg-rich tholeiitic character is shown in the Jensen cationplot (Fig. 2). Trace elements including REE indicate similarities to modern N-MORB compositions (Fig. 3).

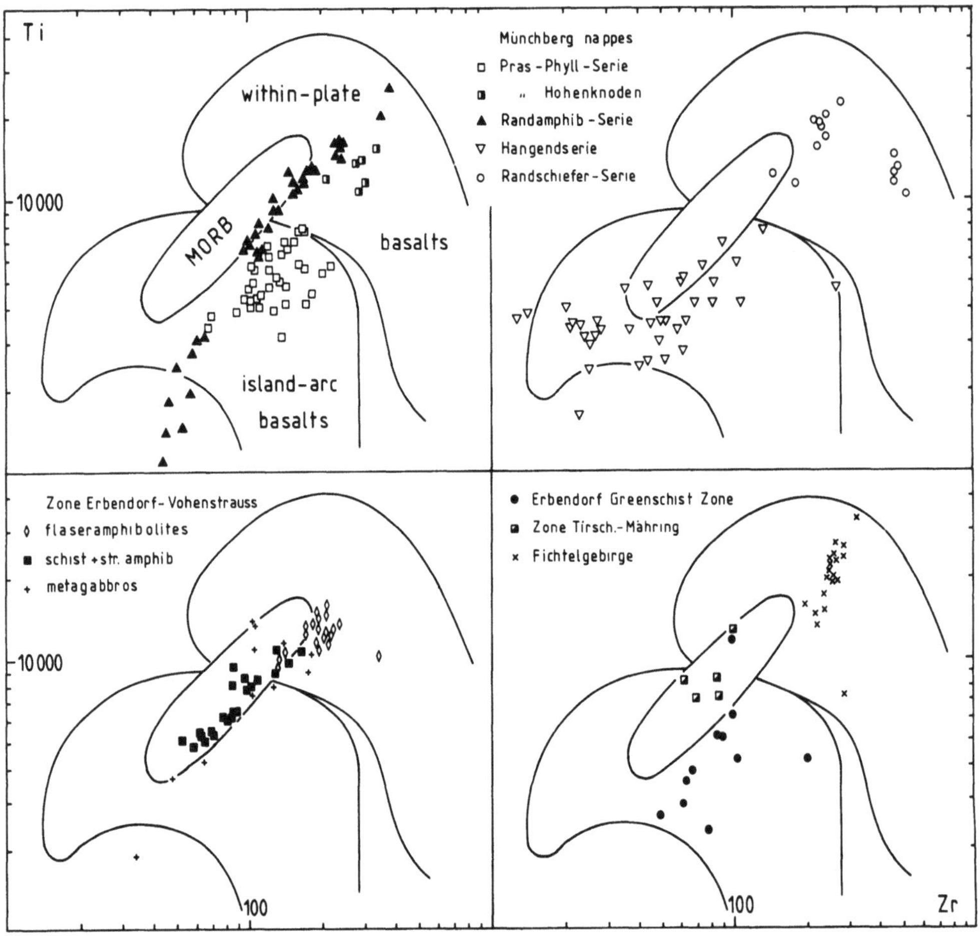

Fig. 3. Comparison of the investigated metabasites with modern basalts from various geotectonic positions in the Ti (ppm) vs Zr (ppm) diagramm (Pearce 1982)

In contrast, the *flaseramphibolites* of the ZEV are clearly enriched in the incompatible trace elements Nb, Ce, P, Zr, Ti and the LREE pointing to a tholeiitic protolith with an E-MORB or within-plate basalt character (Fig. 3). Compared to the schistose and striped amphibolites, the often garnet-bearing flaseramphibolites have generally a higher Fe/Mg ratio (Fig. 2). It should be noted that both amphibolite types show a relatively well-defined regional distribution within the ZEV. We, therefore, do not assume a differentiation relationship between the respective protoliths of the two types. This is in contrast to the situation supposed for the randamphibolites of the Münchberg nappe pile.

The *metagabbros* in the southwest corner of the ZEV are transitional in their geochemical character between the schistose and striped amphibolites and the flaseramphibolites. However, three of the investigated samples (e.g. 3-7 in Table 2) are clearly depleted in incompatible elements and, conformably, have lower Fe/Mg ratios and higher Cr contents (Figs. 2, 3), indicating a more primitive type of gabbroic protolith.

Some of the metabasites in the ZEV are chemically influenced by the intrusion of the Leuchtenberg and Falkenberg granites and/or altered by hydrothermal fluids (Schüssler et al. 1988, Schüssler 1989).

The Fichtelgebirge Crystalline Complex

The metabasites of the Fichtelgebirge form a distinct group which is of alkaline basaltic character (Fig. 1). Modern basalts extruded within plates or at anomalous midocean ridges are similar in composition (Fig. 3). The enrichment of incompatible elements is still stronger than in the flaseramphibolites of the northern ZEV (Fig. 3). The Fichtelgebirge metabasites have no geochemical counterpart in the crystalline basement of northeast Bavaria. However, they agree quite well with the alkaline-basalts of the anchimetamorphic Randschieferserie (Bavarian litho-facies) in the Münchberg nappe pile.

Zone Tirschenreuth-Mähring (ZTM)

The textural similarities between the schistose and striped amphibolites of the ZEV, on the one hand, and of the ZTM and the adjacent Moldanubian s. str., on the other, are matched by a nearly identical geochemical character, conforming to typical N-MORB compositions (Figs. 1, 2, 3).

CONCLUSIONS

Metabasites in each of the individual tectonic units of the Münchberg nappe pile exhibit their own, significant geochemical characteristics. This fact indicates that the respective basaltic protoliths were initially emplaced in different paleogeographial environments. Owing to the absence of age criteria, the relative position of these environments in time and space, prior to the stacking of the Münchberg nappe pile, is so far unknown.

The close geochemical resemblance between the metabasites from the Prasinit-Phyllit-Serie of the Münchberg nappe pile and the Erbendorf Greenschist-Zone is a strong argument for a correlation of the two units which exhibit a similar lithological association and appear to be in an analogous tectonic position.

In contrast, a correlation between the higher, crystalline nappes in the Münchberg pile and the Zone Erbendorf-Vohenstrauss is not encouraged by our results. No geochemical relationship exists between the amphibolites and banded hornblende gneisses of the Hangendserie and the various metabasite types of the ZEV. The randamphibolites, too, find no convincing counterpart in the ZEV. As a consequence we must state that a connection between the higher units of the Münchberg nappe pile and the ZEV cannot be maintained in the straightforward way which has been postulated recently (Weber and Vollbrecht 1986, p. 26).

However, an interestingly close correlation exists between the schistose and striped amphibolites of the Zone Erbendorf-Vohenstrauss, on the one hand, and of the Zone Tirschenreuth-Mähring and the adjacent Moldanubian s. str., on the other. The interpretation of this finding is still open to discussion. Geochemical investigations on metabasites from the Zone Tepla-Domazlice and from the Moldanubian s. str. are in progress and hopefully will provide additional constraints.

The geochemical affinity between the metabasites of the Fichtelgebirge crystalline complex and the alkaline basalts of the anchimetamorphic Ordovician Randschieferserie (Bavarian lithofacies) in the lowermost position of the Münchberg nappe pile should be stressed. However, it would be daring to derive an age relationship from this fact alone.

Acknowledgements. Thanks are due to Ms. Rosemarie Baur (Würzburg), Otto Ewald, Uwe Oppermann, Kurt Pommerenke (Braunschweig), Frank Volker (Karlsruhe), Julian Lodziak, Detlef Requard (Hannover), Jörg Erzinger and Evelyn Zuleger (Gießen) for their assistance in the analytical work. The financial support of Deutsche Forschungsgemeinschaft (grant Ok 2/25 -1 - Ok 2/25-2) is gratefully acknowledged.

REFERENCES

Erzinger J, Heinschild H-J, Stroh A (1984) Bestimmung der seltenen Erden in Gesteinen mit der ICP-AES. In: Welz B (ed) Fortschritte in der atomspektrometrischen Spurenanalytik, 1. Verlag Chemie, Weinheim, pp 251-260

Floyd PA, Winchester JA (1978) Identification and discrimination of altered and metamorphosed volcanic rocks using immobile elements. Chem Geol 21: 291-306

Franke W (1984) Variszischer Deckenbau im Raume der Münchberger Gneismasse - abgeleitet aus der Fazies, Deformation und Metamorphose im umgebenden Paläozoikum. Geotekt Forsch 68: 1-253

Jensen LS (1976) A new cation plot for classifying subalkalic volcanic rocks. Ontario Div Mines MP 66: 1-22

Kraus L (1954) Die geologische Stellung der Prasinit-Phyllit-Serie in der Umrandung der Münchberger Gneismasse. Diss, Univ Würzburg 84 pp (unpublished)

Kreuzer H, Seidel E, Schüssler U, Okrusch M, Lenz K-L, Raschka H (1988) K-Ar geochronology of different tectonic units at the northwestern margin of the Bohemian massif. Tectonophysics (in press)

Matthes S (1951) Die kontaktmetamorphe Überprägung basischer kristalliner Schiefer im Kontakthof des Steinwald-Granits nördlich von Erbendorf in der bayerischen Oberpfalz. Neues Jahrb Mineral Abh 82: 1-92

Matthes S, Olesch M (1986) Polymetamorphic-metasomatic blackwall rocks of the Falkenberg granite contact aureole near Erbendorf, Oberpfalz, Bavaria. Neues Jahrb Mineral Abh 153: 325-362

Matthes S, Seidel E (1977) Die Eklogitvorkommen des kristallinen Grundgebirges in NE-Bayern. X. Bestehen genetische Beziehungen zwischen Eklogit und Meta-Gabbro innerhalb des Münchberger Gneisgebietes? Neues Jahrb Mineral Abh 129: 269-291

Matthes S, Richter P, Schmidt K (1974) Die Eklogitvorkommen des kristallinen Grundgebirges in NE-Bayern. VII. Ergebnisse aus einer Kernbohrung durch den Eklogitkörper des Weißensteins. Neues Jahrb Mineral Abh 120: 270-314

Mielke H, Blümel P, Langer K (1979) Regional low-pressure metamorphism of low and medium grade in metapelites and -psammites of the Fichtelgebirge area, NE-Bavaria. Neues Jahrb Miner Abh 137: 83-112
Oppermann U (1985) Geochemie und Petrographie der Prasinite und Amphibolite am SW-Rand der Münchberger Gneismasse, NE-Bayern. Dipl-Arb, Tech Univ Braunschweig, 140 pp (unpublished)
Pearce JA (1982) Trace element characteristics of lavas from destructive plate boundaries. In: Thorpe RS (ed) Andesites. Wiley, New York, pp 525-548
Peters A (1968) Ein neues Verfahren zur Bestimmung von Eisen(II)oxid in Mineralen und Gesteinen. Neues Jahrb Mineral Monatsh 1968: 209-214
Pommerenke K (1985) Geochemie und Petrographie der Amphibolite im NE-Teil der Münchberger Gneismasse, NE-Bayern. Dipl-Arb, Tech Univ Braunschweig, 135 pp (unpublished)
Puchelt H, Matthes S, Richter P (1978) Aussagemöglichkeiten der seltenen Erden und anderer Spurenelemente für die Eklogite des Münchberger Gneisgebietes. Fortschr Mineral 56: 106-108
Richter P (1984) Wolfram in Graniten Ostbayerns - Versuch einer metallogenetischen Gliederung. Geol Jahrb D 63: 3-22
Richter P, Stettner G (1983) Das Präkambrium am Nordrand der Moldanubischen Region im Raum Tirschenreuth-Mähring (NE-Bayern) und dessen metallogenetische Aspekte. Geol Jahrb D 61: 23-91
Schreyer W (1966) Metamorpher Übergang Saxothuringikum-Moldanubikum östlich Tirschenreuth/Oberpfalz, nachgewiesen durch phasenpetrologische Analyse. Geol Rundsch 55: 491-509
Schüssler U (1987) Petrographie, Geochemie und Metamorphosealter von Metabasiten im KTB-Zielgebiet Oberpfalz, Ostbayern. Diss, Univ Würzburg, 272 pp
Schüssler U (1989) Petrographie, Geochemie und Metamorphosealter von Metabasiten im KTB-Zielgebiet Oberpfalz. Geologica Bavarica (in press)
Schüssler U, Richter P, Okrusch M (1988) Metabasites from the KTB target area Oberpfalz, Bavaria - geochemical characteristics and examples for mobile behaviour of "immobile" elements. Tectonophysics (in press)
Stettner G (1960) Über Bau und Entwicklung der Münchberger Gneismasse. Geol Rundsch 49: 350-375
Stettner G (1975) Zur geologisch-tektonischen Entwicklung des Oberpfälzer Grundgebirges. Aufschluß Sonderband 26: 11-38
Stettner G (1980) Zum geologischen Aufbau des Fichtelgebirges. Aufschluß 31: 391-403
Stosch H-G, Lugmair GW (1987) Geochronology and geochemistry of eclogites from the Münchberg Gneiss Massif, FRG. Terra Cognita 7: 163
Teufel S (1988) Vergleichende U-Pb- und Rb-Sr-Altersbestimmungen an Gesteinen des Übergangsbereichs Saxothuringikum/Moldanubikum, NE-Bayern. Göttinger Arb. Geol. Paläont. 35: 1-87
Voll G (1960) Stoff, Bau und Alter der Grenzzone Moldanubikum/Saxothuringikum in Bayern unter besonderer Berücksichtigung gabbroider, amphibolitischer und kalksilikatführender Gesteine. Geol Jahrb Beih 42: 1-382
Weber K, Vollbrecht A (1986) Kontinentales Tiefbohrprogramm der Bundesrepublik Deutschland, KTB - Ergebnisse der Vorerkundungsarbeiten, Lokation Oberpfalz. 2. KTB-Kolloq, 19.9.-21.9.1986, Seeheim/Odenwald, 186 pp
Wirth R (1978) Geochemie und Petrographie der paläozoischen Magmatite des Frankenwaldes. Diabase, Keratophyre, Pikrite. Diss, Univ Würzburg, 130 pp (unpublished)
Wurm A (1961) Geologie von Bayern. Borntraeger, Berlin, 555 pp

Metallogenic Evolution at the Saxothuringian-Moldanubian Plate Boundary in the Mid-European Variscides

H. Dill

Bundesanstalt für Geowissenschaften und Rohstoffe, Stilleweg 2, 3000 Hannover 51, FRG

Abstract: During the Variscan orogeny and the early platform stage numerous ore mineralizations, some of which have been mined to date (e.g. fluorite) or explored by galleries and shafts (e.g. uranium), evolved in the NE Bavarian basement. These mineralizations have been subdivided into four categories: stratiform/stratabound, thrustbound, granite-related, unconformity-related. The first type shows a bimodal curve in the frequency distribution with a first-order maximum during early Paleozoic and a second-order maximum during Permo-Triassic. Type II is almost exclusively restricted to the Lower Carboniferous, during which the major tectonic movements took place in that region. Type III ore deposits came up at the turn Carboniferous - Permian and type IV commenced subsequent to the peneplanation of the uplifted Variscan basement.

INTRODUCTION

For a long time the Mid-European Variscides have been taken as a folded orogen (Hercyno-type) which strikingly contrasts with the Alpine mountain chains characterized by nappe tectonics and thrusting. During the recent past models on horizontal tectonics were a fresh impetus not least for the understanding of the orogeny in the Mid-European Variscides (see papers presented in this volume and literature cited thereunder) but for the metallogenesis, too. Based on the new geodynamic models in the following a metallogenic study is presented and a subdivision of ore mineralization has been performed. Four distinct types of ore deposition may be defined:

type I: stratiform/stratabound
type II: thrustbound
type III: granite-related
type IV: unconformity-related.

The regional geology, stratigraphy and tectonics are treated in context with these mineralizing processes in as much it is of need for the understanding of the various ore types. For an overall outline of the regional geology and geodynamic evolution the reader is referred to special papers elsewhere in this volume.

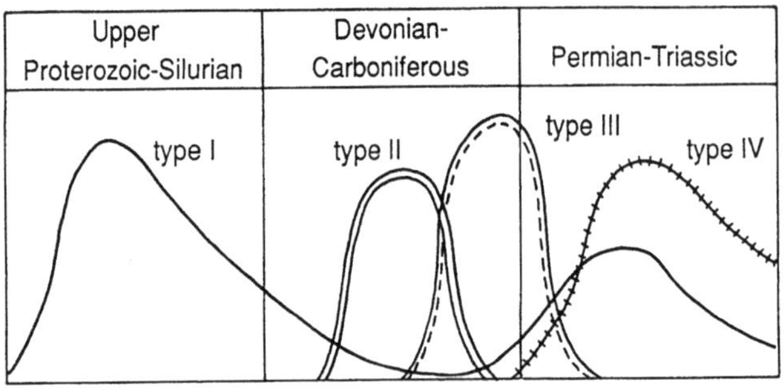

Fig. 1 Frequency distribution of ore types throughout the Variscan orogeny in the NE Bavarian Basement and adjoining areas.

TYPE I - STRATABOUND AND STRATIFORM ORE DEPOSITS

The climax of type I ore mineralization was attained during the initial phases of the evolution of the Mid-European Variscides (Fig. 1). From the L. Proterozoic through the Upper Devonian type I ore mineralizations evolved in that terrain. According to classification schemes proposed for ore deposits of similar kind elsewhere (Sawkins 1984), these stratabound metal concentrations were categorized into "sediment-hosted" and "volcanic-hosted massive sulfides", respectively. The "low grade large tonnage deposits" of the early Paleozoic Graptolite Shales were dealt with separately as "black shale-hosted disseminated ore". The second-most important period, during which sulfides were syndiagenetically accumulated in a continental environment with terrigenous sediments was in the Permo-Carboniferous. Type I Fe depositions evolved from the Middle Ordovician through Upper Devonian.

Sediment-hosted massive sulfides

There are two mining sites in NE Bavaria which were operated for their sediment-hosted "Kieslager deposits", the Bodenmais-Lam-district being situated in the Moldanubian region and the Waldsassen/Bayerland mine lying immediately north of the Erbendorf geosuture in the Saxothuringian Zone.

Numerous studies have centered on the Bodenmais ore deposit and the discussion, on whether syngenetic or epigenetic processes have played the key role in ore deposition, has now shifted towards syngenetic hydrothermal ore deposition (Schmid & Weinelt 1978, Pfeufer 1976) - Fig. 2a.

The sphalerite-pyrrhotite-bearing ore lenses are intercalated into a series of banded cordierite gneisses, cordierite-garnet-sillimanite-bearing metatexites, closely associated with calcsilicates, silicic gneisses and granitic mobilisates (Fig. 2a, b). Both, ore and country rock underwent high-grade

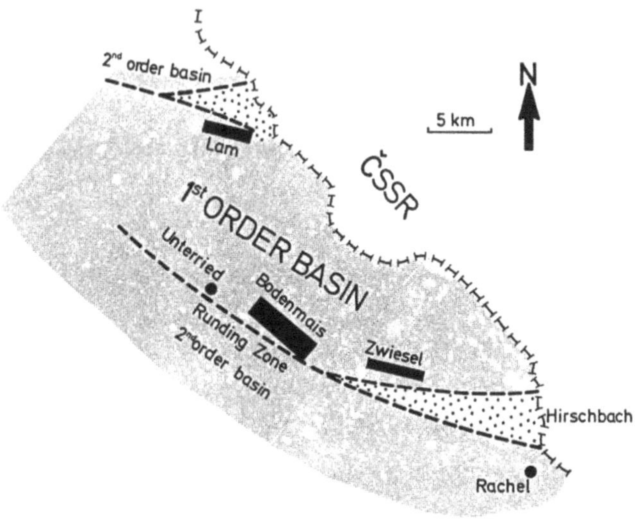

Fig. 2a Sediment-hosted Bodenmais-type Kieslagers (type I).

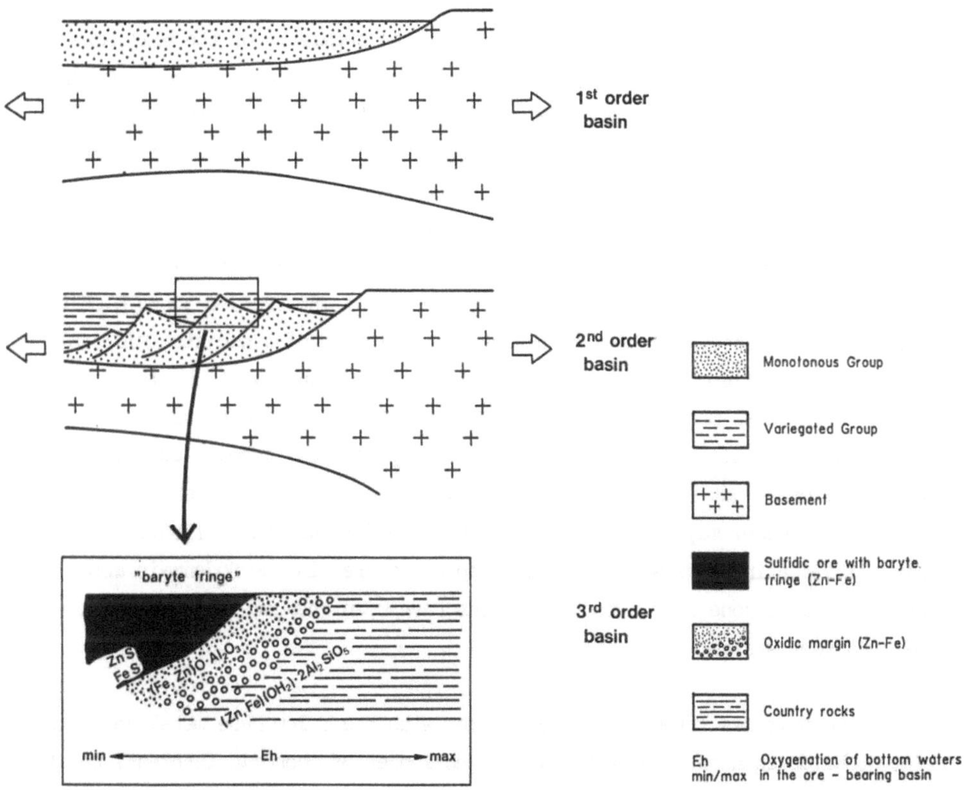

Fig. 2b Model of basin evolution in the Bodenmais area.

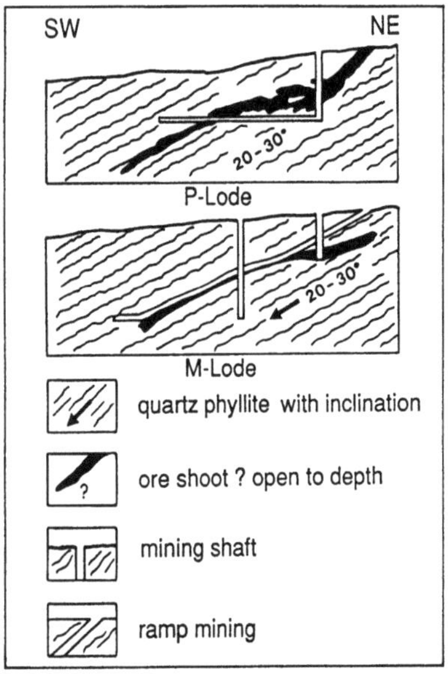

Fig. 2c Cross section through the P- and M-lode of Waldsassen Kieslager (redrawn from Strunz 1953).

metamorphism (Blümel 1978). Inspite of this, the chemistry of the wallrocks (e. g. Ba anomalies), the array of major elements in the ore, proper, (Fe-Zn-Cu-Pb) and the petrography briefly cited above atest to a sediment-hosted rather than volcanic-hosted ore (Dill 1986a, Troll et al. 1987). Subsequently, stratabound ore bodies suffered pervasive re-mobilization triggered by shearing along the Runding mylonite zone (Troll et al. 1987). Of similar type is the Waldsassen "Kieslager" (Wolf 1971, Dill 1985a) which comprises a pyrite("P-lode") and pyrrhotite ore body ("M-lode") interbedded among biotite-chlorite-quartzites, quartz-mica-schists, and biotite mica schists (Fig. 2c). The wall rocks lie close to the andalusite-biotite-boundary of the low-pressure regional metamorphic unit which indicates temperatures of 530° ± 50° C. Igneous activity is manifested by thin tuffaceous beds above and below the ores, but nevertheless the ore is hosted by true metapsammo-pelitic rocks and, hence, may be termed "sediment-hosted" (sensu Large 1980). Magnetite-bearing layers, which were also discovered in the Bodenmais area, and a silified zone in the footwall rocks have been thought to represent an altered vent system.

While the sulfides (from both mining sites with their low base metal contents (e. g. Waldsassen ca. 0.5 wt % Cu)) are no longer of economic interest, the Au contained in these metamorphic rocks was repeatedly the target of exploration activities in the recent past (arsenopyrite from Waldsassen 13 ppm Au - Dill 1985a). The individual ore bodies with pyrite and pyrrhotite are suggested to

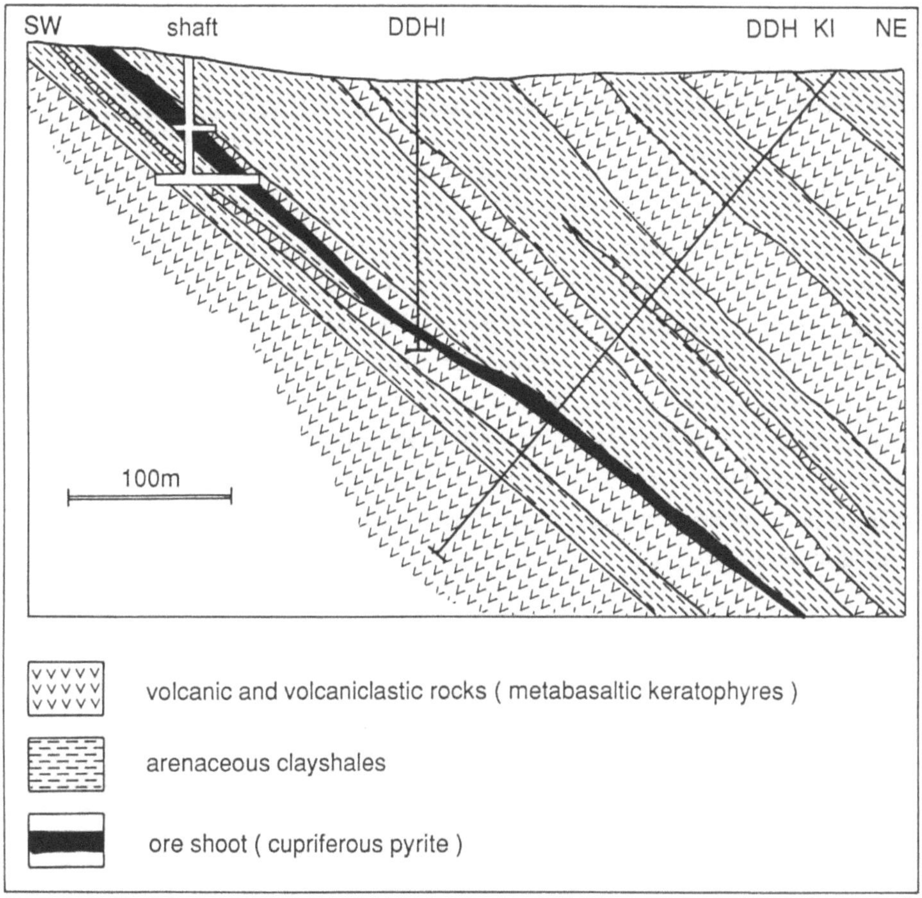

Fig. 3 Volcanic-hosted Kupferberg-type Kieslager (redrawn from Urban & Vache 1972).

have formed in basins of various water depths. According to a model proposed by Plimer and Finlow-Bates (1978) pyrite is inferred to have been precipitated in a relatively shallow water environment, whereas pyrrhotite formed in deeper waters.

Volcanic-hosted massive sulfides

True volcanic-hosted massive sulfides are met in the Kupferberg-Neufang district SW of the Münchberg Gneiss Mass and near Sparneck at its southeastern margin (Fig. 3). In the "prasinite phyllite series" which mainly consists of calc-alkaline volcanics (Okrusch et al. 1986) and the Randschiefer-Series with its alkali basalts, keratophyres and their pyroclastic derivates, stratiform ore bodies of pyrrhotite, pyrite, chalcopyrite, sphalerite and little magnetite have been mined for a long time and re-explored by the end of the 60s. Others, than the aforementioned "kieslager" the "Kupferberg ores" show a conspicuous lack of galena, but an abundance in Cu. Urban and Vache (1972) have stressed the idea on these Cu-rich ores to have been concentrated by supergene processes induced by the uplift of the Variscan basement. This may hardly be maintained when considering the ore textures (e. g. pyrrhotite overlies chalcopyrite and does not display "bird eyes"), mineral association (lack of typical secondary Cu minerals e. g. chalcocite, native Cu). Au is present (2 ppm Au at a maximum - Urban & Vache 1972) though not of economic significance at present. In the adjacent metavolcanic rocks inside the Münchberg Gneiss Mass, the Au contents in Cu-Fe-sulfide-bearing ore shoots, locally containing tellurides merit some attention. It is apparently not a true stratiform volcanic-hosted Au mineralization, rather a preconcentration of Au in a fertile source lithology, from which precious metals became subsequently re-mobilised (see type II ore mineralization) during pervasive fracturation of the MGM.

Disseminated sulfide ore in black shales

There are numerous argillaceous beds in the Mid-European Variscan orogen which bear appreciable amounts of carbon and sulfur to allow syndiagenetic precipitation of pyrite (e. g. Devonian tentaculite shales, Lower Carboniferous sooty shales and roof slates), yet only the Silurian "Lower Graptolite Shales" and Lower Devonian "Upper Graptolite Shales" (Fig. 4) justify denomination as a "low grade large tonnage deposit". Their alum shales, chert and phosphorites contain not only pyrite discernible even with the unarmed eyes, but also chalcopyrite, ruby sphalerite (poor in Fe, rich in Mn and Cd), galena, tetrahedrite and "sooty pitchblende", the last of which mostly encountered along cracks and fissures (Dill 1986b).
Furthermore, Schmid (1980) recorded abnormally high amounts of Eu, Se, Cs, V, Mo and Au from drilling holes spudded in this early Paleozoic black shales. These metalliferous sapropelites formed in a starved basin under poorly oxygenated conditions, whereas the accompanying phosphates derived from coastward upwelling. In the area under investigation (Graefenthal Horst/N Bavaria) no volcanic or volcaniclastic rocks have been mapped so far in these black shale terrains. Consequently, metal-bearing solutions discharged into the reduced bottom waters are certainly not linked with early Paleozoic volcanism. Continuous compaction apparently caused fluids to be expelled from metalliferous Ordovician footwall rocks and fed into these sites of mineralization containing stagnant badly aerated bottom waters. Biogenic reduction of sea water sulfate (Krueger 1973, Dill & Nielsen 1986) supplied the sulfide concentration to precipitate Fe and the base metals. The spread of

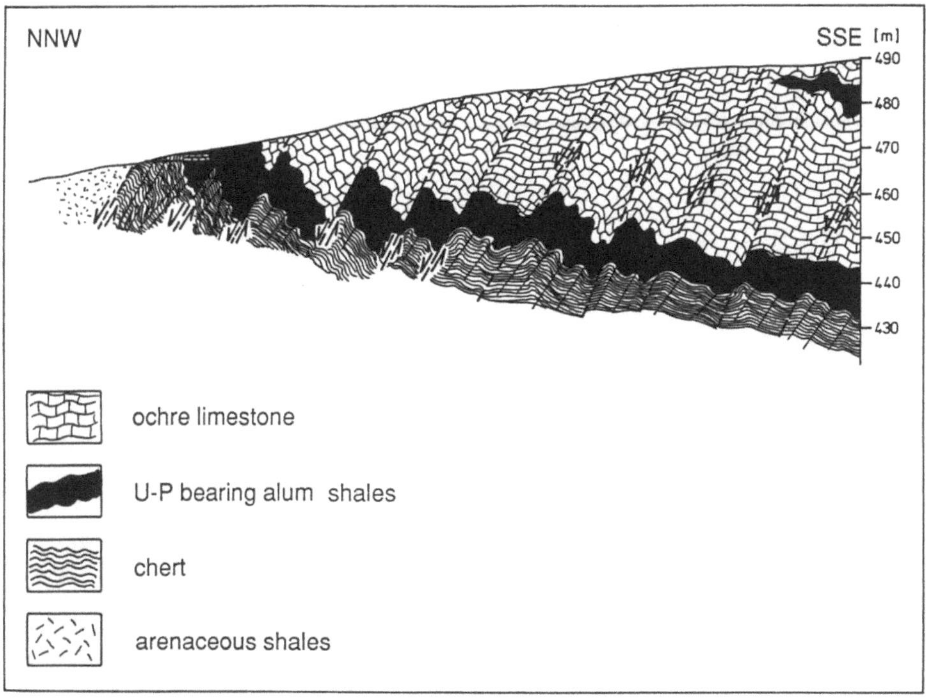

Fig. 4 Disseminated sulfide ores in black shales of the Silurian Graptolite Shales Facies from the Graefenthal Horst near Ludwigsstadt (redrawn from Dill 1985a).

$\delta^{34}S$ is without any debate an additional evidence for excluding any igneous impact on the formation of these pyritiferous black shales.

The oxidic Fe ores

Apart from these Fe-sulfides mentioned for the Silurian Graptolite Shales there are some indications of Fe to be concentrated as $FeCO_3$ in calcareous beds. The euxinic conditions were shortly interrupted during Ludlovian, when the sea level stand was lower and as result of this limestones topped the uplifted swells. The ochre limestone, named on account of its rusty Fe concretions, is a marker of this period (Behrens 1981).

Opposite to this carbonate-hosted Fe mineralization in the ochre limestone, siderite layers with little pyrite, sphalerite and chalcopyrite intercalated among marble horizons ("Wunsiedel Marble") and calc-phyllites were operated for Fe until 1942 (von Horstig & Teuscher 1979) - Fig. 5 - inter alia near Arzberg. It is still a matter of discussion, whether these shelf carbonates represent a carbonate-hosted ore deposit brought about by epigenetic re-

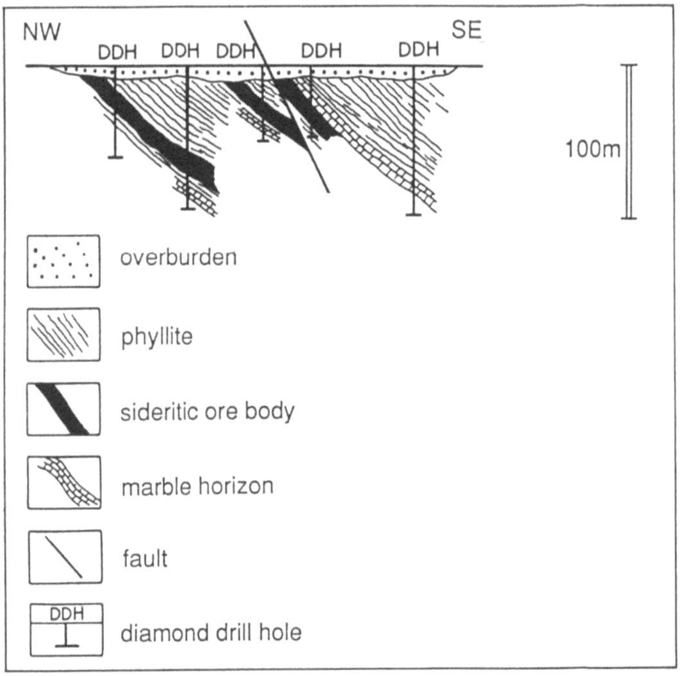

Fig. 5 Carbonated-hosted Fe ores of the Wunsiedel Marble (redrawn from von Horstig & Teuscher 1979).

mobilisation of Fe or a truly syngenetic Fe ore deposit. It seems to be a mixture of the two. Fe was preconcentrated in a fertile near-shore calcareous environment. Subsequently, during the Variscan orogeny siderite with little hematite devolped from this low-metal concentration. Taking into account the lithostatic pressure of 3.5 to 5.5 kbar, constrained by the associated silicate parageneses (Mielke et al. 1980), a temperature of formation of 350° C to 375° C can be calculated for this Fe carbonate using the equation given by Melnik and Siroshtan (1973).
More elevated T values would have led to the release of CO_2 and a replacement of $FeCO_3$ by Fe_3O_4 which has not yet been found here.

A peculiar strataform Fe enrichment occured during middle Ordovician. It contains the full range of Fe ore minerals from Fe silicates (thuringite, aphrosiderite, chamosite), Fe carbonates (siderite, ankerite), Fe oxide (magnetite), to Fe sulfide (pyrite), the abundance in the first ore mineral of which gave its name to this ore type: thuringite ore seam (Dill 1985a). Although the present-day ore textures and assemblages reflect the post-sedimentary diagenetic overprinting and very low grade to low grade regional metamorphism to which the Fe mineralization has been subjected, a tentative description of the environment of deposition can be given. These oolithic Fe ores were laid down in a lagoonal pond coastward of a sand bar, which promoted the high energy water regime in its surrounding and concentration of Fe ooides. It has to be noted, that the pyritiferous facies of these neritic iron-stones closely resembles that of the younger pyritiferous Graptolite shales with respect to sulfur isotope variation. The wide range of isotopically light sulfur common to both pyritiferous rocks and their consistency with regard to their mean $\delta^{34}S$ stress a similar mode of formation by diagenesis and exclude any volcanic-derived S supply.

Like elsewhere in the Variscides, volcano-sedimentary, exhalative hematite ores largely came up during upper Devonian in N Bavaria (Neumann-Redlin et al. 1976). They significantly differ from their Ordovician Fe-bearing predessesors, the thuringite ores, in that the contain much more hematite than magnetite. In turn, little ferroan phyllosilicates and only some pyrite were concentrated marginal to the submarine volcanic cones, which they derived from. These Fe ores are in space and time related to the diabases (= meta-basalts) scattered throughout the Saxothuringian Zone (Fig. 6).

Sandstone-hosted type ore

After the Sudetian movement, the Variscan basement was uplifted and suffered widespread denudation, the debris of which were transported in the narrow troughs embayed into the basement rocks. As a consequence, a new type I ore mineralization evolved. Unlike their early Palaeozoic equivalents Cu, Zn, Pb and U have now subaerially been enriched in an alluvial fan environment during Permo-Carboniferous times (Fig. 7). Among these fan depositions those from pyroclastic and volcaniclastic wet fans seem to be the most prospective host rocks for the above-mentioned elements, since they provide excellent source rock qualities (labile volcanic constituents) combined with good host rock properties (carbonaceous matter as reductant, permeability).

The significance of these embayments (Stockheim, Erbendorf, Weiden, Schmid-gaden basins) as an ore trap decreases from NW towards the SE with the decrease of coal seams and volcaniclastic beds intercalated among these fan depositions.

While the Stockheim basin contains subeconomic stratabound sulfidic pitchblende - carburan mineralization in coal seams of carbonaceous debris flows, the carbargillites of the Schmidgaden basin are giving only a moderate radiometric response (Gudden et al. 1974, Dill et al. 1988) and definite U

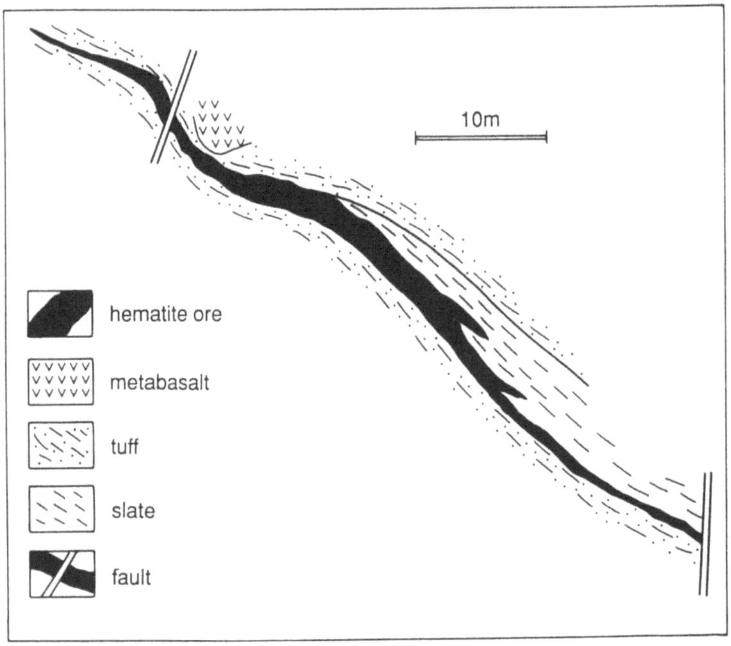

Fig. 6 Exhalative volcanic-hosted hematite ores of the Langenbach mining district ("Lahn-Dill-Ore") (redrawn from Wurm 1961).

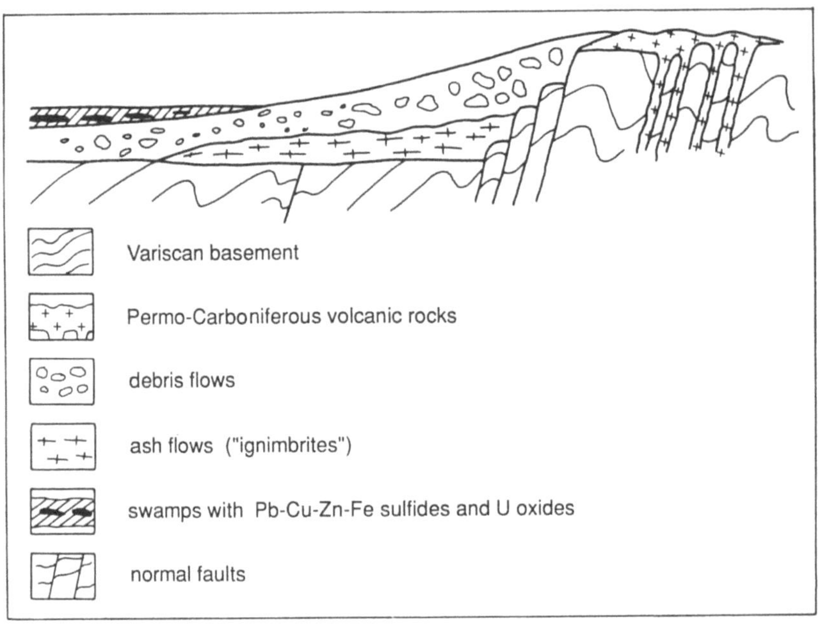

Fig. 7 Sandstone-hosted U-bearing Pb-Cu-Zn-Fe-ore mineralization in the distal fan section of Permo-Carboniferous fans.

minerals do not exist. These ignimbrites, lahars, air fall tuffs merit attention, in specific for their preconcentration of elements (e.g. Ba) which have finally been re-concentrated in veins bound to the Late Variscan unconformity (type IV).

TYPE II THRUSTBOUND ORE DEPOSITS

U and sulfide mineralization bound to bedding-plane-and axial plane parallel fault zones have already been mentioned from the Early Paleozoic Graptolite shales which were squeezed like a kernel into the surrounding brittle country rocks (e.g. ochre limestone, roof slates). Their high carbon content (graphite, semigraphite) acted as an excellent lubricant and facilitated this shearing process. Similar to these black shale-hosted mineralizations, the Au mineralization from the Münchberg Gneiss Mass may also be assigned to the stratabound mobilisates, since ore-bearing structures terminate within the source rocks, proper (source rock = host rock).

Another type of thrustbound ores containing prevalently Sb, As, Au and Hg is hosted by NE-SW-trending shear zones in the environs of the Fichtelgebirge - Erzgebirge Anticline and the Berga Anticline (Fig. 8) (Bachmann 1955, Netschajew & Fritsche 1966, Dill 1985b, Neumann & Tischendorf 1986). The rocks, from which those elements derived do not crop out and can only be

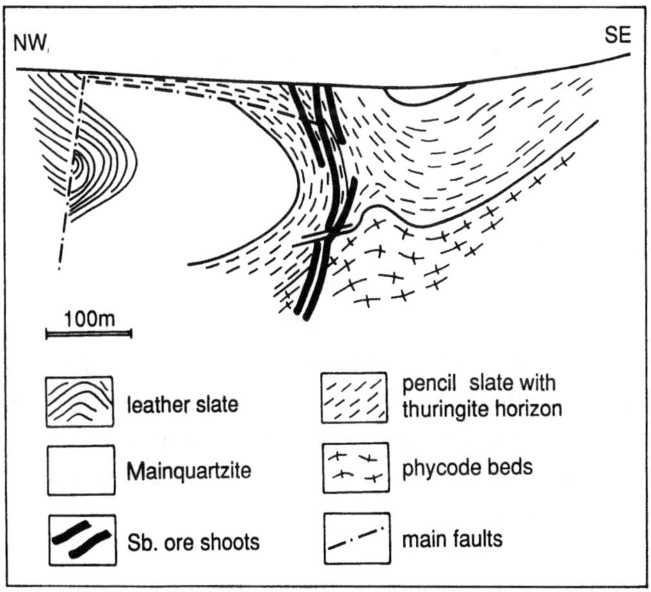

Fig. 8 Thrustbound (type II) stibnite ores from the Berga Anticline (redrawn from Bachmann 1955).

supposed to occur in deeper parts of the host anticlines. The source for Sb, As, Au and Hg is considered in meta-ultra-basic rocks, which crop out in the "Granulite Massif". Hg data are available from serpentinites of this place ($\bar{x}$ = 180 ppb Hg). From reflection seismics sheared slices of similar petrography may be concluded to underly the mineralizing sites at the Berga Anticline.

Two different types of Sb-bearing mineralization evolved in the area under study. Monotonous Sb mineralization with stibnite and traces of native antimony formed distal to the Fichtelgebirge - Erzgebirge anticline. In a proximal position auriferous vein-mineralizations with stibnite and a great variety of sulfantimonides are discernible in the Brandholz - Goldkronach mining sites. In the last of which, W and Sn, typical granite-related elements were discovered in fahlores and bournonite. Not just the trace element variation, but the Pb-isotopes (Hoehndorf & Dill 1986) point to a close relationship to the Late Variscan granites. The "monotonous" Sb mineralizations show a small variation in their Pb/Pb isotope composition. Since the $^{232}Th/^{204}Pb$ model age is higher than assumed in the Stacey-Kramers model, a derivation from a Precambrian source is held to be a reasonable interpretation. Contrary to this, the "polymetallic" Sb-Au-mineralization displays a larger spread due to different host rocks and Variscan granites which apparently contributed to the element composition of these vein mineral assemblages. The story of S derivation in Sb sulfides resembles that for Pb. A derivation of sulfide sulfur from the host rocks may be a plausible explanation for the parallel trend of the marine sulfate age curve and Sb sulfide S-isotope data plotted as a function of host rock (Dill & Nielsen 1986). Variation in time bears a quite constant difference of about 20 ‰ between sulfate of host rock and sulfide $\delta^{34}S$, so that a reduction of host rock sulfate is clearly implied.

TYPE III GRANITE-RELATED ORE DEPOSITS

Posttectonic granites from the NE Bavarian Basement were dated at 320 to 280 Ma (Besang et al. 1976, Wendt et al. 1988). According to Chappel and White's classification scheme (1974) most of them can be defined as S type granites. These granitoids gave rise to the formation of intra- and perigranitic Sn-, W- and U deposits. Moreover there are K-feldspar pegmatites with appreciable amounts of Li-phosphates (Forster & Kummer 1974) bound to these granites in the Oberpfalz Mts., in specific.

Of particular interest are the large pegmatite bodies of Pleystein, Hagendorf and the aplite of the "Silbergrube" near Waidhaus, the last of which is still being operated. Hagendorf, in specific, contains relatively high amounts of phosphate and is in its central part abundant in ore minerals (Fe-rich sphalerite, columbite, hematite, molybdenite, uraninite) indicative of a high temperature of formation.

W ore mineralization with scheelite porphyrablasts in talcized vesuvianite fels of pyrometasomatic origin is spatially related to the older Fichtelgebirge granite in the steatite open pit of Goepfersgruen (Jung & Hoell 1982, Dill 1985a).

Tin mineralization occurs in a great variety of mineral assemblages in the granite (e.g. tin granite) and its immediate surroundings (e.g. Buechig), although it does not attain the economic significance of tin-mineralized granites of the Erzgebirge/GDR, which represents the NE prolongation of the Erzgebirge - Fichtelgebirge anticline. Pitchblende deserves special attention

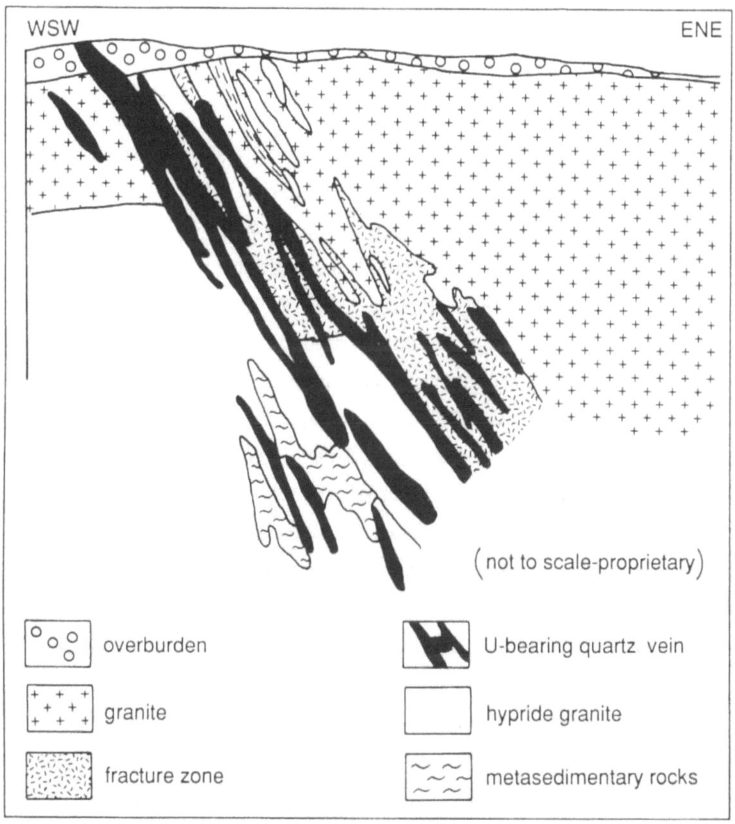

Fig. 9 Cross section through the silicified mineralized structure zone of Grossschloppen U deposit (redrawn from Dill & Kolb 1984).

because it permits a clear determination of the age of formation of these type-III ores. The oldest pitchblende gave an upper intercept in the discordia plot at 295 ± 4 Ma (Carl et al. 1983) and is accompanied by coffinite, brannerite, and various sulfides, arsenides and selenides (Dill 1985a). While the U mineral assemblages bear witness of its close link to the Late Variscan granites, fluorite mineralization shows a wider time interval reaching to Middle Permian. Sm/Nd isotope analyses of fetid flourite point to CaF_2 precipitation at the Permian - Carboniferous boundary (296.6 ± 23.2 to 281.2 ± 22.9 Ma - Leipziger 1986), whereas Rb/Sr isotope analysis of K-feldspars connected with fluorite yielded 204 ± 4 Ma (Lippolt et al. 1985).

TYPE IV UNCONFORMITY-RELATED ORE DEPOSITS

By the end of Upper Carboniferous the uplift of the Variscan basement commenced and a widespread peneplain engendered denudation of this crustal section under tropical to subtropical climates (Falke 1972). It was not just the time of narrow troughs to be subsided into these basement blocks filled with grey and red bimodal clastic and volcaniclastic sediments (see also type I ore deposits) but also the period when fluorite and barite precipitation came into existence.

Intra-Permian and Lower Triassic reworking of these ores lend support to an age of formation of as early as Upper Carboniferous (Reh & Schröder 1974). Supplementary age data of white mica from fault gouge of F-Ba veins e.g. in the Schwarzwald help to confine fluorite-barite mineralization to the Lower Permian through Lower Jurassic (Bonhomme et al. 1983, Mertz et al. 1986). Remobilization of pitchblende II along faults, crossing the main fluorite-barite lode from Nabburg-Wölsendorf constrain the lower limit of $BaSO_4$ - CaF_2 precipitation at 205.9 ± 2.7 Ma (Dill 1987). These age data, yet, do not exclude that younger remobilization of CaF_2 could have taken place in these fluorite-barite deposits. While it may be assumed for the NE Bavarian vein type mineralization, it is proved for the fluorite mineralization from just beyond the border, where fluorite impregnation was encountered in Turonian sandstones of the Jilové Zone/CSSR (Chrt 1982). Ever since controversial ideas have existed on the origin of F-Ba veins. Baumann and Weber (1977) and Weber and Leeder (1982) put forward that these vein mineralizations have derived from a simatic abyssal magmatism controlled by deep-seated NNE-SSW distension zones. Bernard (1975) and Walther (1983) advocated that these ore mineralizations are related to the highly differentiated Late Variscan granites. Dill (1985a) stressed these deposits in the study area to have been spacely linked to the Late Variscan unconformity and originated from crustal fluids.

Current S and Sr-isotope analyses (Hofmann & Baumann 1984, Dill & Carl 1986, Dill & Nielsen 1987) - they shall not be discussed in full detail in this paper; the reader is referred to the publications cited above - discard any hypothesis on these Ba-F-mineralizations to have been vented from a mantle source. By contrast, these results favour an idea of these veins to have been generated by mixing of fluids from "per descensum" and "per ascensum" processes. Fluid inclusion studies yielding low temperatures of formation (T_h: 50 °C to 180 °C) are consistent with this unconformity-model. Considering these type IV unconformity-related vein type deposits attention has to be drawn to the Permo-Carboniferous clastic and volcaniclastic rocks being spread out on the Late Variscan peneplain. They have already been treated as to their role as host for type I ore deposits and now have to be considered as "low metal concentration" or "source rock" for these type IV vein type deposits. Considering them all the wheel has come full circle.

REFERENCES

BACHMANN, M. 1955. Der Antimonerzbergbau in Ostthüringen. Erzmetall 7, 7-14.

BAUMANN, L. & WEBER, W. 1977. Deep faults simatic magmatism and the formation of mineral deposits in Central Europe outside the Alps. In: Metallogeny and plate tectonics in the NE-Mediterranean, Belgrad, 541-551.

BEHRENS, M. 1981. Gräfenthaler Horst und Teuschnitzer Mulde im Fränkischen Schiefergebirge (Exkursion E, "Nördlicher Frankenwald" am 24. April 1980. Jber. Mitt. oberrhein. Ver. N. F. 63, 43-61.

BERNARD, J. H. 1975. Origin and regional extent of the Mesozoic fluorite - barite mineralization of the Bohemian Massif. Vestnik Ustreniko Ust. geol. 50, 357-359.

BESANG, C, HARRE, W., KREUZER, H., LENZ, H., MUELLER, P. & WENDT, I. 1976. Radiometrische Datierung, geochemische und petrographische Untersuchungen der Fichtelgebirgsgranite. Geol. Jb. E8, 3-71.

BLÜMEL, P. 1978. Die Bedeutung kontinuierlicher und diskontinuierlicher Mineralreaktionen als PT-Indikatoren der Metamorphose im prä-permischen Kristallin Süddeutschlands. Z. dt. geol. Ges. 129, 359-375.

BONHOMME, M.G., BÜHMANN, D. & BESUS, Y. 1983. Reliability of K-Ar dating of clays and silifications associated with vein mineralizations in western Europe. Geol. Rdsch. 72, 105-117.

CARL, C., DILL. H., KREUZER. H. & WENDT, I. 1983. U-Pb dating of ores in NE Bavaria. Terra Cognita 1983, 195-196.

CHAPPELL, B.W. & WHITE, A.J.R. 1974. Two contrasting granite types. Pacific Geol. Tokyo 8, 173-174.

CHRT, J. 1982. Fluorite-barite mineralization of the Bohemian Massif. Bull. du BRGM 2, 107-112.

DILL, H. 1985a. Die Vererzung am Westrand der Böhmischen Masse - Metallogenese in einer ensialischen Orogenzone. Geol. Jb. D73, 3-461.

DILL, H. 1985b. Antimoniferous Mineralization from the Mid-European Saxothuringian Zone; mineralogy, geology, geochemistry and ensialic origin. Geol. Rdsch. 74, 447-466.

DILL, H. 1986a. Geochemical facies analysis of Upper Proterozoic metabiolites (Bohemian Massif/F.R. Germany) as guide to metal exploration. In: E.T. Degens, P.A. Meyers, S. Brassel (eds.), Biogeochemistry of Black Shales, Mitt. Geol.-Pal. Inst. Univ. Hamburg 60, 37-53.

DILL, H. 1986b. Metallogenesis of the Early Paleozoic Graptolite Shales from the Graefenthal Horst. Econ. Geol. 81, 889-903.

DILL, H. 1987. Geologic setting and age relation relationship of fluorite-barite mineralization in southern Germany - with special reference to the Late Variscan unconformity. Min. Dep. 23, 16-23.

DILL, H. & KOLB, S. 1984. The Grossschloppen-Hebanz uranium occurrences a prototype of minerlized structure zones associated with desilification and silicification processes (F.R. of Germany, N. Bavaria). In: Fuchs, H.D. (ed.), Vein-type U deposits, IAEA, Wien, 287-303.

DILL, H. & NIELSEN, H. 1986. Carbon-sulphur-iron-variations and sulphur isotope patterns of Silurian Graptolite Shales. Sedimentology 33, 745-755.

DILL, H. & NIELSEN, H. 1987. Geochemical and geological constraints on the formation of unconformity-related baryte deposits of Central Europe. Journ. Geol. Soc. 144, 97-105.

DILL, H. & CARL, C. 1987. Sr Isotope Variation in vein barites from the NE Bavarian basement: relevance for the source of elements and genesis of unconformity-related barite deposits. Mineral. & Petr. 36, 27-39.

DILL, H., TESCHNER, M. & WEHNER, H. 1988. Petrography, inorganic and organic geochemistry of Lower Permian carbonaceous fan sequences ("Brandschiefer Series") - F.R. Germany. Constraints to their paleogeography and assessment of source rock potential. Chem. Geol. 67, 307-325.

FALKE, H. 1972. The paleogeography of the continental Permian in Central-, West- and in part of South Europe. In: FALKE, H (ed.), Rotliegend Essays on European Lower Permian, Brill, Leiden, 281-299.

FORSTER, A. & KUMMER, R. 1974: The pegmatites in the area of Pleystein - Hagendorf/North Eastern Bavaria. Fortschr. d. Mineralogie, 52, Beih. 1, 89-99.

GUDDEN, H., SCHMID, H. & CARDOSO, J. 1974. Uran in Bayern. Report Geol. Surv. of Bavaria, München. 605 pp.

HOEHNDORF, A. & DILL, H. 1986. Lead isotope studies of strata-bound, vein-type, and unconformity-related Pb, Sb, and Bi ore mineralizations from the western edge of the Bohemian Massif (F.R. Germany). Mineral. Dep. 21, 329-336.

HOFMANN, R. & BAUMANN, A. 1984. Preliminary report on the Sr isotopic composition of hydrothermal vein barites in the Federal Republic of Germany. Min. Dep. 19, 166-169.

HORSTIG von, G. & TEUSCHER, E.O. 1979. Die Eisenerze im Alten Gebirge NE-Bayerns. Geol. Jb. D31, 7-47.

JUNG, R. & HOELL, R. 1982. Wolframvorkommen in Nordost-Bayern. Erzmetall 35, 142-147.

KRUEGER, P. 1973. Die Verteilung der Schwefelisotope in sedimentaeren Eisensulfiden unter Beruecksichtigung der Redoxbeziehungen im Bildungsraum. Abh. d. zentr. geol. Inst. 19, 1-170.

LARGE, D.E. 1980. Geological parameters associated with sediment-hosted, submarine exhalative Pb-Zn deposits: An empirical model for mineral exploration. Geol. Jb. D40. 59-129.

LEIPZIGER, K. 1986. Untersuchungen zur Raumschaffung der Fluoritmineralisationen im Woelsendorfer Flußspatrevier. Ph. D. Thesis Univ. Mainz, 185 pp.

LIPPOLT, H.J., MERTZ, D.F. & ZIEHR, H. 1985. The Late Permian Rb-Sr age of K-feldspar from Woelsendorf mineralization (Oberpfalz, F.R. Germany). N. Jb. Miner. Mh. 1985, 49-57.

MERTZ, D.F., LIPPOLT, H.J. & HUCK, K.H. 1986. K-Ar, ^{40}Ar-^{39}Ar and Rb-Sr investigations on the genesis of the Clara vein deposit/Central Schwarzwald. Dt. Geophysik. Ges. Sym. Karlsruhe, p. 235.

MELNIK, Y.P. & SIROSHTAN, R.J. 1973. Physico-chemical conditions of the metamorphism of cherty-iron rocks. UNESCO Earth Sci. 9, 209-216.

NETSCHAJEW, S. & FRITSCHE, E. 1966. Wesenszüge einer Antimonvererzung im Gebiet von Dorfchemnitz-Hormersdorf (Erzgebirge). Zeitschr. f. angew. Geol. 12, 505-511.

NEUMANN-REDLIN, C., WALTHER, H.-W. & ZITZMANN, A. 1976. The iron ore deposits of the Federal Republic of Germany. Iron ore deposits of Europe Vol. I, 165-186.

NEUMANN, W. & TISCHENDORF, G. 1986. Genesis of sulfide mineralization in pyroclasites of Hartmannsdorf, Granulite Massif. Int. Conference on the Metallogeny of the Precambrian. Geol. Surv. Prag, 81-84.

OKRUSCH, M., OPPERMANN, U., RICHTER, P., SCHUESSLER, U., SEIDEL, E. 1986. Vergleichende Untersuchungen an Metabasiten des ostbayerischen Kristallins. KTB-Report, 1-6.

PFEUFER, J. 1976. Das Schwefelerzvorkommen am Silberberg bei Bodenmais (Bayerischer Wald). Erzmetall 29, 340-355.

PLIMER, I.R. & FINLOW BATES, T. 1978. Relationship between primary iron sulfides species sulphur source, depth of formation and age of submarine exhalative sulphide deposits. Mineral. Dep. 13, 399-410.

REH, H. & SCHRÖDER, N. 1974. Erze. In: Hoppe, W. & Seidel, G. (eds.), Geologie von Thüringen. Haack, Gotha/Leipzig, 867-897.

SAWKINS, F.J. 1984. Metal deposits in relation to plate tectonics. Springer, New York, 325 pp.

SCHMID, H. 1980. Geochemische Prospektion auf seltene Elemente in palaeozoischen Schwarzschiefern des Frankenwaldes. Erzmetall 33, 540-545.

SCHMID, H. & WEINELT, W. 1978. Lagerstätten in Bayern. Geol. Bav. 77, 3-160.

STRUNZ, H. 1953. Mineralien und Lagerstätten in Ostbayern; Grube "Bayerland" südlich Waldsassen. Bosse, Regensburg, 72-75.

STRUNZ, H. 1974. Granites and pegmatites in Eastern Bavaria. Fortschr. Miner. 52, 1-32.

TROLL, G., LINHARDT, E. & SKERIES, R. 1987. Petrographic and geochemical studies on country rock of the Bodenmais (Bavaria) sulphide deposit. N. Jb. Geol. Pal. Mh. 1987/12, 726-752.

URBAN, H. & VACHE, R. 1972. Die Kupfererzlagerstaetten von Wirsberg (Oberfranken) im Lichte neuer Aufschluesse. Geol. Bav. 65, 74-106.

WALTHER, H.W. 1983. Über mögliche Ursachen der unterschiedlichen Verbreitung der postvaristischen Baryte und Fluorite im westlichen Mitteleuropa. Z. dt. geol. Ges. 134, 143-151.

WEBER, W. & LEEDER, O. 1982. Structural setting of the late Paleozoic-Mesozoic F-Ba vein deposits of Central Europe. Bull. du BRGM 2, 93-99.

WENDT, I., HÖHNDORF, A., KREUZER, H., MÜLLER, D. & STETTNER, G. 1988: Gesamtgesteins- und Mineraldatierungen der Steinwaldgranite (NE-Bayern). Geol. Jb. E42, 167 - 194.

WOLF, H. 1971. Die Schwefel-Kieslagerstaette "Bayerland" bei Waldsassen in der Oberpfalz und ihr geologischer Rahmen. Acta Albertina Ratisb. 31, 57-100.

WURM, A. 1961. Geologie von Bayern. Borntraeger, Berlin. 555 pp.

Oberpfalz Deep Seismic Reflection Survey and Velocity Studies

J. Schmoll[1], R. Bittner[2], H.-J. Dürbaum[3], T. Heinrichs[4], R. Meißner[2], C. Reichert[3], T. Rühl[5] and H. Wiederhold[3]

[1] Ludwig-Barnay-Str. 9, D-3000 Hannover 1, FRG
[2] Institut f. Geophysik, Olshausenstr. 40, D-2300 Kiel, FRG
[3] Bundesanstalt f. Geowissenschaften u. Rohstoffe/Niedersächsisches Landesamt f. Bodenforschung, Stilleweg 2, D-3000 Hannover 51, FRG
[4] Institut f. Geologie u. Dynamik der Lithosphäre, Goldschmidtstr. 3, D-3400 Göttingen, FRG
[5] Geophysikalisches Institut, Hertzstr. 16 Bau 42, D-7500 Karlsruhe 21, FRG

1 Geological Aims, Planning, Field Investigations

1.1 Purpose of the Investigations

The seismic lines discussed in this paper explore a central part of the Variscan Belt of Europe. This belt is more than 1000 km wide and can be subdivided - in the Central European segment - into several zones, which were first recognized by Kossmat (1927). The zonal boundaries are now regarded as sutures which originated from the closure of several Paleozoic basins. First results on the geophysical-gelogical investigations in the Oberpfalz area are found in DEKORP Research Group (1988).

The DEKORP 4 and 4-Q lines, as well as the closely related KTB 8501 to 8506 lines, were intended to scan the border zone between the Moldanubian Zone and the northward adjacent Saxothuringian Zone (Fig. 1). The Saxothuringian (ST) represents the infill of a basin which opened during the Cambro-Ordovician beyond the rift stage at least into a narrow ocean, and the Moldanubian (MN) contains blocks of pre-Variscan crust (and their Paleozoic cover) which were reactivated by the Variscan orogeny and thrusted toward the NW over the Saxothuringian foreland. Tectonic deformation and co-related metamorphism are polyphase and have produced a highly complex crustal structure For more detailed information see Franke (this Vol.). Special emphasis has been put on the area S of Erbendorf which was under consideration and has since been selected as the site of the Continental Deep Drilling Project KTB.

The long DEKORP 4 profile was intended as a general cross-section from the Teuschnitz syncline in the N to the MN rocks in the S. The KTB 8501 to 8506 lines were arranged in form of a grid to explore the ST/MN boundary region near Erbendorf (i.e., around the KTB drilling site). This array of lines (Fig. 1) represents a first step in obtaining a 3-D view of the target area.

The DEKORP 4-Q line is positioned at a right angle with respect to the DEKORP 4 line near its southern end. It was intended to investigate important tectonic structures such as the Bavarian "Pfahl" and the

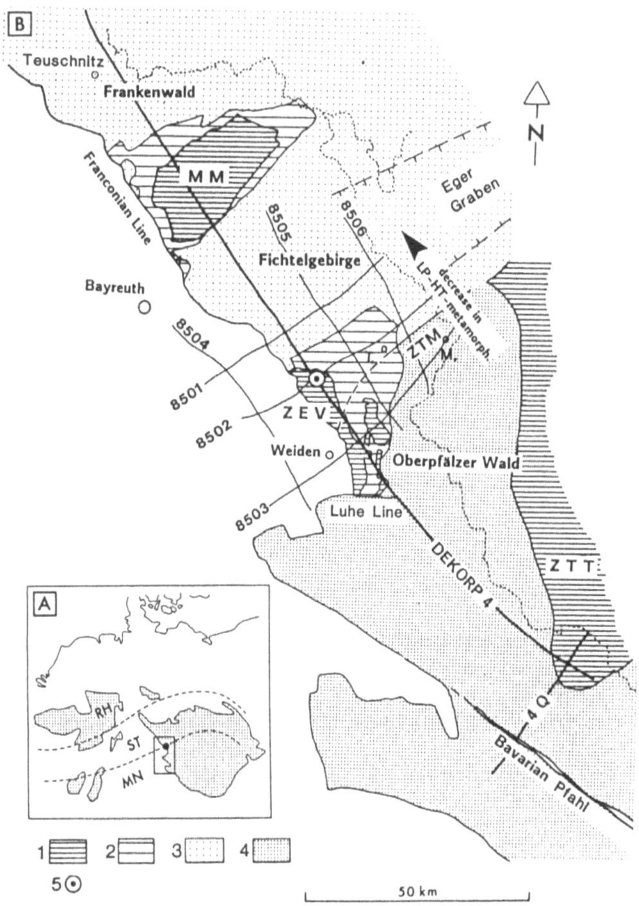

Fig. 1. Geological sketch map of NE Bavaria with the DEKORP-KTB seismic network indicated. A Variscan basement outcrops in Middle Europe with zones after Kossmat (1927). RH Rhenohercynian; ST Saxothuringian; MN Moldanubian. Inset: study area with site of deep drilling (KTB). B Geological map with main tectono-metamorphic units and deep reflection seismic network. 1 MP metamorphic nappes; 2 lower nappes; 3 Saxothuringian; 4 Moldanubian; 5 KTB deep drilling site. Late- to post- Variscan granites are not shown. Reflexion seismic network: heavy lines DEKORP; thin lines KTB. T. Tirschenreuth; M. Mähring; ZTM Zone of Tirschenreuth-Mähring; MM, ZEV, ZTT see list of abbreviations

western margin of the Tepla-Taus-Zone (ZTT), which is marked, in this area, by an important volume of mafic metamorphic rocks (Hoher Bogen).

The KTB 8505 and 8506 lines were arranged to explore the Tirschenreuth/Mähring segment of the ST/MN boundary region. The Erbendorf segment is crossed by the main DEKORP 4 line. KTB 8504 is situated in the Mesozoic foreland, in an area which is probably underlain by a westward continuation of the Erbendorf-Vohenstrauss Zone (ZEV).

6 Abbreviations: DEKORP Deutsches Kontinentales Reflexionsseismisches Programm; KTB Kontinentales Tiefbohrprogramm der Bundesrepublik Deutschland; EB Erbendorf body; FL Franconian Line; HB Hoher Bogen; MM Münchberg Massif; MN Moldanubian; ST Saxothuringian; ZEV zone of Erbendorf-Vohenstrauß; ZTT zone of Tepla-Taus; CDP common depth point; CSG common shot gather; DMO dip moveout; ESP expanding spread; PG geophone position; TWT Two-way traveltime

While the above lines yield cross-sections with respect to the main (Variscan) tectonic trend (SW/NE) in the area, the KTB 8501 to 8503 lines are arranged parallel to the strike. They cut across the NW/SE--directed system of block faults at the SW margin of the Bohemian Massif, the Franconian Line (FL) and associated faults. These faults were expected to produce important steps in the reflection surfaces even at greater depths.

Figures 2 and 3 give predictive geological cross-sections by Franke (1984 unpubl.) for DEKORP 4 and KTB 8502, respectively.

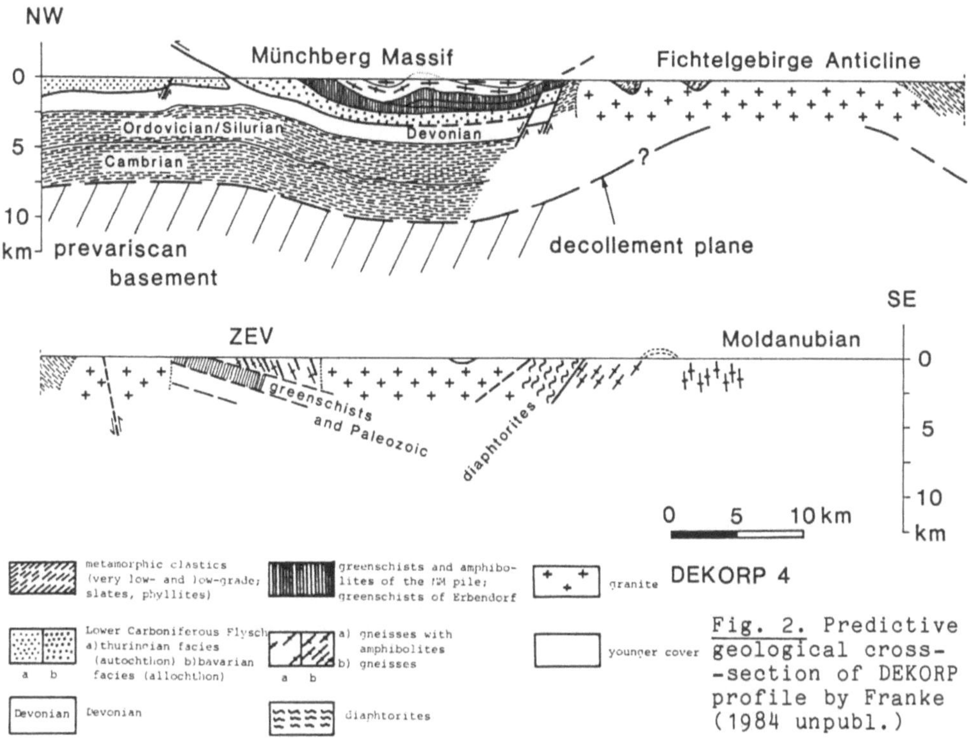

Fig. 2. Predictive geological cross--section of DEKORP 4 profile by Franke (1984 unpubl.)

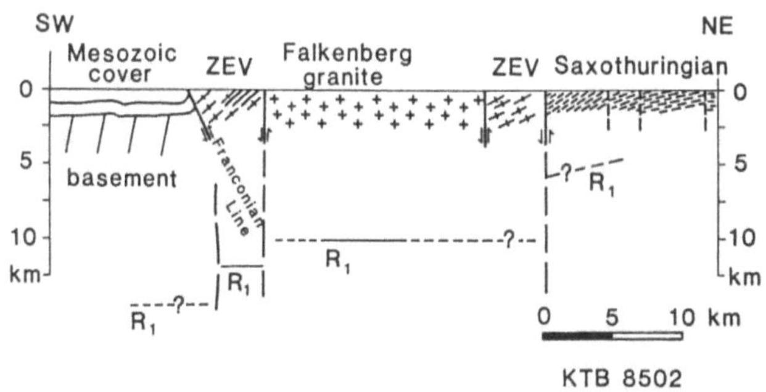

Fig. 3. Predictive geological cross-section of KTB 8502 profile by Franke (1984 unpubl.). R_1 refraction horizon

1.2 Field Technique

The survey of the DEKORP 4 and 4-Q lines was carried out together with that of the six KTB lines in a joint field survey from June to November 1985. Uniform field parameters were chosen to allow a good comparison. A series of wide-angle observations and two expanding spread arrangements were integrated in the survey (DEKORP Research Group 1988). This work was carried out by the contractor PRAKLA-SEISMOS AG in cooperation with crews from universities and geological surveys with their own equipment for the expanding spreads and part of the wide-angle observations. For the accompanying experiments explosives had to be used to provide a suitable energy source whereas for the seismic reflection survey Vibroseis was used. Important data of the Oberpfalz survey 1985 are listed in Table 1. The whole operation was carried out within the framework of the DEKORP project (DEKORP Research Group 1985).

In contrast to the DEKORP 2-South survey (DEKORP Research Group 1985) where Mesozoic and Cenozoic sediments represent the prevailing outcrops, in the Oberpfalz pre-Permian metamorphic and crystalline rocks are present at the surface. This had an important influence on the decision whether explosion seismics (as on DEKORP 2-South) or Vibroseis should be used; due to the difficulty in estimating the shot-hole drilling progress, the Vibroseis method was chosen. Another reason was the very good experience of the Schwarzwald seismic reflection survey in 1984 (Lüschen et al. 1987).

Table 1. Important data of the Oberpfalz survey 1985

Profile	Length	Source parameters for accompanying investigations	
DEKORP 4	186.88 km	Number of shots:	96
DEKORP 4-Q	36.64 km	Drilling rigs:	6
KTB 8501	47.44 km	Drilling depth:	30 m
KTB 8502	50.48 km	Boreholes per shot:	3
KTB 8503	56.80 km	Charge per shot:	90 kg
KTB 8504	50.72 km		
KTB 8505	54.80 km		
KTB 8506	42.60 km		
Total	526.36 km		

The field parameters were derived from experience of the former surveys. Regarding the geometry two facts were decisive: a sufficiently high degree of coverage also in the shallow subsurface should be reached, and an adequately long offset should provide enough normal moveout for an estimation of velocities. These considerations resulted in a compromise: an asymmetric split-spread with 4 km and 12 km spreads, respectively. A list of the parameters used is given in Table 2. The processing was carried out by the DEKORP Processing Center at Clausthal (Bortfeld et al. this Vol.) and, in addition, by several contractors.

Table 2. Field parameters for the near-vertical seismic reflection survey in the Oberpfalz 1985

Method	Vibroseis
Equipment	SERCEL SN 368, telemetric, 200 channels
Sampling rate	4 ms
Sweep	12 - 48 Hz upsweep
Sweep length	20 s
Listening time	12 s
Record length	20 s + 12 s = 32 s
Filters	12 Hz, 18 dB/oct; 89 Hz, 72 dB/oct
Number of geophone groups	200
Geophone group interval	80 m
Geophone pattern	24 fold, in-line
Pattern length	80 m
In-line offset	200 m
Geometry	asymmetric split-spread 4.12 km-0.2 km-VP-0.2 km-12.12 km
Number + type of vibrators	5 VVEA, 16 tons peak force
Vibrator pattern length	145 m/49 m
Vibrator point interval	80 m
Vertical stacking	10 fold
Coverage (ideal)	100 fold
Mean coverage (actual)	75 fold
Datum level	500 m a.s.l.

2 Results of the Near-Vertical Seismic Reflection Survey

All profiles show a great number of reflections in the upper as well as in the lower crust, and altogether they form a good base for a structural interpretation of the crust. The DEKORP 4 profile, which runs perpendicular to the general Variscan strike over a length of about 190 km, represents a key to the structural character of the area under investigation. Therefore, the discussion of this profile will serve as an introduction to the results of the network of the other lines.

2.1 DEKORP 4 Profile, Structural Overview

A line drawing showing the whole migrated DEKORP 4 profile is presented in Fig. 4a. Parts of the migrated record section are reproduced in Figs. 5 through 13; their positions are given in Fig. 4b. In the TWT(two-way-traveltime) range of the first seconds clear and strong reflections allow a convincing correlation with known geological elements (Fig. 2).

In the NW, below the folded Carboniferous of the Frankenwald, strong subhorizontal reflectors at approximately 1 s TWT (i.e., about 2.3 km depth) are observed (Fig. 5). They are interpreted as the boundary between Upper-Devonian (Frasnian) spilites and the overlying late-Devonian and Lower-Carboniferous sediments, as these spilites which are outcropping NE of DEKORP 4 have an appreciably higher acoustic impedance within an otherwise acoustically nearly homogeneous series of anchimetamorphic sediments. Numerous SE-dipping short reflections indicate the

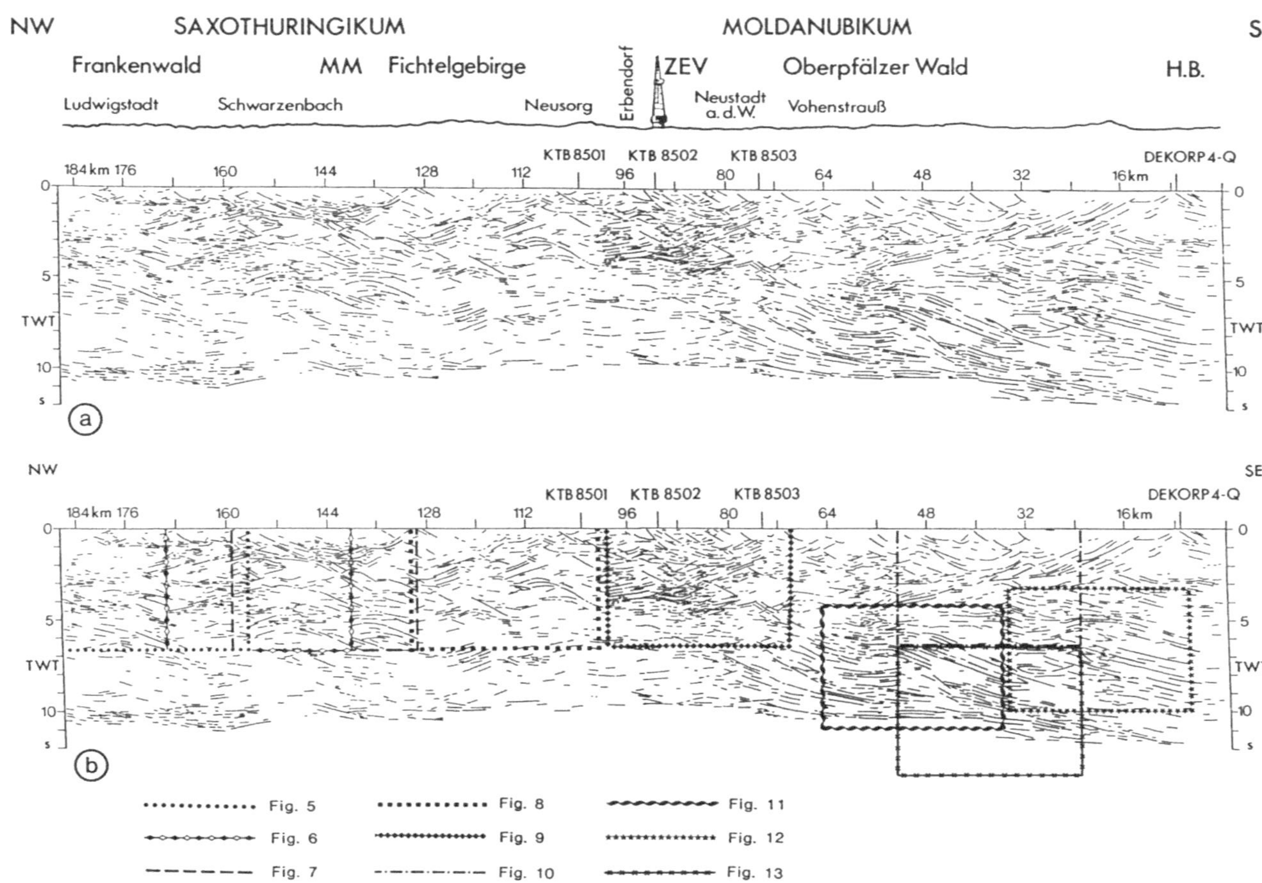

Fig. 4. DEKORP 4 profile: a Line-drawing of the migrated section. b Frames indicate that part of the section which is given in more detail by a following figure

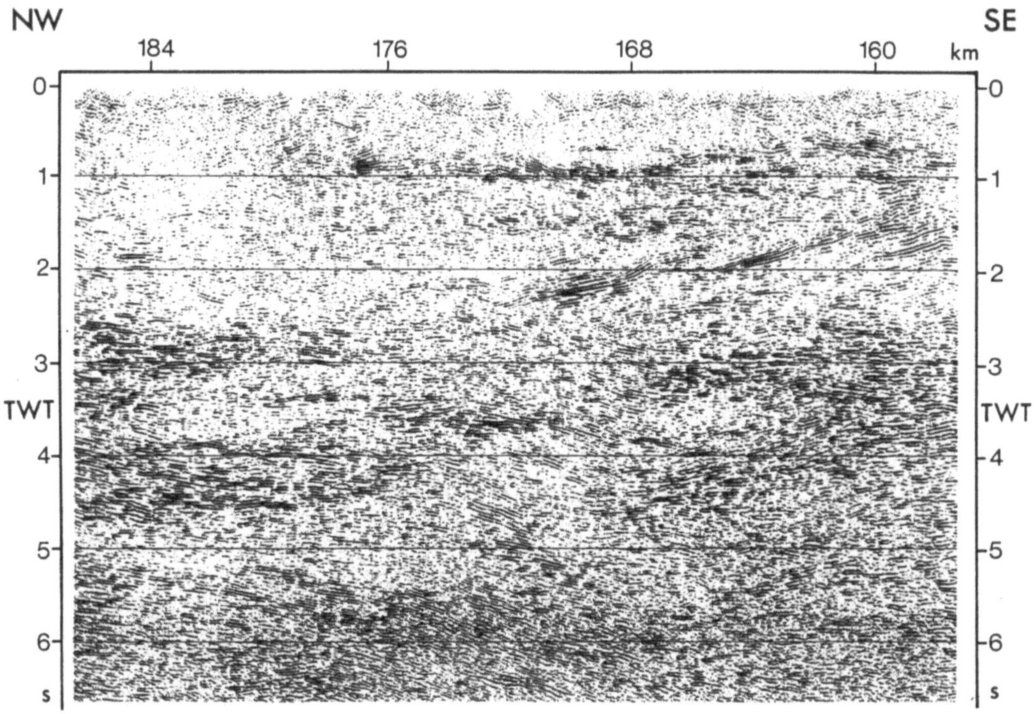

Fig. 5. DEKORP 4 profile; part of migrated section (see Fig. 4b)

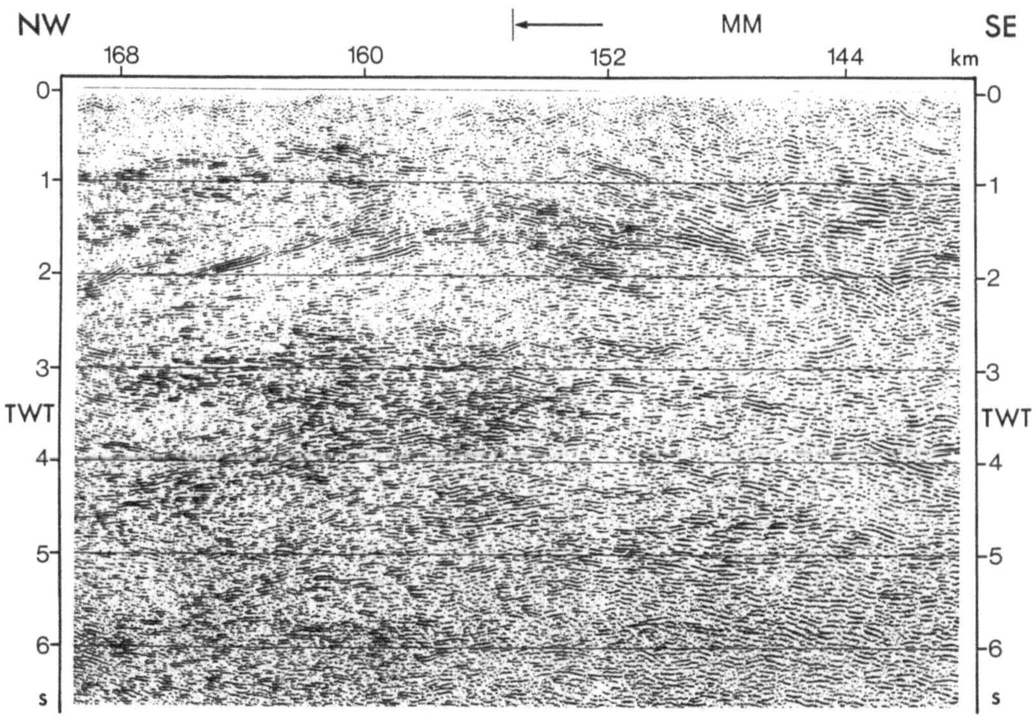

Fig. 6. DEKORP 4 profile; part of migrated section (see Fig. 4b)

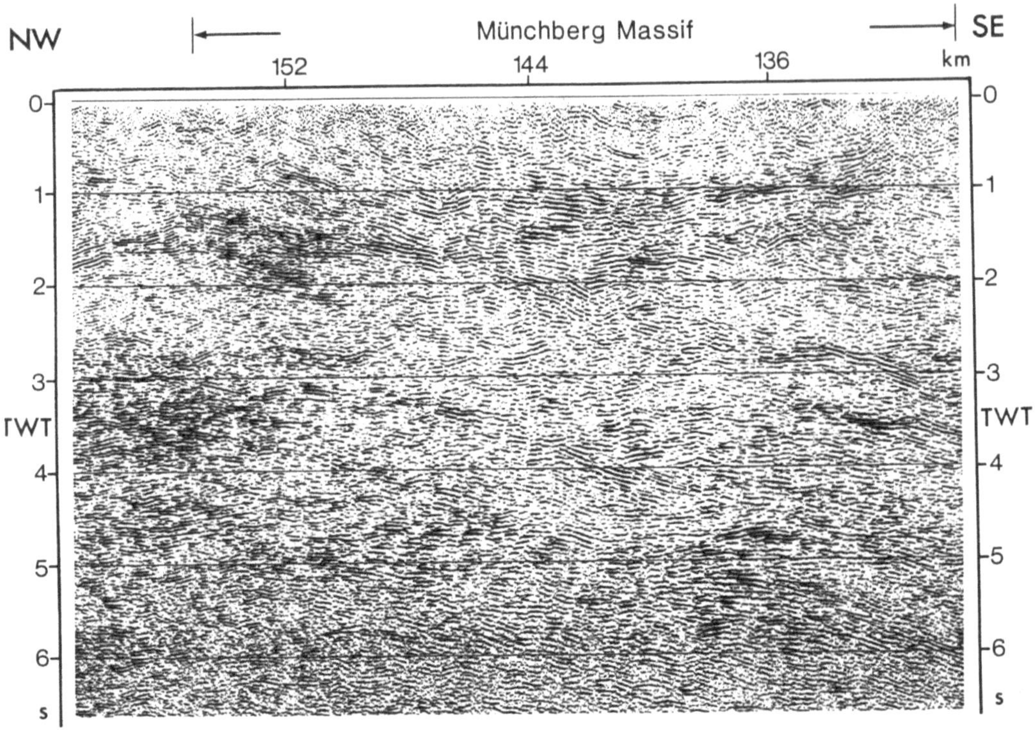

Fig. 7. DEKORP 4 profile; part of migrated section (see Fig. 4b)

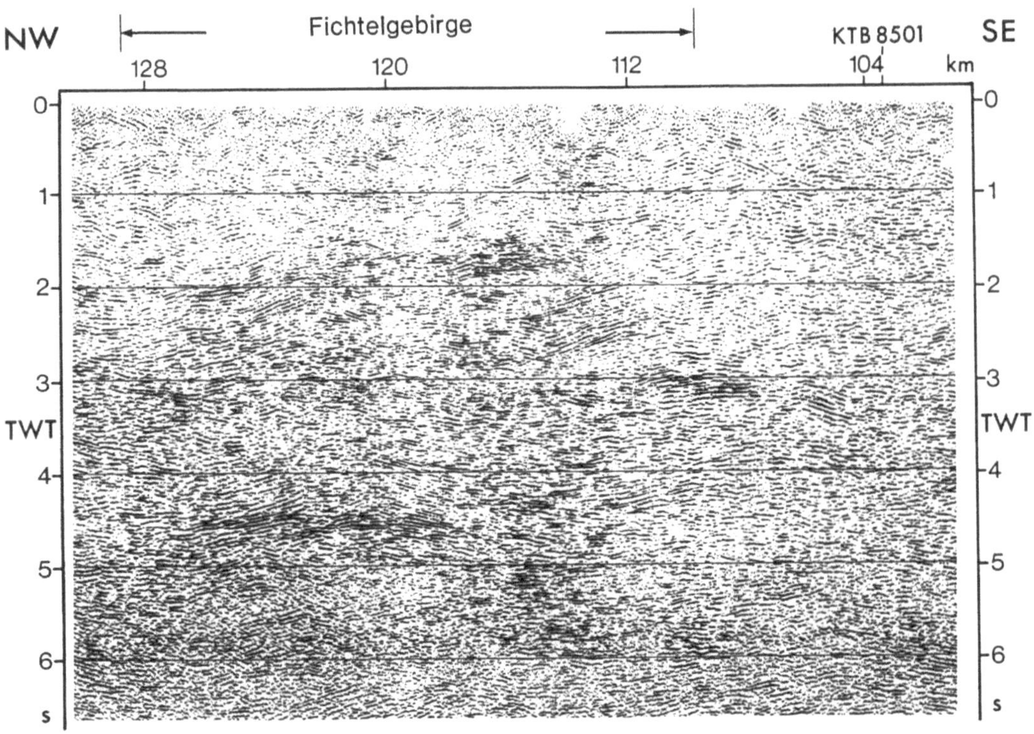

Fig. 8. DEKORP 4 profile; part of migrated section (see Fig. 4b)

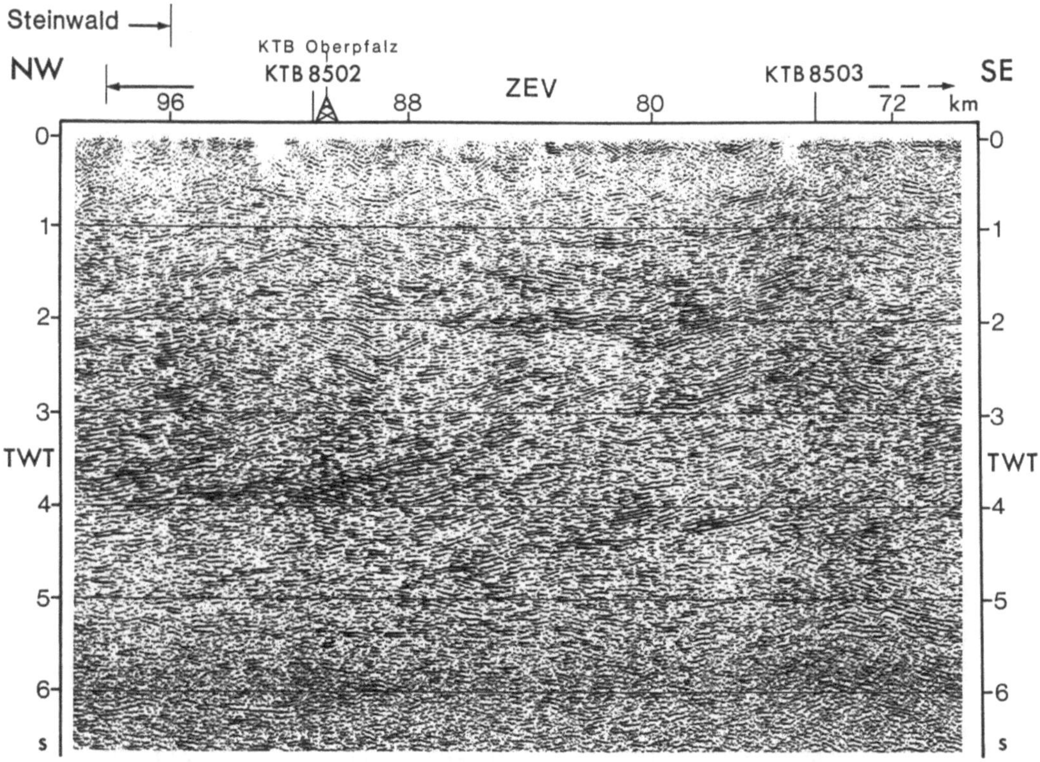

<u>Fig. 9.</u> DEKORP 4 profile; part of migrated section (see Fig. 4b)

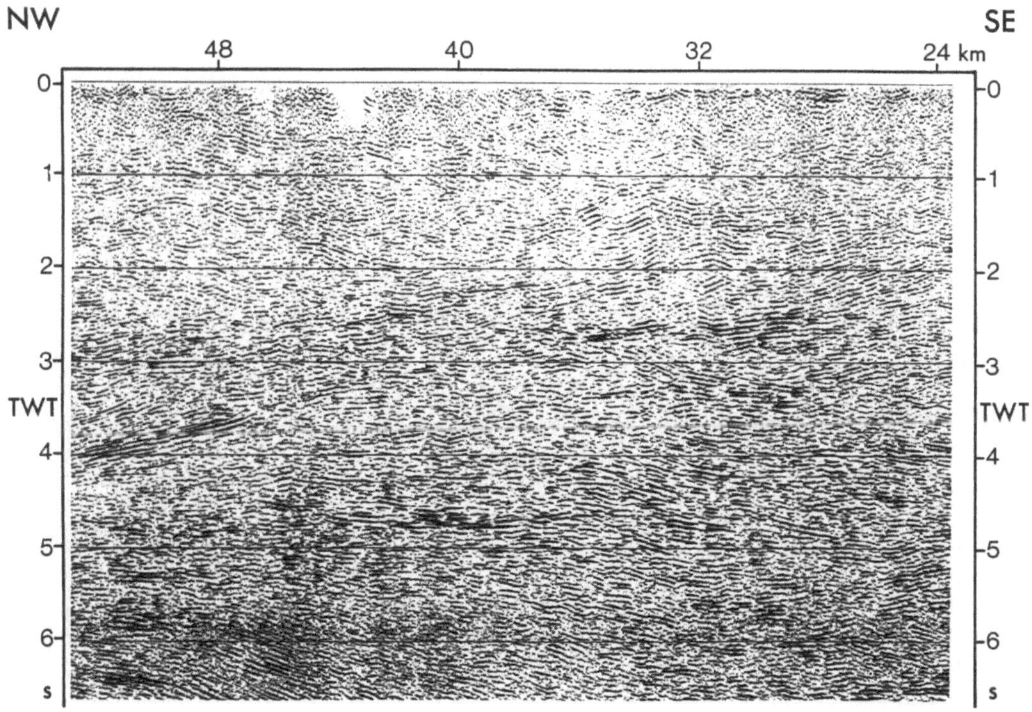

<u>Fig. 10.</u> DEKORP 4 profile; part of migrated section (see Fig. 4b)

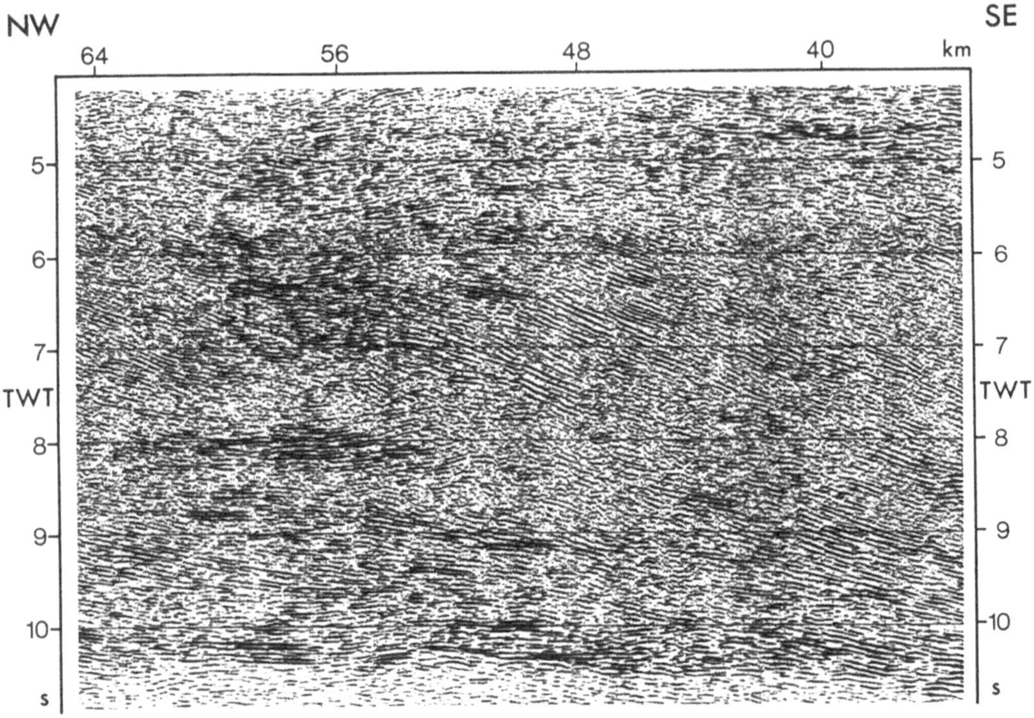

Fig. 11. DEKORP 4 profile; part of migrated section (see Fig. 4b)

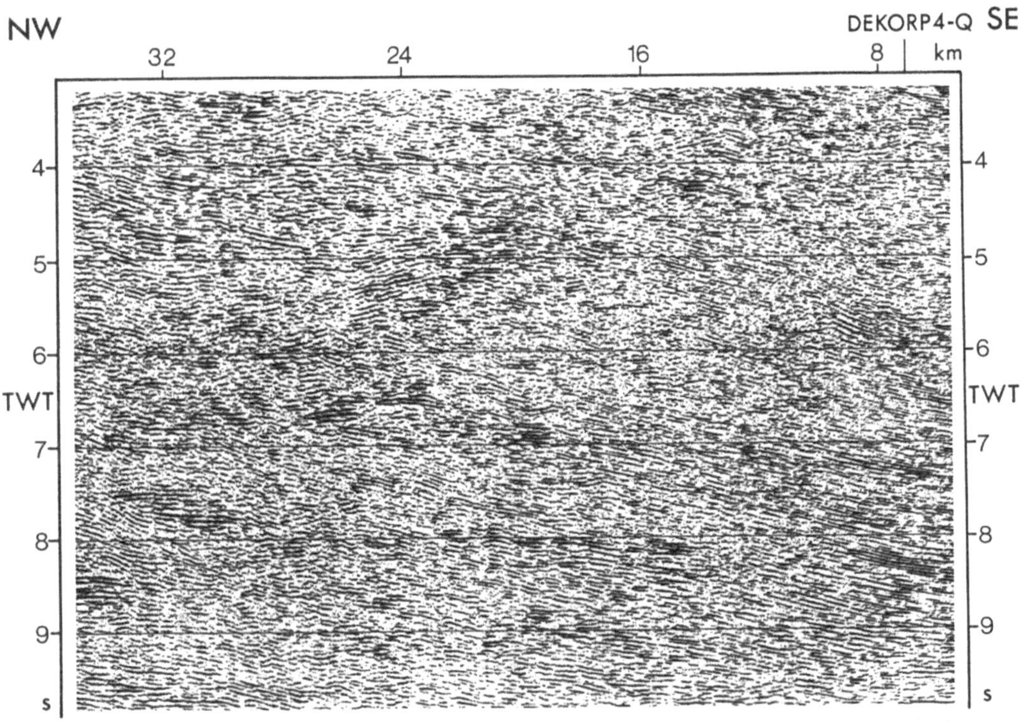

Fig. 12. DEKORP 4 profile; part of migrated section (see Fig. 4b)

tectonic deformation of these sediments. Further to the SE, a bowl-type structure appears in the uppermost crust with its lowest point at about 1.3 s TWT (i.e., about 3.6 km depth, Figs. 6 and 7). It models the base of the Münchberg Massif between km 130 and 155.

Below this shallow domain a wide antiformal structure is seen (about km 144 to 168) between 1.5 and 2.0 s TWT (i.e., about between 4 and 5 km depth) (Fig. 6). Probably, this subsurface structure corresponds to surface structures NE of DEKORP 4, namely the SW plunging Berga and Hirschberg anticlines.

In the Fichtelgebirge, the upper 1.5 s contain only poor reflections, probably due to the lack of lithological contrasts in the granite mountains (Fig. 8). At a level of about 2 s TWT - near km 120 - an antiformal structure is observed as predicted by Franke (1984 unpubl.) (Fig. 2).

Between km 70 and 94, another bowl-shaped structure is observed which corresponds to the nappe complex of the ZEV (Fig. 9). The deepest part near km 82 shows the first good reflections at 1.2 s TWT, corresponding to 3.5 km depth, but similarly arranged reflections can be observed down to 2 s TWT, corresponding to about 6 km depth.

Below the ZEV - down to about 4.0 to 4.5 s - there are a great number of prominent reflections with various dips. They are especially conspicuous between km 84 und 94, where NW dipping and NW rising reflections join in a wedge-type shape.

Between the intersections of DEKORP 4 with KTB 8503 and DEKORP 4-Q the reflections of the uppermost 2 s have varying dips with local antiforms and synforms (Figs. 4a and 10). This is in accordance with the arc-shaped strike of the MN southeast of the Luhe Line which generally dips NE, but there is no detailed local control of the true seismic dip, yet. At the utmost SE end of DEKORP 4 - at the intersection with the DEKORP 4-Q profile - SE-dipping reflections in the first 1.5 s are within the Hoher Bogen (HB) nappe complex, a part of the ZTT.

The general impression of the NW portion of DEKORP 4 is that it contains numerous strong reflections in the upper crust but much less and only weak reflections in the lower crust. This image is in contrast to the SE part of the profile, where numerous strong reflections are observed both in the upper and in the lower crust. The results obtained along the entire DEKORP 4 profile represent a pronounced contrast to all BIRPS lines (12,000 km) around the British Isles (Matthews et al. 1987), the Urach lines (Meissner et al. 1982) and to the Black Forest profiles (Lüschen et al. 1987) with their poorly reflecting upper crust und highly reflecting "laminated" lower crust.

A bunch of strong SE dipping reflectors with remarkable energy appears in the lower crust of the SE portion of the DEKORP 4 profile. They seem to intersect the entire lower crust (Figs. 11 to 13). At this location - at km 24 to 32 - there are two possibilities of interpretation: either the Moho is strongly dipping SE situated 3 km deeper at the SE end of the profile, or the strong reflector at km 24 at 11.7 s TWT belongs to the Moho discontinuity, and there is a step in the Moho with an offset of more than 3 km.

The whole lower crust in the SE portion, especially between km 32 and 64, suggests an image of intensive thrust tectonics with ramp-like structures. The reflections are supposed to originate from changes of the acoustic impedance at numerous shear planes along a tremendous thrust zone. Such changes might be due to mylonitization (Fountain et al. 1984), possible mineralization by lubricating fluids and/or due to imbrications, i.e., the original state of the rock material is highly

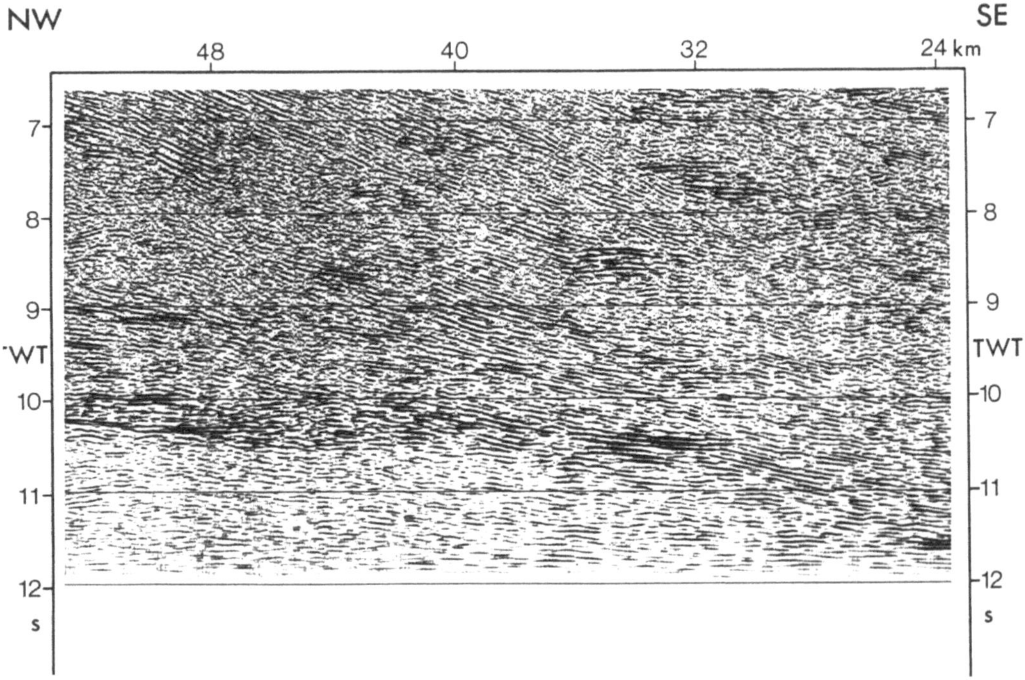

Fig. 13. DEKORP 4 profile; part of migrated section (see Fig. 4b)

modified. Perhaps an additional contribution is provided by formerly subhorizontally layered portions of a laminated lower crust which are now in a different position, more or less parallel to the ramps.

These SE-dipping "thrust reflections" in the SE portion of the profile are restricted to the crust below 4 to 5 s TWT. In the time range between 4 and 5 s TWT they are cut off by subhorizontal reflections which might be interpreted as a master decollement (Fig. 4a). Above this domain strong NW-dipping reflections appear, partly with varying dip. There are several possibilities for interpreting these structures. The NW-dipping reflectors might represent SE-vergent overthrusts, set up in the context of backfolding and development of wedge structures. In part, the reflections might come from outside the vertical plane of the profile and be caused by centroclinal strike of the crustal structures.

2.2 KTB 8501 Through 8506 Profiles, the Suture Zone and the Erbendorf Body

The KTB 8505 profile is the NW/SE running line next to DEKORP 4 in the NE. It is characterized by SE-dipping reflections crossing the entire

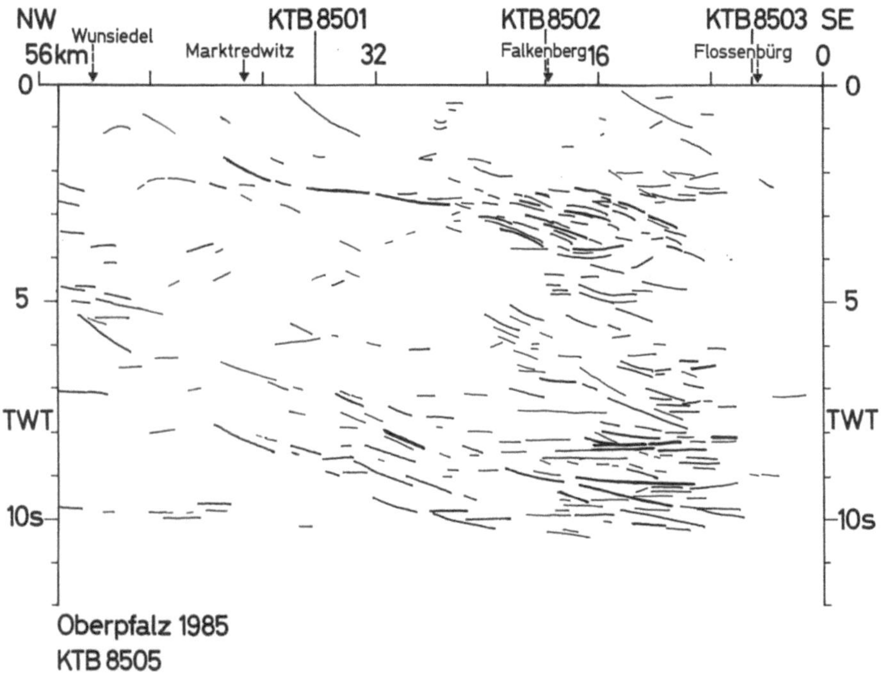

Fig. 14. KTB 8505 profile; line drawing of the migrated section

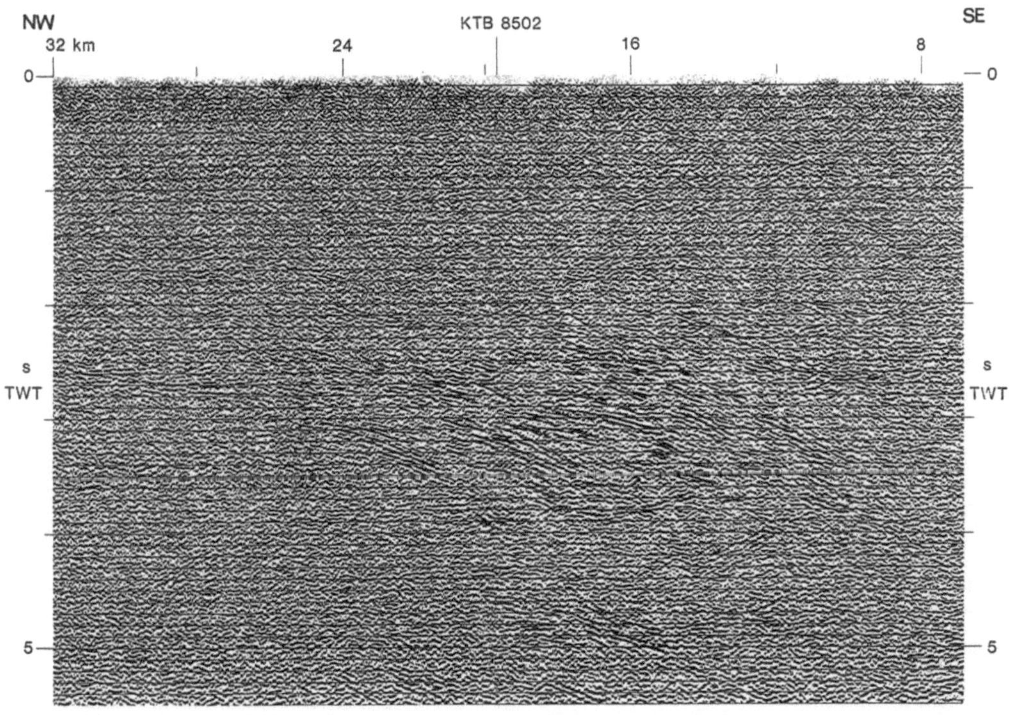

Fig. 15. KTB 8505 profile; part of migrated section with the Erbendorf body

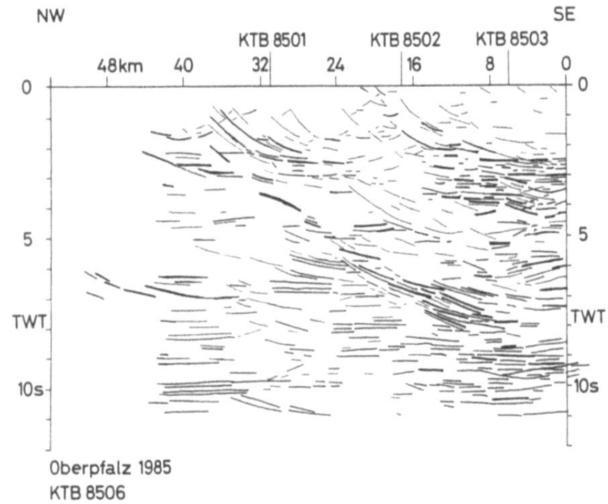

Fig. 16. KTB 8506 profile; line drawing of the migrated section

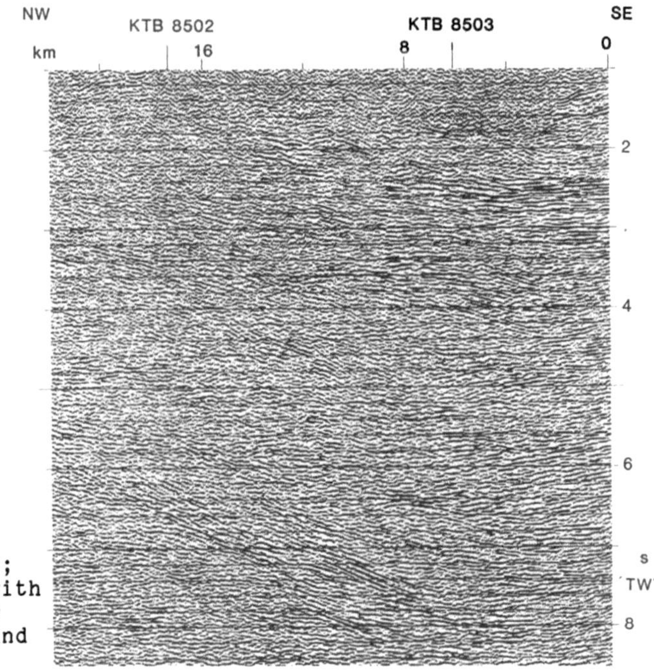

Fig. 17. KTB 8506 profile; part of migrated section with Erbendorf body and dipping reflections in the upper and lower crust

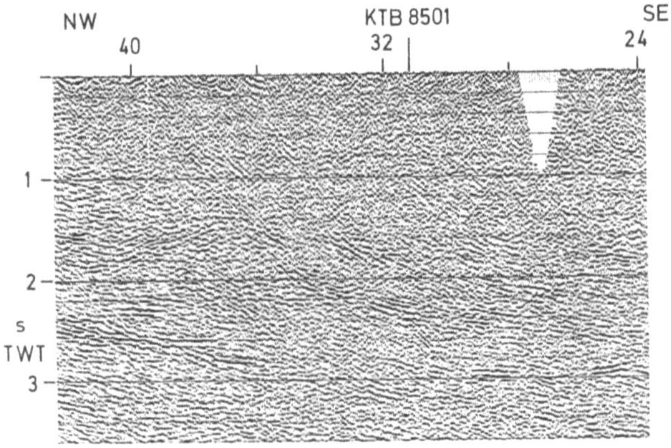

Fig. 18. KTB 8506 profile; part of migrated section with listric shear planes

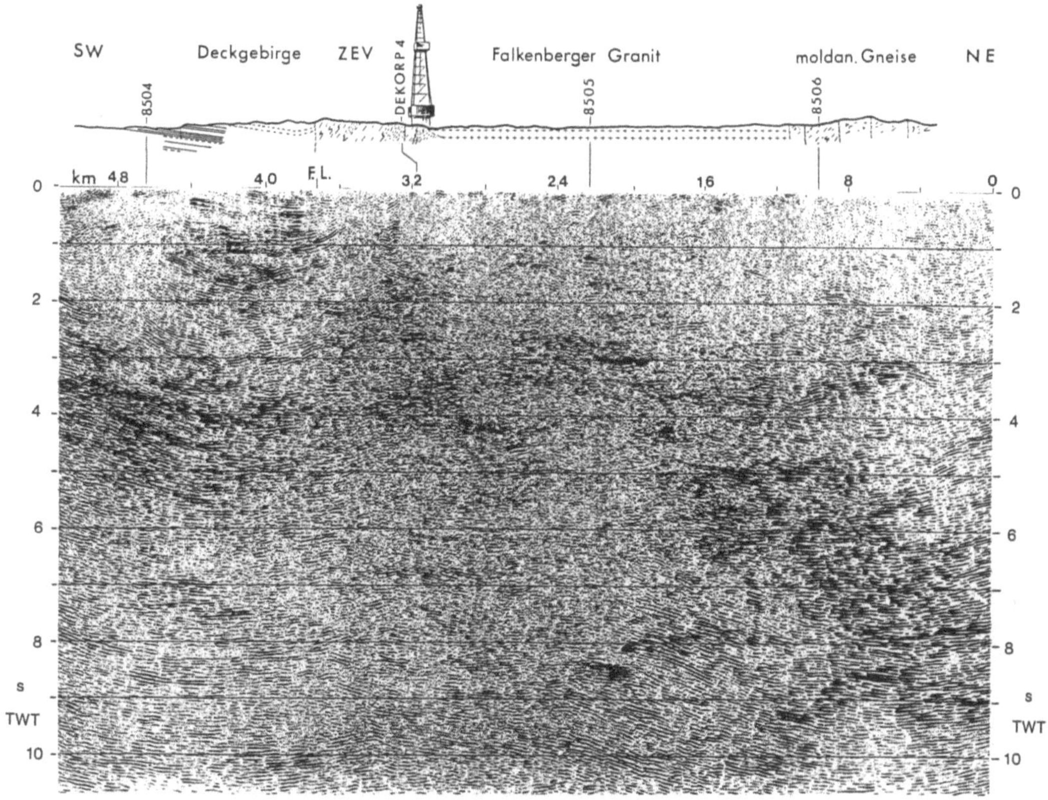

Fig. 19. KTB 8502 profile; migrated section

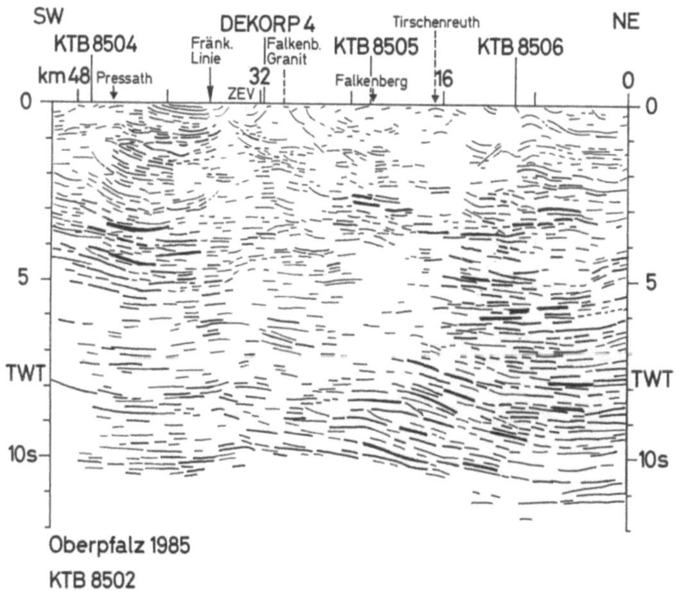

Fig. 20. a KTB 8502 profile; line drawing of the migrated section

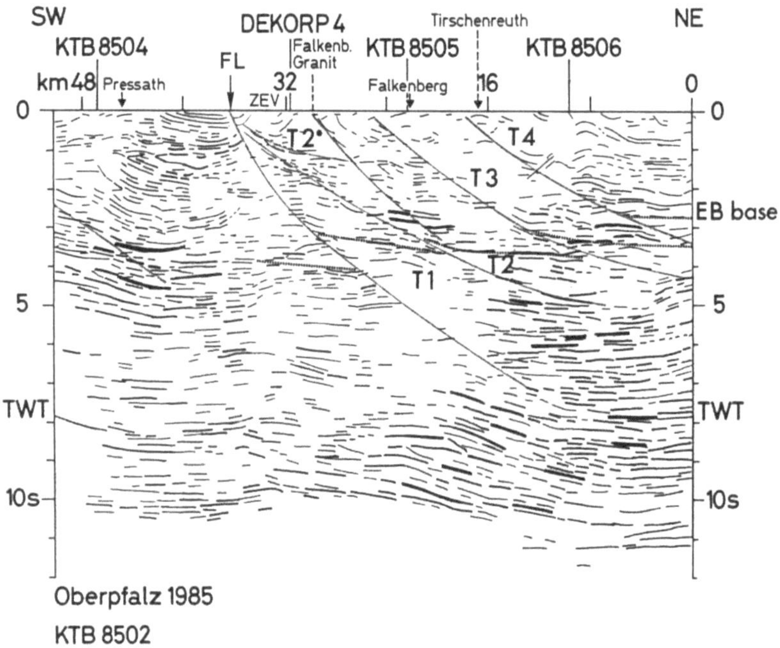

Fig. 20. b the same as a with NE dipping faults indicated (T1 to T4).

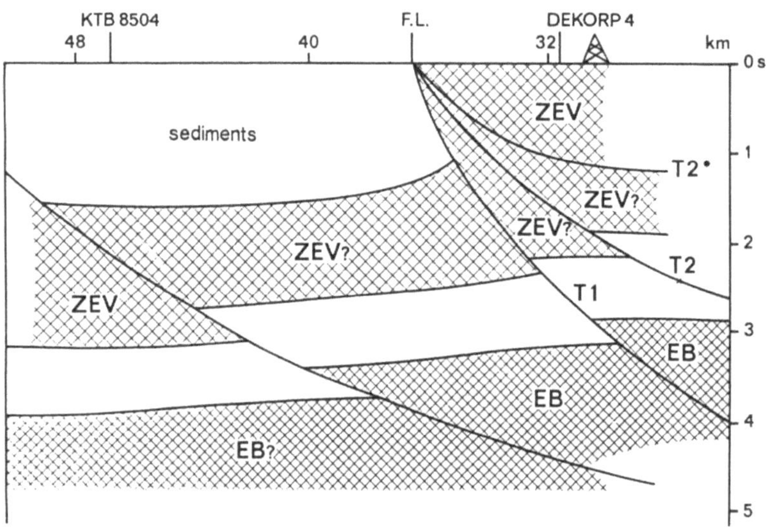

Fig. 20. c Schematic sketch of a fictitious SW/NE profile between the KTB 8502 and 8503 profiles with tentative interpretation of ZEV and EB

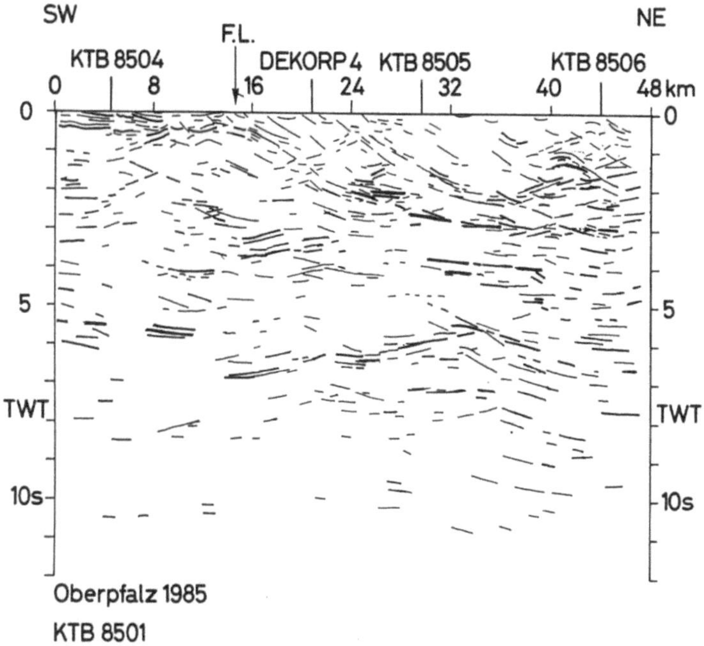

Fig. 21. KTB 8501 profile; line drawing of the migrated section

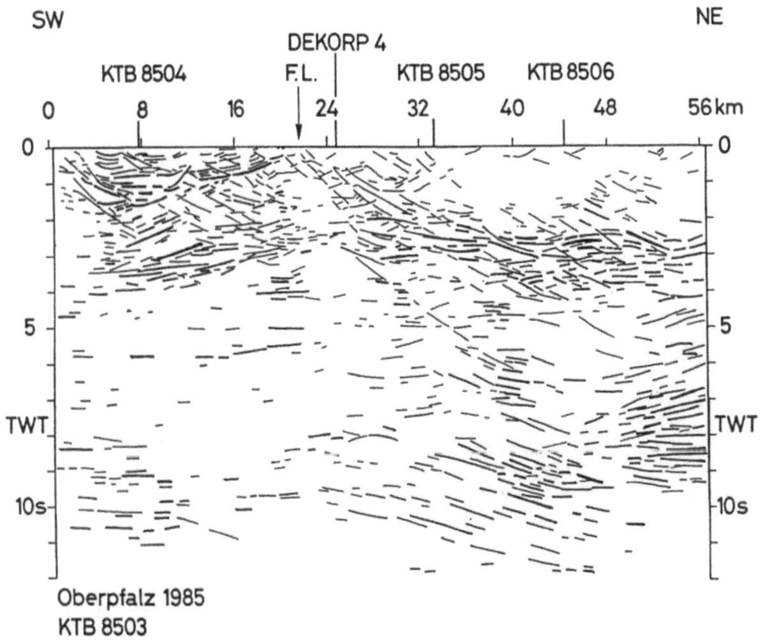

Fig. 22. KTB 8503 profile; line drawing of the migrated section

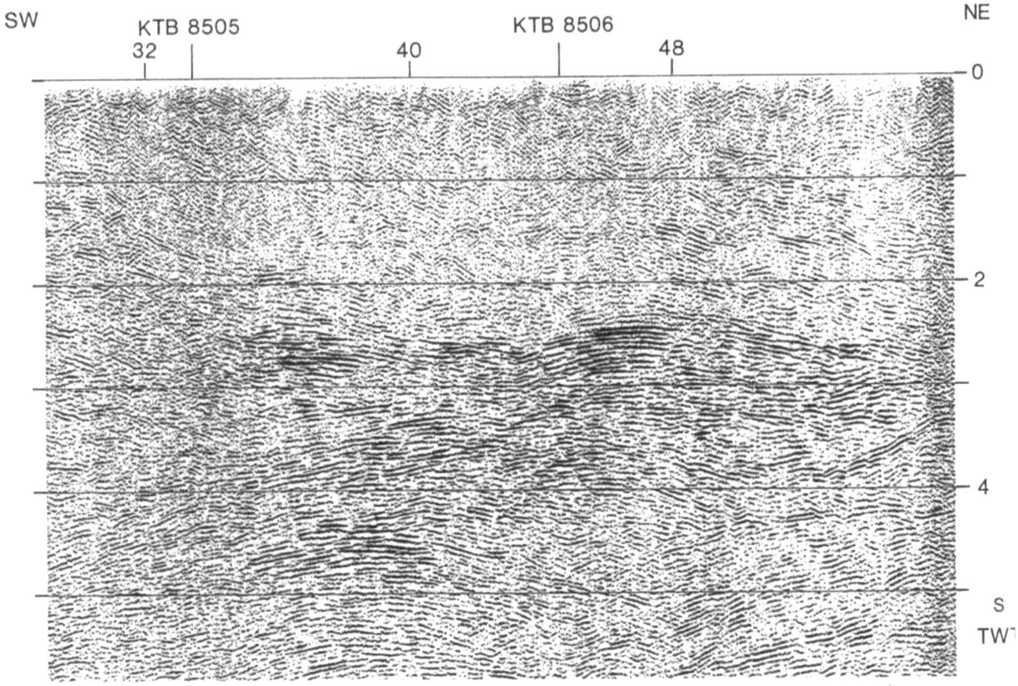

Fig. 23. KTB 8503 profile; part of unmigrated section (coherency filtered)

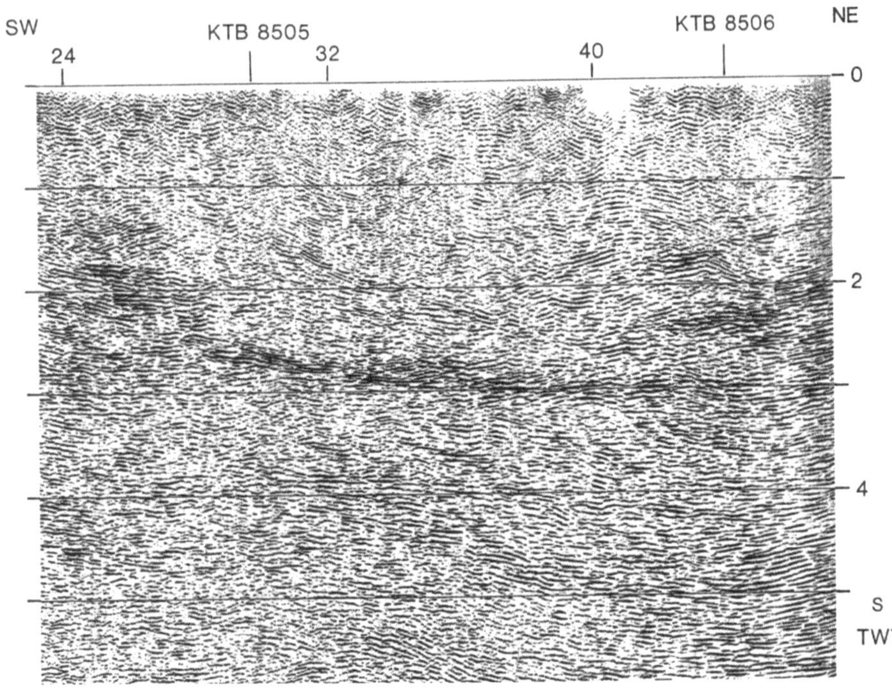

Fig. 24. KTB 8501 profile; part of migrated section (coherency filtered)

crust similar to those in the DEKORP 4 profile (Figs. 14 and 15). In the SE part - between km 12 and 24 and at the same structural position as the highly reflective zone of DEKORP 4's central portion below the ZEV -a pronounced rhombic structure with numerous internal reflections is observed at 2.5 to 4.0 s TWT. This rhombic structure is similar to structures found in zones of compressional tectonics in the foreland of the Alps. In the borehole Perwang 1 such a seismic structure was proven to be an imbricate structure (Kollmann 1977). The seismic image and the tectonic position of this structure on KTB 8505 suggest that we are dealing with an imbricate structure here, too. Its wedge-type NW end passes over into a subhorizontal, very clear reflecting horizon which is interpreted as a decollement plane with a series of curved reflections originating from it with the shape of listric thrust planes. It seems that the SE-dipping extension of the decollement plane touches the basal range of the imbricate structure. Perhaps another thrust plane touches this structure along its top.

The profile next to the NE, running parallel to KTB 8505, is the KTB 8506 profile (Figs. 16 and 17). A great number of reflections are observed. The lower crust is characterized by numerous subhorizontal reflections which might be regarded as laminations. Strong NW-rising aligned reflections intersect nearly the entire crust starting at 9 s TWT in the SE. In the range from 2 to 3 s TWT, a subhorizontal band of reflections in the NW portion of the profile is interpreted as a decollement plane which clearly shows up as the origin of reflecting subsidiary listric thrusts (Fig. 18). At the SE end of the profile a zone of high reflectivity is observed between about 2.5 and 4.5 s TWT; the reflections are mostly subhorizontal, partly SE dipping (Figs. 16 and 17). This feature may represent the imbricate structure, too. However, its shape is not as conspicuous as on the KTB 8505 profile (Fig. 15).

KTB 8502, one of the SW/NE-profiles, shows numerous, often very clear reflections in the entire TWT range observed (Figs. 19 and 20a), whereas on the KTB 8501 and 8503 profiles the lower crust shows comparatively sparse reflections (Figs. 21 and 22). In the range from 2 to 4 s TWT on KTB 8502 and KTB 8503 (Fig. 23) strong reflections are observed, which allows a correlation of the imbricate structure all over the SE part of the seismic grid. This includes the wedge-shaped structure with its high reflectivity on the DEKORP 4 profile below the ZEV. Thus, the image of a continuous imbricate structure extended over a wide area becomes evident. It was named "Erbendorf Body" (EB). It is obvious from the NW/SE-running profiles that the EB overrides a decollement plane exhibiting a ramp-like character. Using the KTB 8501 profile this decollement plane can be mapped over all the NW part of the seismic grid (Figs. 4, 21 and 24).

The basal zone of the EB is depicted in the results of the near-vertical angle seismics structurally similar to those of the wide-angle seismics (DEKORP Research Group 1988). However, a very important additional fact is provided by wide-angle seismics: the basal zone of the EB shows compressional wave velocities of more than 7 km/s. This can be interpreted as lower crust material thrusted upwards along a shear zone. This finding, as well as the picture resulting from the near-vertical angle seismic observations along DEKORP 4 and along the parallel profiles in the NE, suggest that the EB is overriding a decollement plane which is closely related to its origin. The EB basal zone has a slightly antiformal structure. It seems to be aligned with the strong SE dipping reflection band plunging down to the Moho.

Figure 25 shows an attempt to map the base of the EB. The depth contour map was constructed with a mean crustal velocity of 6 km/s. The base of the EB generally rises to the NW. It is divided into blocks, limited by NW/SE striking faults more or less parallel to the FL and dipping to the NE. Caused by these thrusts, the base of the EB rises to the NE to shallower depths.

Figure 26 shows simplified line drawings of the three parallel running profiles DEKORP 4 (only the central portion), KTB 8505 and 8506. The EB is indicated by hatched areas. This figure shows that also the strong SE-dipping reflections appear in all three lines, thus giving evidence for the wide-spread extension of these structural elements.

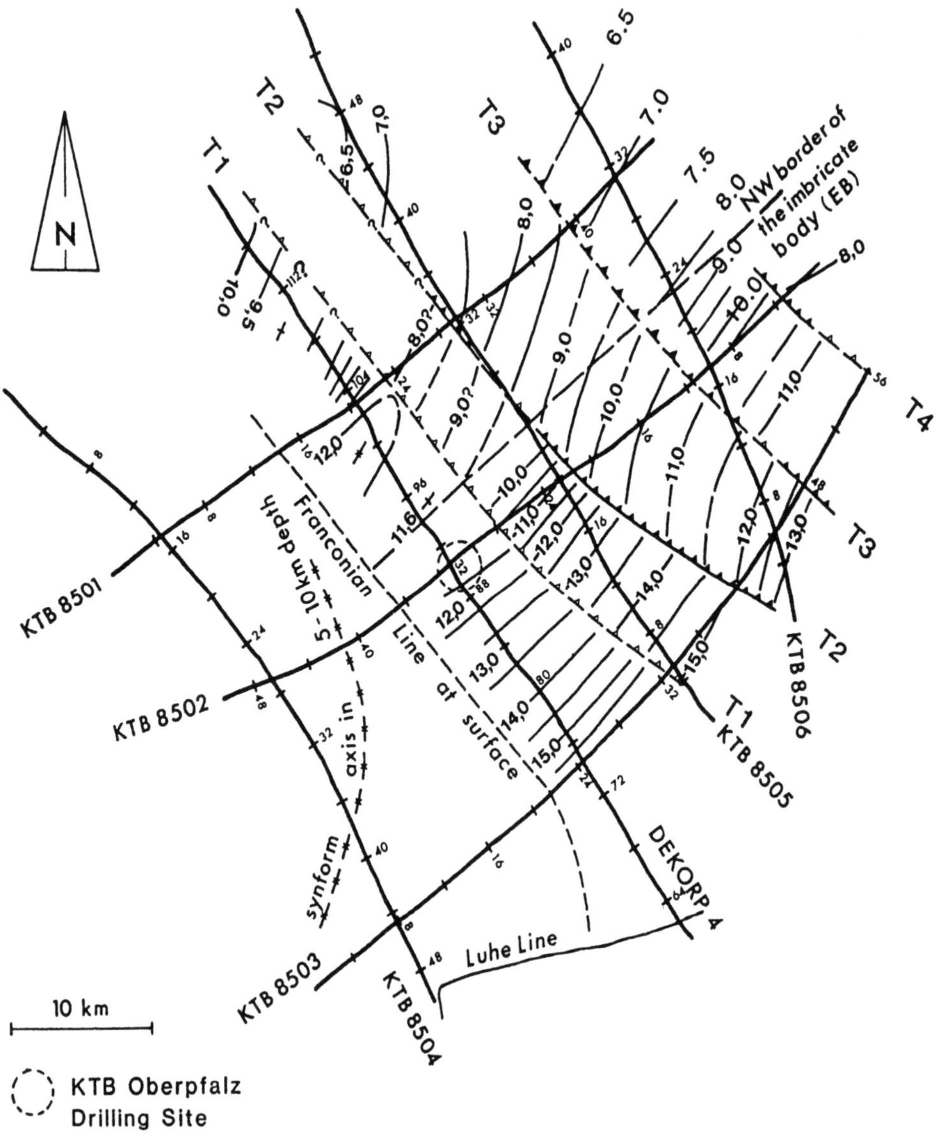

Fig. 25. Depth contour map of the Erbendorf body base and the corresponding decollement plane (depth in km, $V_{average}$ = 6 km/s). T1 through T4 denote thrusts

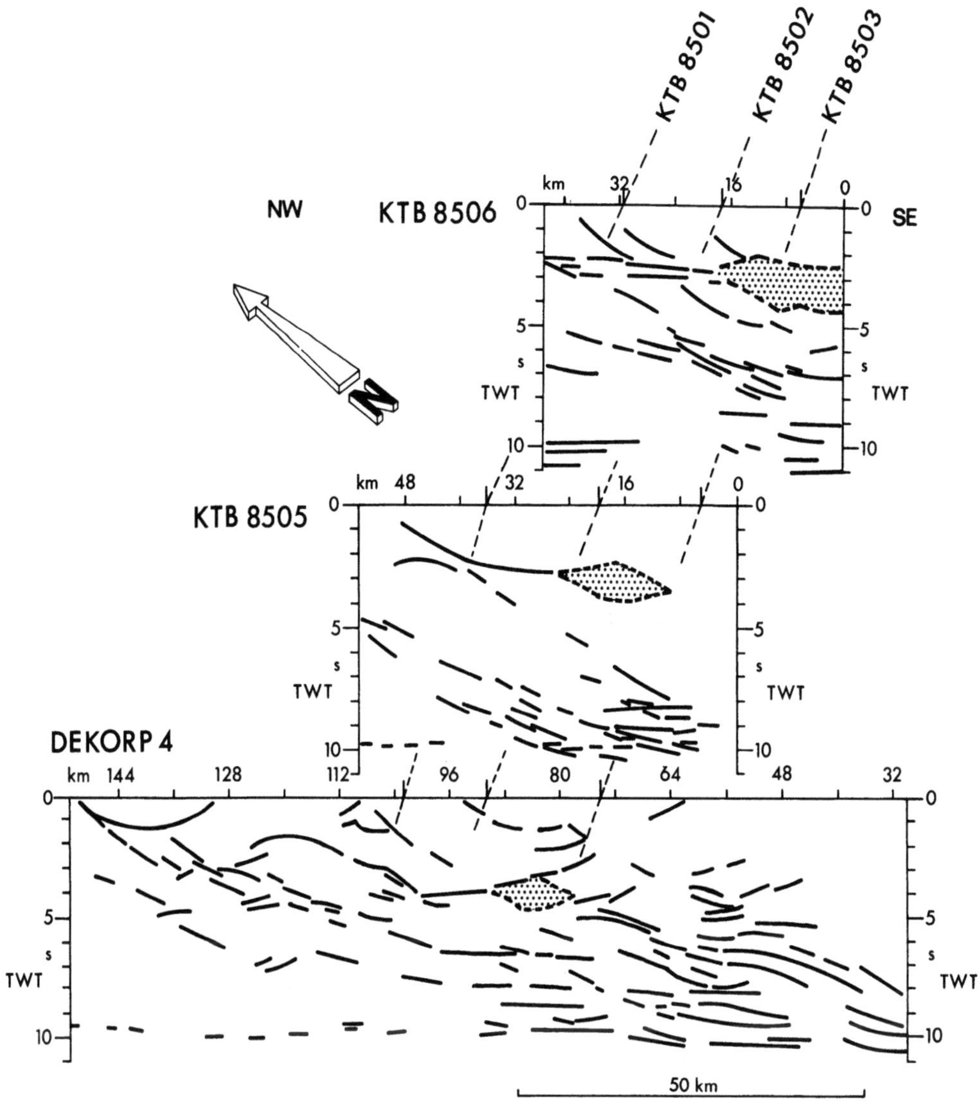

Fig. 26. Arrangement of line drawings with the prominent elements of DEKORP 4, KTB 8505 and 8506 lines. Hatched: Erbendorf body

On the DEKORP 2-South profile (DEKORP Research Group 1985) south of the MN/ST border near Dinkelsbühl a seismic structure had been observed in a tectonically similar position as the EB which was difficult to resolve due to a big cluster of diffractions. Behr (pers. commun.) already at that time interpreted the structure as a complex duplex structure.

The geophysical and geological interpretation led to numerous discussions and investigations. Meanwhile, a much improved migrated seismic section of this part of DEKORP 2-South has been provided by the Seismic Processing Center of BGR, Hanover.

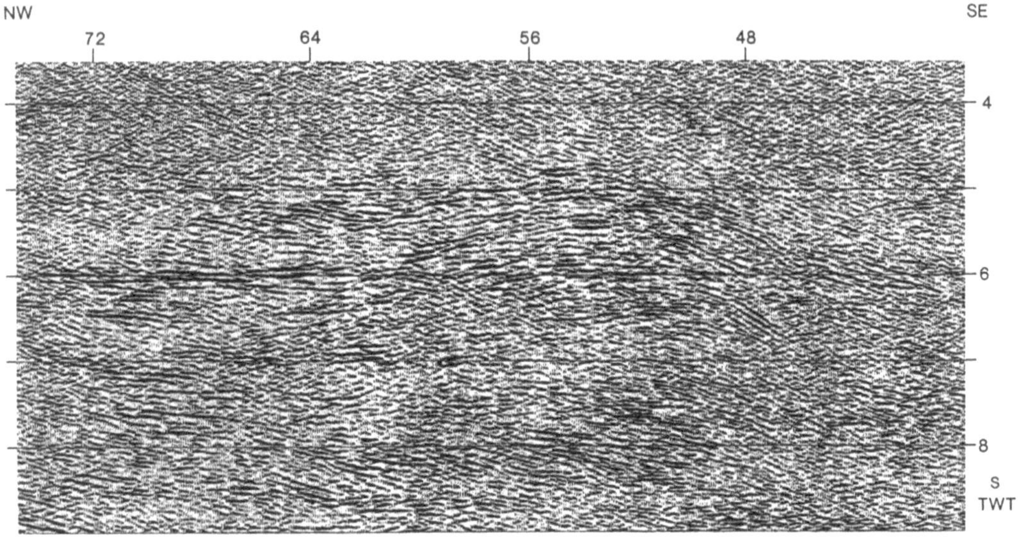

Fig. 27. DEKORP 2-South profile; part of migrated section of the Dinkelsbühl area (coherency filtered)

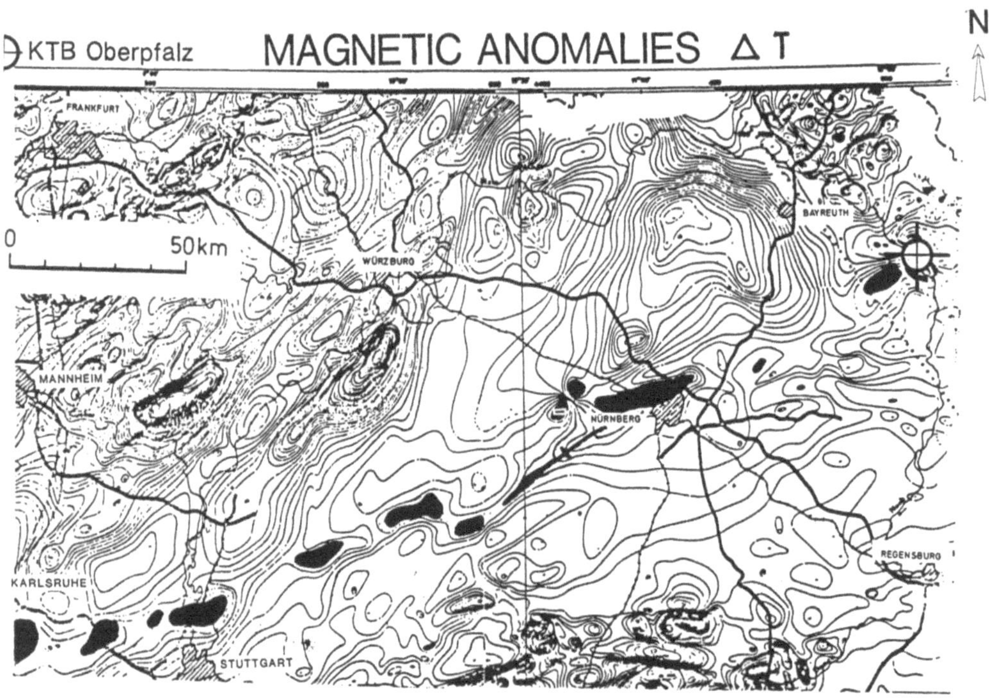

Fig. 28. Anomalies of the total magnetic field (ΔT) in southern Germany; black areas indicate the positive anomalies under discussion (see text) with one exception which is marked by a minus sign

The result (Fig. 27) exhibits a zone with numerous strong reflections at 5 to 6.6 s TWT validating investigations by Kläschen and Bittner (1988) and by Kampfmann (1988), who provided proofs that the seismic data of the Dinkelsbühl structure include much energy reflected from horizons of antiformal shape and with more or less strong dips to the SE. Part of the reflected energy comes from outside the vertical plane of the profile.

The similarity of the Dinkelsbühl structure as shown in Fig. 27 to the EB is striking. This supports the interpretation by Behr and Heinrichs (1987) that the reflecting structure in the Dinkelsbühl area is due to strong compressional shear movements especially in the ramp area of the ST/MN collision zone.

Magnetic results (Bosum pers. commun.) indicate that the observed seismic phenomena along DEKORP 2-South and DEKORP 4, i.e., the EB and the Dinkelsbühl structure, are situated within a zone of prominent anomalies. The map of the total magnetic field of the Federal Republic of Germany (a portion of which is shown in Fig. 28) shows a chain of anomalies between Baden-Baden and Erbendorf, coinciding with the boundary between the ST and the MN. These anomalies strike ENE, clearly deviating from the predominant direction of 45° which is presumably the direction of folding axes within the ST (Fig. 28). Therefore this chain of anomalies should be correlated to the MN/ST suture zone.

An interpretation assuming a subduction of the ST under the MN implies some difficulties. In this case one has to expect the magnetized bodies to be plate-like and dipping towards SSE. Such bodies should show an anomaly with a maximum in the S and a minimum of equal amplitude in the N if they are magnetized parallel to the present field. Moreover, these bodies should be connected here and there with magnetized bodies of the ST infrastructure. None of these effects can be observed.

There is, however, another possible explanation. If the collision of ST and MN has entailed local heating along the suture line, the idea of a production of volumes with a homogeneous magnetization is suggested. This can be effected by creating new grains of ferrimagnetic minerals or by heating existing minerals above their Curie temperature. Processes like that can be the consequence of a metamorphosis by subsidence, by fluid activities or hydrothermal events, respectively, thus creating a "metamorphic-magnetic zone". Such processes could be expected especially around the KTB-location in the area Erbendorf/Windischeschenbach below the ZEV (Pucher 1986).

Thus, these observations might contribute to a better understanding of the reflection seismic observations not only of the lines reported here, but also of the other lines crossing the MN/ST boundary as DEKORP 2-South (DEKORP Research Group 1985).

The EB is cut by several NW/SE trending thrusts (Figs. 20b and 25). They are derived mainly from the interpretation of the KTB 8501 to 8503 profiles. These profiles show in the domain of the first 4 to 5 s TWT, beginning at the FL and following to the NE, a series of NE dipping more or less curved reflections which can be interpreted as overthrusts T1, T2 etc. (Fig. 20b). The same idea is supported by the predominantly subhorizontal reflections in this TWT domain and deeper. These SW vergent listric thrusts may be correlated to the numerous NW/SE striking faults mapped at the surface. Some of these SW vergent overthrusts are outcropping in the surface area of the FL.

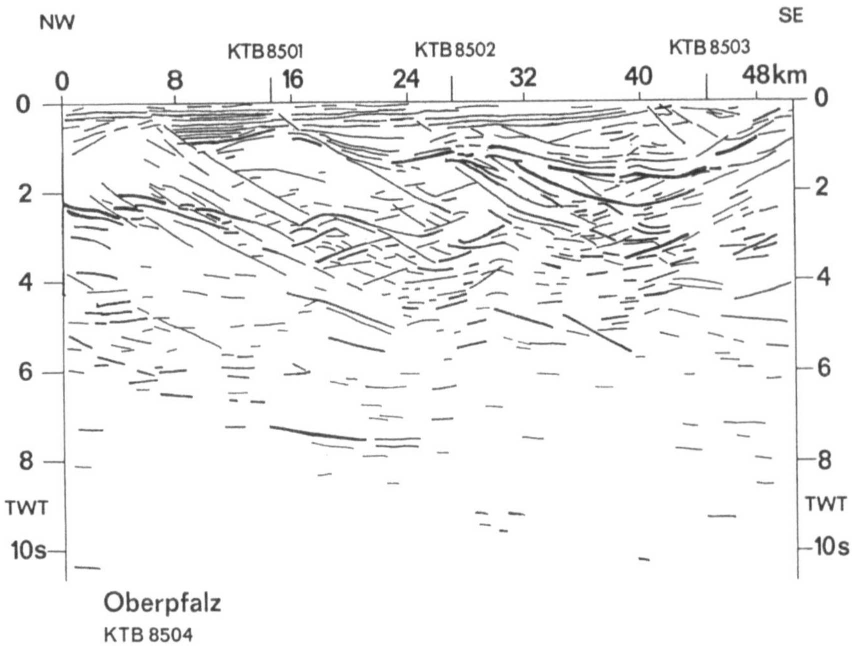

Fig. 29. KTB 8504 profile; line drawing of the migrated section

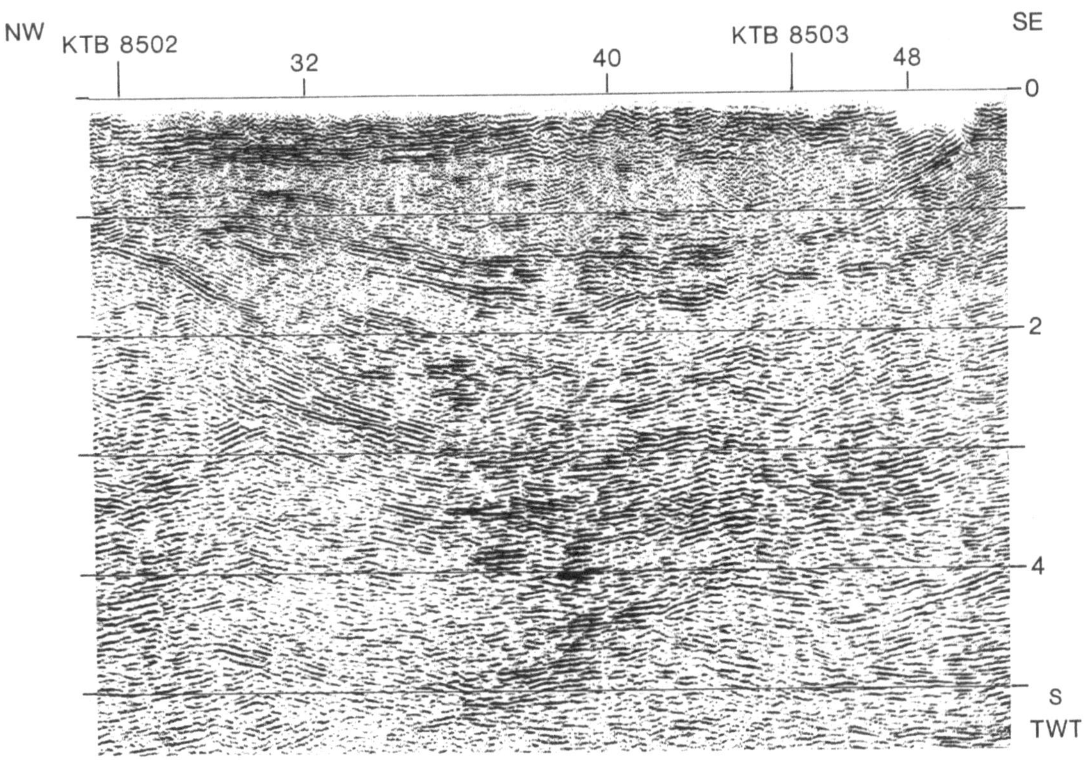

Fig. 30. KTB 8504 profile; part of the migrated section (coherency filtered)

According to the results of geological surface investigations, those of the reflection seismics presented here and gravity surveys (KTB 1986 p.66) the FL fault plane dips to the NE like an overthrust in the upper 2 km of depth. The interpretation of the KTB 8501 through 8503 profiles gives the idea that the FL splits up into several SW vergent overthrusts in depth with decreasing slope to the NE. The geological idea of the FL is a steep, SW inclined normal fault being overturned in the upper part. It is not impossible to assume such a fault plane in the mentioned profiles, according to a lack of reflections. But it is more probable that the strong lateral change of seismic velocities in the overburden at the FL disturbs a correct stacking of common midpoint data, thus yielding a lack of reflections below the surface position of the FL.

Possibly, a larger number of overthrusts contributes to the great throw at the FL at surface, each of them having a smaller amount of throw, respectively. Moreover, the listric geometry of the thrust planes transforms the horizontal component of throw into an increasing vertical component when approaching the surface. On the other hand, the KTB 8502 and 8503 profiles suggest that the FL plunges down to the NE into the lower crust, where it separates two blocks of reflections with different dips and TWT ranges (Figs. 20b and 22).

It seems that the EB in the vertical plane of the DEKORP 4 profile is situated on the downthrow side, i.e. it belongs to foreland domain SW of the FL (Fig. 20). Additionally, on the KTB 8504 profile (Figs. 29 and 30) between the intersections with the KTB 8502 and 8503 profiles the reflections observed in the range of 4 s TWT may originate from the EB wedging out to the NW. Its TWT is in about the same range as at the intersection of the DEKORP 4 and KTB 8502 profiles.

In this context attention should be focussed on the DEKORP 4-Q profile (Fig. 31). In its SW part it intersects an important mylonite zone with the embedded "Bavarian Pfahl", a famous quartz vein. The seismic section shows only a few strong events. One of the most prominent results on this profile is a steeply NE dipping reflection band (true dip angle is about 43°) which is observed over more than 10 km from 0.5 to 2.5 s TWT (Fig. 32). The extrapolation of this reflection band to the surface points exactly to the "Pfahl" outcrop.

It might be that the steeply NE-dipping reflections at 5.5 to 6 s TWT further to the NE on DEKORP 4-Q may be connected to this reflection band, revealing a very great depth range of the "Pfahl". On the other hand, NE-dipping listric reflectors originating from the Pfahl reflection join at about 3 s TWT into a bundle of prominent, horizontal reflectors which correspond to similar horizontal reflections on DEKORP 4 at the intersection of this profile with DEKORP 4-Q. The corresponding horizon, therefore, is subhorizontal and may be related to a deep thrust plane. NE-dipping listric reflectors at km 24 on DEKORP 4-Q may correspond to the SW border of the HB. At the intersection with DEKORP 4, steeply NE-dipping reflectors may be part of another SW vergent overthrust.

It can be concluded that also on the DEKORP 4-Q profile the dominance of SW vergent listric overthrusts is observed and that, similar to the FL, several thrust planes combine at the outcrop of the "Bavarian Pfahl".

An alternative interpretation of the tectonics of the KTB network area is given by Reston and Meissner (1988). They use only the strongest

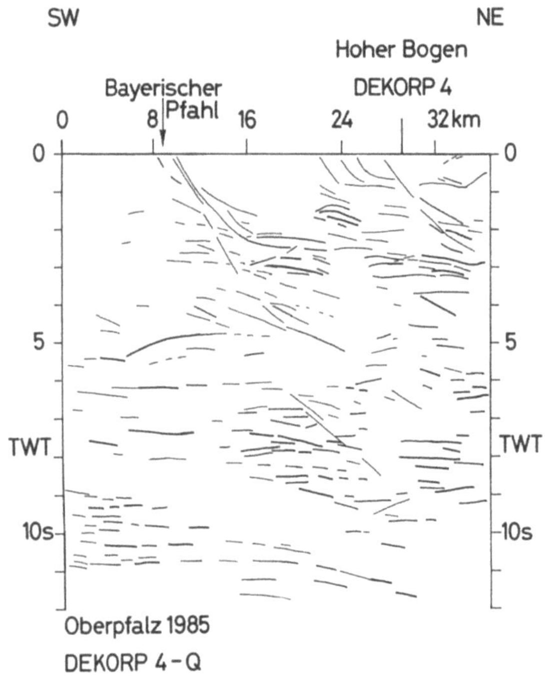

Fig. 31. DEKORP 4-Q profile; line drawing of the migrated section

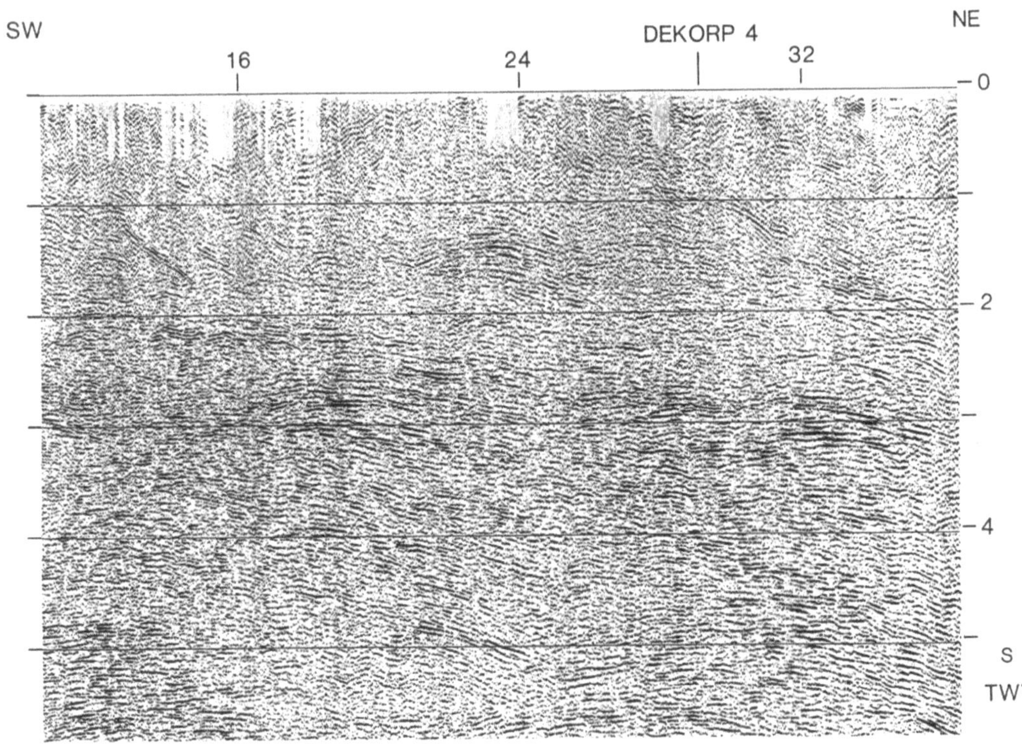

Fig. 32. DEKORP 4-Q profile; part of unmigrated section (coherency filtered)

events and assume the existence of nearly vertical dextral strike slip faults which partly form flower structures.

2.3 Further Results

2.3.1 The Mesozoic Foreland

The grid of seismic reflection profiles was positioned so that information on the subsurface structures would not be obtained only from the outcropping crystalline area E of the FL. The SW portions of the KTB 8501 through 8503 lines and the entire KTB 8504 line are situated W of the FL providing information from below the Mesozoic sediments and tracing the structures observed in the crystalline part.

In this area several reflecting horizons with traveltimes up to 1.6 s are observed from within the Cretaceous, Triassic and Permo-Carbonif-

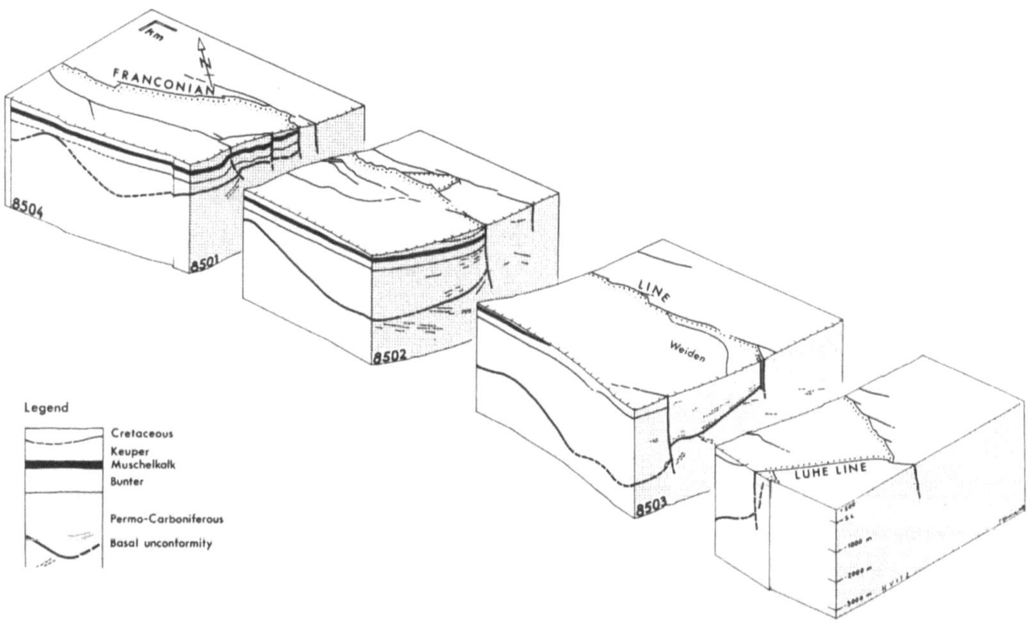

Fig. 33. Block diagram of Permo-Mesozoic foreland basin SW of Franconian Line. Mesozoic structure constrained by depth contour maps and bore holes Wollau and Parkstein (Haunschild and Schröder 1960; Schröder and Siegling 1966; Gudden and Schmid 1985; Emmert 1981). Depth of local unconformity derived from velocity analyses

erous, overlying the basement (Figs. 29 and 30). The maximum thickness of the sediments observed is on the KTB 8504 profile at km 40 (Fig. 29) where the base of the sediments is situated at about 1.7 s TWT corresponding to approximately 3.3 km depth below datum. Figure 33 shows a block diagram of the Permo-Mesozoic foreland basin constrained by the seismic profiles, velocity analyses, and boreholes.

Especially the KTB 8502 (Fig. 34) and 8504 (Fig. 30) profiles show a series of strong reflections SW of the FL in the domain of greater traveltimes. There are several arguments that on the KTB 8504 profile between the intersections with the KTB 8502 and 8503 profiles (km 26 to 48) the bowl-type structural element below the base of the sediments represents the SW continuation of the ZEV, its maximum thickness being situated at about km 40 (from about 1.7 to 2.7 s TWT corresponding to a thickness of 3 km). This interpretation agrees well with the results of the gravity and magnetic investigations, showing in this area between the FL and the Kulmbach-Freihung fault zone a weak and extended maximum which is also interpreted as the SW continuation of the ZEV.

The reflections mentioned above form a synformal structure in connection with the KTB 8502 and 8503 profiles in the depth range of about 3 to 8 km. The axis of this synform is marked in the depth contour map of Fig. 25.

It is probable that below this synformal structure the EB causes the numerous reflections in that traveltime range - especially on the KTB 8502 profile. But a strict correlation of the seismic events is not possible in this area.

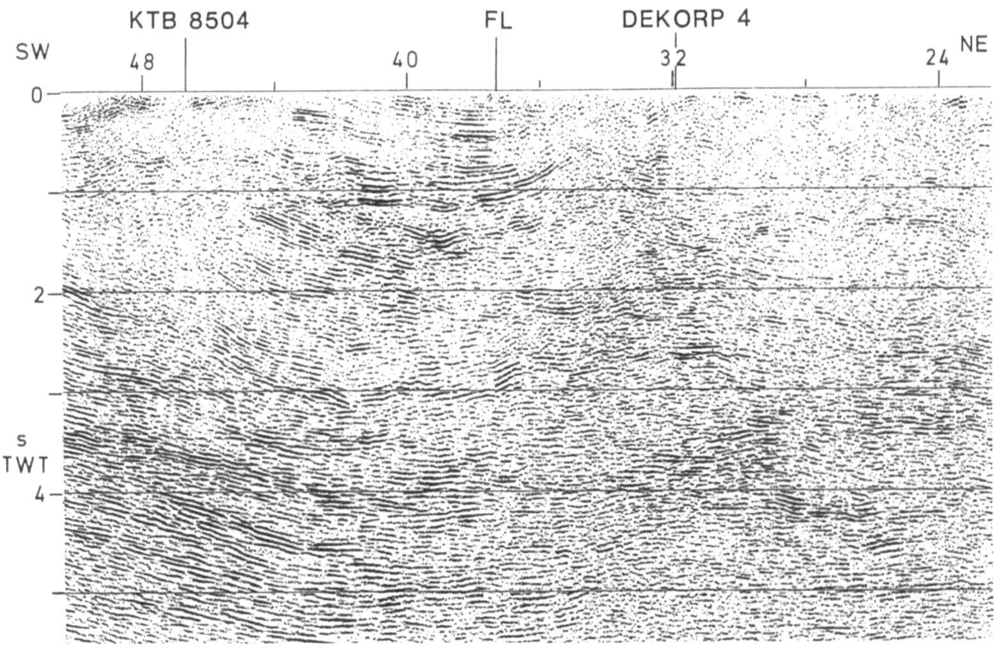

Fig. 34. KTB 8502 profile; part of the migrated section

2.3.2 Seismic Evidence for Nappe Remnants

The DEKORP 4 and 4-Q profiles intersect three nappe remnants (klippen): Münchberg Massif (MM), Zone of Erbendorf-Vohenstrauss (ZEV) and Hoher Bogen (HB) as part of the Tepla-Taus complex (ZTT). The profiles show continuous strong bowl-shaped reflections below the surface position of these nappes. Partly, the reflections are arranged concentrically in different levels. Thus, it can be excluded that these geological units are diapirs or flower structures which have been brought up from greater depth, i.e., their nappe character is strongly supported by the seismic results.

In the domain of the MM (km 130-155) reflections related to the MM and its footwall are observed down to about 1.6 s TWT (about 4.5 km depth) in the center of the synform structure (Fig. 7). This structure is asymmetric; the SE flank rises more or less steeply to the surface at the SE rim of the MM and its adjacent sediments. The NW flank is less steep. Starting from this point at the datum level one can follow an undulating chain of reflections reaching its maximum TWT (about 1 s) near km 140; these reflections should be regarded as base of the nappe pile, implying a maximum thickness of this structure of about 3 km. Deeper reflection horizons down to about 1.6 s TWT crop out at the SE margin of the MM, equivalently, but not at its NW margin. On the contrary, they can be followed nearly continuously to the reflection horizon in the NW part of the DEKORP 4 profile at about 1 s TWT described in the second paragraph of the Section 2.1. Thereby it is concluded that the Carboniferous plunges beneath the nappe pile. The thickness of the Carboniferous reduces towards the SE (Figs. 6 and 7).

In the domain of the ZEV (about km 70-98 on the DEKORP 4 profile, Fig.9) the NW rim of the nappe - at the Erbendorf Line - is connected with SE dipping reflections. Starting from this area, a reflection horizon can be followed more or less continuously towards the SE down to about 0.7 to 0.8 s and a second one to about 1.2 s TWT in the center of the synform (nearly at km 82) and then up to about 0.5 s TWT at km 72. This horizon could be related to the base of the ZEV, implying a maximum thickness of this structure of about 3.7 km. Below that clear bowl-type reflection additional horizons are observed down to about 2 s TWT in the center of the bowl, providing other possible interpretations. Maybe the ZEV base is situated deeper than described above, i.e., at about 1.6 s TWT (about 4.5 to 5.0 km depth), and the deeper reflections must be related to a footwall series similar to the MM situation (greenschists?). Alternatively, this phenomenon could represent a doubling of the ZEV originated by the SW vergent overthrusts T2 and T2* (Figs. 20b, c and 25); reflections from these thrust planes appear as subhorizontal on the DEKORP 4 profile. In this case the total maximum thickness of the ZEV on DEKORP 4 profile amounts to about 6 km at km 80.

The SE part of the ZEV extending to about km 63 (outside Fig. 9) is not as thick as the NW part. This is in accordance with other geologic and geophysical results.

Within the upper part of the ZEV several reflections probably corresponding to intercalations of amphibolites, marbles or other rock material within the gneisses can be observed.

As discussed in Section 2.3.1, it is very probable that the ZEV continues SW of the FL immediately below the sediments of the Permo-Carboniferous. In this area the maximum thickness of the ZEV is estimated to be about 3 km.

The nappe character of the HB is established mainly by the DEKORP 4-Q profile (Figs. 31 and 32); however, only a very small part of this unit in the SW could be covered by this profile. The base of this nappe seems to be situated at about 0.8 s TWT (about 2.5 km depth).

2.3.3 The Moho

The Moho has a variable signature in the area under investigation. In the SE of the DEKORP 4 profile one observes remarkable features at the Moho level. In the left and center parts of Fig. 13 the Moho shows very strong subhorizontal reflections at about 10.0-10.5 s TWT, corresponding to a depth of about 30-32 km. Distinct phases are observed correlating over long distances. At km 36, however, the Moho bends down and dips SE down to about 11.5 s - corresponding to about 35 km depth - where it cannot be followed further due to the restricted record length of 12 s TWT. As mentioned before, there might be a step in the Moho with an offset of more than 3 km in the case of the strong reflector at km 24 at 11.7 s TWT belonging to the Moho discontinuity.

In the central portion of the DEKORP 4 profile the Moho reflections can be seen at about 9.5 s corresponding to about 28.5-29.0 km depth. In the NW of the profile Moho reflections can be located at about 10.8--10.9 s TWT corresponding to about 32-33 km depth. This implies an extended antiform in the central part of the line, the strike of which is unknown. To the NW end of the profile the Moho rises up slightly to about 10.5 s TWT corresponding to about 31-32 km depth.

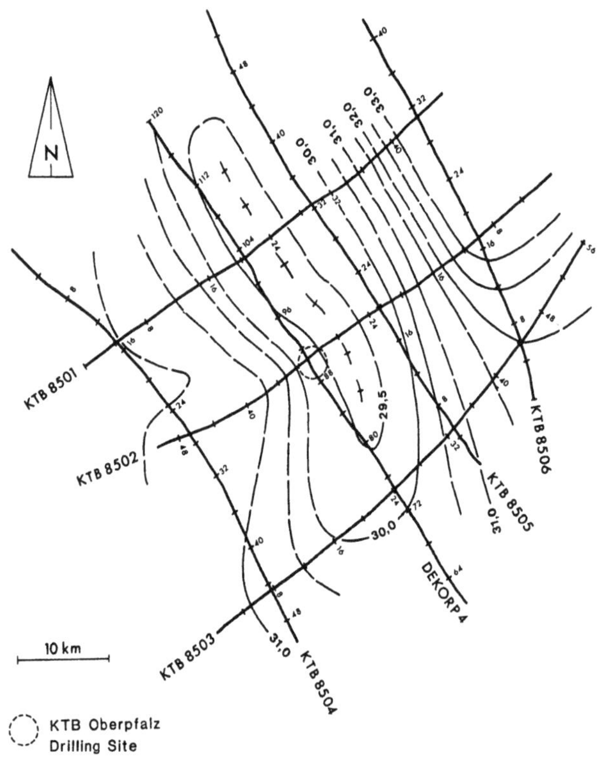

Fig. 35. Depth contour map of the Moho (depth in km, $V_{average}$ = 6 km/s).

A depth contour map of the Moho was established for the small area of the network of the KTB profiles, using a mean crustal velocity of 6.0 km/s (Fig. 35). The dominating NW- to SE-striking antiform axis in the central part may be incorrect, because along the KTB 8501 and 8503 profiles some extrapolations were performed. The strike of the axis changes N of the Luhe Line into a direction similar to the synform axis of Fig. 25.

2.4 Seismic Reasons for the Drilling Location Windisch-Eschenbach

The results of the near-vertical and wide-angle seismic reflection survey in the Oberpfalz area show a strongly structured upper and lower crust. SE-dipping reflectors prevail in the near-vertical survey, some NW-dipping reflectors indicate the complex tectonic movements which must have taken place in this border zone between the ST and MN.

Among the many features observed by seismics and described in the previous paragraphs, the so-called EB and its related structures are the most prominent. It was shown that the EB has an extension of much more than local importance. It is interpreted as a tectonic wedge, probably made up of high-velocity material being thrusted from the MN lower crust up to shallower crustal levels along major thrust systems. The internal imbricate structure of the EB is typical. The structural interpretation along DEKORP 4 gives the following sequence - from top to bottom - at the intersection of DEKORP 4 and KTB 8502, the proposed site of the KTB hole (Fig. 36):

- the ZEV nappe unit
- ST rocks, probably backfolded
- the Erbendorf Body, probably MN lower crust material
- ST rocks again, probably upper crust material subducted below the MN unit
- ST lower crust.

This multifold pattern was an important reason to choose this area as KTB drilling location in the Oberpfalz. By the results of the drilling the validity of the geologic/tectonic interpretation should be established. Moreover, the cause of the seismic reflections can be investigated representing a unique opportunity to study characteristic seismic reflectors of the crystalline crust. As the spacing of the seismic reflection network of the 1985 survey is too coarse due to the character of a reconaissance survey, a 3-D seismic reflection survey will be carried out in order to provide a connection between the features found by drilling and the reflectors observed.

Also other sites have been discussed as possible locations of a KTB-hole in the Oberpfalz, e.g., the intersections of the KTB 8502 line with KTB 8505 or 8506. Also at these locations the EB could have been found. Since the EB rises from W to E, the intersection with the KTB 8506 profile should be the site where the EB could be reached at its shallowest position. But this site has been excluded due to the nearby German/Czechoslovakian border zone and no spatially extended investigations could have been carried out around this place. Also the intersection KTB 8502/KTB 8505 is still too close to the border, at least regarding the necessary additional wide-angle investigations. Moreover, at this location the Falkenberg granite crops out. Its thickness is not very well known. Its base is estimated to be at about 5 to 6 km depth, but no exact information is available. Drilling into many kilometers of granite would not provide interesting geological information.

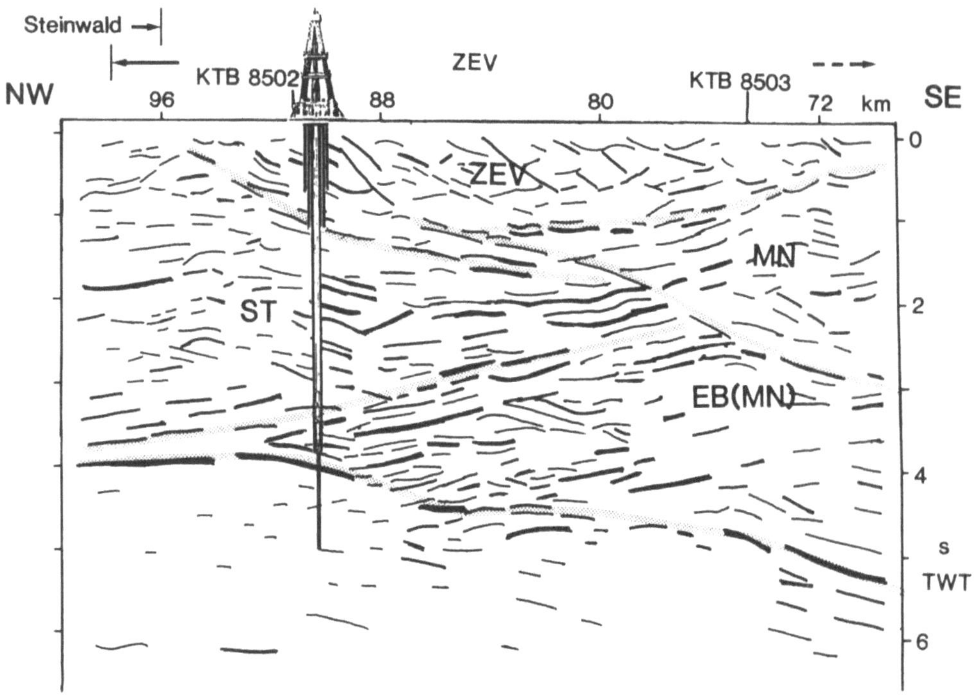

Fig. 36. Central part of the line drawing DEKORP 4 with structural interpretation

On the other hand, the ZEV nappe unit at the intersection DEKORP 4/ KTB 8502 with its multifold components will provide very interesting information also in the uppermost crust. Another reason for this site is the quite unique pattern of the wide-angle seismics. These investigations were carried out with sufficiently high resolution only along the DEKORP 4 line. The extension of the high velocity features shall be studied by two additional wide-angle cross-lines in combination with the 3-D survey in 1989.

When the drilling position in the Oberpfalz area was fixed, it still competed with a suggested site in the Schwarzwald area near the town of Haslach. With respect to the seismic reflection profiles through the Oberpfalz and Schwarzwald sites, a very different character has been observed. This is shown very clearly by Fig. 37a and b which are the result of a line drawing process of the corresponding seismic reflection data. The processing parameters are strictly identical for both data sets. Processing was performed by Wenzel (1986 unpubl.) who kindly provided the figures. The upper crust at the proposed site in the Schwarzwald is nearly void of P-wave reflections except a "bright spot" at about 10.5 km depth. The lower crust shows exceptionally well-developed laminations, the top of which might be reached if drilling were very successful.

In the target area in the Oberpfalz, in contrast, the upper crust exhibits a large number of reflections corresponding to a tectonically very strongly affected crust, whereas the lower crust (beneath the EB) seems to be nearly void of reflections. Thus, there is a reasonable opportunity at this location to study the nature of seismic reflectors from within the crystalline basement.

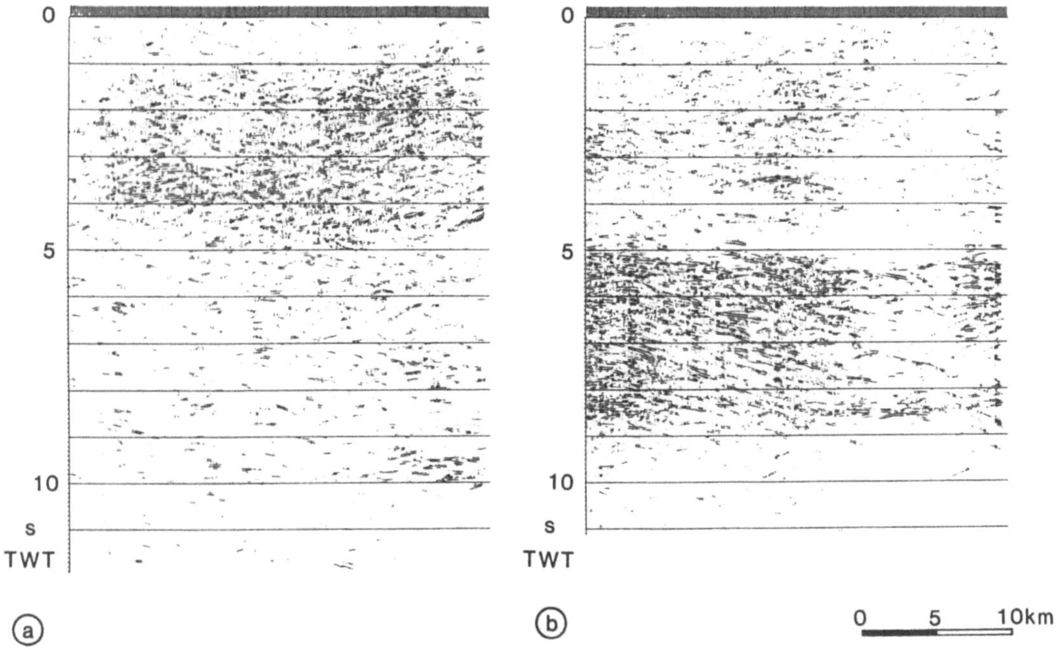

Fig. 37. Automatic line drawing of the central portion of: a DEKORP 4 profile (Oberpfalz) near the town of Erbendorf. b KTB 8401 profile (Schwarzwald) near the town of Haslach. For both profiles identical parameters were used

3 Assessment of Velocities and Relations to Geological Structures

For an assessment of seismic velocities several investigations were performed:

- The velocity distribution of the upper layers was obtained by an inversion of first arivals.
- Two expanding spread experiments have been analyzed.
- Improved stacking velocities were obtained by "dip moveout" procedures.

The results are discussed in the following.

3.1 Velocity Models by First Arrival Analyses

3.1.1 Handling of the Data

For the upper 2 km depth range two-dimensional velocity models were calculated in order to study the relation between physical parameters and the near-surface lithology. For this purpose the traveltime curves of the P-wave's first arrivals of the reflection records were used.

They represent - providing a high degree of coverage - an enormous potential of information about the P-wave velocities of the uppermost depth range. The traveltime curves are inverted by the Slichter method, which is one of several so-called maximum depth methods (Giese 1976). They can yield a rough qualitative image of the velocity structure covered by refracted rays. Slichter (1932) considered the singular transitional case between the normal type of prograde time-distance curves and reversed traveltime curves where all rays are brought to a common focus in time and distance, i.e., the traveltime curve degenerates to a point. Slichter derived the formula

$$z = \frac{x}{\pi} \cosh^{-1}\left(\frac{v_a t}{x}\right)$$

giving a depth value to an apparent velocity v_a obtained by local differentiation of the traveltime curve at offset x and traveltime t. The velocity v_a should apply to a point having coordinates z and x/2, assuming lateral homogeneity. In this way the velocity model is built up by many single velocity values. At the end, the velocity distribution is smoothed by a moving average procedure. Test calculations (Rühl 1987) have shown that the Slichter method results in a good reconstruction of given models and agrees well with results obtained by traveltime tomography. Therefore, this approximation method needing only little computing time has been used during this study of the Oberpfalz data. It should be mentioned that the Slichter model velocities are generally up to 0.2 km/s higher than the expected velocities. Tomographic methods cited above were also carried out in part showing very similar characteristic features and reducing the absolute values of the velocities.

Our investigations revealed the great influence of inhomogeneous near--surface weathered layers on the reconstruction performance. So intensive care must be taken of the elimination of this and other disturbing effects on traveltime curves (Rühl 1987).

By the weathering correction the data are reduced to a reference level that refers to the base of the weathered layer, i.e., to the depth of the first refractor which is used for the correction. For the DEKORP 4, KTB 8505 and 8506 profiles this refractor follows the surface topography at shallow depths and is not marked in the following figures. For the KTB 8501 to 8503 profiles the depth of this reference level is marked.

In the figures (Figs. 38-50), that show the results of these velocity analyses, the following near-surface informations are compiled (from top to bottom): simplified geologic sections, the velocities of the refractor used for weathering corrections, a velocity model with isovelocity lines as function of depth obtained by the Slichter method. It has to be kept in mind, that the top of this model refers to the base of the weathered layer, i.e., it varies along the abscissa. At the bottom the line drawings of the seismic reflection sections down to 2 s TWT are given.

The broken line in the velocity model marks the border of the area within which the velocity calculation is possible. The margins are generally poorly covered, and so there is little information on velocities.

There are small-scale undulations in the velocity models, that certainly have no physical reason but are caused by the approximation procedure. An estimation of the vertical and horizontal resolution of the models gives the values of 200-300 m and 1-2 km respectively. Nevertheless, a correlation of the major structures with geology is attempted.

3.1.2 Results

The results from the NW portion of DEKORP 4, i.e., the transition from Paleozoic metasediments via the Münchberg Klippen to the Fichtelgebirge granites are shown in Fig. 38. A remarkable high of the 6.0 km/s isoline coincides with the amphibolite gneisses of the Münchberg nappe pile. The model shows a low of the 6.0 km/s iso-velocity line in the area of the Carboniferous/Devonian sediments. A sharp transition between the velocity high of the MM and the low following in the N coincides with a small-scale depression of the isolines, possibly caused by a steeply inclined cross fault zone, intersected under an acute angle.

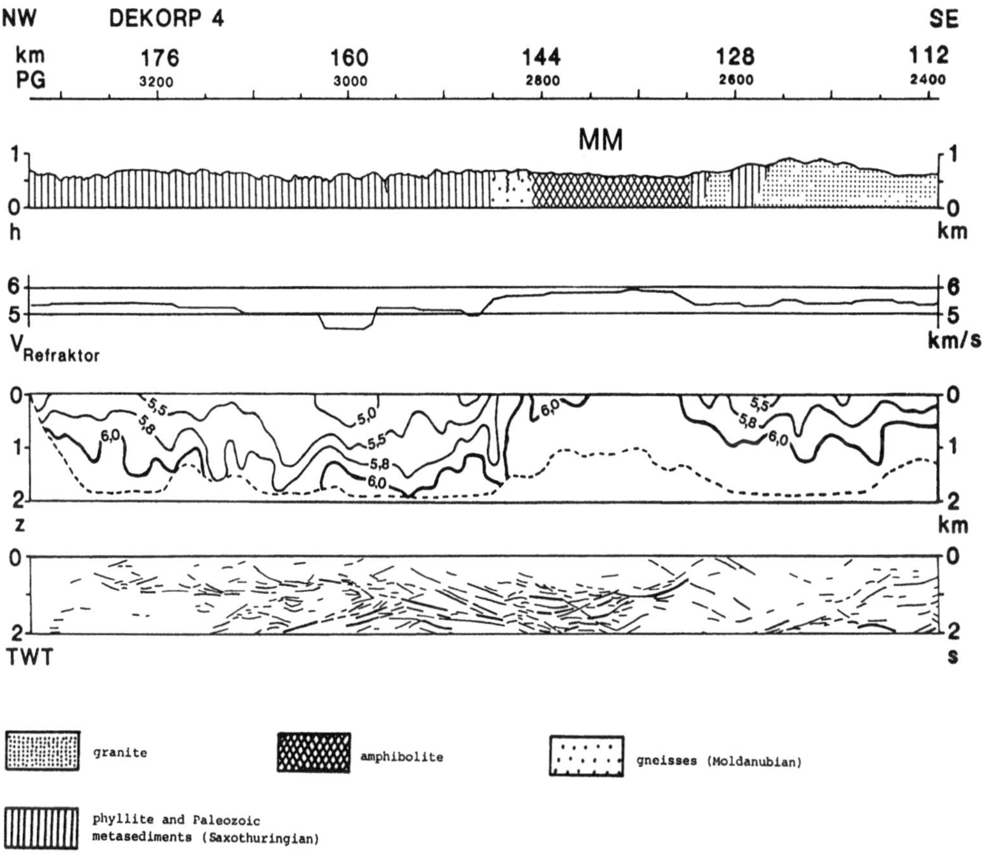

Fig. 38. NW part of DEKORP 4 profile: compilation of surface geology, refractor velocities indicating the seismic velocities at the base of the weathered layer, velocity distribution obtained by first arrival analyses (Slichter method; for reference level see text), line drawing of corresponding part of the near vertical reflection seismic (from top to bottom)

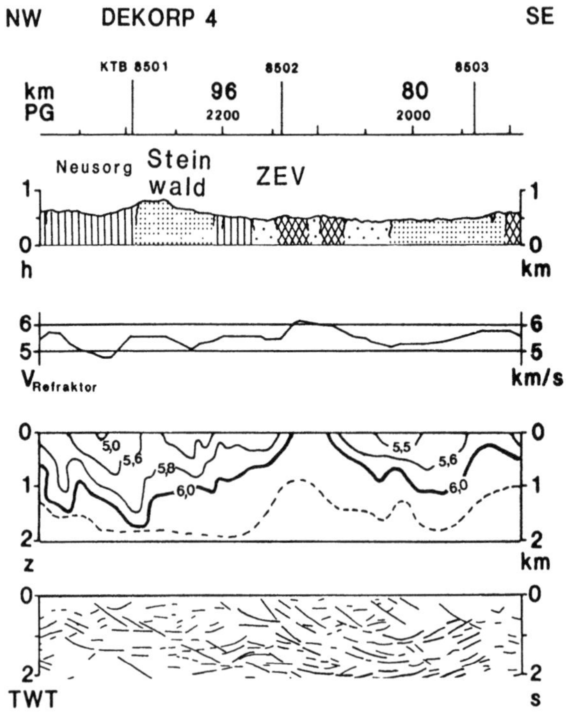

Fig. 39. Central part of DEKORP 4 profile. For detailed description see Fig. 38

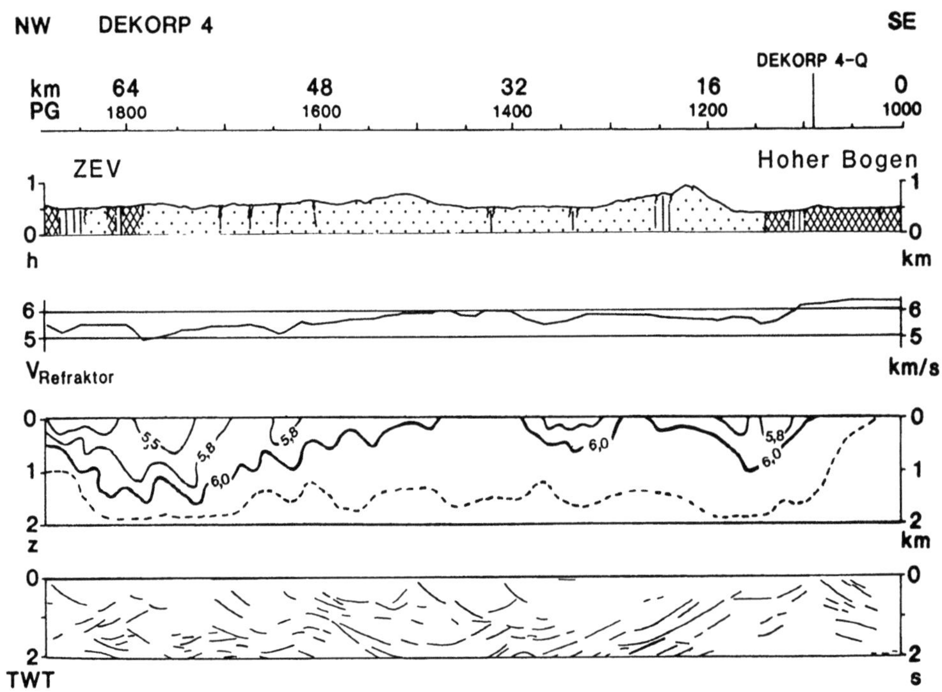

Fig. 40. SE part of DEKORP 4 profile. For detailed description see Fig. 38

The central part of DEKORP 4 (Fig. 39), extending from the Steinwald granite via the Erbendorf-line to the N part of the ZEV, is characterized by a prominent rise of the refractor velocity to more than 6.2 km/s and a corresponding velocity high in the Slichter model. This is confirmed by the massive occurence of amphibolites in the northern ZEV as shown by drilling results in shotholes. A clear velocity low of the refractor at the geophone position PG 2310 and a depression of the iso-velocity lines correlates with the strike fault zone of Neusorg.

The S part of DEKORP 4 (Fig. 40) extends from the S boundary of the ZEV via the widespread occurence of cordierite gneisses to the amphibolite gneisses of the HB. Here again the strong decrease of the refractor velocity down to 4.9 km/s in the diaphtoritic fault zone at the southern edge of the ZEV is remarkable. In the velocity model this zone is marked by a distinct bowl-shaped structure. While the velocity high at the S-end of the profile correlates unambiguously with the amphibolites of the HB, the geological meaning of the velocity high in the center of the cordierite gneisses between PG 1250 and 1450 is not yet understood.

Within the HB complex at the S end of DEKORP 4 the refractor velocities seem to reflect different amphibolite varieties, and a rise of a refractor with 6.3 km/s to less than 50 m below ground may correlate with a narrow zone of suspected eclogites and/or ultramafics within the 'Gabbroic Amphibolite' (Fig. 41). The Slichter model is not reliable in that border zone near the end of the profile due to insufficient data.

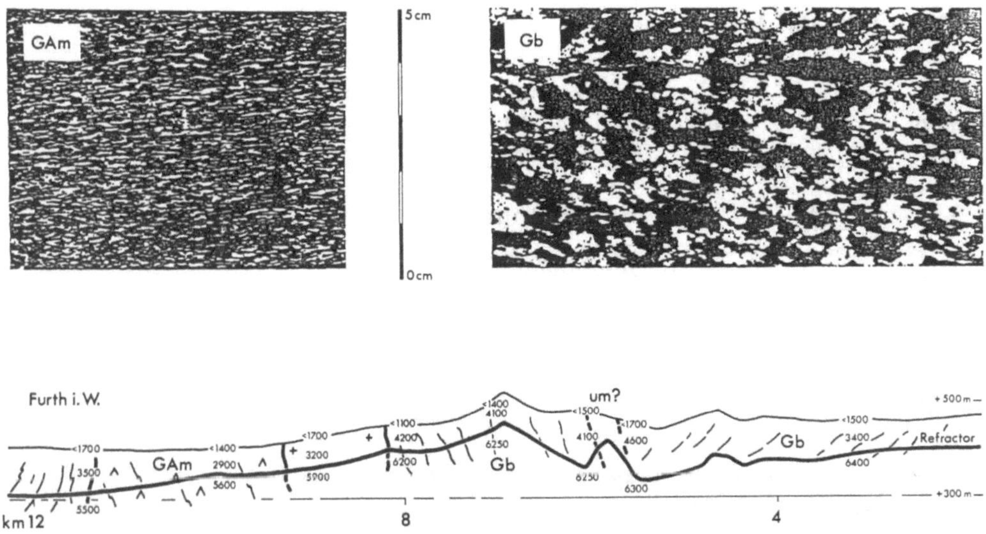

Fig. 41. DEKORP 4, southern end (Hoher Bogen, amphibolite complex). Top characteristic textures of amphibolites; bottom geological section and velocities from first arrivals. Note: refractor velocities near 5600 m/s correspond to schistose amphibolite (GAm) of Hoher Bogen complex, velocities around 6200 m/s reflect gabbroic amphibolite (Gb); um = suspected ultramafics

The intersecting DEKORP 4-Q profile, shown in Fig. 42, reveals in its NE part nearly identical velocity characteristics for the amphibolites of the HB as mentioned in the previous paragraph. The gneisses of the Bavaricum SW of the HB exhibit nearly uniform refractor velocities and Slichter velocities. Especially the Pfahl-mylonites but also other, likewise NE inclined shear zones, are recognized in the reflection line drawing.

The KTB 8505 (Fig. 43) and KTB 8506 (Fig. 44) profiles, both parallel to DEKORP 4, intersect the ST/MN border both comprising a northwestern ST and a southeastern MN portion. On KTB 8505 (Fig. 43) the metasediments of the ST reveal obviously a higher velocity gradient. An explanation is not at hand. The refractor velocity does not show strong lithological differences.

On the KTB 8506 profile (Fig. 44) the MN southeastern part gives a relatively smooth picture with subhorizontal iso-velocity lines, whereas the northwestern ST part offers a pronounced NW vergent saw-tooth pattern, which has a striking similarity to the seismic section. This may be interpreted as an image of a late-Variscan fault- and diaphtoresis-structure and/or an appropriate lithologically differentiated structure.

This saw-tooth pattern in the velocity model comprises a velocity inversion which cannot be detected by the Slichter method in a strict sense as a discontinuous traveltime curve is implied. But it may happen by the picking algorithm that small gaps in the traveltime curve are suppressed using a continuous picking curve instead, thus producing lower velocities and, nevertheless, indicating low velocity zones at a more or less correct position.

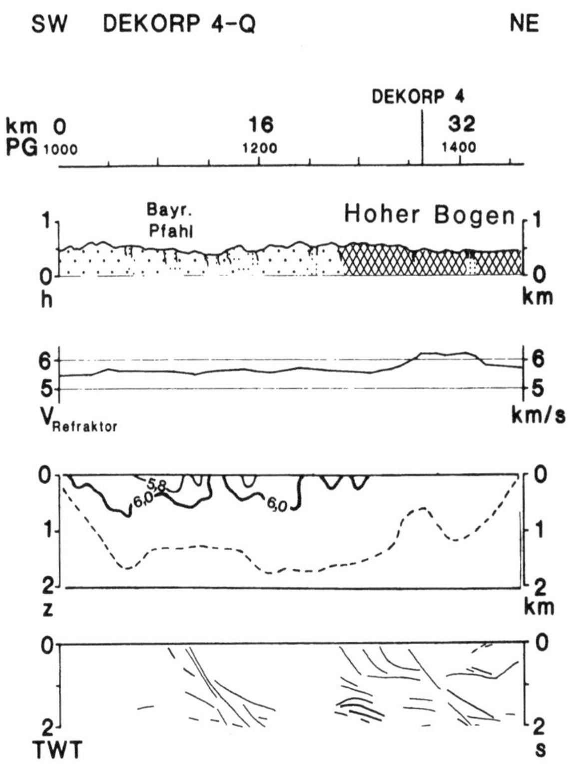

Fig. 42. DEKORP 4-Q profile. For detailed description see Fig. 38

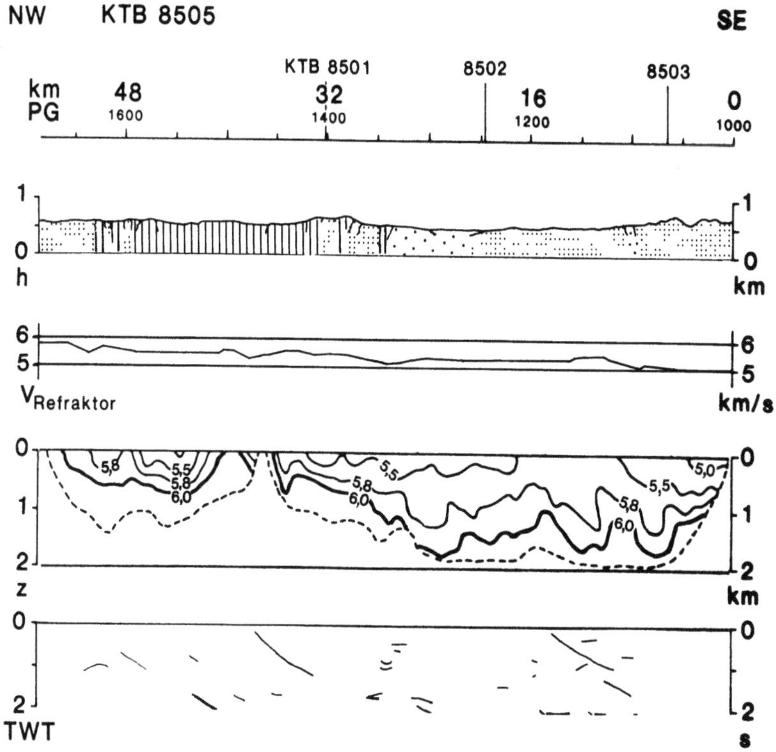

Fig. 43. KTB 8505 profile. For detailed description see Fig. 38

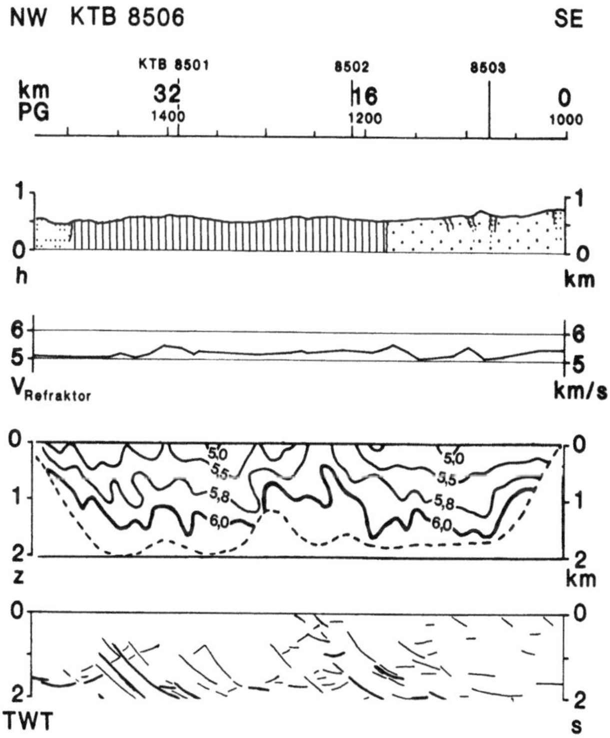

Fig. 44. KTB 8506 profile. For detailed description see Fig. 38

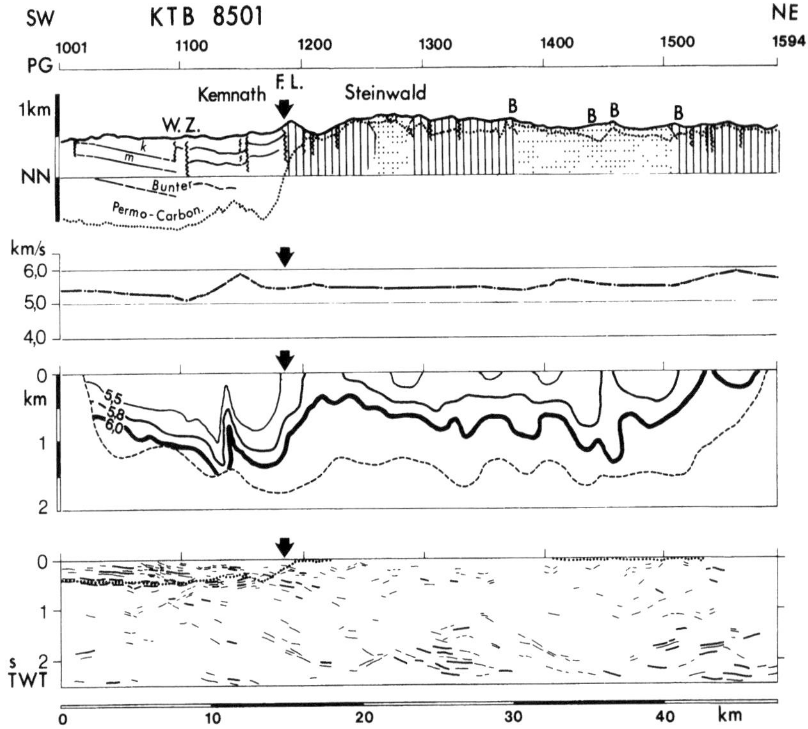

Fig. 45. KTB 8501 profile. For detailed information see Fig. 38. Additionally, the depth of the refractor is indicated by dotted line in the sketch of the surface geology and in the line drawing. W.Z. Weidenberg fault zone, B basalt

The KTB 8501 profile (Fig. 45) is situated in the strike direction of the Variscan belt and perpendicular to the FL, that produces a prominent horizontal velocity gradient. E of the FL the isolines of the granites, embedded in metasediments (phyllites) of the ST, tend to subside. It is questionable whether the opposite steep rise of the isolines at PG 1460 may be combined with a basalt neck. SW of the FL the refractor and thus the reference level of the velocity model - marked in the geological section as dotted line - subsides about 1 km compared with the level NE of the FL, showing a slight high between the FL and the fault zone of Weidenberg. Underneath the SW flank of this high, the velocity model shows a peak that might be related to the Kemnath fault, intersected by the profile under an acute angle, or with a depth continuation of the Weidenberg fault zone inclined very flat to the E.

The Kemnath fault as well as the FL are indicated by delays of the first breaks (Fig. 46). The position of the delays suggests a near vertical attitude of the faults.

The KTB 8502 profile (Fig. 47) intersects the FL farther south in an area where the top of the basement is situated significantly deeper and is not discovered in the refraction section due to the thickness of the Permo-Carboniferous and Cretaceous-Tertiary sediments. The Slichter model exhibits this boundary to be about 1.5 km underneath the first refractor. SW of the FL the first refractor is within the sedimentary section, but NW of it it is directly at the bottom of the weathering

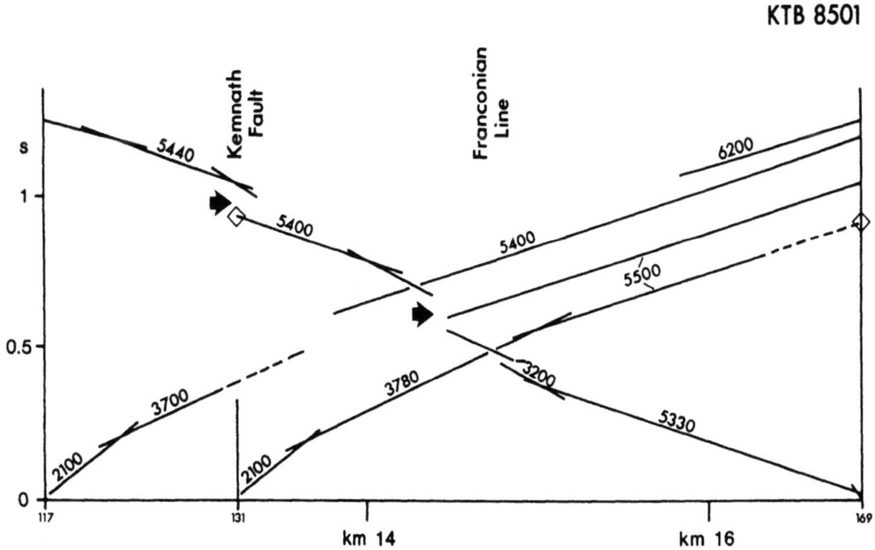

Fig. 46. KTB 8501: Time-distance plot of first arrivals corrected to datum from selected shots across the FL and the Kemnath fault; arrow prominent delay; diamond reciprocal time

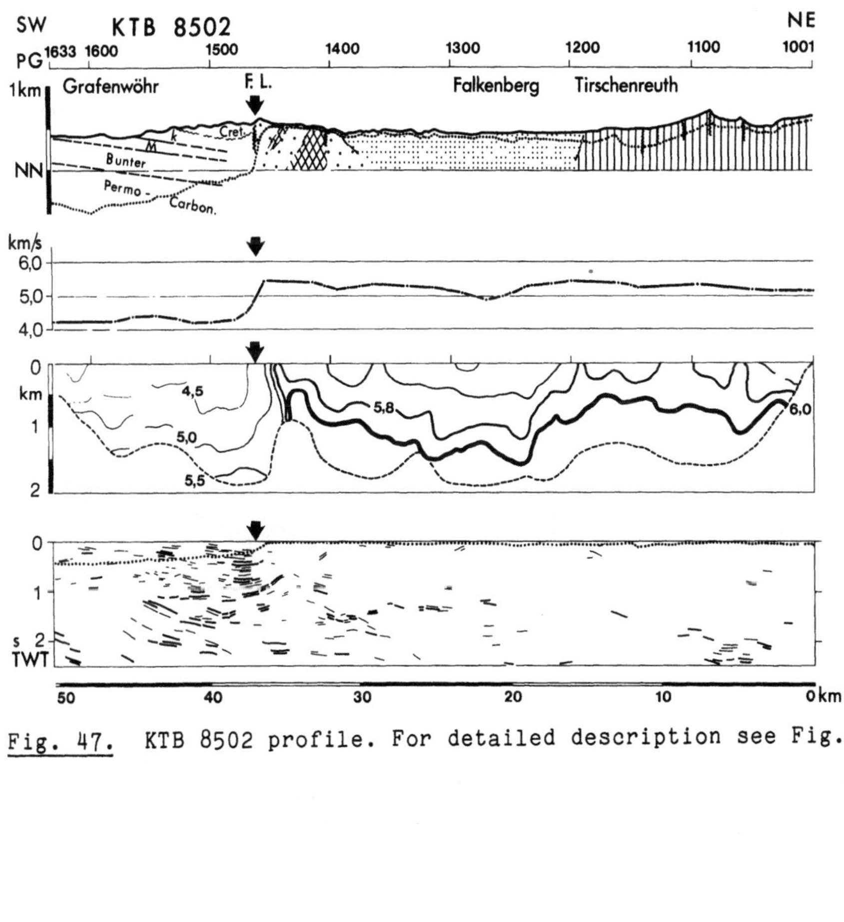

Fig. 47. KTB 8502 profile. For detailed description see Fig. 45

layer. The FL produces again a well-pronounced horizontal velocity gradient in the Slichter model.

Near the E border of the ZEV a moderate velocity high is indicated correlating with the amphibolite gneisses of the ZEV (see also DEKORP 4 profile, central part). Further to the E the elongated body of the Falkenberg granite shows up in the Slichter model as a bowl-shaped velocity structure.

The general structure of the KTB 8503 profile (Fig. 48) is similar to that of KTB 8502. The sedimentary fill of the Weiden trough SW of the FL gives a good reflection picture, thus allowing a comparison with the velocity model. At the E margin of the basin the velocities of the Slichter model are about 5.0 to 5.5 km/s at 1.0 km beneath the reference level. This corresponds to the depth of the crystalline basis at 0.4 s TWT as found in boreholes V10/V13 (Helmkampf et al. 1982) projected onto the profile (Fig. 49a). Slichter's model seems to be reliable in the center part of Fig. 49a, because the iso-velocities agree with the course of the highly reflective basal zone of the sedimentary basin.

This strong reflective band can be interpreted as basal series of the Permo-Carboniferous, containing thin coal seams and well bedded claystone/sandstones in alternating order, as is known from the above mentioned borehole results (Helmkampf et al. 1982). In the W part of the profile the iso-velocity lines rise in contradiction to the reflection section. A possible explanation might be high velocity volcanics in the Lower Permian. The velocity structure of the basement just E of the FL is analogous to that observed on KTB 8502 with a horizontal velocity gradient tilting to the SW.

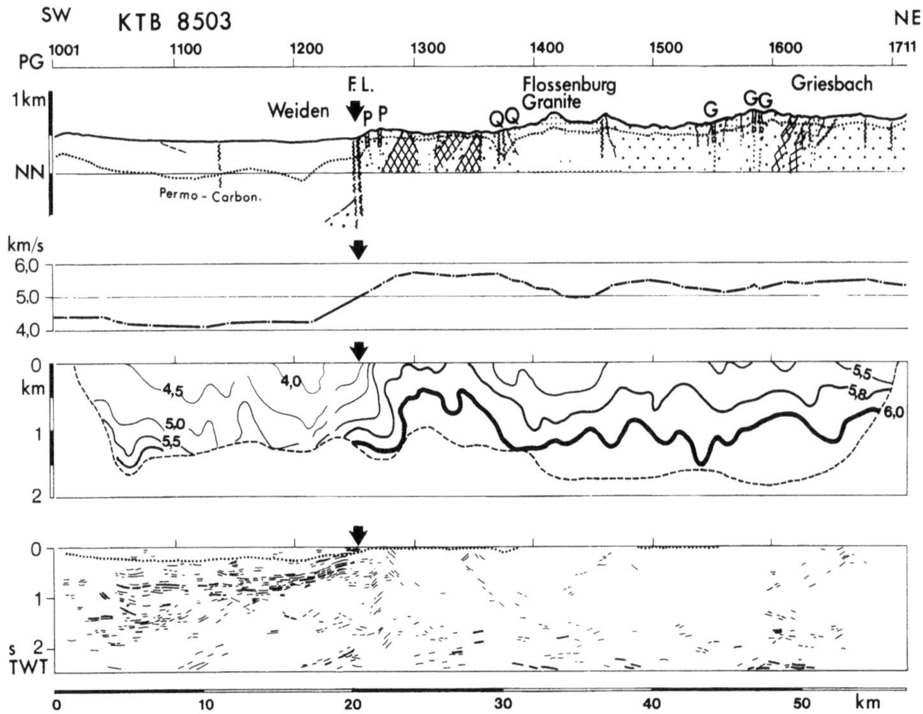

Fig. 48. KTB 8503 profile. For detailed description see Fig. 45. P porphyry; Q quartz vein; G granite.

E of the FL the basement is characterized again by a velocity high in the ZEV and a weak bowl-shaped structure in the Flossenbürg granites. Additional prominent depressions in the velocity model (Fig. 49b), partly also observed in the refraction horizon, correlate - as observed already on other profiles - with silicified fault- and diaphtoresis zones, where the high-grade metamorphic mineral association of the gneisses was transformed by hydration to rock of lesser density.

Figure 50 shows the near-surface velocity analyses of the KTB 8504 profile. The results of one-dimensional selected refraction models are shown along the vertical bars. Data left of the bar result from first arrivals along the KTB 8504 profile; data right of the bar are derived from the intersecting profiles KTB 8501 to 8503. There is also a rough agreement with the Slichter model. Only at the intersection with KTB 8501 the 5 km/s isoline seems to be too deep. The refraction model seems to be more reliable, because it gives a TWT for the basis of the Permian that is consistent with the seismic reflection results on KTB 8504 as well as on KTB 8501. In the S part of the profile the top of the crystalline basement clearly subsides 2 km below the reference level, i.e., close to the validity limit of the Slichter model, confirming the seismic picture.

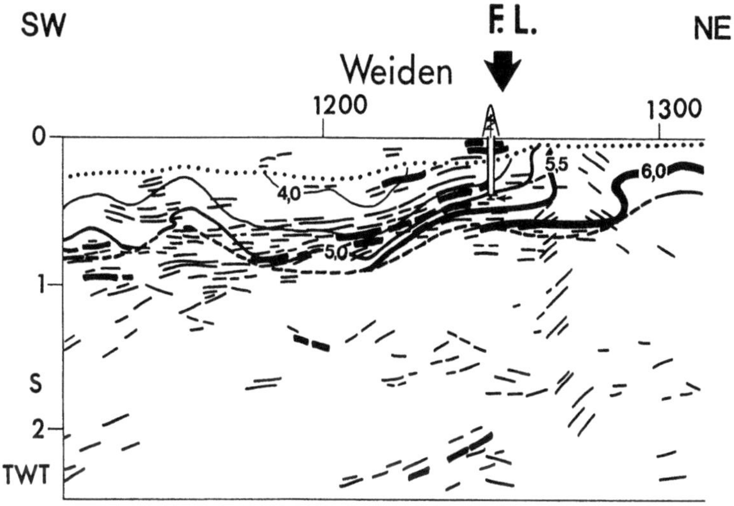

Fig. 49. Parts of KTB 8503 profile: a FL and Mesozoic foreland: Line drawing overlaid by iso-velocity lines obtained by Slichter and refracting horizon (dotted)

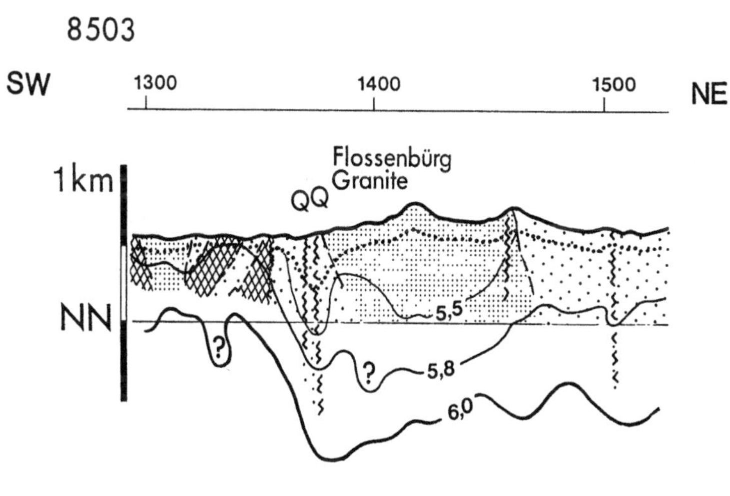

b crystalline basement E of the FL: Surface geology overlaid by iso-velocity lines obtained by Slichter and refracting horizon (dotted). Q quartz vein.

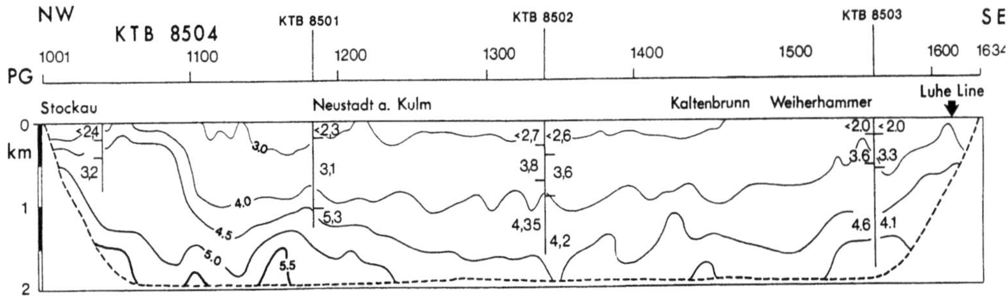

Fig. 50. KTB 8504 profile; iso-velocity lines obtained by Slichter. Selected refraction models are shown by vertical bars; left side of the bar result from first breaks along KTB 8504; right side results from first breaks from intersections with KTB 8501 and 8503 profiles

In Fig. 33 a synthesis of the present state of information by seismic reflection and refraction work about the Permian-Mesozoic foreland basin between Kemnath and Weiden has been given.

3.2 Expanding-Spread Investigations

In addition to the Vibroseis near-vertical reflection survey, two expanding spread experiments with midpoints in the MN and ST zones were performed. A series of 96 borehole shots with charges of 90 kg and 1 km spacing was fired in the NW part of the DEKORP 4 line. The observational scheme is shown in Fig. 51. The recording equipment consisted of 120 channels from the universities of Clausthal, Hamburg and Kiel and from BGR/NLfB (Hanover). The shots were fired in the late afternoons after the daily Vibroseis work.

The data of these experiments were sampled with 4 ms over the record length of 40 s. A bandpass filter of 5 to 32 Hz was used, the dominant signal frequency being around 20 Hz. For the interpretation of the P--waves the record length was reduced to 20 s. Problems for data processing and interpretation were low signal/noise ratios, insufficient statics information, varying frequencies and wave forms (due to different equipments and coupling effects) and the very heterogeneous crust in the vicinity of the boundary between MN and ST. Thus, the chance of destroying signals by stacking of adjacent traces was quite high. Therefore, a special technique was applied using the "energy" chracteristics of traces: squared amplitudes were normalized with regard to the maximum energy per trace. Subsequent mixing and smoothing by use of two moving windows over five traces and 80 ms resulted in the energy sections shown in Fig. 52. First arrivals can be observed very clearly up to a distance of about 60 km in the S part (MN). After some near-vertical reflection bands (P_c) from the lower crust in the offset range from 20 to 55 km, the Moho reflection (P_mP) appears really strong in the near-vertical and in the wide-angle range.

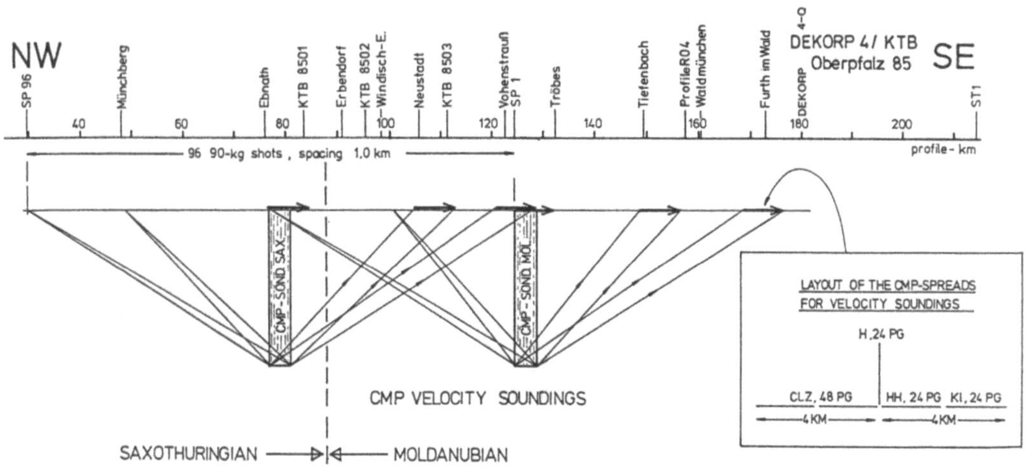

Fig. 51 Observational scheme of the expanding spread experiment, Oberpfalz 1985

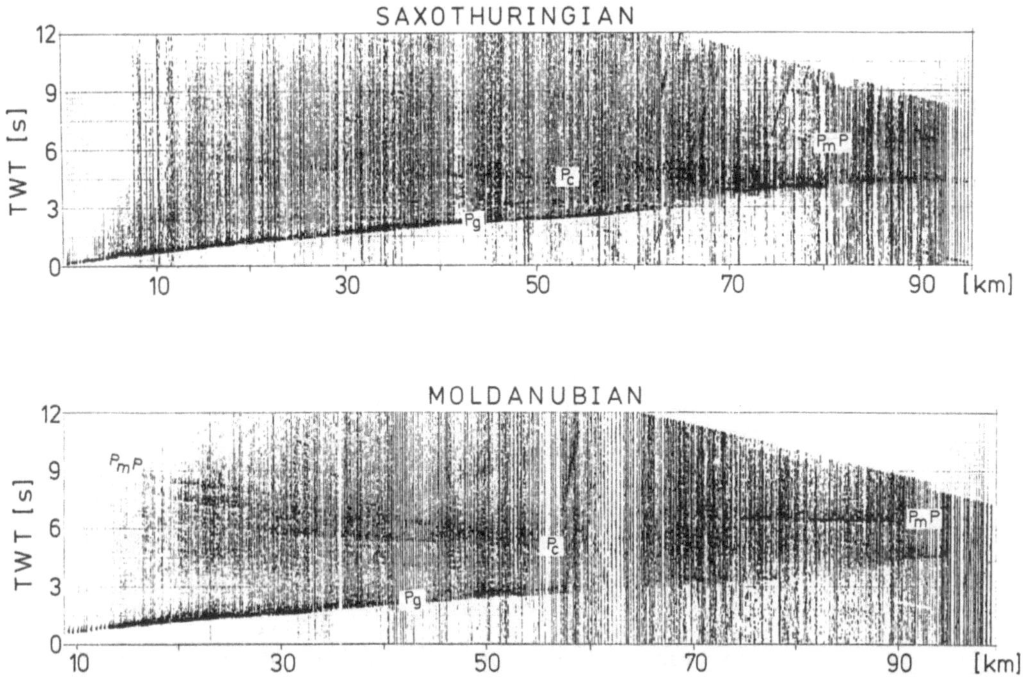

Fig. 52. Energy sections for the Saxothuringian (top) and the Moldanubian (bottom); time coordinate reduced by 8 km/s. For these sections squared amplitudes and a moving window over five traces and 80 ms was used

In the N part (ST) clear first arrivals can be observed up to a distance of 65 km. Here, only one strong reflection band (P_c) appears in the lower crust. In addition, there are weak indications for upper crustal reflections and the P_mP. Using squared amplitudes and moving windows the picking of exact arrival times becomes difficult, Therefore, an "energy-controlled gain" (ECG) was applied to the original traces. Due to this procedure, phases and arrival times are not distorted and a quality improved seismic record is obtained (Bittner et al. 1987).

Figure 53 gives the results of a 1-dimensional interpretation consisting of a direct inversion of first arrivals combined with rms--velocities derived from x^2/t^2 values using the reflections of the entire sections shown in Fig. 52. In order to gain two-dimensional velocity information common shot gathers (CSG) were constructed from single shots along the profile, and two-dimensional ray tracing was used.

Figure 54 shows the interval velocities derived from 1-D and 2-D inversions for the two ESP's. For both ESP's a low velocity zone appears in the depth range of 5 to 10 km which is more pronounced in the MN. Comparing these two velocity depth functions there are no "typical" structures for the MN and ST. A separation into upper and lower crust is not obvious. The mean velocities of the whole crust are identical: 6.25 km/s.

Further investigations resulted in the velocity model shown in Fig. 55 together with a line drawing of nearly the entire DEKORP 4 profile. Areas where ESP rays did not penetrate were completed by velocity

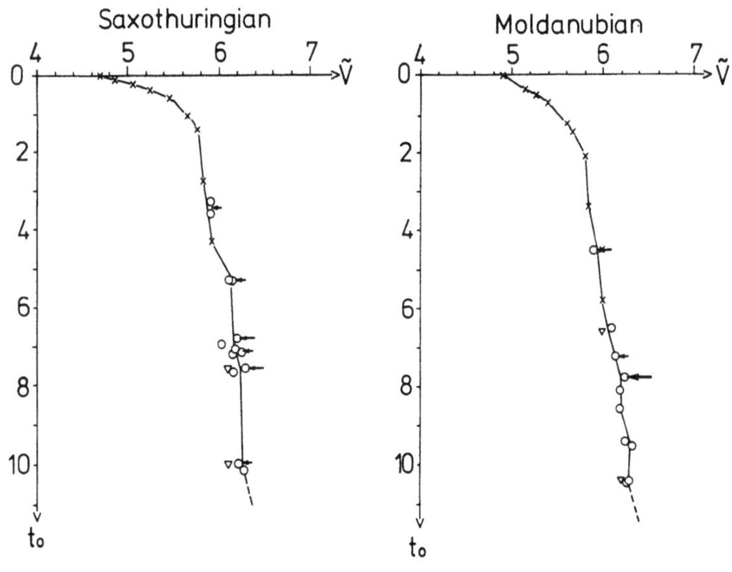

Fig. 53. RMS-velocities derived from x^2/t^2-values using the reflections observed on the energy and "energy controlled gain" sections. Crosses direct inversion of first arrivals (1-D); circles rms-velocities from x^2/t^2-values; triangles optimum stacking velocities from stack tests; arrows dip-corrections, computed from the migrated section of DEKORP 4 (rms-velocities divided by the cosine of the dip-angle)

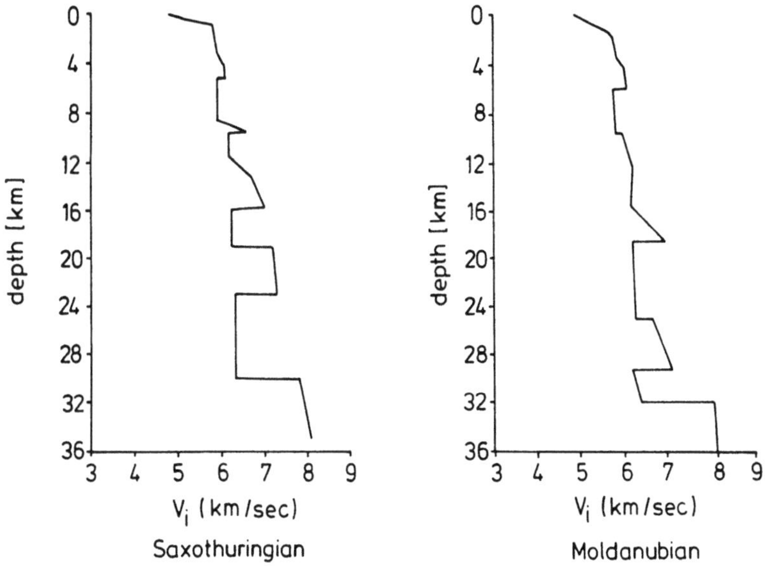

Fig. 54. Computed interval-velocities for the two CMP's (left Saxothuringian; right Moldanubian). 1- and 2-dimensional inversion methods were used

information from wide-angle data (DEKORP Research Group 1988). The velocity model was checked by synthetic seismograms and corrected, so that it fits the observed traveltimes and amplitudes as well (Sörensen 1987).

A comparison with the velocity model derived from wide-angle data (DEKORP Research Group 1988) reveals quite a good structural similarity. Also the velocity values are nearly the same whenever the shape of the particular velocity-depth functions is not identical. For example, the zone of increased velocity starting in the NW at about 22 km (about 7 s TWT) has a shape of a box in the ESP model instead of the spike-type form in the wide-angle model. This might be due to the different ray paths and recording geometry, and, on the other hand, due to the ambiguity of velocity-depth modeling (Berry 1971). The presence of a high velocity zone is necessary to provide coincidence between observed and calculated traveltimes. Its thickness and position is about the same as in the wide-angle seismic model.

3.3 Dip Moveout Velocity Analysis on the KTB 8506 Profile

The main interest in determining velocities from deep seismic data is based on the desire of an optimum seismic section, i.e., a high signal/noise ratio, as can be obtained by stacking of CDP traces. Thus, the determination of stacking velocities had highest priority. Only in the best case (horizontal layers, no lateral velocity changes) has this velocity a geological meaning. If there are reflections from dipping layers the stacking velocity derived from normal moveout is dependent on dip angle: a horizontal layer with high velocity can produce the same moveout as a dipping layer with low velocity. i.e., for an accu-

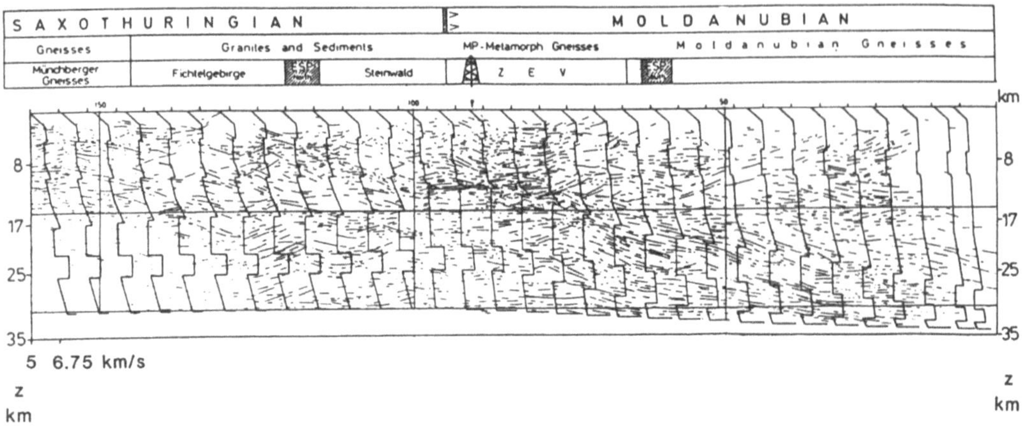

Fig. 55. 2-D velocity model along the DEKORP 4 profile derived essentially from ESP observations with line drawing obtained by the seismic reflection results. The model was completed by data obtained from wide-angle seismic observations

rate stacking of signals from a dipping horizon, a velocity higher than the true velocity above the reflecting layer is required. The stacking becomes difficult or the quality of the stack decreases, if there are reflections with various dips in the same domain of the time section, because then only one stacking velocity, favoring one particular dip can be chosen, i.e., only these reflectors are represented well in the stacked section whose dip was focused by the chosen stacking velocity. Subsequently, the quality of the data migrated after stack is decreased, too: events with low energy might be underrepresented. Following the conventional processing sequence, a great number of stacks with different stacking velocities would be necessary to represent all reflectors with various dip and the same quality. Then, a particular section contains only a part of the entire seismic information. This problem can be overcome by application of the dip moveout (DMO) method. The DMO process removes the influence of dip from stacking velocities and is adequate to a prestack partial migration. The result is a qualitatively better stack as well as dip corrected velocity functions.

The KTB 8506 profile contains several parts where strong horizontal and dipping reflections occur at the same time, thus a DMO velocity analysis was carried out for this profile. While the resulting seismic section does not look much different from that without DMO the velocity--depth functions obtained from this method are more consistent and the velocities itself are better suited and thus more realistic in a physical sense.

The results are demonstrated in Fig. 56. Figure 56a shows the interval velocities derived from stacking velocities used to produce the conventionally stacked section. The areas with velocities equal to and higher than 6.5 km/s and 7.0 km/s, respectively, are indicated by differently

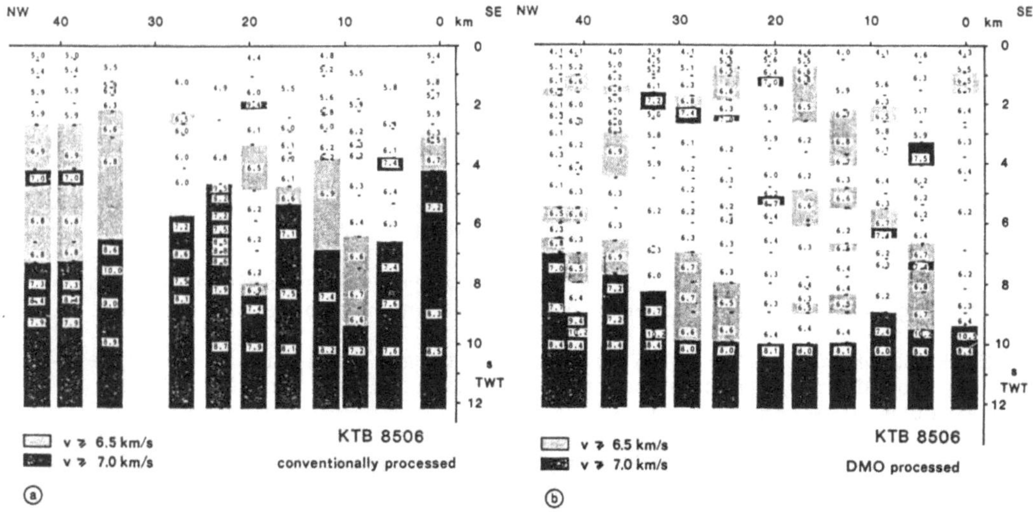

Fig. 56. KTB 8506 profile; interval velocities derived from stacking velocities (a) of the conventionally processed section and (b) of the DMO processed section.

shaded areas. The velocity values were chosen because 6.5 km/s indicates a velocity distinctly higher than the average velocity of the earth's crust in most cases and 7.0 km/s marks velocity values observed normally in the lower crust, i.e., very high velocity values.

In the center part of Fig. 56a unrealistic high velocities occur starting at 5 s TWT. They show an antiformal structure which has no equivalent in the seismic reflection section.

In Fig. 56b the interval velocities obtained by DMO processing are shown. The picture is quite different compared with the velocities from conventional processing: particular velocity zones can be distinguished. High and very high velocities appear:

- in the NW of the profile at about 1.5 to 2 s TWT, dipping weakly to the SE
- in the center of the profile at about 1 s TWT, forming more or less a singularity
- at the SE end of the profile, coinciding with the position of the EB
- at the SE end of the profile below the EB, separated by a small channel of lower velocities
- in the lower crust, corresponding to a normal velocity distribution increasing with depth but showing an upper boundary dipping remarkably to the SE.

Summarizing these features, a fair correlation to the seismic reflection structures is observed; i.e., the SE-dipping tendency is reproduced rather well, and also the shape of the EB seems to be represented. On the other hand, the high velocity singularity in the uppermost part of the center of the profile cannot be interpreted yet, and there are no other seismic indications for such an anomaly.

References

Behr HJ, Heinrichs T (1987) Gelogical interpretation of DEKORP 2-S: A deep seismic reflection profile across the Saxothuringian and possible implications for the Late Variscan structural evolution of Central Europe. Tectonophysics 142: 173-202

Berry MJ (1971) Depth uncertainties from seismic first arrival refraction studies. J Geophys Res 76: 6464-6468

Bittner R, Trappe H, Meissner R (1987) Piggy-back seismic experiments during deep crustal reflection surveys. An Geophphys 5B: 381-388

DEKORP Research Group (1985) First results and preliminary interpretation of deep reflection seismic recordings along profile DEKORP 2-South. J Geophys 57: 137-163

DEKORP Research Group (1988) Results of the DEKORP 4/KTB Oberpfalz deep seismic reflection investigations. J Geophys 62: 69-101

Emmert U (1981) Perm nördlich der Alpen. In: Haunschild H, Jerz H (eds) Erläuterungen zur Geologischen Karte von Bayern 1:500.000. 3. Aufl. Bayer Geol Landesamt, München, pp 34-40

Fountain DM, Hurich CA, Smithson SB (1984) Seismic reflectivity of mylonite zones in the crust. Geology 12: 195-198

Giese P (1976) Depth calculation. In: Giese P, Prodehl C, Stein A (eds) Explosion seismology in central Europe. Springer, Berlin Heidelberg New York Tokyo, pp 141-161

Gudden H, Schmid W (1985) Die Forschungsbohrung Obernsees. Geol Bavarica 88: 5-21

Haunschild W, Schröder B (1960) Geologische Karte von Bayern 1:25.000, Blatt 6237, Grafenwöhr mit Erläuterungen. Bayer Geol Landesamt, München

Helmkampf KE, Kuhlmann J, Kaiser D (1982) Das Rotliegende im Randbereich der Weidener Bucht. Geol Bavarica 83: 167-186

Kampfmann W (1988) A study of diffraction-like events on DEKORP 2-S by Kirchoff theory. J Geophys 62: 163-174

Kläschen D, Bittner R (1988) Laufzeituntersuchungen im CMP-Gather für räumlich verteilte Diffraktoren. 48. Jahrestagung der DGG Köln (abstracts) pp 51

Kollmann K (1977) Die Öl- und Gasexploration der Molassezone Oberösterreichs und Salzburgs aus regional-geologischer Sicht. Erdöl-Erdgas-Z 93 (Sonderausgabe): 36-49

Kossmat F (1927) Gliederung des variszischen Gebirgsbaues. Abh Sächs Geol Landesamt 1: 1-39

KTB 1986 (1986) Kontinentales Tiefbohrprogramm der Bundesrepublik Deutschland, KTB, Ergebnisse der Vorerkundungsarbeiten Lokation Oberpfalz. In: Weber K, Vollbrecht A (eds) 2. KTB-Kolloquium Seeheim/Odenwald, 19-21 Sept 1986

Lüschen E, Wenzel F, Sandmeier KJ, Rühl T, Menges D, Janoth W, Keller F, Söllner W, Stiller M, Thomas R, Krohe A, Stenger R, Fuchs K, Eisbacher G, Wilhelm H (1987) Near-vertical and wide-angle seismic surveys in the Black Forest, SW Germany. J Geophys 62: 1-30

Matthews D and the BIRPS Group (1987) Some unsolved BIRPS problems. Geophys J R Astr Soc 89: 209-216

Meissner R, Bartelsen H, Krey T, Schmoll J (1982) Detecting velocity anomalies in the region of the Urach geothermal anomaly by means of new seismic field arrangements. In: Cermak V, Haenel R (eds) Geothermics and geothermal energy. Schweizerbart'sche Verlagsbuchhandlung, Stuttgart, pp 285-292

Pucher R (1986) Interpretation der magnetischen Anomalie von Erbendorf (Oberpfalz) und dazugehörige gesteinsmagnetische Untersuchungen. Geol Jb E33: 31-52

Reston T, Meissner R (1988) The three-dimensional structure of the Oberpfalz. Tectonophysics (in press)

Rühl T (1987) Tauchwellen-Tomographie für Stripping-Korrekturen und geologische Interpretationen bei den reflexionsseismischen KTB-Profilen im Schwarzwald. Dipl Thesis Univ Karlsruhe

Schröder B, Siegling M (1966) Geologische Karte von Bayern 1:25.000, Blatt 6137, Kemnath mit Erläuterungen. Bayer Geol Landesamt, München

Slichter LB (1932) The theory of the interpretaion of seismic travel-time curves in horizontal structures. Physics 3: 273-295

Sörensen G (1987) Auswertung reflexionsseismischer Messungen auf dem Profil DEKORP 4 in der Oberpfalz. Dipl Thesis Christian-Albrechts--Universität Kiel

Crustal Structure in the Surroundings of the KTB Drill Site as Derived from Refraction and Wide-Angle Seismic Observations

H. Gebrande, M. Bopp, P. Neurieder and T. Schmidt

Institut für Allgemeine und Angewandte Geophysik, Ludwig-Maximilians-Universität München, Theresienstr. 41/IV, 8000 München 2, FRG

1 Introduction

The first knowledge of the seismic crustal structure in the Oberpfalz area came from large-scale refraction surveys carried out during the 1960's and early 1970's utilizing quarryblasts (Giese 1968, 1976; Peters 1974). The pertinent quarries were situated near Böhmischbruck and Mauthaus (Fig. 1). Because of the large spacing between receivers (about 5 km), common in those times, only some rough information on crustal thickness, average velocities, and the existence of a low velocity layer in the middle of the crust could be established. In the light of later investigations (Strößenreuther 1982), a systematic decrease of crustal thickness from 33 km in the Moldanubian (MN) of the southern Oberpfälzer Wald to 30 km in the Saxothuringian (ST) beneath the Fichtelgebirge (Peters 1974) became evident. Moreover, the results indicated an increase of the crustal average velocity from SE to NW.

The former measurements were too wide-meshed for the deduction of more detailed information about the internal structure of the crust. However, the more recent measurements in the southeastern Moldanubian (Miller and Gebrande 1976; Strößenreuther 1982) proved the existence of a high velocity zone in the depth range between 14 and 18 km.

2 Regional Refraction Seismic Investigations in the Oberpfalz

After the establishment of the KTB Project, the observations along the line Mauthaus-Böhmischbruck were resumed in 1981 to obtain a more close-meshed receiver pattern. Explosions from two other quarries near Remmelberg and Tröstau were observed. In 1983 an additional reverse shooting profile was surveyed with underwater explosions as seismic sources (see Fig. 1).

The most surprising result was that an unusal prominent wide-angle reflection appeared in the ST/MN boundary zone near the town of Erbendorf. Figure 2 (top) shows this outstanding feature between distances of 40 and 70 km from the shotpoint Remmelberg and at reduced traveltimes near 1 s with an apparent velocity of about 7.0 km/s. From this section a velocity-depth function was derived (left side of Fig. 3), which shows that near the surface velocity increases rapidly at first - as usual in crystalline - and then more slowly. Under a moderate low velocity zone (v = 5.9 km/s), strongly increasing velocities up to 6.6 km/s follow between depths of 8.5 and 10 km.

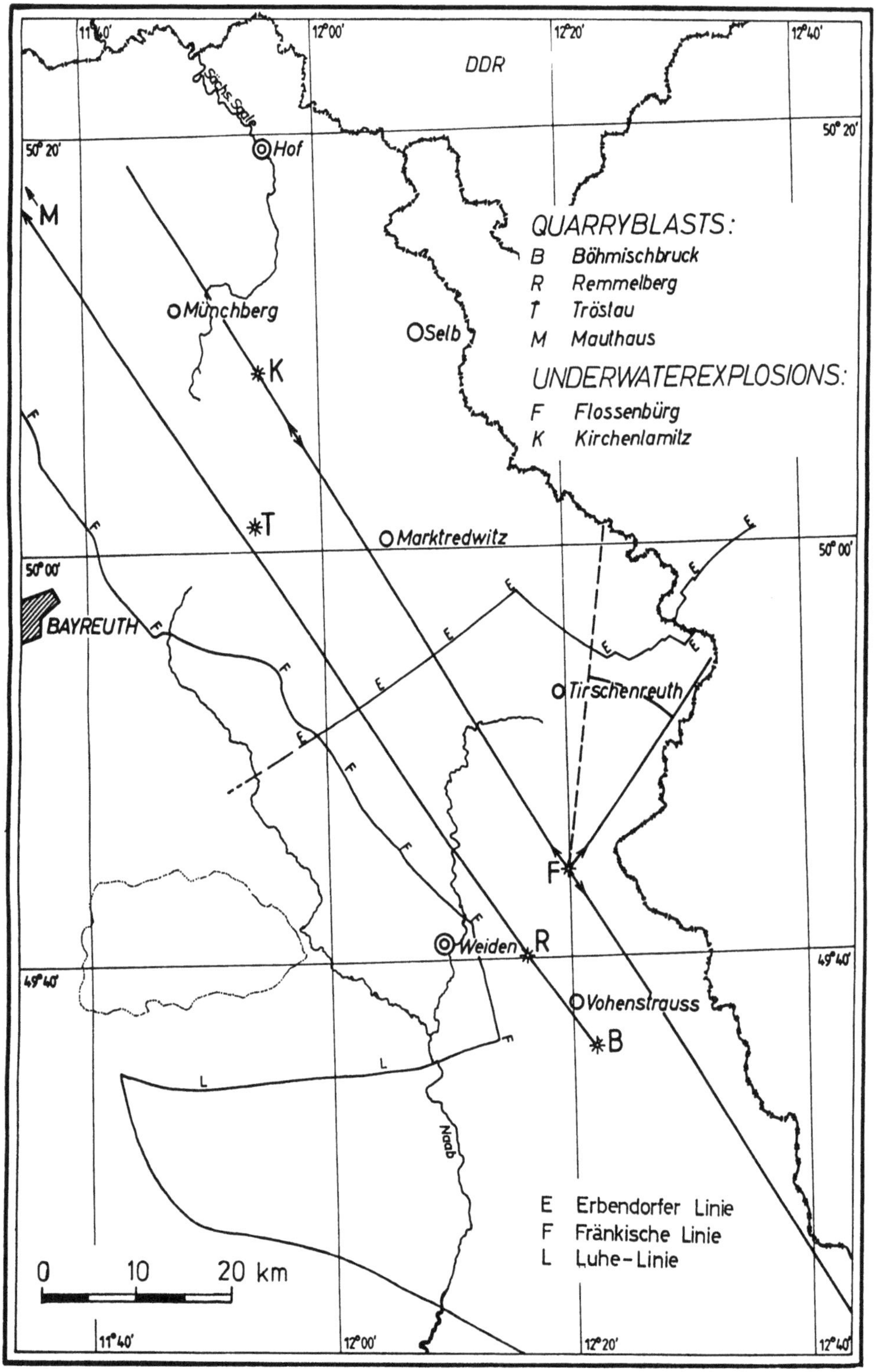

Fig.1. Situation map of refraction profiles surveyed between 1968 and 1983 in the Oberpfalz. Both the lines Böhmischbruck-Mauthaus (**B-M**) and Flossenbürg-Kirchenlamitz (**F-K**) cross the Erbendorfer Linie, a surficial expression of the Moldanubian/Saxothuringian boundary

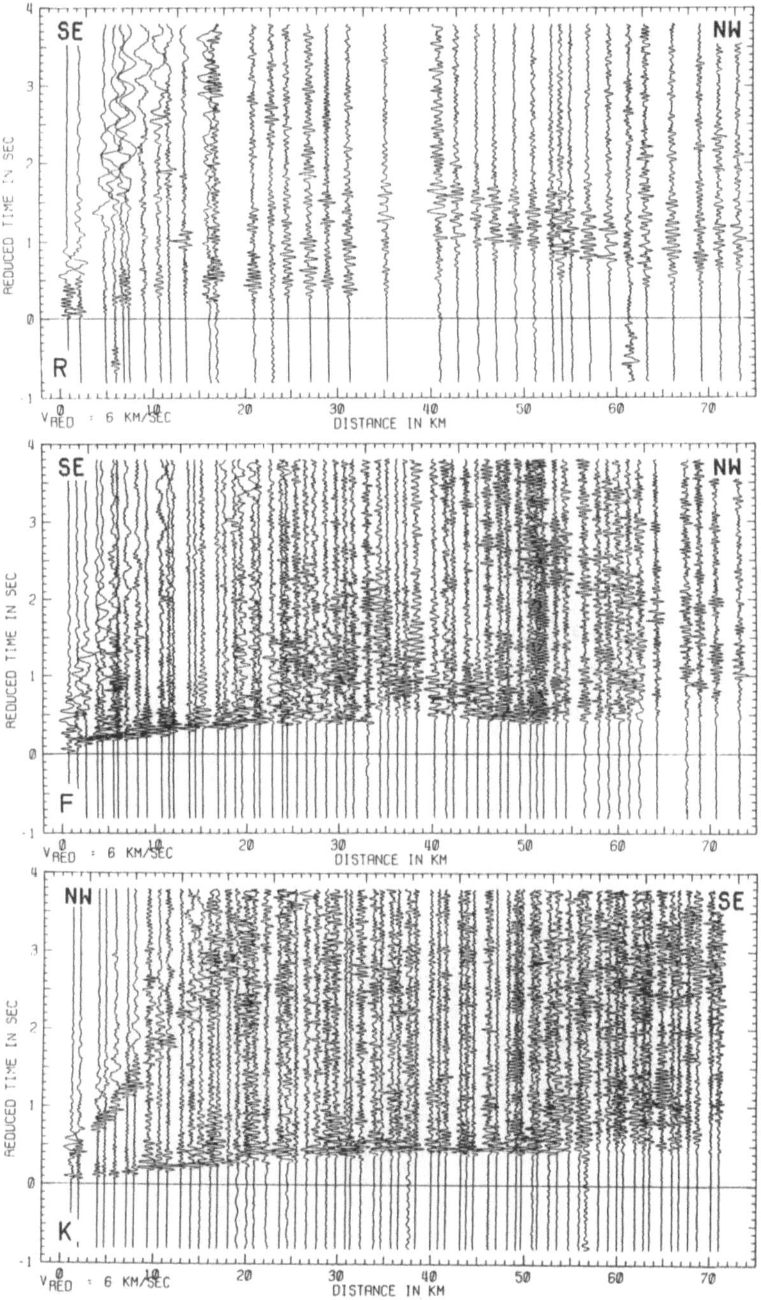

Fig.2. Top Seismogram section from the quarry Remmelberg (**R**) to the NW. A prominent wide-angle reflection appears at reduced traveltimes near 1 s and between distances of 40 to 70 km with an apparent velocity about 7.0 km/s. It bottoms near the town Erbendorf. **Center** Shotpoint Flossenbürg (**F**) observed along the line Flossenbürg-Kirchenlamitz. The strong wide-angle reflections appear at less than 1 s reduced traveltime and within a distance range from 25 to 50 km. **Bottom** Reverse shot Kirchenlamitz (**K**) observed along the line Kirchenlamitz-Flossenbürg. No evident wide-angle reflections can be detected in this section

It was rather exciting to find such high velocities, commonly observed only in the lower crust, in this tier of the crust and thus within the target depth of a potential ultradeep well. This highly reflective high velocity zone was later called Erbendorf body (EB), and it was speculated whether it could be an updoming of the Conrad discontinuity, a mafic intrusion, or an imbricate structure of sialic and basic rocks related to tectonic processes caused by the subduction of the Saxothuringian crust under the Moldanubian crust.

The reverse profile F-K was surveyed to obtain an idea of the NE extent of the Erbendorf body. The seismogram sections (Fig. 2, center and bottom) differ significantly from each other and from the Remmelberg section (Fig. 2, top). Strong wide-angle reflections appear also in the observations of shotpoint F, however, between distances of 25 and 50 km and at lower reduced traveltimes. In the observations of the reverse shot K these wide-angle reflections are missing, or at least no longer distinguishable from first arrivals. Geographically, the Flossenbürg and the Kirchenlamitz v(z)-function correspond to the MN and ST parts of the profile, respectively. In principle they are similar to the Remmelberg v(z)-function, but the low-velocity layer is thinner and pinches out towards the NW and the high-velocity zone is observed already at 6.5 to 8 km depth (Fig.3, right side). In view of the amplitudes, however, it seems to be less well developed than beneath the Remmelberg line.

The most important result of these preliminary investigations was the understanding that in the Oberpfalz area the subsurface - especially in the region of the boundary ST/MN - shows an unusually strong structural heterogeneity and extraordinary high wave velocities in parts of the upper and middle crust. The resolution of structural details had to be left to the superior resolving power of the reflection seismics (DEKORP Research Group 1988). The planning of this survey was based essentially on the results of the investigations discussed above.

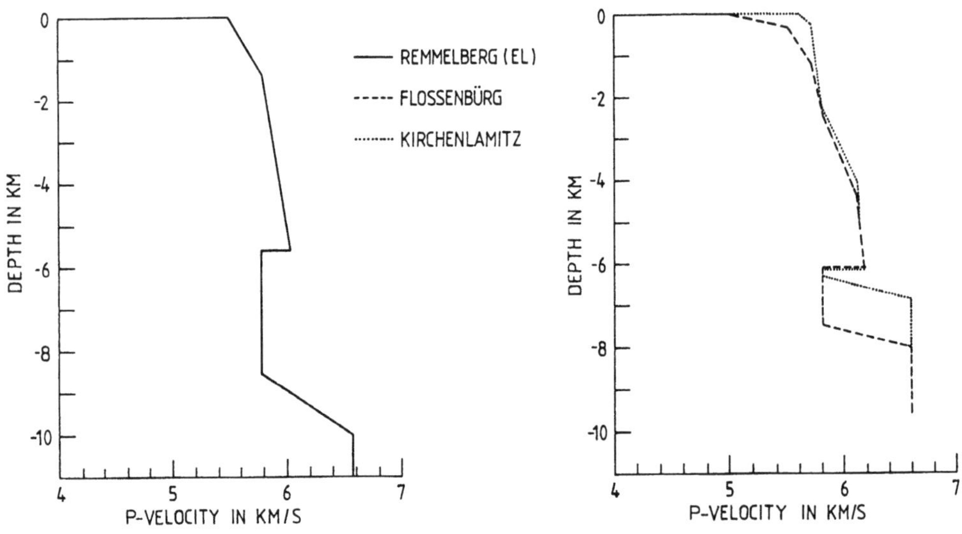

Fig.3. Velocity-depth functions derived from the Remmelberg section (**left**) and from the F and K sections, respectively (**right**)

3 Wide-Angle Reflection Seismics Along DEKORP-4

Originally the DEKORP-4 and KTB reflection survey was planned as shot seismics, and a detailed scheme was worked out to use the explosions for an additional wide-angle survey along most of the lines with the main goals of mapping the Erbendorf body and deriving velocity-depth functions in the wider KTB surroundings. For economic and environmental reasons, however, it was later decided to carry out the reflection survey with the Vibroseis technique and to restrict explosions and wide-angle observations to the line DEKORP-4. Therefore, the extension of the Erbendorf body as a wide-angle reflector and the velocity distribution could be studied in detail only along this line.

3.1 Field Techniques

Ninetysix bore hole shots with charges of 90 kg and 1 km spacing were fired in the northwestern part of the DEKORP-4 line. They were used for three different supplementary programs:

1. for mapping wide-angle reflections from the upper crust by means of the contractor's 200-channel reflection spread operated at 42 to 58 km offset,
2. for mapping lower crustal and Moho wide-angle reflections with a mobile array of 24 3-channel MARS stations at 60 to 90 km offset, and
3. for two expanding spread experiments with the midpoints in the Moldanubian and in the Saxothuringian zone, respectively.

The observational scheme is shown in Fig. 4. Shots moved with 4 km/day from shotpoint 1 near Vohenstrauß towards the northwest, while the contractor's spread and the MARS array (operated by university teams from Berlin, Clausthal, Frankfurt, and München) followed at constant mean shot-to-receiver offset. The expanding spread equipments with a total of 120 channels from the universities Clausthal, Hamburg, and Kiel, and from BGR/NLfB Hanover moved with the same velocity in the opposite direction to keep common midpoints stationary.

The shots were fired in the late afternoons, after the daily Vibroseis work. With some exceptions, especially in dry and hard rocks, their seismic efficiency was sufficient and the noise level at the recording sites was low enough to yield good seismograms up to the maximum distances of 100 km.

3.2 Results of the Wide-Angle Reflection Survey

Figure 5a shows a detailed location map of the shotpoint and the MARS stations. The shot-to-receiver midpoints are plotted in Fig. 5b. They cover 100 km of the DEKORP-4 line and cross the tectonometamorphic boundary between the Saxothuringian (ST) and the Moldanubian (MN) units of the Variscides. The CMP's of the expanding spread observations are indicated by open squares in Fig. 5b. CMP locations are characterized in this chapter by a linear southeasterly increasing CMP-coordinate, the origin of which is the northwestern end of the DEKORP-4 reflection line.

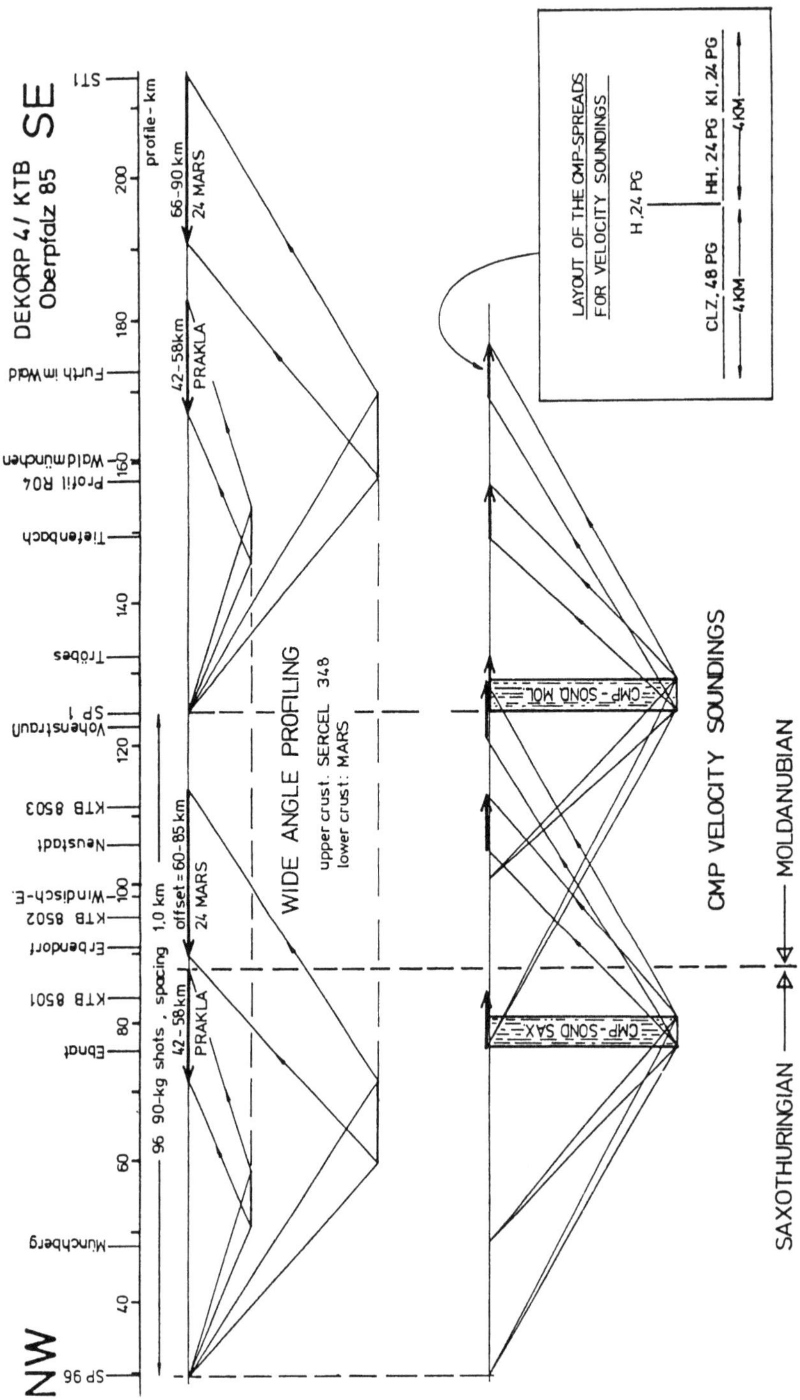

Fig.4. Observational scheme for the accompanying wide-angle reflection seismics along line DEKORP-4. The main goals of the survey were the mapping of wide-angle reflections from the upper crust (contractor's spread) and the lower crust and the Moho (MARS array). The two expanding spreads served for velocity soundings through the entire crust in the Moldanubian and the Saxothuringian domains, respectively

Fig.5a Location map of shotpoints and MARS stations. **b** Locations of shot-to-receiver midpoints (CMP); the CMP's of the expanding spread experiments are indicated by **open squares**. The CMP-coordinate increases towards SE. The KTB location is situated near CMP 96

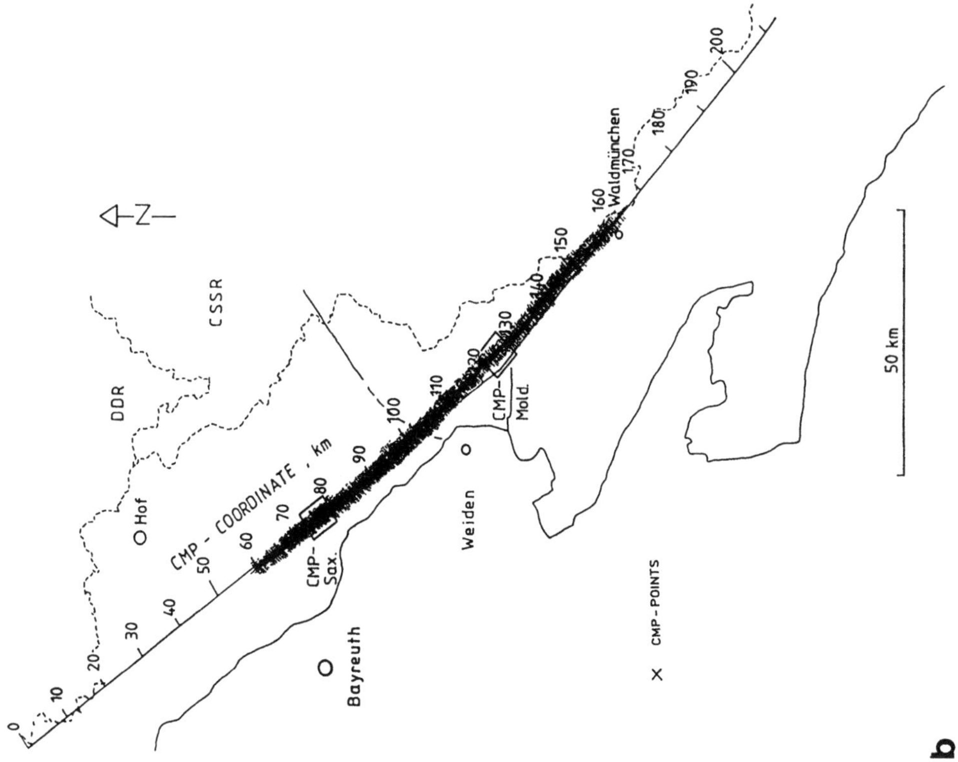

N
DDR
CSSR
O Hof
CMP - COORDINATE , km
Waldmünchen
Bayreuth
Weiden
CMP-Sax.
CMP-Mold.
× CMP-POINTS
50 km

b

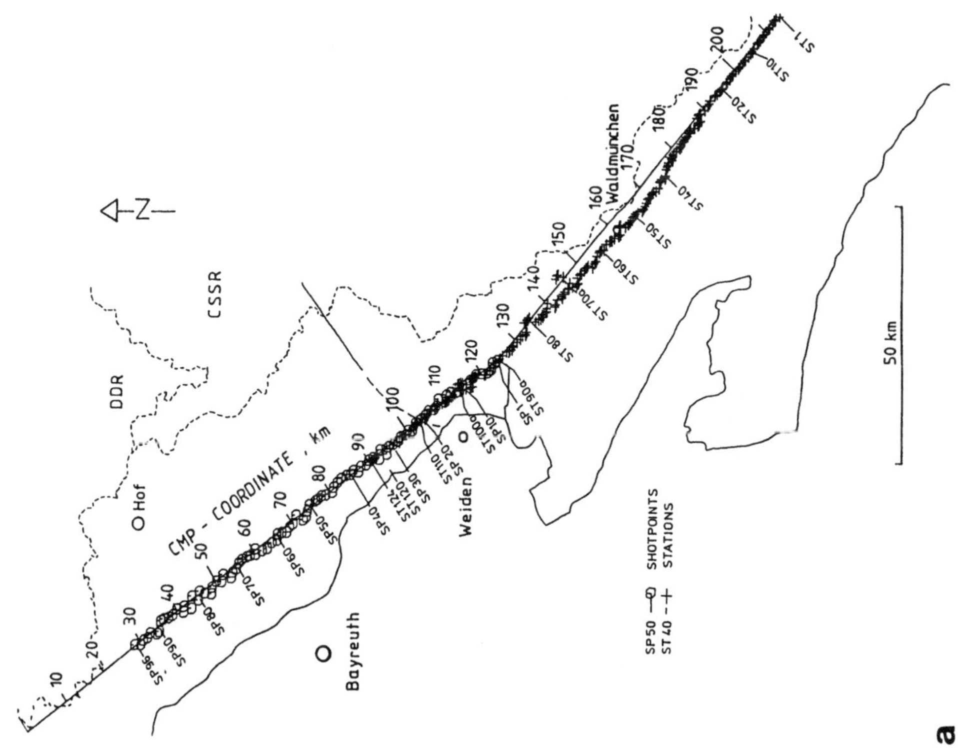

N
DDR
CSSR
O Hof
CMP - COORDINATE , km
Waldmünchen
Bayreuth
Weiden
SP50 SHOTPOINTS
ST40 STATIONS
50 km

a

Due to the multiple coverage, the wide-angle reflections can be processed and presented in different ways. In the following, particular use is made of the common midpoint (CMP) and common-offset (COF) sorting possibilities. CMP sorting suppresses the influence of dip and is therefore particularly suitable for velocity analyses, whereas COF sorting is particularly suitable for structural imaging.

Figure 6 shows three single shot observations in the 42- to 58-km distance range with predominant wave propagation in the ST, in the ST/MN boundary domain, and in the MN. Clear structural changes are obvious, and become even more evident by the CMP sections in Fig. 7a and b. They contain both the SERCEL and the MARS data. In the ST and MN zones proper, the P_g first arrivals are clearly observed up to about 60 km with apparent velocities of about 6 km/s, and are followed by complicated wave groups which cannot be individually resolved using the wide-angle seismics method but which must be interpreted as indications for small-scale heterogeneities in the upper crust. The boundary domain between profile km 80 and 110 (Neusorg-Weiden) deviates from this by two important features:

1. the first arrivals terminate at a distance of 40 to 50 km, which implies a relatively shallow position of a low velocity zone (LVZ) in the upper crust and
2. in the distance range of 40 to 70 km with reduced traveltimes of 1.5 to 0.5 s, an unusually intensive and well-correlatable wide-angle reflection appears (in Fig. 7 center, indicated by P_EP), with apparent velocities up to 8 km/s.

This prominent wide-angle reflection, that had already been found during the regional surveys by means of quarry blasts, is caused by a crustal segment at a depth of 8 to 13 km, the Erbendorf body (EB), that is also characterized in near-vertical angle seismics by an unusually high reflectivity (see also Schmoll et al., this Volume). Its NW-SE extension is best recognizable in the CMP sections (Fig. 7a, b) and the COF sections (Fig. 10). Figure 7a, b gives a selection of six from a total of 46 CMP sections, by which the part of the profile between 60 and 150 km is covered at 2-km intervals (Fig. 5b). The wide-angle image of the EB appears to be most clearly at CMP points 86 to 96; Fig. 7a (bottom) shows a typical example. CMP 84 (Fig. 7a, center) and CMP 98 (Fig. 7b, top) indicate the margins beyond which the EB cannot be traced by wide-angle seismics.

From km 92 onward P_iP reflections from the lower crust appear as distinct wave groups and increase towards SE (in Fig. 7b, center, with reduced traveltimes between 2 and 3 s). Moho reflections (P_MP) are scarce and weak in the ST but grow up when approaching the MN. Between 102 and 106 km they are characterized by apparent velocities up to 10 km/s even in the dip-insensitive CMP sections, indicating Moho undulations and/or strong lateral heterogeneities in the overburden. In the MN proper, the CMP sections show a heterogeneous but not clearly structured upper crust, occasionally good P_iP reflections from the middle crust (top of the lower crust?) and unusually clear Moho reflections (e.g., CMP 130 in Fig. 7b).

Fig.6. Single shot observations at distances between 42 and 58 km with predominant wave propagation in the ST (SP 92 **top**), the MN (SP 25 **bottom**) and the ST/MN boundary domain (SP 59 **center**)

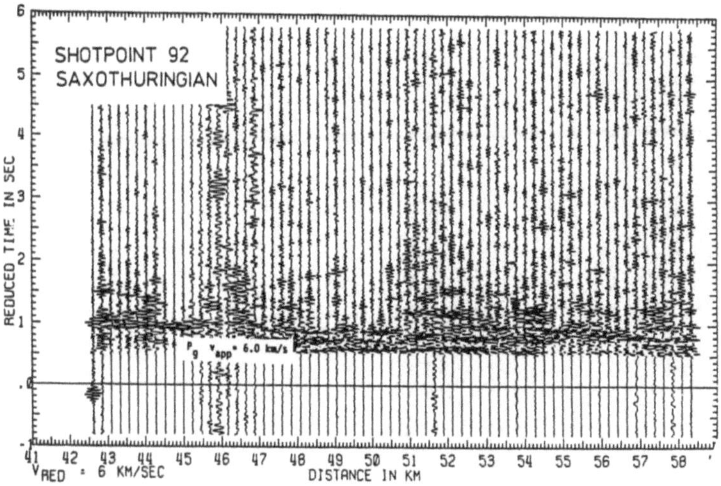
SHOTPOINT 92
SAXOTHURINGIAN
REDUCED TIME IN SEC
P_g v_{app} = 6.0 km/s
V_{RED} = 6 KM/SEC
DISTANCE IN KM

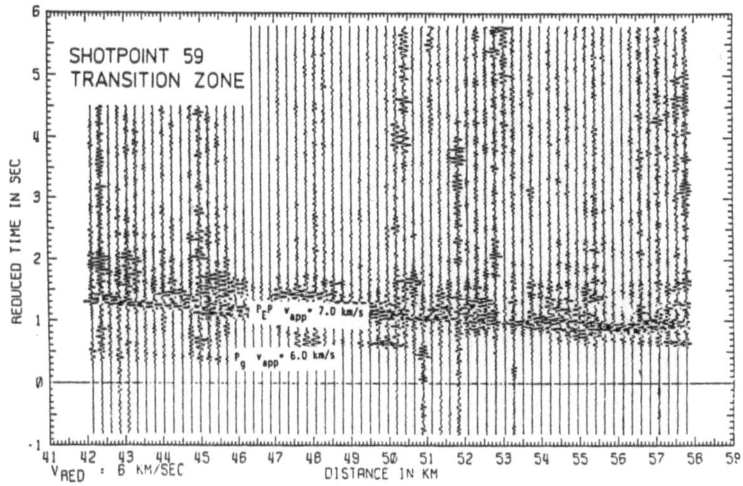
SHOTPOINT 59
TRANSITION ZONE
REDUCED TIME IN SEC
P_cP v_{app} = 7.0 km/s
P_g v_{app} = 6.0 km/s
V_{RED} = 6 KM/SEC
DISTANCE IN KM

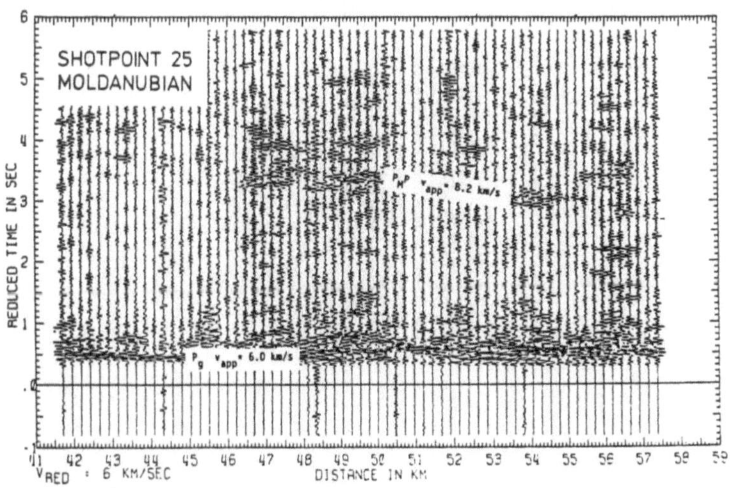
SHOTPOINT 25
MOLDANUBIAN
REDUCED TIME IN SEC
P_MP v_{app} = 8.2 km/s
P_g v_{app} = 6.0 km/s
V_{RED} = 6 KM/SEC
DISTANCE IN KM

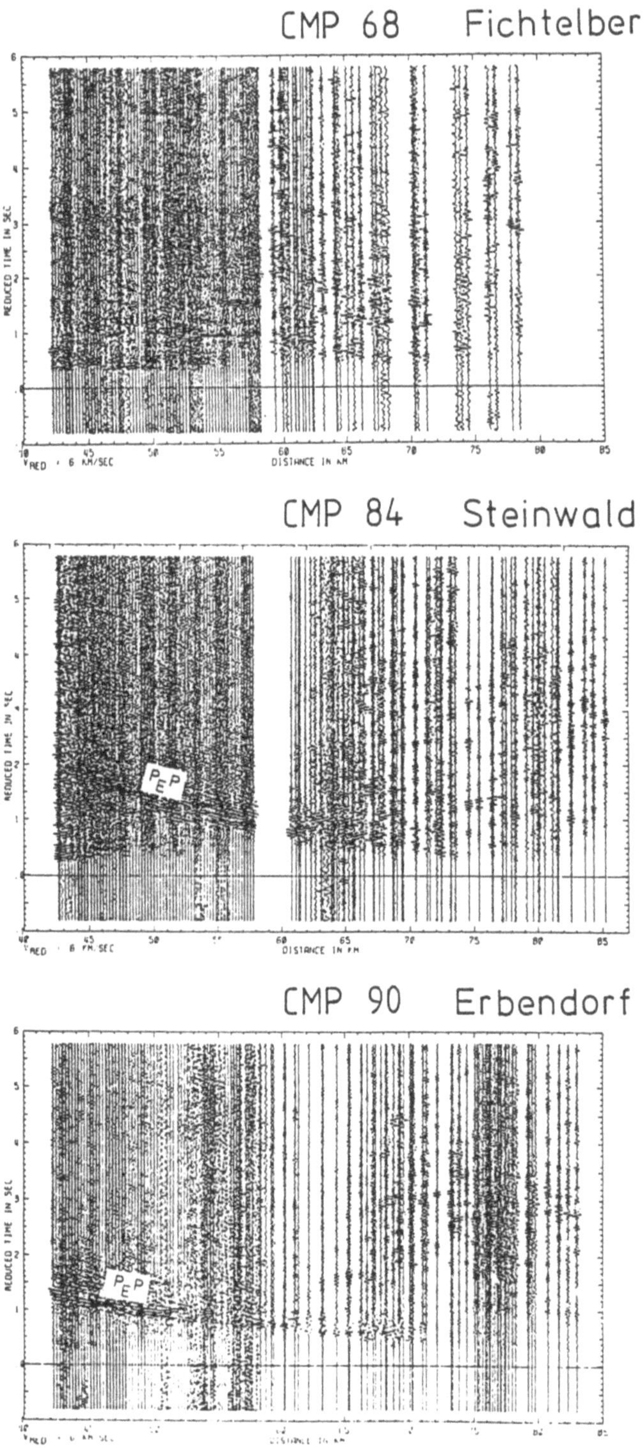

Fig.7. Sequence of six, from 46 available CMP sections between Fichtelgebirge and Oberpfälzer Wald, demonstrating the appearance of

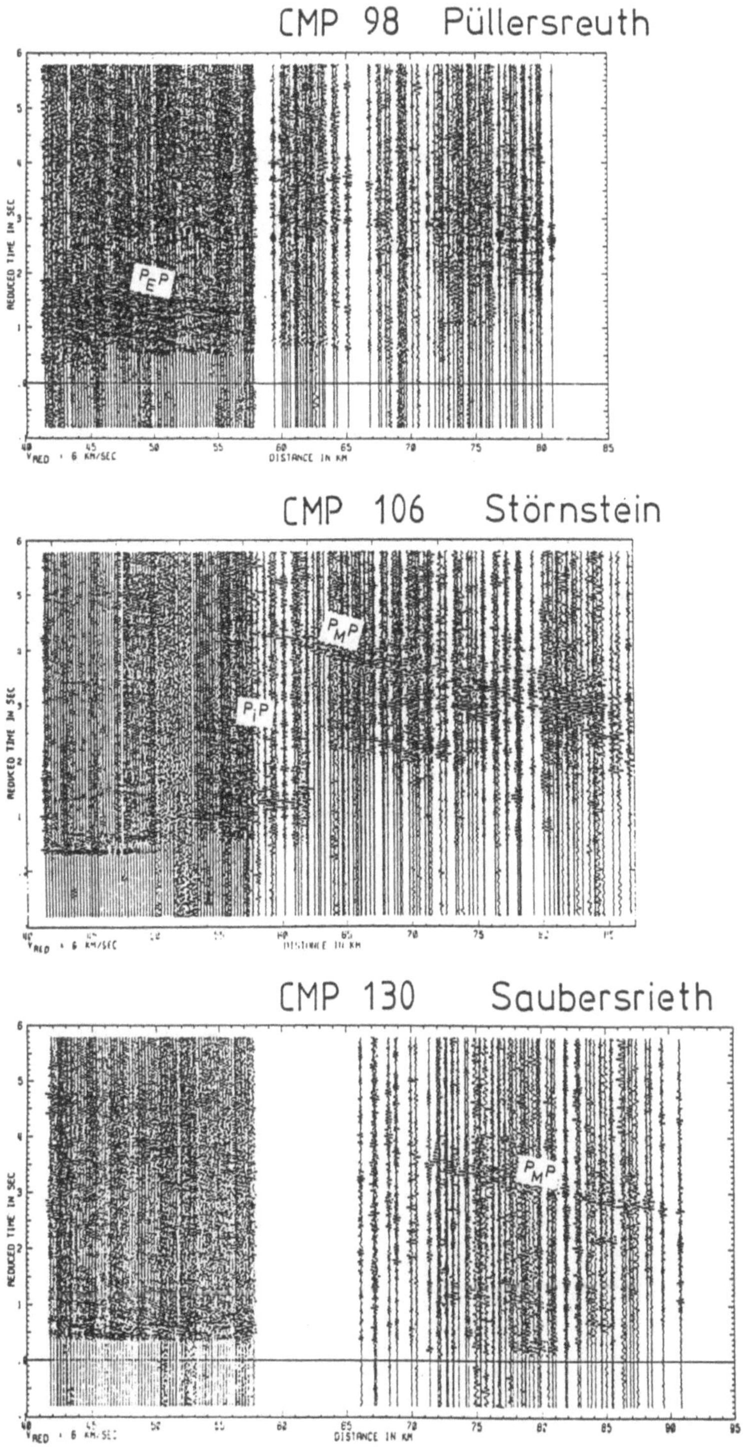

strong wide-angle reflections (P_EP) from the Erbendorf-body in the ST/MN boundary zone (CMP 84-98) and of lower crust (P_iP) and Moho (P_MP) reflections in the MN (CMP 106, 130)

3.2.1 Velocity Analysis

The tightly spaced CMP sections, together with first onsets of the contractor's steep-angle reflection spread, provide an excellent basis for velocity analyses. As shown by Gebrande (1986b), the benefits of the CMP sorting technique are similar in the refraction and wide-angle situations and in the well-known steep-angle reflection case and enable one-dimensional velocity inversions even in the presence of obliquely inclined layers. Alternative variants of the x-t-V-inversion process (Gebrande and Miller 1985; Gebrande 1986b) were applied. This is basically a stripping technique presuming a substratum composed of layers with constant velocities, or constant velocity gradients, or both.
Figure 8a,b shows cuttings of the results for the neighborhood of the KTB location (CMP 96). The considerable scattering of the velocity-depth functions may partially reflect true velocity variations, but probably also the limitations of the method on account of these strong velocity variations. We can, however, rely on the common or slowly changing properties of the velocity distributions and this consistently involves, near the KTB location, a first velocity increase to about 6.4 km/s at 8 to 9 km depth and a second increase to 7.0 km/s or even more at about 12 km depth. These velocities must be attributed to the Erbendorf body.

The determination of a sequence of one-dimensional velocity-depth functions from CMP sections is the first step to the development of a two-dimensional velocity model. A complementary step is seismic imaging by dynamically corrected wide-angle constant offset (COF) sections.

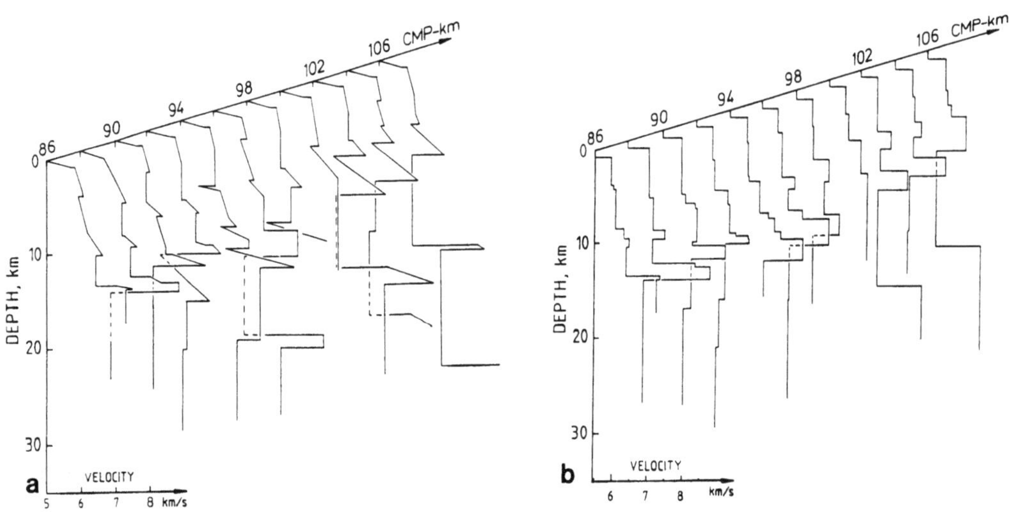

Fig.8. Alternativ CMP velocity analyses in the surroundings of the KTB location by means of the x-t-V inversion: **a** results of a "gradient layer inversion", **b** results of a "generalized Dix inversion". The considerable high velocities at depth around 12 km belong to the lower part of the Erbendorf body.

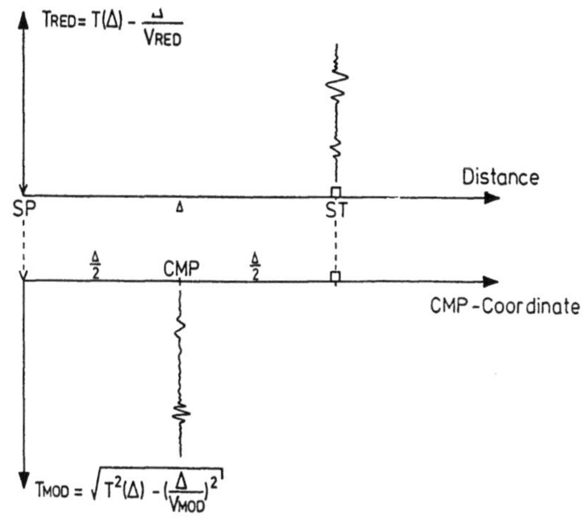

Fig.9. Construction principle of wide-angle COF sections. Seismograms with shot-receiver pairs with equal offset are dynamically corrected and plotted downwards beneath their shot-to-receiver midpoints

3.2.2 COF Sections

The construction principle of COF sections is explained in Fig. 9. Unlike common-shotpoint or CMP sections, which are produced by plotting seismograms in reduced time, T_{red}, against distance upwards, COF sections are generated by plotting seismograms for source-receiver pairs with equal offset in modified time, T_{mod}, against their mid-point coordinate downwards.

In the case of horizontal or moderately inclined layering, wide-angle COF sections simulate zero-offset sections and can immediately be compared with the corresponding results of steep-angle reflection seismics. If correct modification velocities V_{mod} are used, T_{mod} coincides with vertical TWT time. Both sections should image the same structural elements but, due to angle-dependent reflection coefficients, their accentuation may be different. Matching COF sections to zero-offset sections can be an efficient method to estimate average velocities. Modification velocities increasing from 5.80 to 6.25 km/s between 2 and 12 s TWT were found most suitable for the Oberpfalz wide-angle data. Due to the minimum observation distance of 42 km the near-surface domain with less than 2 s TWT cannot be resolved and has therefore been omitted in Figs. 10 and 11.

Figure 10 shows COF sections for 47 and 53 km offset, respectively. In both presentations, which are based on completely independent sets of data, the EB can be recognized as the dominant structure in the upper crust (between profile km 83 and 98 and between 3.5 to 5.0 s modified time). The wide-angle reflections outline an antiformal structure that continues in the NW and SE to deeper crustal levels. The images are not completely identical and deviate also in some detail from the near-vertical reflection seismic image but the general coincidence underlines the significance of the EB phenomenon.

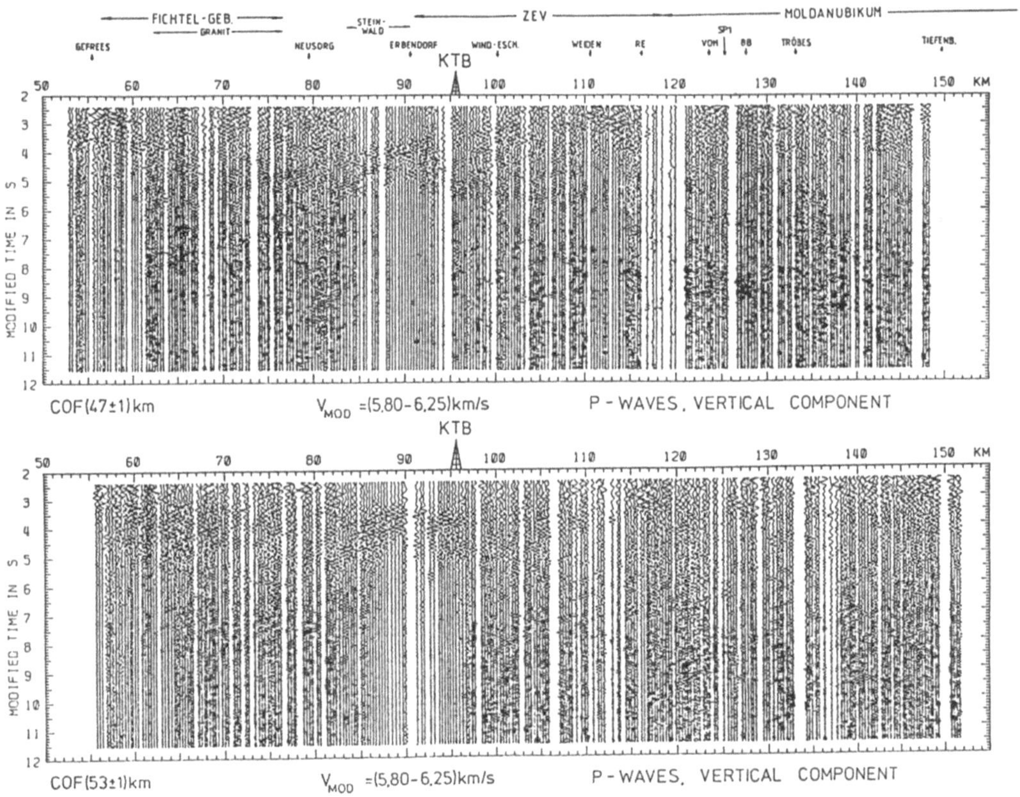

Fig.10. COF sections for 47 km (**top**) and 53 km (**bottom**) offset. The Erbendorf body can be recognized as the dominant structure in the form of an updoming between CMP coordinates 83 and 98 and modified times of 3.5 and 5 s

Fig.11. Imaging the lower crust and the Moho by COF sections with offset 72 km (**top**) and 82 km (**bottom**), respectively. The top of the lower crust appears below the EB at modified times from 7.0 to 7.5 s (about 22 km depth); the Moho rises from the MN towards the ST/MN boundary domain

Fig.12. S-wave imaging of Moho and lower crust by COF 72 and COF 82 (to be compared with P-wave imaging in Fig. 11). The lower crust is hardly recognizable, but the Moho high below the Erbendorf body appears even more clearly than in the P-wave COF sections

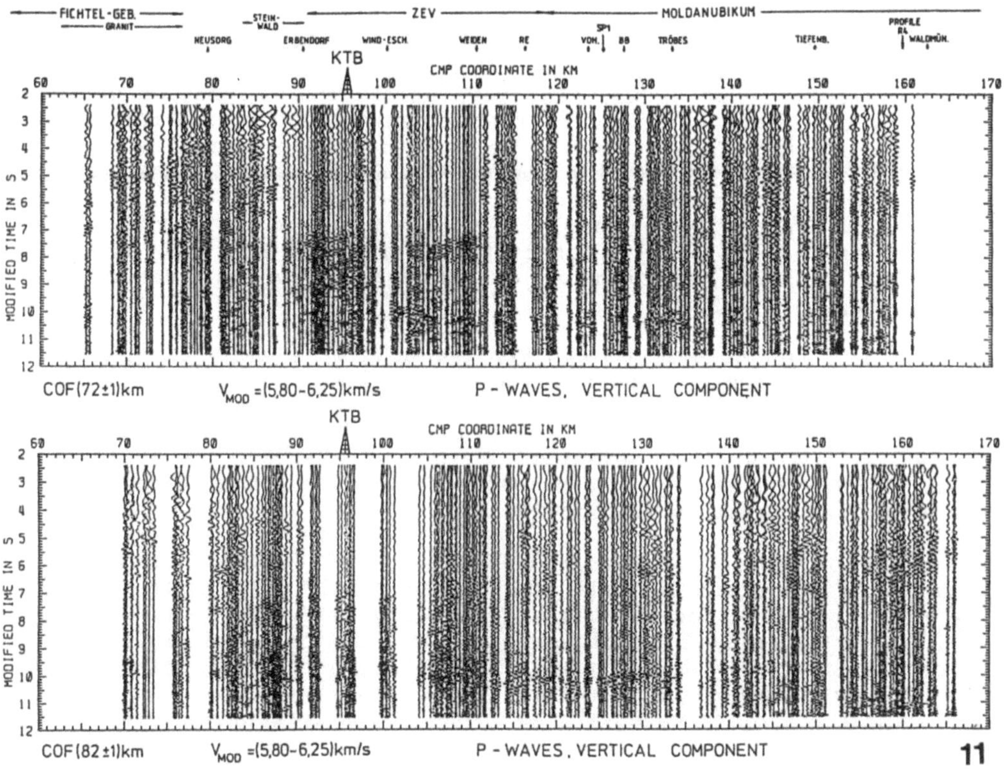
FICHTEL-GEB.
GRANIT
STEIN-WALD
ZEV
MOLDANUBIKUM
NEUSORG
ERBENDORF
WIND.-ESCH.
WEIDEN
RE
SP1
VOH.
BB
TRÖBES
TIEFENB.
PROFILE R4
WALDMÜN.
KTB
CMP COORDINATE IN KM
MODIFIED TIME IN S
COF(72±1)km
V_{MOD} =(5,80-6,25)km/s
P-WAVES, VERTICAL COMPONENT
COF(82±1)km
V_{MOD} =(5,80-6,25)km/s
P-WAVES, VERTICAL COMPONENT

11

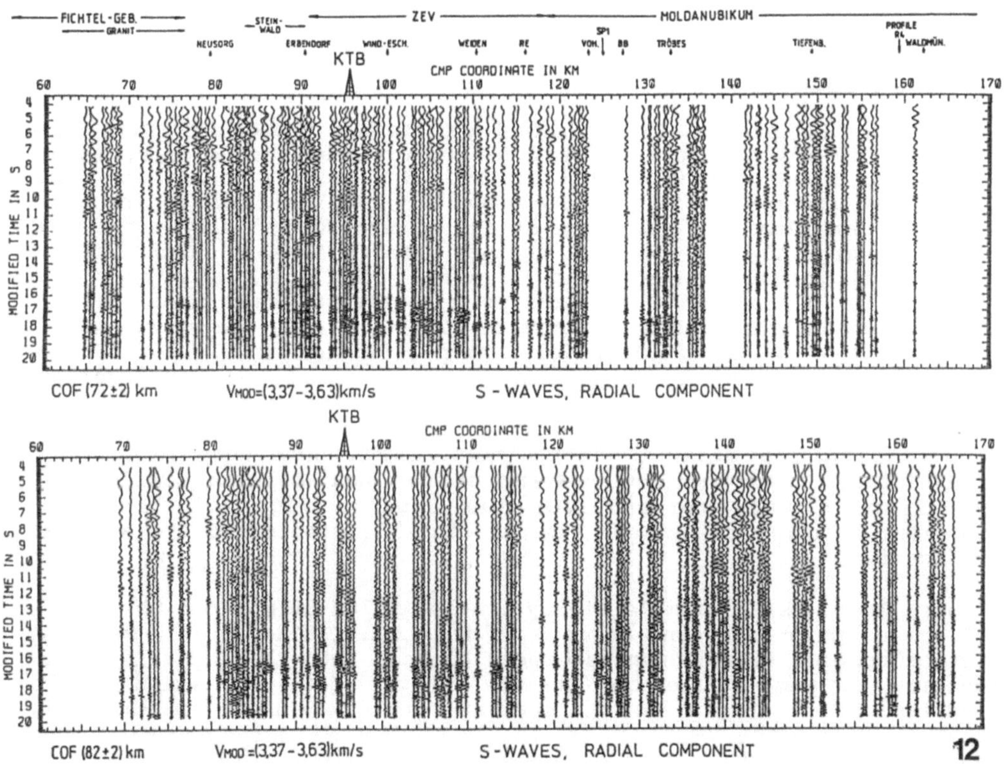
FICHTEL-GEB.
GRANIT
STEIN-WALD
ZEV
MOLDANUBIKUM
NEUSORG
ERBENDORF
WIND.-ESCH.
WEIDEN
RE
SP1
VOH.
BB
TRÖBES
TIEFENB.
PROFILE R4
WALDMÜN.
KTB
CMP COORDINATE IN KM
MODIFIED TIME IN S
COF (72±2) km
V_{MOD}=(3,37-3,63)km/s
S-WAVES, RADIAL COMPONENT
COF (82±2) km
V_{MOD}=(3,37-3,63)km/s
S-WAVES, RADIAL COMPONENT

12

The middle and the lower crust are only inadequately resolved by the small offsets of 47 and 53 km. However, by enlargement of the aperture (COF 72 and 82 km, Fig. 11), structures at greater depths can also be imaged. It appears that there is a further reflector below the EB with T_{mod} = 7.0 to 7.5 s (a depth of approximately 22 km), that presumably marks the top of the lower crust. Accordingly, the EB cannot represent an updoming of the lower crust. If the EB consists of lower crust material - both its characteristic reflection behavior and its velocity advocate this - it must have been brought into its present position by overthrust tectonics (Weber and Vollbrecht, this Volume).

In Fig. 11, particularly at COF 82, the Moho is clearly recognized. It rises from the SW towards the NW. However, the rise is not continuous, but consists rather of small-scale undulations and possible imbrications in the lower crust. Similar structures are imaged by near-vertical seismics. A high of the Moho is indicated below the ST/MN boundary domain, but simultaneously the character of the Moho also becomes more diffuse.

A similar processing of the S-waves (Bopp 1986) likewise shows the rise of the Moho from SE to NW and - surprisingly - the Moho high in the ST/MN boundary domain even more clearly than the P-waves (Fig. 12). On the other hand, it is worth noting that the lower crust which is clearly imaged by the P-waves (Fig. 11) is hardly recognizable in the S-waves. Sandmeier and Wenzel (1986) and Lüschen et al. (1987) have reported similar observations in the Black Forest. This supports the assumption that this could be a general phenomenon of the lower crust. It may be explained by a reduced V_p/V_s-ratio in the middle crust, as found by Strößenreuther (1982) in the southeastern Oberpfalz, and an increase of the V_p/V_s-ratio in the lower crust.

3.2.3 2D Velocity Model

Combining the previously explained structural properties with the results of velocity analyses a first two-dimensional velocity model along the DEKORP-4 line was established. The results of earlier work in the southern Oberpfalz (Strößenreuther 1982) and of independent velocity analyses for the two expanding spread velocity soundings (open squares in Fig. 5b) by Bittner (DEKORP Research Group 1988), were additionally taken into account. The model was improved iteratively by extensive ray-tracing computation based on a further development of the method of Gebrande (1976). Figure 13a shows one of the latest versions of the model, which fits the traveltimes of the most important wave groups in the surroundings of the KTB location within an error of 0.1 - 0.2 s.

Superposing the velocity-depth functions, converted to two-way traveltimes, over the line drawing of the near-vertical reflection seismics is very informative; the result (Fig. 13b) shows a remarkable coherence between the structural elements and the seismic velocities.

Figure 14 gives an example of ray-tracing calculations for the model 25 and a common midpoint about 6 km from the KTB location (CMP 90). The calculated traveltimes have been plotted into the CMP seismogram section to demonstrate the quality of the fit. One obtains the impression that a further improvement should still be possible and that is also the beliefe of the authors. So far, however, any modification which improved the fit for some data subset changed it for the worse in other parts. Therefore, the model presented may be near to some relative optimum but not necessarily the best possible.

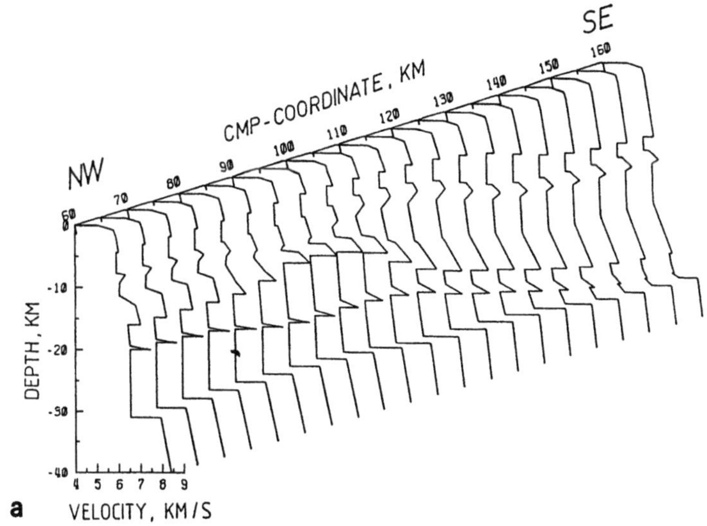

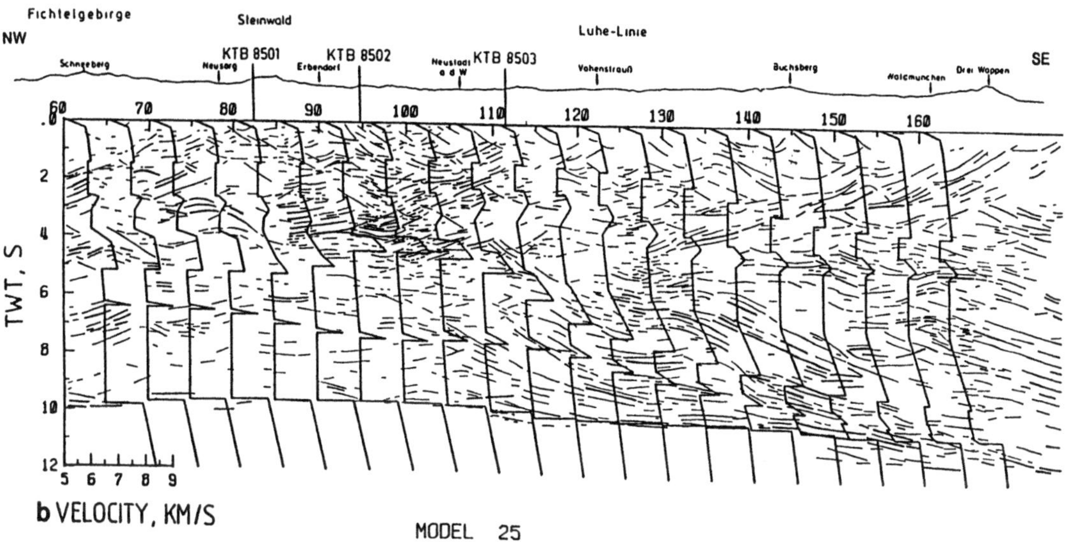

Fig.13a Latest version of the 2D velocity model for the DEKORP-4 line derived by kinematic two-point ray tracing. **b** Superposition of velocity-depth functions (converted to two-way traveltimes) over the line-drawing of the near-vertical reflection seismics demonstrating the coherence between structural elements and seismic velocities

Because of this general weakness of trial and error methods, ray-tracing computations were not further followed up for the time being. Instead, an additional and independent approach to the velocity and imaging problem was undertaken by migration.

3.2.4 Migration of Wide-Angle Data

A detailed comparison of the COF sections (Figs. 10 and 11) with the final migration of steep-angle seismics (Schmoll et al., Fig. 4 and 9 this Volume) certainly shows some encouraging similarities, but also

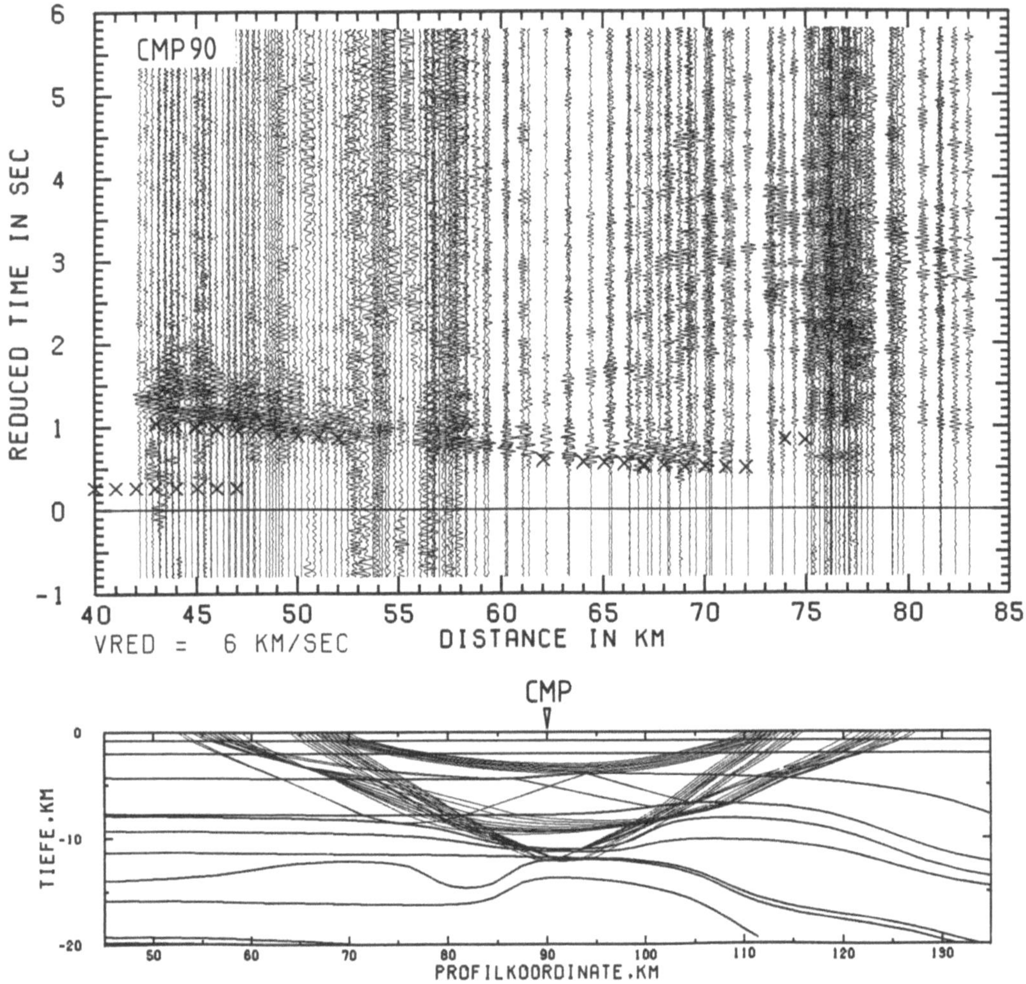

Fig.14. Example of ray-tracing calculations for CMP 90 and model 25. Computed traveltimes have been plotted into the record section

significant differences. The Erbendorf body, for instance, shows its strongest reflections in both types of section at about 3.8 s TWT close to the KTB location, but it appears as a wedge-shaped structure in the steep-angle stack and as an anticline in the wide-angle COF sections. From a purely theoretical viewpoint, such differences seem to be not quite impossible, because steep and wide-angle reflections respond to partially different physical properties. On the other hand, however, the steep-angle image deserves more confidence because of its superior resolution. In addition, the attempts to take over the structural properties of the COF sections straightaway into the 2D velocity model (Figs. 13a,b) were not very successful and point to some distortion of the COF images.

Indeed, a more subtle analysis shows that for large offsets dynamically corrected COF sections represent a more distorted image of the subsurface than the zero-offset sections from near-vertical seismics. Conversely, this means that the process of migration is even more important for wide-angle seismics than it is for near-vertical seismics. Especially on account of the strong lateral heterogeneities in the Oberpfalz area this process should not be omitted. It was therefore necessary to develop a migration method for wide-angle seismics. The common migration techniques of steep-angle seismics could not immediately be adopted, because they are tailored mostly to the zero-offset case and require equidistant input data. Therefore a special 2D-pre-stack migration method was developed, which is essentially a computerized application of classical principles of traveltime migration (Hagedoorn 1954).

The basic concept of the method, called isochrone migration, is illustrated in Fig. 15. An isochrone is the curve or, in the 3D case, the surface of all potential reflection points for rays with the same traveltime from a shotpoint (SP) to a station (ST). For a constant velocity model the isochrones are ellipses with the source and receiver as focal points. The migration is carried out by attributing an instantaneous seismogram amplitude $A(t_i)$ to all points of its corresponding isochrone in a regular subsurface grid and by superimposing all isochrones for all samples and all seismograms of the profile. With sufficient multiplicity of subsurface coverage the amplitudes sum up constructively at the true reflector positions. The method has the following advantages:

- It is a pre-stack depth migration method.
- Pre-sorting of seismograms is not required since the data are processed sequentially sample by sample and seismogram by seismogram; therefore the program runs also on small sized computers.
- Individual shot gathers can be migrated separately and stacked afterwards. To avoid destructive interference due to phase distortion or differing source signals the individual migrations may be stacked by adding their absolute values or their envelopes.

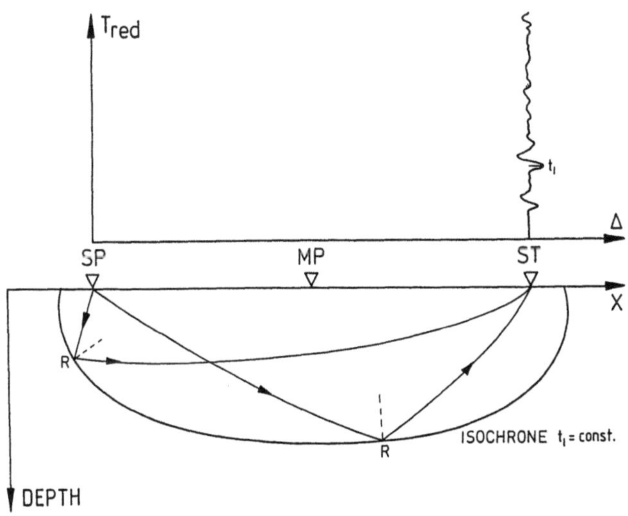

Fig.15. Principle of isochrone migration

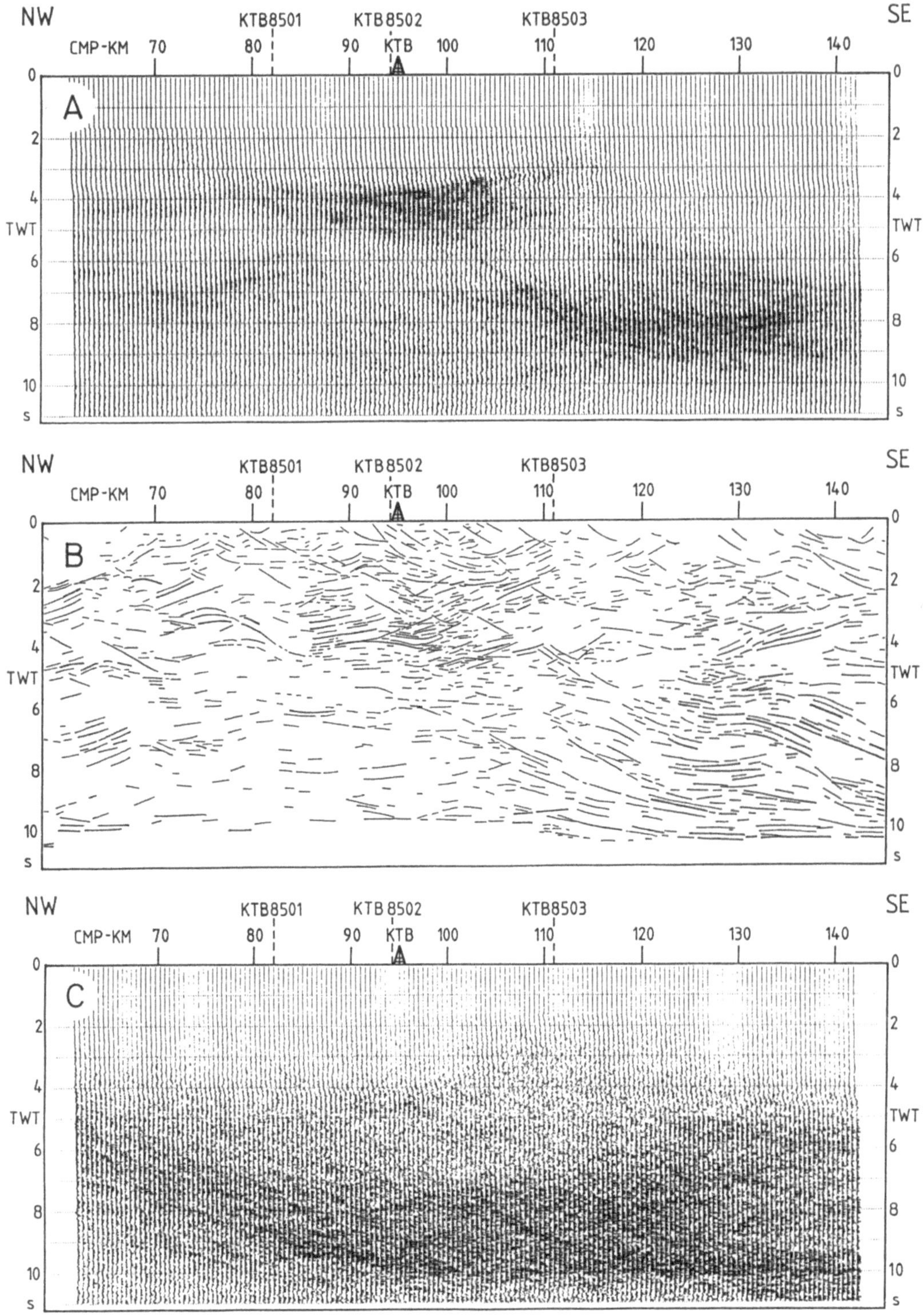
NW
SE
KTB8501
KTB8502
KTB8503
CMP-KM
70
80
90
KTB
100
110
120
130
140
TWT
s
A
B
C

- Unlike with finite difference methods, the migration grid can be confined to the subsurface area of interest. It is not necessary to begin at the earth's surface or beneath source and receiver.

The main shortcoming of the isochrone migration in its present stage of development seems to be its pure ray-geometrical theoretical background. Several tests with synthetic seismograms, however, were very successful, and inspire confidence that the method may also be useful in practice.

The method has been applied in its constant velocity version separately to the data sets focusing on the upper and lower crust (cf. Fig. 4). The results are shown in Fig. 16a and c, respectively. For the sake of comparison with line drawings of the migrated steep-angle section (Fig. 16b) the wide-angle migrations, which are originally depth migrations, were converted into the time domain.

Tentative migrations of a data subset with different constant velocities from 5.9 to 6.1 km/s revealed that the results depend very closely on the chosen velocity. Only within very close limits of the migration velocities (around 5.95 km/s) was convincing resemblance to the migration section of steep-angle seismics obtained. The optimum migration velocity for the upper crust coincides perfectly with the average velocity derived from the 2D model (Fig. 13a,b) down to the depth of the Erbendorf body.

The wide-angle migrations were then extended to the whole data set for the upper crust with the same migration velocity, and Fig. 16a shows the result. Ninetyfive shot gathers with 200 traces each were first independently migrated; then the individual migrations were rectified, stacked, and bandpass filtered from 0.5 to 25 Hz. Due to muting of the first onsets up to reduced times of about 0.8 s, the migration result is only reliable for TWT times greater than about 3.5 s. In the deeper part of Fig. 16a the resolution is limited by the maximum offset of the contractors spread (58 km) which does not include observations of P_MP reflections.

The Erbendorf body appears in the migrated wide-angle section (Fig. 16a) again as a dominant structure in the KTB surroundings at 3.5 to 5 s TWT. But in contrast to the COF sections its image is now wedge-shaped and resembles much more the steep-angle image (Fig. 16b). Surprisingly, the lower crust is also well illuminated by the wide-angle migration in Fig. 16a. The pronounced contrast between its high reflectivity in the Moldanubian and its low reflectivity in the Saxothuringian is remarkable and consistent with the steep-angle results. The apparent transparency of the lower crust beneath the KTB location is obviously due to the shielding effect of the Erbendorf body in the offset range of the contractor's spread (42 - 58 km, cf. Fig. 4). This interpretation is confirmed by the more continuous appearence of the lower crust in the migration result (Fig. 16c) of the MARS spread which enables undershooting of the Erbendorf body and is better tuned to map the lower crust and Moho.

The MARS seismograms in Fig. 16c have been migrated with v_{mig} = 6.1 km/s, that according to the results of velocity analyses and 2D modeling (Fig. 13) seems to be more appropriate for the entire crust.

Fig.16. Comparison of constant-velocity wide-angle migrations of **a** the 42-58 km offset range (contractor's spread, v_{mig} = 5.95 km/s) and **c** the 60-90 km offset range (MARS stations, vertical components, v_{mig} = 6.1 km/s), with **b** the line-drawing of the migrated steep-angle section of line DEKORP-4

The Moho is delineated much more clearly than in the COF sections (Fig. 11) and the top of the lower crust is more distinguished. The Erbendorf body, of course, cannot be portrayed by migration of MARS data, since the observation of the P_EP reflections is limited to smaller offsets.

Structural details displayed by the wide-angle migrations should be interpreted carefully because of possible migration noise. The SE dipping elements in the lower left part of Fig. 16c, for instance, are probably artifacts due to insufficient coverage for CMP coordinates less than 80 km. The SE-dipping structures in the central and right parts deserve more confidence, since they are consistently observed in all three migrated sections (Fig. 16) and there is no deficiency of coverage. Smaller deviations in positioning of the dipping events will probably be eliminated with future use of more realistic depth-dependent migration velocities. Then it will also become possible to condense all wide-angle data in one single migrated section.

4 Discussion and Conclusions

Seismic investigations in the Oberpfalz are a typical example for demonstrating the virtues of close-spaced observations in crustal studies. The Erbendorf body was first discovered accidentally just by recording an additional and favorably situated shotpoint (Remmelberg) along an older reversed refraction line (Böhmischbruck-Mauthaus, Fig. 1) which had not provided any hint to the existence of this outstanding feature; and only by a very close-set observation scheme, both on the source and on the receiver side (Fig. 4), did it become possible to study its extension and its detailed properties along the DEKORP-4 line. These dense and multiply covered observations, with a total of 26,000 seismic traces, provided the data base for an advanced processing of wide-angle reflections.

Several new methods have been developed and applied following the well-approved steps of steep-angle reflection processing: CMP sorting, velocity analysis, simulation of zero-offset sections, and finally migration. Some general conclusions can be drawn from the results:

1. It became evident that the migration process is even more important in wide-angle seismics than it is in steep-angle seismics for a reliable imaging of strong horizontal heterogeneities, as encountered in the Oberpfalz.

2. After migration, wide-angle reflections image essentially the same crustal structures as steep-angle reflections.

3. In spite of their generally weaker structural resolving power, wide-angle reflections may image special structures even more clearly due to their distinctive reflection coefficients; the EB (Fig. 16a) is a good example of this.

4. Wide-angle migration results depend very closely on the average migration velocity and provide similarity with well-processed steep-angle results only within very close velocity limits. Hence, adjusting wide-angle to steep-angle migrations is a very promising method to estimate reliable and physical meaningfull average velocities. This is also true in the case of strong lateral heterogeneities, where direct methods of velocity inversion fail.

As mentioned above, the established average migration velocity (5.95 km/s) down to the depth of the EB is in full agreement with the 2D-model of Figs. 13a,b. Details of this model, however, seem to be capable of improvement. Based on the structural properties shown in Fig. 16, a new sequence of 2D ray-tracing computations has been started, which will most likely lead to an even closer coherence between the velocity structure and the steep-angle section (analogous to Fig. 13b), and to an improved traveltime fit with the original data. It should be kept in mind, however, that a perfect agreement is not to be expected, since a two-dimensional model can never explain all features observed along a one-dimensional seismic line over an intrinsically three-dimensional earth. On the other hand, the three-dimensionality can never be resolved by a single seismic line.

Indeed, there are many indications, e.g., from the refraction lines (cf. Fig. 2) and from the KTB reflection lines (Schmoll et al., this Volume); that lateral heterogeneities perpendicular to the DEKORP-4 line may be of similar importance as in-line heterogeneities. According to present knowledge, it cannot be ruled out that the strong amplitudes of the wide-angle reflections from the Erbendorf body may partly be due to focusing by off-line structures. Future wide-angle measurements planned within an integral program for surveying the local KTB surroundings will take this possibility into account and will be laid out for 3D resolution.

In view of the continental deep-drilling project the main interest focuses on the upper and middle crust. Figure 17 shows a compilation of different p-wave velocity-depth functions obtained for CMP 90 near Erbendorf according to different inversion methods and slightly different traveltime correlations. The scattering of the solutions gives an idea of the uncertainty of present knowledge. At the KTB site the scattering is even larger because of poorer data quality, but the general character of the velocity-depth function seems to be similar to CMP 90 (it should be noted that a special velocity sounding at the KTB site has not yet been carried out). The common features of the velocity-depth functions of Fig. 17 may be expected also at the location of KTB. Beneath a normal increase of velocities, respectively of their smoothed means, down to 3-4 km, and a poorly pronounced low-velocity zone, a marked increase to about 6.3 - 6.4 km/s is expected at approximately 7-8 km depth.

The most spectacular feature is a high-velocity zone at about 12 to 14 km depth which is common to all solutions. This is also true for the "generalized Dix method", which is a special variant of the x-t-V stripping technique and provides interval velocities independent of observed apparent velocities (solid line in Fig. 17). The velocities of up to 7.0 km/s or even more, which are usually observed only in the lower crust, are by no means physically unreasonable at 12-14 km depth. They can be explained by several lithologies, e.g., granulites, gabbros, meta-gabbros, or amphibolites (Gebrande 1982). Sandmeier and Wenzel (pers. commun.) have confirmed by computation of synthetic seismograms that the general type of the velocity-depth function of Fig. 17 may explain not only the traveltimes but also the strong amplitudes of the wide-angle reflections from the Erbendorf body. However, they have also shown that smaller velocities may be sufficient if a lamination of the Erbendorf body with internal velocity contrasts of about 1 km/s is assumed. In any case, such a model requires also rather high velocities if, on the other hand, unreasonably low velocities in the low-velocity lamellae are avoided.

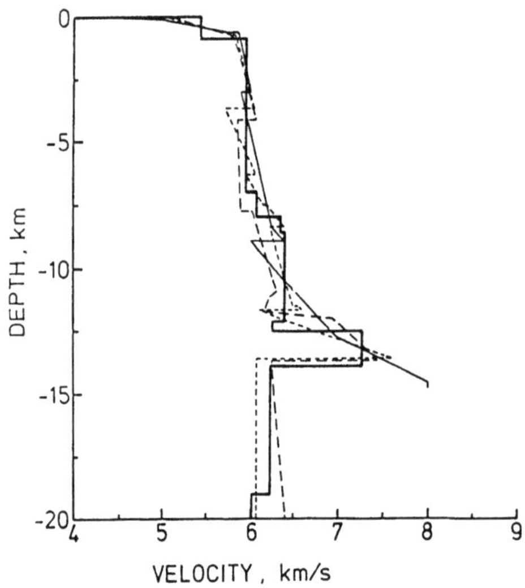

Fig.17. Velocity-depth functions that were obtained for CMP 90 according to different interpretation techniques and for slightly different traveltime correlations. The bandwidth of the different solutions gives an idea of the possibilities and limitations of wide-angle seismics in the upper and middle crust. All solutions show velocities larger than 7 km/s in the lower part of the EB at depth of 12-14 km, but details of layering may be even more complicate

Another argument in favor of a more complicated internal structure of the Erbendorf body comes from gravity. 2D-modeling for a body with the shape of the high-velocity layer (Fig. 13a) and a velocity-related density contrast of 0.3 g/cm^3 (Gebrande 1982) to the surrounding crust with V_p about 6.3 km/s provides a positive gravity anomaly with a maximum of 14 mgal. It seems hardly possible to accommodate such an anomaly to the observed gravity field and to its otherwise known sources (Bücker 1986). A non-two-dimensional structure of the Erbendorf body (see above) or a smaller average density contrast seem to be required. The latter case would imply also a smaller average velocity of the lower Erbendorf body which may be reconciled with the observed high velocities by assuming a lamellae model. It could also more easily explain the ringing waveforms of the Erbendorf reflections in the wide-angle as well as in the steep-angle domain. This aspect, once again, relates the Erbendorf body to the problems of the lower crust. From Fig. 16a-c, however, it is clear that it cannot simply represent an updoming of the lower crust.

Figure 17 is at best a very loose prediction of the velocity-depth function at the KTB location. A detailed prognosis will only be possible on the basis of future measurements and evaluations. Even then, in view of the complex structural properties of the KTB location, drilling will still have the chance to provide some surprises.

Acknowledgments. The funding of the investigations by the German Research Council (DFG) and the Federal Ministry of Science and Technology, Bonn is gratefully acknowledged. We appreciate the continuous support by the DEKORP steering committee and by H. Soffel, director of the Institut für Allgemeine und Angewandte Geophysik, München.
Thanks for excellent cooperation during field work are due to the teams from the Geophysical Institutes of Berlin, Clausthal, Frankfurt, Kiel, Hamburg and München, from the Geological Survey of Lower Saxony and to the Prakla-Seismos crew.
Preprocessing of the contractor's wide-angle data was carried out at the DEKORP Processing Center at Clausthal with valuable support by F. Keller and W. Janoth.
We are indebted to U. Strößenreuther and H. Miller for their earlier contributions to the project.

References

Bopp, M.: Analyse und Interpretation von S-Wellenregistrierungen im Rahmen des Deutschen Reflexionsseismischen Programms DEKORP. Dipl.-Thesis, Ludwig-Maximilians-Universität München, 145 p., 1986

Bücker, C.: Die Anomalien der Schwere im Bereich der Kontinentalen Tiefbohrlokation (KTB) Oberpfalz und ihre Interpretation. Diss. Ludwig-Maximilians-Universtät München, 130 p., 1986

DEKORP Research Group: Results of the DEKORP4-KTB Oberpfalz deep seismic reflection investigations. J.Geophys. **62**, 69-101, 1988

Gebrande H.: A seismic ray tracing method for two-dimensional inhomogeneous media. In: Explosion seismology in Central Europe, Giese, P., Prodehl, C., Stein, A., eds.: 162-167. Berlin: Springer Verlag, 1976

Gebrande H.: Elastic wave velocities and constants of elasticity of rocks and rock-forming minerals. In: Landolt-Börnstein, Geophysik und Weltraumforschung **1b**, Angenheister, G., ed.: 1-99. Berlin-Heidelberg-New York, 1982

Gebrande, H.: Refraktions- und Weitwinkelseismik. In: KTB - Ergebnisse der Vorerkundungsarbeiten Lokation Oberpfalz. 2. KTB-Kolloqium Seeheim/Odenwald 19-21 September, 1986 , Weber K., Vollbrecht A., eds.: 113-135, 1986a

Gebrande, H.: CMP-Refraktionsseismik. In: Seismik auf neuen Wegen. Dresen, et al., eds.: 191-205. Dt. Vereinigung d. Erdölgeol. u. Erdöling. (DVGI), Celle, 1986b

Gebrande, H., Miller H.: Refraktionsseismik. In: Angewandte Geowissenschaften - Band II, Bender, F., ed.: 226-260. Stuttgart: F. Enke Verlag, 1985

Giese, P.: Versuch einer Gliederung der Erdkruste im nördlichen Alpenvorland, in den Ostalpen und in Teilen der Westalpen mit Hilfe charakteristischer Refraktions-Laufzeit-Kurven sowie eine geologische Deutung. Geophysikalische Abhandlungen 1(2), 202 p., Freie Universität Berlin. Berlin: Friedrich Reimer Verlag, 1968

Giese, P.: Results of data generalization. In: Explosion seismology in Central Europe, Giese, P., Prodehl, C., Stein, A., eds.: 339-346. Berlin: Springer Verlag, 1976

Hagedoorn, J.G.: A process of seismic reflection interpretation. Geophys.Prosp. **2**, 85-127, 1954

Lüschen, E. et al.: Near-vertical and wide-angle seismic surveys in the Black Forest, SW Germany. J. Geophys. **62**, 1-30, 1987

Miller, H., Gebrande, H.: Crustal structure in Southeastern Bavaria derived from seismic refraction measurements by ray tracing methods. In: Explosion seismology in Central Europe, Giese, P., Prodehl, C., Stein, A., eds.: 339-346. Berlin: Springer Verlag, 1976

Peters, K.: Ergebnisse der Gravimetrie im Bereich der Münchberger Gneismasse und der Refraktionsseismik längs eines Profils über die Gneismasse. Diss. Ludwig-Maximilians-Universität München, 131 p., 1974

Sandmeier, K.-J., Wenzel, F.: Synthetic seismograms for a complex crustal model. Geophys. Res. Lett. 13, 22-25, 1986

Schmoll, J., Bittner, R., Dürbaum, J., Heinrichs, T., Meißner, R., Reichert, C., Rühl, T., Wiederhold: Oberpfalz Deep Seismic Reflection Survey and Velocity Studies. In: Exploration of the Continental Crust through Drilling IV, Emmermann, R., Wohlenberg, J., eds. Berlin: Springer Verlag, 1988

Strößenreuther, U.: Die Struktur der Erdkruste am Südwestrand der Böhmischen Masse, abgeleitet aus refraktions-seismischen Messungen der Jahre 1970 und 1978/79. Diss. Ludwig-Maximilians-Universität München, 122 p., 1982

Weber, K., Vollbrecht, A.: The Crustal Structure at the Continental Deep Drilling Site Oberpfalz. In: Exploration of the Continental Crust through Drilling IV, Emmermann, R., Wohlenberg, J., eds. Berlin: Springer Verlag, 1988

Processing of Reflection-Seismic Data in the DEKORP Processing Center, Clausthal

Manfred Stiller and Rüdiger Thomas

Institute of Geophysics of the Technical University of Clausthal, Arnold-Sommerfeld-Str. 1, D-3392 Clausthal-Zellerfeld, FRG

1. Introduction

The DEKORP Processing Center (DPC) at the Institute of Geophysics of the Technical University of Clausthal was founded in December 1983 to process deep-seismic, steep-angle data for the DEKORP research project. In September 1984 the DPC was also given the task of processing the geophysical data for the KTB pre-site studies. The DPC was subsequently assigned the Oberpfalz 3-D seismics planned for 1989.

In the meantime, 17 2-D profiles over a total of 1500 profile kilometers have been completed in the Variscan zone. The geographical locations of these profiles are given in Fig. 1a, and Fig. 1b contains values for several survey parameters.

Data processing had already begun during the first field measurements in April 1984. This was only possible through the generous donation of a complete "Phoenix I" computer system (Raytheon RDS 500 and periphery) by Mobil Oil AG, Celle, and through the free-of-charge use of seismic software granted by Seismograph Service Limited, Keston (England). The procurement of a modern processing system capable of handling the extraordinary volume of data had, of course, been planned from the very beginning. Therefore, in May 1986 a "Phoenix DPU" computer (VAX 11/750 and periphery) was installed, and again SSL provided software free of charge for the purpose of research and teaching. Meanwhile, the volume of data has grown to such an extent that both units are being used simultaneously. This is no problem due to the compatibility of the data formats and the similarity of the software.

At the beginning there was, of course, a certain skepsis as to whether a working group at a university would be able to meet the requirements of a project of this type, i.e., producing in a reasonable time results of a quality equal to that of a professional contractor. The success of this experiment, which was the first of its kind in West Germany, was demonstrated by directly comparing the KTB seismic profiles (Oberpfalz, northern Bavaria) processed, due to important deadlines, simultaneously in Clausthal and by various international contractors from the oil industry.

Since DEKORP data processing began, more than 30 gigabytes of seismic data have been converted from field recordings into migrated seismic time sections. This was done initially by four geophysicists and six student assistants, and more recently by

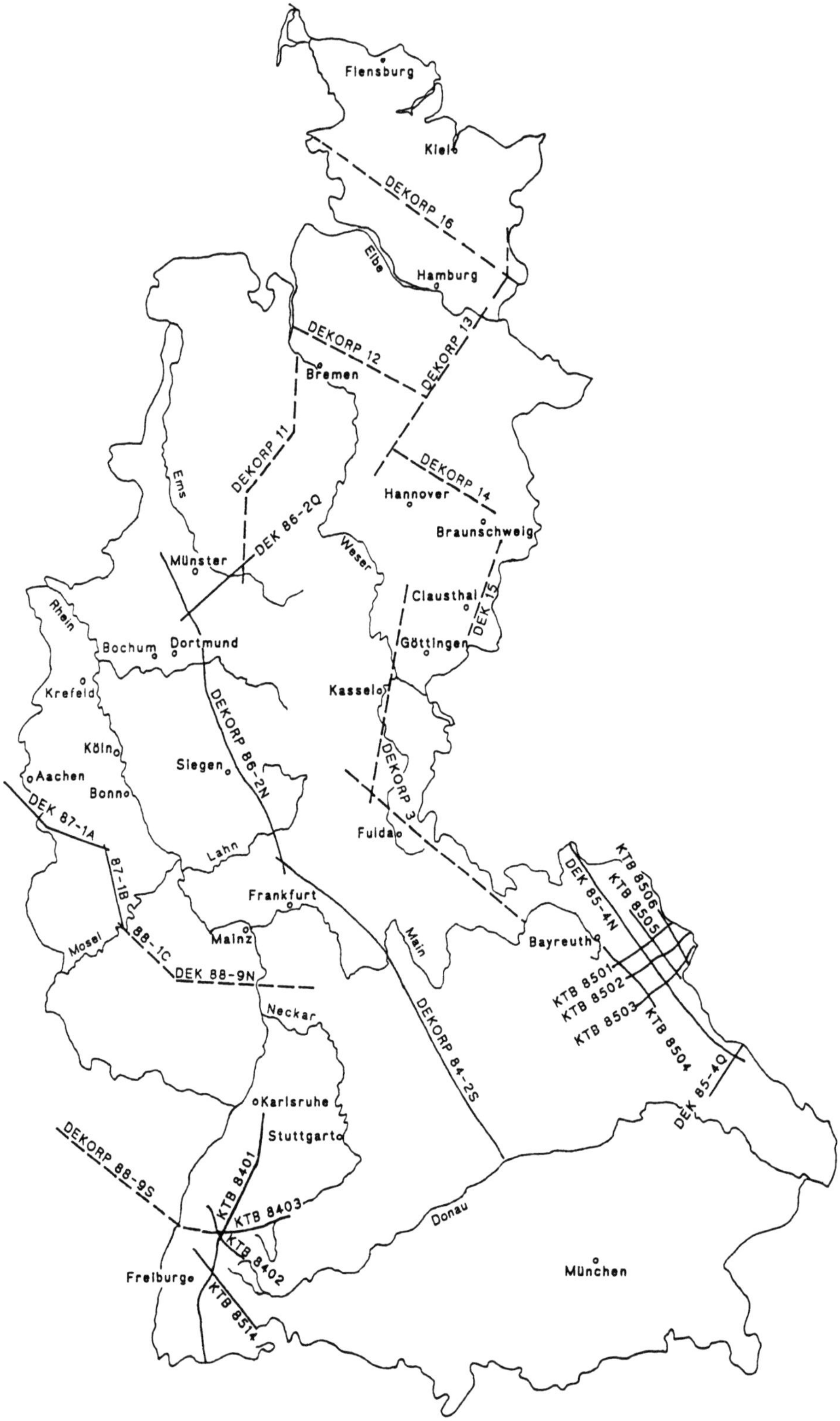

Fig. 1.a Location map of the surveyed (<u>solid lines</u>) and planned (<u>dashed lines</u>) DEKORP/KTB profiles

Profile	Profile length [km]	Record length [s]	Number of shots	Number of channels	Spread (km before/behind shot)	Number of CMP's	CMP-Spacing [m]	Coverage (theor./real)	Data amount [Gigabyte]
DEKORP 84-2S	248	20	784	200	-16/ 0	6253	40	25/ 25	2.92
KTB 8401	168	12	1470	200	- 8/ 8	4070	40	100/ 72	3.29
KTB 8402	45	12	378	200	- 8/ 8	1063	40	100/ 71	0.85
KTB 8403	63	12	574	200	- 8/ 8	1543	40	100/ 74	1.28
KTB 8514	65	12	1078	200	-12/ 4	3262	20	100/ 66	2.41
DEKORP 85-4N	187	12	1646	200	-12/ 4	4647	40	100/ 71	3.68
DEKORP 85-4Q	37	12	316	200	- 4/12	907	40	100/ 70	0.71
KTB 8501	47	12	476	200	- 4/12	1180	40	100/ 81	1.06
KTB 8502	50	12	513	200	- 4/12	1261	40	100/ 81	1.15
KTB 8503	57	12	563	200	- 4/12	1400	40	100/ 80	1.26
KTB 8504	50	12	476	200	-12/ 4	1260	40	100/ 76	1.06
KTB 8505	55	12	500	200	-12/ 4	1364	40	100/ 73	1.12
KTB 8506	42	12	393	200	-12/ 4	1053	40	100/ 75	0.88
DEKORP 86-2N400	160	12	2708	400	- 4/12	7885	20	200/137	12.11
DEKORP 86-2N200	60	12	1102	200	- 4/12	3465	20	100/ 64	2.46
DEKORP 86-2Q	71	12	1126	200	- 4/12	3007	20	100/ 75	2.52
DEKORP 87-1A	93	16	1915	400	-12/ 4	4666	20	200/164	11.41
DEKORP 87-1B	50	16	1049	400	-12/ 4	2506	20	200/167	6.25

Fig. 1.b Characteristic recording parameters of previous DEKORP/KTB profiles

four geophysicists and eleven student assistants. This has been a good opportunity for students to gain practical experience, which is also shown by the great number of theses and oral presentations dealing with the theory and practice of seismic data processing.

The recorded data are excellently suited for studies of the processing and evaluation methods. In contrast to an industrial company, a university-based DEKORP processing center also has the invaluable advantage that the profiles are accessible at any time for further studies under new aspects.

2. Standard Methods for Processing Deep-Seismic Data at the DPC

2.1 Amount and Quality of Data, Processing Sequence

As seen in Fig. 1b deep-seismic profiles cannot be compared directly with profiles surveyed for seismic prospecting. DEKORP and KTB profiles vary in length from about 35 to 250 km, with recording times between 12 and 20 s. An average of nearly 400 km/year have been surveyed.

Moreover, the trace spacings also tend to become progressively smaller in deep seismics. The first DEKORP profile 84-2S had a CMP spacing of 40 m with 25-fold coverage. The recent profiles surveyed in 1987-1988, however, had a CMP spacing of 20 m and 200-fold coverage. Hence, a 13-fold increase in the amount of data obtained per kilometer and per second of recording time was achieved. Despite the use of additional hardware, this increase has repeatedly caused problems in processing time and storage capacity, which will probably continue to be problematic. Considering that the number of surveyed kilometers per year has remained practically constant, the processing capacity has increased at the same rate as the amount of data.

The recording quality of the data is good, with few exceptions. Due to the long profile lengths in deep seismics and the high measuring rate of 4 km or 50 shots per day (for Vibroseis), a high noise level is unavoidable in parts of the seismograms, depending on the recording technique or field conditions. The first quality control is always conducted in the field using monitor playbacks, and information on the unuseable data is made available to the DPC by the field party. However, there may still be noise on some traces in the form of spikes, power-line pickup, and indications of defective or inverted geophones and poor shots. If something of this kind is detected, the noisy traces can be filtered, muted, inverted, or completely eliminated, depending on the nature and amount of noise.

The DPC has access to two complete Phoenix data processing systems, (1) a Raytheon RDS-500 computer (16 bits, 128 KB) with a high-speed Apollo array processor, a disk-drive (80 MB), and four magnetic-tape-drives (800/1600 bits/inch, 75 inches/s) and (2) a VAX 11/750 computer (32 bits, 4 MB) with four disk-drives (three 340 MB and one 690 MB), a high-speed MAP 300 array processor with two disk drives (160 MB each) and five magnetic-tape drives (800/1600/6250 bits/inch, 75 inches/s). A Versatec electrostatic plotter (22 inches wide, resolution of 200 dots/inch) connected with a high-speed rasterizer is at hand. A complete SSC/SSL software package is used with both processing systems for analyzing, processing, and depicting the seismic data. Due to the compatible data formats and comparable routines, simultaneous processing with both machines is possible.

As already mentioned, this software was developed by the SSC/SSL company for seismic prospecting and was made available as a standard processing package for research and teaching. Initially, there were frequent problems resulting from the fact that several program parameters had not been dimensioned for and tested with the large amounts of data that are usual with deep seismics. Hence, the DPC also has the important task of modifying and expanding the programs so that they meet the requirements for deep seismics as well. This often requires much time and patience, which usually goes unmentioned, although it is of fundamental importance for the success of the data processing described in the following.

Nonetheless, the pure processing time (i.e., the time needed to determine the processing parameters plus machine time) is much greater than in seismic exploration. It averages ca. 1 week/10 km depending on the length of the profile. A very efficient, three--phase processing sequence was developed soon after the foundation of the DPC. This sequence involves production of a single-fold section, raw stack, and final stack. This is done so that geologists and mineralogists can evaluate the first results directly after measurement, as well as to meet DPC's own processing requirements.

A single-fold section is generated in the processing center directly from shot gathers, i.e., after demultiplexing the data (Section 2.2), but before CMP sorting. With a geophone spread of 16 km, each shot covers a subsurface range of 8 km and therefore shots spaced 8 km apart must be edited and joined together. Thus, in the case of the KTB lines, for which the shots were placed at 80 m intervals, every hundredth shot is to be selected. After filtering, scaling, and rough dynamic corrections, an initial interpretation of these sections can be made, because dominant events are already visible in the unstacked shots. For further processing, information on where to place the analysis points for the raw stack is also taken from these seismograms.

Compromising between processing time and quality (only a few velocity and muting functions are used), the aim of this first raw stack of all traces is to obtain as soon as possible an overview of the total profile and a basis for a raw migration. All major geological units, structures, and boundaries can now be recognized and evaluated for an initial fundamental interpretation. Proceeding on this basis, a large number of analyses are then performed to determine the individual processing parameters for the very time-consuming final stack.

Additional processes (e.g., determining and applying residual static corrections) involve extensive calculations for this final stack and produce, together with optimally processed input data, clearly better resolution of all, even small and near-surface events. Figure 2 shows a complete scheme for the final seismic processing as used at the DPC. Based on numerous tests and modifications, this flowchart has been proven to be the most appropriate for deep seismics using the Vibroseis method.

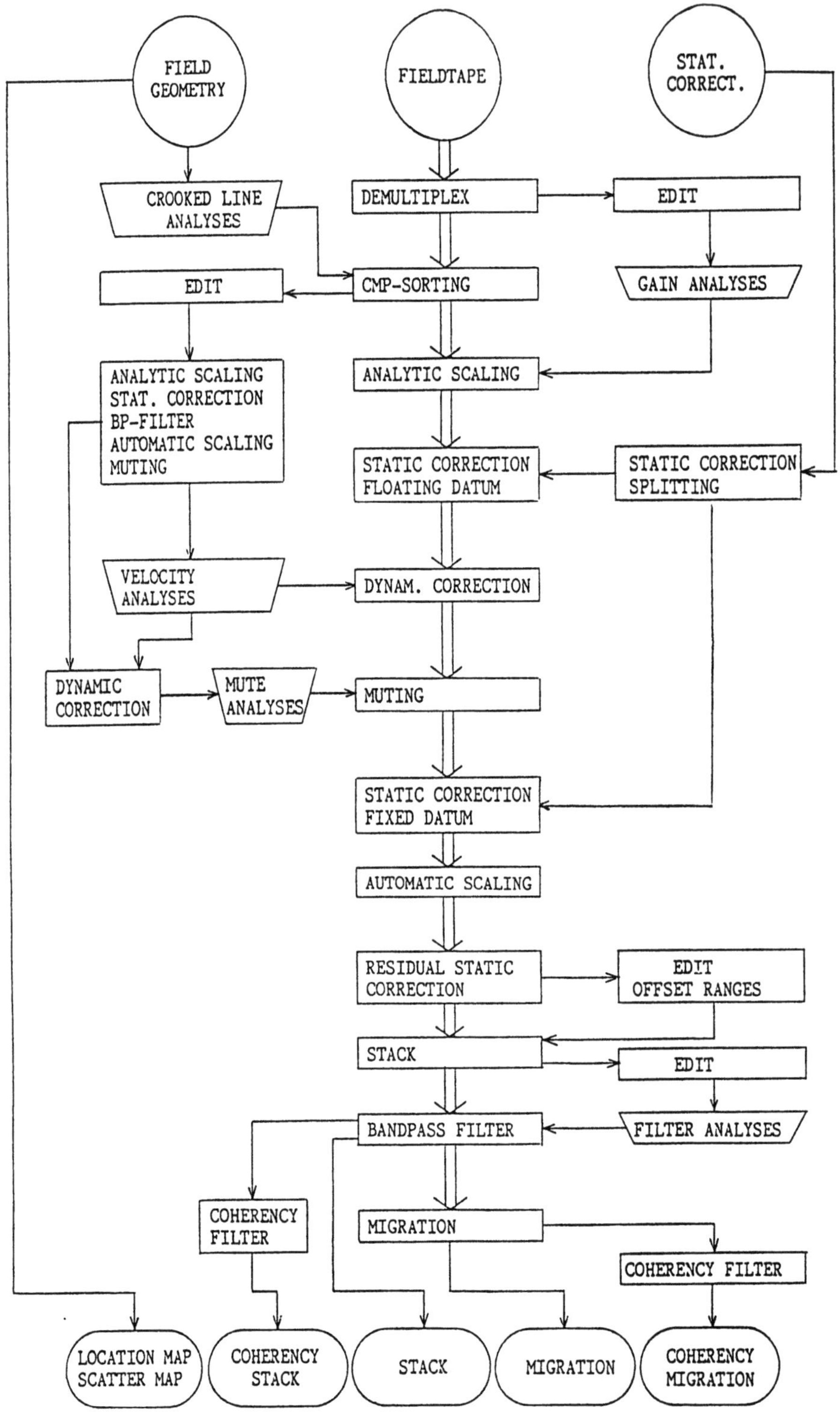

Fig. 2. Generalized flowchart for deep-seismic vibroseis processing

The position of the respective steps in the total sequence is shown. The circles at the top of the figure symbolize the magnetic tapes supplied by the contractor. The trapezoids to the left and right indicate the analyses necessary for the individual processing steps, while the rectangles represent the calculation processes. The ovals at the bottom indicate the plots made available for geological interpretation. The central sequence marked by broad arrows represents the flow of all the data from the field tapes to the final sections. Each of the processing steps will be presented in the following chapters, and their significance for deep-seismic data processing is explained.

2.2 Demultiplexing, CMP Sorting, and Analytical Scaling

The digital field systems common today record seismic signals with a time-multiplex method, i.e., for defined points in time the instantaneous signal amplitudes for all geophones are recorded on magnetic tape (for deep seismics the sampling rate is usually 4 ms). Because all steps in the processing center are done trace by trace, the samples have to be rearranged into a trace-sequential format in an initial step, called demultiplexing, resulting in so-called common-shotpoint gathers (CSP). Simultaneously, the true signal amplitudes are restored. This is necessary owing to the automatic, time-dependent gain control in the recording device, which is done to accommodate the high dynamics of the ground motion (up to 100 dB).

A CSP for profile KTB 8514 Black Forest is shown in Fig. 3a. Due to the enormous differences in amplitude from the beginning to the end of the unprocessed seismogram, almost only the first arrivals and the strong air waves can be recognized.

Seismic reflection profiles are generally recorded on the basis of multifold coverage to improve the signal/noise ratio, i.e., one and the same subsurface point is repeatedly encountered at various angles of incidence by different shots. The signals for one reflection element, recorded on different traces, differ from one another only by their traveltime, whereas the qualitative information is theoretically the same despite the various wave paths - a necessary supposition for the stacking process.

Thus, the demultiplexed data in CSP form must be rearranged by sorting into so-called common-midpoint gathers (CMP) in order to be processed further. The corresponding sorting instructions are dependent on the recording configuration and the absolute coordinates of the source and receiver positions, which must be derived from the basic field parameters made available to the DPC by the contractor.

Curves and sharp bends (up to 200 m/km) in the long profiles and the impossibility of setting up an absolutely straight shot--point line under the conditions of accessibility in Central Europe, especially with the Vibroseis method (lateral offsets of up to 1000 m), nearly always lead to the problem that the theoretical subsurface reflection points (i.e., the points halfway between the respective shot and geophone) are not located directly below the geophone line, but are scattered more or less around this line. If, however, the lines of reflection and measurement do not coincide, an ideal CMP line must be defined in the processing center using the so-called "crooked-line" sorting process.

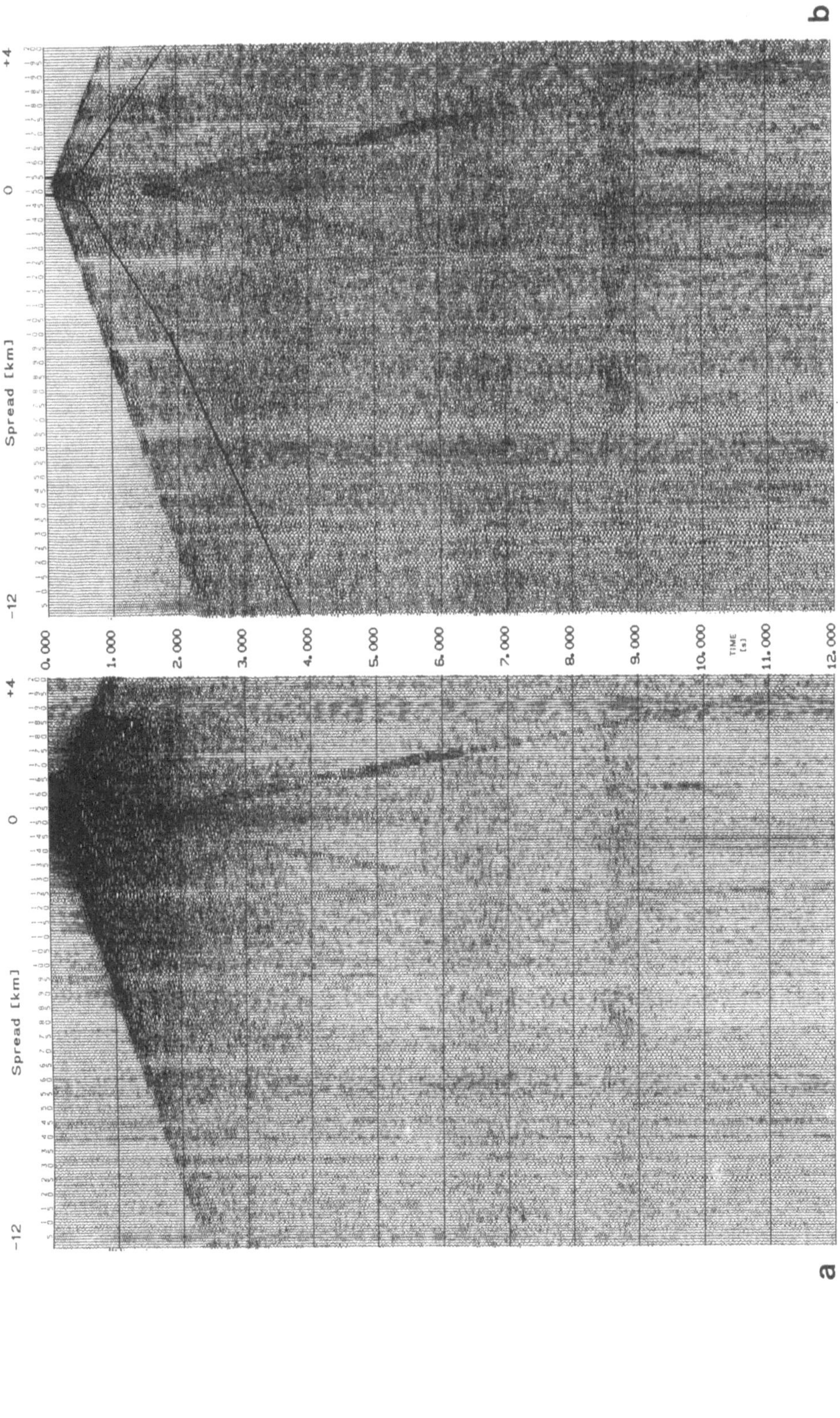

Spread [km]
-12
0
+4
0.000
1.000
2.000
3.000
4.000
5.000
6.000
7.000
8.000
9.000
10.000
TIME [s]
11.000
12.000
a
b

The first step involves creating so-called scattergrams, which represent the distribution of the theoretical subsurface reflection points. In a long profile like DEKORP 86-2N, for example, there are ca. 1.2 million individual scatter points. In the second step, a reflection line is then calculated which represents the best fit through this cloud of points. The essential parameters are the permissible curvature of the defined line and the permissible deviation of individual points from this line. Figure 4a shows the positions of some of the geophones and vibrators, and Fig. 4b the resulting scattergram with the calculated CMP line (part of profile KTB 8506).

The systematic decrease in energy with increasing traveltime (Fig. 3a) can be attributed to essentially two causes. First, the energy per area decreases owing to the expansion of the wave front (spherical divergence). This decrease is due to the fact that the amplitude is inversely proportional to the distance traveled; in a homogeneous medium the amplitude is proportional to $1/T$ as well, where T is traveltime. Second, during propagation a portion of the seismic energy is irreversibly transformed into heat (absorption). This process is an exponential function of the distance; in the case of a homogeneous medium, the amplitude is also proportional to $\exp(-\alpha * T)$, where α is an absorption parameter. Correcting the recorded amplitude values by $k * T * \exp(\alpha * T)$, where k is a constant, appears to be suitable for compensating for these two effects. The result can be seen in Fig. 3b.

This process of analytical scaling, in contrast to automatic scaling (AGC), has the decisive advantage that the values of the applied correction curve can be stored as a small number of parameters and thus the original amplitudes can be recovered at any time.

Because in conventional data processing automatic scaling is indispensable before stacking, amplitude equalization with T^2 up to ca. 3 s has proven to be sufficient for obtaining a rather constant energy level as a basis for subsequent AGC. However, no automatic scaling is allowed when processing with true amplitudes. In that case, to determine appropriate values of α, extensive gain analyses are necessary to take into account all vertical and horizontal changes in the energy curves (see Section 3.4).

2.3 Static Corrections, Dynamic Corrections, and Muting

The recorded traveltimes are not only dependent on the structures in the deeper subsurface, but also on near-surface irregularities in velocity, the variable thickness of the weathering layer, the differing drilling depths (in explosion seismics), and the topography of the earth's surface. These influences can be controlled by static, i.e., traveltime-independent, corrections.

←

Fig. 3.a Unprocessed common shotpoint gather of Black Forest line KTB 8514 (FF 523; location 1399; km 32)

Fig. 3.b Same as Fig. 3a, but after analytical scaling; muting curve given by solid line

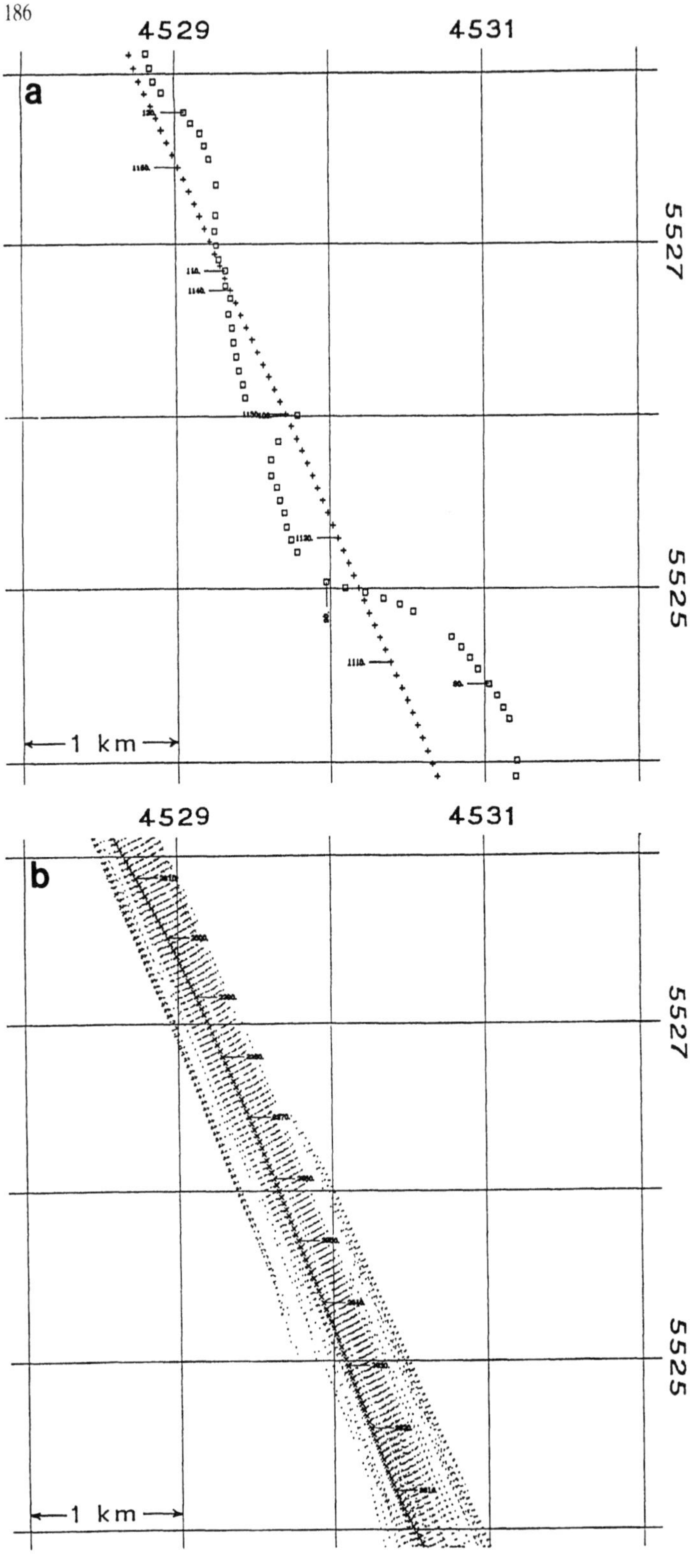

Fig. 4.a Positions of geophones (<u>pluses</u>) and vibrators (<u>squares</u>), segment of Oberpfalz line KTB 8506 (km 8.2-12.7)

Fig. 4.b Scattergram for subsurface points of surface layout in Fig. 4a; <u>line of crosses</u> is CMP processing line

The measured traveltimes are corrected by static shifts so that after that all receiver and source positions are theoretically located at a uniform reference level below the aforementioned irregularities.

The correction group of the field party evaluated the first arrivals using the monitor playbacks. Offset gaps were closed with special near-line measurements. The basic static correction values obtained in this way for each source and receiver were given to the DPC.

Because the static correction values scatter greatly for profiles in areas with a rough topography, it is advantageous to first correct all traces to a variable reference level (floating datum). Therefore, a block-static value is determined for each CMP gather, this value being equal to the mean of the total static corrections of all traces of this CMP. Then, before dynamic corrections, only a static shift corresponding to the deviation of each individual trace from this intermediate datum is applied resulting in a much smaller time delay. The advantage here is that the appropriate stacking velocity is then much more realistic and its values are much less scattered along the profile.

The block-static value, which is constant within each of the gathers, does not necessarily have to be applied before stacking. However, transition to a constant datum directly following the dynamic correction is recommendable for determining the residual static corrections.

It follows from the principle of multifold coverage that each subsurface reflection point is associated with several different shotpoint-receiver pairs. The different travel paths result in different traveltimes. The hyperbolic curvature (normal moveout, NMO) of the traveltime curves must be corrected dynamically, i.e., dependent on offset, time, and velocity, so that related events appear simultaneously, which is absolutely necessary for stacking. The offset of each trace is known. Thus, for each traveltime only the accompanying stacking velocity has to be determined by velocity analyses at discrete points along the profile.

This process of determining appropriate velocity-time functions is very problematic because common correction procedures assume horizontal or slightly dipping, well-reflecting horizons, which is seldom the case with complex geological structures in deep seismics. Therefore, Section 3.1 is devoted to the determination of stacking velocity as the most critical processing parameter.

In addition to steep-angle reflections, the seismograms also contain first arrivals, direct waves, refractions, as well as wide-angle events. These undesirable signals must be eliminated as a function of source-receiver distance (offset) before further processing. The suppression of the unuseable zones by zeroing all samples above a user-defined curve is referred to as "initial muting". To prevent jumps at the end of a muted zone, generation of a transition zone by tapering is indispensable. Figure 3b shows a typical muting curve in a CSP.

Long profiles over different geological units require frequent checking and spatial and temporal variation of the muting curves because optimum stacking is dependent on this to a great degree. The muting parameters must be frequently changed along the profiles and often lie outside the boundaries common in seismic prospecting (up to 6 s with 12 km offset). Especially for short traveltimes, not only high coverage, but also the good match between a few coherent traces yield better results.

2.4 Automatic Scaling, Residual Static Corrections, and Stacking

In the above discussion only the energy of each individual trace has been considered. It is just as decisive for a good stack that all traces have the same average energy. Deviations from the average energy are caused, for example, by differences in the coupling of the geophones and vibrators to the ground, as well as the changing nature of the ground. To eliminate these effects, only CMP traces of similar type and quality may be used in the summation process for meeting the basic stacking requirements.

In conventional seismic data processing, correcting the systematic decrease in energy with time by an analytical gain function is followed by automatic amplitude control. This is done to level out all residual horizontal and vertical variations in energy. Therefore, a time window of user-defined length is moved sample by sample over the trace, and a factor is calculated from the RMS (root mean square) energy. The central sample of each window is then divided by this factor so that the mean energy in every trace window is subsequently the same for all traces.

Stacking is substantially improved by using residual static corrections because the basic static values determined in the field are not accurate enough, as experience has shown. Thus, subsequent corrections are necessary, which are done at the DPC using two independent programs for calculating residual static correction values.

Surface-consistent static values for all sources and receivers are calculated with the first method, which assumes that the near-surface influences, such as topography and the thickness and velocity of the weathering layer, produce constant time delays regardless of the wave path; i.e., a constant time shift is assigned to each given location at the surface.

For this purpose three relative shifts are calculated for each trace. This is done by cross-correlating three different model traces with all traces of the respective CMP gather to which the basic static and dynamic corrections have already been applied. The first model trace is the preceding stacked CMP. The second is that trace of the current CMP gather which produced the greatest maximum during the first cross-correlation pass. The third model trace is the current stacked CMP, which has been corrected by the better of the two residual static sets of values calculated in the first two cross-correlation passes. Finally, that set of values that appears to be the most reliable of the three sets can serve as a basis for calculating the final residual static correction values.

Subsequently, the shot and geophone static values are calculated by the Gauss-Seidel iteration method. They are the solution of a system of simultaneous equations in the form of a matrix containing the coefficients of all source and receiver locations and an accompanying vector describing the relative static shifts. The number of iterations can be selected, and on the basis of these source and receiver corrections, a new system of equations can be defined for better convergence and solved in a second run. Finally, the residual static correction is split into a portion for the shot and a portion for the geophone. The residual static correction component for each specific source and for each specific receiver is always the same, and the total residual static correction value for each CMP trace is made up of the sum of the respective components (like for the basic static corrections).

In practice, better results are obtained when, on the one hand, the correlation windows (temporally and spatially interpolated) are placed only over the zones of good reflectivity and, on the other hand, only traces with offsets smaller than 4 km are used for calculating the geophone and shot portions. This is done to eliminate any effects of horizontal, instead of vertical variations. Moreover, any systematic velocity errors arising during the preceding dynamic correction are less significant for small offsets.

A final stack without residual static corrections is, of course, done beforehand to check the quality of the velocities used for the dynamic corrections and to compare the changes with the raw stack on one hand, and with the final stack corrected by residual statics on the other.

Figure 5 shows a portion of profile KTB 8502, which in (a) is stacked without surface-consistent residual static corrections and in (b) with them.

Determination of subsurface-oriented residual static correction values is the second method for improving the stacking amplitudes. In the DPC it is used after application of surface-consistent residual static correction values to produce increased coherency from top to bottom within each CMP by slightly shifting the individual traces with respect to each other.

The amount of shift of the individual CMP traces is determined by cross-correlating them with a pilot trace. This pilot trace is obtained by weighted stacking a small, user-defined number of adjacent CMPs onto the middle CMP. A graduated weighting in percent of all CMPs involved in the pilot trace is done so that the influence of each CMP increases with decreasing distance from the middle CMP.

The number of adjacent CMPs used is dependent on the geological structures appearing in the profile. When the number of stacked CMPs is too large, there is the danger that small-scale structures may be blurred. Furthermore, steep dips can lead to the obliteration of reflections already on the pilot trace and thus cannot be taken into account in the correlations.

In contrast to the surface-consistent method, it is recommendable when using the subsurface-oriented method to take the total trace length as the correlation window. By doing so, the residual static corrections are not too strongly influenced by any single event. If this leads to poorer results in certain traveltime ranges, separate residual static values for the upper and lower traveltime range can also be calculated in two runs and applied to the respective windows. In this way, time delays that are not

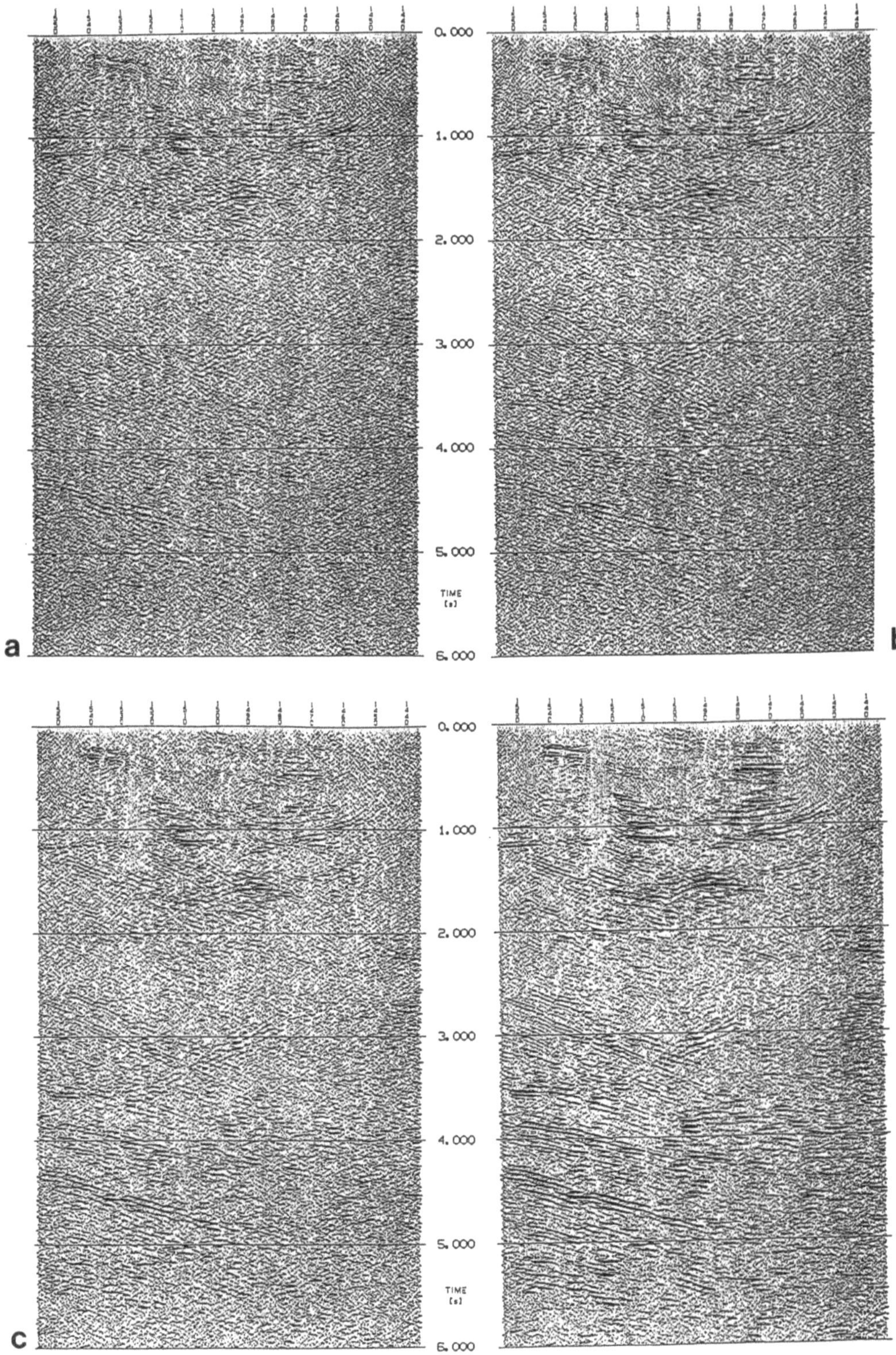
0.000
1.000
2.000
3.000
4.000
5.000
TIME
(s)
6.000
a
b
c
d

purely of static nature, but are the products of complex subsurface structures, can be taken into account within certain limits. This technique is useful particularly for the long traveltimes in deep seismics.

Figure 5c shows the stacking result using both surface- and subsurface-oriented residual statics. The improvement over the version with only basic static corrections (Fig. 5b) is obvious.

In a direct comparison of the two methods it can be concluded that with the first method true static values for shot and geophone positions can be calculated using the relative time delays between the model trace and the CMP traces, taking each source and receiver location into consideration. In the second procedure, however, the time delay of each individual CMP trace relative to the pilot trace is used directly as the static value without any consideration of the source and receiver location. In both cases the maximum permissible residual static correction values must naturally be limited to realistic values on the order of about half of the signal period.

Now that all necessary static and dynamic corrections have been completed, all traces of each CMP gather may be stacked, i.e., horizontally added and then divided by the number of traces. In this way a theoretical improvement in the signal/noise ratio is obtained proportional to the root of the multiplicity of coverage. By producing a stacked zero-offset time section, the amount of data for each profile is reduced to ca. one-eightieth (for the recording configuration of the DEKORP/KTB measurements in the Black Forest and Oberpfalz).

Experience in deep-seismic data processing has shown that, owing to the normally very great inhomogeneity of the recorded data, there must be sufficiently large subsurface coverage. Yet, the stacking result will not necessarily improve with increasing degree of coverage. Because this observation is difficult to explain or generalize, Section 3.2 is devoted to a more detailed study on this.

The instantaneous coverage increases along the time axis due to offset-dependent muting. The signals also appear to have a random component. These two observations apparently cause the mean stacking amplitudes in the upper time ranges to be greater than in the lower ones when divided only by the respective instantaneous coverage. The same is true for those CMPs with decreasing coverage toward the margins of the section. Dividing not by the coverage itself but by the 0.6th power of the same is an appropriate way of compensating for this effect. This is an

Fig. 5.a Part of Oberpfalz line KTB 8502 (km 35-44.5), stack without residual static corrections

Fig. 5.b Same as Fig. 5a, but additionally with surface--consistent residual static corrections

Fig. 5.c Same as Fig. 5b, but additionally with subsurface-oriented residual static corrections

Fig. 5.d Same as Fig. 5c, but additionally with coherency filtering

empirical value which ensures that in the muting range where the coverage is small, the divisor is larger than that in the deeper traveltime range, where full coverage is attained. When this modified compensation is done during stacking, no further automatic scaling is necessary afterwards.

2.5 Frequency Filtering, Coherency Filtering, and Migration

By cutting out or attenuating undesirable noise frequencies, filtering (mostly bandpass filtering) allows the desired signal to be accentuated and a clearer, interpretable seismic image to be obtained. Because it is linear, the filter process can also be done after stacking.

Frequency spectra are calculated for various points of the section to determine the spatial and temporal variations in the filtering parameters. In addition, different bandpasses are applied to groups of 51 CMPs that are representative for a rather large profile range. The high- and low-pass boundaries are determined by shifting a narrow pass band of constant width stepwise over the entire frequency range. In critical cases, varying the high cut-off frequency while maintaining a constant low cut-off (and vice versa) also provides reliable information. In nearly all DEKORP/KTB profiles the maximum signal frequency range is already fixed by the sweep range (12-48 Hz). Differing surface-wave intensities and changing absorption characteristics necessitate frequent control of the filter parameters. Section 3.3 deals with how drastically the signal frequencies can change over short distances.

When the amplitude of monofrequency noise (e.g., 50 Hz from power lines, 16 2/3 Hz from the electric railroad overhead lines) is not too high, steep-sided notch filters can recover masked signals.

A repeated improvement of the signal/noise ratio, and thus a clearer appearance and a better interpretability, is obtained by coherency filtering. This procedure is applied mainly to stacked sections and is aimed at intensifying previously existing signals persisting over several traces and at suppressing incoherent noise. The way in which this works is briefly discussed in the following:

First, trial stacks of a specific number of adjacent traces along lines of various dip yield cumulative traces with differing amplitudes. The trace length is divided into a user-defined number of windows. For each window, the cumulative trace with the highest amplitude is selected. Then, the trace segment chosen for each window is combined with those selected for the other windows into a single coherency trace. Next this trace is added to the middle one of the input traces, and the result is a coherency-filtered trace. Finally, the coherency-filtered section is obtained by proceeding one trace further, and repeating this process for the same number of adjacent traces until all traces have been processed. This method is thus able to accentuate certain alignments (dominant dips) within the time window being considered, whereby it must be emphasized that to obtain good results, it is important to choose reasonable values for the parameters.

The largest positive and negative angles of dip to be considered are dependent on the steepest inclinations being present; the temporal changes in inclination are decisive for the length of the window. The number of traces for which the coherency is determined is dependent on the mean number of traces over which the signals persist. The weighting factor for the addition of original and coherency traces is selected on the basis of the intensity of the reflections already present in the original section.

The example in Fig. 5d shows (in comparison with Fig. 5c) the improvements obtained with appropriate coherency filtering. In Section 3.3 several further practical aspects of coherency filtering are explained.

Data processing at the DPC concludes with the migration of the zero-offset stacked sections. The purpose of this step is to maintain a spatially true picture of the subsurface, i.e., the energy distributed on diffraction hyperbolas is placed to the apices of these hyperbolas, reflection elements appear with the true inclination at the right places, so-called "smiles" become synclinal structures, and all distortions are eliminated.

Basic problems with migration are the determination of appropriate migration velocities, the proper treatment of steeply inclined structures, limiting of the level of noise, obtaining tolerable processing times, and the necessity of splitting of long profiles into several segments due to limited storage capacity.

In standard seismic processing, finite-difference (FD) migration is commonly used because it permits variable velocities in all directions and produces the best signal/noise ratio. The disadvantage of long processing times can generally be compensated for by resampling the data (changing the sampling interval from 4 to 8 ms).

However, the limits of the FD migration are soon reached with the presence of steeper dips resulting in blurring due to frequency--dispersion effects. In these cases, (1) resampling the data changing the sampling interval from 4 to 2 ms, (2) decreasing the spacing by inserting interpolated traces, or (3) using smaller downward continuation steps can help. Yet these three suggestions have the disadvantage that they drastically increase processing time, so that other migration methods could be more reasonable. Figure 6 shows the effect of FD migration on a part of profile DEKORP 85-4Q (area of the Bavarian Pfahl): (a) the unmigrated stack, (b) the FD migration with resampling at 8 ms, and (c) the FD migration with resampling at 2 ms.

Frequency-wavenumber migration (f-k) and summation migration (Kirchhoff) are alternative migration methods. Theoretically, f-k migration of all dips can be done free of dispersion, but it becomes inaccurate in the presence of variations in velocity. Thus, it is unsuitable for the final processing of whole sections and is limited to special, small-scale applications. However, because f-k migration can be done very rapidly, it is an excellent method for determining migration velocities when diffractions are present. For this purpose the section is f-k migrated testwise with different, constant velocities, whereby at different points and in different depth zones the diffraction hyperbolas contract

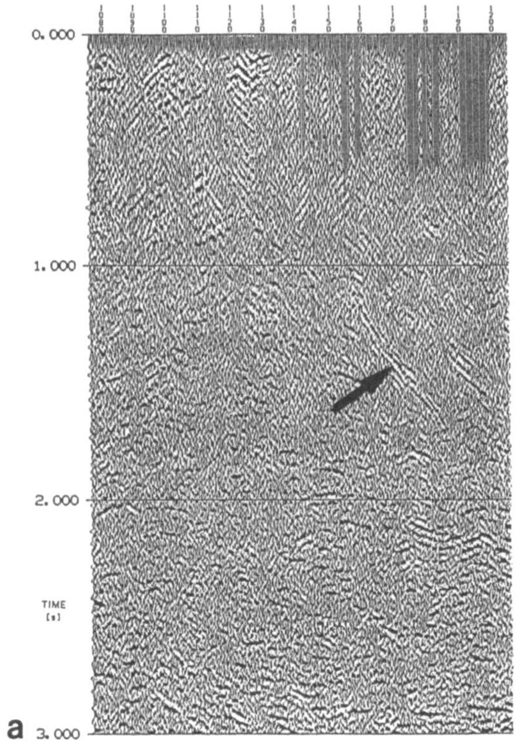

Fig. 6.a Segment of Oberpfalz line DEKORP 85-4Q (km 6-16.5), unmigrated stack

Fig. 6.b FD migration (resampling to 8 ms) of Fig. 6a

Fig. 6.c FD migration (resampling to 2 ms) of Fig. 6a

Fig. 6.d f-k migration (resampling to 8 ms) of Fig. 6a

Fig. 6.e Kirchhoff migration (resampling to 8 ms) of Fig. 6a

to their apex at different velocities. A migration-velocity model for a subsequent FD migration can be constructed on the basis of these f-k migrations.

The Kirchhoff migration is able to migrate all dips correctly with acceptable processing time and also allows variable velocities. However, it requires highly noise-free input data; if this is not the case, the signal/noise ratio tends to decrease (Fig. 6e).

In areas lacking diffractions, smoothed and, if need be, dip--corrected (i.e., reduced) stacking velocities can be used as reference values for the migration velocities so that migration can be done without previous, extensive testing (e.g., for raw stacks).

All migrations (particularly for short profiles) are more or less strongly disturbed by superimposed marginal effects. The amplitudes of the traces at the margins are smeared on semicircles whose radii correspond to the two-way traveltimes.

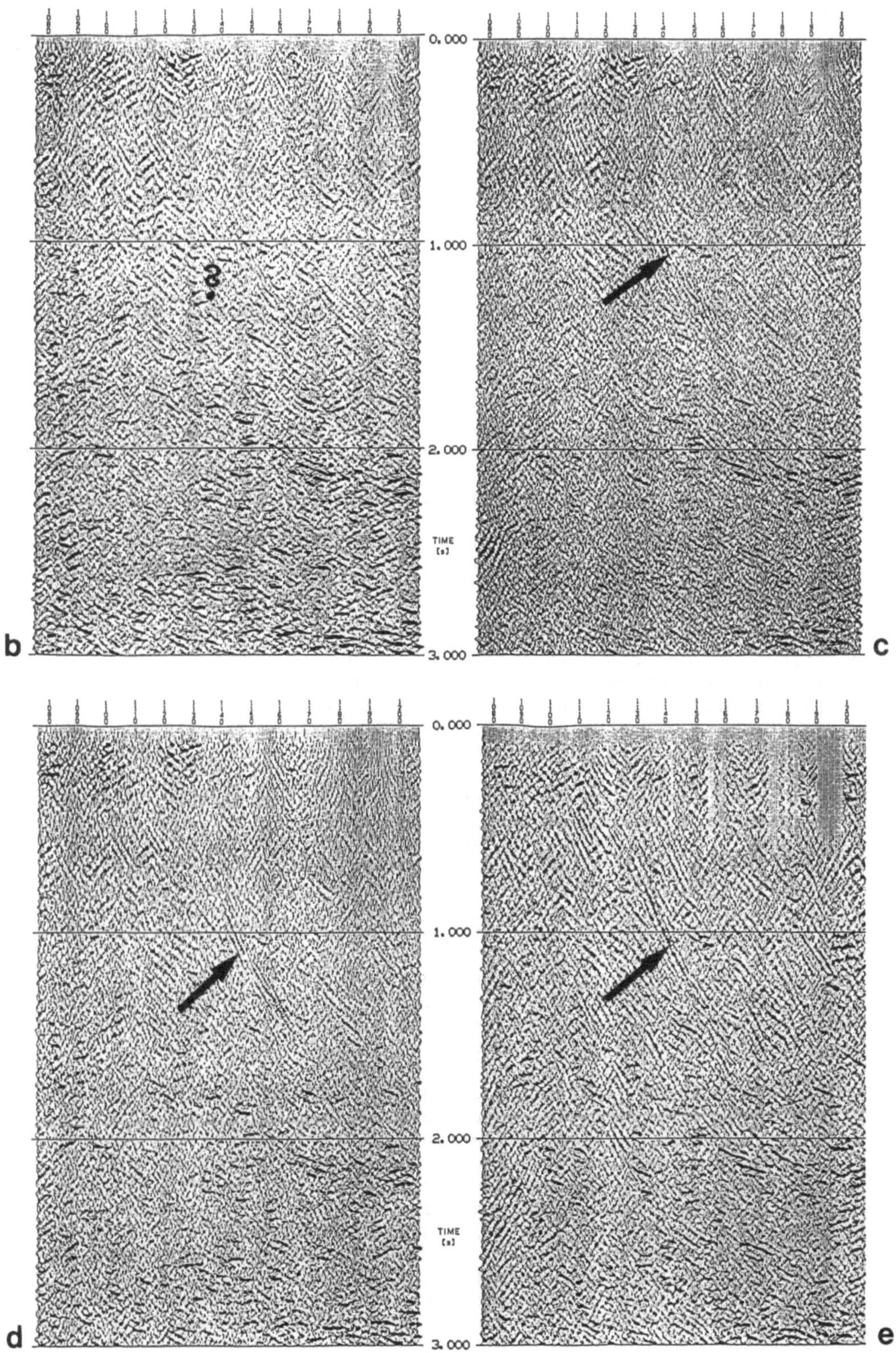
0.000
1.000
2.000
TIME [s]
3.000
b
c
d
e

This is a fundamental problem, whose effects can be reduced by supplementing the profile with a sufficient number of zero traces and, if need be, by horizontally tapering the outer data traces. However, this problem can never be eliminated completely.

Coherency-filtered, stacked sections should not be used as input sections for migration, otherwise "smiles" can occur due to the overly great differences in dynamics. On the other hand, coherency filtering can always be used following migration when it is certain that the noise effects of the migration will not be intensified.

3. Special Features of the Deep-Seismic Data Processing at the DPC

3.1 Determining the Stacking Velocities

Determining appropriate stacking velocities is of decisive importance for seismic processing because it has a considerable influence on all subsequent steps (e.g., residual static corrections, horizontal stacking, migration). This is true for both prospecting and deep seismics. However, due to the frequent absence of strong and continuous reflectors in the crystalline crust, finding useable velocities for dynamic correction with standard methods of analysis is often difficult.

It is first necessary to exactly define the term "stacking velocity": As described in Section 2.3, the curvature of the traveltime hyperbolas must be eliminated by time-dependent, normal moveout corrections - on the basis of multifold coverage. The correction velocity that produces the maximum stacking amplitude in the zero-offset section at a time T_0 is designated by the stacking velocity v_s for that time T_0. This is far from always being a value on the order of magnitude of rock-specific velocities or those determined by seismic refraction measurements. It is very important that these correction velocities are not be used as reference values for geological interpretation. Only under very restricted conditions (i.e., horizontal or nearly horizontal stratification, homogeneous bedding, good impedance contrast) can the interval velocities be estimated. These conditions are often met in seismic prospecting, usually in sedimentary rocks. In deep seismics, calculation directly back to layer velocities results in errors that are too large, due to the presence of velocity gradients and complicated structures, and thus should be omitted in nearly all cases.

In deep seismics a combination of several methods of analysis is generally necessary to determine appropriate stacking velocities so that the disadvantages of some methods are compensated for by the advantages of others. This also allows the results of one method to be used to double-check the results of the other methods. Depending on the number of ambiguities in the stacking maxima, the simultaneous use of up to five different methods of analysis may be necessary for deep seismics because no one method alone is able to supply all the needed information on stacking velocity. At least two different methods of analysis are routinely performed to obtain the velocities for the final stacks of the DEKORP/KTB lines.

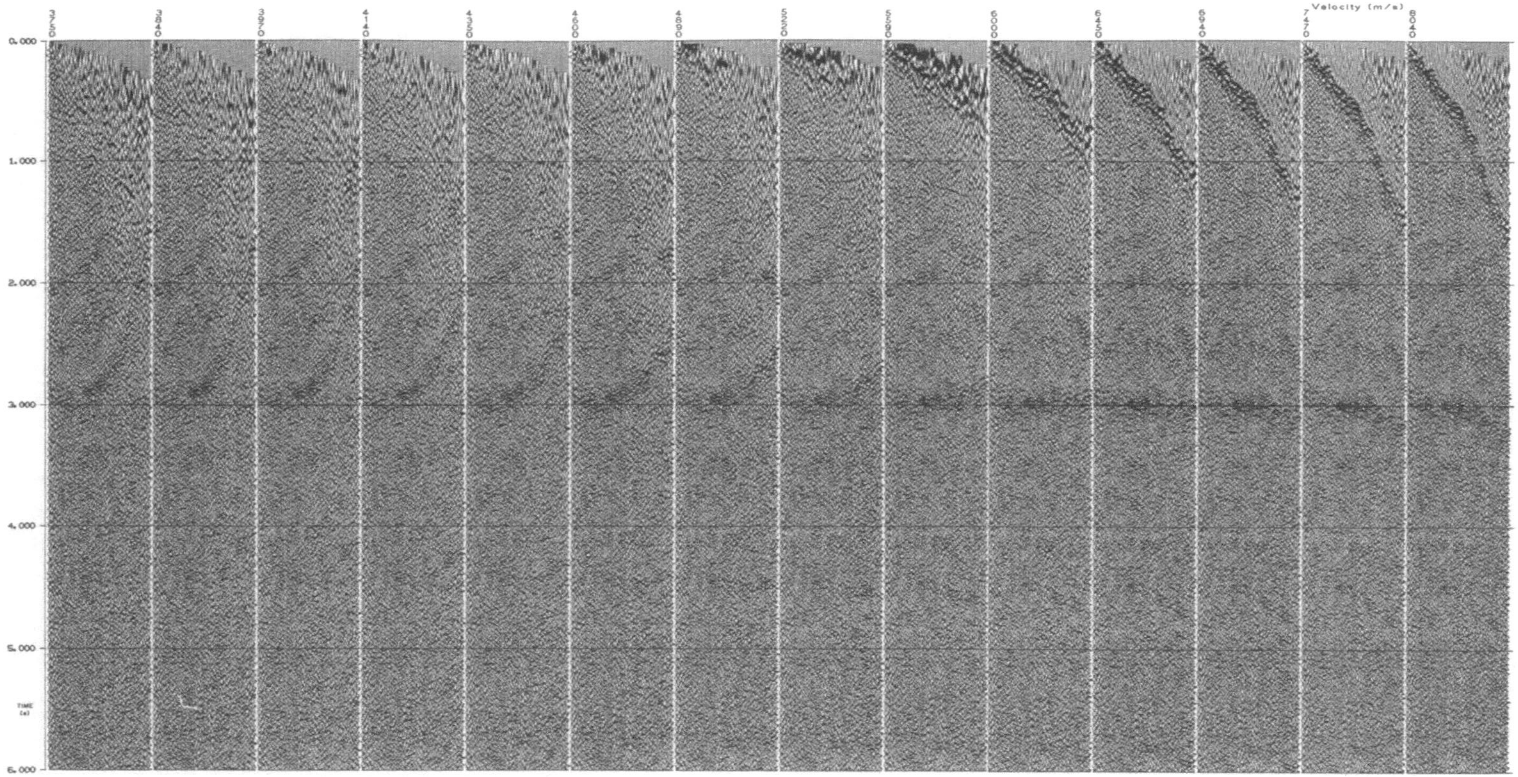

Fig. 7.a Single-CMP velocity analysis (unmuted) of CMP 2935 of the Oberpfalz line KTB 8501 (km 37.5)

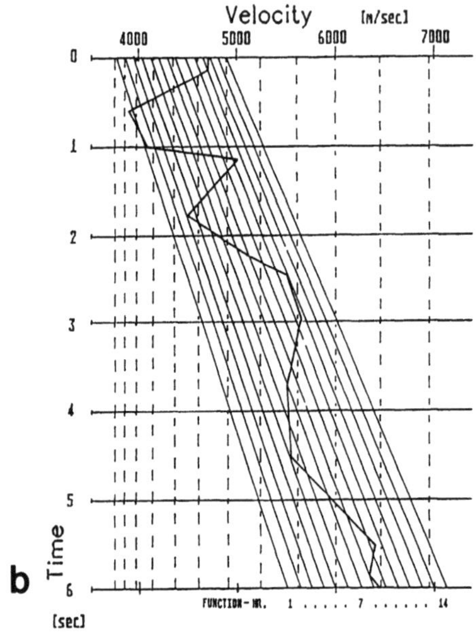

Fig. 7.b Comparison of resolution using constant velocities (dashed lines) and velocity functions (solid lines); picked velocity curve is added (cf. Fig. 7d, e)

Fig. 7.c Constant-velocity stacking analysis of CMPs 2920-2940 from Oberpfalz line KTB 8501 (km 37-37.8) with picked velocity curve

Fig. 7.d Same as Fig. 7c, but with velocity-function stacking analysis

Fig. 7.e Same as Fig. 7d, but velocity determination with stacking-semblance analysis of the CMPs 2926-2934 (km 37.2-37.5)

In the following discussion, all methods used at the DPC for determining appropriate stacking velocities will be presented briefly using a CMP range from profile KTB 8501 (Fig. 9b, profile kilometer 37.5) as an example. The limitations of all methods will be discussed. The methods are the single CMP analysis, constant-velocity stacking analysis, velocity-function stacking analysis, semblance analysis, and stacking analysis of large areas. The pronounced reflector at ca. 3 s can serve as orientation and for comparison of the individual methods.

If various constant velocities are applied to a single unstacked CMP gather, the stacking velocity for a specific point in time is assumed to be correct when the hyperbola curvature vanishes at that time. The velocity function in which all hyperbolas have been flattened will yield the best stacking result. This method has the advantage that analyses can be carried out quickly and that information on each trace of the CMP gather and thus the contribution of the different offset ranges can be recognized. However, a disadvantage is that inclined horizons lead to less normal moveout and consequently cannot be differentiated from horizontal reflectors of higher velocity. In addition, the stacking process, which improves the signal/noise ratio, is lacking as well, so only the stacking velocities of prominent reflectors (Fig. 7a) can be determined easily.

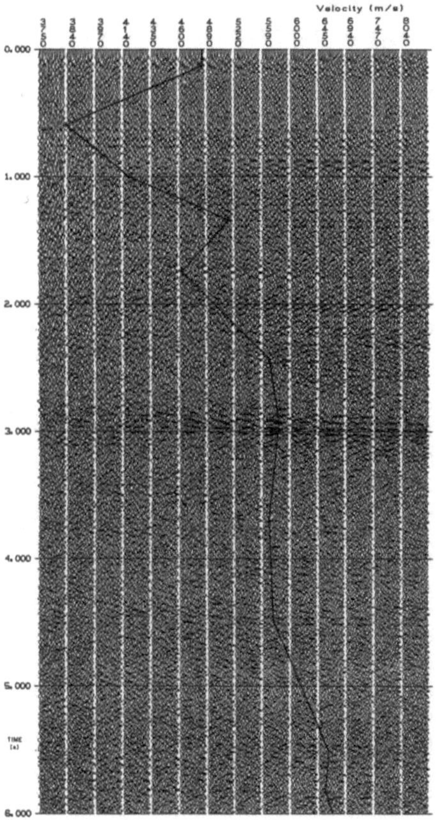

Velocity (m/s)
TIME (s)

c

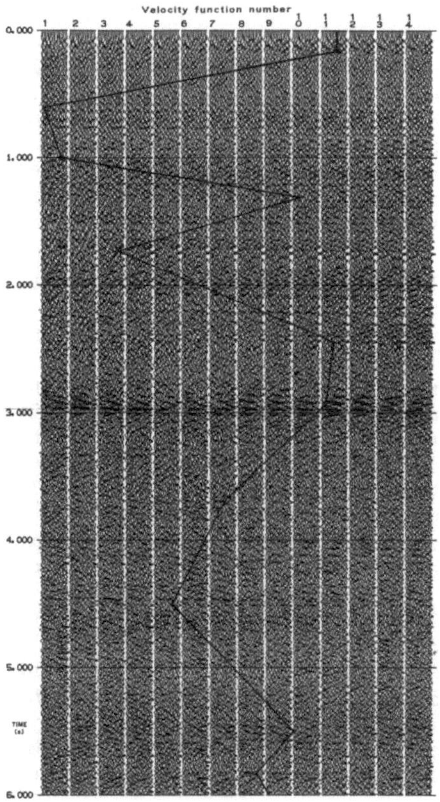

Velocity function number
TIME (s)

d

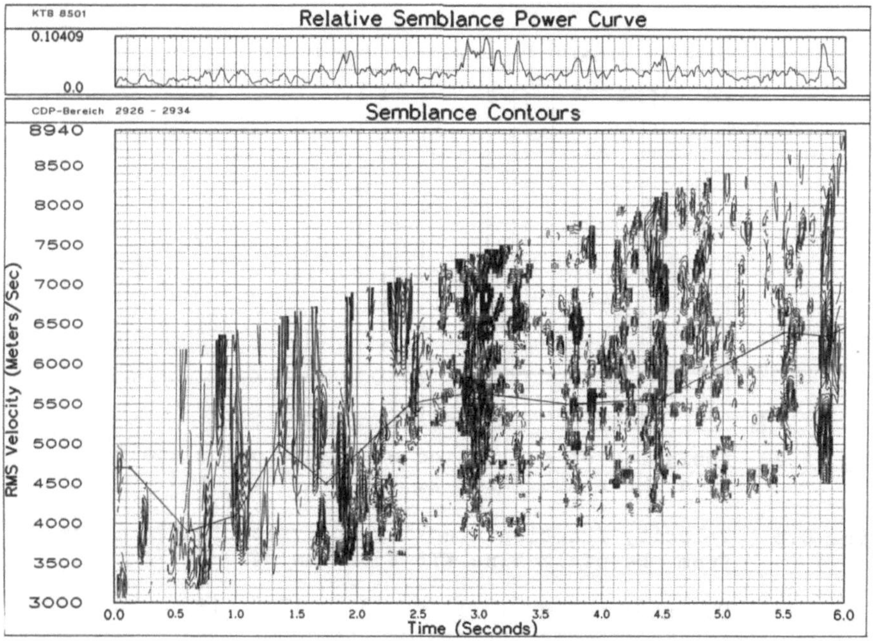

KTB 8501
Relative Semblance Power Curve
0.10409
0.0
CDP-Bereich 2926 - 2934
Semblance Contours
RMS Velocity (Meters/Sec)
Time (Seconds)

e

None of the disadvantages of using a single unstacked CMP gather occur when a group of several stacked CMPs (typically 21 at the DPC) is analyzed after application of various constant velocities. When this is done, the maximum stacking amplitude for each point in time can be determined as a function of velocity. A good stack can be obtained only when all traveltime hyperbolas in each CMP gather at every point in time have been flattened completely (Fig. 7). The already long processing times caused by having to consider a group of CMPs and subsequently stacking them is of particular consequence because the traces are also corrected dynamically with very high velocities for short traveltimes and with very low velocities for long traveltimes.

The use of unrealistic velocities can be avoided by producing trial stacks not with various constant velocities, but with various velocity-time functions that vary around a model function (e.g., based on raw stack analyses). When this is done, higher resolution is obtainable because much denser scanning is attained using the same number of trial stacks (Fig. 7b). Nevertheless, the velocities producing the maximum stacking amplitudes are much more difficult to pick because every function has a different velocity at different times. The correction velocity for any particular time is taken from a large table containing the complete velocity-time test field. Figure 7d shows an excerpt of an analysis done by this method.

In the first three methods presented, the corrected and, depending on the method used, stacked data itself was analyzed. Semblance analysis, on the other hand, deals with the depiction of isolines which represent the stacking energy as a function of the chosen correction velocity range. This method can also be applied to a small group of CMPs in which all traces with the same offset have previously been stacked. The selected number of

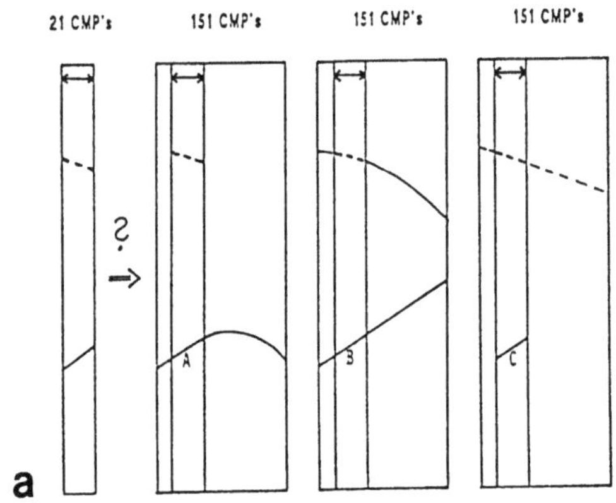

Fig. 8.a Ambiguity of evaluation when a velocity analysis of a CMP range which is too small is used

Fig. 8.b Velocity function stacking analysis of CMPs 2866-3016 from Oberpfalz line KTB 8501 (km 34.3-40.7)

Function number	11	15	17	18	20	24
Velocity at 3 s	5100	5820	6180	6360	6720	7440 m/s

Velocity function number
1 1
1 5
1 7
1 8
2 0
2 4
0.000
1.000
2.000
3.000
4.000
5.000
6.000
TIME (s)

b

participating CMPs (dependent on the dips occurring) should not be too large; this is to avoid misstacking during common offset summation. Nine CMPs were used for the semblance analysis shown in Fig. 7e. Due to the continuous variation in velocity, semblance analysis is excellently suited for supplementing one of the previously described methods because the minimum and maximum stacking amplitudes are easier to locate in the centers of the resulting "islands".

One basic phenomenon occurring in all methods of analysis is the simultaneous appearance of various events which cannot be NMO--corrected unambiguously with only one velocity. In this case it often cannot be decided which signal should be preferred when analyzing only a small group of CMPs. Then, the logical consequence is that a rather large segment of the profile has to be analyzed dynamically to determine those stacking velocities that are connected with plausible large-scale structural elements.

Figure 8a shows these problems schematically: a test stack over a narrow CMP range shows two dipping events. It cannot be discerned here whether the strong element is (A) the flank of a diffraction, (B) a portion of an inclined reflection, or (C) an isolated reflection element which just happens to be situated in the analyzed range. Likewise, the weak element can certainly be a portion of a continuous (weak or interrupted) horizon.

Figure 8b shows in a very illustrative way the changes in the appearance of the reflective zone at ca. 3 s when 151 CMPs are stacked with different dynamic corrections. On the one hand, different events are seen in any one time zone for different velocities. On the other hand, even slight changes in velocity can lead to a distinct change in the total appearance. Despite the wealth of information contained in the velocity analysis over such a wide range, the evaluation is very complicated in practice. Firstly, the velocity within a range of 151 CMPs is, of course, not horizontally constant (and this with varying significance for different time periods). Secondly, arbitrary horizontal and vertical jumps in velocity are only permissible within certain mathematical limits (occurrence of imaginary interval velocities) so that not all events can be picked optimally. Nonetheless, any range of only 21 CMPs never has the interpretative significance as the total range of 151 CMPs. Thus, this type of representation always offers additional interpretative help.

Velocity analyses for the final stack of profile KTB 8501 were done and evaluated with all five of the described methods. A comparison of the raw stack (Fig. 9a) with the final stack (Fig. 9b) clearly shows the improvements which, for the most part, were obtained through optimized dynamic corrections, attained using combined techniques of velocity analysis. Furthermore, the following processes such as residual static corrections and coherency filtering are all the more effective after optimum dynamic correction.

The remainder of this chapter will now present another practical example in which the relationship between stacking velocity and dip of a horizon is very clearly shown.

With the same shot-geophone configuration, inclined bedding produces shorter traveltimes for greater offsets than horizontal bedding does. On the other hand, these shorter traveltimes result in higher apparent velocities, with which the hyperbola curvature must be corrected to obtain an optimum stack. If different dips occur at the same depth, different stacking maxima are then obtainable for the same traveltime T_0 - but at different velocities.

During site selection for the KTB the bedding velocities had to be determined as exactly as possible at the intersection between the KTB 8502 and DEKORP 85-4N profiles in the Oberpfalz region, and subsequently all structures had to be identified correctly. For this purpose, special velocity analyses were done for each of the two key zones of the sections. For these analyses 200 CMPs (KTB 8502) and 400 CMPs (DEK 85-4N) were used for the test stacks with a set of constant velocities.

The results for two velocity values selected for the problem mentioned above are shown in Fig. 10. Figure 10a presents for the core zone of profile KTB 8502 the result of a trial stack with a velocity of 5050 m/s, and Fig. 10b the result of a trial stack with a velocity of 7200 m/s. It can clearly be recognized that the stack with the lower velocity accentuates mostly horizontal elements, whereas the much higher velocity brings out an inclined horizon.

Presuming that other influences can be ignored and that dipping strata are the main reason for this effect, the lower velocity could serve as the mean true velocity. An apparent dip of ca. 32° for the event between 0.5 and 2.5 s can be taken from the section. However, the true dip may be obtained directly from the profile after migration and conversion to a 1:1 depth section. In this case a true dip of 40° can be calculated with the information on hand.

Because the angle of dip α influences the velocity by $1/\cos(\alpha)$, a stacking velocity of 7000 m/s is obtained for the inclined horizon, which is confirmed by the stacking result with the high velocity.

Such model calculations are naturally to be used only with great caution. For example, it must be made sure that a seismic element identified as being horizontal does not belong to an inclined, geological element with a strike parallel to the profile. At the same time, in 2-D seismics, bedding does not always strike normal to the profile, so that only an apparent angle of dip may be obtained.

To deal with these problems, which occur often particularly in deep seismics, the velocity analyses for the final stacks of two profiles in which steeply inclined bedding was of interest (DEKORP 85-4N, KTB 8514) were evaluated in a way that the horizontal elements could be accentuated in a first stack and the inclined elements in a second.

Fig. 9.a Raw stack of Oberpfalz line KTB 8501 based on conventional velocity analyses

Fig. 9.b Final stack of Fig. 9a with improvements obtained by combined velocity analysis; the analyses of Figs. 7a, c, d, e, and 8b lie in the range of locations 1430-1510 (km 34.3-40.7)

SW
NO
0.000
1.000
2.000
3.000
4.000
5.000
6.000
7.000
8.000
9.000
10.000
11.000
12.000

a

SW
NO
0.000
1.000
2.000
3.000
4.000
5.000
6.000
7.000
8.000
9.000
10.000
11.000
12.000

b

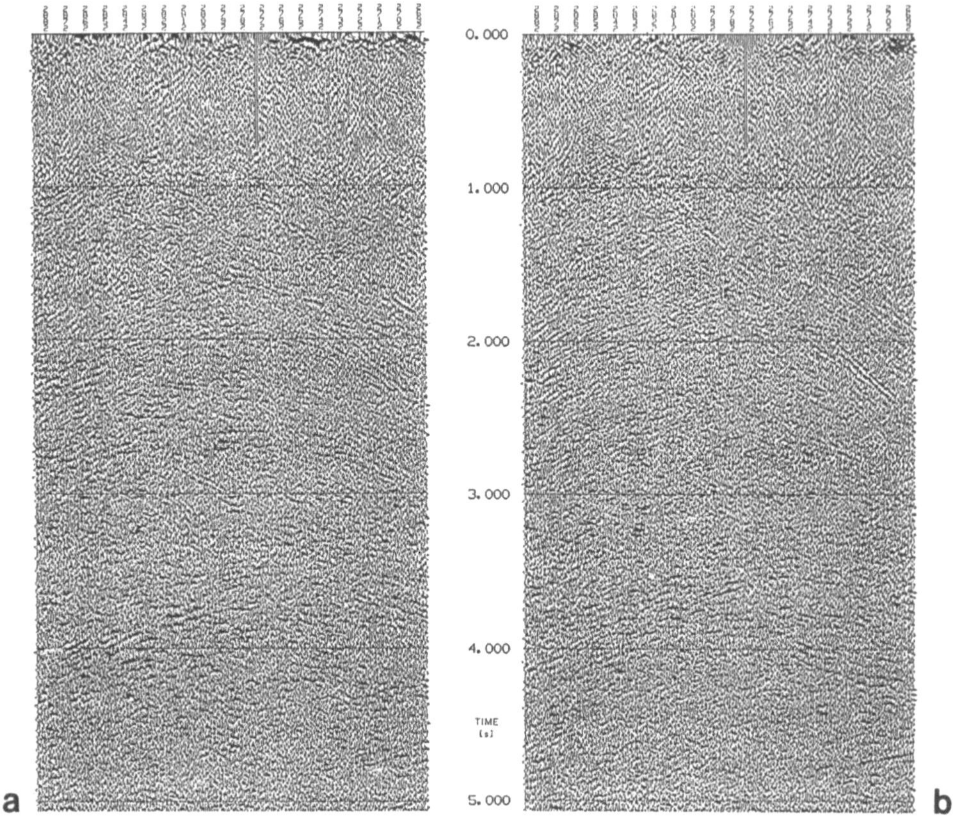

Fig. 10.a Constant-velocity stack with v_s = 5050 m/s for a part of Oberpfalz line KTB 8502 (km 27.2-35.5)

Fig. 10.b Same as Fig. 10a, but v_s = 7200 m/s

This makeshift way of producing several sections must be used until methods for considering all inclinations in one stacked section, which are currently being developed (dip moveout, automatic dynamic corrections), are ready for routine application in standard processing.

2.2 Offset-Restricted Stacking

To be able to identify horizontal velocity variations undoubtedly present within a 16-km-long spread, offset-restricted trial stacks with constant velocities were done for the central part of profile DEK 85-4N (locations 2050-2250), in addition to the velocity analyses already described. Only traces from user--selected offset ranges were considered in test stacks for a total of 30 constant velocities between 4000 and 7625 m/s. This was done for the following five offset ranges: two analyses with 40-fold coverage each over a partial spread of 8 km (i.e., offsets from -4 to +4 km and from -12 to -4 km), two analyses with 20-fold coverage each over a partial spread of 4 km (i.e., offsets from -4 to 0 km and from 0 to +4 km), and one analysis with 80-fold coverage over the total offset range from -12 to +4 km.

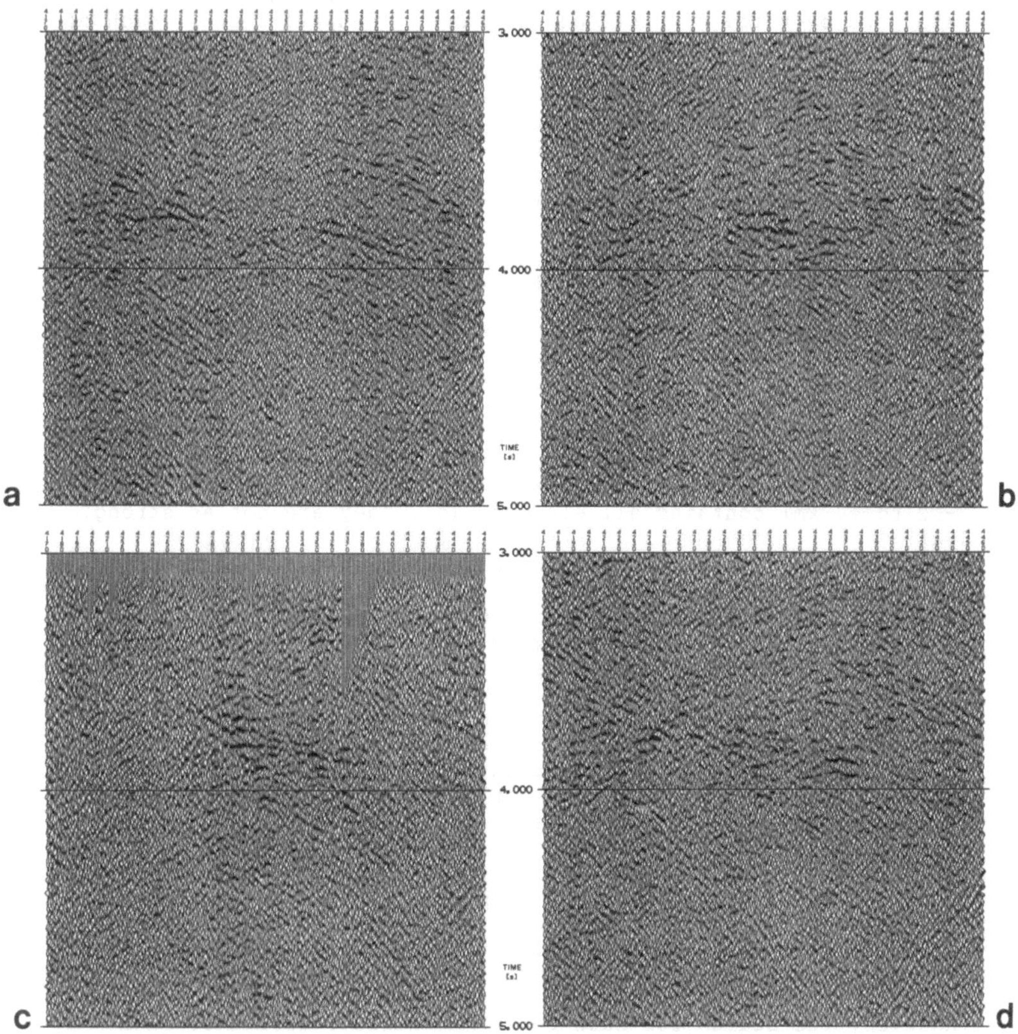

Fig. 11.a Offset-restricted stack of a segment of Oberpfalz line KTB 8502 (offset range of from -4 to 0 km)

Fig. 11.b Same as Fig. 11a, but offset range of from -8 to -4 km

Fig. 11.c Same as Fig. 11a, but offset range of from -12 to -8 km

Fig. 11.d Same as Fig. 11a, but stack of all traces (offset range of from -12 to +4 km)

It was found that similar stacking results could be attained over a wide velocity range for all offset ranges. Consequently, a detailed analysis differentiating lateral changes in velocity was not possible. This conclusion was supported by three other offset-restricted velocity analyses (-4 to +4 km, +4 to +12 km, -4 to +12 km) along the intersecting profile KTB 8502 (locations 1350-1450).

Because an evaluation of the velocity that takes into consideration the offset effects was neither practical nor expedient, all traces in the center of profile DEK 85-4N were corrected dynamically with a velocity model obtained from analyses over the complete offset range. Then ten stacks, each of different offset ranges, were done to try to recognize the effect of the offset alone, i.e., independent of velocity effects. The stacks were as follows: four stacks with 20-fold coverage over a partial spread of 4 km (0 to +4 km, -4 to 0 km, -8 to -4 km, and -12 to -8 km), three stacks with 20-fold coverage over 8 km (-4 to +4 km, -8 to 0 km, and -12 to -4 km), two stacks with 60-fold coverage over 12 km (-8 to +4 km, -12 to 0 km), and one stack over the total offset range from -12 to +4 km.

An important result of the comparison of several of these stacks (Fig. 11) is that all offset ranges can contribute to the final result to different extents, and that the signal/noise ratio does not necessarily have to improve with increasing coverage. Thus, the information obtained from the stack with only short (a) source-receiver distances (offsets of from -4 to 0 km) differs greatly from those with medium (b) or large (c) shot-geophone distances (offsets of from -8 to -4 km and -12 to -8 km). Accordingly, although the stack of all traces (offsets of from -12 to +4 km) contains all elements (d), they are not as strong.

Studies on unstacked CMP gathers also show that traveltime hyperbolas frequently begin or end abruptly with certain offsets, and they are uniformly continuous with other offsets. Accordingly, the greatest stacking amplitudes are found sometimes on traces with small offsets, sometimes with greater offsets, and sometimes with all offsets. Even in the latter case the information contained in the near-shot traces can differ from that in the traces with great offsets. Except for the general inhomogeneities in deep-seismic data previously described, there are no concrete and conclusive explanations for these phenomena. In the examples given here, small-scale, near-surface heterogeneities could have led to breaks in the horizon at 3.8 s in the stack only with small offsets. In contrast, rays for which geophone and vibrator were positioned outside the disturbed zone do not traverse the inhomogeneous zones at all (stack only with large offsets) and thus the information of the reflections from the deeper subsurface arrives undisturbed (undershooting).

However, the search for such possible causes requires so much time and yields such contradictory results that it is simply not practical in routine data processing. This is particularly true because the parameters of nearly all aspects of data processing (e.g., crooked-line sorting, eliminating disturbed traces, finding appropriate correction velocities, determining residual static corrections, selecting suitable migration methods) are strongly influenced by anything deviating from the theoretical concept of multiple coverage. Strictly speaking, if certain causes for deviations from the model assumptions have been recognized, many processing parameters (from the beginning) still have to be double-checked and, if need be, changed. In view of the aforementioned amount of data and processing time this is practically impossible even in an optimally equipped processing center.

Nevertheless, in practice, at the DPC, special processing methods - adapted to special situations - are used, e.g., stacks with various velocity-time models or, in cases of offset problems, stacks with certain shot-geophone distances. The entire profile KTB 8502 was also processed with restricted offset range (-4 to +4 km). Particularly the great variety of processing versions offers a solid basis for well-founded interpretations which - above and beyond the known geological conceptions - attempt to incorporate all information on the subsurface contained in the measured data.

Summarizing the problems on the offset dependency of amplitudes in CMP gathers, the following, extremely important conclusion can be drawn: A strong reflection in the zero-offset stacked section is generally an indication of large reflection coefficients for the subsurface. In contrast, the lack of distinct reflections does not necessarily indicate low reflectivity in the subsurface, but can also, for many reasons, be caused by not meeting the basic conditions of stacking (information of the same type and intensity on all traces of a CMP gather).

This is basically true for all branches of practical reflection seismics, but does appear to play a much more decisive role in deep seismics than in seismic prospecting. Therefore, this effect should never be neglected in any statement concerning the continuity of horizons and existence of transparent zones, and every geological interpretation that is essentially based on such an assumption must be double-checked carefully.

3.3 Improving the Signal/Noise Ratio

Second to stacking, frequency filtering is another important step in improving the signal/noise ratio. As already indicated in Section 2.5, profiles of great lengths (up to 250 km), which traverse quite different geological structures, and long recording times (up to 20 s), which reach vertically down into the mantle, necessitate frequent checking of the filtering parameters and require considerable spatial and temporal variation of these parameters.

The question of whether the signal signature and the related frequencies contained in the recorded seismic signals can be correlated with the changes in the surface geology is of special interest. The KTB Black Forest lines 8402 and 8403 will be used to illustrate this.

A Figure in another part of this book[1] shows the location of those profiles and an overview of the surface geology of the Black Forest. KTB 8402 is the NW-SE-oriented profile. To investigate the above question this area can be roughly divided into two parts: Gneisses dominate in the north, the Triberg granite crops out in the south. KTB 8403 traverses the Black Forest from west to east. In the east there is a 200- to 300-m--thick cover of Buntsandstein and Muschelkalk. Towards the west, the profile crosses the boundary to the Triberg granite, and finally the gneisses of profile KTB 8402 are encountered.

Instead of regarding the entire area, only the frequencies of the signals recorded for the Buntsandstein are compared to those recorded for the gneiss. To do this several filter tests were performed and frequency spectra were calculated for different time windows.

1 Fig. 1 in Lüschen, E. et al., Near Vertical and Wide-Angle Seismic Surveys in the Black Forest, SW Germany

Figure 12a shows such a filter test for profile KTB 8403 between locations 1102 and 1121 in the area of Buntsandstein cover. At the DPC the filter test in the various profile segments is normally calculated for a range of 51 CMPs. To be able to present the results in a clear way only a representative segment of 20 CMPs was chosen.

The left-most panel shows the unfiltered data, the following panels were filtered with various bandpasses shifted over the frequency range. The transmission width is 5 Hz in each case, and the corresponding values for each panel are given at the top. The possible frequency range is 12-48 Hz due to the sweep used.

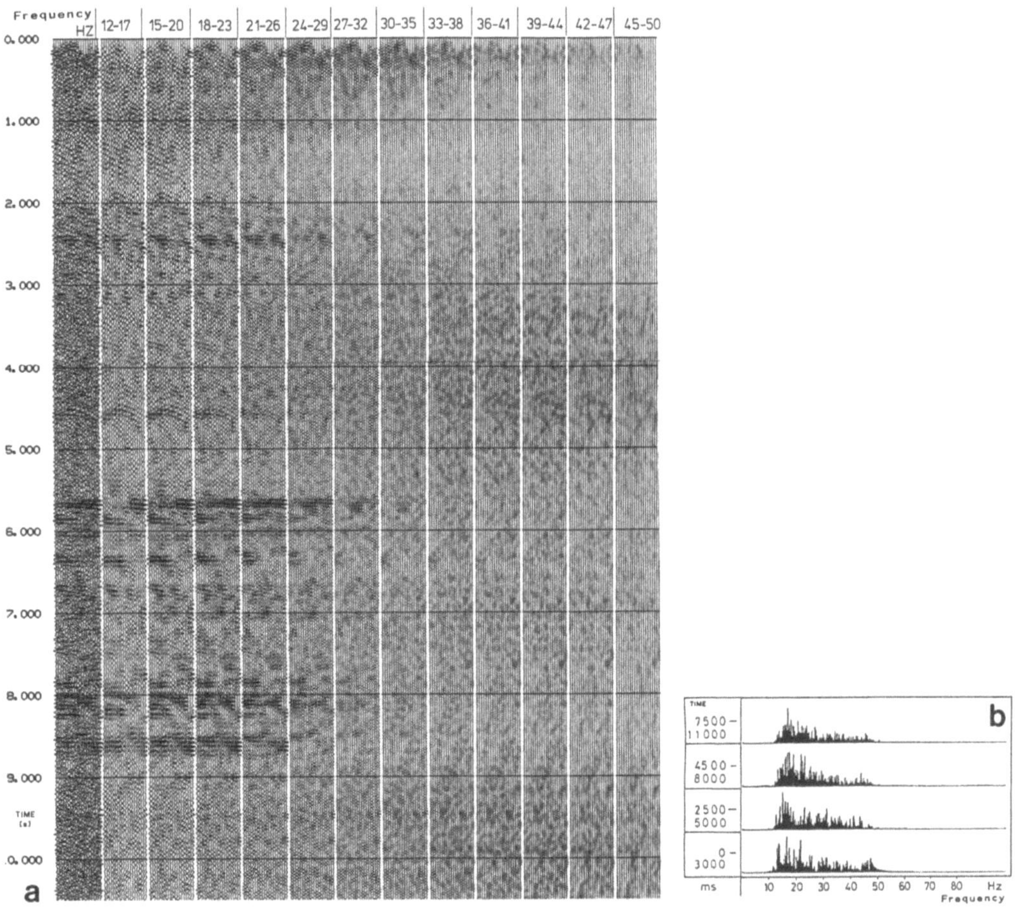

Fig. 12.a Filter test on a part of Black Forest line KTB 8403 (km 8.2-9.7), Buntsandstein

Fig. 12.b Frequency spectra in different time windows for Fig. 12a

Supplementary information is given by the frequency spectra calculated for four windows (Fig. 12b). It must be noted that each window is normalized to its maximum value separately. Consequently, direct comparison of amplitudes from window to window is not possible. Thus, only the information within each individual window can be evaluated quantitatively.

Distinct signal frequencies of up to 25 Hz at the most can be taken from the amplitude spectra. The rest is so homogeneous that it can be regarded as noise. The main frequencies contributing the most to the signal shape lie around 18 Hz. The filter test confirms the very low frequency content, showing main frequencies at 18 to 20 Hz. Thus, the Buntsandstein cover must be very absorbent.

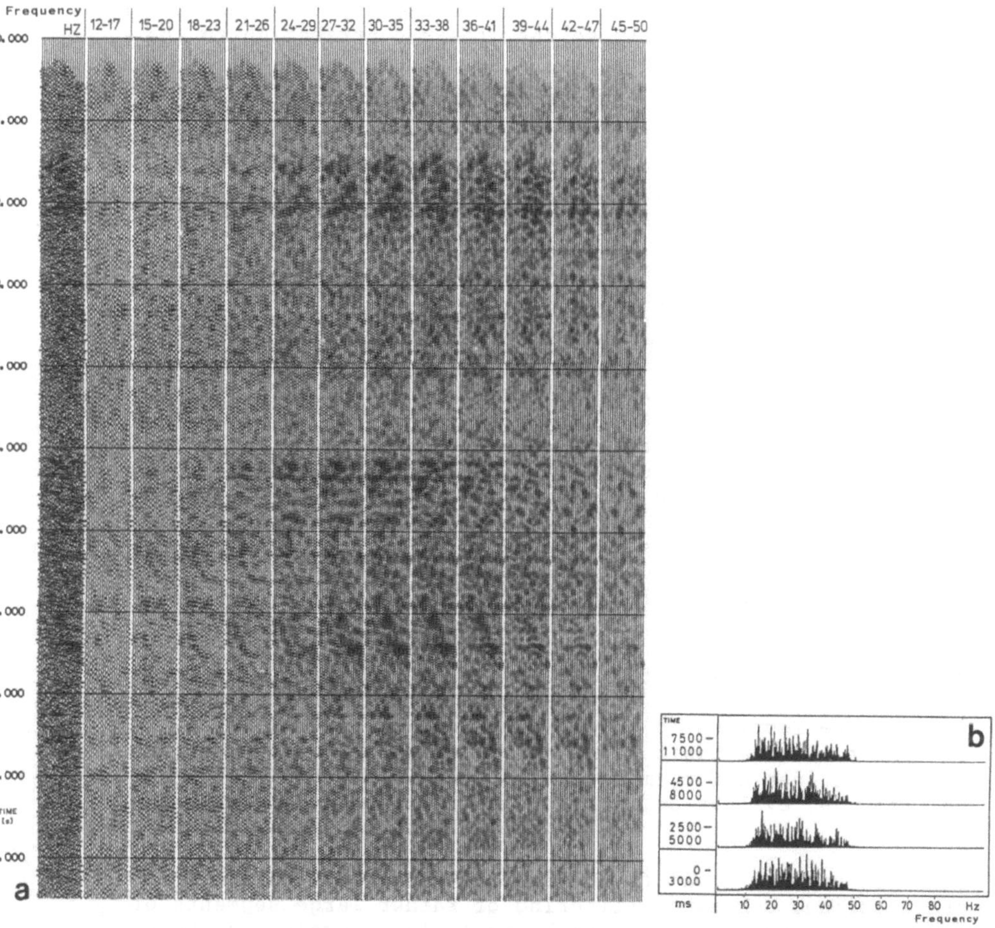

Fig. 13.a Filter test on a part of Black Forest line KTB 8402 (km 10.5-12), gneiss

Fig. 13.b Frequency spectra in different time windows for Fig. 13a

In contrast, Fig. 13a and b shows the relationships for gneiss (KTB 8402, locations 1131-1150). Useable portions of the signals from 15 to 40 Hz can be taken from the frequency spectra. The filter test confirms this as well. Furthermore, it allows more exact statements to be made about the main frequencies, which lie between 25 and 35 Hz, and thus are clearly higher than in the first case. Hence, these higher frequencies have been absorbed to a lesser extent in the gneiss. When filtering, such great differences in frequency content must be considered if all information contained in a stacked section is to be worked out clearly.

A further processing step, which contributes essentially to improving the signal/noise ratio and has proved its value for deep seismics, is coherency filtering, the theoretical background of which was already treated in Section 2.5.

The question of what significance a coherency-filtered section has in comparison with an only stacked section will now be dealt with. For this purpose Fig. 14a shows profile KTB 8504 which in part (a) is stacked and frequency-filtered and in part (b) is additionally coherency-filtered. The coherency trace and stacked trace are weighted equally so that their contribution to the final result is the same.

A direct comparison clearly shows that many reflections can be more easily recognized in the coherency-filtered section, whereas they can only be presumed to exist in the noncoherency-filtered stack. As to the question of what value a coherency-filtered section has in geological interpretation, the differing aims of seismic prospecting and deep seismics must be discussed. In seismics for oil and gas exploration, small-scale faults or phase changes in reflections, for example, are of especially great significance. At the same time the profiles are limited to a few kilometers and are often laid for answering very specific questions.

In contrast, an overview of profile lengths of up to 250 km involving traveltimes of 12 to 20 s is required for evaluating a deep-seismic section. At the beginning of an interpretation, the primary attention is directed at continuous reflections, large--scale fault zones, and geological boundaries, e.g., those between the Saxothuringian and Moldanubian zones in the area of the KTB site in the Oberpfalz region. Questions concerning such large areas can be answered much more easily and clearly using a coherency-filtered section, as demonstrated in the depicted example (Fig. 14a, b).

Moreover, coherency filtering permits the interpreter who is looking for small-scale structures to carry over persisting coherent reflections to the noncoherency-filtered section. This is done to form a basic framework of good reflections for the subsequent search for small-scale structures. This means that a well-founded evaluation is much easier with the help of both sections.

There is another area of application: For a publication or oral presentation, coherency filtering of rather large segments of sections can be useful for emphasizing large-scale structures, which are described, by clear and contrasty illustrations. Line drawings alone or already evaluated stacks are mostly used, which robs the critical reader or listener of the chance to form his own opinion. Even with coherency-filtered sections that are greatly reduced in size, diagrams can be produced in which the essential structures are still recognizable, allowing individual interpretation.

3.4 Processing with True Amplitudes

In addition to conventional stacks with automatic, time-dependent amplitude control, processing with the best possible true amplitudes (TA processing) was also done for profile DEK 84-2S and for three of the Black Forest profiles (KTB 8401, 8402, and 8403). This chapter will introduce the problems of this type of processing, the characteristic differences, and the respective advantages and disadvantages of both processing methods.

The differences in energy between the first arrivals and the deep reflections are extremely great in a seismic-reflection field record. With good ground coupling, they have a dynamic range of up to 100 dB, corresponding to 100,000-fold smaller amplitudes at the end of the record than at the beginning. In addition to this decrease in energy with traveltime, there are also spatial variations in energy, above all in land seismics. On the one hand, physical and geological factors are responsible for these differences in amplitudes (e.g., spherical wave propagation, absorption, reflection and transmission, diffraction, noise, etc.). On the other hand, there are also factors that depend on the recording situation, such as differing shot charges and borehole depths in explosion seismics, differences in vibrator coupling and surface conditions in Vibroseis, and in both methods the differences in geophone coupling.

All of these influences must be taken into account in well--planned, data-dependent processing when a migratable and interpretable stacked section is the desired product. There are three main methods that can be used for correcting amplitudes, i.e., multiplication with a constant factor, the analytical or programmed gain function, and automatic gain control (AGC) with or without variable window length and window shift. These methods are listed according to their efficiency with regard to amplitude correction; i.e., AGC changes the relative amplitudes within a trace the most, the gain curve changes them very evenly, and multiplication with a constant does not change them at all.

As already described in the foregoing, in conventional seismic data processing an AGC is nearly always done following an empirically determined analytical gain function, eliminating divergence and absorption effects. This is done to intensify all weak amplitudes and to attenuate all stronger ones. In this way a legible, balanced seismogram image is produced in which all correlatable reflections are emphasized equally, which is the advantage. The disadvantage, however, is the loss of the true reflection amplitudes.

Fig. 14.a Stack of Oberpfalz profile KTB 8504 without coherency filtering

Fig. 14.b Same as Fig. 14a, but with coherency filtering

a

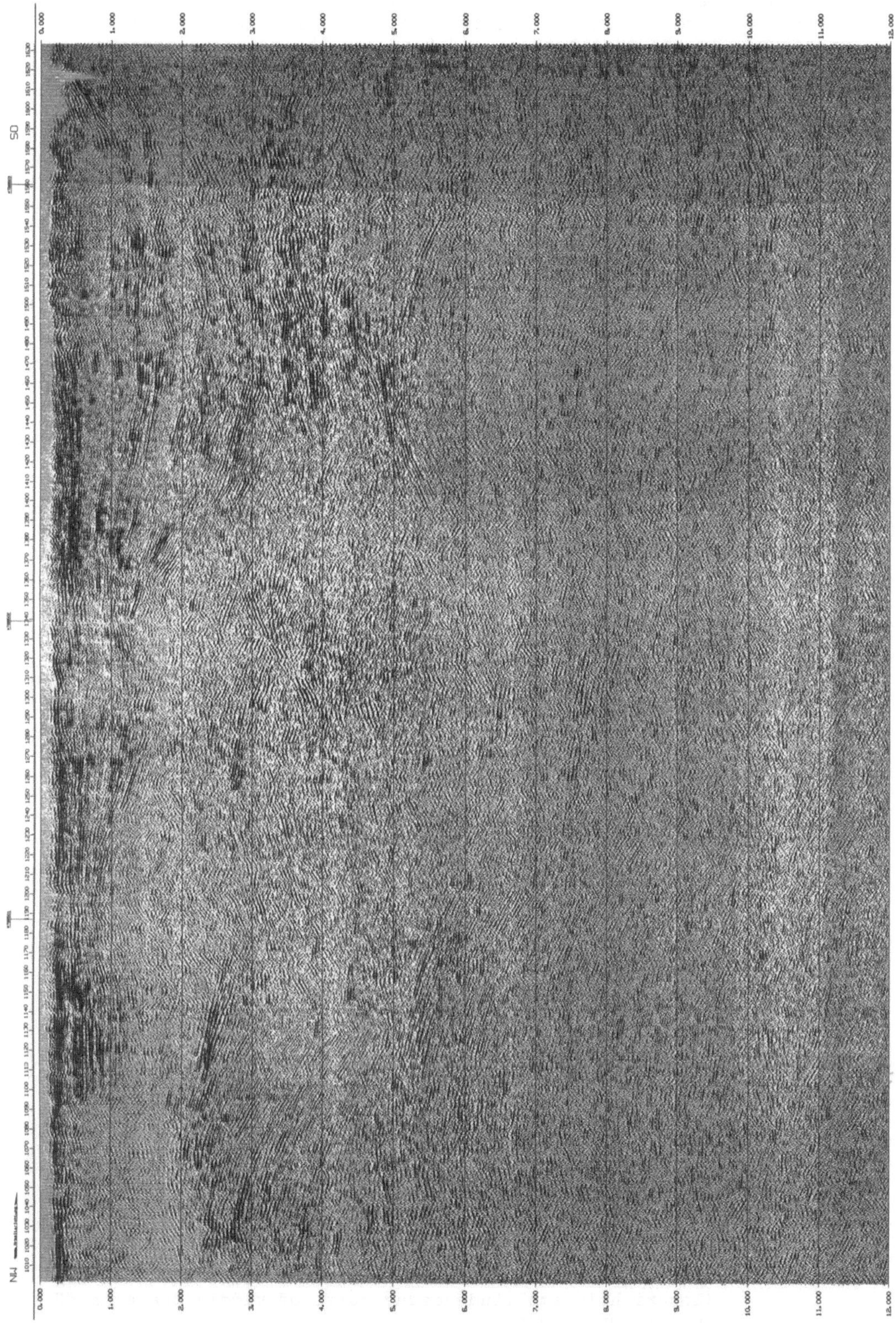

b

For this reason, at the DPC a method has been developed, after numerous attempts, which incorporates all necessary amplitude corrections and yet leaves the reflectivity information itself unchanged. As the fundamental condition for the stacking process, it must be ensured with this method that all traces of a CMP gather are equivalent, so that some traces do not contribute more to the stacking amplitude than others - a great problem for all TA techniques.

In the following discussion, the effects of the various necessary processes will be elucidated using single shots so that the ensuing problems and relationships can be more clearly explained. Energy analyses of common-shot point data (200 regularly spaced, sequential traces from one source) can be evaluated much more reliably than those of CMP-sorted data (ca. 80 irregularly spaced, sequential traces from various sources). The amplitude correction can also be done after CMP sorting, because only single-trace processes are involved here. Thus, the flowchart in Fig. 2 is still valid, apart from the omission of AGC.

Figure 15a shows two adjacent, unprocessed Vibroseis field records of profile KTB 8401 (FF 1036, 1037; locations 2215, 2216) directly after demultiplexing. Their differing total energies and the occurrence of zones within the spread with more or less high amplitudes and several clipped traces are immediately obvious. These effects are presumably due to differing source and receiver couplings, field and surface conditions, and disturbed locations or geophones.

The first process is a so-called horizontal trace equalization. For this, the RMS energy for each trace is first determined for a time window that is not too small and in which a relatively constant energy level is expected. Then, each trace is multiplied by an individual constant factor so that the RMS energy in the user-defined time window is then the same for all traces. In this case, the range of 5 - 10 s was selected for normalization, i.e., beyond the greatest decrease in energy and the influence of any surface waves. Now, the two shots can be compared with one another for the first time because source and receiver influences were compensated for, as far as possible, without changing the relative amplitudes within each trace (Fig. 15b).

In addition, the general amplitude decay of the central trace of the record at the left is shown logarithmically with respect to time. It is strongest in the first three seconds, i.e., ca. -60 dB, and is thereafter relatively constant. The causes for this were described in detail in Section 2.2. Figure 15c shows the shot record on the left of Fig. 15b, but after traveltime--proportional, spherical divergence correction; in the lower time range the amplitude level rises by about 15 dB.

Fig. 15.a Two common shotpoint gathers of Black Forest profile KTB 8401 (FF 1036, 1037; locations 2215, 2216; km 97.2)

Fig. 15.b Same as Fig. 15a, but with horizontal trace equalization (logarithmic amplitude-decay curve of middle trace in dB in the upper left)

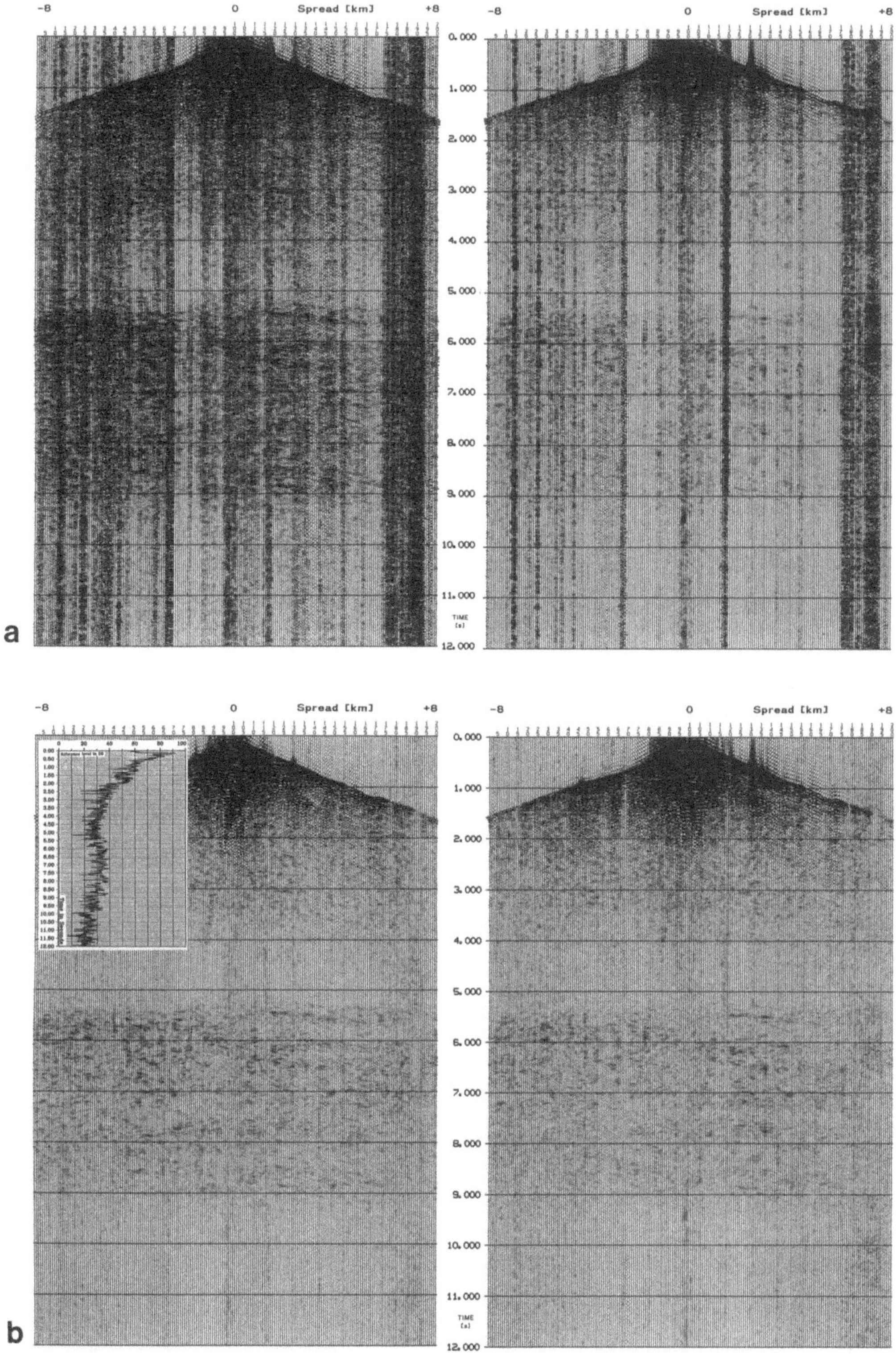
-8
0
Spread [km]
+8
0.000
1.000
2.000
3.000
4.000
5.000
6.000
7.000
8.000
9.000
10.000
11.000
12.000
TIME [s]
a
b

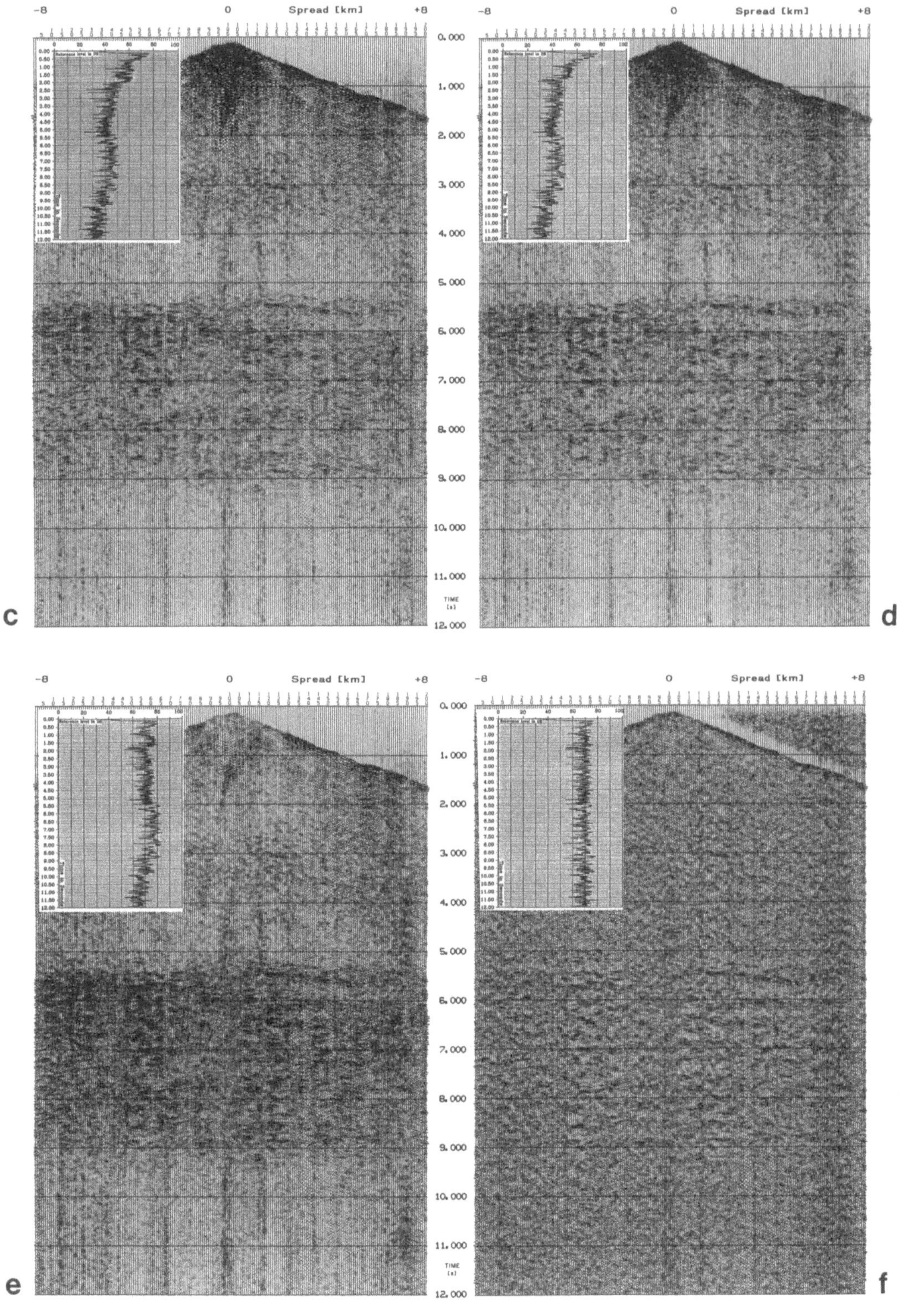
-8
0
Spread [km]
+8
0.000
1.000
2.000
3.000
4.000
5.000
6.000
7.000
8.000
9.000
10.000
11.000
TIME [s]
12.000
c
d
e
f

Before anything can be said about the amounts of reflection or absorption, any high-energy surface waves should always be eliminated with appropriate high-pass filtering (Fig. 15d). This portion amounts to ca. 5 dB, and the protrusion in the energy curve at ca. 1.7 s is eliminated. The reflections remain unchanged, but are now seen much more prominently.

The residual amplitude corrections of ca. 30 dB now remaining can be interpreted as an absorption effect. There are two methods for eliminating this effect, i.e., an analytical and a programmed gain function. With the former the attempt is made to fit a curve with the form $\exp(\alpha * T)$ to the data. With the latter the desired correction curve is digitized as time/dB pairs. The latter method was employed for the TA processing of DEK 84-2S, the former for the Black Forest profiles. After determining an appropriate value of α the record has finally been properly prepared, as shown in Fig. 15e, for TA stacking, because the true reflection amplitudes have, for the most part, remained unchanged. The energy jump at 5.3 s, where the sudden appearance of increased reflectivity in the lower crust is very obvious, makes this particularly evident. It amounts to about 6 dB (corresponding to a doubling of the amplitudes) both before and after the amplitude correction.

For comparison, the picture in Fig. 15f was produced by conventional AGC processing. The equalizing effects of the automatic regulation ensure that the strong reflections are attenuated and the weak ones are intensified. When this is done, the energy jump is no longer present and the dynamics are collectively smoothed out; weak reflections are more easily recognizable because they are brought up to the level of the strong ones. But AGC cannot, of course, distinguish between useful signals and noise. Therefore, "automatic noise control" can also result as an undesirable side-effect.

TA processing is characterized by the exclusive application of smooth correction curves to prevent any irregular distortion of the relative reflection amplitudes [compare the decrease in energy to be corrected (A) and the correction curve for it (B) in Fig. 16a]. AGC, in contrast, produces a very irregular correction curve (Fig. 16b) so that the energy values can be brought to a very uniform level, instead of retaining the surely more correct energy jump.

Now that the ways the individual processes work have been explained using single records, a few comments must still be made on the efforts involved in TA stacking. Whereas AGC allows the traces to be corrected, with practically no checking, to an uniform amplitude level for stacking, all intermediate results in TA stacking must

---

Fig. 15.c Left CSP from Fig. 15b, but additionally with spherical divergence correction

Fig. 15.d Same as Fig. 15c, but additionally with high-pass filtering

Fig. 15.e Same as Fig. 15d, but additionally with absorption correction (corresponding to the preparation for a true-amplitude processing)

Fig. 15.f Same as Fig. 15e, but with automatic gain control (corresponding to the preparation for a conventional stack)

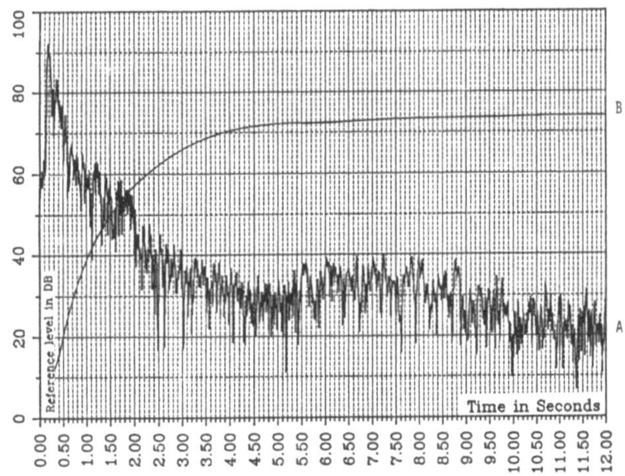

Fig. 16.a Amplitude-decay curve from Fig. 15b (A) and correction curve determined by gain analysis (B) for true-amplitude processing

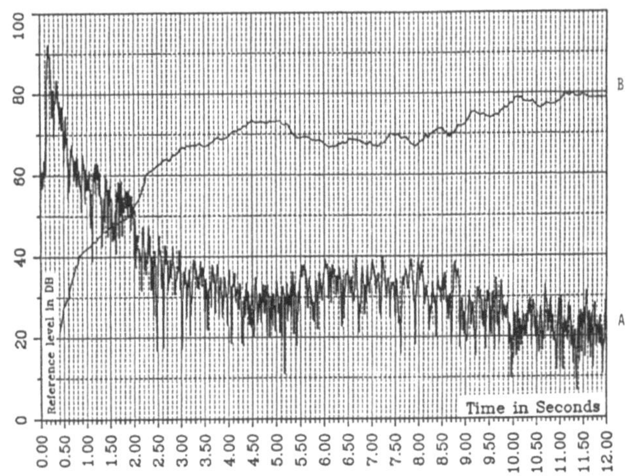

Fig. 16.b Same as Fig. 16a, but correction curve determined automatically with AGC for conventional processing

be checked frequently because the parameters (frequency contents, surface waves, absorption, etc.) vary greatly both horizontally and vertically. Energy analyses for closely spaced, representative section ranges form the basis for the necessary spatial and temporal interpolations of the correction curves.

Two Figures in another part of this book[2] show a comparison of a conventional AGC stack with a TA stack for a part of the line KTB 8401 (nearly the same CMP range as for the single records in

2 Fig. 3 and 4 in Lüschen, E. et al., Near Vertical and Wide-Angle Seismic Surveys in the Black Forest, SW Germany

Fig. 15). The strong reflections from the laminated lower crust and the "Haslach diffractor" are much more prominent in the TA version, whereas in the AGC version even the weak reflections in the first seconds are visible, although the contrast has decreased substantially.

Another disadvantage of TA stacking is also evident: the residual noise in the uppermost time ranges. It is impossible to check all traces to be stacked due to their sheer numbers, i.e., several hundred thousand. Only by averaging the conditions within certain zones and interpolating between analyses is it possible to determine all the needed processing parameters. Thus it happens repeatedly that individual traces dominate in the first seconds of the stack where the coverage is low but the decrease in energy is the greatest.

Thus, both processing methods have their advantages and disadvantages. The AGC version is better suited for recognizing and following up subsurface structures because it depicts all reflecting horizons uniformly regardless of their true reflection amplitudes. The TA version, on the other hand, is more appropriate roughly estimating reflection and absorption parameters. Therefore, simultaneous and comparative evaluation with both versions offers additional, valuable help in geological interpretation.

4. Summary and Outlook

A detailed summary of the deep-seismic data processing concept for the DEKORP/KTB reflection lines was presented in the foregoing chapters. A generalized program flowchart ("from the field tape to the evaluatable seismic section") served as a guide for introducing all necessary processing steps and their parameters, and their respective position and significance within the total scheme were explained. Yet, this scheme cannot be regarded to be rigid and must be frequently modified depending on the data (omitting, adding, or exchanging processing steps). The numerous diagrams should aid in illustrating the effects.

In recent years, experience and routine have been gained at the DPC in processing deep-seismic data. However, this should not obscure the fact that data of this type have numerous typical characteristics which require particular care when selecting the parameters for most processing steps. Profile lengths of up to 250 km over very diverse geological regions and recording times of up to 20 s down into the earth's mantle necessitate constant double-checking and well-planned spatial and temporal variation and interpolation of all processing parameters. On the other hand, both the recording- and subsurface-dependent inhomogeneities in the data are typical of nearly all lines.

Special problems were singled out, particularly in Section 3, with which it was shown how consistent quality in the results can be ensured through conscious application of analysis techniques and two-dimensional variation of the processing parameters.

Furthermore, new processes are continually being tested for their applicability in deep seismics and, if need be, modified or newly developed (frequency-wave number filtering, zero phase deconvolution, slant stack, automatic line drawing, dip moveout, true amplitude migration, etc.). When necessary, three-dimensionality is also considered from the very beginning of the development of a process so that full advantage can be taken of it for the planned KTB 3-D seismic surveys in the Oberpfalz region in 1989.

Electrical Resistivity Studies in the Vicinity of the KTB Drill Site, Oberpfalz

Volker Haak

Institute of Meteorology and Geophysics of the J.-W. Goethe-University, Feldbergstr. 45, 6000 Frankfurt, FRG

1 FOREWORD

Measurements of the electrical resistivity distribution are becoming more and more popular in studies of the Earth's crust. Nevertheless, causes and interpretations still often remain a matter of speculation. The deep drilling project therefore offers a unique possibility to answer problems as to the existence in particular of low resistivity zones in the crust. It was the aim of the measurements described here to "forecast" the existence of such low resistivity zones in the vicinity of the drilling site. Since the final location of the drilling site was decided after most of the measurements had been concluded, it turned out that no electrical depth sounding was done at or even within a few kilometers of the drilling site. Nevertheless these results are, hopefully, representative for the drilling site.

This report is made by the spokesman for a large group of scientists from many universities. The following persons and universities have performed the measurements and the data interpretation presented here:

<u>Aachen</u>, RWTH: G. Voigt, C. Wendel, J. Wohlenberg;
<u>Berlin</u>, Freie Universität: Th. Blümecke, V. Haak, M. Martinez, U. Seidel, V. Rath, P. Rüdinger;
<u>Berlin</u>, Technische Universität: H. Brasse, F. Leonhardt;
<u>Braunschweig</u>, Technische Universität: Ch. Drews, H.-M. Maurer;
<u>Braunschweig</u>, GEOMETRA: K. Bahr, U. Czarkowski, U. Jensen;
<u>Köln</u>, Universität: H.N. Eilenz, T. Hanstein, C.H. Kalle, K.-M. Strack;
<u>München</u>, Universität: A. Berktold, H. Dittus, U. Teufel.

2 INTRODUCTION

The electrical resistivity of rocks of the continental crust varies by more than six orders of magnitude and may thus serve to distinguish the various types of rocks. These quite dramatic changes are caused by

- the content of fluids
- the concentration of solutes
- the content of carbon.

It is necessary for these conductivity-enhancing factors to have a high or almost complete electrical connectivity in order to produce a measurable effect.

Other important factors controlling the electrical resistivity of the crust are:

- temperature above 500 °C
- melt.

Since these two factors do not play a dominant role in a a passive continental crust, electrical conductivity anomalies may well be due to anormal abundances of carbon, fluids, or solutes. For a long time these factors did not seem to be interesting aspects for crustal studies. However, many recent studies have shown conspicuous aspects of the distribution of the electrical conductivity of the continental crust which make it necessary to reconsider the role that carbon and water play in the crust:

- the low resistivities of the lower crust
- low resistivity "layers" in the upper crust.

These properties seem to exist independently of the type of crust, whether tectonically active or passive, whether young or old, but with various structures from one crustal type to another. An early presentation of the crustal resistivity was given by Angenheister (1962, presented by Kertz (1964), Fig. 6), who summarized the general "agreement" that the resistivity of the continental crust is high, at least 1000 ohm m. Blohm et al.(1977) and Van Zijl et al. (1977) were among the first who detected definitively the low resistivities even of the old (shield) crust. Jödicke et al. (1983) could even follow the dipping course of such low resistivity layers in the Rhenish massif for more than 100 kilometers. This latter study seems to approach very closely to a possible explanation for the cause or the nature of the low resistivities, since there could in this case exist a correlation with the tectonics pointing at graphite or fluids. Possibly graphite could play an important role, but still Shankland and Anders'(1983) statistical investigation and their conclusion that fluids may be the cause for the decrease in resistivity must be kept in mind, in particular when considering the obviously high correlation between low resistivity and high seismic reflectivity zones in the lower continental crust (see Thompson et al. 1983). Thus, although the actual cause of these low resistivity layers is still a matter of speculation it seems already obvious that the low resistivities should signify distinct stages in the evolution of the continental crust:

- the low resistivity of shear zones
- a vanishing effective pressure which opens pathways for fluids, thus lowering the resistivities
- deeply buried paleo-reliefs with deposits of biogenic black shales.

The important point for future research is to understand which conductivity-increasing factors are inherent in tectonic and petrologic processes. The consideration of such models is essential for the understanding and application of all geoelectrical models. Haak and Hutton (1987) and Jones (1987) have discussed such a correlation of electrical structures and tectonic/petrographic models. It is a most important aspect of the Continental Deep Drilling Projects (KTB) to drill through such a major high conductivity structure in order to understand its nature. The interpretation of conductivity structures in continental rift systems becomes even more complicated, since temperature and melts can also play an important role in increasing the electrical conductivity.

3 EXPERIMENTAL TECHNIQUES

Six various electromagnetic methods were applied in the course of the presite study. Since the non-electrician can hardly see the differences between the methods, and thus might believe that there exists gross redundancy, the different methods with their particular sensitivities to different aspects of the electrical conductivity distribution will be briefly outlined. Indeed, as no perfect electrical method exists , it is necessary to apply as many methods as available and also to incorporate new methods (e.g. LOTEM), even if their response to great depths of investigation in a high resistivity crystalline crust is not yet completely known.

Magnetotellurics (MT) measures the natural, time-varying electrical and magnetic field in the period range 10-10,000 s (corresponding to middle crust to upper mantle). Processed quantities are the apparent resistivities and phase differences as function of period. At each site four apparent resistivities and four phases are obtained. The interpretation is simple in the case of 1-D resistivity, where all four values are equal. In the Oberpfalz the actually observed 3-D resistivity structure prohibits any simple interpretation. The greatest disadvantage of MT is the static shift of the apparent resistivity curve, claiming too great depths and resistivities for the crust.

Medium-magnetotelluric MMT, same as MT, but measures in the range 100 Hz to 100 s (upper crust).

Audio-magnetotelluric AMT: same as MT, but measures in the range 1000 Hz to 3 Hz (uppermost crust, to about 5 km).

Controlled source audio magnetotelluric (CSAMT = AAMT for active audiomagnetotelluric) measures single frequency sinusoidal signals artificially induced by a current dipole successively in the period range from 0.001 to 10 s. The advantage is the effective minimization of technical noise by stacking. The disadvantage is the limitation of the depth of penetration by the "Near Field Effect" observable by the 45° increase of the apparent resistivities with period and the corresponding zero phase. The depth of investigation is a little less than in AMT. The effect of strong lateral variations of the electrical resistivity on the apparent resistivities and phases has not yet been considered in model interpretations.

Time domain electromagnetic method (LOTEM) measures the transient decay of the vertical magnetic field component induced by a current dipole, as in the CSAMT method. Advantages are the highly effective minimization of noise by selected stacking and digital signal processing analogous to seismics and in particular the nonexisting or small static shift effect of the apparent resistivities, as suffered by all MT methods. The influence of strong lateral variations of the electrical resistivities on the measured quantities is also not yet fully understood and may affect the 1-D model interpretations. LOTEM is a very new technique, with little experience as yet in studies of the deeper crust of high electrical resistivity. The survey in the Oberpfalz was actually the first one where large depth penetration (10 km) was obtained with a "low power" source (60 A).

Schlumberger direct current method (DCR) is the classical method for measuring the electrical resistivity particular in sedimentary region. The application in crystalline environment with strong lateral changes of electrical resistivity poses serious problems in interpretation, in particular for depths greater than a few hundreds of meters.

Beside the differences in the experimental techniques and differences in the depth range of investigation, the methods have different sensitivities to the vertical resistivity structure. While the natural methods use horizontally directed current systems, the active methods use also vertically directed current components. In the case of anisotropic or small-scale inhomogeneous resistivity structures, these differences may lead to different resistivities and their corresponding depths. Of particular interest is therefore the joint model interpretation of the various measurements at one site. This will be a task for the future.

4 MEASUREMENTS OF THE ELECTRICAL CONDUCTIVITY IN THE OBERPFALZ AREA

Three periods of measurement have taken place in the whole Oberpfalz area according to the time, place and methods involved. They are thus of varying importance to the KTB project.

In the first period before about March 1986, MT, AAMT, and DCR measurements were performed on a "large scale" with widely spaced stations, mostly east of the Franconian Line. These measurements have not direct relevance to the geological structure of the drilling area except for a very few measurements, which will be discussed later.

In the second period after March 1986, very intensive field measurements (AMT, AAMT, MMT, LOTEM) started in the area around the then still unknown place which was later to become the actual drilling site (Fig. 1). These measurements and the successive interpretations became more important for the selection of the drilling site. They will be discussed in this report.

After the final decision of the drilling site a third period was started in a square 1 x 1 km immediately around the pre- and main drilling site. The data of these measurements have not yet been processed. The actual drilling site is represented in Fig. 1 by the black diamond.

→

Fig. 1. Position map with explanation: Geological map with the position of the sites where the electrical resistivity has been measured. The diamond marks the position of the drilling site

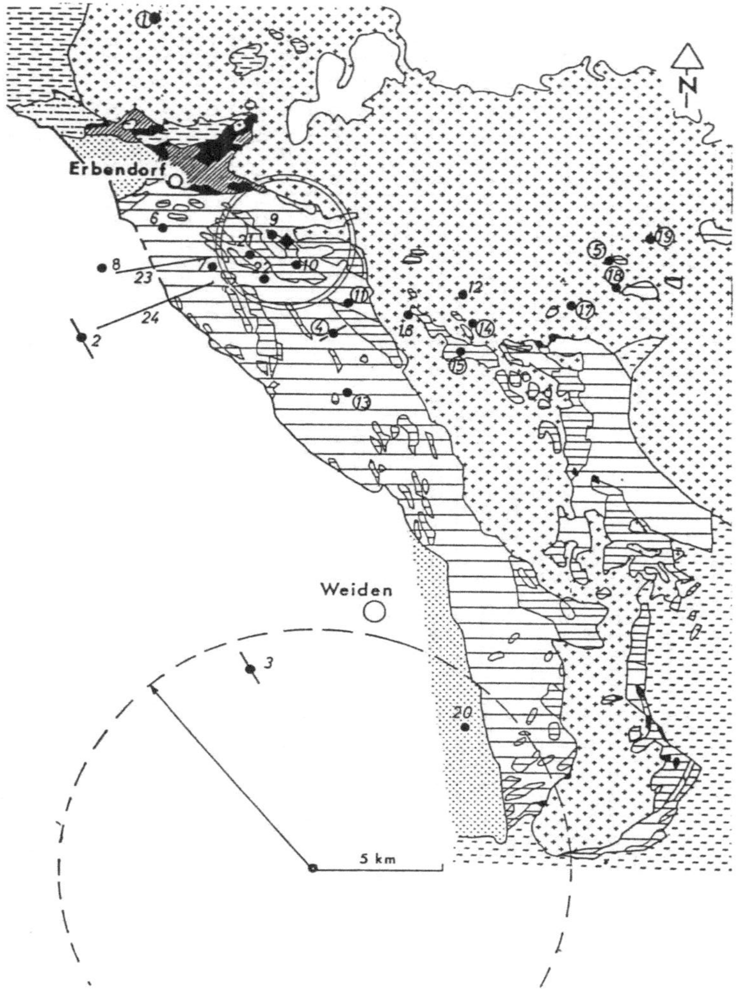

MESSPUNKTE ZUR ELEKTRISCHEN LEITFÄHIGKEIT IN DER UMGEBUNG DES LOKATIONS-VORSCHLAGES OBERPFALZ

Nr.	Projekt-Bezeichn.	........METHODEN					....
01	WAL		MMT	AMT	AAMT		
02	PRE, WAE	MT		AMT			
03	TIP	MT		AMT			
04	PUE	MT	MMT	AMT	AAMT	LOTEM	DCR
05	BEI	MT	MMT	AMT	AAMT	LOTEM	
06	N4			AMT	AAMT		DCR
07	N5, Bühl			AMT	AAMT	LOTEM	DCR
08	N6, BUE7				AAMT		DCR
09	6					LOTEM	
10	5					LOTEM	
11	4					LOTEM	
12	BAU				AAMT		
13	21					LOTEM	
14	16					LOTEM	
15	18					LOTEM	
16	19					LOTEM	
17	15					LOTEM	
18	9					LOTEM	
19	8					LOTEM	
20	S1			AMT	AAMT		DCR
21	GEI			AMT			
22	WIL			AMT			
23	PROFIL BUE			AMT			
23	PROFIL MUE			AMT			

This report tries to outline the most important results of several field campaigns in the second period. They were performed by working groups of various universities. In most cases the processing and interpretation of the data are part of university theses, some of which are still in progress.

5 MAGNETOTELLURIC STUDIES

The MT "survey" by the Free University of Berlin started in 1983 and covered the whole Oberpfalz area, including Saxothuringicum and Moldanubicum, with about 20 sites (Blümecke, 1984). Only three of these stations are situated in the area presented by Fig. 1. The interpretation of these 20 MT stations is in the present state still problematic, since the obviously existing but still unknown static shifts and the large scatter of the processed apparent resistivities and phases prohibit any simple 1- or 2-D modeling. The only "understandable" data have been measured at the stations included in Fig. 1, nos. 2, 3 and 4. The data display a clear 2-D-resistivity structure with strike direction parallel to the Franconian Line which separates the sedimentary part in the West from the metamorphic rocks (gneisses and amphibolites) in the east. Station no. 2, PRE, will be discussed in more detail. Apparent resistivities and phases of the main impedances are presented in Fig. 2, which are already rotated in a coordinate system of the strike direction of the Franconian Line (FL). Both apparent resistivity curves and both phase curves indicate clearly a very good conductor beneath the also low resistive sediments. In the first step, one of the two curves of the apparent resistivities can be used for a 1-D model; the question is which one. According to theory and experience, the "E-polarization curve" should be selected for this purpose. At first glance the identification of E- and B-polarization seems to be a simple matter, i.e., the major resistivity curve with the electric field component parallel to the Franconian Line should be the E-polarized apparent resistivity. However, unfortunately, the actually observed geomagnetic induction vector at this site also points to the south, parallel to the Franconian Line. Theoretically, the induction vector should be at right angles to the Franconian Line. Since the induction arrows also of most of the MT sites not shown here point more or less to the south (Blümecke 1984), they obviously indicate a more regional EW-striking conductivity anomaly (Berktold 1987). It was therefore decided that the major apparent resistivity curve represents the locally valid E-polarization.

This apparent resistivity curve and the corresponding phase curve was therefore inverted by the Monte Carlo method (Martinez 1987), with the result shown in Fig. 3. A large group of resistivity depth models can fit the same data, with the important aspect common to all models that there exists a low resistivity layer at about 15 km depth, sandwiched between high resistivity material.

The question was whether this layer continues to the east, crossing the Franconian Line. The corresponding E-polarized apparent resistivity curve at station no. 4, just across the Franconian Line, displayed a constant value of 10 ohm m from 50 s up to several 100 s, which did not contradict the existence of such a low resistivity

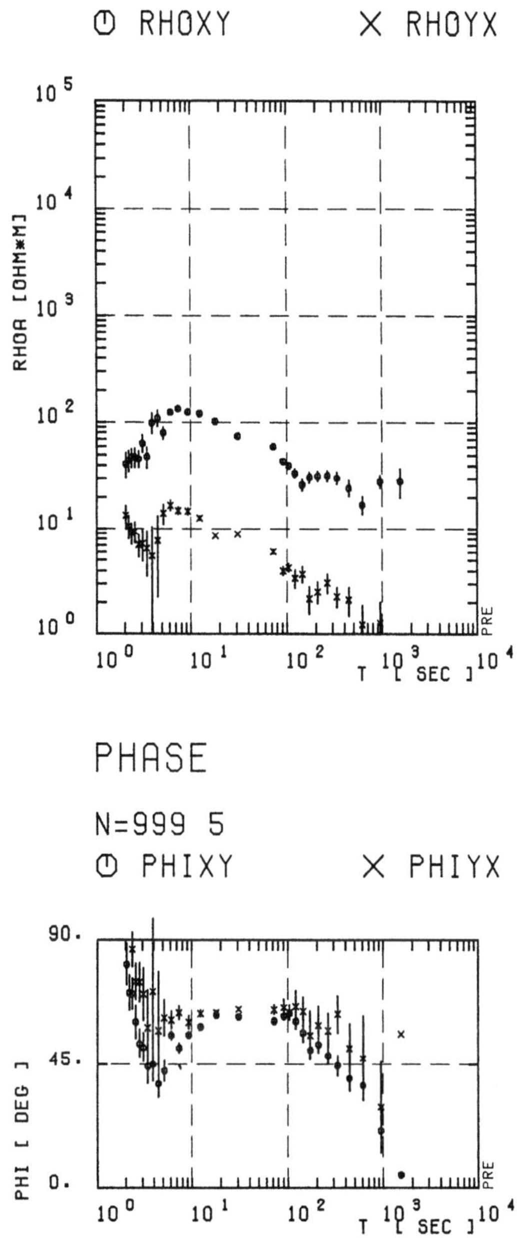

Fig. 2. Apparent resistivities and phases measured by MT at site no. 2. The major resistivity curve represents the E-polarization with the E-field parallel to the strike of the Frankonian line

layer. It seemed impossible, however, to infer any definitive depth of this layer. For this purpose it was necessary to measure apparent resistivities at shorter periods. This was a task for the MMT method.

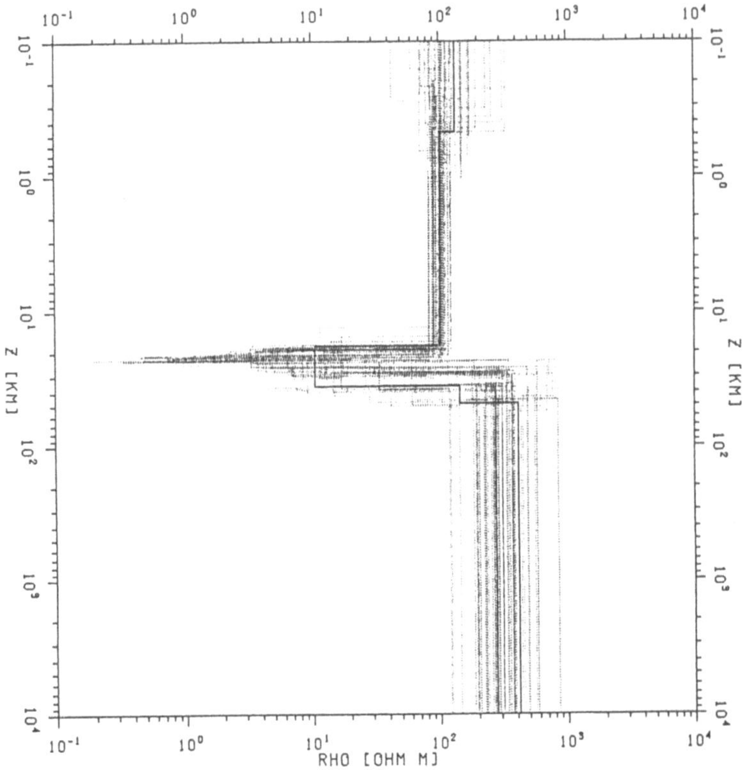

Fig. 3. Models of the resistivity depth function calculated by the Monte Carlo method from the E-polarized apparent resistivity and phase curve of Fig. 2

6 MMT-STUDIES

The measurements were performed by GEOMETRA (1986) at three sites within a few days. The resulting curves for stations no. 4 and 5 are shown in Figs. 4 and 5. They correspond to the E-polarization. At station 1 the data quality was too poor and is not displayed here. The 1-D-inversion (M. Martinez) yielded the resistivity depth function shown. At station 4, the low resistivity layer appears at only 3 km depth, at station 5 at 10 km depth. These values will be discussed later, together with the results obtained by the LOTEM method.

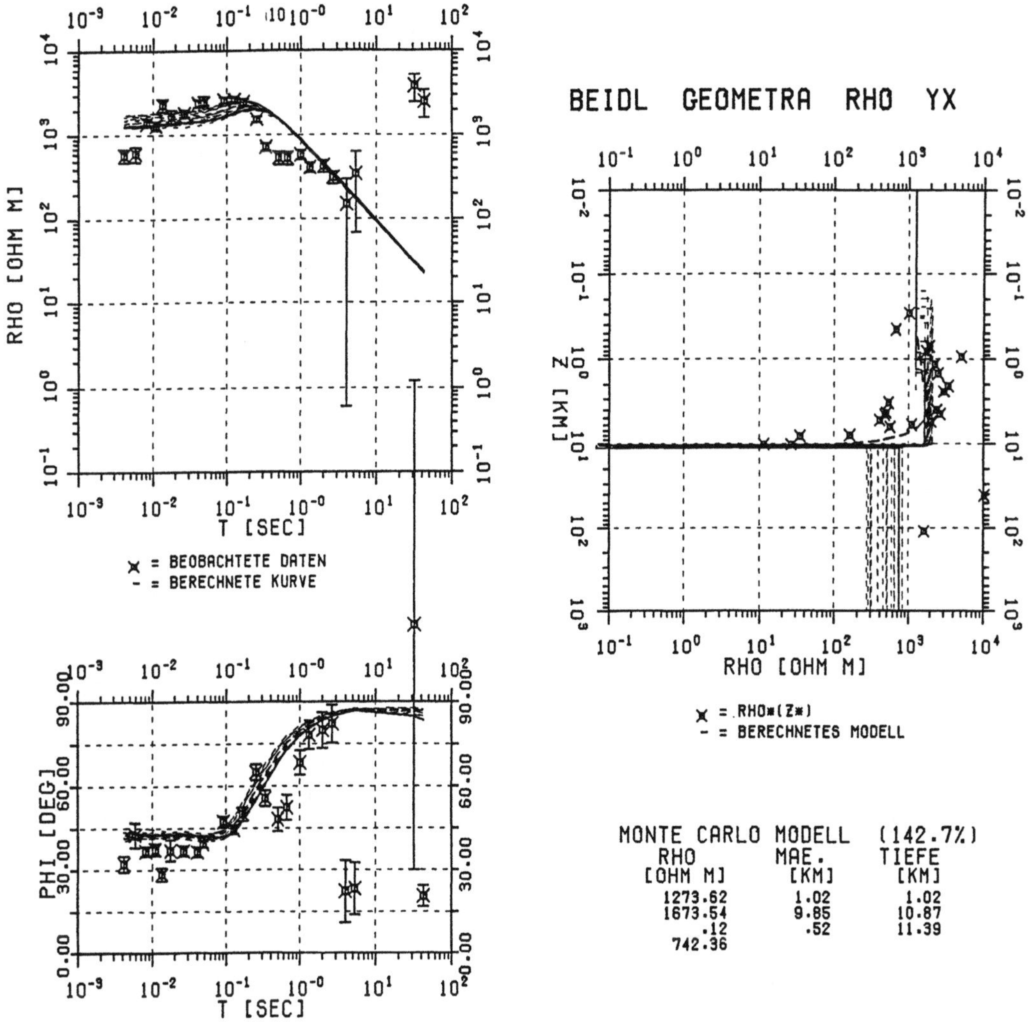

Fig. 4. Apparent resistivities and phases for site no. 5, with the resulting resistivity depth function, measured by MMT

7 THE LOTEM-EXPERIMENT

The measurements and interpretations were performed by the University of Köln (LOTEM-group; 1986a,b). At about 20 sites (see Fig. 1) the vertical component of the transient magnetic field was recorded by induction coils. The source was a dipole feeding square-shaped electrical current pulses into the ground.

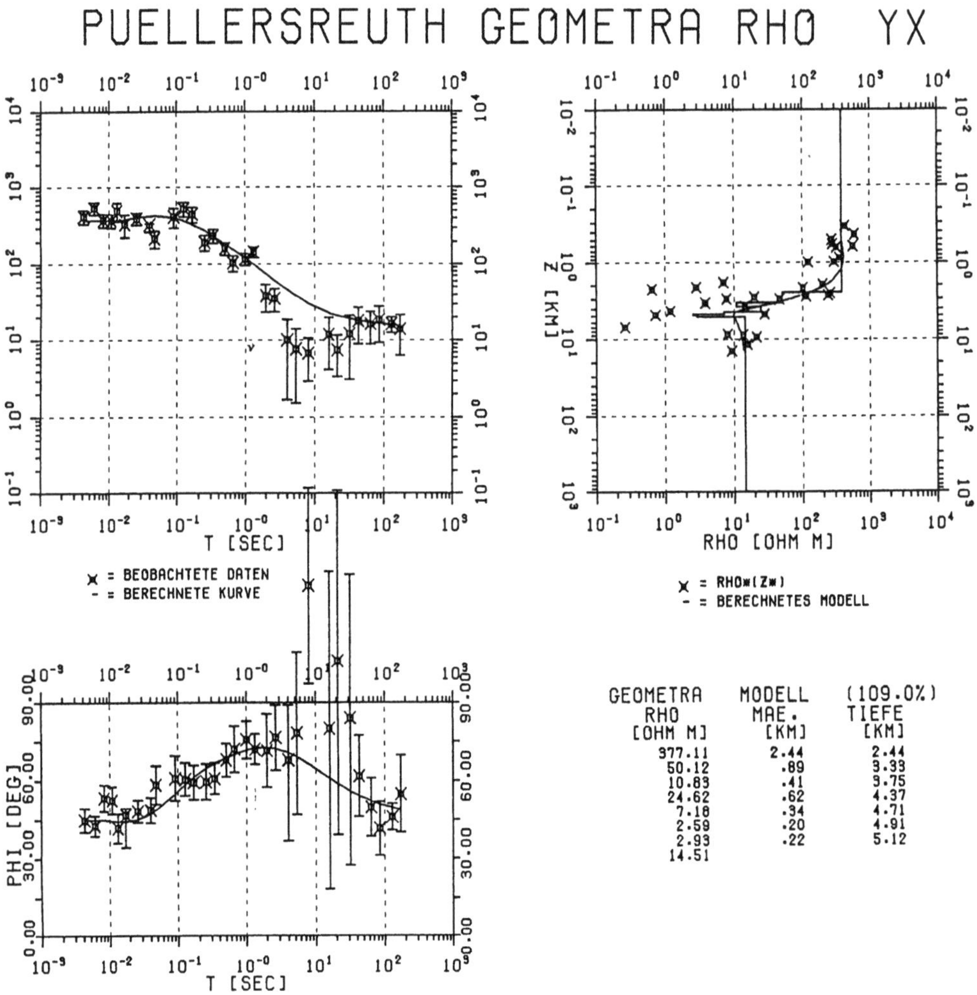

GEOMETRA RHO [OHM M]	MODELL MAE. [KM]	(109.0%) TIEFE [KM]
377.11	2.44	2.44
50.12	.89	3.33
10.83	.41	3.75
24.62	.62	4.37
7.18	.34	4.71
2.59	.20	4.91
2.93	.22	5.12
14.51		

Fig. 5. Apparent resistivity and phase curve with the resulting resistivity depth function at site no. 4, measured by MMT

Considerable data improvement was achieved by a specially selected stacking method and prestack digital data processing techniques similar to those used in reflection seismics (Hanstein et al. 1986). The quality of the final signals was rather good. A prominent aspect was the extreme short duration time of the transient signal, see Fig. 6, where a typical signal is compared with the signal measured in the Black Forest. About half of the sites displayed reversals of the transient signal, which could not be interpreted quantitatively.

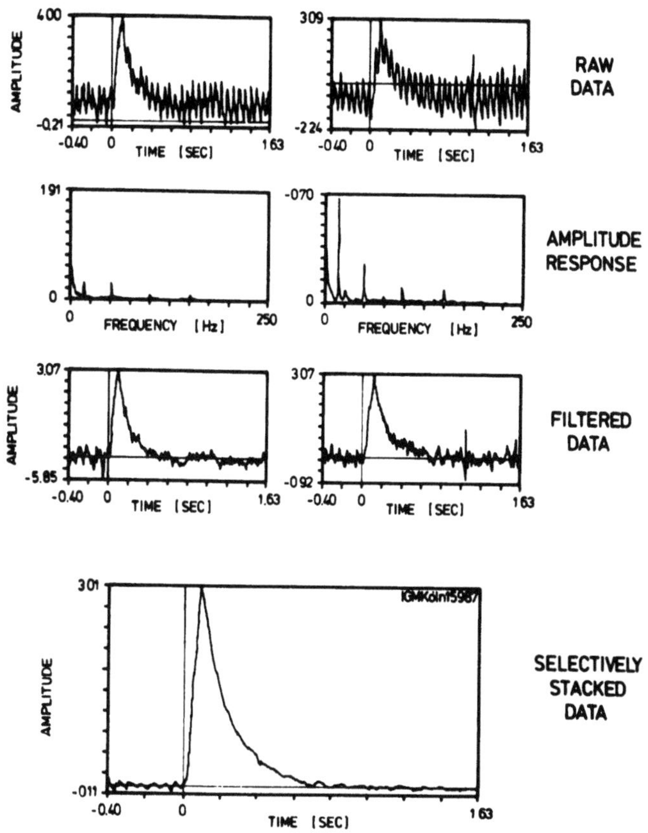

Fig. 6. Typical time functions of the vertical component of the transient magnetic field in the Oberpfalz at station no. 12. The very short duration of the signal indicates the high resistivity of the upper crust

For display only the signals (without reversals) have been transformed into the "early-time" and "late-time" apparent resistivity curves. Purely numerical reasons prohibited a deconvolution of these apparent resistivities with the response function of the equipment. In particular, the early time curve therefore does not represent the approximate true resistivity of the uppermost part of the crust (Fig. 7), and is thus called the resistivity transform, which is the convolution of the apparent resistivity with the system response accounting for analog electronics and the source ramp function. The 1-D inversion uses the original voltage decay curve and convolves in the subsequent step the synthetic calculated apparent resistivity curve with the system response. Figure 7 shows the resulting 1-D model at one site. A low resistivity layer appears clearly at 10 km depth. The corresponding theoretically calculated model curve (1000 m) is shown at the left, now convolved with the response function, for comparison with the "measured" apparent resistivities. Model curves 250 m and 500 m may well demonstrate the sensitivity of the model curve to the variation of the thickness and the conductivity-depth product of the low resistivity layer at 10 km depth.

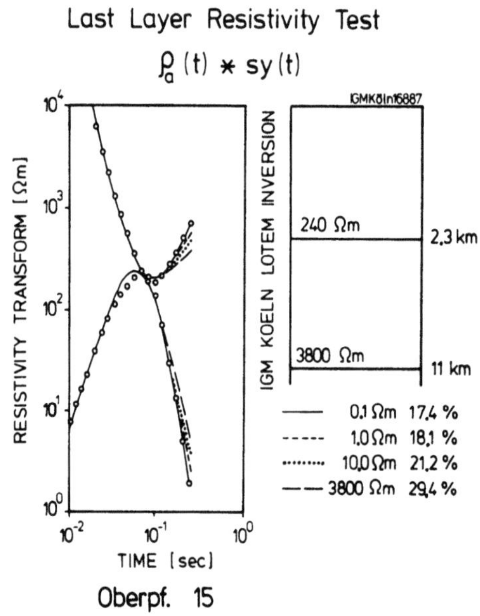

Fig. 7. Early and Late Time apparent resistivities measured by the LOTEM method. The deconvolution with the response function of the equipment was not applied in this presentation. At the right the resulting resistivity model is displayed, with three options of the thickness of the low resistivity layer, where the continuous line (1000 m) is fitting the data

The inversion of the data at all sites without those where reversals had been observed was, however, not unique, but yielded models with low resistivity layers also at depths shallower than 10 km. However, this "layer" varied its depth from site to site quite erratically. This problem of non-uniqueness inherent often almost all geophysical inversion techniques can often be solved with a piece of a-priori knowledge: from a quite general point of view, such a low resistivity "layer" cannot jump from place to place within 1 or 2 km distance, either from the point of view of geology or from the point of view of the limited resolution of the applied method itself. It should be a purely mathematical effect. With the at least reasonable assumption that this low resistivity layer should be at approximately the same depth at each site WITHOUT prescribing the depth itself the repeated inversion of the same data yielded indeed a common depth for all sites which turned out to be close to 10 km. (Fig. 8)

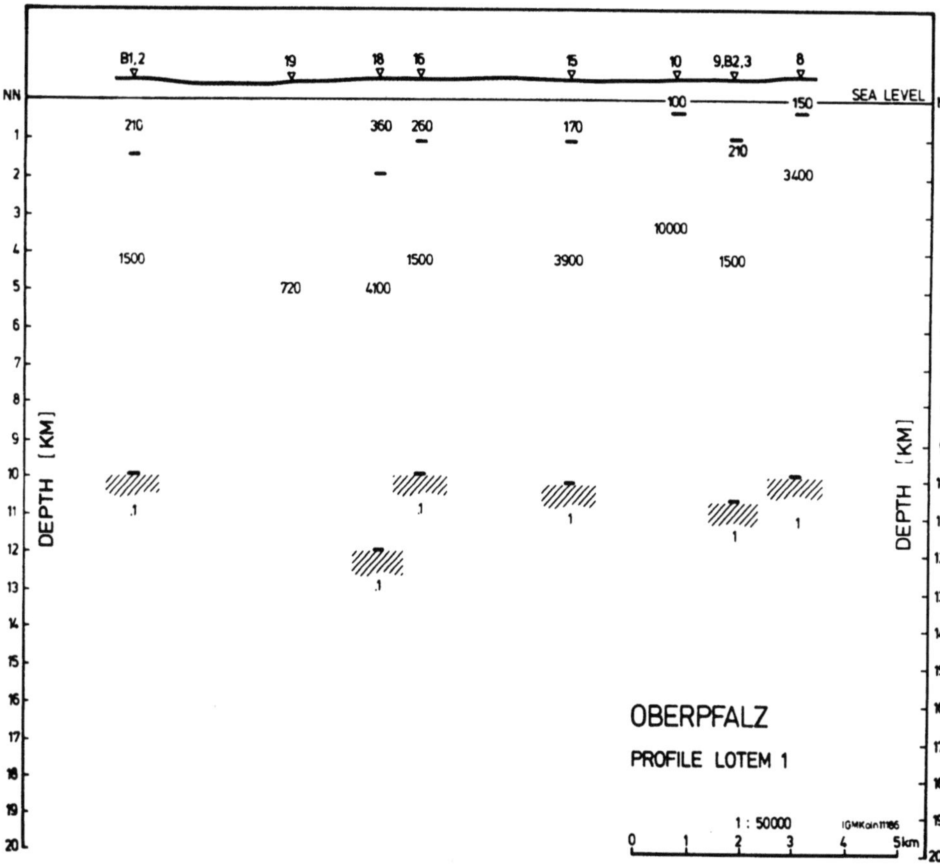

Fig. 8. Synoptic presentation of the LOTEM results

8 AMT-AAMT-DCR STUDIES

The AAMT study was performed by the Technical University of Braunschweig (Drews 1986) and already during the first measuring period yielded unexpected results at the sites situated in the area of Fig. 1. At stations 6 and 7 (Figs. 9 and 10), the prominent feature was the existence of very low resistivity at a depth of about 300 m beneath the highly resistive gneisses (Drews 1986). These unexpected results were later confirmed by a rather intensive AMT study along the profiles 23 and 24, executed by the Technical University of Berlin (Leonhardt and Brasse 1986) (Fig. 11). The DCR study, performed by the Technical University of Aachen (Wohlenberg et al. 1986) confirmed the high resistivity of the uppermost (300 m) gneisses in this area. Further AAMT measurements at sites no. 4, 5 and between them yielded a continuous high resistivity (greater than 1000 ohm) of the upper 5 km (Fig. 1).

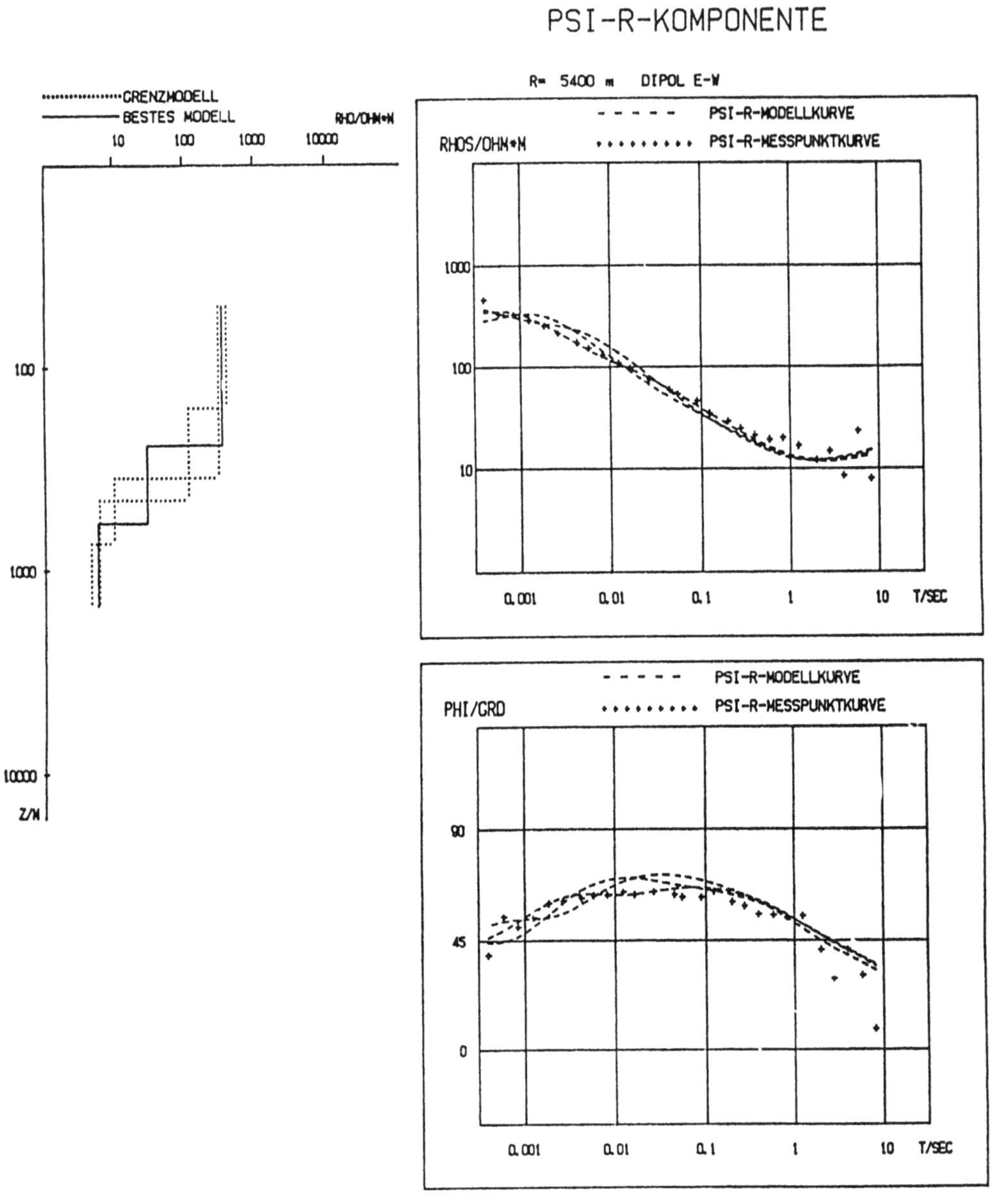

RHO/OHM*M	BESTES MODELL :	.307E+03	.278E+02	.540E+01
	GRENZMODELL :	.358E+03	.107E+03	.587E+01
	GRENZMODELL :	.274E+03	.101E+02	.507E+01
D/M	BESTES MODELL :	.245E+03	.344E+03	
	GRENZMODELL :	.159E+03	.290E+03	
	GRENZMODELL :	.338E+03	.392E+03	

Fig. 9. Apparent resistivity and phase measured by the AAMT method at site no. 7 with the resulting resistivity model. The prominent aspect is the very low resistive layer starting at about 300 m depth

N10
Beidl-Schoenthan

PSI-R-KOMPONENTE

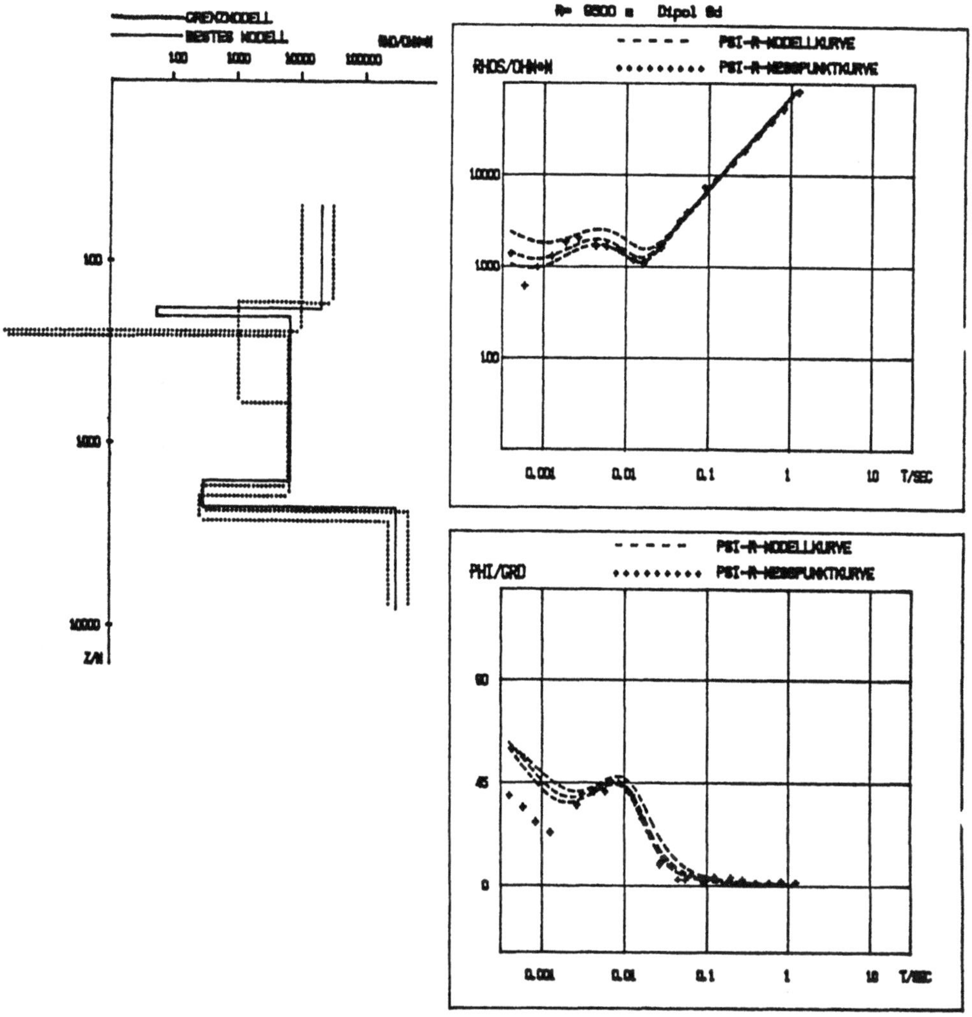

	BESTES MODELL :	.187E+05	.585E+02	.813E+04	.277E+03	.286E+03
RHO/OHM*M	GRENZMODELL :	.280E+05	.107E+04	.829E+04	.277E+03	.209E+03
	GRENZMODELL :	.908E+04	.258E+00	.585E+04	.277E+03	.481E+03
D/M	BESTES MODELL :	.188E+03	.200E+02	.142E+04	.680E+03	
	GRENZMODELL :	.180E+03	.420E+03	.198E+04	.829E+03	
	GRENZMODELL :	.288E+03	.887E-01	.148E+04	.882E+03	

Fig. 10. Apparent resistivity and phase measured by the AAMT method at site no. 5, with the resulting resistivity mode. The resistivities of the upper 5 km are very high. The low resistivity layer at 10 km depth (Figs. 4 and 7) does not show up since the depth of penetration of this method was too small

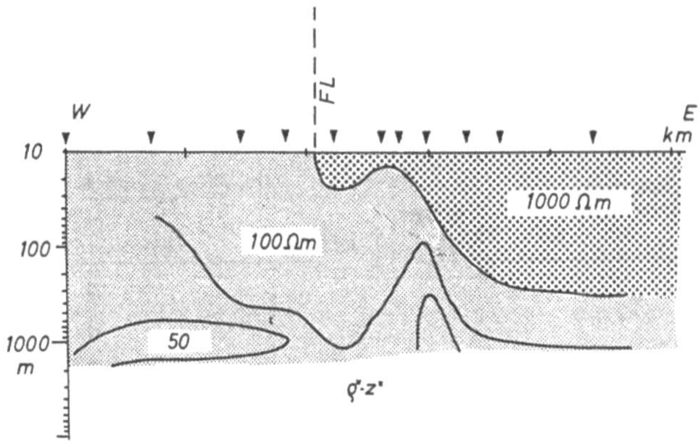

Fig. 11. Cross-section of the rho*-z* model along profile no. 23 (see Fig. 1). It shows low resistive sediments (100 ohm m) to the west of the Frankonian Line ("FL") and a low resistivity layer beneath the high resistive gneisses to the East of the FL. This low resistivity layer reaches probably to the drilling site about 3 km further to the right

The AMT study at about 20 sites started in the sediments W of the Franconian Line crossing the Frankonian Line. It revealed the continuous low resistivity "layer" along both profiles, see Figs. 10 and 1. Presented are the rho*-z* values which should be understood as a close approximation to the true resistivity-depth values. At the present state of knowledge this low resistivity layer continues even to the actual drilling site where a transition to high resistivities occurs in a still unclear fashion. It is also unclear how far to the south this low resistivity reaches, probably not as far as site no. 4. The nature of the low resistivity is unknown.

9 DISCUSSION AND CONCLUSION

The electrical resistivity studies in the area of Fig. 1 obtained two important results: A shallow conductor along profiles 23 and 24, and a deep conductor at 10 km depth.

The shallow conductor, observed both by AMT and AAMT at ab out 20 sites, exists at a depth of about 300 down to at least 1000 m. The resistivities according to AAMT are well below 10 ohm m, according to AMT of about 50 to 100 ohm m. The values do not differ much from the values of sedimentary resistivities but they display an obvious anisotropic resistivity with the lower value in the vertical direction opposite to the case of undisturbed sediments. Very recent Self-Potential (SP) measurements detected a large SP anomaly at the drilling site, where presumably this good conductor ends. It is therefore possible that graphite could play a certain role in this

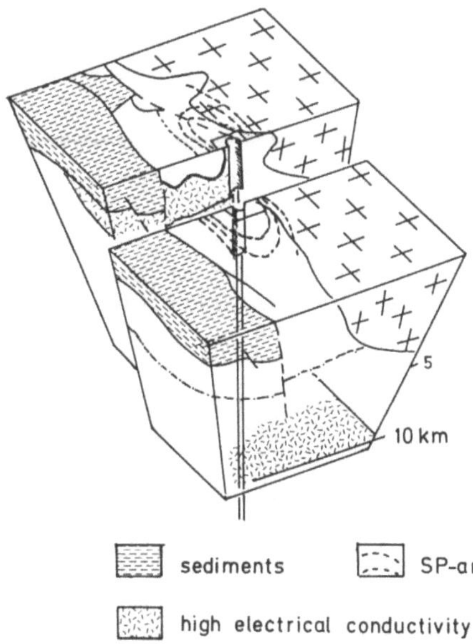

Fig. 12. An updated bloc model of the main results of the geoelectrical investigations around the KTB site: There exists a shallow ($\sim$ 1 km deep) low resistivity layer and a deeper ($\sim$10 km) low resistivity layer. The shallower layer reaches from the Franconian Line to the KTB-site

case, as well as in decreasing considerably the resistivities and in causing the SP anomaly. On the other hand, also strongly weathered rocks filled with saline water could at least decrease the resistivities.

The deeper conductor at 10 km depth is still more enigmatic. The depth of 10 km has been confirmed by MMT and LOTEM at site no. 5. The shallower depth at site no. 4, detected by MMT and MT, and the greater depth at site no. 2, as obtained by MT, may be due to static shift problems since LOTEM, which is not influenced by these problems, yields a continuous depth for this layer at nine sites (see circles in Fig. 1). The NS extent, in particular towards the drilling site, is an open question. The nature of the low resistivity is also an open question, in particular when the extremely low value of 0.1 ohm m, as determined with LOTEM, is considered.

At present a still larger number of different electrical methods at a number of densely spaced sites are operating in the vicinity of the drilling site. From the point of view of the electrical studies, the drilling results are awaited with a great deal of interest.

REFERENCES

Angenheister G (1962) Prot Erdmagnet Tiefensondierung. Kassel

Blohm EK, Worzyk P, Scriba H (1977) Geoelectrical deep sounding in Southern Africa using the Cabora Bassa power line. J Geophys 43:665-679

Blümecke T (1984a) AMT Untersuchungen in Nordbayern. Protokoll Koll Elektromagn Tiefenforschung, Grafrath/Obb, Haak V, Homilius J (eds) Freie Univ Berlin, pp 59-68

Blümecke T (1984b) MT Untersuchungen zur Tiefbohrlokation Oberfpalz. Protokoll Koll Elektromagn Tiefenforschung, Grafrath/Obb, Haak V, Homilius J (eds) Freie Univ Berlin

Drews C (1986) Ermittlung der Tiefenverteilung der elektrischen Leitfähigkeit in der Oberpfalz mit Hilfe der Aktiven Audiomagnetotellurik. Dipl Arbeit Braunschweig

GEOMETRA (1986) Bericht über die MMT-Oberpfalz-Messungen

Haak V, Hutton R (1986) Electrical resistivity in continental lower crust. In: Dawson JB et al. (eds) The Nature of the lower continental crust. Geol Soc Spec Publ No 24, pp 35-49

Hanstein T, Eilenz H, Strack KM (1986) Einige Aspekte der Aufbereitung von LOTEM Daten. Protokoll Koll Elektromagn Tiefenforschung, Lerbach bei Köln, Haak V, Homilius J (eds) Univ Frankfurt and NLfB Hannover

Jones AG (1987) MT and reflection: an essential combination. Geophys JR Astr Soc 89:7-18

Jödicke H, Untiedt J, Olgemann W, Schulte L, Wagenitz V (1983) Electrical conductivity structure of the Crust and Upper Mantle beneath the Rhenish Shield. In: Fuchs K et al. (eds) Plateau Uplift, pp 288-302

Kertz W (1964) The conductivity anomaly in the upper mantle found in Europe. J Geomagn Geoelectr vol XV, no 4, pp 185-182

LOTEM-group (1986a) Report on the LOTEM measurements Oberpfalz

LOTEM-group (1986b) Poster. 2 KTB Kolloquium Seeheim

Shankland TJ, Ander ME (1983) Electrical conductivity, temperatures and fluids in the lower crust. J Geophys Res 88: 9475-9484

Strack KM (1985) Das Transient-Elektromagnetische Tiefensondierungsverfahren, angewandt auf die Kohlenwasserstoff- und Geothermie-Exploration. Mitt Inst Geophys Meteorol Univ Köln, Heft 42

Teufel U, Dittus HJ, Berktold A (1986) Audiomagnetotellurik, Magnetotellurik und erdmagnetische Tiefensondierung im mittleren und nördlichen Schwarzwald. Protokoll Koll Elektromagn Tiefenforschung, Lerbach bei Köln, Haak V, Homilius J (eds) Univ Frankfurt and NLfB Hannover

Tezkan B (1986) Erdmagnetische und magnetotellurische Untersuchungen auf hochohmigen Kristallinstrukturen des Hochschwarzwaldes und des Bayerischen Waldes bei Passau. Univ Göttingen

Thompson BG, Nekut A, Kuckes AF (1983) Deep crustal electromagnetic sounding in the Georgian Piedmont. J Geophys Res 88:9461-9473

Untiedt J (1986) Die elektrische Leitfähigkeit in der Erdkruste in den Zielgebieten Oberpfalz und Schwarzwald - Zusammenfassung der Ergebnisse der elektromagnetischen Voruntersuchungen. Protokoll Koll Elektromagn Tiefenforschung, Lerbach bei Köln Haak V, Homilius J (eds) Univ Frankfurt und NLfB Hannover

Van Zijl JSV et al. (1977) Electrical studies in the deep crust in various tectonic provinces in southern Africa. In: Heacock JG (ed) AGU Monogr 20, pp 470-500

Wohlenberg J, Voigt G, Wendel C (1986) Bericht zu den Voruntersuchungen der KTB: Geoelektrische Tiefensondierung in der Oberpfalz. Lehr- und Forschungsgebiet für Angew Geophys, RWTH Aachen

Petrography and Petrology of the KTB Location Schwarzwald

Wolfhard Wimmenauer, Hans Klein, Hiltrud Müller and Rüdiger Stenger

Mineralogisch-petrographisches Institut der Universität, Albertstr. 23b, D-7800 Freiburg i. Br., FRG

Abstract. The metamorphic massif of the Central Schwarzwald has been chosen as a possible site of a deep continental research drilling. Apart from the outstanding results of the geophysical and geological investigations carried out in that region, the petrographic conditions and petrological aspects known so far, cause far-reaching expectations from the materials encountered in the drilling. The target area is a typical section of the pre-Variscan metamorphic basement with predominant paragneisses, orthogneisses, and various specialized intercalations. The paragneiss complex, of Proterozoic age, has been divided in several lithostratigraphic units, which can partly be correlated with certain members of the Proterozoic sequence in Bohemia. Numerous geochemical analyses of certain paragneiss sequences and their statistical evaluation are useful means for further distinctions and correlations.- A high-pressure phase of the metamorphism is clearly expressed in many eclogites and accompanying intermediate and acid (former supracrustal) rocks. Several narrow belts with mylonitic granulites are interpreted as deep-seated shear zones of a pre-Variscan tectonic event. In the Haslach area, a granulite zone seems to mark the border between a gneiss unit with HP relics and another one where those are lacking.- The problem raised by the tonalitic-trondhjemitic-granodioritic orthogneisses is whether they were intrusive (originating from the lower crust) or subautochthonous formations by anatexis of their paragneiss environment.- For the completion of the general view, some paragraphs on the metamorphic units north and south of the Central Schwarzwald are briefly discussed.

1 GENERAL VIEW

Regarding the metamorphic evolution of the Schwarzwald, four principal geological domains can be distinguished. They comprise Variscan and pre-Variscan rock sequences of magmatic and sedimentary origin. From north to south, the following units are exposed (Fig. 1):

1. The metamorphic sequence of Baden-Baden - Gaggenau, most probably of older Paleozoic age. This unit belongs to the Saxothuringian zone of the Variscan orogen. It appears to be thrusted under the Moldanubian basement along a plane striking WSW-ENE, which continues to the southwest as the line of Lalaye - Lubine in the Vosges.
2. The gneiss massif of the Central Schwarzwald (plus a few smaller gneiss fragments in the Northern Schwarzwald): pre-Variscan gneisses and migmatites with metabasitic intercalations. This unit extends for nearly 100 km from north to south; its maximum E-W extension is about 40 km. Gneisses and migmatites are intruded by Variscan granites, particularly in the north (granites of Forbach, Bühlertal, Raumünzach, Seebach, and Oberkirch) and in the east (granite of Triberg). In the

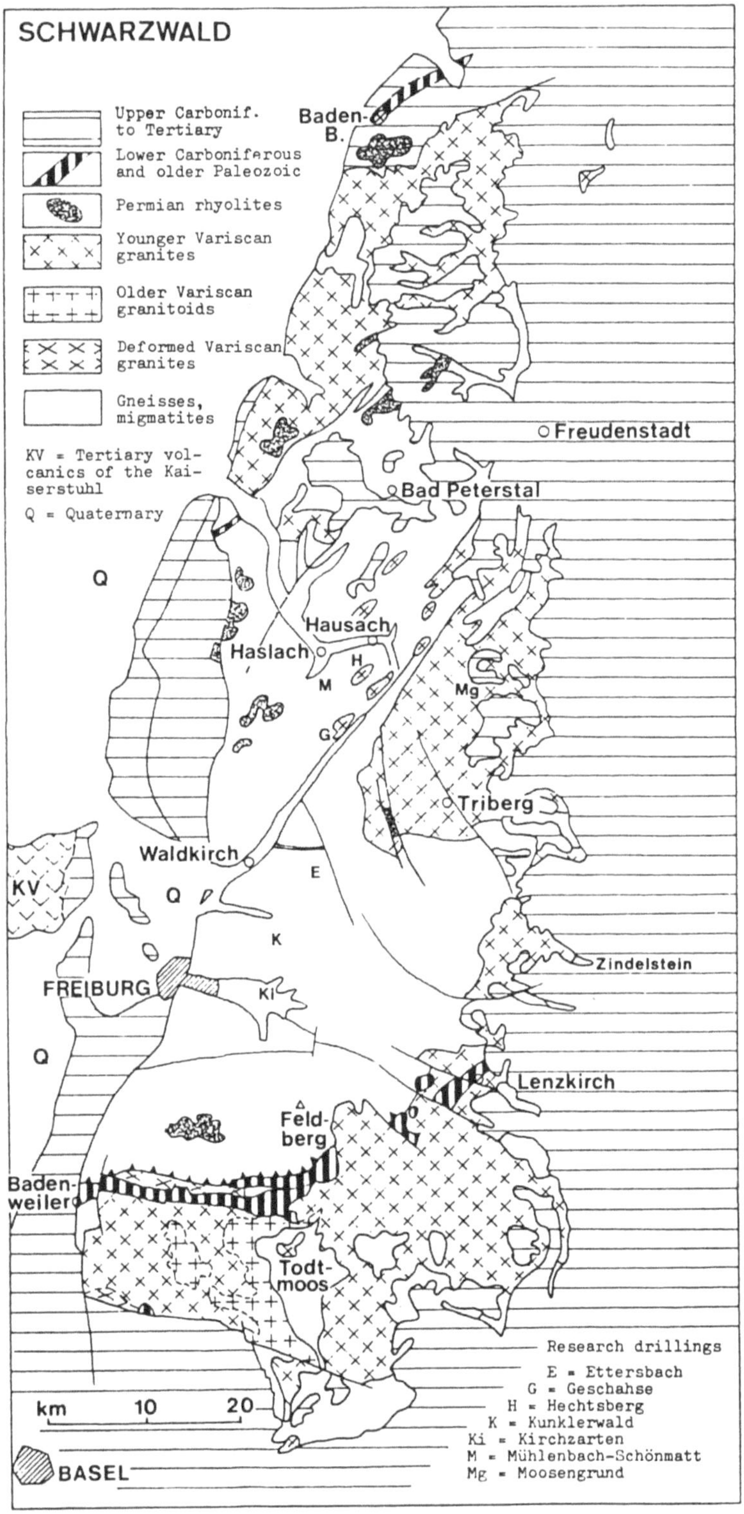
SCHWARZWALD
Upper Carbonif. to Tertiary
Lower Carboniferous and older Paleozoic
Permian rhyolites
Younger Variscan granites
Older Variscan granitoids
Deformed Variscan granites
Gneisses, migmatites
KV = Tertiary volcanics of the Kaiserstuhl
Q = Quaternary
Baden-B.
Freudenstadt
Bad Peterstal
Hausach
Haslach
H
M
Mg
G
Triberg
Waldkirch
E
KV
Q
K
Zindelstein
FREIBURG
Ki
Q
Lenzkirch
Feldberg
Badenweiler
Todtmoos
km 10 20
BASEL
Research drillings
E = Ettersbach
G = Geschahse
H = Hechtsberg
K = Kunklerwald
Ki = Kirchzarten
M = Mühlenbach-Schönmatt
Mg = Moosengrund

extreme south, an early Variscan in-situ granitization strongly modified the pre-existing gneiss structures and compositions ('Randanatexit' of Altherr and Maass 1977).
3. The Paleozoic of the Badenweiler-Lenzkirch zone, 50 km in length and up to 3 km wide. This unit contains clastic sediments of Upper Devonian to Lower Carboniferous age, acid and intermediate volcanics associated with them, and a sequence of metamorphic sediments probably of older Paleozoic age, the so-called 'Alte Schiefer' between Schönau and Lenzkirch. In the western part of the zone, a large shear body of metagranite (Münsterhalden granite, 20 km long and up to 2 km wide) is intercalated between the Paleozoic and the gneisses, respectively, within the Paleozoic sequence. The whole Badenweiler-Lenzkirch unit is thrusted under the gneisses along a W-E to SW-NE striking plane ('Südschwarzwälder Hauptbewegungszone'. A general view of the tectonic situation of that zone and the structure of the Schwarzwald as a whole has recently been given by Krohe and Eisbacher (1988).
4. The metamorphic units of the Southern Schwarzwald: gneisses, migmatites, and subordinate schists (of probably older Paleozoic age). A zone of early Variscan in-situ granitization is known as 'Wehra-Wiesetal diatexite'. Gneisses of Schwarzwald type have also been met with in several deep drillings in northern Switzerland. Several large Variscan granite plutons have intruded the pre-Variscan metamorphic units; the youngest of them even breached the Paleozoic zone of Badenweiler-Lenzkirch and the thrust structures mentioned above.

2 THE METAMORPHIC PALEOZOIC OF BADEN-BADEN

The metamorphic rocks at the northern termination of the Schwarzwald are exposed in several small, disconnected areas between Baden-Baden in the SW and Gaggenau in the NE (Sittig 1965, 1966). The contact of this Saxothuringian unit with the Moldanubian part of the Schwarzwald is hidden under Carboniferous and Permian sediments, but the approximate position of that limit can be inferred from the occurrences of gneiss near Neuweier in the SW and in the town of Gaggenau (Hummelberg quarry) in the eastern part of the zone. In addition, the Saxothuringian area is distinguished by marked gravimetric and magnetic anomalies, whose southeastern limits coincide with the supposed borderline in the underground.

The Paleozoic metasediments and associated metavolcanic rocks are subdivided into two main units:

1. The northern zone with low-grade metapelites, metagreywackes, and metamorphic basic tuffs or tuffites. Characteristic mineral associations are:
Quartz + feldspars (albite to oligoclase) + muscovite + chlorite + biotite in the metapelites and
Plagioclase (albite, oligoclase to andesine) + actinolite + epidote + chlorite ± magnetite ± quartz in the metagreywackes and the more basic rock types.
Dolomitic marble occurs in two small exposures west of Gaggenau. A narrow (intrusive) body of massive uralite-monzonite is intercalated between the metasediments and metabasites north of Ebersteinburg. Members of the same unit occur, modified by contact metamorphism, at a few places right in the city of Baden-Baden near the Friesenberg granite, a Variscan intrusion in the metamorphic Paleozoic.

Fig. 1 Schematic geological map of the Schwarzwald

2. The southern zone with medium-grade metapelites and metabasites. The following parageneses have been observed in the metapelites (mica schists):
Quartz + garnet + biotite + plagioclase ± tourmaline,
Quartz + garnet + biotite + muscovite + plagioclase + kyanite ± andalusite ± tourmaline,
Quartz + garnet + biotite + muscovite + plagioclase + K-feldspar (metablastic) + kyanite ± tourmaline.
The plagioclases are mostly oligoclase; the K-feldspars contain up to 20 mol-% albite.

The contact metamorphism generated parageneses with quartz, plagioclase, biotite, muscovite, andalusite, cordierite, and fibrolitic sillimanite. The rare metabasites in Baden-Baden are biotite- or garnet-bearing amphibolites.

From the above parageneses and the application of various petrological thermometers and barometers, the following conditions of metamorphism appear to be most probable (Müller 1988b):

1. Metapelites and metabasites of the northern zone: 420 - 520°C, 0.2-0.4 GPa;
2. Metapelites of the southern zone: 490 - 560°C, 0.4 - 0.6 GPa;
3. Contact-metamorphic rocks in Baden-Baden: 550 - 660°C, 0.2 GPa.

The regional metamorphism was effectuated during the collision of the Saxothuringian and the Moldanubian plates and most probably prior to the intrusion of the Friesenberg granite and the contact metamorphism connected with it. A younger Variscan tectonic phase also affected this granite and is probably responsible for the present position of the low- and medium-grade units in the same crustal level.

Geochemically, the low-grade metapelites are distinguished from those of the medium-grade zone by lower contents of Al, Mn, Ga, Rb, Sr, La, Y, Nb, and Th, while the metagreywackes of the same units have significantly different averages of Mg, Zn, Sr, and Zr (Müller 1988b). With the medium-grade metapelites, only three (K, Cu, Ba) of 29 analyzed elements are significantly different from the average of the Villé schists in the northern Vosges, so these two units appear to be the next candidates for a correlation on geochemical grounds.

As to the causes of the prominent gravimetric and magnetic anomalies NE of Baden-Baden, a larger intrusion of a heavy igneous rock such as the uralite monzonite (D = 2,81) may be responsible for the gravity high, while the magnetite-bearing basic layers of the northern zone do not seem to fulfill the requirements imposed by the observed anomaly (Plaumann 1985, Pucher 1986).

3 THE GNEISSES AND MIGMATITES OF THE CENTRAL SCHWARZWALD

3.1 Occurrence and Tectonics

The gneisses and migmatites of this unit form a continuous massif between the master fault of the Rhine graben in the west, the Variscan granites in the north and east and the thrust zone of Badenweiler-Lenzkirch in the south. The major tectonic features of the Schwarzwald have recently been described by Krohe and Eisbacher (1988). With respect to the tectonic phenomena seen in the gneiss exposures, Kleinschmidt et al. (1986) distinguished at least six phases of deformation, four of which can clearly be correlated with certain metamorphic events. The D3 deformation generated the ubiquitous, pervasive foliation of the gneisses; it is most strongly expressed in the blastomylonitic structures of the granulites (see section 3.2.2). D4 accompanied and followed the widespread formation of anatectic veins (metatects), while D5 (early Variscan) produced linear structures, s-c mylonites, shear bands, and microcrenulations. D6 is a late Variscan cataclastic deformation in certain zones of pronounced extension. D5 and D6 implicate increasing retrograde transformation of the earlier high- and medium-grade metamorphic parageneses.

3.2 Petrography, Geochemistry, and Origin

3.2.1 The Paragneisses and their Intercalations (Table 1)

Paragneisses with quartz, plagioclase, and biotite as main constituents are the most widespread metamorphic rocks in the Central Schwarzwald. Varieties with sillimanite and/or cordierite are also very frequent. The chemical compositions of those gneisses indicate greywackes and pelites as protoliths. Quartz-rich paragneisses (from orthoquartzitic sandstones) and orthoclase-rich gneisses (from arkoses) are much rarer.

Table 1. Chemical analyses of characteristic gneisses from the Kinzigtal region

	1	2	3	4	5	6	7	8	9	10	11	12
SiO_2	64.09	65.56	61.60	90.50	61.90	68.08	64.79	62.87	75.06	72.37	64.91	60.20
TiO_2	0.80	0.74	0.71	0.29	0.78	0.38	0.56	0.73	0.06	0.25	0.74	0.82
Al_2O_3	17.82	15.69	18.34	4.05	18.40	16.22	17.14	16.51	13.64	14.68	15.69	17.71
Fe_2O_3	1.08	1.14	6.44[a]	0.31	2.18	0.54	0.87	1.89	0.16	0.50	1.14	1.90
FeO	4.12	4.79	-	0.54	3.54	2.90	2.90	4.01	0.78	2.09	4.79	5.05
MnO	0.06	0.09	0.09	0.05	0.06	0.06	0.08	0.11	0.08	0.07	0.09	0.11
MgO	2.00	2.51	2.87	0.37	2.75	1.53	1.91	2.69	0.23	0.65	2.51	3.91
CaO	0.56	2.19	3.56	0.10	0.25	2.45	3.66	4.69	0.41	1.49	2.19	3.35
Na_2O	2.13	3.83	3.56	0.79	1.59	4.12	3.89	3.46	4.06	3.72	3.83	3.22
K_2O	4.48	2.56	2.63	1.19	4.74	2.75	2.06	2.41	4.48	3.61	2.56	2.40
P_2O_5	0.13	0.18	0.18	0.06	0.10	0.12	0.21	0.19	0.17	0.10	0.18	0.27
L.o.i.[b]	3.34	1.22	n.d.	0.46	3.37	1.44	1.43	1.19	0.55	0.58	1.22	1.39
Total	100.61	100.50	(99.99)	99.11	99.66	100.59	100.61	100.24	99.68	100.11	99.85	100.15
Rb	177	113	96	n.d.	n.d.	85	73	73	136	101	83	95
Sr	167	257	204	n.d.	n.d.	193	240	421	42	114	223	175
Ba	350	896	455	n.d.	n.d.	453	472	658	259	555	813	410
Zr	235	170	176	n.d.	n.d.	92	198	139	55	120	201	149

1: Metapelitic paragneiss, 0.8 km NNW Mitteltal (Zell a.H.); 2: Metagreywacke paragneiss, Heiligendobel (Haslach); 3: Average of 22 pelitic metagreywacke paragneisses, drilling Hechtsberg (Haslach); 4: Metaquartzite gneiss, Im Grün (Zell a.H.): lithostratigraphic unit of Welschensteinach-Oberharmersbach; 5: Graphite-bearing metapelite gneiss, Kinzig bridge near Steinach (Zell a.H.); lithostratigraphic unit of Welschensteinach-Oberharmersbach; with 0.49% C. 6: Granodioritic orthogneiss, Einetwald north of Steinach (Zell a.H.); 7: Trondhjemitic orthogneiss, Stimmel NE of Urenkopf (Haslach); 8: Tonalitic orthogneiss, west of Prechtaler Schanzen (Hornberg); 9: Leucocratic gneiss (leptynite), near Heizenberg, Welschbollenbach (Zell a.H.); 10: Blastomylonitic granulite, acid layer, Salmensbach (Haslach); 11: Blastomylonitic granulite, intermediate (metagreywacke) layer, Salmensbach (Haslach); 12: Kinzigite, Strohbach near Gengenbach (Zell a.H.).

Names in parentheses: designations of the topographic maps 1:25000.

a = total iron as Fe_2O_3 b = loss on ignition

Garnet- and cordierite-bearing paragneisses are known as kinzigites. Hornblende-bearing paragneisses are transitional to amphibolites (see section 3.2.3). Other mineral constituents of petrogenetic relevance are staurolite (rare), kyanite (rare), graphite, and sulphides. Muscovite (sericite), chlorite, and calcite are products of younger retromorphic processes. In rare lime-silicate rocks, there appear diopside, Ca-Al-garnet, and wollastonite along with amphibole, plagioclase, and quartz. Marbles are extremely rare.

Acid leptynites, mostly of K-rhyolitic composition, are frequent intercalations of the paragneisses in certain areas. Wimmenauer (1980, 1984) defined several lithostratigraphic associations of paragneisses with leptynites in the Central and Southern Schwarzwald (e.g. associations of Todtmoos, Bad Peterstal and Nordrach). Another assemblage of leptynites with paragneisses and amphibolites occurs along the southern and southeastern margin of the gneiss massif close to the thrust zone of Badenweiler-Lenzkirch. These leptynites are different from those in the interior of the massif by their pronouncedly sodic character and a repeated intercalation of amphibolite. The most prominent exposure of such a leptynite-amphibolite association is the quarry No. 1 at Zindelstein. Here, more than hundred layers of leptynite (mm to dm thick) alternate with amphibolite layers of similar thickness. The association has much in common with the 'association leptyno-amphibolique' of the Massif Central in France (Wimmenauer and Lim, 1988). Between Sulzburg and Lenzkirch, the leptynites were the preferred substratum to the in-situ granitization mentioned in section 1. Certain leucocratic, magnetite-bearing gneisses between Belchen and Gisiboden engender a marked magnetic anomaly.

In the course of the reconnaissance studies for a deep research drilling, Hirschmann (1985, 1986) developed further the lithostratigraphic knowledge of the Central Schwarzwald. He distinguishes, besides the orthogneisses, six units in the rank of 'formations', listed below from the base to the top.

1. Formation of Welschensteinach-Oberharmersbach: predominant metapelites, intercalations of metaquartzite, and graphite-bearing gneiss.
2. Transitional formation, similar to (1), with additional, but subordinate leptynites, amphibolites, and calc-silicate rocks.
3. Leptynite formation, including Wimmenauer's assemblages of Nordrach, Bad Peterstal-Mauren, and perhaps Zindelstein. K-leptynites predominant in the Nordrach unit; K-Na-leptynites in Bad Peterstal-Mauren; multiple alternation of Na-leptynites with amphibolites in Zindelstein.
4. Sulzburg-Vöhrenbach formation: paragneisses, leptynites (predominantly sodic), amphibolites. This unit forms the southern and southeastern margin of the Central Schwarzwald metamorphic basement.
5. Formation with amphibolites. Metapelitic and metagreywacke gneisses with numerous, mostly small lenses of amphibolites (often eclogitogenic). Metaquartzites, leptynites, leucocratic gneisses, and calc-silicate rocks occur in minor quantities.
6. Formation with metagreywackes. Large areas with metapelitic paragneisses and frequent metagreywacke intercalations; some amphibolites and leptynites (the Monotonous Group of Wimmenauer 1980).

The stratigraphic sequence of these 'formations' could be derived from several less disturbed sections in the field, but very often their relations have been strongly altered by repeated tectonic dislocations, particularly by the so-called 'ductile shear zones' (pre-Variscan) and the major Variscan disturbances. Hirschmann (1985, 1986) estimates the thickness of the formation with metagreywackes (6) to 1000 to 2000 m, while the thickness of the other units varies between 500 and 1500 m.

Since the distance between the Central Schwarzwald and the Bohemian Massif is more than 300 km, an unequivocal correspondence of the Precambrian stratigraphic units cannot be supposed immediately. Still, the 'Leptynite Formation' may be regarded as the most characteristic member of the sequence, whose equivalent in Bohemia is one of the most persistent features in the Lower Moldanubian. It belongs to the Klet Group in the sense of Zoubek (1982). According to Hirschmann, neither the Krumlov Group (above the Klet Group) nor the Monotonous Group (beneath it) can be recognized in the Schwarzwald. Therefore, the whole of the 'formations' defined so far could represent a very thick and differentiated equivalent of the Moldanubian Klet Group. The Leptynite Formation plays the dominant role as a 'leader horizon' in such a correlation.

More details of the field observations and a thorough discussion of the stratigraphic and tectonic aspects by G. Hirschmann are expected in near future.

Another approach to obtaining knowledge of the conditions of the Precambrian sedimentation and volcanism is the geochemical study of the

Table 2. Average chemical analyses of metagreywacke gneisses from the Central and Southern Schwarzwald (Müller 1988b)

	1		2		3		4	
SiO_2	61.60 ±	3.16	60.55 ±	4.27	68.97 ±	2.03	65.11 ±	2.70
TiO_2	0.71	0.14	0.98	0.30	0.50	0.18	0.23	0.28
Al_2O_3	18.34	0.91	17.42	0.96	15.66	0.67	16.00	1.12
Fe_2O_3	6.44	0.79	8.03	2.88	4.54	1.15	5.84	1.12
MnO	0.09	0.02	0.11	0.04	0.08	0.03	0.09	0.03
MgO	2.87	0.97	3.44	0.91	1.41	0.69	2.63	0.87
CaO	3.56	1.50	3.42	1.63	2.31	0.53	3.27	1.33
Na_2O	3.56	0.54	3.07	0.55	3.74	0.46	3.29	0.47
K_2O	2.63	0.51	2.77	0.55	2.62	0.57	2.74	1.28
P_2O_5	0.18	0.05	0.20	0.06	0.16	0.08	0.21	0.15
Rb	96 ±	25	109 ±	26	70 ±	15	111 ±	56
Sr	204	61	230	77	222	78	363	214
Ba	455	132	587	174	613	179	1079	868
Sc	17	4	21	8	11	3	16	5
Y	25	6	35	12	31	5	34	12
La	20	9	19	8	21	6	25	15
Ga	21	2	16	1	15	1	19	3
Pb	27	21	20	14	18	4	21	10
Zr	176	38	164	39	166	29	201	80
Hf	5	1	6	1	5	1	6	2
Nb	9	4	15	10	9	3	10	5
V	93	35	165	52	71	35	102	31
Cr	78	84	80	64	53	41	74	50
Ni	10	8	23	23	14	15	33	31
Co	13	4	15	4	10	4	14	4
Cu	9	7	18	17	10	10	22	8
Zn	111	59	98	24	53	14	73	24
Th	11	4	11	4	12	3	21	10
U	5	2	3	1	3	1	1	1

1: Drilling Hechtsberg near Haslach (n = 22); 2: Drilling Ettersbach near Simonswald (n = 9); 3: Drilling Kunklerwald near Glottertal (n = 14); 4: Southern Schwarzwald (n = 21).

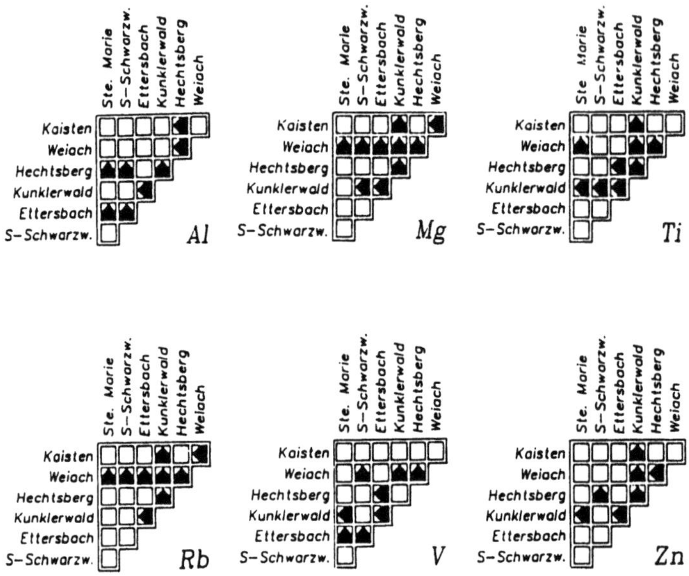

Fig. 2 Examples of highly significant differences between paragneiss series from the Schwarzwald and northern Switzerland (Kaisten, Weiach). Black symbols indicate differences above the 99% significance level. The 'roof tops' point towards the unit with the lower element content. From Müller (1988).

respective rocks (Table 2, Fig. 2). Analyses of metasediments of the Schwarzwald, Vosges, and northern Switzerland gneiss basement and their statistical processing revealed many highly significant differences between rock sequences from different areas (Müller and Wimmenauer 1986, Müller 1988b). The results obtained so far confirm the model of a lithostratigraphic division of the Precambrian in the Schwarzwald and emphasize a remarkable difference between the metasedimentary units north and south of the Badenweiler-Lenzkirch thrust zone. On the other hand, closer relations between the gneisses of Sainte-Marie-aux-Mines (Vosges) and the Central Schwarzwald are suggested.

3.2.2 Granulites and Related Rocks

Granulites and related rocks occur in several distinct zones of the Central Schwarzwald. The first zone extends discontinuously over 10 km in the St. Peter-Glottertal-Kandel area (Groschopf et al. 1980). It has been penetrated with an apparent thickness of 250 m in the Kunklerwald research drilling (1985). A second zone of more than 20 km in length has been recognized between the Elz and Kinzig valleys. The 'granulitic' rocks are intercalations within common paragneisses; they have been subjected, together with the latter, to a regional HT-LP metamorphism (amphibolite facies metamorphism and anatexis, see section 3.2.7). The best preserved granulites are strongly foliated, compositionally banded rocks with acid (granitoid) and intermediate (metagreywacke) layers. The paragenesis quartz + mesoperthite + plagioclase (antiperthitic) + kyanite + garnet + rutile is taken as a criterion for a MP-HT metamorphism. The texture of the most relictic rocks is blastomylonitic with felspar 'Augen' and strongly elongated quartz grains

('Plattenquarze'). Hacker (1986) studied the development of the granulite fabrics in great detail; while the MP deformational phase is expressed in discoidal and platy quartz, the younger anatectic and metablastic overprint resulted in a general recrystallization and disordering of the oriented structures. Apparently, these zones of ductile (HT) shearing were particularly susceptible to feldspar metablastesis. Later deformational phases (D4, D5, see section 3.1) could again modify the fabrics by renewed shearing and re-orientation.

Genetically, the granulite horizons can be interpreted as fragments of the deeper crust, transported upwards on major thrust planes (Behr et al. 1986) or as in-situ formations on those very same planes under appropriate conditions (cf. Fuchs and Scharbert 1979). In any case, fragments of metagabbro and ultrabasites otherwise unknown in the environment emphasize the tectonic transport along those zones. In the Elz-Kinzig valley granulite zone, the granulites are locally seen as phacoliths in a gneiss and metablastite surrounding and eclogite-amphibolite lenses are associated with them.

3.2.3 Metabasites (Table 3)

Small lenses of metabasites (a few meters to 1 km maximum in length) are very frequent in the gneisses of the Central Schwarzwald. At least two main types can be distinguished petrographically:

1. Eclogitogenic amphibolites and rare relict eclogites; they occur preferably in the monotonous metagreywacke and metapelitic paragneisses (see section 3.2.1).
2. Common hornblende-plagioclase amphibolites in various environments (paragneiss, orthogneiss, alternating with leptynite).

The petrographic phenomena of the eclogitogenic amphibolites are very complex (Klein and Wimmenauer 1984). Characteristic reaction textures are:

1. Omphacite --> diopside + plagioclase symplectite --> diablastic hornblende + plagioclase;
2. Garnet --> hornblende + plagioclase (kelyphite);
3. Garnet + quartz --> plagioclase + orthopyroxene;
4. Kyanite --> corundum (± spinel) + plagioclase;
5. HP-mica --> oligoclase (and/or orthoclase) + corundum (or sillimanite) + Fe-phlogopite.

The compositions of the relict HP phases and the reactions indicate an evolution from a high-pressure stage (≈ 1.5 GPa, 600 - 700° C) through a granulite stage to an amphibolite stage, the latter being characteristic for the surrounding gneisses and migmatites (exceptions see section 3.2.4).

The 'common' amphibolites consist predominantly of hornblende and plagioclase; quartz, biotite, diopside and rarely garnet may also be present. In the largest amphibolite body of the Schwarzwald, Urenkopf near Haslach, there occur ultrabasic differentiates (spinel- and plagioclase-bearing hornblendites). Low-grade alterations with actinolite, epidote, chlorite, sericite, prehnite, albite, and calcite are very widespread in all of the metabasite varieties.

Geochemical studies on the ultrabasites and metabasites of the Central Schwarzwald by Eckhardt et al. (1986) indicate an island-arc-tholeiitic character of most of the amphibolites of Hirschmann's 'Formation with amphibolites', while Büsch et al. (1985) emphasize the low-potassium-tholeiitic affinities of the amphibolites associated with leptynites in

Table 3. Chemical analyses of characteristic metabasites and accompanying HP-rocks from the Kinzigtal region

	1	2	3	4	5	6	7	8
SiO_2	51.80	49.88	47.20	63.00	67.41	55.29	47.47	43.50
TiO_2	0.42	0.93	1.02	0.67	0.61	0.84	0.15	0.10
Al_2O_3	6.77	10.49	19.99	17.31	15.31	20.92	25.80	18.87
Fe_2O_3	2.53	1.64	1.40	1.26	1.63	2.40	0.16	3.08
FeO	6.14	8.71	6.35	3.99	2.94	4.19	3.06	5.40
MnO	0.16	0.24	0.10	0.10	0.13	0.14	0.07	0.13
MgO	17.81	13.12	7.61	2.45	1.87	4.97	4.81	15.55
CaO	13.23	11.00	11.18	3.70	2.90	4.18	15.83	10.32
Na_2O	0.66	1.67	2.12	4.26	4.41	4.81	1.49	0.75
K_2O	0.05	0.38	0.86	1.03	0.67	0.41	0.42	0.78
P_2O_5	0.03	0.16	0.08	0.13	0.19	0.15	0.02	0.02
L.o.i.[a]	1.51	1.59	1.37	1.82	1.74	1.35	1.40	1.76
Total	101.08	99.81	99.28	99.72	99.81	99.65	100.68	100.26
V	137	257	243	111	93	123	108	56
Ni	94	123	12	23	10	10	25	60
Cr	1369	842	166	68	36	37	480	246
Rb	10	10	26	35	21	14	10	20
Sr	104	169	295	312	204	47	276	114
Ba	81	85	162	355	216	124	22	58
Zr	46	28	46	126	148	275	29	20

1. Garnet pyroxenite (gn, cpx, opx, hbl), Hausach (Hornberg);
2: Eclogite amphibolite, Mühlenbach (Haslach);
3: Eclogite amphibolite, Am Thurm near Hausach (Hornberg);
4: Intermediate former HP-rock, associated with 3, Am Thurm (Hornberg);
5: Acid former HP-rock (metagreywacke), associated with 3 and 4;
6: Intermediate garnet-gedrite-albite amphibolite, Sulzbach valley (Haslach);
7: Diopside-bearing amphibolite, Hohebannstein (Haslach);
8: Spinel-bearing hornblendite, Hohebannstein (Haslach).

Names in parentheses: designations of the topographic maps 1:25 000.

a = loss on ignition

Zindelstein. The rare ultrabasites in the Central and Southern Schwarzwald originate from slightly depleted mantle lherzolites and harzburgites.

3.2.4 Intermediate and Acid High-Pressure Rocks (Table 3)

Some of the eclogite-amphibolite bodies in the Central Schwarzwald are interlayered with or accompanied by intermediate and acid metamorphic rocks, which also show relics of high-pressure parageneses (Wimmenauer and Adam 1985). The variety of their chemical analyses comprises andesitic, trondhjemitic, keratophyric, greywacke, and pelitic compositions. Garnet and kyanite occur as HP-mineral relics, but certain textures indicate the former existence of other HP phases, e.g.

1. Jadeite + quartz --> albite + little amphibole;
2. Mg-Ca-Al garnet --> gedrite + plagioclase (± biotite, iron ore);
3. Omphacite --> diopside + plagioclase (symplectite);
4. HP mica --> feldspars + corundum (or sillimanite) + Fe-phlogopite.

The observations indicate that at least parts of the gneiss massif, including rocks of supracrustal origin, were subjected to high-pressure conditions together with their basic intercalations (the present eclogites, Fig. 3).

Another noteworthy phenomenon are quartz veins within some of the eclogites, which contain pseudomorphs after a HP mica, as mentioned above, up to 5 cm large, plus rutile, zoisite, and black apatite.

3.2.5 Ultramafitites

Ultramafitites of mantle origin occur as rare and very small bodies in different gneiss environments. Most of them are serpentinized spinel lherzolites. In one of the occurrences (Hohfelsen near Bernau, close to the Badenweiler-Lenzkirch thrust zone), the transformation of 'primary' garnet to spinel has been observed. Baatz (1988) estimates a minimum pressure of 2 GPa for the earliest recognizable parageneses in the ultramafitites. After Werling (1986), a HP-HT mantle paragenesis was re-equilibrated at temperatures between 800 and 1100°C and pressures below 1 GPa (garnet-spinel-opx-cpx peridotite). The kelyphitization of garnet and the formation of Mg-chlorite, orthoamphibole, and serpentine occurred under amphibolite facies and lower conditions.

3.2.6 The Orthogneisses of the TTG Suite (Table 1)

Orthogneisses of tonalitic, trondhjemitic, and granodioritic composition are particularly frequent in the Central Schwarzwald (Elz and Kinzig valley areas). They form lenses and belts of up to 10 km in length and a few kilometers in width, mostly concordant with the surrounding paragneisses. The contacts with the latter are rarely clear-cut; mostly, gradations in composition and structure or repeated alternations of the two main gneiss types are observed. These phenomena have been interpreted by Rein (1952) and Mehnert (1953) as expressions of a migmatitic interpenetration of the plutonic protoliths of the orthogneisses proper and their metasedimentary country rocks. The question is still open, whether the plutonites were intrusive or in-situ formations by anatectic mobilization of their immediate environment (anatexis I of the earlier authors).

The petrography of the orthogneisses is rather simple. Plagioclase, quartz, biotite, and orthoclase are the predominant constituents of the granodioritic types. In the trondhjemitic and tonalitic orthogneisses, orthoclase is only a minor constituent or absent. Green hornblende is characteristic of the tonalitic orthogneisses. The anorthite contents of the plagioclases vary between 25 and 45%. Occasionally, garnet occurs along with biotite, but not with amphibole. Allanite is a very characteristic accessory mineral in the 'amphogneisses', that is, in the rocks transitional between typical ortho- and paragneiss.

The 'Flaser' texture, a distinguishing mark of the orthogneisses, seems to be inherited from an earlier (plutonic) coarse-grained fabric. In many cases, a cataclastic comminution of the minerals seems to have preceded the present granoblastic texture.

Aplitic, more fine-grained orthogneisses are frequent in the Kinzig valley area; they occur as deformed dikes or veins in the common orthogneisses or as larger bodies associated with them. Garnet and sillimanite are minor mineral constituents of these meta-aplites.

3.2.7 The Migmatites

Migmatites of anatectic origin are very widespread throughout the Central Schwarzwald, particularly in its southern part (Schauinsland-Feldberg area). Mehnert (1953-1963) described these rocks in great detail and interpreted them as products of partial melting of para- and orthogneisses (anatexis). The conditions of their formation were studied experimentally by Winkler and von Platen (1961), Steuhl (1962), Mehnert et al. (1973), Mehnert and Büsch (1982). The petrographic phenomena vary greatly depending on the type of gneiss affected by those processes, the degree of partial melting, and the tectonic movements contemporaneous with the anatexis.

The most likely P-T conditions during the climax of the anatexis are 0.3 to 0.4 GPa and 650 to 720°C; somewhat higher temperatures should be assumed for the mobilization of hornblende-tonalitic metatects.
The parageneses plagioclase + quartz + biotite + cordierite ± orthoclase and plagioclase + hornblende + quartz are most characteristic for the migmatites. Biotite and sillimanite, regularly separated from each other by cordierite, are often observed in the restites of metatexites. Garnet occurs only in special intercalations, such as kinzigites (Büsch et al. 1980).

Generally, the anatexis seems to have taken place in a closed system, still, Schneider and Büsch (1980) assume migration of K, Rb, and Ba to some extent, because the budget of these elements between the paleosome and the anatectic products is not balanced. Introduction of potassium is also taken into consideration for certain K-feldspar metablastites along the southern margin of the gneiss massif immediately north of the Badenweiler-Lenzkirch thrust zone and near Offenburg, there, especially, known as 'durbachites' (Morche 1979).

An anatectic overprint of early Variscan age transformed the sodic leptynites (see section 3.2.1) on the southern border zone to granitoid rocks ('Randanatexit' of Altherr and Maass 1977, 'Gisiboden-Verband' of Werling and Altherr 1986). The granitoids are remarkably richer in K than the relict lepynites and leucocratic gneisses of the same zone. Paragneisses and hornblende gneisses are also affected by K-feldspar metablastesis. K-Ar datings of hornblendes from such rocks gave model ages between 328 and 342 Ma (Werling and Altherr 1986). With the approach to the thrust zone of Badenweiler-Lenzkirch, the gneisses, the metablastic and anatectic rocks are increasingly mylonitized, but their amphibolite-facies parageneses are partly preserved.

4 THE BADENWEILER-LENZKIRCH ZONE

Geology, petrography, and tectonic history of the Badenweiler-Lenzkirch zone were the subjects of many studies in the past (e.g. Sittig 1969, 1981; Altherr and Maass 1977; Burgath 1972; Burgath and Maass 1973; Maass 1981; Fluck et al. 1980). In a recent note, Werling and Altherr (1986) have drawn up a picture of the metamorphic evolution in the central part of that zone (surroundings of Bernau; see also the Geological Map 1 : 25 000, sheet Feldberg by Wimmenauer and Schreiner 1981). In this area, the zone consists of at least three main units of different lithology and metamorphic status:

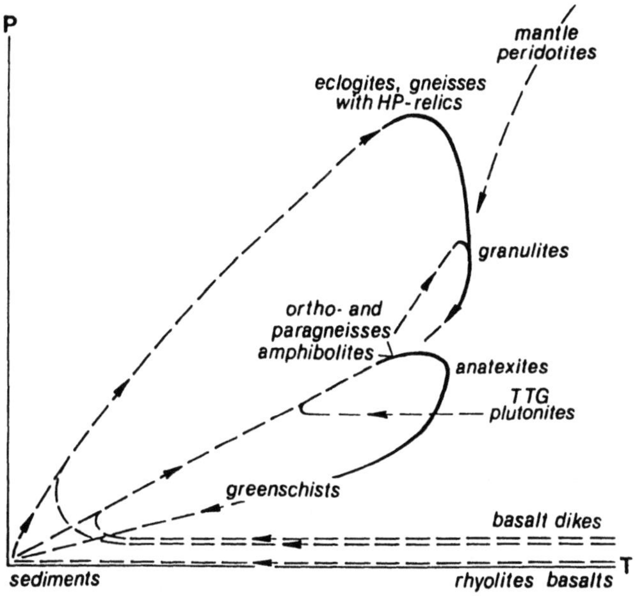

Fig. 3 Schematic diagram of the metamorphic paths of the gneisses and their intercalations in the Central Schwarzwald (not to scale). P and T data are available for the sections shown by heavy lines; broken lines indicate inferred sections of the evolution.
Three main branches of the metamorphic evolution are assumed:
1. Sedimentary and basaltic protoliths were subjected to a HP metamorphism up to 1.5 MPa and temperatures of 650 to 700°C. The retrograde evolution through a granulite, an amphibolite, and, locally, a greenschist stage are well documented.
2. Sedimentary and volcanic (acid and basic) protoliths have been metamorphosed in the LP amphibolite facies to common gneisses, leptynites, and amphibolites. In particular, deep-seated shear zones, the conditions of the granulite facies were reached (shown as an extra loop arising from the gneisses towards higher P and T conditions). The plutonites of the TTG suite took place within the paragneisses earlier to the main regional metamorphic event. All of the gneisses, including those of the HP branch, were more or less intensely affected by the regional anatexis (T approximately 700°C, P less than 0.5 GPa).
3. The rare mantle peridotites bring along relict parageneses of very high P and T; they underwent retrograde transformations of various degrees, particularly serpentinization.

1. The assemblage of Geschwend-Sengalenkopf: metapsammites, metaconglomerates, nodular schists with cordierite. Muscovite, quartz, garnet, andalusite, cordierite, and biotite characterize the degree of metamorphism.
2. The assemblage of Bernau: metapsammites with quartz, albite-rich plagioclase, muscovite, and chlorite; epidote, chlorite, adularia, and rare actinolite have been observed in quartz veins. The contact-metamorphic effects from the Bärhalde and St. Blasien granites produced cordierite nodules (altered), pale biotite, and a weak blastic crystallization of plagioclase.
3. The sedimentary and volcanic rocks of the Upper Devonian and Lower Carboniferous: siltites, greywackes, pelites, siliceous sediments; 'porphyries' of varying composition (granitic, trachytic according to

Burgath 1972). Although the sediments and volcanics are usually considered as non-metamorphic, certain 'metamorphic' minerals seen in narrow veins (chlorite, epidote, actinolite, adularia, pyrite) indicate conditions exeeding those of common diagenesis. Farther to the west in that zone, basaltic-andesitic pyroclastites habe been transformed to spilites (Burgath 1972).

4. Between Schönau and the Blössling mountain (west of Bernau), the southern border of the Paleozoic zone is made up of metapsammites and greenschists, the so-called 'Südrandschuppen', which again show a higher grade of metamorphism (actinolite, chlorite, albite-rich plagioclase, and epidote in the basic, metavolcanic members of the assemblage; Baatz 1976).

5 THE GNEISSES AND MIGMATITES OF THE SOUTHERN SCHWARZWALD

Gneisses and migmatites occur in several 'rafts' between the large Variscan plutons (areas of Badenweiler, Todtmoos, Urberg, Wutach and Steina valleys) and in a more continuous zone in the extreme south of the Schwarzwald basement outcrop. Gneisses belonging to the same zone have also been met with in several deep drillings in northern Switzerland (e.g. Kaisten AG and Weiach TG). On the first approximation, the gneisses of the Southern Schwarzwald seem to have the same characteristics and history as those of the Central Schwarzwald. Still, there are some remarkable differences between the two units, which even emphasize more the role of the Badenweiler-Lenzkirch zone as an important dividing line. The main points of difference between the Central and the Southern Schwarzwald are listed in Table 4. Among the most interesting rocks in the Southern Schwarzwald, the metagabbros and meta-anorthosites have to be mentioned. Innumerable blocks of decimeter to hectometer size occur as inclusions within the 'Wehra-Wiesetal diatexite', a zone of Variscan in-situ granitization. From the wide dispersion of those rocks and the fragmental nature of the individual bodies it may be concluded that one or a few intrusions of gabbroic and related rocks have been disrupted tectonically; their fragments were scattered over a distance of more than 20 km.

Eclogites of crustal origin, as those of the Central Schwarzwald, do not occur in the southern unit. A few bodies of mantle peridotite are associated with the leptynites of the Todtmoos and Wutach areas.

These and some other differences evoked in Table 4 indicate that two blocks of pre-Variscan basement, widely apart from each other before, have been approached to within a few kilometers distance along the Badenweiler-Lenzkirch tectonic zone.

The serpentinites of the Southern Schwarzwald are interpreted as mantle fragments; they are classified as alpinotype. Slightly depleted lherzolites consist of olivine, orthopyroxene, clinopyroxene, and spinel as 'primary' minerals. The thermobarometric evaluation of the compositions of these phases indicates a pressure of 2.0 to 2.5 GPa and a temperature of roughly 900°C for their final equilibration (Stenger et al. 1987, Baatz 1988).

A concise review of the geology of the Southern Schwarzwald from the Precambrian up to the Quaternary has recently been given by Huber and Huber-Aleffi (1984). In addition to the general view, the book contains detailed results on the phenomena of ductile and brittle deformation and fracturing in this area. The studies were executed on behalf of the 'Nationale Genossenschaft für die Lagerung radioaktiver Abfälle' (NAGRA) in Baden (Switzerland).

Table 4: Major differences between the metamorphic basements of the Central and the Southern Schwarzwald (after Stenger et al. 1988)

Central Schwarzwald	**Southern Schwarzwald**
Widespread metapelitic and metagreywacke paragneisses	Metapelitic and metagreywacke paragneisses, with pyroxene gneisses and calc-silicate rocks in the units of Murgtal and Kaisten (AG)
K-rich leptynites in the center; Na-rich leptynites along the southern margin	K-rich leptynites widespread
Tonalitic-trondhjemitic orthogneisses prominent	Lacking
Crustal eclogites and eclogite-amphibolites frequent	Lacking
Some garnet-spinel peridotites	Spinel peridotites only
Metagabbros very rare	Metagabbros and meta-anorthosites, fragments of layered intrusions
Pronounced low-velocity zone	Fading out
Pronounced lamination of the lower crust	Weak lamination of the lower crust
Higher values of Bouguer gravity	Lower values of Bouguer gravity

6 PETROGRAPHY OF THE KTB TARGET AREA AROUND HASLACH IM KINZIGTAL

An area of 10 x 10 km surface around the projected drilling site has been mapped in detail by H. Klein and R. Stenger (1987, Fig. 4). The following geological units are distinguished on the map:

1. Metagreywacke and metapelitic paragneisses: plagioclase-quartz-biotite gneisses, partly with sillimanite and, more rarely, orthoclase, cordierite, or garnet. Subordinate intercalations are metaquartzite gneisses, graphite-bearing gneisses, hornblende gneisses, and calc-silicate rocks.
2. Leucocratic gneisses and leptynites with K-feldspar, plagioclase, quartz, little biotite, garnet, and sillimanite. They form minor conformable intercalations in other gneisses, but also areas of considerable extension north and southeast of Haslach.
3. Alternation of ortho- and paragneisses (partly 'heterogene Mischgneise' of the earlier authors).
4. Orthogneisses of granodioritic, trondhjemitic, and tonalitic composition (plagioclase, quartz, biotite ± orthoclase or hornblende, little garnet in some cases).

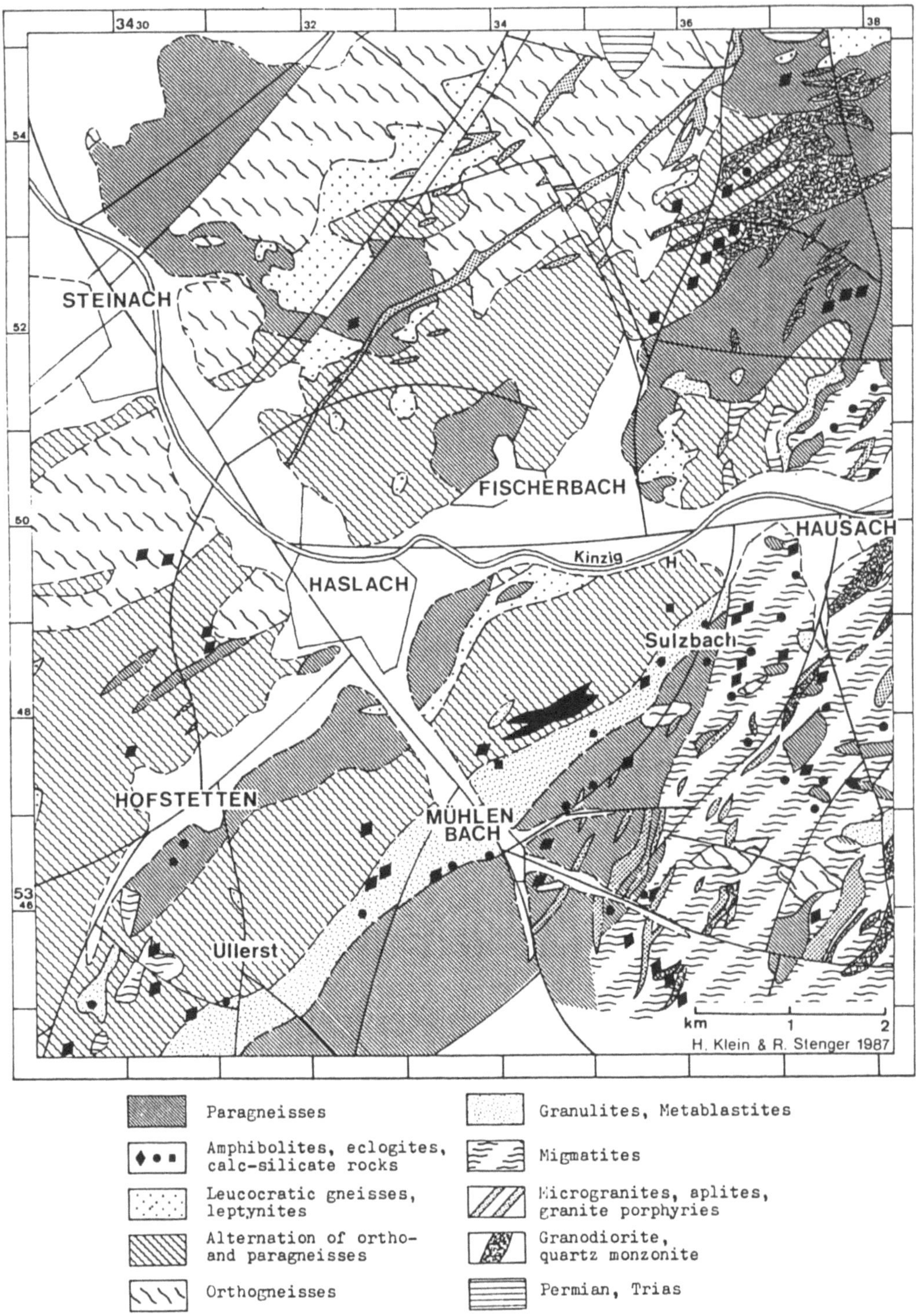

Fig. 4 Petrographic map of the Kinzigtal area around Haslach. Fracture tectonics after Kleinschmidt, Flöttmann, and Gallus (in von Gehlen et al. 1986). H = site of the Hechtsberg research drilling.

5. Blastomylonitic granulites in a prominent SW-NE striking zone in the southeastern section of the map. Characteristic relic parageneses are quartz + mesoperthite + plagioclase (partly antiperthitic) + garnet + kyanite + rutile. The typical granulites show compositional banding and often a sort of 'Augen' texture with porphyroclasts of feldspar. The granulite relics are widely transformed to metablastic gneisses with a prominent granular texture (Kinzigitgneise of the ancient Geological Map 1 : 25 000).
6. Eclogite amphibolites occur within the same zone and also in the common paragneiss and migmatite environment.
7. Lenses and layers of common amphibolite are found within the paragneisses, the orthogneisses, the orthogneiss-paragneiss alternation, and the migmatites. The largest amphibolite body on the Urenkopf comprises metagabbroic, meta-anorthositic, and hornblenditic varieties.
8. The Variscan plutonic activity is represented by many dikes and stocklike bodies of microgranite (mostly aplitic), granodiorite and quartz monzonite (the 'Glimmersyenite' of the earlier authors), and porphyritic microgranites ('Granitporphyre' of the Geological Map 1 : 25 000).

Among the paragneiss assemblages of the Southern Schwarzwald, that of the Murgtal excels by its frequent intercalations of pyroxene gneiss, calc-silicate rocks, and rare marble (Krützfeld 1980). The same assemblage has also been penetrated for several hundred meters in the drilling of Kaisten (AG) in northern Switzerland. The drilling Weiach (TG) encountered a gneiss suite which shows quite significant geochemical differences towards all other sequences studied so far (particularly high contents in so-called basaltophile elements).
Regarding the metamorphic conditions in the Southern Schwarzwald, it is noteworthy that the high-pressure metamorphism of crustal rocks known from the Central Schwarzwald (see sections 3.2.3 and 3.2.4) cannot be observed. In some of the leptynites near Todtmoos, relics of a granulite stage (orthopyroxene, garnet, and kyanite in acid rocks) are preserved. The predominant status of the leptynites and gneisses is that of the amphibolite facies; anatectic phenomena are widespread and lead, in the Wehra-Wiesetal diatexite, to the formation of coarse-grained granitoid rocks. On the other hand, very fine-grained leptynites of igneous and sedimentary origin occur as intercalations in the same zone in the Wehra valley ('Gersbacher Schiefer', Lämmlin 1981, Vogt 1983).

The area recently mapped is a typical section of the pre-Variscan basement of the Central Schwarzwald, since most of the rock types described in sections 3.2.2 - 3.2.7 are represented. The different metamorphic units are arranged in SW-NE striking bands from a few hundred meters up to two kilometers in width. More irregular dispositions are seen between Steinach and Fischerbach, where lobate areas of leucocratic gneiss and paragneiss interfere with the general pattern, and in the east and southeast, where the anatectic overprint gave rise to the formation of larger volumes of migmatites.
The ductile shear zone with blastomylonitic granulites and metablastites (between Ullerst and Hausach) separates a section with eclogite amphibolites in the southeast from another one with common amphibolites in the northwest. Thus, the zone seems to mark the border between two units of the pre-Variscan basement with different metamorphic features. Although it is quite well expressed in the field and on the map, it is not easy to define the structural relations of those two units and the shear zone proper. According to Behr et al. (1986) and Kleinschmidt et al. (1986), the ductile shear zones of the Central Schwarzwald are important elements of a pre-Variscan thrust event; along the shear zone

mapped in the Kinzigtal area, a nappe of metamorphic rocks with eclogites and other HP rocks has been thrusted over metamorphic rocks of MP and LP origin (with common amphibolites), possibly in a northward direction. The granulite parageneses and blastomylonitic textures may have been acquired during the process of ductile shearing (D3 deformation, see section 3.1). They have been strongly overprinted by the deformations D4 (and the concomitant anatectic processes), and D5. Late fracture tectonics again disturbed the earlier architecture.

REFERENCES

Altherr, R. and Maass, R. (1977): Metamorphite am Südrand der zentralschwarzwälder Gneismasse zwischen Geschwend und Bernau.- Neues Jahrb. Geol. Palaeontol., Abh., **154**, 129-154, Stuttgart.

Baatz, K. (1976): Petrographie und Geochemie der Südrandschuppen der Zone Badenweiler-Lenzkirch.- Unpubl. Thesis (Diplomarb.), Freiburg i.Br., 67 pp.

Baatz, K. (1988): Ultramafitite in Schwarzwald und Vogesen.- Thesis Freiburg i.Br., 191 + 209 pp.

Behr, H.J., Hacker, W., Heinrichs, T., and Hirschmann, G. (1986): Contribution to: von Gehlen, K. et al. (1986).

Burgath, K. (1972): Der unterkarbonische Vulkanismus des Schwarzwaldes.- Thesis Freiburg i.Br., 350 pp.

Burgath, K. and Maass, R. (1973): Die variszische Entwicklung im südlichen Schwarzwald.- C. R. 7ème Congr. Stratigr. Géol. Carbonifère, **2**, 195-209, Krefeld.

Büsch, W., Schneider, G., and Mehnert, K.R. (1974): Initial melting at grain boundaries, II.- Neues Jahrb. Mineral., Mh., **1974**, 345-370, Stuttgart.

Büsch, W, Matthes, S., Mehnert, K.R., and Schubert, W. (1980): Zur genetischen Deutung der Kinzigite im Schwarzwald und Odenwald.- Neues Jahrb. Mineral., Abh., **137**, 233-256, Stuttgart.

Büsch, W., Hofer, A., and Spohn, A. (1985): Tholeiitische Amphibolite des Schwarzwaldes: eine Vergesellschaftung von MORB und IAT?.- Neues Jahrb. Mineral., Abh., **152**, 2344, Stuttgart.

Eckhardt, F.J., Burgath, K., and Mohr, M. (1986): Contribution to: von Gehlen, K. et al. (1986).

Fluck, P., Maass, R., and von Raumer, J.F. (1980): The Variscan units east and west of the Rhinegraben.- Colloq. C 6, 26. Congr. Internat. Géol., 112-131, Paris.

Fuchs, G. and Scharbert, H.G. (1979): Kleinere Granulitvorkommen im niederösterreichischen Moldanubikum und ihre Bedeutung für die Granulitgenese.- Verh. Geol. Bundesanst., **1979**, 29-49, Wien.

von Gehlen, K., Kleinschmidt, G., Stenger, R., Wilhelm, H., and Wimmenauer, W., (editors, 1986): Ergebnisse der Vorerkundungsarbeiten, Lokation Schwarzwald.- 2nd KTB Colloq., 160 pp., Seeheim.

Groschopf, R. and Schreiner, A. (1980): Geol. Karte Baden-Würtemberg 1:25 000, **7913** (Freiburg i.Br. NO), with explanations, 112 pp., Stuttgart.

Hacker, W. (1986): Contribution to: von Gehlen, K. et al. (1986).

Hirschmann, G. (1985): Lithostratigraphic units in the gneisses of the Central Schwarzwald.- 2nd Internat. Symp. Observ. Contin. Crust through Drilling, Excursion Guide Schwarzwald, 8-15, Seeheim.

Huber, M. and Huber-Aleffi, A. (1984): Das Kristallin des Südschwarzwaldes.- NAGRA Techn. Ber., **84-30**, 226 pp., Baden (CH).

Jenkner, B. (1986): Geothermievorbohrungen im Mittleren Schwarzwald für das Kontinentale Tiefbohrprogramm der Bundesrepublik Deutschland.- Zwischenbericht des Geol. Landesamtes Baden-Württemberg, 23 pp., Freiburg i.Br. 1986.

Kessler, G. and Leiber, J. (1980): Geol. Karte Baden-Württemberg 1:25 000, **7813** (Emmendingen), with explanations, 151 pp., Stuttgart.

Klein, H. and Stenger, R. (1987): Geologisch-petrographische Kartierung im mittleren Kinzigtal (Zentralschwarzwald).- Jahresh. Geol. Landesamtes Baden-Württemberg, **29**, 47-59, Freiburg i.Br.

Klein, H. and Wimmenauer, W. (1984): Eclogites and their retrograde transformation in the Schwarzwald.- Neues Jahrb. Mineral., Monatsh., **1984**, 25-38, Stuttgart.

Kleinschmidt, G., Flöttmann, T., and Gallus, B. (1986): Contribution to: von Gehlen, K. et al. (1986).

Krohe, A. and Eisbacher, G.H. (1988): Oblique crustal detachment in the Variscan Schwarzwald, Southwestern Germany.- Geol. Rundsch., **77**, 25-43, Stuttgart.

Krützfeld, B. (1980): Geländeaufnahme des Gesteinsprofils im Murgtal zwischen Hottingen und Murg.- Unpubl. Thesis (Diplomarb.), 91 pp., Freiburg i.Br.

Lämmlin, I. (1981): Petrographische und geochemische Untersuchungen im Gneisgebiet des Südschwarzwaldes.- Thesis Freiburg i.Br., 157 pp.

Lippolt, H.J., Kober, B., Hradetzky, H., and Mertz, D. (1986): Contribution to: von Gehlen, K. et al. (1986).

Maass, R. (1981): The Variscan Black Forest.- Geol. Mijnbouw, **60**, 137-143, Amsterdam.

Mehnert, K.R. (1953 - 1962): Petrographie und Abfolge der Granitisation im Schwarzwald.- Neues Jahrb. Mineral., Abh., **85**, 59-140, **90**, 39-90, **98**, 208-247, **99**, 161-199, Stuttgart.

Mehnert, K.R. and Büsch, W. (1982): The initial stage of migmatite formation.- Neues Jahrb. Mineral., Abh., **145**, 211-238, Stuttgart.

Mehnert, K.R., Büsch, W., and Schneider, G. (1973): Initial melting at grain boundaries of quartz and feldspar in gneisses and granulites.- Neues Jahrb. Mineral., Monatsh., **1973**, 165-183, Stuttgart.

Morche, W. (1979): Petrographische und geochemische Untersuchungen am Durbachit des Schwarzwaldes.- Unpubl. Thesis (Diplomarb.), 131 pp., Freiburg i.Br.

Müller, H. and Wimmenauer, W. (1986): Geochemische Korrelation moldanubischer und saxothuringischer Metasedimentserien aus Schwarzwald, Vogesen und der Nordschweiz.- Fortschr. Miner., **64**, 127, Stuttgart.

Müller, H. (1988a): Geochemistry of metasediments in the Hercynian crust of Schwarzwald, Vosges, and northern Switzerland.- Tectonophysics **157**, 97-108, Amsterdam.

Müller, H. (1988b): Geochemie prävariskischer Metasedimenteinheiten aus Schwarzwald, Vogesen und der Nordschweiz unter besonderer Berücksichtigung der Schiefer von Baden-Baden.- Thesis Freiburg i.Br., 205+96 pp.

Plaumann, S. (1985): Schweremessungen im Nordschwarzwald (Raum Baden-Baden - Gaggenau).- Ber. Niedersächs. Landesamtes Bodenforsch,. Archiv No. 97981, Hannover.

Pucher, R. (1986): Interpretation der magnetischen Anomalie bei Baden-Baden.- Geol. Jahrb., **E 33**, 53-64, Hannover.

Rehfeld, U. (1983): A relictic staurolite-kyanite mineral paragenesis in the paragneisses of the Central Schwarzwald.- Neues Jahrb. Mineral., Abh., **146**, 170-181, Stuttgart.

Rein, G. (1952): Der Werdegang des Orthits in der magmatischen und metamorphen Abfolge des mittleren Schwarzwaldes.- Neues Jahrb. Mineral., Abh., **84**, 365-435, Stuttgart.

Schädel, K. and Stober, I. (1988): Neue Beobachtungen zur Gneis- und Granittektonik im mittleren Schwarzwald.- Geol. Jahrb., **A 107**, 1-34, Hannover.

Schneider, G. and Büsch, W. (1980): Zur Geochemie von Migmatiten am Beispiel von Gesteinen des Schwarzwaldes.- Berliner Geowiss. Abh., Ser. A, **33**, 1-33, Berlin.

Sittig, E. (1965): Der geologische Bau des variszischen Sockels nordöstlich von Baden-Baden.- Oberrhein. Geol. Abh., **14**, 167-207, Karlsruhe.

Sittig, E. (1966): Das metamorphe Altpaläozoikum des Nordschwarzwaldes.- Jahresber. Mitt. Oberrhein. Geol. Ver., N. F. **48**, 121-131, Stuttgart.

Sittig, E. (1969): Zur geologischen Charakterisierung des Moldanubikums am Oberrhein (Schwarzwald). Oberrhein. Geol. Abh., **18**, 119-161, Karlsruhe.

Sittig, E. (1981): Evidence for wrench-faulting within the Paleozoic Badenweiler-Lenzkirch zone (Southern Schwarzwald, W-Germany).- Neues Jahrb. Geol. Paläontol., Monatsh., **1981**, 431-448, Stuttgart.

Stenger, R., Baatz, K., Klein, H., and Wimmenauer, W. (1987): Metamorphic evolution of the pre-Hercynian basement of the Schwarzwald (FRG).- Terra cognita, **7**, 167, Strasbourg.

Stenger, R., Baatz, K., Klein, H., and Wimmenauer, W. (1988): Metamorphic evolution of the pre-Hercynian basement of the Schwarzwald (FRG).-

Tectonophysics, **157**, 117-122, Amsterdam.

Steuhl, H.H. (1962): Die experimentelle Metamorphose und Anatexis eines Parabiotitgneises aus dem Schwarzwald.- Chem. Erde, **21**, 413-449, Jena.

Vogt, U. (1983): Gesteine des Grundgebirges nordöstlich von Wehr (südlichster Schwarzwald).- Unpubl. Thesis (Diplomarb.), 85 pp., Freiburg i.Br.

Werling, E. (1986): Metamorphose und Tektonik in der Badenweiler-Lenzkirch-Zone im Raum Bernau.- Unpubl. Thesis (Diplomarb.), 121 pp., Karlsruhe.

Werling, E. and Altherr, R. (1986): Die metamorphe und tektonische Entwicklung des Südrandes der zentralschwarzwälder Gneismasse bei Bernau.- Abstr. Sympos. Strukturgeol., Kristallingeol., Tektonik, 77-80, Tübingen.

Wimmenauer, W. (1980): Lithology of the Precambrian in the Schwarzwald. An interim report.- Neues Jahrb. Mineral., Monatsh., **1980**, 364-372, Stuttgart.

Wimmenauer, W. (1984): Das prävariskische Kristallin des Schwarzwaldes.- Fortschr. Mineral., **62**, 69-86, Stuttgart.

Wimmenauer, W. and Adam, A. (1985): High-pressure phenomena in acid rocks accompanying eclogites in the Schwarzwald (SW Germany).- Terra cognita, **5**, 426, Strasbourg.

Wimmenauer, W. and Lim, S.K. (1988): L'association leptyno-amphibolique de la Forêt-Noire (R.F.A.).- Bull. Soc. Géol. France, 8ème Sér., **4**, 35-41, Paris.

Wimmenauer, W. and Schreiner, A. (1981): Geol. Karte Baden-Württemberg 1:25 000, **8114** (Feldberg), with explanations, 134 pp, Stuttgart.

Wimmenauer, W. and Stenger, R. (1988): Acid and intermediate HP-metamorphic rocks in the Schwarzwald (Fed. Rep. Germany).- Tectonophysics, **157**, 109-116, Amsterdam.

Winkler, H.G.F. and von Platen, H. (1961): Experimentelle Gesteinsmetamorphose V: Experimentelle anatektische Schmelzen und ihre petrogenetische Bedeutung.- Geochim. Cosmochim. Acta, **24**, 250-259, Oxford.

Zoubek, V. (1982): Über den Stand der Untersuchungen im Moldanubikum und die Beziehungen Moldanubikum-Oberproterozoikum (Brioverien).- Z. Angew. Geol., **28**, 305-315, Berlin.

Structural and Basement Evolution in the Central Schwarzwald Gneiss Complex

T. Flöttmann and G. Kleinschmidt

Geologisch-Paläontologisches Institut, J. W. Goethe-Universität, Senckenberganlage 32, D-6000 Frankfurt am Main, FRG

INTRODUCTION

The Central Schwarzwald Gneiss Complex (CGC) belongs to the highly metamorphic core ("Moldanubian") of the European Variscan orogenic belt. The prevailing lithologies are metasediments of possibly Precambrian depositional age (Wimmenauer 1980). Intercalated orthogneisses are abundant (Fig. 1). The dominant metamorphism took place under upper amphibolite facies conditions, locally reaching anatexis. Relics of eclogites and granulites occur as isolated but scattered pods within the paragneisses or form distinct chains within the basement (Fig. 1). The various lithologies underwent temporarily separate structural and metamorphic histories in space and time. As a consequence of the dominant low pressure/high temperature (LP-HT) metamorphism the segments of different crustal provenance were welded together and their previous tectonometamorphic history was somewhat blurred.

Structural analysis of metasediments, granulites and eclogites and their respective PT-paths are utilized to elucidate some aspects of the CGC-basement evolution.

STRUCTURAL ANALYSIS OF DIFFERENT LITHOTECTONIC UNITS

Structural Analysis of the Paragneisses

The first deformational event (D_1, Fig. 1) in the metapelitic rocks is recorded by relic quartz veins representing a first schistosity (s_1), which formed under greenschist facies conditions (Voll 1960).

During the second deformation (D_2) the first schistosity is folded isoclinally by B_2-folds (see Fig. 2).

The third deformation (D_3) refolds the B_2-folds homoaxially. The s_3-foliation, marked by a conspicuous separation of light (plagioclase + quartz) and dark (biotite) layers, is the dominant planar structural element in most gneisses. Shear along s_3-planes causes asymmetrical thickening and disruption of earlier fold hinges, locally leading to the formation of rootless intrafolial folds.

Due to the fourth deformation D_4, the dominant foliation (s_3) and the older structures are refolded by open, partly upright folds of dm to m scale (Fig. 2).

A fifth deformation (D_5) leads to crustal scale shearing within the basement (Kleinschmidt et al. 1985; Flöttmann 1988; Krohe & Eisbacher 1988). Deformation commences on gently NW dipping shear planes that evolve from and cut through B_4-folds. These planes are often coated with a mobilisate constisting of plagioclase, K-feldspar and quartz. On shear planes that have undergone marked noncoaxial deformation, synkinematic fibrolite grows at the expense of feldspars and biotite. The final deformational product is a mylonite consisting of alternating streaks of fibrolite and platy ribbon quartzes (HT-mylonites, Fig. 3). Within these zones B_4-hinges rotate into near parallelism with the stretching lineation formed by elongate fibrolites and biotites on the mylonite planes. The lineation represents

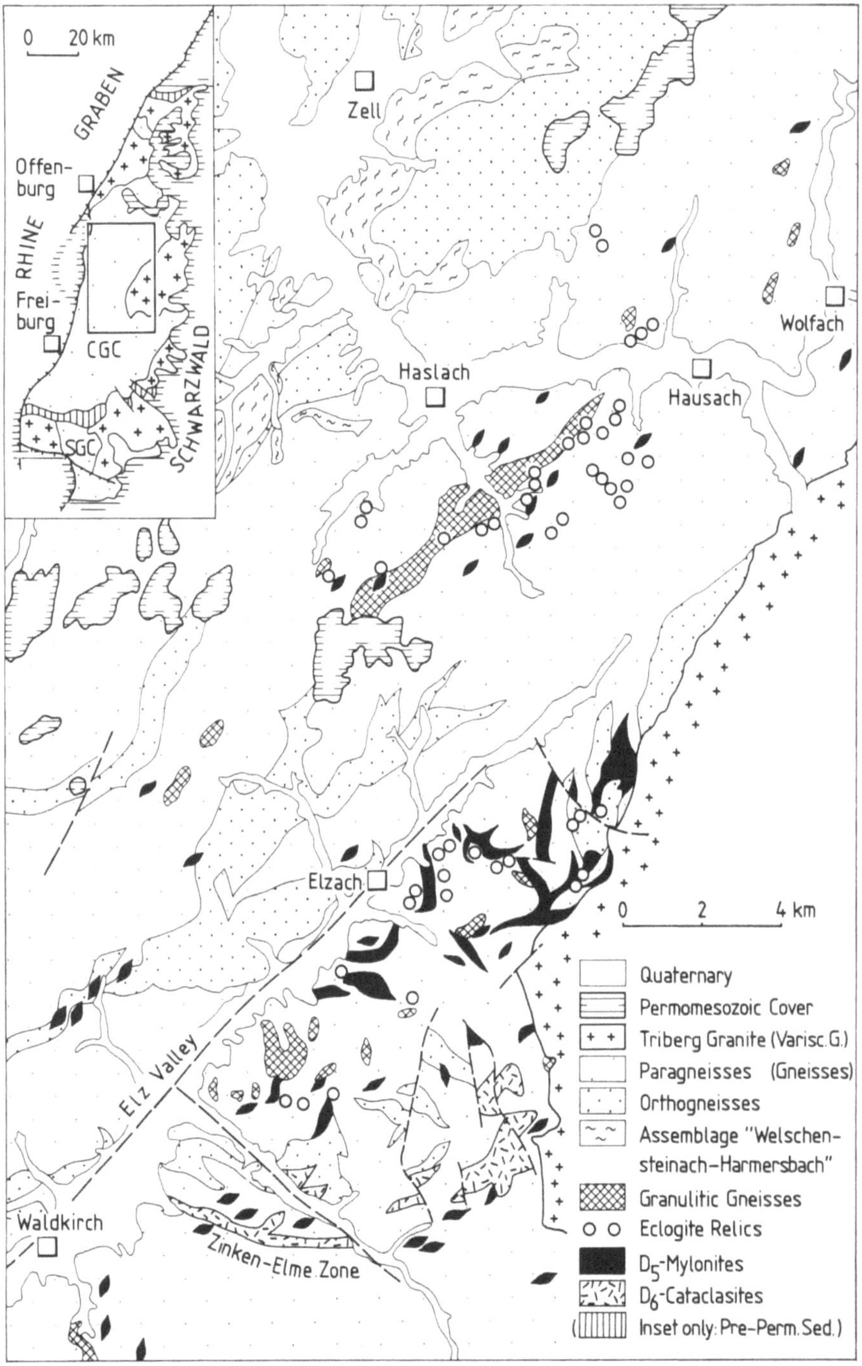

Fig. 1. Simplified map of the major geological units

the x-direction, and the mylonitic foliation represents the xy plane of the finite strain ellipsoid (x > y > z). Grain boundary diffusion was the dominant quartz deformation mechanism during this stage (Flöttmann 1988).

During later phases of the same deformation (D_5) the dominant deformation mechanism of quartz switched to crystal-plastic deformation. This is indicated by extensive quartz-ribbon formation as a consequence of progressive subgrain rotation ("rotation recrystallization" Burg 1986), grain boundary migration and undulose extinction of subgrains (MT-LT mylonites, Fig. 4). Feldspars deform brittlely. Andalusite reacts syndeformationally to form white mica. These features are particularly well developed in sheared lower Carboniferous (Lippolt et al. 1986) granitic dykes. Microfabrics such as s-c structures, non-conjugate ecc cleavages (compare Fig. 3, 4) and porphyroclast-systems etc. are used for a kinematic analysis of the D_5-Deformation (Simpson & Schmid 1983; Passchier & Simpson 1986; Platt & Vissers 1980).

South-southeastward thrusting dominated during the first deformational stages. South of the Elz-Valley the thrust planes dip 10-40° towards the NW. W of the Triberg Granite thrust motions follow planes that dip 20-50° towards the W. Thrusts consist either exclusively of fibrolite (HT) mylonites or, locally, show lower temperature mylonites as well. North of the Elz-Valley the deformation is less penetrative but follows generally the same trend.

Strike slip movements are concentrated in the narrow Zinken-Elme Zone. Ductilely deformed granites and gneisses show dextral sense of shear along steeply N-dipping (50-85°), narrow mylonite planes. Geometrical relationships and stress tensor analysis indicate that thrusts and strike-slip movements did not occur simultaneously. At least the early thrusting phases clearly predate strike-slip movements (Flöttmann 1988).

Regional crustal extension results in a widespread cataclastic deformation, D_6, in the CGC (Flöttmann et al. 1986; Kleinschmidt et al. 1985, Krohe & Eisbacher 1988). Deformational conditions are below quartz-recrystallization threshold temperature. Prehnite, epidote and chlorite are stable during deformation.

North of the Elz-Valley cataclasis concentrates on narrow SW-NE striking low angle normal faults (sensu Davis 1983). Tectonic transport is directed towards the SE, indicating NW-SE extension of the crust. South of the Elz-Valley cataclasis occurs on a regional scale (compare Fig. 1). In the Zinken-Elme Zone (Groschopf 1973) cataclastic shear zones reactivate the steep anisotropy, tectonic transport is directed toward the W-NW. The crustal extension coincides temporarily and spatially with intrusions of large volumes of granites (Flöttmann et al. 1986; Krohe & Eisbacher 1988).

Structural Analysis of the Granulites

The rocks here referred to as "granulites" are characterized by the lack of primary hydrous minerals. A commonly developed mineral paragenesis is: kyanite + mesoperthite + garnet + quartz. Orthopyroxene which, according to Winkler (1979), is indicative of "granulite facies" metamorphism is absent. Here the usage of the term granulite follows the workers of the "Granulitgebirge" in Saxony (e.g. Behr 1961, cum lit.).

The oldest and most prominent structural feature of the granulites is a planar mylonitic foliation consisting of layers with relic feldspar prophyroclasts in a matrix of recrystallized feldspars alternating with layers of quartz discs and platy ribbons. Stretched and recrystallized feldspar porphyroclasts and micas grown in garnet pressure shadows show overwhelmingly symmetrical orientation with respect to the dominant mylonitic foliation. Feldspar strain has strongly oblate character (Flöttmann 1988). The geometry of folded veins in the granulites indicates that irrotational shortening deformation geometry was dominant.

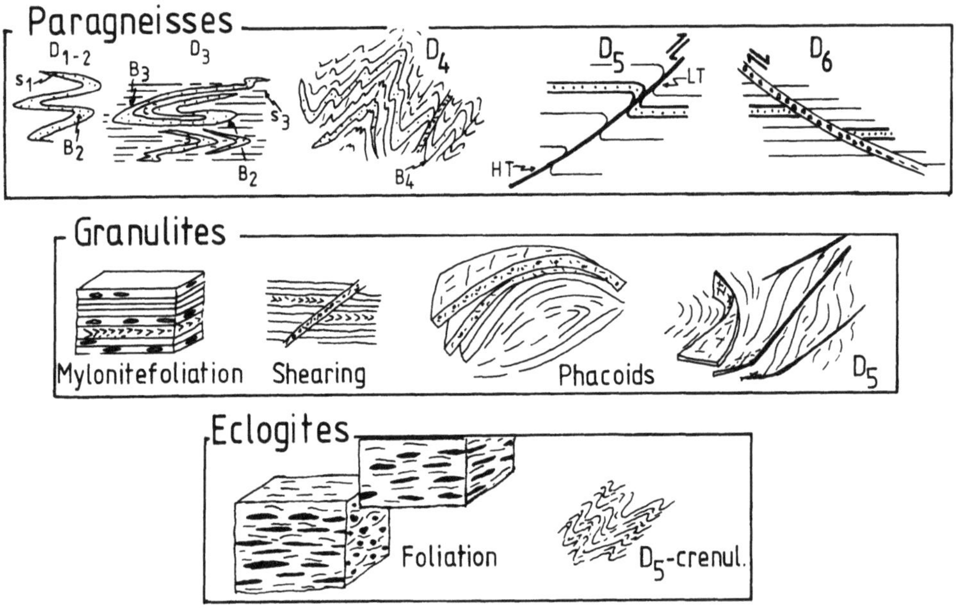

Fig. 2. Schematic sketches illustrating the structural development in the different lithotectonic units. Paragneisses are initially folded homoaxially (D_1–D_3). The dominant planar foliation develops from s_3, which is refolded by D_4. Crustal scale shearing and low angle normal faulting with cataclasis are the final deformations. In the granulites firstly a planar mylonite foliation develops (oblate strain), succeeded by retrograde shearing and phacoid formation. Eclogites partly undergo prolate strains (rodding), some types are deformed by the granulite-mylonitisation as well. Notice that both, eclogites and granulites are deformed together with their paragneiss country rock from D_5 onwards.

Kyanite is usually aligned in the granulite foliation, however some kyanite overgrows quartz ribbons. Kyanite usually reacts to form coronas of corundum ± anorthite and spinel (hercynite), which again reacts to form rims of cordierite in the presence of sillimanite (Klein & Wimmenauer 1984; Eigenfeld 1952).

The reorientation of the granulite foliation is highly irregular due to intense secondary (post-mylonitic) shearing and folding of the original mylonites. On outcrop scale however, granulites form conspicuous bulbuous, phacoid shaped bodies with diametres up to ten metres (Fig. 2). Although highly variable in detail, most granulite phacoids have some features in common: The core is often made up of a granoblastic variety, showing relics of disc quartzes in a matrix of feldspar. These "core granulites" (Behr 1961) are wrapped by strongly foliated, mylonitic types. The preferred orientation gradually fades toward the phacoid margins due to extensive metablastic feldspar growth as a consequence of the dominant LP-HT metamorphism.

D_5-shear zones with typical fibrolites cut through the granulite outcrops and their metablastic outer shell.

Fig. 3. HT-mylonite. Quartz ribbons alternating with fibrolite seams containing biotite relics. Notice spaced ecc-cleavage indicating dextral sense of shear. Scale bar: 0,1 mm.

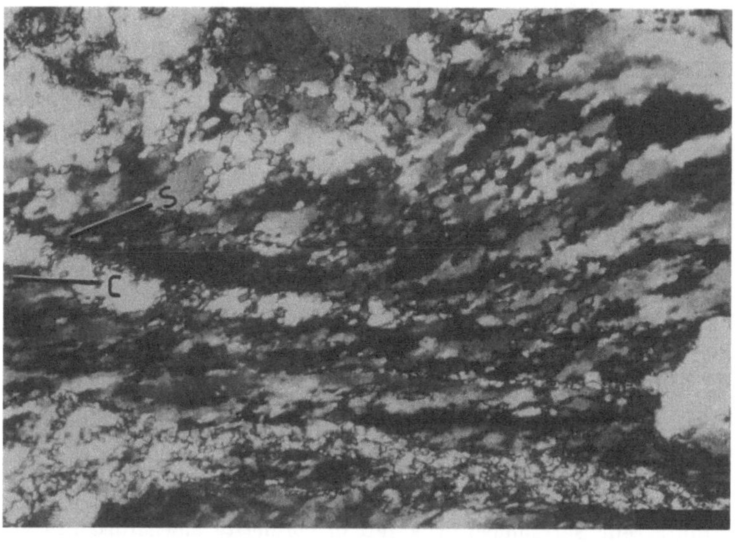

Fig. 4. MT-mylonite. Asymmetrically aligned, dynamically recovering and recrystallising quartz ribbons form s- and c-planes, indicating dextral sense of shear. Scale bar: 0.1 mm

Structural Analysis of "Eclogites"

Due to their stiff rheology only few eclogites are deformed penetratively. Eclogite inclusions in granulites show chocolat tablet boudinage (Ramsay 1967) as an effect of granulite flattening (oblate strain). Apart from this some eclogites underwent earlier and independent deformation as indicated by strong rodding (prolate strain, comp. Fig. 2). Internally the rods consist of unstrained, recrystallized diopside and plagioclase. Eclogites within D_5-shear zones develop a microcrenulation.

PT-PATHS

PT-Path in Paragneisses

The dominant metamorphic imprint within the paragneisses is the LP-HT metamorphism. Mineral paragenesis allowing a reconstruction of the prograde PT-path are preserved in the so called Assemblage of Welschensteinach-Harmersbach (Wimmenauer 1980, see Fig. 2).

Andalusite relics are cut through by a planar anisotropy, which on the other hand is cut off by the regional schistostiy s_3. The older fabric is believed to represent remainders of s_2. Since s_2 deforms andalusite, the latter must have grown previous to the D_2-deformation. Staurolite, growing from primary muscovite in the presence of biotite, overgrows s_2. The staurolite forming reaction most probably is:

$$Chl + Mus = Stt + Bio + SiO + H_2O \qquad \text{(Hoschek 1969).}$$

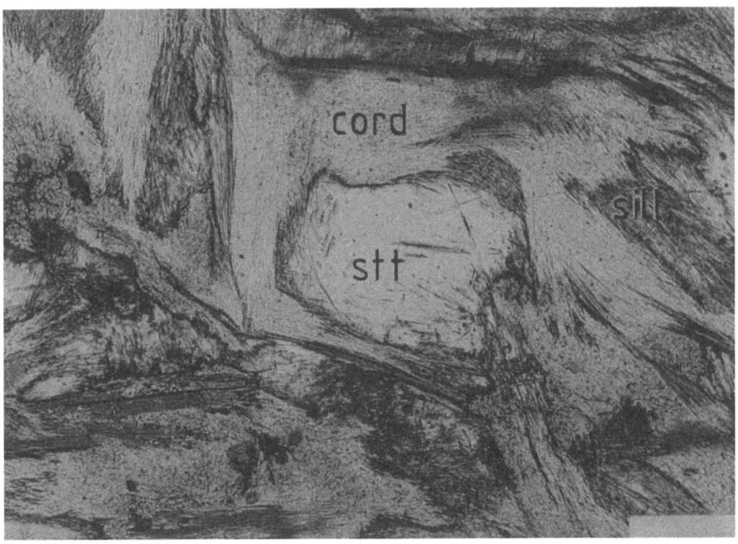

Fig. 5. Staurolite relic surrounded by a felt of cordierite and sillimanite needles (high relief). Scale bar: 0.1 mm.

Kyanite also grows from muscovite and forms an assemblage with staurolite. Staurolite decomposes according to the following reactions (compare Fig. 5):

Stt + Mus + Qz = Sill + Bio + Qz (Hoschek 1969)

and

Stt + Qz = Cor + Sill + H_2O (Richardson 1968).

Finally, one sample shows the classical "high grade in" reaction:

Mus + Qz = Sill + Kspar + H_2O (Winkler 1979).

These reactions are interpreted to indicate the following metamorphic history (Fig. 6): Andalusite is a relic of prograde metamorphism. The intersection of the "staurolite in-isograd" with the kyanite-sillimanite boundary (Holdaway 1971) indicates shallow medium pressure metamorphism of approximately. 5 kb minimum pressure at 500-600°C (MP-MT metamorphism). The staurolite out reactions indicate successive decompression to 4 resp 2-3 kb. Unlike other rock types, this particular one lacks anatectic phenomena, thus the maximum recorded temperatures are around 600-680°C. For most rocks with anatectic conditions the temperature range is around 650-720°C at 3-4 kb (compare Stenger et al. 1989).

PT-Path in Granulites

Granulite metamorphism (Fig. 6) is characterized by the paragenesis mesoperthite + kyanite + garnet. The PT-conditions of metamorphism are confined by the intersection of the reaction curve:

Qz + Mus + Naspar = Kspar + Al_2SiO_5 + L (Thompson 1982)

with the line of kyanite-sillimanite coexistence (Holdaway 1971). The PT-range obtained is 8 kb minimum pressure at temperatures of 700-800°C, which is generally in accordance with the results of other workers (Petrakakis 1986; Stenger et al. 1989). The retrograde reaction of hercynite in the presence of sillimanite to form cordierite,

Herc + Sill + Qz = Cor (Richardson 1968),

indicates PT-conditions of 700°C at pressures of 3-4 kb. Since no muscovite occurs, temperatures are believed to be on the product side of the muscovite out isograd. The PT-conditions outlined by the latter reaction coincide conspicuously with those of the LP-HP metamorphism in the paragneisses.

PT-Path of the Eclogites

The PT-conditions of the peak eclogite metamorphism (Fig. 6) and their retrogressive metamorphism have been worked out in detail by Klein & Wimmenauer (1984). They found 3 distinct metamorphic stages (see also Wimmenauer & Stenger 1989):

- The jadeite component in omphacite gives PT-conditions of a minimum of 12-15 kb at a temperature of about 600°C for the eclogite stage proper (HP-MT).
- A second stage of metamorphism took place at slightly reduced pressures but higher temperatures ($\approx$ 700°C).
- Finally, the eclogites reach the LP-HT metamorphic stage.

PT-DEFORMATION PATH

The PT-conditions of the various deformations are based on the following observations and arguments:

PT-Conditions During Paragneiss Deformations:

D_1 (greenschist stage):
Relic quartz veins (s_1) may represent remainders of a prograde greenschist facies deformation.

D_2 (medium pressure, MP-MT-stage):
Isoclinal folding of s_1, s_2 deforms older andalusite relics, kyanite and staurolite indicate medium pressure metamorphic conditions during D_2.

D_3 (amphibolite facies, LP-HT-stage):
Homoaxial, asymmetrical refolding of B_2. A regional foliation, s_3, is marked by separation of dark and light components, anatectic melts occur preferably along small shear zones.

D_4 (LP-HT-stage):
Refolding of s_3 and older structures. LP-HT-metamorphism outlasts folding as indicated by feldspar-metablastesis.

D_5 (retrograde shearing, HT and MT mylonites):
Compressive retrograde shearing, initially in the stability field of sillimanite: $T_{(min)}$ = 500°C at 4 kb (lower pressures would result in higher temperatures). Later deformation took place under reduced temperatures (T around or above 300°C, indicated by quartz recrystallisation behaviour).

D_6 (LT-retrograde cataclasis):
Extensional retrograde shearing. Quartz reacts brittlely, mineral paragenesis indicate prehnite-pumpellyite facies conditions with temperatures predominantly below 300°C.

PT Conditions During Granulite Deformation

For the Saxony granulite complex Lister & Dornsiepen (1982) give PT-conditions of deformation of app. 6 kb at 600-700°C. Behr (1980) gives an estimate of 5-7 kb at 600-800°C. The fact that granulite ribbon grains are overgrown by kyanite may indicate that pressures in the CGC were as high as 8 kb at least during initial granulite mylonitisation. At lower pressures sillimanite would be the stable Al_2SiO_5-phase under the given metamorphic conditions.

DISCUSSION

Structural analysis and metamorphic evolution reveal two distinct episodes of basement evolution in the CGC, pre- resp. post-dating the dominant LP-HT metamorphism (see Fig. 6). Eclogites, granulites and paragneisses follow different PT and deformation paths before the onset of LP-HT metamorphism. Radiometric age data for the older metamorphic events are not evidently consistent: According to U/Pb zircon dating the orthogneiss protolith intrusion took place app. 520 ± 15 Ma b.p. (Todt & Büsch 1981). The same method gives an age of 480 Ma for the LP-HT metamorphism (Steiger et al. 1973; Lippolt et al 1986). However, U/Pb data for the HP eclogite metamorphism indicate roughly the same (490 Ma) or even younger ages of 410 Ma (Schleicher & Kramm 1986). The latter age was also found in the granulites (Lippolt et al. 1986). Despite of the conflicting absolute radiometric age data, geological evidence indicates the following relative timing sequence of events: The oldest metamorphism is documented in the HP-eclogites, which are at least partly affected by the younger granulite deformation. Eclogites and granulites are overprinted by LP-HT metamorphism. The entire basement was deformed by SE directed thrusting after the high temperature metamorphism. No geological reasons necessitate a major hiatus between peak metamorphism and retrograde thrusting. Because the thrusts cut through lower Carboniferous granites (340 Ma; Lippolt et al. 1986) and the oldest post-tectonic granites yield ages of 325 Ma (Wendt et al. 1970; Lippolt & Rittmann 1984), D_5 as well as D_6 deformations were clearly of Carboniferous age.

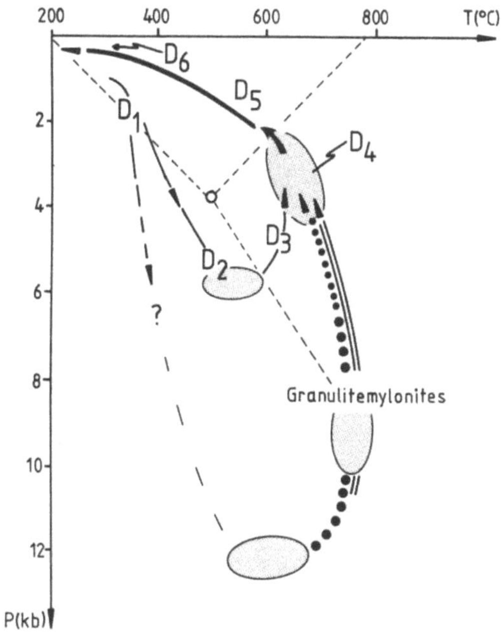

Fig. 6. PT-deformation paths. Shaded areas indicate periods of metamorphism. Paragneiss path (continuous line) shows initial deformation under greenschist facies conditions. A second metamorphism followed under increased pressure (MP-MT). The dominant foliation s_3 and subsequent B_4-folds formed under prevailing LP-HT conditions. The eclogites (dotted line) show relics of a HP-MT stage before reaching the MP-HT granulite metamorphic conditions (granulite path: parallel lines). The different units were amalgamated during LP-HT metamorphism, and subsequently deformed together. Notice that tectonometamorphic history of the paragneisses is possibly more complex since portions of the "felsic" country rock could have followed the prograde eclogite loop (dashed line). For further explanation see text. Al_2SiO_5 stability fields according to Holdaway (1971).

Scattered eclogite relics in the paragneisses could indicate that larger volumes even of "felsic" country-rocks suffered HP-metamorphism, as was suggested by Heinrich (1982) for parts of the Western Alps. Rocks of lower crustal provenance occur within thrusts in other parts of the Variscan chain (Burg et al. 1984). Applying these concepts would automatically imply that the prograde tectonothermal evolution of paragneisses that suffered HP-metamorphism would differ considerably from the evolution of those rocks that underwent only shallow to medium pressure metamophism (e. g. Assemblage of Welschensteinach-Harmersbach).

The prolate and highly symmetrical granulite deformation could be the result of lower crustal stretching as discussed by Weber (1984). The highly variable phacoid shape of the granulite outcrops is interpreted to be due to passive granulite deformation between lower crustal segments during early convergent motions that took place prior to or during LP-HT metamorphism. Continued convergence initially led to thrusting towards the SE and later to dextral strike-slip movements. Isostatic uplift of the tectonically thickened crust resulted in low angle normal faulting. The overall extensional regime facilitates the intrusion of large volumes of granites.

REFERENCES

Behr H J (1961) Beiträge zur petrographischen und tektonischen Analyse des Sächsischen Granulitgebirges. Freiberger Forschungsh. C119: 1-118

Behr H J (1980) Polyphase shear zones in the granulite belts along the margins of the Bohemian massiv. J Struc Geol 2: 249-254

Burg J P (1986) Quartz shape fabric variation and c-axis fabrics in a ribbon-mylonite: arguments for an oscillating foliation. J Struc Geol 8(2): 123-131

Burg J P, Leyreloup A, Marchand J, Matte Ph (1984) Inverted metamorphism zonation and large scale thrusting in the Variscan Belt: an example in the French Massiv Central. In Hutton D H W, Sanderson D J (eds) Variscan Tectonics of the North Atlantic Region. Geol Soc Lond Spec Publ 14: 33-46

Davis G H (1983) Shear-zone model for the origin of metamorphic core complexes. Geology 11(6): 342-347

Eigenfeld R (1952) Über Paragneise granulitischer Prägung im Raume Kandel-Glottertal des Schwarzwaldes nordöstlich Freiburg i.Br. N J Miner Abh 84: 241-280

Flöttmann T, Gallus B, Kleinschmidt G (1986) Variskische Kataklase im Mittleren Schwarzwald. N Jb Geol Paläont Mh 1986(8): 459-466

Flöttmann T (1988) Strukturentwicklung, P-T-Pfade und Deformationsprozesse im zentralschwarzwälder Gneiskomplex. Frankfurter geowiss Arb Serie A Bd 6 p 206

Groschopf R (1973) Eine weitere Zone prä-permischer Sedimente im Schwarzwald. Jh geol LA Bad Württ 15: 247-250

Heinrich C A (1982) Kyanite-Eclogite to Amphibolite Facies Evolution of Hydrous Mafic and Pelitic Rocks, Adula Nappe, Central Alps. Contrib Miner Petrol 81: 30-38

Holdaway M J (1971) Stability of Andalusite and the Aluminium Silicate Phase Diagram. Amer J Sci 271: 97-131

Hoschek G (1969) The stability of Staurolite and Chloritoid and their Significance in Metamorphism of Pelitic Rocks. Contrib Miner Petrol 22: 208-232

Klein H, Wimmenauer W (1984) Eclogites and their retrograde transformation in the Schwarzwald (Fed. Rep. Germany). N Jb Miner Mh 1984(1): 25-38

Kleinschmidt G, Flöttmann T, Gallus B (1985) Structural analysis, central Schwarzwald. 2nd. Int. Symp on Observation Continental Crust through Drilling, Seeheim 4.-6.Oct. 1985: 62

Krohe A, Eisbacher G H (1988) Oblique crustal detachment in the Variscan Schwarzwald, southwestern Germany. Geol Rundschau 77(1): 25-43

Lippolt H J, Kober B, Hradetzky H, Mertz D F (1986) Schwarzwald, Geochronologische Studien an Metamorphiten und Mineralisation. Kontinentales Tiefbohrprogramm der BRD 2nd KTB Koll 19.9.- 21.9.1986 Seeheim, Poster Programm: 41

Lippolt H J, Rittmann K L (1984) Die jüngere variskische Geschichte der Granite des SE-Schwarzwaldes - $^{40}Ar/^{39}Ar$-Untersuchungen an Glimmern. Fortschr Miner 62(Beih.1): 134-135

Lister G S, Dornsiepen U. F (1982): Fabric transition in the Saxony granulite terrain. J Struc Geol 4(1): 681-692

Passchier C W, Simpson C (1986) Porphyroclast systems as kinematic indicators. J Struc Geol 8(8): 831-843
Petrakakis K (1986) Metamorphoseentwicklung in der südlichen Bunten Serie am Beispiel einiger Gneise, Moldanubikum, Niederösterreich. TMPM 35: 243-259
Platt J P, Vissiers R L M (1980) Extensional structures in anisotropic rocks. J Struc Geol 2(4): 379-410
Ramsay J G (1967) Folding and fracturing of rocks. McGraw Hill, New York, p 568
Richardson S W (1968) Staurolite Stability in a Part of the System Fe-Al-Si-OH. J Petrol 9(3): 467-488
Schleicher H, Kramm U (1986) Altersbestimmungen nach der Uran/Blei-Methode an mittel- bis hochdruckmetamorphen Gesteinen des Schwarzwaldes. 76. Jahrestagg Geol Verein Gießen, Abstr: 69
Simpson C, Schmid S M (1983) An evaluation of criteria to deduce the sense of movement in sheared rocks. Geol Soc Amer Bull 94: 1281-1288
Steiger R H, Bär M T, Büsch W (1973) The Zircon Age of an Anatectic Rock in the Central Schwarzwald. Fortschr Miner 50 (Beih. 3): 131-132
Stenger R, Baatz K, Klein H, Wimmenauer W (1989) Metamorphic evolution of the pre-Hercynian basement of the Schwarzwald (Federal Republic of Germany). Tectonophysics 157: 117-121
Thompson A B (1982) Dehydration melting of pelitic rocks and the generation of H_2O-undersaturated granitic liquids. Amer J Sci 282: 1567-1595
Todt W A, Büsch, W (1981) U/Pb investigations on zircons from pre-variscan gneisses. I. A study from the Schwarzwald, West Germany. Geochim Cosmochim Acta 45: 1789-1801
Voll G (1960) New work on Petrofabrics. Liverp Manch Geol J 2(3): 503-567
Weber K (1984) Variscan events: Early Paleozoic continental rift metamorphism and late Paleozoic crustal shortening. In: Hutton D H W, Sanderson D J (eds): Variscan tectonics of the North Atlantic region. Geol Soc Lond Spec Publ 14: 3-22
Wendt I, Lenz H, Harre W, Schoell M (1970) Total rock and mineral ages of granites from the southern Schwarzwald, Germany. Eclog Geol Helv 63(1): 365-370
Wimmenauer W (1980) Lithology of the Precambrian in the Schwarzwald. An interim report. N Jb Miner Mh 1980(8): 364-372
Wimmenauer W, Stenger R (1989) Acid and intermediate HP metamorphic rocks in the Schwarzwald (Federal Rebublic of Germany). Tectonophysics 157: 109-116
Winkler H G F (1979) Petrogenesis of Metamorphic Rocks, 5th edn. Springer, Berlin Heidelberg New York p 348

Ore and Mineral Deposits of the Schwarzwald

K. von Gehlen

Institut für Geochemie, Petrologie und Lagerstättenkunde, J. W. Goethe-Universität, D-6000 Frankfurt am Main, FRG

1 INTRODUCTION

The Schwarzwald (Black Forest) is one of the old mining regions of Germany. Indications of mining by the Romans were found at Badenweiler (Kirchheimer 1976). The oldest document proving mining activities dates from 1028 (Metz 1957c, Schneiderhöhn 1958b). Several towns were originally medieval miners` towns: Münster, Todtnau, Sulzburg, Prinzbach (Metz 1957c), and Neubulach (Metz 1977). Part of the medieval wealth especially of Freiburg (im Breisgau) came from silver mining profits.

The (former) mining potential of the Schwarzwald can be estimated from the calculations by H.W.Walther (1981) to have been about 200,000 to 300,000 t of Pb + Zn. The Schauinsland mine near Freiburg has produced a total of 80,000 t of Zn>Pb>>Ag metal from about 192,000 tons of concentrates extracted from about 1,200,000 tons of raw ore during the time from 1901 to 1953 (Schürenberg 1957). The fluorite and barite potential of the Schwarzwald is of the order of 10 million t (extrapolated from Huck & H.W.Walther 1984). For the important Schwarzwald mines and mineralization districts, see Fig. 1.

Today, underground mining is going on at three places with fluorite (plus barite) as the main product: from north to south, at the Käfersteige mine near Pforzheim, at the Clara mine near Wolfach, and at the Gottesehre mine near St.Blasien (soon to be closed). The proposed Schwarzwald KTB drilling site near Haslach is less than 1 km from several smaller vein deposits of the Kinzig valley mining district which were once mined (Fig. 2; Bliedtner and Martin 1986).

For more recent reviews of mineral deposits in the Schwarzwald, mostly presenting only parts of them either geographically or regarding types of deposits, see, e.g., Schneiderhöhn (1941: 512-519 etc.), Metz et al. (1957), Kirchheimer (1957), Metz (1959, 1977, 1980), Wimmenauer (1975, 1979a, 1980a, 1985), H.W.Walther (1984: 192-196, 219-222), H.W.Walther et al. (1986: 188-191, 276-278), or Bliedtner and Martin (1986). Similarities between mineralizations in the Vosges and the Schwarzwald were pointed out by Agard et al. (1975b) and Fluck and Lougnon (1984).

2 TYPES OF MINERAL CONCENTRATIONS IN THE SCHWARZWALD FORMED UP TO VARISCAN (HERCYNIAN) TIMES

There are several events during the geological history of the Schwarzwald during which interesting mineral concentrations formed more or less effectively. The most important events were:

1. Precambrian sedimentation of the starting material of the present paragneisses and most migmatites;

2. Early Paleozoic intrusion of granites, the starting material of the present orthogneisses;

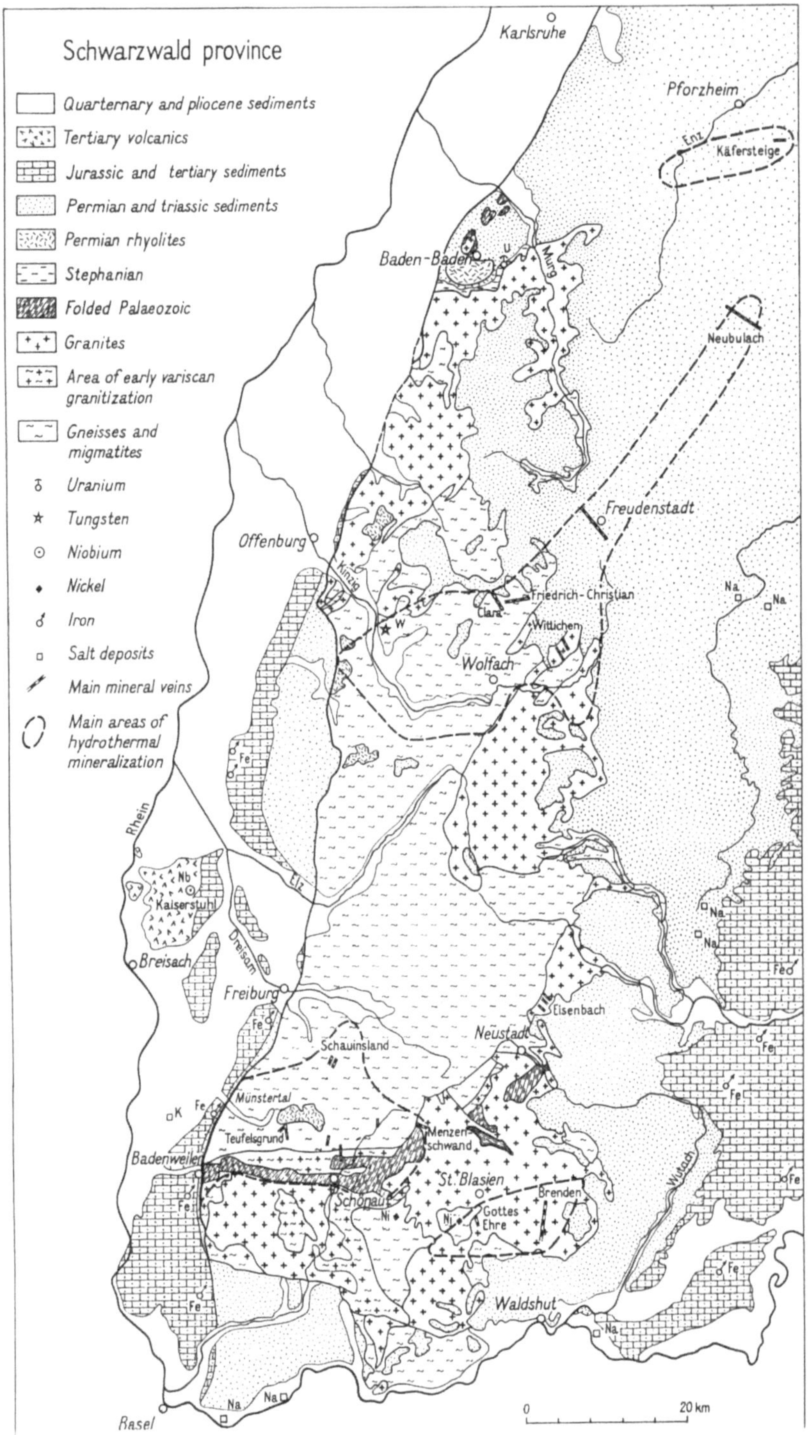

Fig. 1. Important ore and mineral districts in the Schwarzwald (H.W. Walther et al. 1986: 189)

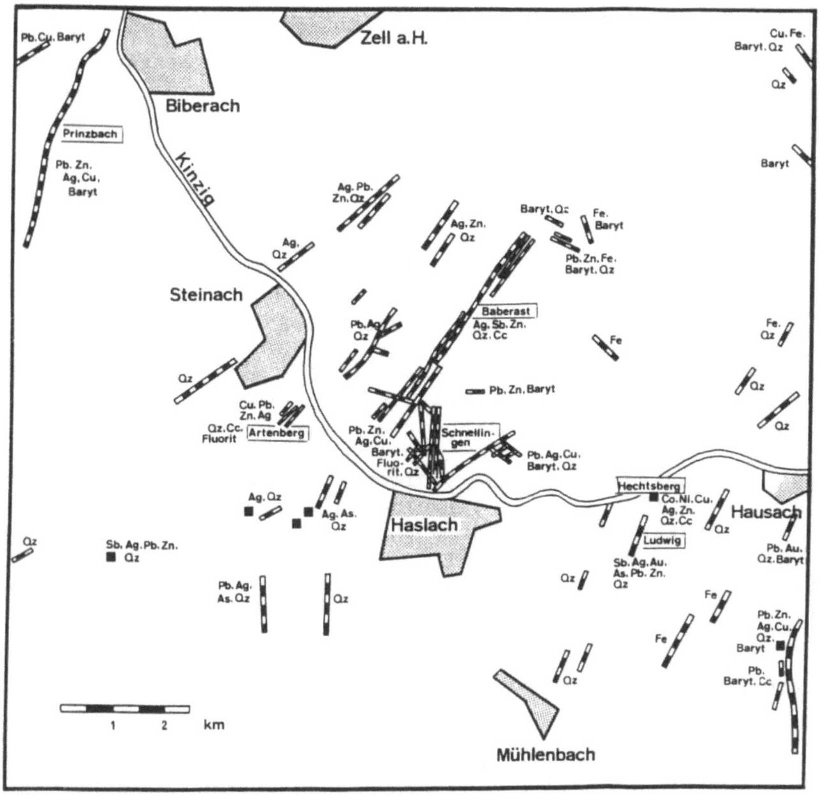

Fig. 2. Hydrothermal vein deposits around Haslach/Schwarzwald (von Gehlen et al. 1986b: 134, Fig. 6.1, by M.Martin)

3. Metamorphism and migmatization (anatexis);

4. Variscan (= Hercynian) magmatic activity (predominantly granitic);

5. Post-Variscan hydrothermal convection leading to vein-type mineralization.

Almost the entire Schwarzwald basement belongs to the Moldanubian zone of the Variscan orogen; only the northernmost part near Baden-Baden consists of Saxothuringian rocks. In the Schwarzwald there are small E-W trending grabens filled with Paleozoic sediments and volcanics, the most important being the so-called Kulmzone of Badenweiler-Schönau-Lenzkirch (see Fig. 1). Compressional tectonics were dominant in the southern Schwarzwald during the Late Variscan ("Kulmzone"), while in the central Schwarzwald, extension is dominant (Behr and others). The existence of thrusts within the metamorphic basement has been postulated (e.g., Behr et al. 1986).

2.1 Precambrian Sedimentation

The Moldanubian Schwarzwald predominantly consists of rocks comparable to the Monotonous Series of the Bohemian Massif and the Vosges (Wimmenauer 1980b, 1984). These are mainly former greywackes and pelites with a sedimentation model age of 700 M.a. (Lippolt 1986: 299) containing only minor mineral concentrations of interest. Small bodies of amphibolites and eclogites indicate former basaltic rocks within the dominating sediments.

Graphite is a minor constituent of kinzigite (more than 4 % graphite was found at the type locality near Schenkenzell, quoted by Osann 1927, p. 28). Graphite contents of 1-2 % have been found locally in paragneisses near Haslach (R.Metz, personal communication). Osann suggested its formation from organic matter and pointed out that graphite does not occur in orthogneisses.

Strontium isotope studies have revealed that most of the strontium found in post-Variscan vein barites, fluorites and calcites must have come from basement rocks such as gneisses and amphibolites (Lippolt 1984, R.Hofmann & Baumann 1984, von Gehlen et al. 1984; see Sect. 5.3). This could also apply to (parts of) the barium in the vein barites.

It is possible that less monotonous rocks which show up in certain parts of the Schwarzwald (Wimmenauer 1980b, 1984) may form a larger percentage of the uppermost kilometers of a deep drill hole. Minor CO_2 in fluid inclusions in rocks may indicate the existence of C-bearing rocks below (Behr; cf. J.Walther and Althaus 1986).

The tungsten maxima in stream sediments shown in the Geochemical Atlas (Fauth et al. 1985, pp.64-65) especially around Nordrach, Eisenbach, and Untermünstertal mostly seem to be caused by W-bearing veins (see Sect. 2.4). Some minor scheelite has also been found in veinlets in amphibolites (K.-P. Burgath and F.-J. Eckhardt, personal communication).

2.2 Metabasites and Meta-ultramafitites with Ni-Fe-Cu Sulfides

Lenses of metabasites to meta-ultramafitites south of the Kulmzone carry minor Ni-pyrrhotite ores at Horbach-Wittenschwand (partly serpentinized hornblende-peridotite to pyroxenite, Metz 1980: 435) and at Todtmoos-Mättle (norite, Metz 1980: 446). According to Metz and Rein (1958: 64) they have been metamorphosed and partly migmatized and are therefore older than the anatexis. Schneiderhöhn (1958a) reported investigations by D. Hoenes who found only 0.35 % Ni and 0.2 % Cu, respectively, 0.1 % Ni and 0.07 % Cu at Horbach, and 0.X % Ni at Todtmoos. At Horbach, about 130 t of Ni metal have been recovered from 1848 to 1880 (Metz and Rein 1958, pp.64-66). Metz (1980, pp.418-447) gave a very detailed account of both places and their mining history. At both localities, the ores have been mined out. The deposits have been listed among the possible sources of platinum group metals found in gravels of the river Rhine (Goldenberg 1988).

2.3 Metamorphism and Anatexis (Migmatization) of the Gneisses

Several phases of metamorphism, with anatexis locally reached during the LP-HT stage about 490 M.a. ago, did not create ore and mineral deposits. Only very minor sulfide and oxide minerals in restites formed during anatexis were observed (Wimmenauer 1951).

Sillimanite may locally be enriched (Kandel, St.Peter-St.Märgen, Metz 1959; near Prinzbach/Biberach, R.Metz, personal communication). Eigenfeld-Mende (1948) first studied special metamorphic reactions in metabasites north of the Kulmzone. During the "first" anatexis with intrusion of acid magmas forming the later orthogneisses, an important mineral formed was Th-bearing orthite (Rein 1952).

2.4 Paleozoic Sedimentation and Volcanism

Stratabound traces of gold in Lower Carboniferous coarse-grained sediments of the Kulmzone east of Badenweiler are being investigated by Krützfeld. Accompanying fine-grained minerals isolated from the sediments are pyrite, arsenopyrite, sphalerite, and galena. There are apparently no gangue minerals. The origin is still problematic (Krützfeld 1985; cf. also Sect.4, D).

Scheelite occurrences (with quartz and adularia) in veinlets in the western Kulmzone seem to be related to Carboniferous pyroxene andesites (Burgath, personal communication; cf. Sect.4, D, H).

The sediment-hosted uranium enrichment in Upper Carboniferous strata at Müllenbach near Baden-Baden was interpreted by Kneuper et al. (1977) and others as the product of synsedimentary (partly "roll-front-type") mineralization of Carboniferous age (see also Wimmenauer et al. 1980). However, recent studies led Brockamp and Zuther (1983; Brockamp et al. 1985, 1987) to the assumption of a Jurassic epigenetic hydrothermal formation process (see also Sect. 5.4). Tapfer (1987) stressed that the distribution and high concentrations of Cu,Zn,Pb,S are not typical of roll-front-type uranium deposits. Apparently, a predominantly synsedimentary mineralization was overprinted by a hydrothermal event in Jurassic times and partly recrystallized, as Clauer (in Brockamp et al. 1985) determined K/Ar ages around 150 and 190 M.a.

2.5 Mineral Concentrations in Variscan Granites

Cassiterite has been found in traces in schlieren of the Triberg granite and the Bärhalde granite (Osann 1927: 75; R. Metz and W. Wimmenauer, personal communication). Some granite areas show slightly increased tungsten (W) contents, according to the stream sediment map in the Geochemical Atlas (Fauth et al. 1985, pp. 64-65).

Small pitchblende-bearing veinlets together with bismuth and other minerals, often oxidation products, have been reported from the Triberg granite (e.g., Walenta 1980) which is a typical "fertile" Sn-W-U granite (Fritsche and Föhse 1981).

3 VEIN-TYPE DEPOSITS OF VARISCAN AGE

The great problem in recognizing Variscan vein-type mineralizations in the Schwarzwald was and still is that there must be mineralizations of different ages (discussed, e.g., by von Gehlen et al. 1970; von Gehlen 1987):

1. There are radiometric Variscan ages (Sect. 3.3) and relations to Variscan granites (Sects. 2.5, 3.1);

2. There are Tertiary mineralizations in the Rhinegraben faults (Sect. 4, K);

3. Veins (mostly barite-rich) in the Triassic sedimentary cover must be Triassic or younger in age;

4. For the majority of vein-type deposits in the basement, no clear age indications were available, so they could be of Variscan or Tertiary age or of any age in between.

Formerly (e.g., Schneiderhöhn 1941 and others), most deposits of group 4 were interpreted as Late Variscan. For veins of group 3 in the sedimentary cover, Schneiderhöhn developed his "secondary hydrothermal" interpretation, postulating redeposition of parts of earlier mineralizations (in the basement) at higher levels.

In 1983, Bonhomme et al. obtained Liassic K-Ar ages for illites in wallrock of Clara veins (see, e.g., discussion in von Gehlen 1987, pp. 88-94; cf. Lippolt 1984, Lippolt et al. 1986 a,b,c). Another important observation was that by Huck (1984) indicating Lower Triassic ages for the Clara fluorite and barite stages (see Sect. 5.4). Last but not least, Behr and Horn (1984; Behr et al. 1987) found strong differences between the compositions of fluids in fluid inclusions in Variscan (in the Rhenohercynian) and those in Post-Variscan vein minerals. Fluid inclusions may therefore apparently be used for a discrimination. Most observations in Schwarzwald veins so far place them in the Post-Variscan group (cf. Sect. 5.1).

Therefore, only those deposits are listed in this chapter for which there are strong arguments for a Variscan age.

3.1 Wolframite-Bearing Pneumatolytic Veins

Walenta (1967; Walenta et al. 1970) found wolframite and also some scheelite in pneumatolytic quartz-tourmaline veins connected with the Nordrach granite. The main region of wolframite occurrences is restricted to the Rossgrabeneck-Schwaigwald area between Biberach and Nordrach. Accessories are hematite and traces of rutile, arsenopyrite, pyrite, pyrrhotite, cassiterite, and bismuth (cf. Sect. 4, I). Scheelite (+ calcite, fluorite, pyrite) in a granite porphyry near Badenweiler was found by Schatz (1970).

3.2 Ag-Bi-Co-Cu-U Veins Around Wittichen

A genetic connection between the Wittichen granite (the northern part of the Triberg granite) and its Ag-Co-Bi(-U) and Cu-Bi mineral-bearing numerous barite veins has been postulated (cf. Sect. 2.5). Early U/Pb age determinations of Wittichen pitchblende gave 235 ± 5 M.a. (Kirchheimer 1957); although this corresponds to the Middle Triassic (Muschelkalk), the real age would probably come out older if redetermined and recalculated, as in the case of Menzenschwand (see Sect. 3.3), and might be Carboniferous. For details on these veins, which were in part mined for various raw materials for a long time, see e.g., Mehnert (1949), Wimmenauer (1952), Kirchheimer (1957: 92-94), Walenta (1972), Bliedtner and Martin (1986: 484-569); cf. Section 4, Q.

3.3 U-Mineralization in Hematite-Bearing Quartz Veins at Menzenschwand

At this locality, U/Pb investigations on pitchblende indicated a Variscan age for at least this early mineral: Menzenschwand material gave between 190 and 325 M.a. (model ages ca. 230 to 250 M.a., Wendt et al. 1974). The latter ages were recognized as discordant and a primary age of 310 ± 3.5 M.a. = Upper Carboniferous was derived (Wendt et al. 1979). For a description of the deposit see Bültemann (1965) and Metz (1980, pp.411-418); cf. Section 4, M.

Except for the U-minerals, the veins are similar (Schauer 1962, Wimmenauer 1980a) to the hematite-bearing "silicification zones" of Faisi and von Gehlen (1957) in the southwestern Schwarzwald (see Sect. 4, H). These may be of Variscan age according to a paleomagnetic study (von Gehlen and Schmitz 1981) but could also be Post-Variscan because of their field relations to the Münstertal fluorite veins (Faisi and von Gehlen 1957) or because of possibly erroneous paleomagnetic "ages" due to block tilting. Scheelite seems to indicate a Carboniferous age (Burgath and Mohr 1984; see Sect. 4, H).

At Menzenschwand, apparently only the oldest pitchblende-bearing phase is definitely of Variscan age. Fluid inclusion studies (Sect. 5.1) may solve the problem.

4 VEIN-TYPE DEPOSITS OF POST-VARISCAN AGE

As the major proportion of the Schwarzwald vein-type deposits apparently is of Post-Variscan age, vein classifications used in the literature are listed in this chapter. Maus and Renk (1981) suggested a division of the apomagmatic veins into ten groups on the basis of the dominant gangue minerals and stressed the importance of complex redeposition processes in most veins. They also pointed out that some veins show numerous oxidation minerals while others do not, this difference being only partly explained so far.

Schürenberg (1957) distinguished a number of groups of vein-type hydrohydrothermal deposits in the southern Schwarzwald:

A. Quartz-fluorite veins with galena and sphalerite (type Schindler; Münstertal-Wiesetal). They exhibit the most complete succession of ore and nonmetallic vein minerals in the southern Schwarzwald (see Schürenberg 1950 for the Teufelsgrund and Schindler veins; von Gehlen 1955, Sehlke 1956, Wecht 1955). The mineralization took place in three main phases, the second and third of which partially repeating the first. Toward the east, the successions become poorer in minerals, e.g., the quartz-arsenopyrite-pyrite assemblage of the first phase disappears. From Wieden, also wolframite (Walenta 1967), an Ag-As and an Ag-Sb association (Grundmann 1971) and cinnabar (Otto and Schürenberg 1974) were reported. For tectonic reasons, the Wieden and Todtnau veins should be of Mesozoic or Tertiary age (Sperling 1972).

B. Quartz-fluorite veins with galena (type Ruprechtsgangzug; St. Blasien and surroundings, including the Gottesehre vein at Urberg; Otto 1967).

C. Quartz veins with pyrite, arsenopyrite, and tetraedrite (type Wildsbach; St. Ulrich and Sulzburg).

D. Quartz veins with pyrite, arsenopyrite, and antimonite (type Münstergrund; St. Ulrich, Sulzburg; Walenta 1957, 1958; Hofherr 1984). The Sulzburg WNW-ESE vein system of about 5-km length occurs in gneisses but mostly in sediments of the Kulmzone, here apparently following a coal-rich horizon (cf. Sect. 2.4), with the main phase (Pb-Zn-Sb) not influenced by the (Tertiary?) Schweighof fault (Hofherr 1984). The quartz veins with traces also of bismuth, molybdenite and gold in the Kulmzone region, called group 1 by Maus and Renk (1981) probably belong here. Burgath and Mohr (1984) mentioned quartz-antimonite and gold mineralizations between type H veins (see below) just north of the Carboniferous volcanites of the "Kulmzone".

In antimonite-bearing veins of the Adlersbach valley near Hausach near Hausach (Kinzig valley, central Schwarzwald), Walenta (1957) found traces of gold.

E. Quartz-calcite-barite veins with sphalerite and galena (type Schauinsland), including the important Schauinsland deposit: e.g., Kreutzwald (1936), Schürenberg (1957, pp. 118-164). Mittelstädt (1987) performed a geochemical study.

F. Veins with the simpler succession quartz-barite-sphalerite-galena (type Kropbach; Münstertal district).

G. Quartz-siderite-barite veins with chalcopyrite and sphalerite (type Riggenbach; Münstertal district).

H. Quartz veins with barite, hematite, pyrite (and sometimes fluorite), generally striking WNW-ESE (type Tirolergrund). There are continuous gradations into sometimes wide silicification zones (Faisi and von Gehlen 1957). The pitchblende-bearing veins of Menzenschwand (see type M. and discussion in Sect. 3.3) are, except for the uranium mineral phase, similar to this group. Veins of this group carrying some pyrite and chalcopyrite may have been the target of mining at Gersbach, Wehra valley (Metz 1980, pp.449-455). Maus and Renk (1981) mentioned scheelite besides hematite in "stengelquarz" veins (= type H). Burgath and Mohr (1984) found two generations of scheelite in type H veins within and north of the Kulmzone volcanics and suggested a genetic connection with these Carboniferous volcanics.

J. Unusual veins with, in addition, Cu sulfides (Rotenbach; Wecht 1955) or Co-U minerals (Kobaltgrube near Sulzburg; Wimmenauer 1955a).

K. (Symbol first used by Metz 1957b): Mineralizations within the Tertiary Rhinegraben faults with quartz, barite, fluorite, and minor sulfides, especially galena (Metz 1957b; Badenweiler: Kirchheimer 1957: 42-48; Metz 1959: 227-229; Weiler near Lahr: Kirchheimer 1957: 57-74; Walenta and Wimmenauer 1961; Bliedtner and Martin 1986: 178-183).

Schürenberg extended his classification to other parts of the Schwarzwald in R.Hofmann & Schürenberg (1979, published Oct. 1979; to be abbreviated below as H&S). Unfortunately, Wimmenauer (1980a, published March 1980) independently published another extension of the Schürenberg system (repeated in Wimmenauer 1985 and H.W.Walther et al. 1986: 190-191; to be abbreviated as W80). The symbols in W80 mostly overlap the older ones of Metz (1957b; for type K) and of H&S (for types L to P). There is even a third extension of the Schürenberg system by Martin (in an introductory chapter of Bliedtner and Martin 1986: 23; to be abbreviated as B&M) using the letters L to R for central Schwarzwald veins in still another way.

The solution to this dilemma cannot be another new classification. The system of R.Hofmann and Schürenberg has priority.

I. (Used by W80 only, therefore undisputed; in B&M misquoted for type J, see above): Quartz-tourmaline veins with wolframite, scheelite and bismuth (Rossgrabeneck near Zell am Harmersbach; Walenta et al. 1970; discussed in Sect. 3.1: most probably of Variscan age).

J, K. (See above; symbols already used in 1957).

L. (After H&S: 34; = type P of W80): Quartz and quartz-barite veins, partly with iron and manganese oxides (e.g., Eisenbach, Faisi 1951, Wimmenauer 1979b; U-bearing, Kirchheimer 1953, Wimmenauer 1955b, von Gehlen 1958; cf. type V).

M. (After H&S: 35; = part of type H of W80): Quartz-fluorite-barite-hematite mineralization with pitchblende at Menzenschwand (similar to H except for the uranium minerals, W80; for strontium isotope ratios see R.Hofmann 1985; discussed in Sect. 3.3: partly Variscan).

Central Schwarzwald:

N. (After H&S: 37; = part of type N of W80): "Precious quartz veins" of the central Schwarzwald, without barite, resembling the "precious quartz formation" of Freiberg/Saxony (Ursula near Welschensteinach, after Vogelgesang 1865; Walenta 1957: 35-40; B&M: 267-270; cf. type P).

O. (After H&S: 37; = part of type N of W80): Silver-rich calcite vein Wenzel, Frohnbach valley (Metz 1957a; Walenta 1957: 55-58; B&M: 440-456).

P. (After H&S: 37, 39; = part of type N of W80): "Precious" lead veins around Hausach and Wolfach, e.g., Fortuna, Gelbach valley, and Gabriel, Einbach valley; (W80:) veins of the Pb-Zn and related types in the central and lower Kinzig valley (Metz 1957a, B&M); some of the deposits resembling the precious quartz formation of Freiberg in Saxony (W80): Ludwig and Friedrich near Hausach (after Vogelgesang 1865; Walenta 1957: 40-54; cf. type N).

Q. (after H&S: 37, 39; = type K of W80): Co-Ni-Bi-Ag-U veins with barite, fluorite, calcite, and quartz (Wittichen district; B&M: 484-569; discussed in Sect. 3.2: probably at least in part of Variscan age). The Daniel vein near Wittichen (in L of W80) contains Cu and Bi sulfides (Mehnert 1949; B&M: 551-557; cf. type U).

R. (After H&S: 37, 41; = part of types L and M of W80): Quartz- and/or fluorite-bearing barite-Cu-Pb veins at Schapbach and Wildschapbach; (W80:) barite veins, mostly devoid of or poor in ore minerals. The most important example is that of the Clara veins at Oberwolfach, which cut gneisses as well as the overlying Lower Triassic sandstone (Huck 1984, Huck in B&M: 366-399). Diffuse hydrothermal mineralization in the lowermost Buntsandstein (su1 and su2) around the outcrop is connected with the first major phase in the veins (Maag 1987). For traces of, e.g., secondary gold among the many minerals known see Walenta (1987). A vein of fluorite and quartz (mostly pseudomorphous after barite) with galena and chalcopyrite is the Friedrich-Christian vein at Wildschapbach (Laufhütte 1954; Metz 1957a; = M of W80), "the most important Pb deposit" of the central Schwarzwald together with the Herrensegen vein (Metz et al. 1957: 261; B&M: 336-355).

S. (After H&S: 37, 49): Barite veins with hematite and manganese oxides in the central Schwarzwald (Harmersbach, Ohlsbach; B&M: 40, 63-75).

Northern Schwarzwald (Metz 1977):

T. (After H&S: 49; = part of type L of W80): Quartz-fluorite veins with traces of sulfides in the Triassic sedimentary cover of the northern Schwarzwald, including the important Käfersteige vein near Pforzheim (Renk 1975) with fluorite >> barite and minor quartz.

U. (After H&S: 49): Barite veins with fahlore in the Triassic sedimentary cover of the northern Schwarzwald (e.g., Freudenstadt; Neubulach, Metz 1977: 218-243, with Cu-Bi sulfides, Kluth 1965, cf. type Q; introducing "secondary-hydrothermal" ideas: Schneiderhöhn 1949).

V. (After H&S: 49): Barite veins with Fe-Mn minerals in the Triassic sedimentary cover of the northern Schwarzwald, especially at Neuenbürg-Pforzheim (Metz 1977, Wimmenauer 1979b, Joachim and Smykatz-Kloss 1985; cf. type L).

5 FORMATION OF THE SCHWARZWALD POST-VARISCAN VEIN-TYPE DEPOSITS

The ore-petrological investigation of drill cores of several geothermal drill holes in the Schwarzwald by Friedrich et al. (1986) revealed only minor sulfidic and oxidic ore minerals. In the two holes close to the Haslach KTB locality, however, the number and quantity of sulfide minerals was somewhat higher. Pyrite, marcasite, pyrrhotite, arsenopyrite, chalcopyrite, sphalerite, galena, bornite, and chalcocite were observed, often together with Fe-Ti oxides. The composition of some of these minerals in KTB cores could give indications of temperatures and sulfur and oxygen fugacities during formation and eventually of the proximity to ore deposits.

5.1 Fluid Inclusion Studies

The fluid systems of the Post-Variscan mineralizations -- as preserved in primary fluid inclusions -- differ strongly not only from those in Variscan vein minerals (Behr and Horn 1984), but also from the rock fluids (Behr et al. 1986, KTB Report; for a summary of the Schwarzwald reports, see von Gehlen et al. 1986b, pp. 125-138; cf. also J.Walther and Althaus 1986). The temperatures of vein mineral formation have been rather low, $< 200\,^{\circ}C$ after fluid inclusion studies by Behr et al. (1986, KTB Report). In addition, salinities were much higher in the vein mineral fluids than in the rock fluids. Only locally, rock fluids similar to the fluids in vein minerals (with NaCl, $CaCl_2$) occur which can be explained by retrograde decomposition of plagioclase (J.Walther and Althaus 1986, KTB Report).

The fluids in Post-Variscan vein minerals show the following ranges (Behr et al. 1986, KTB Report, from more than 2000 measurements; Te = eutectic temperature, Tm = melting temperature of ice, Th = homogenization temperature):

Te = -51 to -76 °C (high content of $CaCl_2$);
Tm = -10 to -44 °C (salinity > 23.3 wt% NaCl equivalent);
Th = <+40 to +180 °C (maximum +120 to +180 °C).

Almost all fluid inclusions are two-phase inclusions (only in barite: one-phase), they homogenize into the liquid phase and have densities of 1.1 g cm^{-3} and more. Because of the rather thin cover (maximum 1-2 km) and therefore low pressure during mineralization, Th almost corresponds to the temperature of formation. Values outside the ranges above were only found in the Schnellingen veins near Haslach (Behr et al. 1986, KTB Report).

Behr & Gerler (1987) concluded that the Post-Variscan vein mineralization must be derived from saline basinal brines influenced by marine conditions. These brines mixed with ascending fluids of low salinity and somewhat higher temperatures.

Hydrothermal alteration of granites (and gneisses; Huck 1984) may have supplied Sr, Ba and Pb into the fluids (Simon and Hoefs 1987; cf. also Murad 1970, 1974). On the other hand, there is an enrichment of Ba, Ag and Cu in the country rocks directly adjacent to the "Clara" veins (Zirngast 1978; for an early trace element study, see Degens 1956). A number of alteration studies has been performed in the Schwarzwald; there were certainly several phases of alteration (Huck 1984). Fluid inclusions in the Triberg granite and its alteration were studied by Homann (1986).

During early phases of the formation of the Clara mineralization, rare conditions must have existed, because the rare hydrothermal mineral sellaite MgF_2 formed (Maus et al. 1979).

Primary and remobilized fluorites may be distinguished by means of rare-earth element patterns (Möller et al. 1982).

5.2 Radiometric Age Studies

An important group of genetic arguments comes from radiometric ages determined on sheet silicates (illite, sericite) which are assumed to have been formed at the time of the mineralizations (K/Ar, $^{40}Ar/^{39}Ar$ and Rb/Sr data; Bonhomme et al. 1983, Brockamp et al. 1985, Mertz et al. 1986, Lippolt et al. 1986). Permian to Upper Jurassic ages have been obtained [for a compilation, see von Gehlen 1987, fig.4; the number of Upper Jurassic (Malm) ages has increased meanwhile: Mertz 1987]. The general problems in radiometric dating of mineral veins which are generally poor in suitable minerals (except for the occurrence of uranium minerals, cf. Sects. 3.2 and 3.3) were discussed in detail by Lippolt (1984, 1986).

Apparently, as suggested by several authors, the opening of the Atlantic Ocean in Jurassic times has caused extension also in the Schwarzwald. This may have been accompanied by a thermal event (Brockamp et al. 1987).

Fissure fillings of anhydrite and gypsum in geothermal boreholes in the central Schwarzwald are even younger. They contain one-phase fluid inclusions of very low salinity (< 1 wt% NaCl equivalent). Temperatures of formation must have been < 40 °C which indicates subrecent formation. Sulfur isotope ratios are different from those of sulfates from the other veins (von Gehlen et al. 1962, 1986a; for ranges of S and Sr isotope ratios in and around the Schwarzwald, see compilation and discussion by von Gehlen 1987: 97-100).

5.3 Lead and Strontium Isotope Studies

Lippolt et al. summarized their Pb and Sr isotope results in a poster (1986b).

Pb isotope ratios in galenas from eight central Schwarzwald mineralizations vary in wider limits than those in basement rocks (Lippolt et al. 1986, KTB Report). Ore lead is definitely more radiogenic in comparison. Ore lead from the Clara/Friedrich-Christian vein system is identical to that of the "Bleiglanzbank" (galena-bearing horizon) of Upper Triassic (Keuper) age in the sedimentary cover. This may have genetic implications (Lippolt et al. 1983; 1986, KTB Report; see Fig. 3 and Sect. 5.4). There are several minor stratabound Pb-Zn etc. mineralizations in various horizons of the Triassic sedimentary cover in SW Germany (cf., e.g., Amstutz and Fontbote 1985; B.Hofmann 1979).

According to Lippolt et al. (1986, KTB Report), the lead of the vein galenas cannot be derived from the earth's mantle. Also lead mobilization from the adjacent country rocks can be ruled out. Remobilization must have played an important role, supplying lead from different sources. The primary source of the central Schwarzwald ore lead must have been the basement rocks; the Pb isotope ratios are typical of lead from the upper crust.

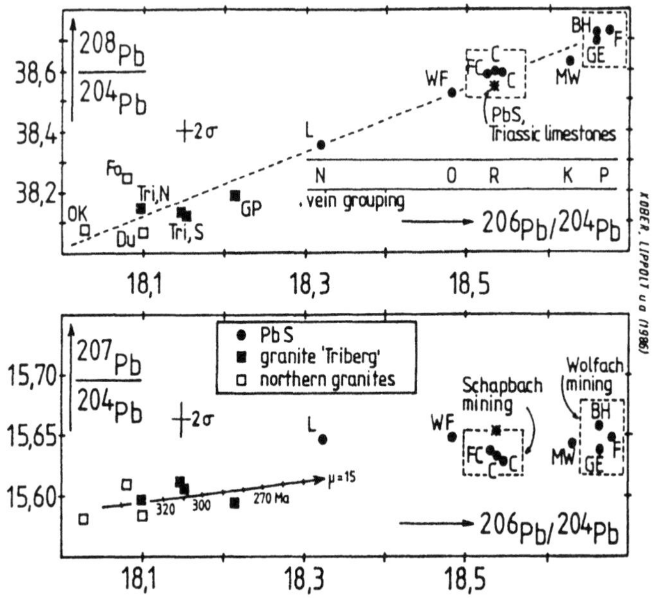

Fig. 3. Pb isotope ratios in galenas from vein mineralizations of the central Schwarzwald (black circles), in K-feldspars from Variscan plutonites of the central (black squares) and northern Schwarzwald (white squares), and average of Upper Triassic "Bleiglanzbank" galenas (stars) (Lippolt et al. 1986, Fig. 2; also Fig. 6.4 by Lippolt et al. in von Gehlen et al. 1986b: 136). -- Vein groups O, R, K, P as in Section 4 (after H&S); mines: L Ludwig im Adlersbach; WF Wenzel im Frohnbach; C Clara; FC Friedrich-Christian; MW Michael im Weiler; BH Bernhard im Hauserbach; GE Gabriel im Einbach; F Fortuna im Gelbach

The Sr isotope ratios (see compilation in von Gehlen 1987, Table 2) in vein barites, fluorites and calcites from the "Clara" and "Friedrich-Christian" veins range from 0.7111 to 0.7175. This means that no contributions from the earth's mantle or from granitic magmas can be recognized (Lippolt et al. 1986, KTB Report; see also Dill and Nielsen 1987). Evaporites from the sedimentary cover and Tertiary volcanics cannot be the Sr source. The strontium must be derived from the metamorphic rocks (mostly gneisses) of the Schwarzwald basement (von Gehlen et al. 1984; Fig. 5 of von Gehlen 1987; cf. Sect. 2.1). Barites throughout the Schwarzwald have comparable $SrSO_4$ contents (R.Hofmann 1979) and rather similar Sr isotope ratios and should therefore be genetically related (R.Hofmann and Baumann 1984; Lippolt et al. 1986, KTB Report).

5.4 Conclusions

A report on the state of knowledge of the processes which apparently have formed the great number of Post-Variscan veins in the Schwarzwald and its vicinity was recently given by von Gehlen (1987). Incorporated were many results of geochemical and other studies during the last years, as well as e.g., the most important observations by Huck (1984) during his mapping of the singular exposures of about 10-km length of veins at various levels of the Clara mine underground, i.e. singular in comparison to the often poor to non-existent vein outcrops at other Schwarzwald localities left behind by former mining.

The following "presently best" genetic interpretation was given by the author: "Important recent results by several authors seem to indicate that most vein-type mineralizations are of Mesozoic age and have been precipitated from far-reaching "formation water" circulation systems which brought components from above -- in part saline -- as well as from the basement. Locally, relations to stratabound mineralizations in Triassic sediments are likely. There are neither indications of mantle sources nor of magmatic activity related to the mineralizations" (von Gehlen 1987, p.87, partly following Behr and Gerler 1987).

Post-Variscan movements, opening fissures which were filled by vein minerals, must have occurred almost continuously from the Upper Permian (?) to the Upper Jurassic. However, the movements were not strong because they can hardly be recognized in the Triassic to Jurassic cover. Tectonic arguments indicate Post-Variscan fissure opening also in the basement (Sperling 1972). An important observation is that the fluorite veins of the Clara turn almost horizontal just below the Lower Triassic (Buntsandstein) sedimentary cover (Huck 1984, p.44; reproduced in von Gehlen 1987, Fig.3). A halo of diffuse hydrothermal mineralization up to the su2 member of the Buntsandstein corresponding to the first major phase in the veins (Maag 1987) seems to definitely indicate a Lower Buntsandstein age for this phase. The younger barite veins at the Clara reach higher levels in the stratigraphic column. These "geological" ages contrast with Jurassic radiometric ages (Mertz 1987; see Sect. 5.2) which may be caused by an additional Jurassic thermal event as shown by Brockamp et al. (1985, 1987) at Müllenbach (Sect. 2.4).

5.5 Outlook

A KTB drill hole at the Haslach site would cross the temperature range of formation especially of the vein-type mineralizations and would therefore permit direct investigation of the concentration processes by rock and fluid studies. In the 300 °C temperature range, rocks would presently be in the process of fluid inclusion mobilization (J.Walther and Althaus 1986) which could be studied in situ. Likewise, the process of healing of fluid-filled cracks forming secondary fluid inclusions could be studied in situ as it occurs down to about 150 °C-200 °C; already at intermediate depths, the KTB hole would expose regions with active retrograde mineral transformation and transport of dissolved components (J.Walther and Althaus 1986). Rock alteration, fluid and mineralogical studies then possible in situ would be of fundamental importance in the recognition of -- so far only postulated -- mineral and ore concentration processes.

Acknowledgements: I am grateful to K.-P. Burgath, F.-J. Eckhardt, H.W. Walther and especially to R. Metz and W. Wimmenauer for information on unpublished results.

REFERENCES

Agard J, Fluck P, Weil R, Wimmenauer W (1975a) Geologie des gites minéraux des Vosges et des regions limitrophes. Mém BRGM 87: 1-189

Agard J, Fluck P, Wimmenauer W, Weil R (1975b) Comparaison des minéralisations des Vosges et de la Foret Noire. Mém BRGM 87: 123-130

Amstutz GC, Fontbote L (1985) Correlative observations on the lead--zinc-(barite-fluorite) occurrences in the Germanic Triassic lithofacies of Central Europe. Geol Jahrb D 70: 179-211

Behr HJ, Gerler J (1987) Inclusions of sedimentary brines in post--Variscan mineralizations in the Federal Republic of Germany -- a study by neutron activation analysis. Chem Geol 81: 65-77

Behr HJ, Horn EE (1984) Unterscheidungskriterien für Mineralisationen des varistischen und postvaristischen Zyklus, die aus der Analyse fluider Einschlsse gewinnbar sind. Schriftenr GDMB 41: 255-269

Behr HJ, Hacker W, Heinrichs T, Hirschmann G (1986) Tektonische Entwicklung des Schwarzwaldes. 2nd KTB Colloq, Seeheim, Poster vol, p 61

Behr HJ, Horn EE, Frentzel-Beyme K, Reutel Chr (1987) Fluid inclusion characteristics of the Variscan and post-Variscan mineralizing fluids in the Federal Republic of Germany. Chem Geol 61: 273-285

Bliedtner M, Martin M (1986) Erz- und Minerallagerstätten des Mittleren Schwarzwaldes. Geol Landesamt Baden-Wuerttemberg, Freiburg i Br, 782 pp

Bonhomme MG, Bühmann D, Besnus Y (1983) Reliability of K-Ar dating of clays and silicifications associated with vein mineralizations in Western Europe. Geol Rundsch 72: 105-117

Brockamp O, Zuther M (1983) Das Uranvorkommen Müllenbach/Baden-Baden, eine epigenetisch-hydrothermale Imprägnationslagerstätte in Sedimenten des Oberkarbon (Teil II: Das Nebengestein). Neues Jahrb Mineral Abh 148: 22-33.

Brockamp O, Zuther M, Clauer N (1985) Alter und Genese der Uranlagerstätte Müllenbach (Baden-Baden). Fortschr Mineral 65, Beih 1: 34

Brockamp O, Zuther M, Clauer N (1987) Epigenetic-hydrothermal origin of the sediment-hosted Müllenbach uranium deposit, Baden-Baden, W-Germany. Monogr Ser Mineral Deposits 27: 87-98

Bültemann HW (1965) Die Uranvorkommen Ellweiler in Rheinland-Pfalz und Menzenschwand im Schwarzwald. Erzmetall 18: 79-83

Burgath K-P, Mohr M (1984) Quarz-Hämatit-Fluorit-Mineralisationen mit seltenen Übergemengteilen im Schwarzwald. Unpubl Colloq Intraformationale Lagerstättenbildung, Bonn, Oct 1984, p 11

Degens E (1956) Geochemische Untersuchungen im Nebengestein fluor--barytischer Co-Ni-Bi-Ag-U-Erzgänge des mittleren Schwarzwaldes. Glueckauf 92: 842-852

Dill H, Nielsen H (1987) Geochemical and geological constraints on the formation of unconformity-related vein baryte deposits of Central Europe. J Geol Soc 144: 97-105

Eigenfeld-Mende I (1948) Metamorphe Umwandlungserscheinungen an Metabasiten des Sdschwarzwaldes. Mitt Bad Geol Landesamt, NF 1: 1-111

Faisi S (1951) Die Eisen- und Manganerzgänge von Eisenbach (südöstlicher Schwarzwald) und ihre tektonische Stellung. Neues Jahrb Mineral Abh 83: 53-150

Faisi S, Gehlen K von (1957) Verkieselungszonen und ihre Bedeutung für die varistische Vererzung im südwestlichen Schwarzwald. Neues Jahrb Mineral Abh 91: 351-360

Fauth H, Hindel R, Siewers U, Zinner J (1985) Geochemischer Atlas Bundesrepublik Deutschland. Verteilung von Schwermetallen in Wässern und Bachsedimenten. BGR, Hannover. Schweizerbart, Stuttgart, 79 pp (multicolored maps)

Fluck P, Lougnon J (1984) Les grands traits de la métallogenie des Vosges. Fortschr Mineral 62, Beih 2: 53-65

Friedrich G, Keyssner S, Vogt JH, Vogtmann J (1986) Schwarzwald, Erzpetrologische-lagerstättenkundliche Untersuchungen. 2nd KTB Colloq Seeheim, Poster vol, p 44
Fritsche R, Föhse H (1981) Uranprospektion im Triberger Granit. Aufschluss 32: 369-377
Gehlen K von (1955) Gesteine und Blei-Zink-führende Flußspatgänge zwischen Feldberg und Belchen. II. Die Flußspatgänge von Wieden und ihre tektonische Stellung. Neues Jahrb Mineral Abh 88: 15-54
Gehlen K von (1958) U- und W-haltiger Hollandit und Kryptomelan von Eisenbach (südöstlicher Schwarzwald). Neues Jahrb Mineral Monatsh 1958: 73-84
Gehlen K von (1987) Formation of Pb-Zn-F-Ba mineralizations in SW Germany: a status report. Fortschr Mineral 65: 87-113
Gehlen K von, Schmitz U (1981) Palomagnetische Altersbestimmungen an hämatit-führenden Verkieselungszonen im Südschwarzwald. Fortschr Mineral 59, Beih 1: 50-52
Gehlen K von, Nielsen H, Ricke W (1962) S-Isotopen-Verhältnisse in Baryt und Sulfiden aus hydrothermalen Gängen im Schwarzwald und jüngeren Barytgängen in Süddeutschland und ihre genetische Bedeutung. Geochim Cosmochim Acta 26: 1189-1207
Gehlen K von, Zeino H, Murad E, Samman A (1970) Neue Ergebnisse zur Genese der Vererzung im südlichen Schwarzwald. Z Dtsch Geol Ges 121: 75-78
Gehlen K von, Baumann A, Hofmann R, Grauert B, Nielsen H (1984) Strontium-Isotopie in Barytgängen des Schwarzwaldes. Fortschr Mineral 62, Beih 1: 70-71
Gehlen K von, Grauert B, Nielsen H (1986a) REE minerals in southern Schwarzwald veins and isotope studies on gypsum from the central Schwarzwald, F.R. Germany. Neues Jahrb Mineral Monatsh 1986: 393-399
Gehlen K von, Kleinschmidt G, Stenger R, Wilhelm H, Wimmenauer W (eds) (1986b) Kontinentales Tiefbohrprogramm der Bundesrepublik Deutschland KTB, Ergebnisse der Vorerkundungsarbeiten Lokation Schwarzwald. 2nd KTB Colloq, Seeheim, I 160 pp.
Goldenberg G (1988) Platinmetalle im Rheinsand. Aufschluss 39: 57-64
Grundmann G (1971) Neues über Silbererze vom Tannenbodengang bei Wieden (Sdschwarzwald). Aufschluss 22: 195-200
Hofherr IM (1984) Eine Antimon-Mineralisation südlich von Sulzburg im Südschwarzwald. Ber Naturforsch Ges Freiburg im Breisgau 74: 49-71
Hofmann B (1979) Blei-, Zink-, Kupfer- und Arsenvererzungen im Wellengebirge (unterer Muschelkalk, Trias) am südlichen und östlichen Schwarzwaldrand. Mitt Naturforsch Ges Schaffhausen 31: 157-196
Hofmann R (1979) Die Entwicklung der Abscheidungen in den gangförmigen, hydrothermalen Barytvorkommen Mitteleuropas. Monogr Ser Mineral Deposits 17: 81-214
Hofmann R (1985) Strontium-Isotopie der jüngeren Abscheidungen in der Uranlagerstätte Menzenschwand und Vergleiche mit anderen Lagerstätten der weiteren Umgebung. Fortschr Mineral 63, Beih 1: 101
Hofmann R, Baumann A (1984) Preliminary report on the Sr isotopic composition of hydrothermal vein barites in the Federal Republic of Germany. Mineral Deposita 19: 166-169
Hofmann R, Schürenberg H (1979) Geochemische Untersuchungen gangförmiger Barytvorkommen in Deutschland. Monogr Ser Mineral Deposits 17: 1-80
Homann KD (1986) Genese und hydrothermale Überprägung des Triberg Granites; Schwarzwald, Granite. 2nd KTB Colloq, Seeheim, Poster vol, p 45
Huck K-H (1984) Die Beziehungen zwischen Tektonik und Paragenese unter Berücksichtigung geochemischer Kriterien in der Fluß- und Schwerspatlagerstätte "Clara" bei Oberwolfach/ Schwarzwald. Thes, Univ Heidelberg, 177 pp

Huck K-H, Walther HW (1984) Zur wirtschaftlichen Bedeutung der post--varistischen Mineralisation in Mitteleuropa. Schriftenr GDMB 41: 405-418

Joachim H, Smykatz-Kloss W (1985) Die manganhaltigen Brauneisen-Baryt-Gänge des Neuenbürger Reviers (nördlicher Schwarzwald), BRD. Chem Erde 44: 311-339

Kirchheimer F (1953) Weitere Untersuchungen über das Vorkommen von Uran im Schwarzwald. Abh Geol Landesamt Baden-Wuerttemberg 1: 1-60

Kirchheimer F (1957) Bericht über das Vorkommen von Uran in Baden-Württemberg. Abh Geol Landesamt Baden-Wuerttemberg 2: 1-127

Kirchheimer F (1976) Bericht über Spuren römerzeitlichen Bergbaus in Baden-Württemberg. Aufschluss 27: 361-371

Kluth GC (1965) Die Kupfer-Wismut-Erze des mittleren Schwarzwaldes und ihre Geschichte. Thes, Univ Heidelberg, 199 pp

Kneuper G, List KA, Maus HJ (1977) Geologie und Genese der Uranmineralisation des Oostroges im Nordschwarzwald. Erzmetall 30: 522-526

Kreutzwald A (1936) Die Blei-Zinkerzgänge am Schauinsland im südwestlichen Schwarzwald. Neues Jahrb Mineral Geol Palaeontol Beil 70 A: 234-334

Krützfeld B (1985) Beobachtungen an Vererzungen in Sedimenten der Zone Badenweiler-Lenzkirch. Aufschluss 36: 261-265

Laufhütte D (1954) Der Bleiglanzgang Friedrich-Christian im Wildschapbachtal. Thes, Univ Freiburg i Brsg, 95 pp

Lippolt HJ (1984) Experimentelle Datierungen von postvaristischen Mineralisationen, Möglichkeiten und Grenzen. Schriftenr GDMB 41: 305-359

Lippolt HJ (1986) Die Retentivität isotopischer Chronometer, Minerale als Stoppuhren geologischer Prozesse. Heidelberger Geowiss Abh 6: 285-302

Lippolt HJ, Schorn U, Pidgeon RT (1983) Genetic implications of new lead isotope measurements on Schwarzwald vein and Upper Triassic sediment galenas. Geol Rundsch 72: 77-104

Lippolt HJ, Kober B, Hradetzky H, Mertz DF (1986a) Schwarzwald, Geochronologische Studien an Metamorphiten und Mineralisation. 2nd KTB Colloq, Seeheim, Poster vol, p 41

Lippolt HJ, Kober B, Mertz DF, Keuthen M (1986b) Schwarzwald, Pb- und Sr-Isotopengeochemie von Mineralisation und Nebengesteinen. 2nd KTB Colloq, Seeheim, Poster vol, p 42

Lippolt HJ, Mertz DF, Huck K-H (1986c) The genesis of the Clara and Friedrich-Christian vein deposits/Central Schwarzwald (FRG): evidence from Rb-Sr, Sr87/86, K-Ar and Ar40/Ar39 investigations. Terra Cognita 6: 228

Maag WH (1987) Buntsandsteinsedimente des mittleren Schwarzwaldes unter Beeinflussung hydrothermaler Ereignisse. Heidelb Geowiss Abh 8: 157-159 160-162

Maus H, Renk A (1981) Der kristalline Schwarzwald -- seine Gesteine und Lagerstätten. Aufschluss 32: 323-332

Maus H, Gundlach H, Podufal P (1979) Über den Sellait (MgF_2) der Grube Clara, Oberwolfach, Mittlerer Schwarzwald. Neues Jahrb Mineral Abh 136: 10-25

Mehnert KR (1949) Die Kupfer-Wismut-Lagerstätte Daniel bei Wittichen (mittl. Schwarzwald). Neues Jahrb Mineral Geol Palaeontol Monatsh 1949 A: 217-260

Mertz DF (1987) Isotopengeochemische und mineralogische Untersuchungen an postvariscischen hydrothermalen Silikaten. Thes, Univ Heidelberg, 207 pp

Mertz DF, Lippolt HJ, Huck K-H (1986) K-Ar, $^{40}Ar/^{39}Ar$ and Rb-Sr investigations on the genesis of the Clara vein deposit/central Schwarzwald. Abstr DGG/ILP Meet Karlsruhe, p 235

Metz R (1957a) Die Blei-Zinkerzgänge im mittleren Schwarzwald nördlich des Elztales. Beih Geol Jahrb 29: 189-202

Metz R (1957b) Die Erzführung der Schwarzwald-Randverwerfungen. Beih Geol Jahrb 29: 203-207
Metz R (1957c) Die Geschichte des Blei-Silber-Zinkerzbergbaus im Schwarzwald. Beih Geol Jahrb 29: 208-256
Metz R (1959) Einige Gesteins- und Mineralfundpunkte im Schwarzwald in der Umgebung von Freiburg. Aufschluss 10: 205-230
Metz R (1977) Mineralogisch-landeskundliche Wanderungen im Nordschwarzwald, besonders in dessen alten Bergbaurevieren, 2nd edn. Schauenburg, Lahr/Schwarzwald, 632 pp, 1 map
Metz R (1980) Geologische Landeskunde des Hotzenwalds mit Exkursionen, besonders in dessen alten Bergbaugebieten. Schauenburg, Lahr/ Schwarzwald, 1117 pp, 4 maps
Metz R, Rein G (1958) Erläuterungen zur Geologisch-petrographischen Übersichtskarte des Südschwarzwaldes, 1:50 000. Schauenburg, Lahr/ Schwarzwald, 126 pp, 1 map
Metz R, Richter M, Schürenberg H (1957) Die Blei-Zink-Erzgänge des Schwarzwaldes. Beih Geol Jahrb 29: 277 pp
Mittelstädt R (1987) Flüssigkeitseinschluß- und Schwefelisotopenuntersuchungen an Mineralen der Blei-Zink Lagerstätte "Schauinsland" im südlichen Schwarzwald. Dipl Thes, Univ Göttingen, 119 pp
Möller P, Maus H, Gundlach H (1982) Die Entwicklung von Flußspatmineralisationen im Bereich des Schwarzwaldes. Jahresh Geol Landesamt Baden-Wuerttemberg 24: 35-70
Murad E (1970) Geochemie der "gangbildenden" Elemente Blei, Zink, Barium und Fluor in Eruptivgesteinen des südlichen Schwarzwaldes. Thes, Univ Frankfurt a M, 77 pp
Murad E (1974) Hydrothermal alteration of granitic rocks and its possible bearing on the genesis of mineral deposits in the southern Black Forest, Germany. Econ Geol 69: 532-544
Osann A (1927) Die Mineralien Badens. Schweizerbart, Stuttgart, VIII 239 pp
Otto J (1967) Der Fluoritgang der Grube "Gottesehre" bei Urberg, Kr. Säckingen (Südschwarzwald). Jahresh Geol Landesamt Baden-Wuerttemberg 9: 25-50
Otto J, Schürenberg H (1974) Neue Mineralfunde im südlichen Schwarzwald. Aufschluss 25: 205-211
Rein G (1952) Der Werdegang des Orthits in der magmatischen und metamorphen Abfolge des mittleren Schwarzwaldes. Neues Jahrb Mineral Abh 84: 365-435
Renk A (1975) Der Flußspatgang der Grube "Käfersteige" bei Pforzheim (Nordschwarzwald). Dipl Thes, Freiburg i Br, 82 pp
Schatz RH (1970) Neue Mineralfunde im Steinbruch Schweighof bei Badenweiler im SW-Schwarzwald. Aufschluss 21: 103-104
Schauer H (1962) Uranerzführende Gänge im Bärhaldegranit (Südschwarzwald). Neues Jahrb Mineral Monatsh 1962: 121-124
Schneiderhöhn H (1941) Lehrbuch der Erzlagerstättenkunde, Band I. Die Lagerstätten der magmatischen Abfolge. Fischer, Jena, XXIV 858 pp
Schneiderhöhn H (1949) Schwerspatgänge und pseudomorphe Quarzgänge in Westdeutschland. Neues Jahrb Mineral Geol Palaeontol Monatsh 1949 A: 191-202
Schneiderhöhn H (1958a) Die Erzlagerstätten der Erde, Band I. Die Erzlagerstätten der Frühkristallisation. Fischer, Stuttgart, XIII 315 pp
Schneiderhöhn H (1958b) Die Entwicklung der Kenntnis der Erzlagerstätten im Schwarzwald. Z Dtsch Geol Ges 110: 17-21
Schürenberg H (1950) Die Erzgänge Teufelsgrund und Schindler im Untermünstertal und ihr quantitativer Mineralgehalt. Neues Jahrb Mineral Abh 81: 123-182
Schürenberg H (1957) Die Blei-Zink-Erzgänge des Südschwarzwaldes. Beih Geol Jahrb 29: 33-188

Sehlke K (1956) Gesteine und Blei-Zink-führende Flußspatgänge zwischen Feldberg und Belchen im Hochschwarzwald. IV. Erzführende Flußspatgänge der Umgebung von Todtnau (Südschwarzwald). Neues Jahrb Mineral Abh 89: 258-280

Simon K, Hoefs J (1987) Effects of meteoric water interaction on Hercynian granites from the Südschwarzwald, southwest Germany. Chem Geol 61: 253-261

Sperling H (1972) Zur Altersstellung der Wiedener und Brandenberger Flußspatgänge (Südschwarzwald). Mineral Deposita 7: 351-359

Tapfer M (1987) Geochemical characteristics of non-strata-bound uranium mineralization of the Müllenbach deposit (Northern Schwarzwald, W.Germany). Monogr Ser Mineral Deposits 27: 99-106

Vogelgesang W (1865) Geognostisch-bergmännische Beschreibung des Kinzigthaler Bergbaus. Beitr Statist Großherzogt Baden 21

Walenta K (1957) Die antimonerzführenden Gänge des Schwarzwaldes. Jahresh Geol Landesamt Baden-Wuerttemberg 2: 13-67

Walenta K (1958) Die antimonerzführenden Gänge bei St.Trudpert im Münstertal und bei St.Ulrich (südlicher Schwarzwald). Jahresh Geol Landesamt Baden-Wuerttemberg 3: 53-79

Walenta K (1967) Scheelit und Wolframit von neuen Fundorten im Schwarzwald. Jahresh Geol Landesamt Baden-Wuerttemberg 9: 51-58

Walenta K (1972) Die Sekundärmineralien der Co-Ni-Ag-Bi-U-Erzgänge im Gebiet von Wittichen im mittleren Schwarzwald. Aufschluss 23: 279-329

Walenta K (1980) Neue Uranvorkommen im Gebiet von Schramberg und Triberg und ihr Mineralbestand. Aufschluss 31: 141-150

Walenta K (1987) Gold von der Grube Clara. Aufschluss 38: 40-44

Walenta K, Wimmenauer W (1961) Der Mineralbestand des Michelsganges im Weiler bei Lahr (Schwarzwald). Jahresh Geol Landesamt Baden-Wuerttemberg 4: 7-38

Walenta K, Sawatzki G, Dayal R (1970) Die Wolframerzvorkommen im Gebiet des Nordracher Granitmassivs und seiner Umgebung. Jahresh Geol Landesamt Baden-Wuerttemberg 12: 207-226

Walther HW (1981) Quantitative regionale Metallogenese als Beitrag zur Frage: Wo sind die Metalle der Zukunft? Erzmetall 34: 432-438

Walther HW (1984) La metallogénie de la Republique Fédérale d'Allemagne. In: Laffitte P, Emberger A (eds) Mémoire explicatif de la carte métallogenique de l'Europe et des pays limitrophes, 1:2 500 000. 1. Synthese par zone redactionelle. UNESCO Sci Terre 17: 187-235

Walther HW, with 26 co-authors (1986) Federal Republic of Germany. In: Dunning FW, Evans AM (eds) Mineral deposits of Europe, vol 3 Central Europe. Instn Mining Metall Mineral Soc, London, pp 175-301

Walther J, Althaus E (1986) Schwarzwald, Fluide Einschlüsse. 2nd KTB Colloq, Seeheim, Poster vol, p 46

Wecht EHP (1955) Gesteine und Blei-Zink-führende Flußspatgänge zwischen Feldberg und Belchen im Hochschwarzwald. III. Gesteine und Erzgänge am Silberberg im Feldberggebiet, Hochschwarzwald. Neues Jahrb Mineral Abh 88: 273-308

Wendt I, Lenz H, Höhndorf A (1974) Das Alter des Bärhalde-Granites (Schwarzwald) und der Uranlagerstätte Menzenschwand. Geol Jahrb E 2: 131-143

Wendt I, Lenz H, Höhndorf A, Bültemann H, Bültemann W-D (1979) Das Alter der Pechblende der Lagerstätte Menzenschwand, Schwarzwald. Z Dtsch Geol Ges 130: 619-626

Wimmenauer W (1951) Erzmikroskopische Beobachtungen an Gesteinen des schwarzwälder metamorphen Grundgebirges. Mitteilungsbl Bad Geol Landesanst 1950: 14-20

Wimmenauer W (1952) Mikroskopische Untersuchungen an Uranerzen aus dem mittleren Schwarzwald. Mitteilungsbl Bad Geol Landesanst 1951: 75-83

Wimmenauer W (1955a) Der Mineralbestand der uranhaltigen Erze aus der Kobaltgrube "Segen Gottes" bei Sulzburg (südlicher Schwarzwald). Jahresh Geol Landesamt Baden-Wuerttemberg 1: 61-71
Wimmenauer W (1955b) Über den Mineralbestand der uranhaltigen Manganerze von Eisenbach und einiger anderer "Psilomelane" des Schwarzwaldes. Jahresh Geol Landesamt Baden-Wuerttemberg 1: 72-78
Wimmenauer W (1975) Caractères généraux et métallogenie des gites de la Foret Noire; Index des gites minéraux de la Foret Noire et de ses environs. In: Mém BRGM 87: 121-123 152-161
Wimmenauer W (1979a) Paragenese of hydrothermal ore and mineral deposits of the Schwarzwald (FRG). Freiberger Forschungsh C 345: 105-114
Wimmenauer W (1979b) Die Eisenerzgänge im Schwarzwald. Geol Jahrb D 31: 49-63
Wimmenauer W (1980a) The mineral deposits of the Schwarzwald. Erzmetall 33: 150-152
Wimmenauer W (1980b) Lithology of the Precambrian in the Schwarzwald. An interim report. Neues Jahrb Mineral Monatsh 1980: 364-372
Wimmenauer W (1984) Das prävaristische Kristallin im Schwarzwald. Fortschr Mineral 62 Beih 2: 69-86
Wimmenauer W (1985) Hydrothermal vein deposits of the Schwarzwald. 2nd Int Symp Observ Cont Crust through Drilling, Seeheim, Excursion Guide Schwarzwald, pp 45-54
Wimmenauer W, Hauptmann A, Huber C, Hurrle H, Lim SK, Martin M, Otto J, Renk A (1980) Introduction to the Schwarzwald Excursion. 8th Int Geochem Explor Symp, Hannover, Excursion Guide, C 4, 24 pp
Zeino H (1967) Die Genese der hydrothermalen Mineralisation im südwestlichen Schwarzwald. Thes, Univ Frankfurt a M, 117 pp
Zirngast M (1978) Zur Geologie und Geochemie der Schwerspatlagerstätte "Clara" bei Wolfach (mittlerer Schwarzwald). Thes, Univ Braunschweig, 164 pp

Near-Vertical and Wide-Angle Seismic Surveys in the Schwarzwald*

E. Lüschen[1], F. Wenzel[1], K.-J. Sandmeier[1], D. Menges[1], Th. Rühl[1], M. Stiller[2], W. Janoth[2], F. Keller[2], W. Söllner[2], R. Thomas[2], A. Krohe[3], R. Stenger[4], K. Fuchs[1], H. Wilhelm[1], and G. Eisbacher[3]

[1] Geophysikalisches Institut, Universität Karlsruhe, Hertzstr. 16, 7500 Karlsruhe 21, FRG
[2] Institut für Geophysik, Universität Clausthal, Arnold-Sommerfeld-Str. 1, 3392 Clausthal-Zellerfeld, FRG
[3] Geologisches Institut, Universität Karlsruhe, Kaiserstr. 12, 7500 Karlsruhe, FRG
[4] Mineralogisch-petrographisches Institut, Universität Freiburg, Albertstr. 23b, 7800 Freiburg, FRG
*reprint of J. Geophys. (1987) 62: 1-30

Abstract

A unified seismic exploration program, consisting of 345 km of deep reflection profiling, a 200 km refraction profile, an expanding spread profile and near surface high-resolution measurements, revealed a strongly differentiated crust beneath the Black Forest. The highly reflective lower crust contains numerous horizontal and dipping reflectors at depths of 13-14 km down to the crust-mantle boundary (Moho). The Moho appears as a flat and horizontal first order discontinuity at a relatively shallow level of 25-27 km above a transparent upper mantle. In the seismic model based on near-vertical and wide-angle data the lower crust consists of lamellae with an average thickness of about 100 m and velocity contrasts increasing with depth. The upper crust is characterized by a discontinuous pattern of mostly dipping reflectors which are related to Hercynian overthrusting and accretion and to late-Hercynian extensional faulting. A bright spot at 9.5 km depth is interpreted to be due to low-velocity material. The lower part of the upper crust appears as a relatively transparent zone which is also identified as a low-velocity zone situated directly above the laminated lower crust.

Key words: Reflection seismology - Deep-crustal reflection profiling - Refraction seismology - Hercynian crustal evolution

1 Introduction

Continental deep drilling is one of the key projects of the International Lithosphere Program (ILP) which is aimed at unravelling the dynamics and evolution of the lithosphere. The contribution of the Federal Republic of Germany consists of the continental deep drilling programm KTB (Kontinentales Tiefbohrprogramm, Althaus et al., 1984) and of the continental seismic reflection program DEKORP (Deutsches Kontinentales Reflexionsseismik Programm, Bortfeld et al., 1985).

Deep drilling interest concentrated on two potential sites, both located within Hercynian crystalline basement complexes - the Oberpfalz (Upper Palatinate) in eastern Bavaria and the Schwarzwald (Black Forest) in southwestern Germany. In both areas comprehensive efforts have been made to apply all kinds of geoscience investigations (Alfred-Wegener-Stiftung, 1985) for reconnaissance studies. Reflection measurements were organized within the framework of the DEKORP. In this paper we describe the contribution of seismic methods to the understanding of structural and physical properties of the Hercynian crust beneath the Black Forest in connection with other geophysical sounding methods. The seismic survey consisted of four types of investigations:

A Deep reflection profiling (near-vertical incidence, section 3)

B Deep refraction profiling (long range profiles, section 4, 4.1)

C Expanding spread profile (ESP, section 4, 4.2)

D High resolution methods (near-surface, borehole measurements, section 7)

This article deals with the programs A, C and D and with a combined reflection-refraction interpretation, using other geophysical data as well. Details of the refraction experiment B can be found in Gajewski and Prodehl (1987).

The reflection network was planned to get regional structural control by a 170 km long NS-profile and spatial control of the central Black Forest region by three shorter intersecting profiles. The NS profile 8401 follows approximately the morphological axis of the Black Forest crossing the Central Schwarzwald Gneiss complex (CSGC) which is bounded in the west by the Tertiary Rhinegraben, in the north by the southward dipping Saxothuringian-Moldanubian suture zone, in the south by the NW-dipping "Badenweiler-Lenzkirch" crustal thrust zone, and in the east by gently eastward dipping Mesozoic sediments (Fig. 1). Two other profiles (8402, 8403) are focussed on the proposed drilling area. A fourth supplementary line (8514) was planned to study the tectonic significance and lateral extent of the "Badenweiler-

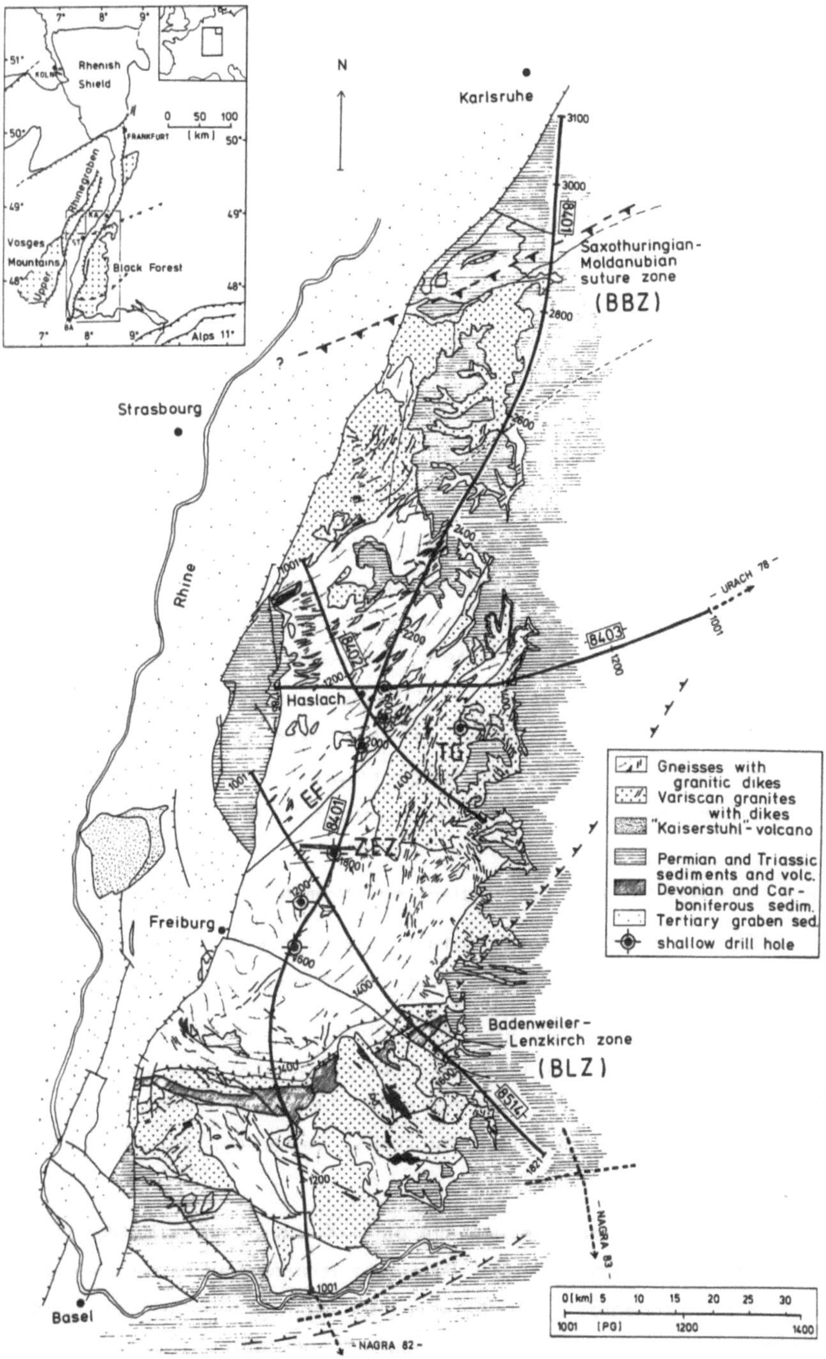

Figure 1: Geological-tectonic map of seismic reflection profiles 8401, 8402, 8403 and 8514. The Central Schwarzwald Gneiss Complex is bounded by two Hercynian thrust zones, the Baden-Baden zone (BBL) in the north (Saxothuringian-Moldanubian suture zone) and the Badenweiler-Lenzkirch zone (BLZ) in the south. ZEZ = Zinken-Elme zone, TG = Triberg Granite, EF = Elztal Fault, PG = Geophone location.

Lenzkirch"-zone. The profile net is connected with reflection profiles in the south (Finckh et al., 1986) and in the east (Bartelsen et al., 1982, Walther et al., 1986).

The field and processing parameters were chosen to provide good resolution throughout the entire crust. In order to link the crustal seismic sections to surface structures for their geologic-petrological interpretation, we performed near surface high resolution measurements at selected sites and a systematic evaluation and tomographic inversion of first arrivals of the reflection recordings. From the interpretation of the refraction data an outline of the two-dimensional velocity structure of the Black Forest emerged. Additional velocity information of the target area and further constraints on the physical nature of the crust were established by intensive one- and two-dimensional seismic modelling of reflection and refraction data and by an Expanding Spread Profile (ESP) focussed on the Central Black Forest.

In the past, intensive geophysical studies in the region concentrated mainly on the Rhinegraben. Results of these studies in particular of the seismic investigations are documented in e.g. Illies and Mueller (1970), Illies and Fuchs (1974), Edel et al. (1975), Prodehl et al. (1976), Illies (1981), and Zucca (1984). Fuchs et al. (1987) opened a new phase of discussion on the properties and the evolution of the lower crust beneath the Rhinegraben system, relating seismic reflection results in the Rhinegraben proper (Dohr, 1970) with corresponding results described in this paper.

2 Geological and petrological environment

The Black Forest is the uplifted eastern shoulder of the Rhinegraben rift, where rocks of the Hercynian basement complex of Central Europe are exposed.

The structural zonation of the European Hercynian fold belt is described in Behr et al. (1984). It is subdivided in two external zones and one internal crystalline zone. The external Rhenohercynian and Saxothuringian zones consist of Paleozoic sedimentary rocks, volcanics, and a minor proportion of crystalline blocks.

The Black Forest and Vosges are part of the internal Moldanubian zone which consists of Precambrian and Paleozoic terranes accreted during the Acadian/ Hercynian orogenic cycle between the Southern continent Gondwana and a Northern continent N America-Europe. Thrust belts formed north and south of the internal crystalline zone. The northern belt (Saxothuringian zone, northern part of the Armorican Massif) is characterized by NW-vergence of the major structures whereas the southern belt (Moravian zone, Southern Central Massif, Southern Armorican Massif) shows SE- and S-directed vergence.

During an early stage of the Acadian/Hercynian orogenic cycle oceanic crust was subducted along both mobile belts. This event was followed by continent-continent collision in Carboniferous time (Lorenz and Nicholls, 1984; Matte, 1986). The later stage is characterized by complex wrench faulting and extensional tectonics of Basin-and-Range type.

The crystalline basement of the Black Forest consists of high-grade gneisses and migmatites intruded by Hercynian granitoids. The Paleozoic Badenweiler-Lenzkirch zone separates two crystalline blocks: the Central Schwarzwald Gneiss Complex (CSGC) and the Southern Schwarzwald Complex (SSC) which differ significantly both in their petrological and geophysical characteristics.

The CSGC consists of intensely deformed and transposed metapsammitic and metapelitic gneisses of Precambrian age.

The high-grade gneisses and migmatites of the basement complex (CSGC) display a polyphase metamorphic evolution. Numerous small bodies of eclogites (or eclogitic amphibolites), ultramafics (both spinel- and garnet-bearing serpentinites and pyroxenites) and granulitic gneisses indicate older high and medium pressure metamorphic events (Klein and Wimmenauer, 1984; Wimmenauer and Adam, 1985).

The Hercynian tectonic evolution of the CSGC can be documented along two NE to ENE trending belts of strongly deformed Paleozoic sediments and volcanics: the Baden-Baden zone in the north and the Badenweiler-Lenzkirch zone in the south. These zones represent thrust faults opposed to each other with northwestward and southeastward vergences. The Baden-Baden zone is regarded as the boundary between the Moldanubian and the Saxothuringian zone of the Hercynian fold belt. Northwestward directed tectonic transport is indicated by transposed foliations in weakly metamorphosed sediments which dip to the SE beneath higher grade micaschists and gneisses. Overthrusting was followed by dextral strike-slip motions along this zone. The Badenweiler-Lenzkirch zone separates the Central Schwarzwald Gneiss Complex from the Southern Schwarzwald Complex. Southeast-directed tectonic transport is indicated by a 2-4 km wide, strongly deformed W-E striking zone of Paleozoic sediments, metasediments, volcanics, and mylonitic gneisses which dip to the NW beneath the CSGC. Sporadic fossils indicate that sedimentation occurred between around 370-335 Ma. In this time span the sedimentary environment changed from deep water deposition to shallow marine and continental sedimentation (Sittig, 1969). This may be the result of a considerable tectonic shortening in this time span. The mylonites and inverse metamorphic isograds demonstrate overthrusting of the hanging wall CSGC over the sedimentary sequence to the SE. As in the Baden-Baden zone, thrust movements were followed by strike-slip motions. Within the CSGC further convergent zones exist (e.g. Zinken-Elme zone).

Hercynian crustal evolution culminated with the emplacement of several suites of granitic plutons mostly of S-type character, the oldest of which show penetrative mylonitic/cataclastic deformation (370-335 Ma). The suite of the syn- to post-tectonic granites covers a time span of more than 70 Ma (370-290 Ma). Structural and geochemical investigations reveal the evolutionary trends both in time and space of the various granite suites (Emmermann, 1977). Ignimbrites and rhyolites of Permian age mark the final stage of the Hercynian magmatic activity (Lippolt et al., 1983).

Late Carboniferous to Permian intramontane troughs filled with clastic sedimentary rocks and extrusive magmatic activity indicate a change in tectonic style towards extension of the Hercynian basement during this time. Within the basement, the pattern of late Hercynian extension is indicated by porphyric and lamprophyric dikes and by post-metamorphic cataclastic deformation within the CSGC, the SSC and within the granite plutons.

In Mesozoic time subsidence of platform type is characterized by non-marine redbeds and shallow marine epicontinental deposits.

Due to the stress regime in the foreland of the Alpine collision zone a system of grabens formed in the early Tertiary. Subsidence in the Rhinegraben started at 45-40 Ma. Uplift of the graben shoulders of more than 2000 m caused exposure of the Hercynian basement. Related to the rift event, the Moho rose to extremely shallow depths by updoming in the southern Rhinegraben where a depth of 24 km is shown by intensive seismic refraction studies (Edel et al., 1975).

A specific goal of the seismic sounding was to clarify the role and extent of Hercynian thrust and extension tectonics for the entire Black Forest region and in particular for the central Black Forest area.

3 Seismic reflection profiling and structural image

For economic reasons and difficult terrain and permitting conditions we decided to use the Vibroseis technique, bearing in mind the convincing results of the COCORP (USA) and ECORS (France) programs for crustal reflection profiling. Explosive sources were used in additional wide-angle measurements with equipment of various research institutions in order to get velocity control (section 4).

The general goal of the geophysical reconnaissance program was to collect all kinds of data for a long N-S crustal section. Consequently the N-S reflection profile was regarded as a key profile to study the regional geologic and tectonic setting of the Black Forest. Since COCORP started deep crustal reflection profiling in 1975, it has been demonstrated that profiles of at least

100 to 200 km length are necessary to identify the structural-tectonic style of a crustal segment. The profile 8401 therefore follows the morphological axis of the Black Forest along 170 km crossing the Central Gneiss Complex and the adjacent thrust zones: "Baden-Baden" in the north and "Badenweiler-Lenzkirch" in the south (Fig. 1). At the southern end the line joins the reflection network of northern Switzerland (Finkh et al., 1986). For three-dimensional structural control the central target area was covered by two intersecting profiles generating a triangle of 3-4 km side length together with line 8401. Profile 8402 is about 45 km long and traverses the crystalline basement almost perpendicularly to the Hercynian strike. Profile 8403 is a prolongation of the Urach profile U 1 (Bartelsen et al., 1982) forming a 120 km E-W-transect, roughly perpendicular to the Rhinegraben. It will be extended into France crossing the Rhinegraben and the Vosges mountains in 1988 as a joint ECORS-DEKORP program.

These three profiles 8401-8403 with a total length of 278 km were accomplished in a three-month campaign in 1984. A fourth profile of 65 km length (8514) was recorded in order to investigate prominent reflectors of profile 8401 in the upper crust in greater detail, particularly the northward dipping Paleozoic "Badenweiler-Lenzkirch" thrust zone (BLZ) and its implications for the proposed drill site near Haslach (Fig. 1). This profile was recorded in November 1985.

3.1 Field survey

The field work was performed by a Vibroseis crew of the company PRAKLA-SEISMOS AG, Hannover. The routine oil-and-gas exploration crew was supplemented by additional vibrators, trucks, and personnel recruited from the University of Karlsruhe who assisted in the weathering survey, in static corrections, geological mapping and in other crew operations. The total crew consisted of 57 persons and about 30 trucks, all operating within a range up to 40 km.

The field parameters (Table 1) were chosen to meet the following conditions: a multifold horizontal coverage of about 100 (as estimated from a comparative Vibroseis-explosion test on profile DEKORP 2-S, Bortfeld et al., 1985), a CDP-spacing of 40 m and a daily progress of 4-5 km, i.e. a 200-channel recording system along a 16 km spread with 80 m source and geophone group spacing had to be utilized. A symmetric split-spread with 200 m in-line offset was chosen to increase the shallow coverage and resolution. Upper crustal resolution and velocity control by normal move-out require different spread configurations. We decided in favour of structure resolution (split-spread) while velocity control was achieved separately by an expanding-spread profile

Table 1. Acquisition parameters

Source	5 vibrators VVDA (14 t) on profiles 8401, 8402, 8403 and VVEA (19 t) on profile 8514	
	spacing between vibrators	20 m (8514: 10 m)
	moveup between sweeps	6 m (8514: 2 m)
	total source array length	146 m (8514: 48 m)
	vibrator point (VP) spacing	80 m (8514: 40 m)
	signal	upsweep 12–48 Hz, logarithmic, duration 20 s
	no. of sweeps per VP	12 (8514: 5)
	in-line offset	2 × 200 m
	max. off-line offset	700 m
Receiver	no. of geophone groups	200
	group spacing	80 m
	spread length	16.24 km
	spread	split spread (8514: asymmetric)
	no. of geophones (10 Hz) per group	24
	pattern	in-line, weighting uniform
	spacing between geophones	3.5 m
	total array length per group	80 m
Recording	equipment SERCEL 348 (8514: SERCEL 368), telemetric, automatic noise reduction before correlation, correlator-stacker CS 2502	
	traces	200
	correlation before vertical stack, sampling rate	4 ms
	total recording time (uncorrelated)	32 s
	correlated record length	12 s
	vertical stacking of correlated records	12-fold (8514: 5)
	preamplification	$2^7 = 42$ dB
	low-cut filter	12.5 Hz
	high-cut filter	62.5 Hz
Coverage	reflection	max. 100-fold
	refraction	max. 200-fold (8514: 400-fold)
	CDP spacing	40 m (8514: 20 m)

(section 4.2) and by a refraction experiment also along the main N-S line (Gajewski and Prodehl, 1987, and section 4.1). A source pattern of five vibrators and 146 m length with 10-12 fold vertical stacking yielded reliable suppression of groundroll noise. A 12-48 Hz upsweep of 20 s length was chosen based on a startup test. The received signal was filtered and digitized at each geophone group consisting of an 80 m in-line array of 24 geophones and a remote box. The data were then multiplexed and telemetrically transmitted to the recording truck by a 16 km long cable. Automatic noise rejection and full precision correlation were applied before 12-fold vertical stacking and storage on magnetic tapes.

In contrast to the explosion technique, the Vibroseis method allows choice of bandwith of the emitted signal. In practice, however, vibrator-ground interaction may influence the spectral content of the seismic signal, depending on mechanical properties of the vibrator system and elastic properties of the ground. During the Vibroseis survey daily similarity tests were performed under different geological and coupling conditions. The baseplate signal can be regarded as seismic source signal (Lerwill, 1981). The amplitude spectra of the similarity recordings were compared with the spectra of the field· recordings aquired at the same vibrator locations. In

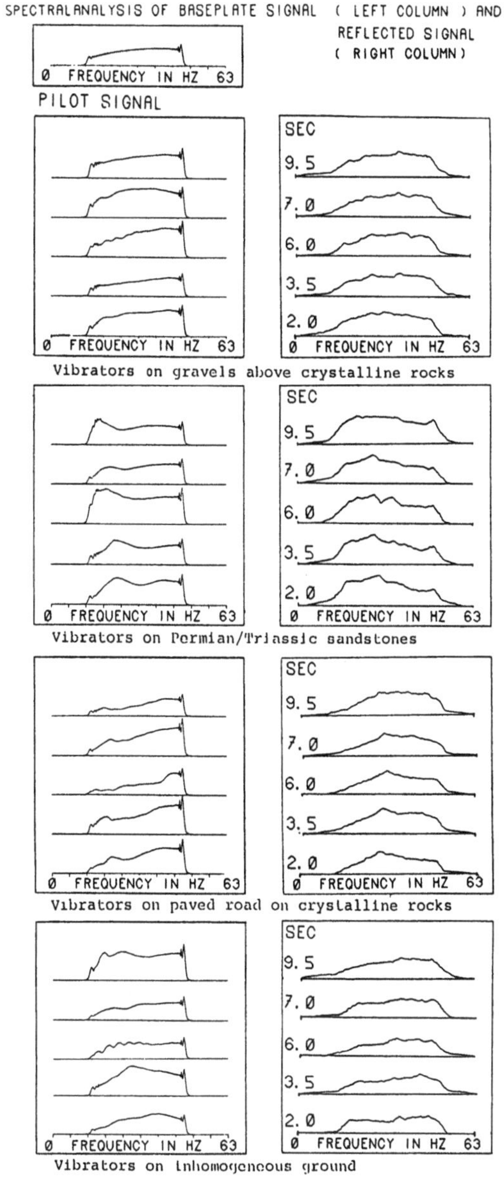

Figure 2: Spectral analysis of baseplate signals of five vibrators from similarity tests (left panels) and of corresponding seismic records (right panels, sum spectra of high S/N ratio traces from five 1 s time sections starting at TWT indicated lefthand).
Amplitudes are normalized, scales are linear. Pilot signal (upsweep, 12-48 Hz, 20 s length) on top.

Fig. 2 four examples are shown to demonstrate the ground-dependent differences between amplitude spectra of the pilot signal (12-48 Hz), baseplate signals of five vibrators, and of 5 seismogram portions of 0.5 s length at different recording times.

The upper example shows an ideal broadband source spectrum which is similar to the pilot signal showing the normal situation in the crystalline area of the Black Forest. In the other three examples of Fig. 2 resonant frequencies appear, which in most cases are low above thick Permian and Triassic sediments and high on paved roads. The studies of Vibroseis source characteristics showed that mainly broadband seismic signals were emitted resulting from favourable vibrator-ground interactions.

A detailed study of vibrator-ground interactions has been carried out by Schnell (1987) who considered the influence of the road and the near-surface geology.

3.2 Data processing

The data processing for Black Forest profiles was carried out at the DEKORP PROCESSING CENTER, Institut für Geophysik, Technische Universität Clausthal. Details and problems of the processing sequence have been discussed by Bortfeld et al. (1985) regarding the DEKORP profile 2-South. True-amplitude and crooked-line processing were applied routinely and post-stack coherency-filtering optionally. The high signal/noise ratio even in the common source gathers is shown in Fig. 3.

Common midpoint (CMP) sorting took into account crooked line geometry. A lateral deviation of vibratorpoints up to 800 m due to terrain conditions and a slightly bent receiver line required a crooked line processing, for which a processing line through the centre of the midpoint distribution was defined.

Stacking-velocity analysis was based on visual inspection of constant velocity stacks of groups of 20-50 CMPs using 28 velocities in the range of 4000-9760 m/s. Often the picking of optimum velocities was hampered by the lack of continuous reflectors in the upper crust. There the quality of stacked sections could often be enhanced by unconventional, high stacking velocities (see also section 6).

Figure 3: Examples of common source gathers (profile 8401, VP 2178) in three different display and scaling modes. Left: original field recording without further processing, middle: AGC (automatic gain control)-scaled record after application of static corrections, right: TA (true amplitude) record. TA includes horizontal trace equalization and amplitudes are multiplied between 0 and 3 s TWT with $k\,T\,e^{aT}$ (k = const., T = traveltime, a = constant of absorption) to compensate for spherical divergence and absorption. The values have been assessed empirically in order to obtain a balanced seismogram. Note the strong reflections from the lower crust beginning at 5 s TWT (two way traveltime) even in single-fold seismograms. Length of spread: 16.24 km.

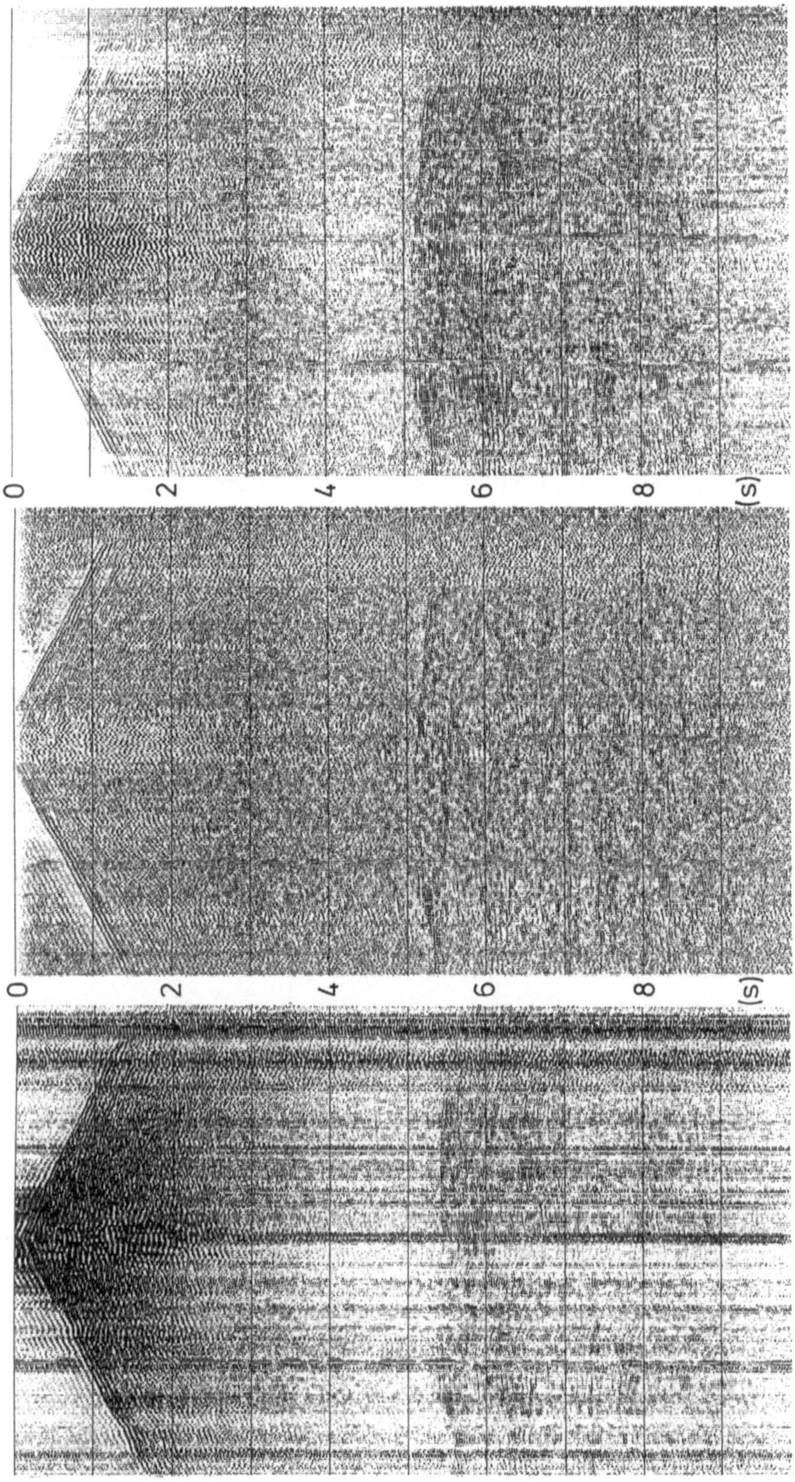
0
2
4
6
8
(s)
0
2
4
6
8
(s)

Figs. 4, 5, 6 illustrate the different character of imaging by AGC (automatic gain control)-, TA (true amplitude)-stacking and FD (finite difference) migration and their specific advantages, chosen from the central part of profile 8401.

For interpretation purposes all processing modes were taken into account in order to exploit the special advantages of each section. The stacked sections reveal a better signal/noise ratio and often a better readability than the migrated sections. The AGC-stacked data make reflections visible mostly by their coherence, even the weaker elements whose amplitudes are below the ambient noise level. The TA-stacked data allow a good estimate of the real amplitude differences and reflection coefficients. On the other hand they show a rather heterogeneous quality within the first two seconds TWT due to varying coupling conditions at the surface.

Migration corrects for structural distortions by focussing diffractions and moving dipping events to their real location. Migration often produces artificial effects known as "smiles", especially in the deeper parts of the sections (Warner, 1987), therefore three-dimensional control by the intersecting profiles is essential for a reliable interpretation of the data.

Migration velocities were based on smoothed stack velocities. In addition to finite difference-migration, constant velocity migrations have been applied on selected portions. A high amplitude event at a midcrustal level, which shows clear diffractions at each side, was used as a test object for migration efficiency. Based on imaging-time variation (De Vries and Berkhout, 1984) migrated sections for different migration velocities were generated in one

Figure 4: Example for a conventional AGC stack from the central part of line 8401. Note the balanced energy which is determined from a moving time window of 500 ms. Frequency filtering, bandpass characterized by lower pass - stop band frequency/upper stop - pass band frequency: 0-2.4 s TWT: 14-20/37-48 Hz, 2.4-4.4 s TWT: 11-18/37-48 Hz, 4.4-12 s TWT: 8-15/37-48 Hz.
Length of section: 22.4 km.

Figure 5: TA stack corresponding to the section of Fig. 4. This section gives a better estimation of true reflection energy, but is of rather heterogeneous quality within the first two seconds TWT due to varying coupling conditions of detectors and sources and varying ambient noise. A horizontal scaling is included normalizing the trace energy between 3.5 and 12 s TWT.

Figure 6: Finite difference migration based on the AGC stack of Fig. 4. The curved reflector pattern in the deeper part of the section is partly due to sideswipe. The majority of these elements correspond to true dipping reflections controlled spatially by the intersecting lines 8402 and 8403.

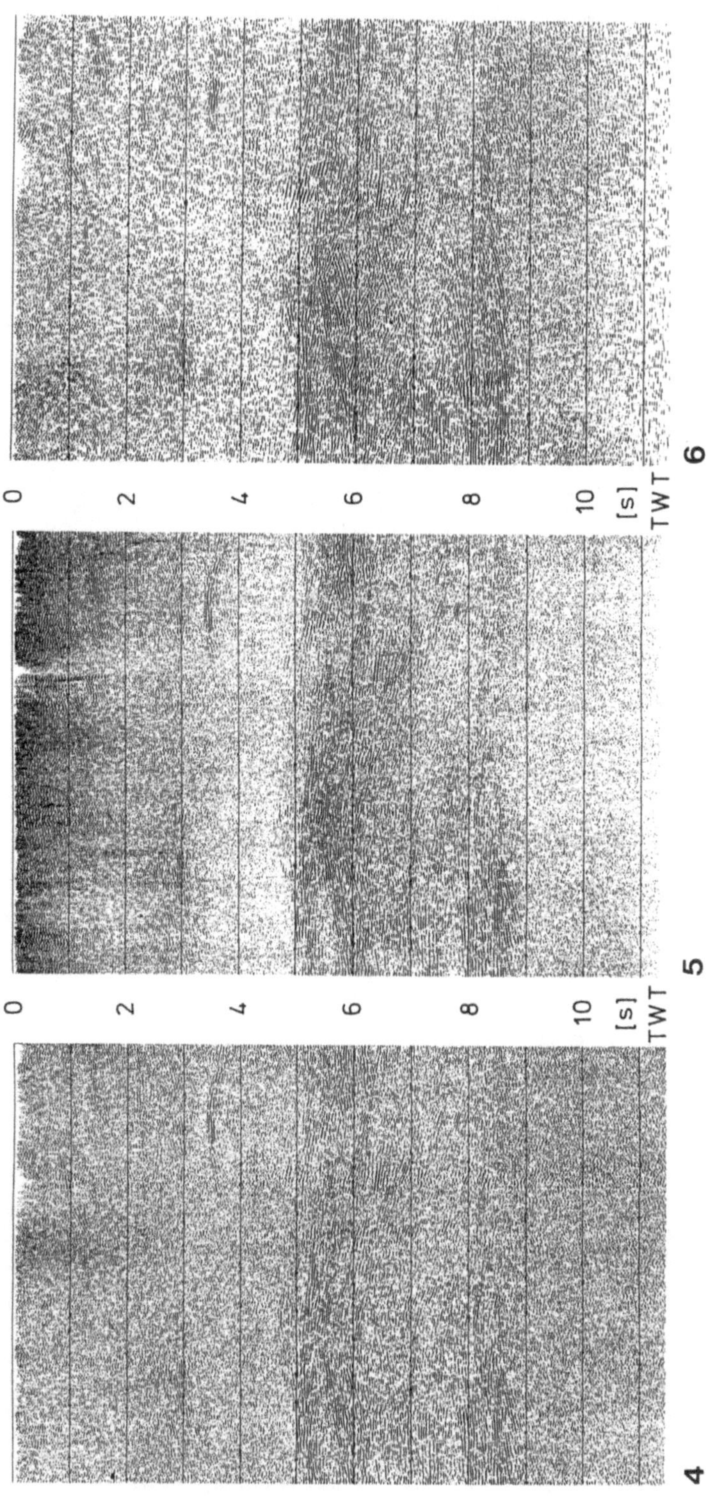
0
2
4
6
8
10
[s]
TWT
4
5
6

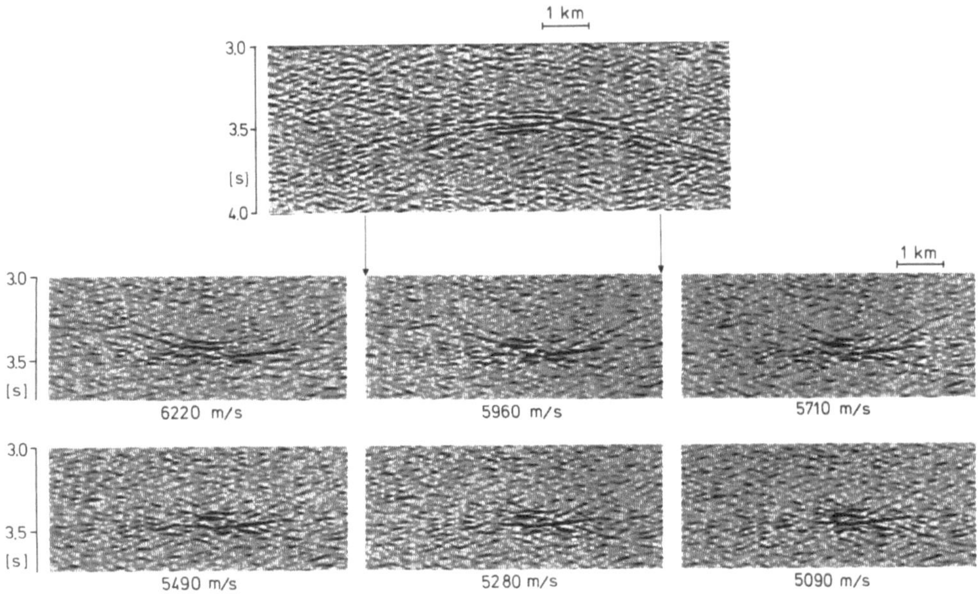

Figure 7: Constant velocity migrations of a high amplitude event beneath Haslach (Bright Spot) (Location 2070 of line 8401). Using the criterion of optimal focussing the migration velocity of approx. 5700 m/s may be regarded as maximum acceptable value.

single, recursive downward continuation process (Fig. 7). The variation between 6220 and 5090 m/s shows that the migration process for these data is rather insensitive to velocity variations. Using the criterion of optimal focussing we determined 5700 m/s as maximum velocity for this particular data set.

3.3 Presentation of data

Fig. 8 shows a compilation of structural information obtained by the complete profile net, based on migrated sections presented as simplified line drawings.

Each line or reflection event does not necessarily represent a first-order discontinuity. Laminated zones with a thickness below the seismic wavelength increase the reflectivity by constructive interference of multiple internal reflections in certain frequency bands (Fuchs, 1969). The crustal structure

Figure 8: Line drawings for profiles 8401, 8402, 8403 and 8514, based on migrated sections. Lines correspond to coherent phases. The depth scale at the righthand side of profile 8401 is derived from wide-angle data. BBL = Baden-Baden zone, BLZ = Badenweiler-Lenzkirch zone, EF = Elztal Fault, TG = Triberg Granite, ZEZ = Zinken-Elme zone, PG = Geophone location.

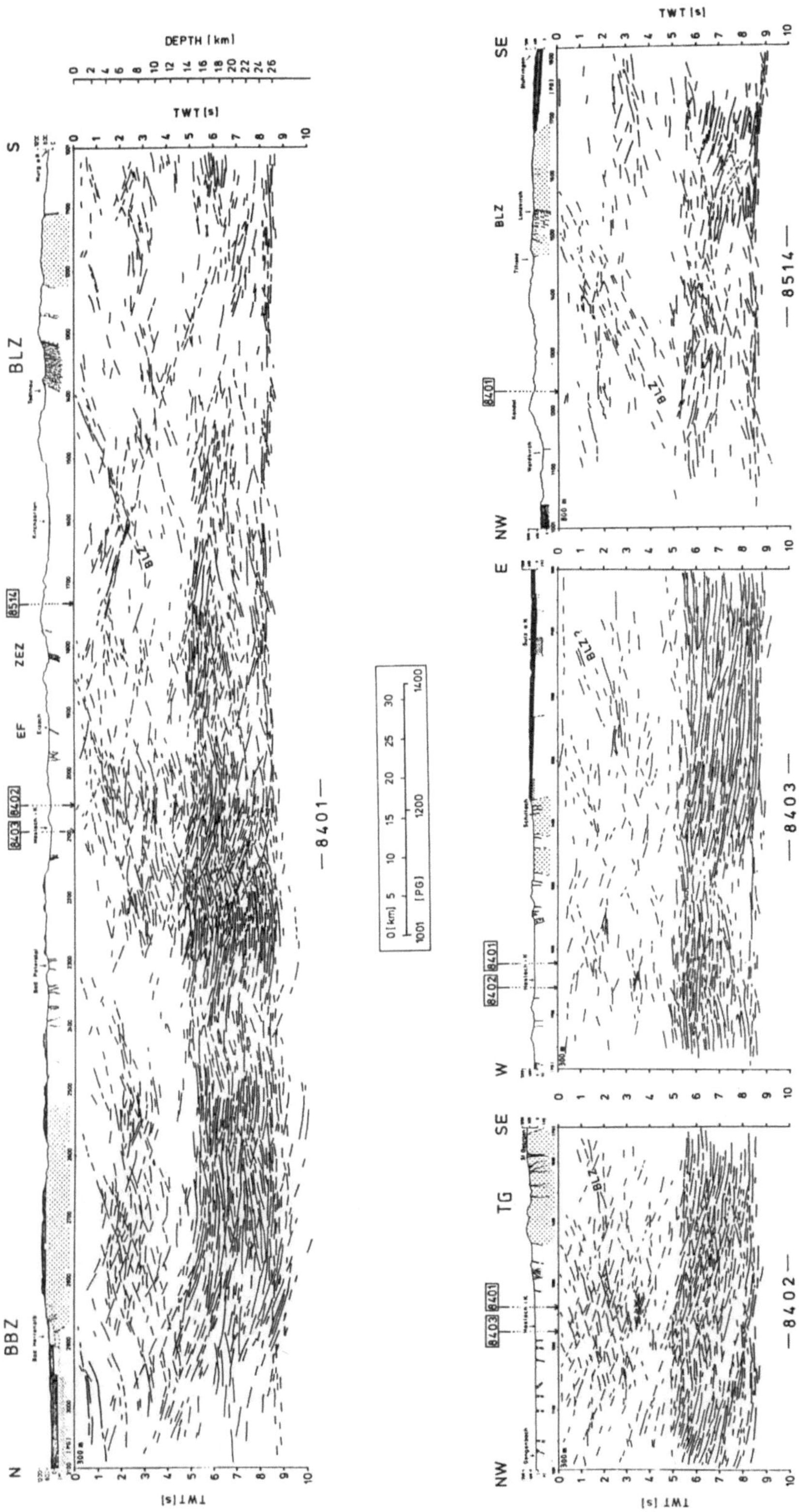
N
BBZ
BLZ
S
ZEZ
EF
8403
8402
8514
DEPTH [km]
TWT [s]
— 8401 —
0 [km] 5 10 15 20 25 30
1001 [PG] 1200 1400
NW
TG
SE
8401
BLZ
— 8402 —
W
E
BLZ ?
— 8403 —
— 8514 —

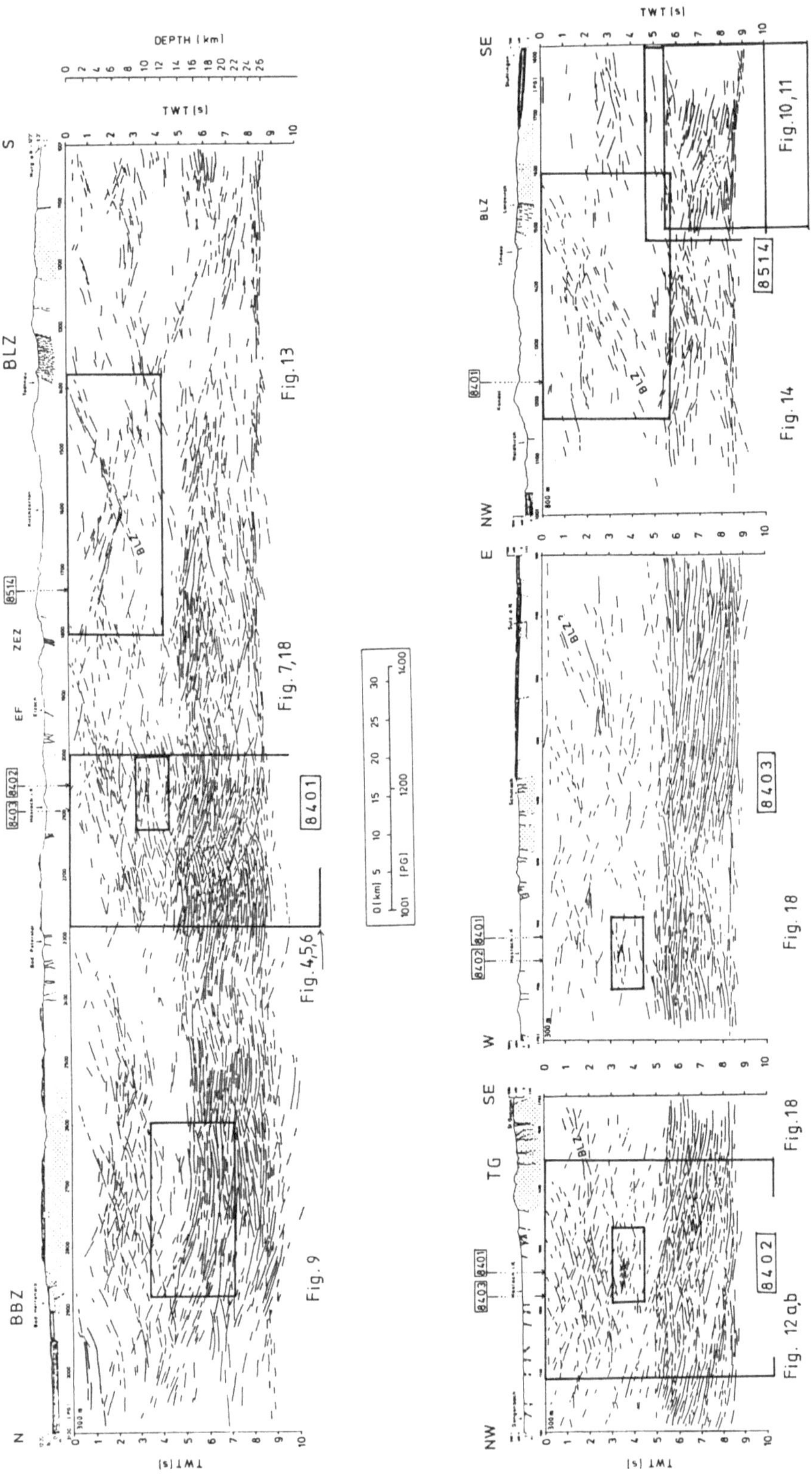
S
BLZ
ZEZ
EF
BBZ
N
DEPTH (km)
TWT (s)
Fig.13
Fig.7,18
Fig.4,5,6
Fig.9
8401
8514
8402
8403
0 (km) 5 10 15 20 25 30
1001 (PG) 1200 1400
NW
SE
BLZ
Fig.10,11
Fig.14
W
E
Fig.18
TG
Fig. 12a,b

image is dominated by a strong differentiation into a highly reflective lower crust of laminated character between 5 and 9 s TWT, and an upper crust of lower reflection density. Fig. 9 shows a selected seismogram close-up from the TA stacked profile 8401, which demonstrates this contrast. The lower part of the upper crust all over the profile net is characterized by a mostly transparent zone. The top of the laminated lower crust was found at an almost constant level of 5 s TWT corresponding to 14 km depth with small undulations. The reflective character of the lower crust breaks down abruptly at about 8.5 s TWT corresponding to a depth of 25.5 km with slight undulations and a relative updoming beneath the Kirchzarten basin (location No 1600, 8401). This observation leads us to define the crust-mantle boundary by the deepest continuous reflectors, which coincides with the conventional Moho defined by refraction seismics (compare section 4.1 and 8). The upper mantle is the most transparent part in the seismic sections. No significant reflection signals could be identified within the last three seconds of the recording time; maximum time was 12 s corresponding to a maximum depth of approximately 36-38 km. Fig. 10 shows an example from profile 8514 which demonstrates the abrupt change of the reflectivity pattern at the Moho.

The lower crust is widely characterized by horizontally layered reflections, which are sometimes consistent over 10 to 20 km, with some discordant patterns of dipping elements. Smiles appear at some deeper locations, in the

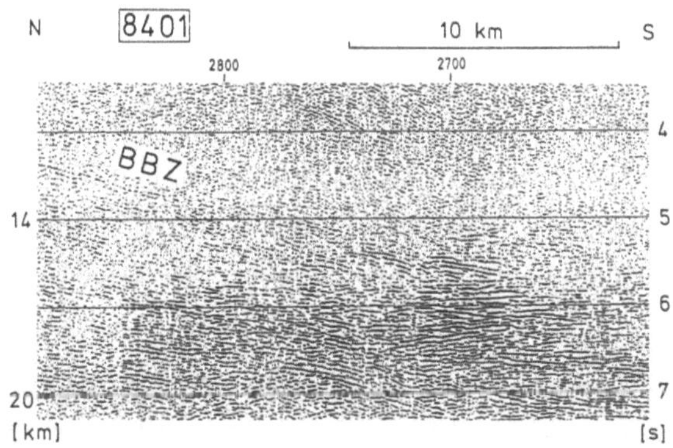

Figure 9: Section of the TA stack from the northern part of line 8401 showing strong differentiation between the highly reflective lower crust beginning at about 5 s TWT and the transparent middle crust. The dipping element on the left (4-5 s TWT) may be related to the Baden-Baden thrust zone.

Figure 8a: Same as Figure 8. Positions of seismogram sections presented in Figs. 4-7, 9-14 are marked.

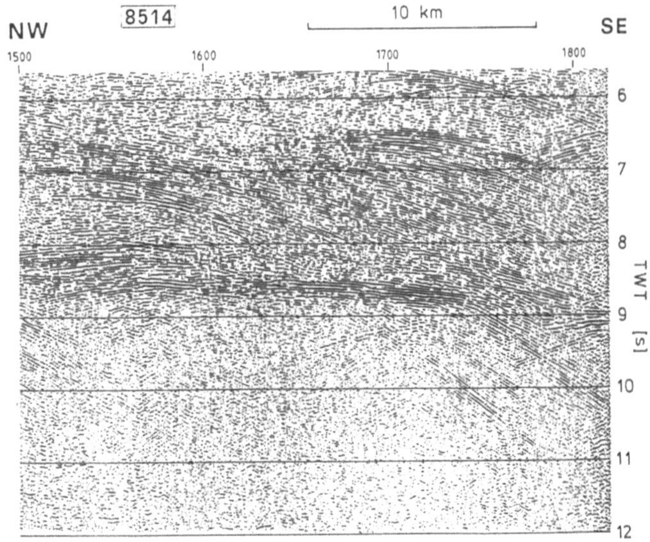

Figure 10: Stack section from the southeastern part of line 8514. Note the strong reflections in the lower crust, in particular the strong continuous reflection from the crust-mantle boundary. This section is dominated by diffractions, even for more than 9 s TWT, having their apices in the lower crust. Diffractions originating at the crust-mantle boundary (e.g. on the right side) indicate vertical steps or small scale undulations.

central part of profile 8401 (Fig. 6) and in the southeast part of profile 8514 (Fig. 11). Locally the lower crust also contains transparent patterns above the Moho, for instance in the southern part of profile 8401, at location 1500 of profile 8403 and in the northwestern part of profile 8514. Between locations 2300 and 2500 of profile 8401 (Fig. 8) the reflection energy fades out throughout the whole crust. This is probably caused by scattering of seismic energy due to an accumulation of vertically arranged fractures and dikes in the upper crust, rather than due to disturbances during data acquisition. The laminated lower crust was intensively modelled (section 5) in order to derive physical constraints for petrological interpretations.

The upper crust contains mostly dipping, often cross-cutting and high amplitude events of local character. At the northern end of profile 8401 the Rotliegend-trough appears in the seismic section with some internal formations and vertical faults. In this sedimentary basin the frequency content of the seismic signals is reduced so that the laminated character of the lower crust cannot be resolved. Seismic stripping and long term statics (compare section 3.4.1) can improve the data quality beneath the Permian sedimentary basin. In the middle and lower crust at the northern end of line 8401 a pattern of southward dipping reflectors is clearly visible (Figs. 8, 9); only a few reflectors penetrate the horizontally layered crust-mantle boundary. This may

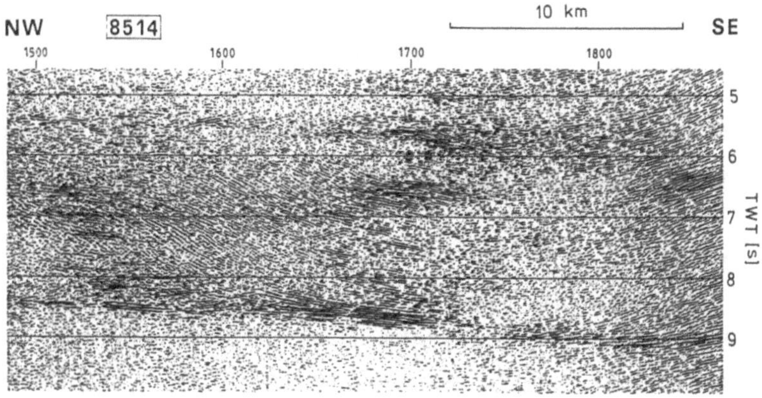

Figure 11: Migration of stack section from Fig. 10. Note that deep diffractions are focussed mostly at their source structures. Also migration produces artificial, curved elements (6-8 s TWT).

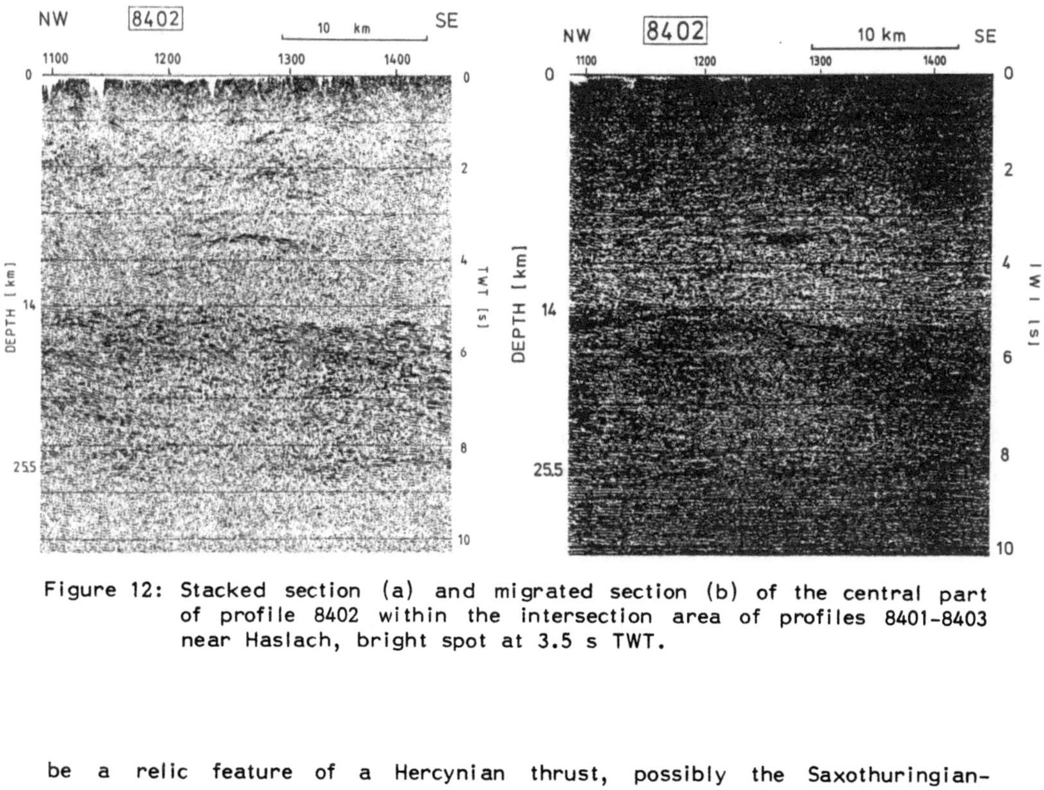

Figure 12: Stacked section (a) and migrated section (b) of the central part of profile 8402 within the intersection area of profiles 8401-8403 near Haslach, bright spot at 3.5 s TWT.

be a relic feature of a Hercynian thrust, possibly the Saxothuringian-Moldanubian suture zone of the Hercynian belt, also called Baden-Baden zone (BBZ, section 2). Further to the south, particularly in the Central Schwarzwald Gneiss Complex a criss-cross pattern of low-angle reflectors prevails in the upper crust (compare seismogram close-ups in Figs. 4 and 5 for profile 8401 and Fig. 12a,b for profile 8402). In this area cataclastic shear zones dipping to the SE are found at the surface, which are identified as reflectors by high resolution seismics (section 3.4.2). These SE-dipping reflectors of the upper crust are interpreted as late-Hercynian extensional faults. On the other hand, north- and northwest-dipping reflectors of the criss-cross pattern can be correlated with mylonitic, ductile deformation zones associated with intraplate convergence and thrust zones of Hercynian age in the southern Black Forest (Zinken-Elme zone, ZEZ and Badenweiler-Lenzkirch zone, BLZ).

The BLZ is a relic of a major Hercynian thrust and can be traced clearly on the profiles 8401 and 8514. It dips N to NW down to a depth of 10-12 km (Fig. 13 from profile 8401, Fig. 14 from profile 8514) and intersects or is intersected by another reflector. This Glottertal reflector cannot be traced clearly to the surface. It shows the strongest reflection amplitudes observed in the upper crust of the Black Forest. From the profiles 8401 and 8514 its strike is determined to be parallel to 8514 and its dip to be to the SSW. Probably it is an extension fault related to corresponding shear zones of normal fault character and Hercynian origin which were mapped at the surface. Recently collected seismicity data demonstrate that this reflector represents an active fault (Bonjer et al., 1986), indicating that the present stress field is released along ancient structures. Hence brittle deformations may be responsible for the pronounced Glottertal-reflections.

The criss-cross pattern of low-angle reflectors reveals that low-angle convergent and divergent structures contribute a much greater deal to the architecture of the crust than previously assumed. In accordance with geological findings (sections 2, 7.2) we believe that thrusts penetrating the entire crust are overprinted by low-angle normal faults in the upper crust and by a strong horizontal 'lamination' in the lower crust, including the crust-mantle boundary. Alternatively, such an irregular pattern could be influenced by strong shallow velocity inhomogenities which tend to smear out individual reflections into a complex pattern (Peddy et al., 1986). However, the persistence and continuity of lower crustal reflectors indicate that a strong influence of distortions in the upper crust can be excluded.

3.4 Special investigations

3.4.1 Velocity structure from first arrivals

First arrivals and subsequent refracted arrivals as well as surface waves are systematically eliminated by the muting procedure in standard reflection data processing. This causes a lack of coverage and structural information in the topmost part of the seismic section, mainly within the first half second TWT corresponding to a depth of about one kilometre. To compensate for this loss of information, the first arrivals were evaluated separately to determine in detail the lateral variation of velocity structure for the upper two kilometres.

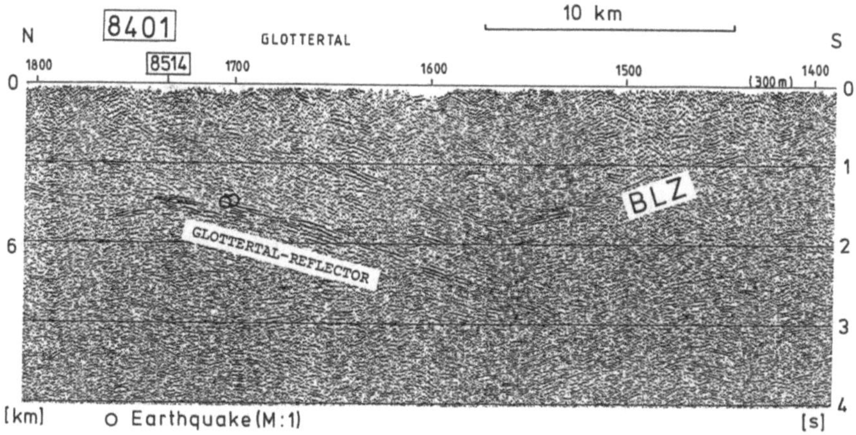

Figure 13: Upper crustal reflections from a southern part of the stacked line 8401. The northward dipping elements on the righthand side are connected with the Badenweiler-Lenzkirch thrust zone (BLZ). The southward dipping prominent Glottertal-reflector is interpreted as a low-angle normal fault. An earthquake sequence is located within this element (Bonjer et al., 1986). Precision of Location is $\pm$100 m. Focal mechanism is of shallow thrust type. Three-dimensional control is provided by the intersecting line 8514 (compare Fig. 14).

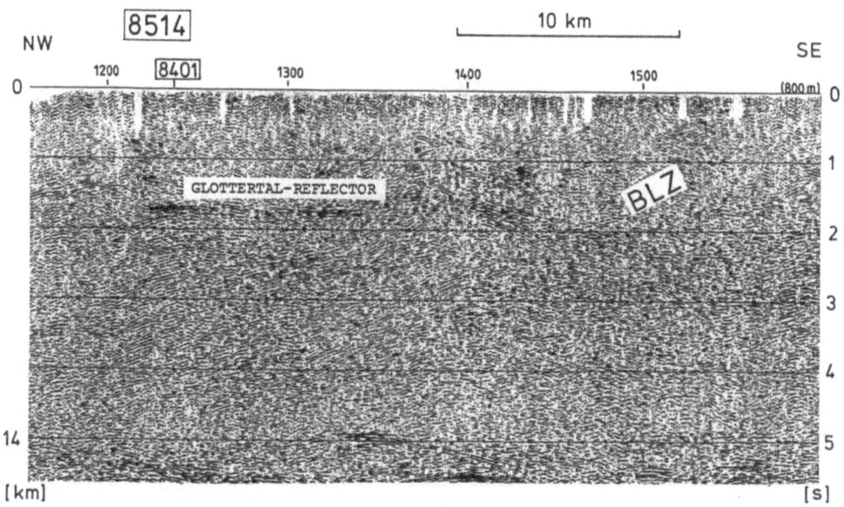

Figure 14: Part of the stacked profile 8514. The Glottertal and Badenweiler-Lenzkirch-elements are delineated in a NW-SE direction allowing a reconstruction of dip and strike with the aid of profile 8401. The Glottertal reflector dips to SW and strikes NW-SE.
The surface trace of the BLZ is curved (compare location map Fig. 1). Its dip of about 30° can be followed to a depth of about 12 km.

The refraction technique benefits from the extremely high data redundancy that reaches a maximum coverage of nearly 200 for each subsurface point.

The first arrivals of selected high-gain field records were picked manually and transferred to traveltime-distance plots. The "Plus-Minus"-Method of Hagedoorn (1959) was applied graphically to determine the laterally varying refractor velocity of the crystalline basement and its depth. Short refraction lines (weathering surveys) were carried out additionally at spacings of about 2 km to obtain information about the thickness and the velocity of the weathering zone and to control the intercept times and to calibrate the first arrivals of the Vibroseis data, since an in-line offset of 200 m of the Vibroseis split-spread geometry had to be bridged.

Basic static corrections were computed based on the results of the "Plus-Minus"-method, topographic data and the interpolated results of the weathering survey. Statistical improvement of these static corrections (Fromm, 1984) was obtained by automatic evaluation of first arrivals. Short-period perturbations on the traveltime curves were transformed into delay times of the weathering layer, which were added to the basic corrections. Static corrections included an offset correction and were computed for a reference level of 300 m above mean sea level (profile 8514: 800 m).

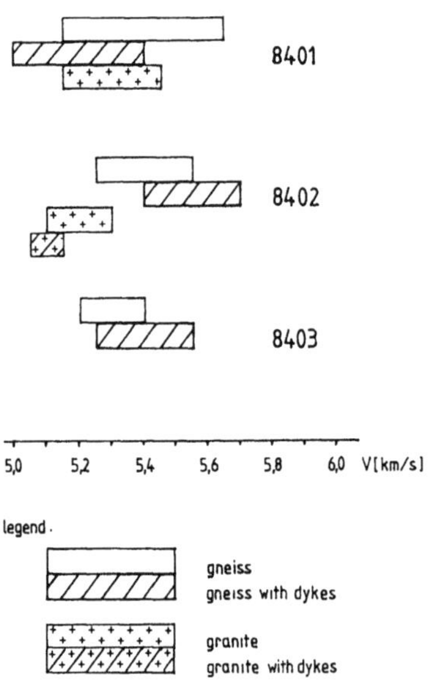

Figure 15: Representative basement velocities for zone beneath weathering layer derived from first arrivals of Vibroseis field recordings. Length of bars corresponds to standard deviation of the mean values.

Although surface rocks are composed of crystalline basement, gneiss or granite with occasional cover of a few meters of Quaternary deposits, the evaluation of the refraction data clearly show a considerable thickness of low velocity material varying between 2000 and 3500 m/s and with thicknesses between 50 and 300 m (see upper parts of sections in Fig. 16). This is probably due to lateral variations of rock weathering. The refractor velocity of the "solid" basement varies between 5000 and 5700 m/s. In Fig. 15 these velocities are related to different petrological assemblages which are documented by abundant exposures. Generally, granites tend to have lower velocities than gneisses. If associated with granitic dikes, both units show a relatively wide scattering of velocities. This might be explained by an azimuthal velocity dependence on the strike directions of the dikes with respect to the orientation of the profiles.

To investigate the velocity structure of the uppermost crust in greater detail the complete set of first arrivals was picked by using the single channel algorithm (Ketelsen et al., 1983). This data set provides a maximum 200-fold coverage which is on the average reduced to about 70-100-fold because of missing vibrator stations and missing picked values. The traveltime-distance curves were inverted by different techniques (Giese, 1976, Slichter, 1932) yielding velocity-depth functions. As the most powerful and economical method, the maximum depth estimation-method of Slichter (1932), making use of the triple data set comprising distance, traveltime and apparent velocity was used for preference.

Computations with synthetic, even laterally varying models, have demonstrated the reliability of this method. The high redundancy of velocity-depth data was used to smooth the velocity structure by averaging. Its reference level is the topography of the seismic basement (Fig. 16). A CMP-sorting-technique (Gebrande, 1986) has been applied but gave no further improvement. Another approach to control the velocity structure is the application of iterative tomographic methods (e.g. Gorden et al., 1970; Censor, 1983) which provide a velocity image directly from traveltime. The results corresponds to those obtained by the Slichter-method with only slight improvements (Rühl, 1987).

The velocity sections of Fig. 16 (lower parts) display considerable lateral variation of the velocity structure ranging from 4000 to 6000 m/s. The lower velocities clearly correlate with a Permian and Triassic sedimentary cover in the northern part of profile 8401 and in the eastern part of profile 8403. Some conspicuous correlations between known petrological complexes are obvious. On profile 8401 higher velocity material updomes in the core of the central complex of gneisses and anatexites. The section of profile 8402 reflects a velocity difference between the Triberg granite (5200 $\pm$100 m/s) and

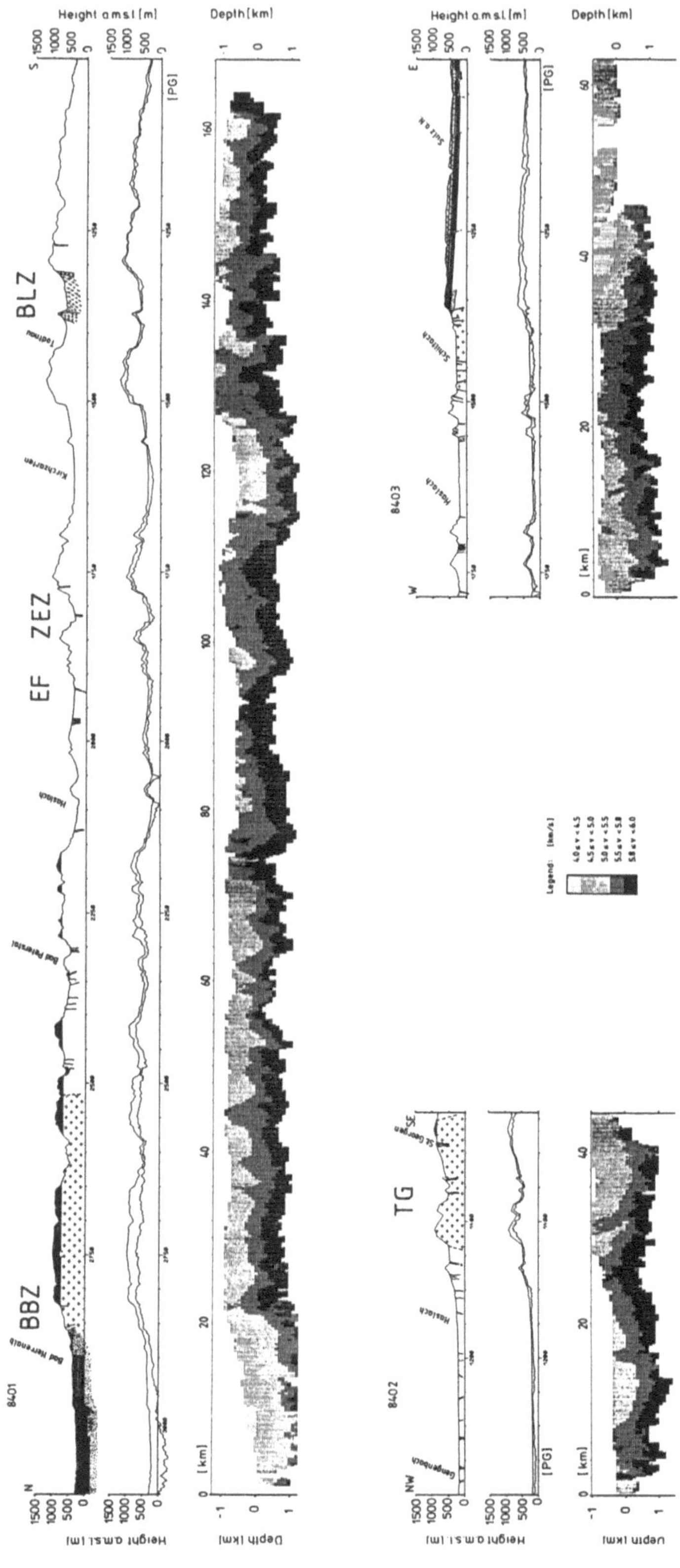
8401
N
S
BBZ
TG
EF
ZEZ
BLZ
Bad Herrenalb
Bad Peterstal
Haslach
Kirchzarten
Todtnau
Height a.m.s.l. [m]
Depth [km]
[km]
[PG]
8402
NW
SE
Gengenbach
St. Georgen
8403
W
E
Schiltach
Sulz a.N.
Legend: [km/s]

the gneisses (5400 ±150 m/s, compare Fig. 15) and a synclinai basin within the gneisses. High near-surface velocities at km 5 and 25 of profile 8402, km 25 of profile 8403, km 92 and 108 of profile 8401 correlate with belts of granulitic gneisses and mylonites (Hacker and Hirschmann, 1986).

In addition to geological interpretations, the assessment of long term statics offers an important practical application. Based on the velocity model of Fig. 16 long term effects of the order of 20-30 ms comparable to a seismic period were estimated within a CMP gather using a deeper reference level. These statics are not yet applied on the data presented in this paper.

3.4.2 Seismic probing of near-surface crystalline rocks

One important but indirect approach to explain the nature of deep crustal reflectors is sophisticated modelling that estimates the physical parameters as described in section 5.

As another approach, in-situ probing of seismic reflectors in crystalline rocks has been attained through geophysical logging tools and vertical seismic profiling (VSP) in a series of boreholes, and through their correlation with structures and lithologies of the cored rocks. The borehole studies were complemented by near surface reflection and refraction profiles of high resolution in order to link deep seismic sections to known near-surface structures and lithology. Out of a great variety of seismic imaging tools and widespread resolution capabilities applied at different sites in the Central Black Forest, two are presented in this paper.

For the investigations a car-mounted high-pressure pneumatic hammer was used at 20 m intervals for generating compressional (P) and horizontally polarized shearwaves (SH). 24 receiver arrays of vertical geophones or alternatively, horizontal geophones for SH-recording, were deployed at 10 m intervals. Recording with a 24-channel digital signal-enhancement unit deployed in off-end geometry yielded a 6-fold coverage with CMP spacing of 5 m. The daily progress was 600-800 m.

←

Figure 16: Evaluation of the Vibroseis - first arrivals for profiles 8401, 8402 and 8403. Upper sections: surface geology, crosses: granites, white: gneisses (with dikes), other signatures: Permian rocks and Paleozoic sediments of the Badenweiler-Lenzkirch zone.
Intermediate sections: thickness of the weathering zone (heterogeneous, velocities of 2000-3500 m/s); upper line topography, lower line seismic basement;
Lower sections: two-dimensional velocity structure of the basement, for velocities see legend.
Vertical exaggeration of topography is 4.

First arrivals and refraction lines yield P-velocities of 600 m/s (SH: 340 m/s) in the upper few meters and basement velocities increasing from 2800 to 4200 m/s (SH: 1700-2000 m/s) in the upper 50 m. In P- and SH-reflection sections near Haslach an interrupted reflection can be traced over the complete sections dipping 30-40° to the east. This reflection can be traced upward into a cataclastic zone related to a low-angle normal fault. This fault thus could be followed from the outcrop to a depth of about 700 m by this shallow reflection survey. Comparison with the deep reflection section 8402 (section 3.3, Fig. 12) shows a continuation of this fault-reflector to a system of reflectors which, dipping to the SE, penetrate the entire upper crust. We therefore interprete these and similar reflectors as indicators of late-Hercynian extension (compare section 7.2). However, in general the shallow seismic sections show weak amplitudes and little consistency of reflectors, especially when using SH-waves. This may be a consequence of complicated ductile folding and metamorphic transposition on a scale less than the seismic wavelength (P: 40-100 m, SH: 40-130 m), which might also account for the relative transparency and complexity of the entire upper crust.

Fig. 17 shows an example for a data compilation for the well "Hechtsberg" (300 m depth) near Haslach. The original VSP record section displays the downgoing direct wave and the tube wave. The averaged P-wave velocity decreases from top to bottom from 5400 to 5000 to 4300 m/s which is confirmed by the sonic log. This corresponds to a petrographic change (decrease of mafic constituents, decrease of degree of foliation) from orthogneisses to paragneisses. The upgoing wavefield delineates a sequence of reflections, which can be correlated with sudden velocity breaks in the sonic log, and with structural, and petrographic variations revealed by the core evaluation. At 290 m depth a strong reflection corresponds to a cataclastic shear zone which is interpreted as a late-Hercynian extensional fault zone. It is probably connected to a similar zone outcropping about 500 m north of the borehole.

3.4.3 Haslach "bright spot"

An isolated reflection event beneath Haslach at about 9.5 km depth, which looks like a bright spot (Sheriff and Geldart, 1983) is located beneath the crossing area of three profiles near Haslach (compare Fig. 12). Therefore good three-dimensional control exists for stacked as well as migrated sections. They are presented in Fig. 18 together with a position map indicating the lateral extent of 3-5 km. The reflection amplitudes observed in true amplitude sections are comparable to those of prominent reflections from the lower crust.

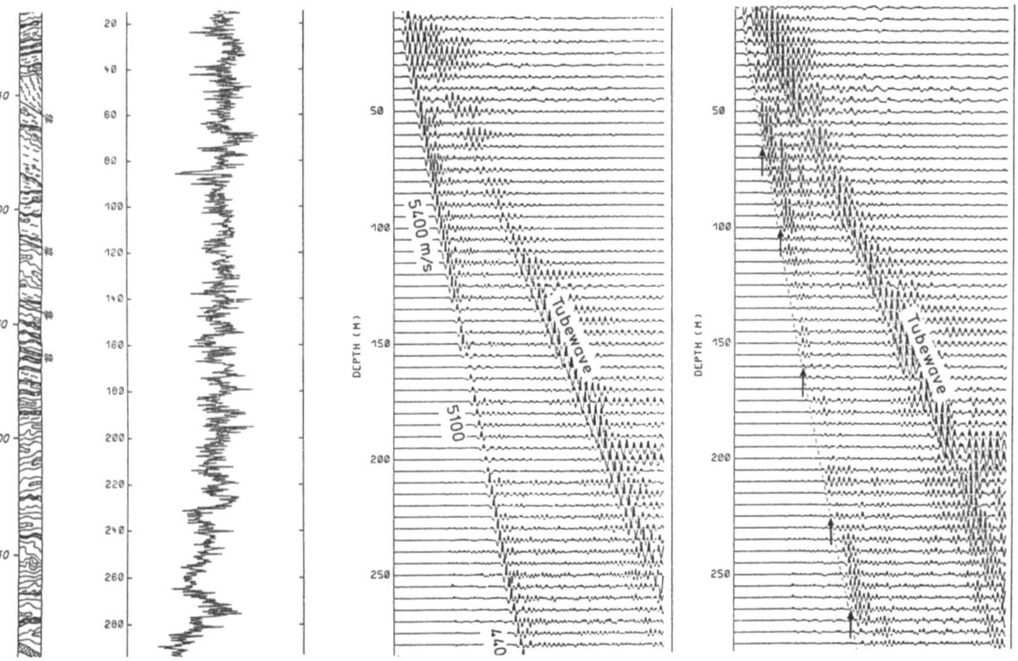

Figure 17: Borehole data of site "Hechtsberg" from left to right:
- Structures and lithology derived from cores (simplified). Different gneisses and shear zones indicated.
- Sonic log (digitized from analog data of Vogelsang, 1986).
- Vertical seismic profiling (original data, vertical component). Clearly recognizable is the direct wave (labelled by corresponding velocities) and the tube wave. Note modified time scale.
- Upgoing wave field after application of a f-k-filter. Note strong reflections particularly from a cataclastic shear zone (depth 280 m). Arrows indicate reflections.

In view of the strong reflections from the bright spot even in single-fold seismograms (source-point gathers) we tried to determine the polarity of the reflection coefficient. Single-fold uncorrected seismograms from source-point gathers were plotted in wiggle-area mode in two ways. Positive and negative values respectively above a certain threshold value were blackened and then compared to each other (Fig. 19). In traces with a high signal/noise ratio we may identify visually a symmetrical form of a Klauder wavelet consisting of a main lobe and two side lobes. This waveform represents a negative polarity of the reflection. The inspection of a variety of source-point gathers shows that wherever a Klauder wavelet can be recognized, its polarity is negative. This observation indicates that if the reflection is from a single interface then it is caused by material of lower impedance than the surrounding

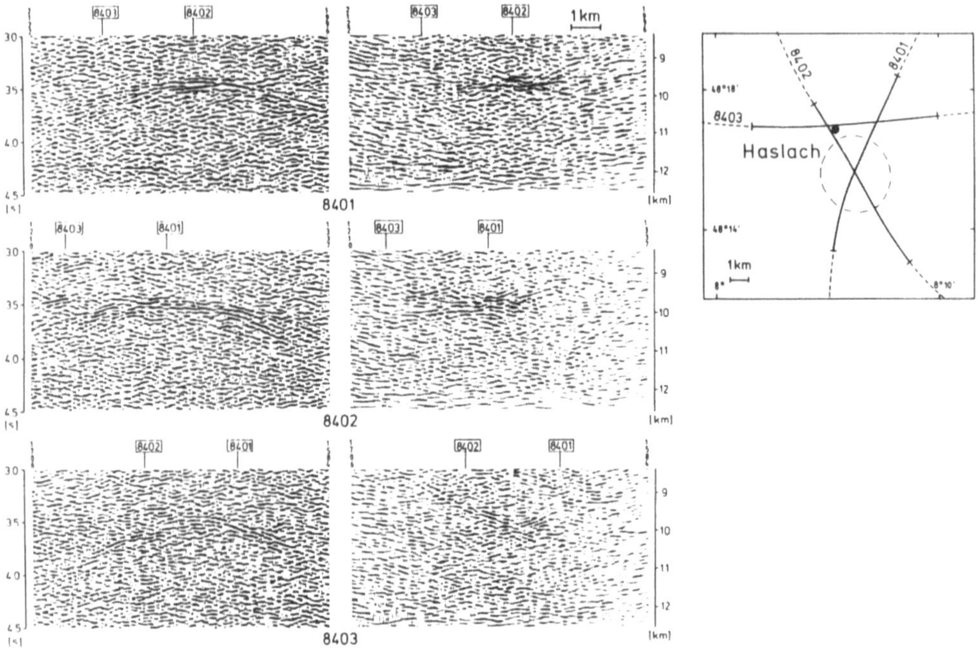

Figure 18: Haslach "Bright Spot" represented by stacked (left) and migrated sections (right) of profiles 8401, 8402, 8403. Migration velocity 5700 m/s. Position sketch shows location and extent of the high amplitude event (dashed oval). Point marks village Haslach.

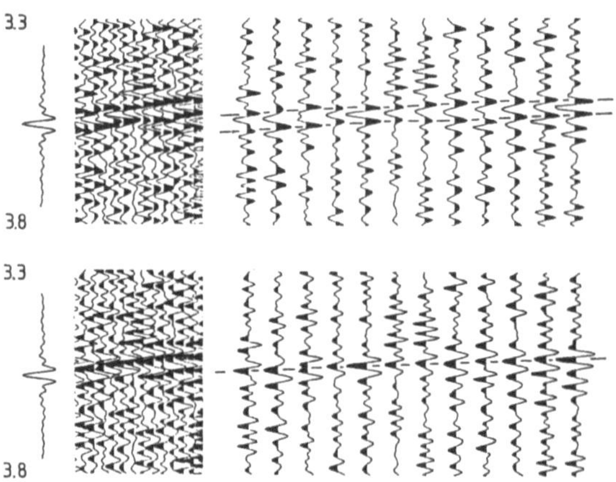

Figure 19: Single fold seismograms (line 8402) of the Haslach "Bright Spot" between 3.3 and 3.8 s TWT. 12 traces are plotted with different spacings. In the upper part the positive values are blackened, in the lower part the negative ones. The theoretical Klauder wavelet with negative polarity is shown on the lefthand side. Positive values correspond to compression at the surface. The signals exhibit negative polarity because maximum correlation appears to the left.

medium. A velocity of 5.4-5.5 km/s of the surrounding medium is obtained from wide-angle experiments (see section 4.1). Therefore a velocity clearly below 5.4 km/s must be associated with a material causing high amplitude reflections with negative polarity. A possible explanation might be a concentration of trapped fluid. Its origin could be related to dehydration of Hercynian-age overthrusting of "wet" sediments. A similar bright spot appears on section 8514 (Location 1350, 5 s TWT, compare Fig. 14). The strong reflection amplitudes could result from a contact between impervious rocks and rocks with fluid-filled porosity. Lower crust or mantle derived gases (CO_2, H_2O) may be trapped in addition or alternatively to fluids from underthrusted sediments.

Similar bright spots were found in recent COCORP surveys (Brown et al., 1986). They are typically at midcrustal levels and can be of variable origin. Morton and Sleep (1985) inferred a magma chamber beneath an actively spreading backarc basin from similar phenomena.

4 Velocity control by refraction and wide-angle reflection measurements

Whereas near-vertical reflection profiling provides detailed images of crustal structure the reliability of the resulting velocity information is usually poor. For instance, the analysis of stacking velocities alone gives no hint of a low-velocity zone in the middle crust. Therefore additional wide-angle measurements are required to determine the velocity needed for time-depth conversion and petrological interpretations. A gross two-dimensional velocity distribution was derived from refraction profiles. Expanding spread soundings, focussed on the potential drilling site were performed to deliver a high precision velocity-depth distribution. In addition a wide-angle Vibroseis experiment undershooting the Rhinegraben yielded velocity control for the adjacent graben proper and served to prepare a feasibility study for the Rhinegraben experiment in 1988 (Damotte et al., 1987).

4.1 Refraction profile

In August 1984 the Black Forest was studied by a seismic refraction survey along a 240 km NS profile from its northern end to the Swiss molasse basin (Gajewski and Prodehl, 1987; Fig. 20). Five shots, offset by 30 to 40 km were recorded with stations spaced between 1.5 and 2 km apart. The geometry was designed to resolve a two-dimensional velocity distribution. The velocity model, displayed in Fig. 32, is essentially based on the correlation of the wave P_G diving through the uppermost crust, reflections P_L from the top of the lower crust, and P_M from the crust-mantle boundary. The model was successively improved by synthetic seismograms based on two-dimensional asymptotic ray theory (Gajewski and Prodehl, 1987).

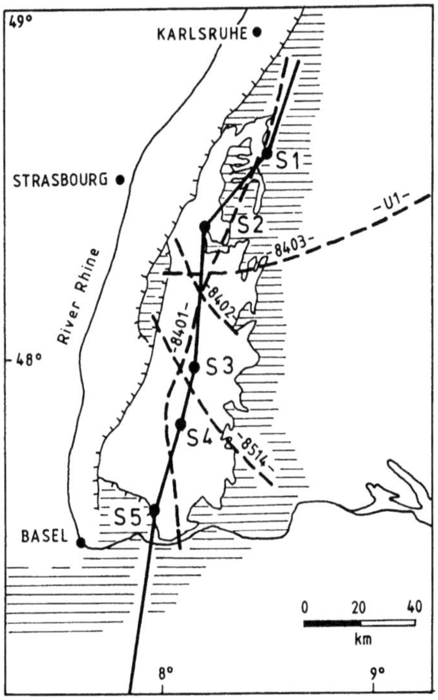

Figure 20: Location map showing reflection profiles and the NS refraction profile. Black circles are shotpoints observed to both sides with receiver spacings of 1.5-2 km (from Gajewski and Prodehl, 1987).

The main finding is a low-velocity zone (5.4-5.5 km/s) which is pronounced in the northern and central Black Forest and which fades out to the south. It is located directly on top of the lower crust which has an average velocity of 6.6 km/s. The top of the lower crust is at a depth of 14 to 15 km beneath the central Black Forest. In this model the crust-mantle boundary appears as a first-order discontinuity and does not show any major depth variations north of the Swiss molasse basin.

The refraction line and the NS reflection profile 8401 practically coincide. Structural features observed in near-vertical reflection sections and velocity information inferred from wide-angle measurements were jointly evaluated.

4.2 Expanding spread profile

From the results of the refraction and reflection profiles the region around Haslach appeared as typical normal Moldanubian crust right in the center between two large thrust zones (Baden-Baden-zone in the north and Badenweiler-Lenzkirch-zone in the south). It was therefore decided to explore this crustal section and its velocity structure by an expanding spread profile (ESP). The principle goals of this experiment were:

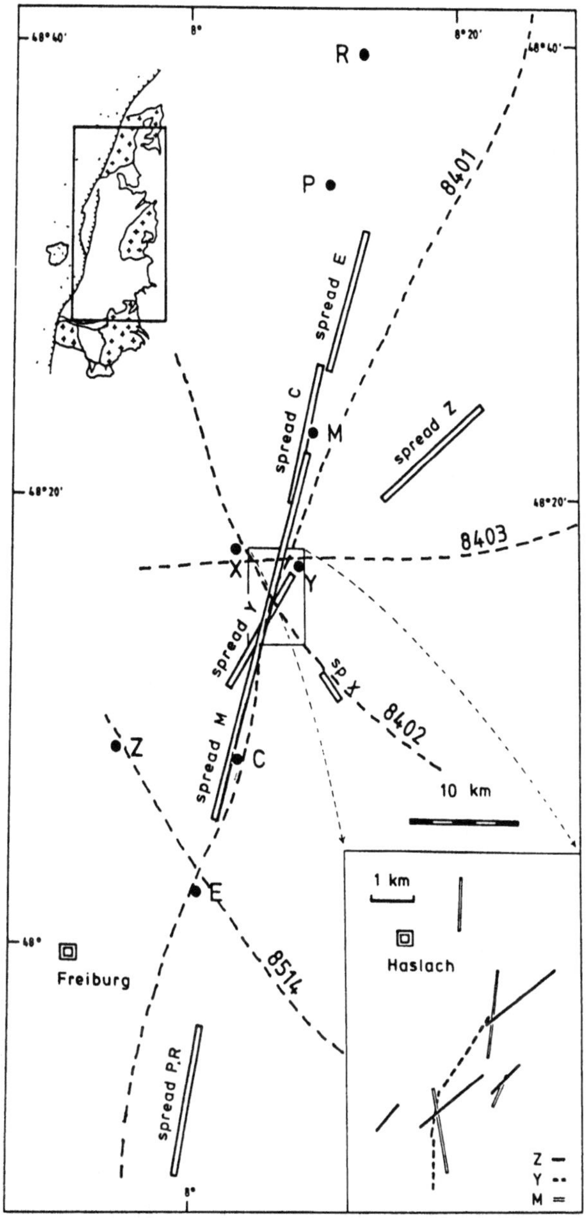

Figure 21: Schematic map of the "Expanding Spread"-profile with shotpoints and corresponding geophone layouts consisting of four independent digital recording units. The inset shows the blown up central region with true midpoint configurations for shots M, Y and Z covering the lateral extent of the Haslach "Bright Spot". The inset shows also location of reflection profiles.

1) to investigate the nature of the bright spot at about 9.5 km depth (see section 3.4.3)

2) to determine the average velocities in the upper crust and the entire crust for precise time-depth conversions

3) to investigate the wide-angle characteristics of the lamellar lower crustal reflections.

Figure 22: Section of shotpoint Y observed by three multichannel digital recording units in the offset range of 5 to 13 km.
Time scale is unreduced. P_G first arrivals from crystalline diving wave, followed by upper crustal reflections of local character,
P_B bright spot phase. These signals are missing at greater offsets (compare Fig. 23).
P_L first of lower crustal reflections.
P_M latest reflections which correlate with the Moho-phase from refraction survey and from modelling.
Amplitudes are normalized for each trace.

Our intent was to get high resolution observations in the wide angle range, comparable to the resolution of the near-vertical recordings. Small explosive charges of 10-50 kg were used as sources, distributed in patterns of 30-50 shallow drillholes (depth 2-3 m) in crystalline rock. This arrangement produced a broadband signal with dominant frequencies between 20 and 30 Hz. Shot R in the north (Fig. 21) was fired in a similar arrangement and with 60 kg charge in a small pond. The recording spread consisted of 144 channels spaced 80 m apart enabling exact phase correlation whereas in the refraction experiment (section 4.1) a spacing of 1.5-2 km was used.

The field experiment was performed in October 1985. The locations and corresponding geophone spreads together with common mid-point ranges of some selected recordings (shots M, Y and Z) are given in Fig. 21. In total eight different shotpoints were chosen for data evaluation. The recordings from shotpoints Y and Z are shown in Figs. 22 and 23. In both seismogram sections a clear P_G-phase from the first arrivals can be identified. Phases reflected from the bright spot at about 9.5 km depth (phase P_B) can only be correlated in the record section of shot Y in the near vertical range. It is followed by a band of strong reflections from the laminated lower crust beginning with phase P_L caused by the same laminated structure seen in near-vertical sections. With the small geophone spacing of 80 m it is possible to correlate individual phases over 400 to 1500 m. Consequently, the lateral extent of single lamellae is of the order 200 to 800 m (see also section 5.2). By contrast, in the refraction profile with a geophone spacing of 1.5-2 km these small-scale reflecting structures cannot be resolved. Therefore a correlation in widely spaced refraction measurements is subject to spatial aliasing simulating large-scale structures.

From the geometric configuration of the expanding spread profile the average velocity down to the first lamellae of the lower crust is determined using the T^2-X^2-diagram for the combined sections of the shots M, Y and Z (Fig. 24). The slope of the solid straight line fitting the beginning of lower crustal reflections yields an average velocity of 5.62 km/s. The error bounds are

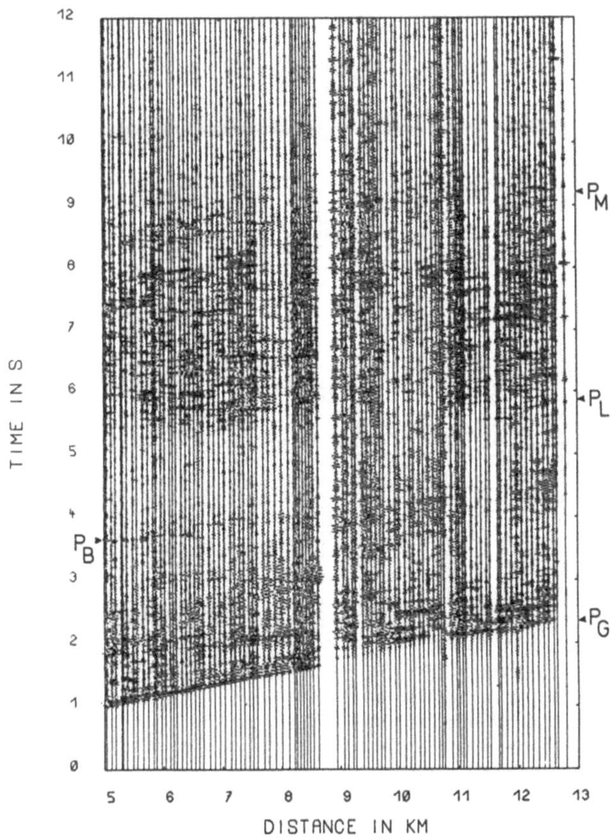

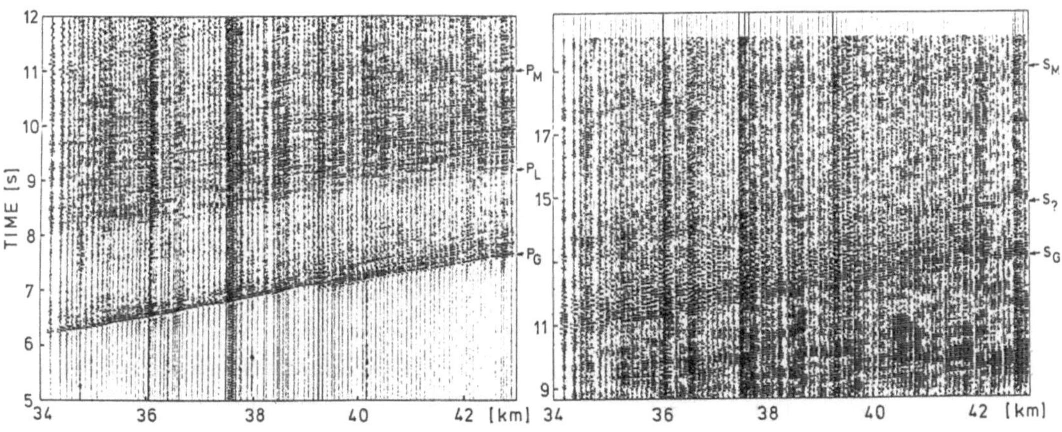

Figure 23: Section from shotpoint Z with offset range of 34 to 43 km. The left part shows compressional wave arrivals P_G, P_L, P_M (compare figure 22). The righthand part is from the same recording with a time scale multiplied by SQRT (3). Coherent phases can be compared with corresponding P-phases. S_G and S_M are clearly identified as shearwaves. The $S_?$-phase can be alternatively interpreted as S_L or as a converted phase (S_MP)from the crust-mantle boundary.

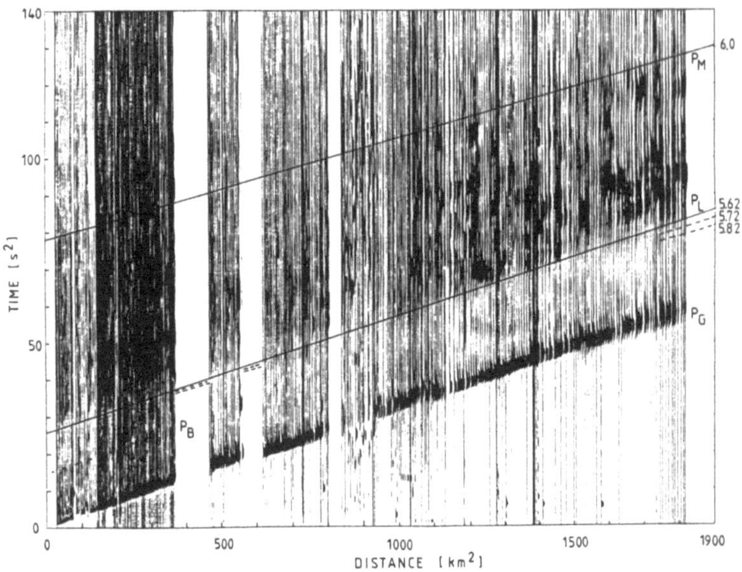

Figure 24: T^2-X^2 presentation of Expanding Spread data for shots M, Y, Z and corresponding spreads arranged according to distance from shotpoints within 0 to 43 km offset range. The seismograms are displayed as energy traces.
Solid straight lines yield RMS-velocities (values in km/s at the righthand side) for the upper crust and for the whole crust respectively. They are determined by the vertical TWT, i.e. t_o, fixed by the near vertical reflection profiles within ± 0.2 s for the whole Haslach region and by the first and last reflected energy from the lower crust in this diagram. The sensitivity of this procedure is indicated by dashed lines with greater values and fixed time t_o.

not more than ± 2 % taking vertical two-way traveltimes and traveltimes of reflections at individual common midpoints from the original data set into account. The sensitivity of this procedure is demonstrated by the dashed lines in Fig. 24 corresponding to a velocity increase of 0.1 and 0.2 km/s. This approach is only valid for a homogeneous medium. In a stratified medium this is an approximate procedure and velocity and depth do not necessarily coincide with the true RMS values. For the upper crust we studied the possible variations by comparing the true and the apparent RMS velocities for a variety of realistic models. We never found deviations larger than 0.02 km/s. The average velocity of 5.62 km/s ± 2 % for the upper crust confirms the results of modelling the refraction data (section 5.1), especially the reality of the low-velocity zone between 7 and 14 km depth.

The average velocity down to the crust-mantle boundary defined by the last reflected energy corresponding to phase P_M is estimated to be 6.0 km/s ± 4 %

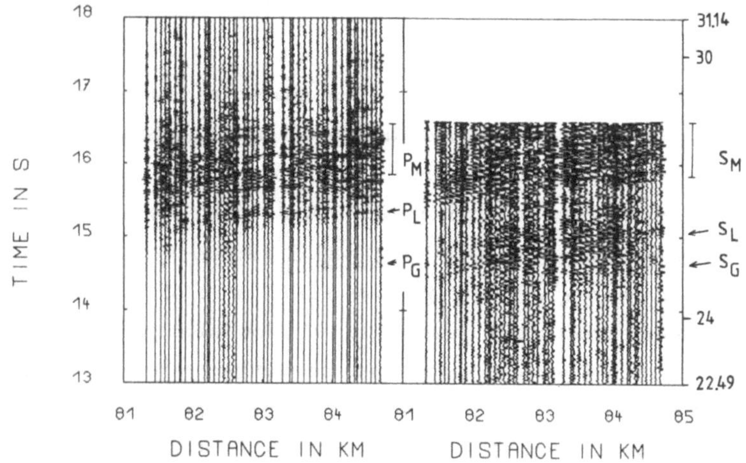

Figure 25: Section from shotpoint R with a selected offset range of 81 to 85 km. P-waves in the left part, S-waves on the right side. The righthand part is from the same recording with a time scale multiplied by SQRT (3). Coherent phases can be compared with corresponding P-phases. S_L arrivals are faster than P_L with respect to a ratio of 1.73. This indicates that the mean value of the V_p/V_s-ratio (or the Poisson ratio) for the upper crust is less than 1.73.

giving an interval velocity of about 6.5 km/s for the lower crust in agreement with the refraction modelling (section 5).

In the ESP-sections shearwave phases from the upper crust and the crust-mantle boundary can be correlated and compared with the corresponding P-phases in the undercritical as well as in the supercritical range. Two seismogram sections for different distance ranges have been subdivided into a P- and a S-wave section in Figs. 23 and 25. The time-scale of the S-wave section is amplified by SQRT (3) corresponding to a Poissons ratio of 0.25. In Fig. 23 the P-wave section shows the characteristic response of the crust in the upper (P_G-phase) and lower part (phase P_L and P_M characterizing top and bottom of the laminated lower crust). In the S-wave section the diving wave from the upper crust (S_G) has the same velocity in the amplified time scale as the corresponding P-phase. Therefore Poissons ratio is normal in the penetrated part of the upper crust. A distinct onset from the Moho (S_M) can also be correlated yielding the same Poissons ratio for the whole crust whereas the lower crustal S-wave reflections cannot be identified unequivocally. The phase $S_?$ can be interpreted in two ways:

a) it represents the converted P-S phase from the Moho.

b) it characterizes the S-wave reflection from the top of the lower crust which would imply a strong reduction of Poissons ratio in the low velocity zone in order to explain the time-shift with respect to the P-wave reflection.

A similar comparison at supercritical offsets shows the same results in this range (Fig. 25). In contrast to the situation in the undercritical range the traveltime differences between converted P-S phases from the Moho and the shearwave reflections from the top of the lower crust are large enough to allow for the unequivocal identification of the S_L-phase. P_L and S_L exhibit a significant time difference with an average V_P/V_S ratio of only 1.67 down to the top of the lower crust. With respect to this low mean value the velocity ratio in the P-wave low velocity zone must be further reduced regarding the value 1.73 in the first 6-7 km depth obtained from shotpoint Z. For the whole crust a V_p/V_s-ratio of 1.73 as a mean value is inferred from P_M and S_M phases. In order to compensate for the reduction within the P-wave low velocity zone, an increase of the V_p/V_s-ratio in the lower crust must be assumed. These observations are based on shotpoints Z (Fig. 23), P and R (Fig. 25). The other shotpoints did not generate enough shearwave energy to provide reliable correlations on vertical component recordings. These qualitative results need further justification by a controlled shear wave experiment scheduled for 1987.

During the KTB reflection campaign a wide-angle Vibroseis experiment was carried out as joint ECORS/DEKORP operation as a feasibility study concerning a future seismic reflection survey of the Rhinegraben rift. Details concerning data acquisition, communication, data processing and influence of near-surface geology are given in Damotte et al. (1987). In Fig. 26 the data of this Rhinegraben undershooting test are compared with expanding spread records of the Black Forest in the offset range 70 to 85 km. The arrivals P of the lower crust beneath the Rhinegraben are about 0.3 s later than from beneath the Black Forest, both phases being propagated on a purely crystalline path. The duration of the lower crustal reverberations amounts to about 1 s in the Rhinegraben compared to 2 s in the Black Forest, indicating a thickness ratio of about 1 : 2 in the lower crust. These observations agree very well with reflection seismic findings of Dohr (1970). In his section obtained in the central graben near Rastatt south of Karlsruhe the difference in TWT between the top of the lamellae at about 7 s and the bottom of the sedimentary fill near 2 s is just 5 s, as is the TWT through the crystalline basement from the surface to the top of the lamellae in this KTB reflection data in the Black Forest. One possible interpretation is that the entire brittle upper crust of the graben subsided without appreciable change in thickness during Tertiary graben formation. A comprehensive discussion of consequences of this hypothesis for crustal development can be found in Fuchs et al. (1987).

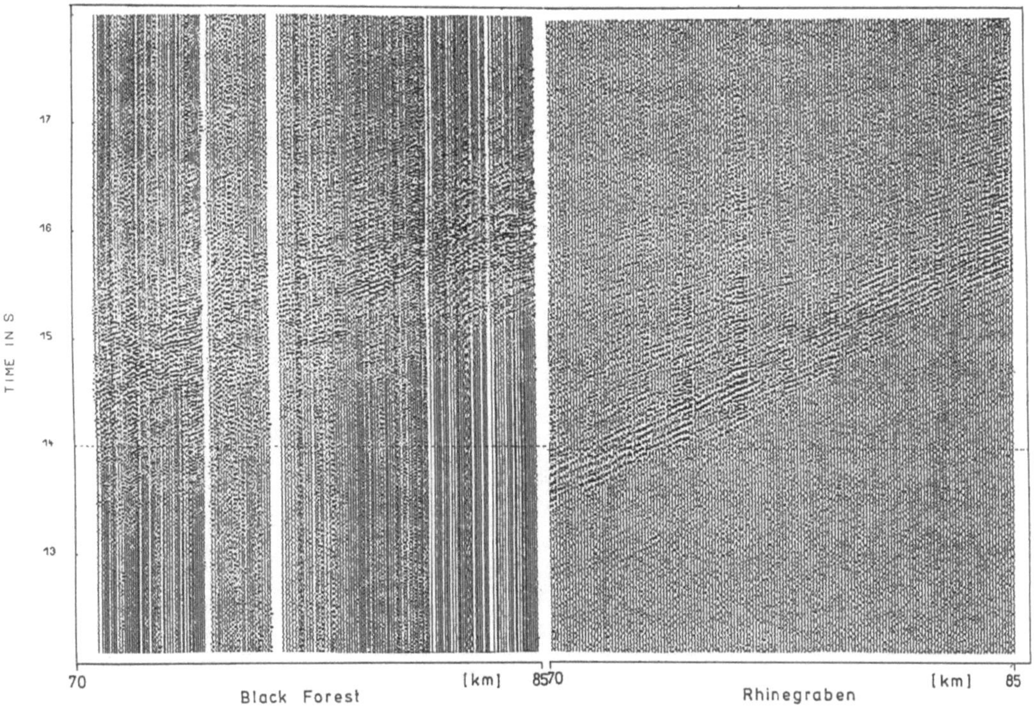

Figure 26: Comparison of wide angle seismograms from the Expanding Spread-profile in the Black Forest (left panel shots R, P and corresponding spreads) and the Rhinegraben undershooting (right panel, vibrators in the Black Forest, recording spread in the Vosges Mountains on WE profile at 48°15' latitude). Time scales not reduced.

Probably the most outstanding feature is the different energy distribution in Fig. 26. In the Black Forest the greatest amplitudes are coming from lamellae near the crust-mantle boundary whereas in the Rhinegraben most of the energy is concentrated in the earlier reflections from the lower crust. This indicates stronger velocity contrasts of the upper part of the lower crust in the Rhinegraben than beneath the Black Forest.

5 Seismic modelling

Important constraints on the physical properties of the earth's crust are provided by determining compressional and shear wave velocities from refraction seismic data. Gajewski and Prodehl (1987) derived a two-dimensional model by ray-tracing methods and asymptotic ray theory (see section 4.1). Here we focus on the central Black Forest, modelling the record section of shotpoint S3 (Figs. 22 and 30) by the Reflectivity Method (RM) of Fuchs and Mueller (1971). Several circumstances are favourable in this respect. The shot was detonated in crystalline basement rocks and generated a broad band source signal with frequencies up to 25 Hz. In the two-dimensional velocity model of Gajewski and Prodehl (1987) and in the structural image of the Vibroseis survey the crustal region north of S3 appears to be sufficiently laterally homogeneous so that the RM can be applied. The computations are performed on a CDC CYBER 205 so extremely fast that we do not face any practical limits concerning vertical complexity, frequency content and number of models (Sandmeier and Wenzel, 1986). Since the reflection profile 8401 almost coincides with the refraction line we can compare results of two-dimensional modelling of source-point gathers with those of the RM.

The interpretation of refraction and reflection data sets of the same region with different sophisticated modelling techniques yields a detailed image of the crustal velocity structure of the central Black Forest.

5.1 Modelling of wide-angle seismograms

As the modelling procedure has been described in detail in Sandmeier and Wenzel (1986), the results are only summarized here. An attempt was made to match not only particular phases such as those from the upper and lower crust (P_G, P_L), the Moho reflection (P_M), and shear wave onsets but the entire wavefield in the observational parameter space of offset and traveltime. We do not consider the data set to be composed of several distinct phases and some kind of noise, but rather we realize that the long reverberations are mainly caused by the lamination of the lower crust and that their amplitude and frequency behaviour reflects properties of the lamination. This point of view turns out to be substantial because the classic idea of phase correlation inherently contains a high frequency approximation and breaks down if it is applied to a structure layered on a scale less than a dominant wavelength. However, this is the case in the lower crust of the central Black Forest. We elaborate this point further in section 6. In order to find error bounds for our model the parameter space consisting of compressional and shear velocity and attenuation as functions of depth was systematically sampled. About 60 different models were calculated for this purpose.

Figure 27: Reflectivity modelling of the refraction profile S3-North. Observed data in the upper part, synthetic seismograms in the lower part (time scale reduced with 8 km/s).
Inset shows corresponding velocity-depth function. The splitting of P_G-phases is due to surface multiples.

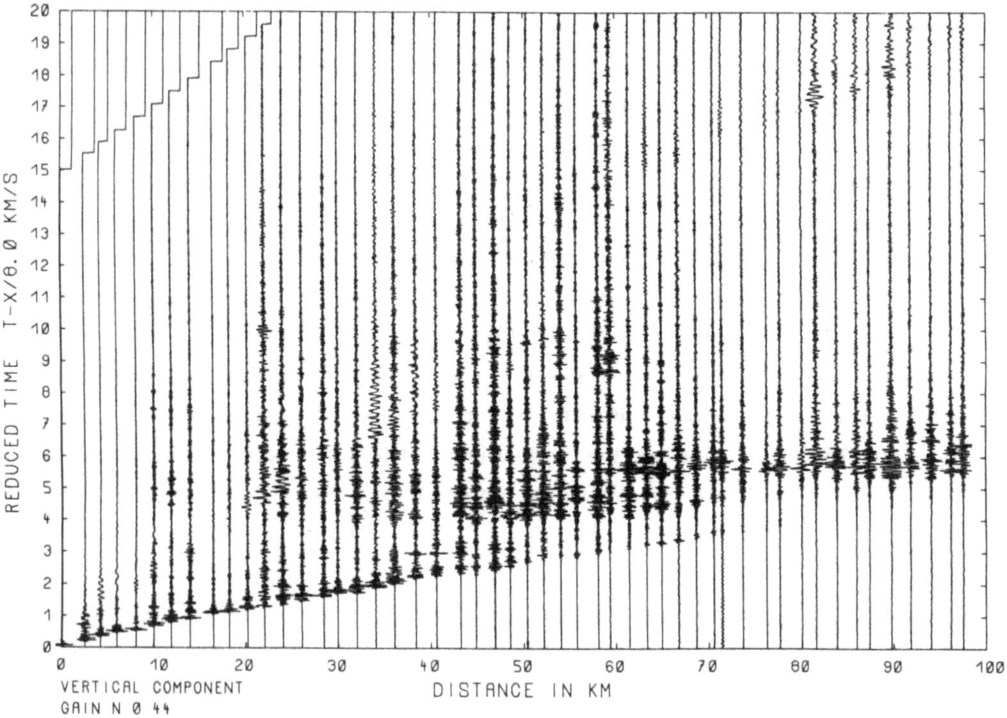
REDUCED TIME T-X/6.0 KM/S
DISTANCE IN KM
VERTICAL COMPONENT
GAIN N 0 44

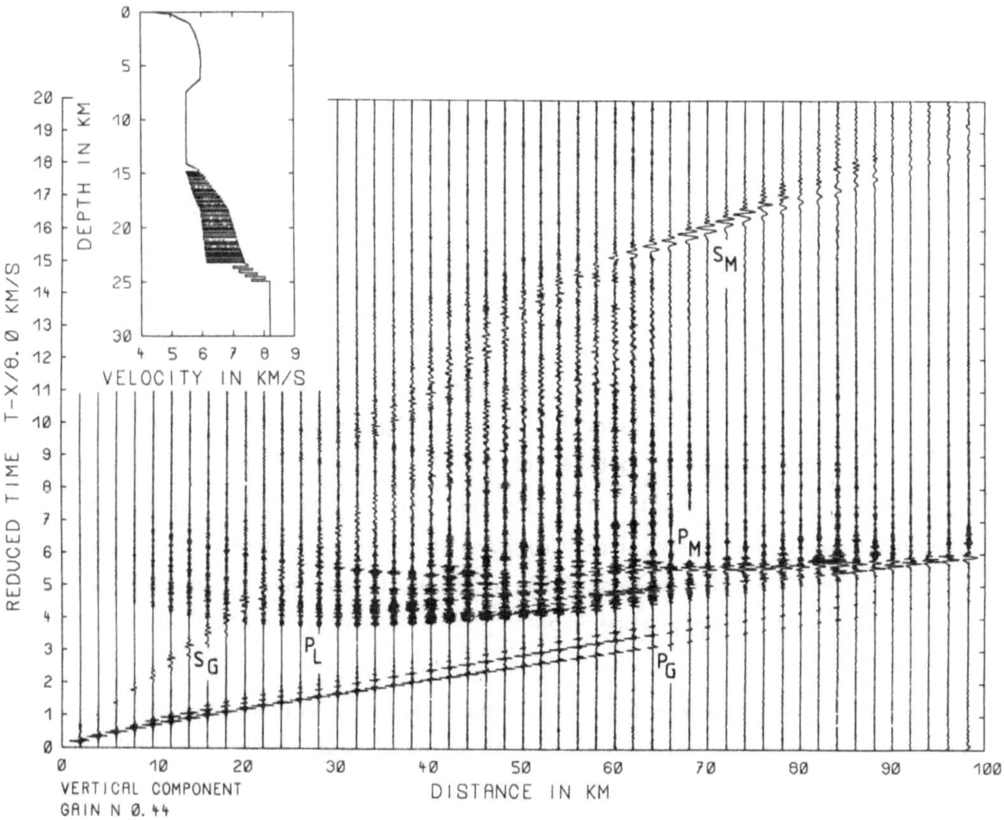
DEPTH IN KM
VELOCITY IN KM/S
REDUCED TIME T-X/6.0 KM/S
S_M
P_M
S_G
P_L
P_G
DISTANCE IN KM
VERTICAL COMPONENT
GAIN N 0.44

Fig. 27 shows the data of shotpoint S3, the best model we found and the corresponding synthetic seismogram section. The ratio of compressional to shear velocities is assumed to be SQRT (3). This value does not contradict the interpretation of shear velocities based on ESP data because no shearwave signals from the lower crust were observed in the section of S3. The density is given by RHO = 0.252 + 0.3788 $\cdot$ V_P . The quality factor Q amounts to 400 in the upper, 1000 in the middle and 2000 in the lower crust. The main results can be summarized as follows: There is a pronounced low velocity channel in the upper crust with a reduction in seismic velocity from 6.0 to 5.5 km/s. It is located directly above the laminated lower crust. The thickness of the lamellae is randomly varied with an average value of 120 $\pm$ 30 m. The average local velocity in the lower crust increases from 6.0 to 6.8 km/s. The magnitude of the reflection coefficients increases from top to bottom. The crust-mantle boundary is modelled as a steplike transition zone consisting of first order discontinuities. The mean V_p/V_s-ratio has been proved to be normal.

In Fig. 28 (panel (b) to (f)) synthetic seismograms for 5 of about 60 models are presented in order to give an impression of the sensitivity of the synthetics in reaction to model variations. Panel (a) displays the data at offsets between 50 and 80 km reduced with 8 km/s. The main phases labeled P_G, P_L, P_M in Fig. 27 are clearly recognizeable in this range. In (b) the synthetic seismograms of the best solution are displaced. In (c) the average thickness of the lamellae is increased to 200 instead of 120 m. The reverberations from the lower crust change significantly. They are small at small offsets and become larger beyond 75 km and the amplitudes around 70 km grow compared to those of the Moho reflection. A similar obvious misfit with the data appears if the average thickness is decreased to 50 m (d). The amplitudes of the P_L wavetrain become much smaller relative to P_G and they almost disappear between 65 and 75 km. In the model of Fig. 27 the lamellae ride on a gradient structure of 0.21 s^{-1} between 15 km and 18.5 km and 0.07 s^{-1} between 18.5 km and 23.5 km depth. The variation of the gradient is not arbitrary. If it is replaced by a constant gradient the P_L reverberations get only gradually weaker between 60 and 75 km as (e) demonstrates whereas the double gradient results in a sharper drop in amplitude at 65 km, in agreement with the observation (compare (a)). An evidently unsatisfactory synthetic record section arises if the gradient is assumed to be zero so that the lamellae are simply superimposed on a constant velocity of 6.7 km/s (compare (f)).

In order to test the significance of the frequency content for modelling a complex structure we have applied a 15 cps low-pass filter (Fig. 29) on the

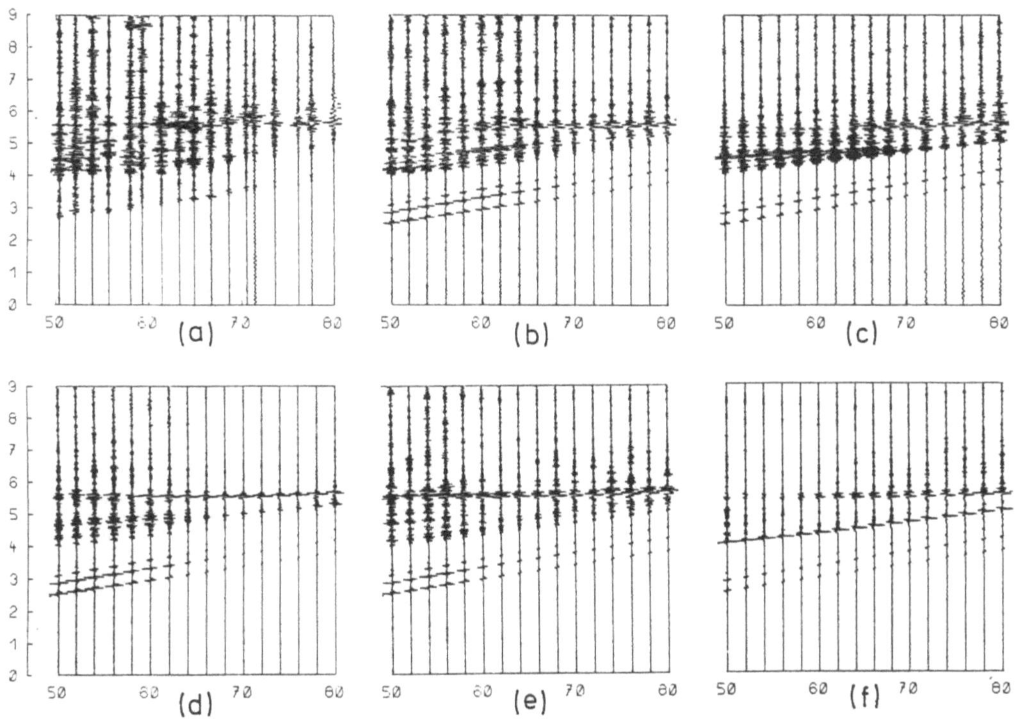

Figure 28: Synthetic seismogram sections for offset range from 50 to 80 km.
a) observed data (same as in Fig. 27 upper part)
b) synthetic data of final model (same as in Fig. 27 lower part), average lamellae thickness 120 m
c) synthetic data corresponding to modified model with lamellae of average thickness of 200 m
d) thickness of lamellae: 50 m
e) single lower crustal velocity gradient instead of two gradients
f) no gradient superimposed on lamellae.
Average crustal velocities are the same in b) to f).

particular seismogram sections of Fig. 28. In this case it is not possible to draw the same conclusions from modelling the data as in the high-frequency case. Apart from panel (f) the response from the lower crust is remarkably diminished and from the real data of panel (a) we would not expect a laminated structure with high reflectivity above the crust-mantle boundary. It is not clear which model fits the data best although the characteristics described above for the high-frequency source remain the same qualitatively. This example shows that physical properties of a complex structure (e.g. a laminated zone as proposed for the lower crust of the Black Forest) can only be inferred from data with sufficiently high frequencies. As in most of the earlier refraction profiles low frequencies around 10 Hz are dominant, their interpretation generally cannot reveal the complexity of the crust but represents only an average over the prevailing wavelength.

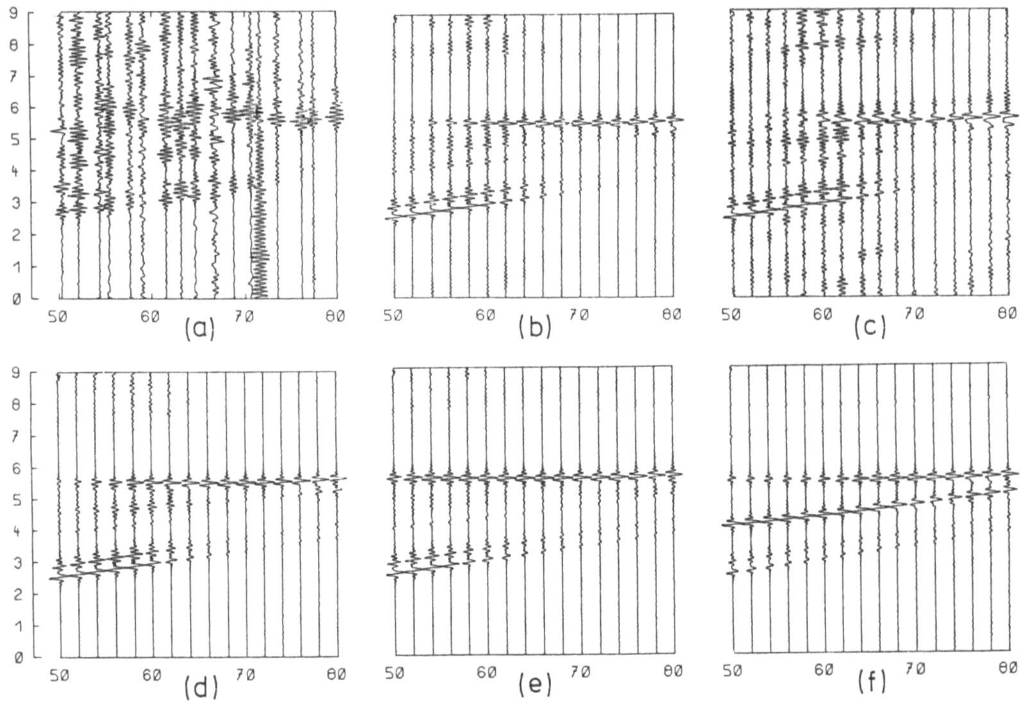

Figure 29: Observed and synthetic seismograms corresponding to same models as in Fig. 28, but low-pass filtered with 15 Hz cut off. Note that lower crustal reverberations are not clearly recognizable.

5.2 Modelling of near-vertical seismograms

A shortcoming of the RM is that the medium has to be laterally homogeneous. Despite of the fact that velocities in the central Black Forest are only slowly varying laterally we know from reflection sections that the lower crust contains numerous reflecting elements with apparent lengths in the kilometre range and that it is therefore laterally variable on this scale. We focussed our interest on two problems: 1) What kind of two-dimensional structure has to be assumed to provide reflection seismograms as observed in the central Black Forest? 2) How do those models compare with the one-dimensional model derived with the RM? For this purpose we employed a program that solves the two-dimensional acoustic wave equation numerically (Wenzel et al., 1987). It follows the approach of Gazdag (1981) and Kosloff and Baysal (1982) who calculate the spatial derivatives in the wave equation with Fourier transform methods. With this algorithm which is implemented on the CDC CYBER 205 the seismic response of a crustal block of 30 x 30 km, a size large enough to synthesize source-point gathers, can be calculated. Steep travel paths of lower crustal reflections are assumed in order to prevent conversion so that the equations of elasticity degenerate into acoustic equations. On the other hand full wave solutions are required because the fact that the lower crust apparently contains many features in the range of a wavelength or less prohibits high-frequency approximations.

The utilisation of the two-dimensional acoustic wave equation does not allow a quantitative comparison of synthetic seismograms and data because a line source is assumed instead of a point source and only velocity variations (i.e. compressibility variations) are considered while density remains constant. However, a qualitative comparison based on the abundance of reflection events, their lateral extent and consistence is possible.

The lower crust is modelled with randomly distributed lamellae of variable length, thickness, and concentration. A single lamella is generated by imbedding a high-velocity rectangle within a lower background velocity. Fig. 30 shows an observed source-point gather and a synthetic one containing only lower crustal reflections. Here we used lamellae with a thickness of 120 m and lengths between 1 and 2 km. The number of high-velocity lamellae is such that the total area filled with lower velocity material equals the area with higher values. If we deviate significantly from this selection of parameters the overall appearance of the synthetics loses its similarity with the data in the sense that reflections become laterally too consistent or too numerous. The wavelet is a sweep autocorrelation function with frequencies between 10 and 35 Hz. The data have been filtered with the same frequency window to allow comparison. Details of the numerical method and of the results are given in Wenzel et al. (1987). Here we only present one example of the results: A structure containing randomly distributed lamellae shows a reflection response with similar characteristics as the data if the parameters mentioned before are used. Along vertical cross sections through the two-dimensional model we get sequences of alternating high and low-velocity layers with an average thickness of 120 m. This is in agreement with the result derived by modelling of refraction data with the RM.

6 Compilation of velocity information

Following our concept of unified interpretation of all accessible data we analyse observed and modelled velocities in terms of their compatibility. Fig. 31 shows the available velocity information derived for the central Black Forest near Haslach. Physical velocities are displayed as local and root mean square values (RMS) respectively. Solid lines represent these compressional wave velocities inferred from modelling of refraction seismograms with the Reflectivity Method (RM). Dashed lines correspond to refraction interpretations of the same record section based on Asymptotic Ray Theory (ART). For details

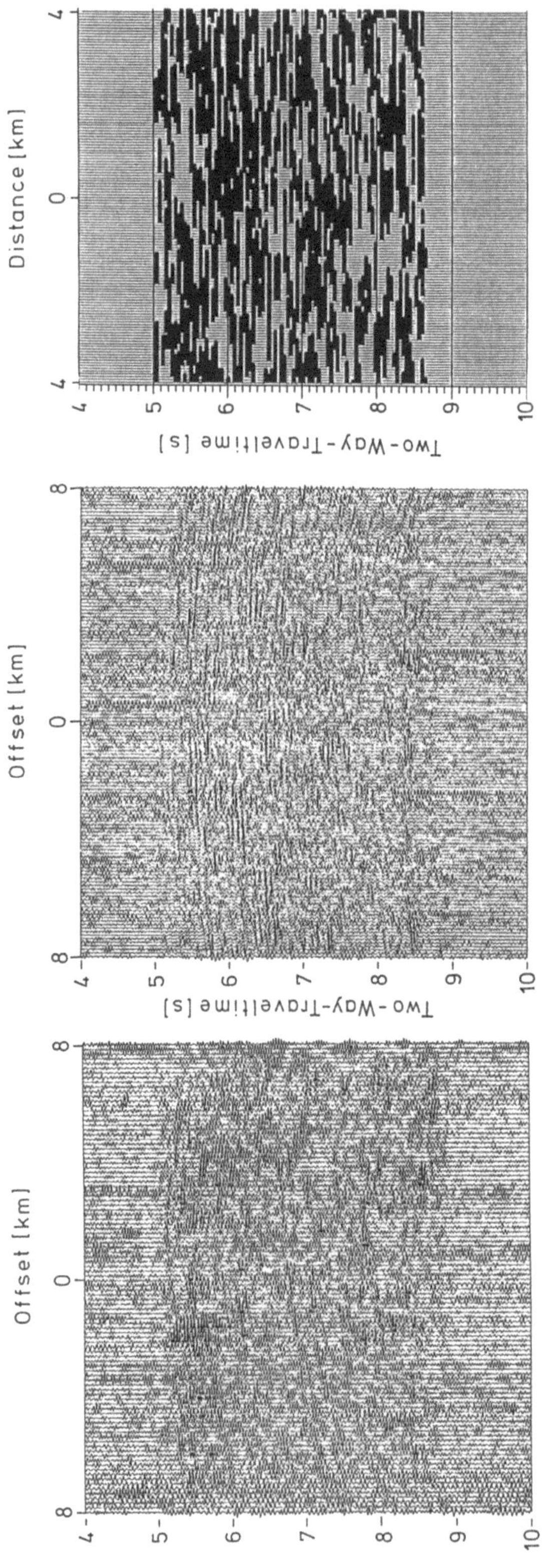

Distance [km]
Offset [km]
Offset [km]
Two-Way-Traveltime [s]
Two-Way-Traveltime [s]

see section 4.1 and 5. Dotted lines represent stacking velocities used for processing the data of the reflection profiles 8401, 8402, 8403 near Haslach.

Physical velocities

Two velocity depth functions were independently derived from the 1984 refraction experiment using RM and ART for the interpretation. Both models coincide in the upper 7 km and in the thickness of the low velocity channel where the velocities only slightly differ. Consequently the position of the top of the lower crust is in agreement. A striking difference appears in the lower crust. For its upper part a relatively high velocity (6.5 km/s) is derived by Gajewski and Prodehl (1987) from a phase correlated at the outermost parts of the profile and interpreted as a diving wave. This phase cannot be clearly recognized in the profile of shotpoint S3 displayed in Fig. 27 because of its restricted offset range. Better evidence is attained in the sections of adjacent shotpoints S2 and S4 which have a larger offset. The details of the lower crustal RM model are essentially constrained by the data in the offset range between 40 and 90 km. Our RM model does not reproduce the diving wave from the lower crust. It could only be generated in the synthetic seismograms by increasing the velocities between 14 and 18 km depth up to 6.5 km/s. However, in this case a much poorer fit of the data in terms of amplitude and frequency characteristics in the offset range up to 90 km would result. Consequently we conclude that an interpretation of the record sections in the frame of a one-dimensional model is incomplete. The remaining discrepancies must be attributed to lateral variations of depth and magnitude of velocity contrasts.

Both interpretations of the data agree in the RMS velocities throughout the whole crust and in the depth of the crust-mantle boundary at about 25 km as a first order discontinuity. The major features of the velocity depth profile presented here are qualitatively in good agreement with previously derived models, see Mueller et al. (1969), Mueller (1977) and Deichmann and Ansorge (1983).

Figure 30: Two-dimensional modelling of a common vibrator-point gather. Observed data are shown in the lefthand panel from 4 to 10 s TWT. Synthetic seismograms are displayed in the central panel, the corresponding model in the righthand panel. Dark areas represent high velocity lamallae imbedded in a lower background velocity. Special emphasis was placed on the lamellar structure of the lower crust. The lamella distribution with depth is varied laterally. Lamellae have randomly distributed horizontal extents of 1 to 2 km. See text for further explanations.

Stacking velocities

Representative stacking velocities for the central Black Forest are displayed as dotted lines in the left panel of Fig. 31. They show considerable variation in the range of 1 km/s at all depths. In order to examine their compatibility with refraction velocities one has to be aware of their meaning and accuracy. Stacking velocities are formal velocities used to produce a stacked section with optimum contrasting structures; consequently they must be considered processing velocities. In our case 20-50 CDPs were first stacked with trial velocities, then displayed, and finally the best section was chosen by

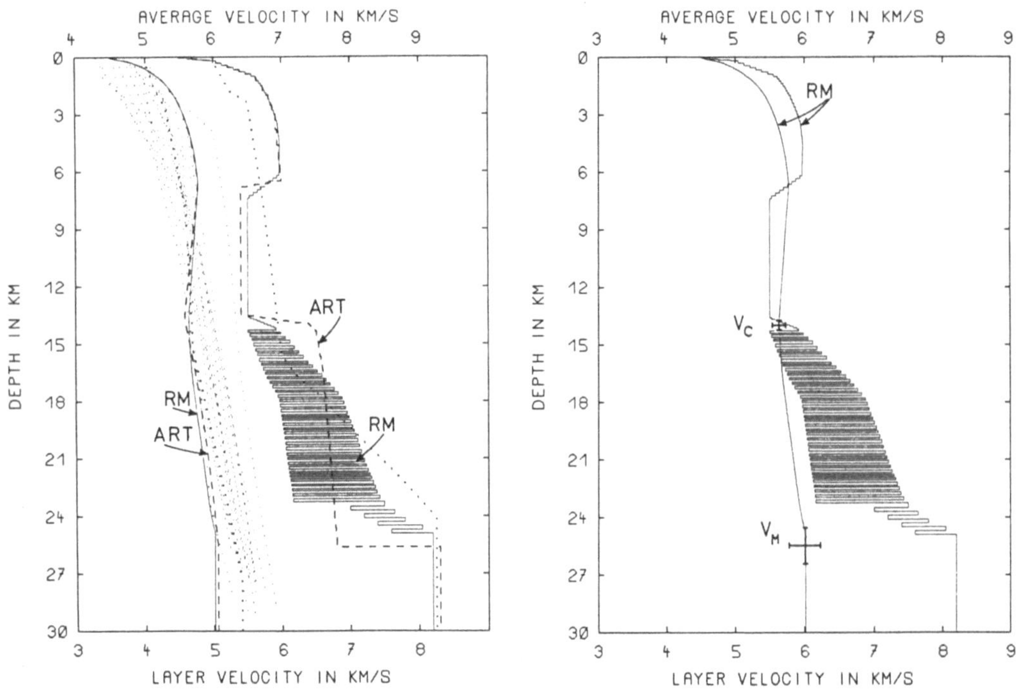

Figure 31: Velocity compilation for the Central Black Forest (near Haslach). left panel: local velocities from ray tracing (ART) and reflectivity (RM) modelling (dashed and solid lines respectively, scale at the bottom). Corresponding RMS-velocities are displaced by one unit to the left (scale on top, dashed and solid lines) together with stacking velocities of the reflection survey (dotted lines). Thick dots represent interval velocities derived from a typical stacking velocity function.
right panel: comparison of RMS velocities of the final RM-model and ESP derived values of V_C = 5.62 km/s $\pm$2 % for the upper crust and of V_M = 6 km/s $\pm$4 % for the entire crust (compare Fig. 24). Corresponding depth uncertainties are $\pm$2 % and $\pm$4 % respectively. For completeness the interval velocities of the RM model are shown at the same scale.

individual judgement. Under certain circumstances stacking velocities are closely related to physical velocities. If a horizontal reflector is overlain by a laterally homogeneous medium and if the recording offsets are within the hyperbolic range the best stack should arise if the stacking velocity is the physical RMS velocity (Dix, 1955). The accuracy of its determination is limited however (Hajnal and Sereda, 1981). Dipping reflectors and point diffractors have optimum stacking velocities larger than the physical ones (Dinstel, 1971). As both events and diffractions are observed in the lower crust of the central Black Forest we attribute the systematic deviation of the stacking velocity from the physical RMS value to these features. Similar observations are reported from the DEKORP 2-South reflection profile (Bortfeld, et al. 1985). Further discrepancies are introduced if lateral velocity inhomogeneities along a CDP are present. These effects have been studied e.g. by Miller (1974) and Blackburn (1980) with respect to sedimentary regions. However, in the crystalline environment of the Black Forest the physical velocities do not change dramatically. Another point is that CDPs very often do not show consistent reflectors in the crystalline basement. Therefore an interpretation of stacking velocities of the type familiar from exploration experience in sedimentary regions must be regarded with caution. If, however, the stacking process is focussed on a consistent reflection, e.g. from the Moho using a recording spread of more than 20 km, a reliable interpretation of stacking velocities in terms of physical velocities is possible. This has been demonstrated on the 1978 Urach reflection profile (Bartelsen et al., 1982).

The stacking velocities are biased towards low values in the upper crust and towards high values in the lower crust (left panel of Fig. 31). We cannot yet present a straightforward explanation of this fact.

Comparing stacking and refraction velocities we have to consider that the stacking velocity involves vertical, the refraction velocity mostly horizontal ray paths. Discrepancies could then be caused by seismic anisotropy (Winterstein, 1986).

7 Compilation of geophysical data and geological-petrological interpretation

7.1 Data

The deployment of a variety of geophysical sounding tools on a regional scale gives a synchronous view of crustal properties through different imaging techniques. Physical anomalies and gradients may be tested for their compatibility and consistency. In Fig. 32 structural information from the

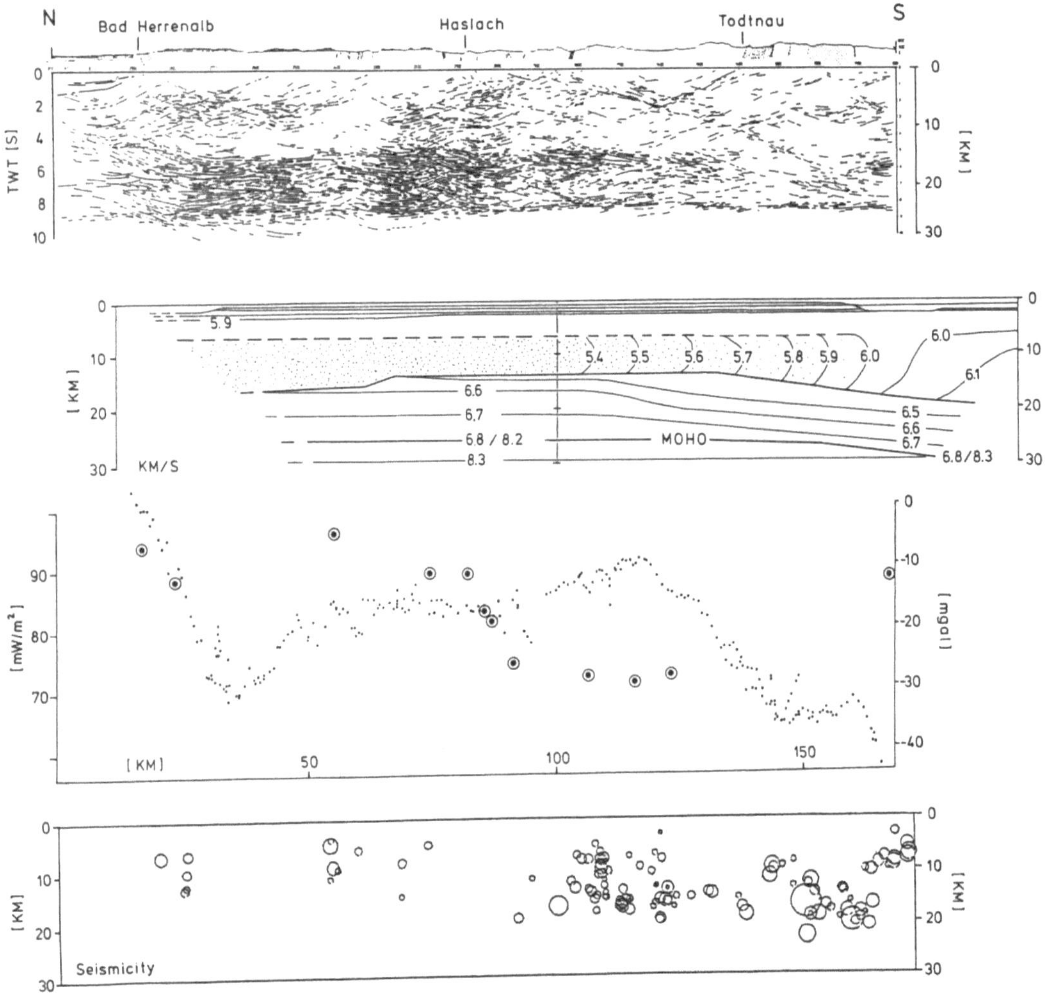

Figure 32: Compilation of geophysical data for the N-S profile. From top to bottom: structural data from the Vibroseis reflection survey (line drawing of migrated line 8401), velocity model from the refraction survey, Bouguer gravity anomaly (thin dots) - heat flow (thick dots), seismicity distribution (magnitudes between 1 and 5; projected within a strip of ±10 km width). For references see text.

Vibroseis reflection profile 8401 (from Fig. 8), the refraction seismic model from Gajewski and Prodehl (1987), the Bouguer gravity anomaly along the seismic profile 8401 (Götze et al., 1986), the heat flow distribution derived from borehole measurements (Stiefel et al., 1985) and the seismicity distribution (Bonjer and Apopei, personal communication) are displayed. Additional information obtained on electrical conductivity from magnetotelluric and electromagnetic sounding methods has been presented by Berktold et al. (1985) and by the LOTEM Working Group (1986).

One of the most exciting questions concerns the compatibility of crustal discontinuities defined by refraction and reflection methods, i.e. the compatibility of wide-angle and near-vertical measurements. Wide-angle refraction ray paths are sensitive to large-scale vertical and horizontal velocity gradients. Near-vertical incidence of elastic waves more accurately images the vertical sequences of fine structures in terms of their reflectivity at a resolution within the scale of the seismic wavelength. Gajewski and Prodehl (1987) transformed their refraction velocity-depth model to a velocity-vertical traveltime (TWT) model and compared it to the near vertical (TWT) reflection section 8401. They found coincidence of the Moho (velocity step from 6.8 to 8.2 km/s) with the deepest accumulation of consistent near-vertical reflections, and coincidence of the Conrad discontinuity (midcrustal velocity step from 5.4-6.0 to 6.5 km/s) with the upper level of the highly reflective lower crust. An equivalent approach is a depth conversion of the reflection section using the refraction velocity model. Both presume identical horizontal and vertical velocities and are not conclusive in case of seismic anisotropy. We have preserved the vertical traveltime scale on the reflection sections because of its objective measured character.

A more reliable test for compability of reflecting structures in the near-vertical, with first or second order velocity discontinuities in the wide-angle range beneath a fixed surface location, is an expanding spread experiment. In our case the most important crustal discontinuities, the boundary between upper and lower crust and the crust-mantle boundary, are documented continuously from the near-vertical to the wide-angle range in the Haslach region (section 4.2). High frequency signals and a detector spacing of 80 m enabled a continuous phase correlation and showed coincidence of these discontinuities. This is confirmed by near-vertical and wide-angle modelling presented in section 5.

Another very striking coincidence is observed between the upper crustal low-velocity zone and a zone of relative transparency in the reflection sections. The base of the low-velocity zone also coincides with the first reflections from the lower crust. Gajewski et al. (1987) determined that the regional extent of the low-velocity zone agrees with the exposed crystalline area of the Black Forest. To the south the low-velocity zone fades out accompanied by a progressively weaker reflectivity of the lower crust. The coincidence of the low-velocity zone and relatively transparent zone appears to be considered as a special local phenomenon. Otherwise, if correlated with the transparent zone in the EW profile 8403, the low-velocity zone would extend further to the east, as far as the region of the Urach geothermal anomaly (Bartelsen et al., 1982).

The top of the low-velocity zone is neither clearly marked by the reflectivity distribution nor by wide-angle reflections in the expanding spread profile. It may be of transitional character as well as of undulatory form following the reflection pattern in the upper crust. Its depth, and its interval velocity of 5.4-5.5 km/s, are constrained first by the velocity-depth function inferred from the first arrivals P_G which show an increase of velocity up to 6 km/s at 6-7 km depth, and second by the average velocity of 5.62 km/s between the surface and the lower crust as inferred from the ESP as well as from the bulk of wide-angle and refraction data.

Strong lateral gravity gradients in the north and in the south may be related to the thrust belts "Baden-Baden" and "Badenweiler-Lenzkirch". The gravity minimum in the South coincides with a decrease of reflectivity in the lower crust and the fading out of the low-velocity zone in the upper crust.

Two and three-dimensional gravity modelling has been performed by Götze et al. (1986). The velocity model of Fig. 32 has been transformed to a 2D-density model using an empirical velocity-density relationship and inversion techniques. The computed gravity (Fig. 33) fits the observed curve satisfactorily. In the central Black Forest a slight misfit might suggest an additional influence of low-density material in the upper or middle crust. The fading out of the low-velocity/density zone in the south and the subsidence of the lower crustal isolines is fairly well matched by the computed gravity and the observed data.

In contrast to the velocity-density model a structure-density model has been tested for an accretion hypothesis of underthrusted Paleozoic metasediments, which is supported by structural imaging by the reflection seismics. By optimizing structures and densities a complete fit could be forced. Gravity modelling may constrain structural models if the results are inconsistent with the data, but consistence is not a general proof of the models. Severe uncertainties may arise from three-dimensional effects of the Rhinegraben which strikes nearly parallel to the gravity survey.

A special phenomenon of interest is the strong Bouguer maximum near the "Baden-Baden" zone in the north known as "Kraichgau"- anomaly (Fig. 33, left side). The gravity high could be related to a shallow body of mafic-ultramafic material within the Saxothuringian-Moldanubian suture zone. The suture zone is delineated clearly by a strong lateral gravity gradient (compare gravity map by Gerke, 1957). Fountain and Salisbury (1981) reported similar gravity anomalies in other continental suture zones. Just north of the Baden-Baden suture zone the reflective character of the lower crust seems to continue into the "Kraichgau"-area, revealed by reflections at 5.2 s, 7 s and 9 s (Demnati and Dohr, 1965). This indicates that it

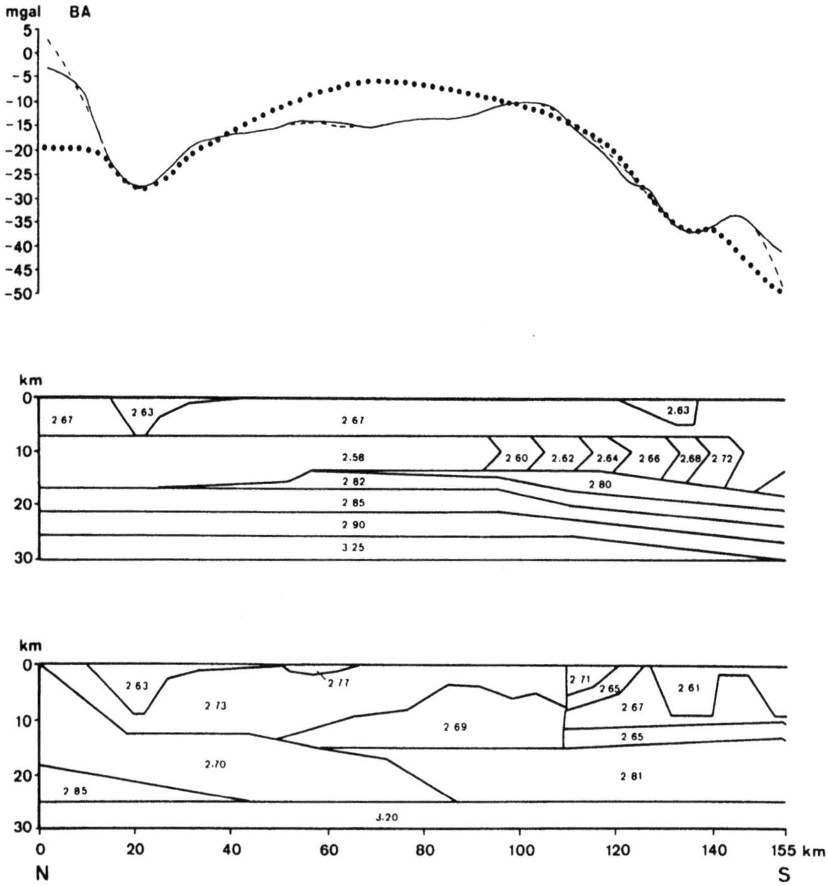

Figure 33: Gravity modelling.
Upper part: observed (---) and computed (.... model A, —— model B) Bouguer anomaly.
Lower part:
A: velocity model converted to a density model using an empirical relationship. Lower crustal densities obtained by an inversion technique. Granite massifs are added for completeness.
B: Structure model based on a hypothetical geological-tectonic model.

Numbers are densities in g/cm³.

originated after the Hercynian suturing provided that the "lamination" represents the same geological phenomenon both north and south of the Baden-Baden zone. The widespread appearance of the lamination in large areas in Europe suggests nonlocal geological and physical processes.

Gravity and geomagnetic mapping in the Central Black Forest reveal relatively small anomalies with SW-NE trends coincident with the strike of the Hercynian tectonic fabric. Granite bodies such as the "Triberg"-granite are clearly displayed by local gravity minima. The maximum depths of the granite bodies are estimated to be about 4-8 km calculated for granite densities of 2.62 g/cm^3 and gneiss densities of 2.73 g/cm^3 determined by weighing core samples (Plaumann et al., 1986). This density decrease of 4 % corresponds to a velocity decrease of the same amount (compare section 3.4.1). Reflections from the expected base of the granite bodies are widely scattered so that corresponding correlations are ambiguous (compare northern part of profile 8401, locations 2500-2900, Fig. 8). The basal contact between granites and gneisses must be of discontinuous or scattered geometry compared to a seismic wavelength, or it does not represent a contrast of impedance at that depth.

The surface heat flow (Stiefel et al., 1985) is relatively high (70-95 mW/m^2) and shows strong lateral variations. This may be due to increased basal heat flow from the upper mantle as a consequence of the Rhinegraben rift formation with increased convective and conductive heat transport still active today.

From the rheological point of view, the crust may be subdivided into an upper brittle and a lower ductile part (Chen and Molnar, 1983; Meissner and Strehlau, 1982). The brittle-ductile transition is expected to be represented by the deepest earthquake foci. In the northern part of the cross section (Fig. 32, lower section) the seismicity is low and maximum hypocentral depth correlates with the reflection-refraction defined boundary between the upper and lower crust. In the southern part the seismic activity increases and the foci reach depths of about 20 km. The deepest earthquakes are located within the laminated lower crust.

Magnetotelluric studies revealed a high-conductivity layer (650 s) at mid-crustal depths, which however cannot be correlated to either the low-velocity zone or to the laminated lower crust, and a pronounced conductivity anisotropy (factor of anisotropy 5) with maximum conductivity in Hercynian direction (NE-SW) (Berktold et al., 1985, Schmucker and Tezkan, 1987). Recently performed transient electromagnetic soundings in the central Black Forest near Haslach yielded another high-conductivity layer with about 5 Ohm-m dipping to the NW between 9 and 6 km depth (LOTEM Working Group, 1986) correlating with the top of the low-velocity layer.

7.2 Discussion and conclusions

Jones and Nur (1984) and Smithson et al. (1986) have shown by specific field studies, laboratory data of physical properties and seismic modelling, that cataclastic fault zones and ductile mylonite zones can be regarded as candidates for upper crustal reflectors. The seismic structural image of the upper crust in this view is considered as marker of tectonic events. Fig. 35 reflects the basic tectonic interpretation of the seismic sections in the Black

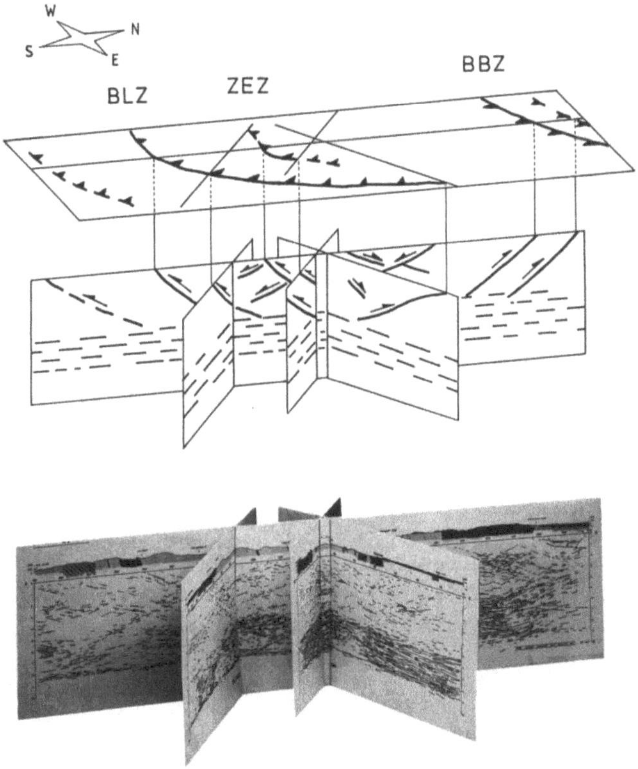

Figure 34: Three-dimensional structure diagram provided by line drawings of the net of four reflection profiles (lower part). View is from ESE. Interpretational fence diagram showing Hercynian thrust and Late Hercynian extension tectonics (upper part). BLZ = "Badenweiler-Lenzkirch" zone, ZEZ = "Zinken-Elme" zone, BBZ = "Baden-Baden" zone.

Forest. The upper crust is characterized by a rather heterogeneous seismic image. Reflectors are much less continuous than in the lower crust and reveal an increased complexity of structures at the scale of the seismic wavelength (150-300 m). Dominant dipping reflectors seem to die out at a depth of about 8-10 km. In the Black Forest this might be related to a change of the reflective character of a fault zone due to a change of deformation style from a brittle regime with non-cohesive gouge, breccias, and random-fabric cataclasites to mylonites. Alternatively an overprint by recent fluid activity could be taken into account resulting in a homogenization of seismic structures caused by alternations in mineralogy and petrology. Prominent reflectors in the upper crust in the Black Forest are interpreted in terms of Hercynian tectonic processes. The convergent zones of the Baden-Baden suture zone (BBZ) in the north and the Badenweiler-Lenzkirch zone (BLZ) in the south are visible in the seismic reflection profiles with bands of inclined reflectors. The reflecting band related to the Baden-Baden zone dips 40° to the south and reaches into the deeper crust. On profiles 8401 and 8514 the Badenweiler-Lenzkirch zone can be traced from the surface to a depth of more than 12 km as a number of strongly reflecting elements dipping NW. NW-dipping structures in the eastern section of profiles 8402 and 8403 indicate the continuation of the Badenweiler-Lenzkirch structure beneath the Mesozoic sedimentary cover to the NE. Other NW-dipping reflectors south of the Elztal fault reach a depth of at least 5 km and are interpreted as belonging to another intraplate convergent zone similar to the Badenweiler-Lenzkirch area ("Zinken-Elme"-zone, ZEZ). These zones BBZ, BLZ and ZEZ constitute thrust faults opposed to each other with northwestward and southeastward vergencies (compare section 2).

Low-angle normal faults characterized by cataclastic deformation in the Central Black Forest are identified as reflectors by near-surface high-resolution measurements (section 3.4.2) and are correlated with several SE- and NW-dipping reflectors in this area (see profile 8402). These reflectors, often cross-cutting the above described reflector pattern, are related to extensional tectonic processes during the late Hercynian and are regarded as markers of the change of the tectonic style from compressional to extensional character.

The lower part of the upper crust (e.g. 7-14 km) is seismically characterized by a P-wave low-velocity zone, a reduced V_P/V_S-ratio (but no shearwave velocity inversion) and a zone of relative transparency. This coincides with a zone of high electrical conductivity (Fig. 35). Model interpretations of the nature of low-velocity zones in reactivated continental crust may include granitic layers (Mueller, 1977), underthrusting of sedimentary rocks, and fluid overpressure related to metamorphic or tectonic processes (e.g. Fyfe and Kerrich, 1985). Dehydration reactions during metamorphic processes can reduce the effective pressure at a given crustal level leading to the formation of new microcracks and thereby lowering the P-wave velocity (but increasing the S-wave velocity; Kern 1982). Velocity inversions can also be triggered by phase transitions in the solid state (quartz) in regions of high heat flow (Kern, 1982).

A bright spot situated in the low-velocity channel is probably caused by a body of lower impedance which might be taken as evidence for a fluid accumulation (section 3.4.3). These completely different observations favour a major role of fluids in a porous or fractured medium determining physical properties and triggering petrologic processes in addition to temperature and pressure. In the Kola deep borehole a low-velocity zone between 4.5 and 9 km depth is characterized by the presence of mineralizing fluids in connection

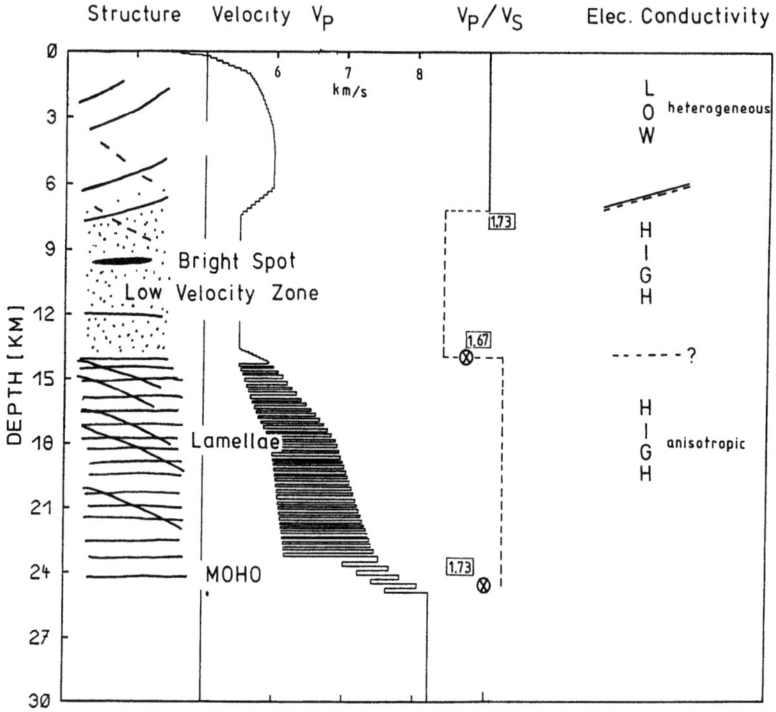

Figure 35: One-dimensional presentation of geophysical targets in the Central Black Forest (Haslach region).
From left to right: Structures, P-wave-velocities, V_p/V_s-ratio (numbers indicate average values for the upper crust and for the entire crust respectively), electrical conductivity (qualitative values). See text for quantitive values. The precision of the upper boundary of the high conductivity layer is of the order of 10-15 %.
Poisson's ratios are 0.25 (V_p/V_s = 1.73), 0.22 (1.67) as mean values for the entire crust and for the upper crust respectively. The dashed line is approximate distribution of local values.

with hydrofracturing, increase in porosity and strong textural variations of the Precambrian rocks (Kremenetsky and Ovchinnikov, 1986). Fuchs et al. (1987) propose a similar mechanism involving dehydration of parts of the lower crust by decompression. The generation of microcracks by fluid overpressure may explain the combined effects of both low-velocity and high electrical conductivity. Also, a layer of decreased velocity and increased conductivity has been found in crystalline rocks of the borehole Urach 3 (Swabian Alb) displaying high microcrack density and strong hydrothermal alteration (Stenger, 1982). The same relationships have been found in the NAGRA borehole "Böttstein" (NAGRA, 1985).

The lower crust is characterized by a strong and laterally consistent lamination with alternating high and low P-velocities differing by about 10 %

and a vertical layering on a scale of 100 m. The horizontal extent of single reflecting elements is in the range of a few hundred meters. The average V_P/V_S-ratio in the lower crust is slightly increased above the crustal mean value of SQRT (3) (compare section 4.2). Seismic lamination of the lower crust is a common feature in Hercynian and Caledonian consolidated continental crust (Meissner et al., 1983; Bois et al., 1986; Bortfeld et al., 1985, Matthews and Cheadle, 1986). Lithological interpretations involve compositional or metamorphic layering, underplating, magmatic differentiation, partial melting, fluid enrichment, penetrative ductile deformation (mylonitic banding) or anisotropy.

A great amount of experimental data on velocities of presumed lower crustal rocks has been published in recent years (Christensen, 1979; Christensen and Fountain, 1975; Fountain, 1976; Kern, 1982; Kern and Schenk, 1985 and others). Information on the composition of the lower crust is obtained from outcropping rock series or by analysis of xenoliths in volcanic host rocks. Exposures representing complete lower crustal sections in Europe (Ivrea zone/Western Alps; Calabria/Southern Italy) reveal granulite facies rocks of alternating mafic to acidic composition as major constituents (Fountain and Salisbury, 1981; Schenk, 1984). Measurements of compressional and shear wave velocities in these rocks at lower crustal PT-conditions (Fountain, 1976; Kern and Schenk, 1985) have demonstrated the following relations:

- Both V_p and V_s increase with increasing amount of pyroxene, amphibole, garnet and sillimanite.
- Poisson's ratio is high with high feldspar content and low with high quartz content.
- Strong anisotropy of seismic velocity is caused by preferred lattice orientation of minerals and by metamorphic/compositional layering.
- Metapelitic granulite-facies rocks of intermediate silica and high alumina content have velocities equivalent to mafic granulites (V_p = 6.8 - 7.5 km/s).

Using geophysical and geological data from the Ivrea zone, Hale and Thompson (1982) generated synthetic seismograms with interfering reflections from a stack of thin layers similar to those observed in our reflection data. Blundell and Raynaud (1986) suggested alternatively a contribution by sideswipes from a single undulatory surface. Although these lateral effects cannot be completely excluded, a major influence is ruled out by spatial control provided by the intersecting lines.

Anisotropy may be an important factor with respect to the strongly reflecting character of the lower crust. Experimental data on the elastic properties of mylonitic rocks (with anisotropy up to 20 %) demonstrate that deep crustal reflections can be explained by extensive ductile deformation of rocks, provided the mylonitic zone is thick enough relative to the seismic wavelength (Jones and Nur, 1984). Late Hercynian extension of the crust in the Black Forest which produced cataclastic fault zones in the upper crust could be accompanied by ductile deformation in the lower crust thus producing the laminated seismic structure.

Comparative near-vertical and wide-angle observations provide additional hints on the Rhinegraben rift evolution. Fuchs et al. (1987) stressed the identical thickness of about 14 km of the crystalline upper crust beneath both graben and shoulders, based on near vertical reflections from the lower crust. They concluded that the upper crust could have subsided about 4 km, without considerable deformation, into the lower crust. Wide-angle observations (section 4.2) indicate a reduced thickness and stronger impedance contrasts of the lamellar structure in the graben proper. These observations favour an origin of the lamellae prior to the beginning of Rhinegraben subsidence and uplift in the Eocene which have modified the preexisting lamallae. In the reflection profiles 8402 and 8514 (see Figs. 10 and 12a), the image of lower crustal lamination is characterized by more widespread diffraction legs and shorter reflection elements than in profiles 8401 and 8403. This may be due to the presence of line diffractors or elongated lamellae with Hercynian strike perpendicular to profiles 8402 and 8514. This observation is consistent with profiles U1 and U2 of the Urach Geothermal project (Bartelsen et al., 1982) and DEKORP 2 South (Bortfeld et al., 1985) and may be regarded as indication of an origin of lamellae during late Hercynian or Mesozoic extension. Another relationship exists probably to the conductivity anisotropy with preferred Hercynian strike (Berktold et al., 1985). The lower crustal lamination found by various authors from BIRPS-, ECORS- and DEKORP-groups cited above as typical for Hercynian Central and Western Europe, may have its counterpart in the Mesozoic rift margin of the eastern North American coastal plain (Cook et al., 1983).

The crust-mantle boundary has a highly variable appearence in reflection profiles, particularly in those of the COCORP, due to different tectonic regimes, thus reflecting different origins (Oliver, 1982). Its formation might be triggered by thermal events and could represent the final stage of an advanced differentiation process (Meissner and Wever, 1986). Strongly reflecting crust-mantle boundaries often seem to be associated with rifted areas. Klemperer et al. (1986) studied pronounced, mostly horizontally layered Moho-reflections in the Basin and Range and found indications favouring an origin by magmatism and extension.

The Moho beneath the Black Forest is developed throughout the entire study area as the deepest horizontally layered and laterally consistent accumulation of high amplitude reflectors at a relatively shallow depth of 25-27 km. Increasing amplitudes at greater source-receiver offsets indicate that sharp discontinuities or lamallae are superimposed on a velocity gradient zone of the type determined by Deichmann and Ansorge (1983) for the eastern border of the Black Forest by a refraction survey (Fig. 30). Continuity and truncation relationships with dipping reflectors suggest a younger than Hercynian age. The Moho as well as horizontal lamellae in the lower crust seem to have overprinted dipping crustal reflectors which delineate thrust zones of Hercynian age (BB-zone and BL-zone).

The age of the Moho is constrained on one hand by these truncations and by its continuity across suture zones and on the other hand by small vertical offsets especially in the southeastern part of line 8514 which are clearly recognized by overlapping diffractions with apices within the Moho (Fig. 10). If these offsets are markers of tectonic events disrupting a generally rather smooth Moho, they are related to the Alpine orogeny; this favours a Moho age of Late Hercynian to Mesozoic time.

Fig. 35 summarizes the relevant geophysical information for the Haslach region. From the structural image of the near-vertical reflection survey several relics of thrusting and normal faulting can be expected in the upper crust above a very pronounced low-velocity zone coinciding with a relatively transparent zone at 7-8 km depth. A local object of special interest is a bright spot at about 9.5 km depth.

A series of coincident geophysical observations raises crucial geo-logic-petrological questions concerning recent metamorphic processes and hydrothermal mineralization. The upper crustal low-velocity zone coincides with a transparent zone where dipping reflectors die out, with a zone of decreased Poissons-ratio (meaning that the S-wave velocity is not decreased as much as the P-wave velocity) and with a level of high electrical conductivity. These crustal properties seem to be related to recent fluid activity in a porous or fractured medium, to increased basal heat flow and to the brittle-ductile transition. The origin of the geophysical anomalies and gradients is strongly associated with crustal structure, composition and processes.

Acknowledgments

The funding of the KTB reconnaissance program (project RG 8314 6) by the Federal Ministry of Science and Technology, Bonn, is gratefully acknowledged. We thank especially H. Schwanitz and his crew from PRAKLA-SEISMOS AG,

Hannover, for excellent performance during the Vibroseis reflection survey and much helpful cooperation. The DEKORP-Processing Center at Clausthal, directed by R.K. Bortfeld, using hard- and software from Mobil Oil AG, Celle, and Seismograph Service LTD, London, performed the standard processing. The Geological Survey of Lower Saxony and the DEKORP management, Hannover, A. Hahn, H.-J. Dürbaum, Ch. Reichert and J. Schmoll provided administrative and consultative service. The THOR Geophysikalische Prospektion GmbH, Kiel, provided their pneumatic hammer device and specialists and the University of Kiel, R. Meissner and H. Stümpel, the recording equipment for the high resolution surveys. We thank the university teams from the Geophysical Institutes of Zürich, Hamburg, Kiel, Clausthal and Karlsruhe for their enthusiastic participation in the ESP and other experiments. R. Vees (Clausthal) organized pond shot R of the ESP. We are indebted to the entire interdisciplinary Black Forest Working Group for many fruitful and encouraging discussions, in particular we appreciate the contributions of W. Wimmenauer, K. Schädel, J. Jenkner, B. Jenkner, H. Klein (Freiburg), G. Kleinschmidt, Th. Flöttmann, B. Gallus (Frankfurt), H.-J. Behr (Göttingen), K.-M. Strack (Köln), H.-J. Götze, J. Schmalfeldt (Clausthal), A. Stiefel, K.-P. Bonjer, I. Apopei, J. Ebel, D. Gajewski, St. Holbrook, H.-P. Schnell, W. Wälde, P. Blümling, J. Mechie, C. Prodehl (Karlsruhe). A. Stiefel provided the heat flow data, and K.-P. Bonjer and I. Apopei the seismicity data of Fig. 32. Special thanks are given to D. Gajewski and C. Prodehl for continuous discussions and for their seismic refraction data and results and to K.-M. Strack for the LOTEM contribution.

We appreciate very much the continuous support by the local authorities and individuals, forestry offices, the Baden-Württemberg mines office and Geological Survey (Freiburg) and the public media. They all created an animating environment for geoscientific research in the Black Forest. The mayor and the municipal offices of Haslach provided buildings, which served as headquarters for our activities for two years.

We owe our thanks to S.L. Klemperer (BIRPS) for carefull and critical reading of the manuscript, for many helpfull and stimulating suggestions concerning its content and presentation. We wish to thank G. Müller for his support in editing and publishing this paper.

We thank Mrs. S. Stuhlert for typing the manuscript.

Contribution No 339 of the Geophysical Institute, Karlsruhe.

Contribution No 207 of Special Research Program SFB 108 "Stress and Stress Release in the Lithosphere" of Karlsruhe University.

References

Alfred-Wegener-Stiftung: 2nd International Symposium on Observation of the Continental Crust through Drilling, Abstracts, Seeheim, Federal Republic of Germany, Oct. 4-6, 1985, Bonn, 96 p., 1985.

Althaus, E., Behr, H.J., Eder, F.W., Goerlich, F., Maronde, D. and Ziegler, W.: Kontinentales Tiefbohrprogramm ("KTB") (Continental Deep Drilling Program) of the Federal Republic of Germany - Advances and Status 1984, Terra Cognita, 4, 389-397, 1984.

Bartelsen, H., Lüschen, E., Krey, Th., Meissner, R., Schmoll, J. and Walther, Ch.: The Combined Seismic Reflection-Refraction Investigation of the Urach Geothermal Anomaly, in: The Urach Geothermal Project (ed. R. Haenel), Schweizerbart'sche Verlagsbuchhandlung, Stuttgart, 247-262, 1982.

Behr, H.-J., Engel, W., Franke, W., Giese, P. and Weber, K.: The Variscan belt in Central Europe: main structures, geodynamic implications, open questions. In: H.J. Zwart, H.-J. Behr and J. Oliver (eds.), Appalachian and Hercynian Fold Belts. Tectonophysics, 109, 15-40, 1984.

Berktold, A., Musmann, G. Tezkan, B., Wohlenberg, J.: Electrical conductivity studies, Schwarzwald Working Group. - Abstract, 2nd International Symposium on Observation of the Continental Crust trough Drilling, Alfred-Wegener-Stiftung, Bonn, October 4-6, 1985, p. 70, 1985.

Blackburn, G.: Errors in stacking velocity - True velocity conversion over complex geologic situations, Geophysics, 45, 1465-1488, 1980.

Blundell, D.J. and Raynaud, B.: Modeling Lower Crust Reflections Observed on BIRPS Profiles, in: Reflection Seismology: A Global Perspective (eds.: Barazangi, M. and Brown, L.), Geodynamics Series Volume 13, American Geophysical Union, Washington, D.C., p. 287-295, 1986.

Bois, C., Cazes, M., Damotte, B., Galdéano, A., Hirn, A., Mascle, A., Matte, P., Raoult, J.F. and Torreilles, G.: Deep Seismic Profiling of the Crust in Northern France: The ECORS Project, in: Reflection Seismology: A Global Perspective, (eds.: M. Barazangi, L. Brown) Geodynamics Series Volume 13, American Geophysical Union, Washington, D.C., p. 21-29, 1986.

Bonjer, K.-P., Apopei, I., Lüschen, E., Fuchs, K. and Sandmeier, K.-J.: The Glottertal - 1986 Micro-Earthquake Sequence - Evidence of Brittle Deformation in the Upper Crustal Reflector, Abstract of Poster Program, 2. KTB Kolloquium 19.9.-21.9.1986 Seeheim, p. 54, 1986.

Bortfeld, R.K., Gowin, J., Stiller, M., Baier, B., Behr, H.J., Heinrichs, T., Dürbaum, H.J., Hahn, A., Reichert, C., Schmoll, J., Dohr, G., Meissner, R., Bittner, R., Milkereit, B., Gebrande, H.: First results and preliminary interpretation of deep-reflection seismic recordings along profile DEKORP 2-South, J. Geophys., 57, 137-163, 1985.

Brown, L., Barazangi, M., Kaufman, S. and Oliver, J.: The First Decade of COCORP: 1974-1984, in: Reflection Seismology: A Global Perspective, (eds.: M. Barazangi, L. Brown) Geodynamics Series Volume 13, American Geophysical Union, Washington, D.C., p. 107-120, 1986.

Censor, Y: Finite Series-Expansion Reconstruction Methods, Proc. IEEE, 71, 409-419, 1983.

Chen, W.-P. and Molnar, P.: Focal depths of intercontinental and intraplate earthquakes and their implications for the thermal and mechanical properties of the lithosphere, J. Geophys. Res., 88, 4183-4214, 1983.

Christensen, N.I.: Compressional wave velocities in rocks at high temperatures and pressures. Critical thermal gradients, and crustal low-velocity zones. J. Geophys. Res., 84, 6849-6857, 1979.

Christensen, N.I. and Fountain, D.M.: Constitution of the lower crust based on experimental studies of seismic velocities in granulite. Geol. Soc. Amer. Bull., 86, 227-236, 1975.

Cook, F.A., Brown, L.D., Kaufman, S. and Oliver, J.E.: The COCORP seismic reflection traverse across the southern Appalachians, American Association of Petroleum Geologists Studies in Geology 14, 61 p., 1983.

Damotte, B., Fuchs, K., Lüschen, E., Wenzel, F., Schlich, R., and Torreilles, G.: Wide angle Vibroseis test across the Rhinegraben, Geophys. J.R. astr. Soc., 89, 313-318, 1987.

Deichmann, N. and Ansorge, J.: Evidence for Lamination in the Lower Continental Crust Beneath the Black Forest (Southwestern Germany), J. Geophys., 52, 109-118, 1983.

Demnati, A. and Dohr, G.: Reflexionsseismische Tiefensondierungen im Bereich des Oberrheingrabens und des Kraichgaues, Zeitschrift für Geophysik, 31, 229-245, 1965.

De Vries, D. and Berkhout, A.J.: Influence of Velocity Errors on the Focusing Aspects of Migration, Geophysical Prospecting 32, 629-648, 1984.

Dix, C.H.: Seismic velocities from surface measurements, Geophysics 20, 68-86, 1955.

Dinstel, W.L.: Velocity spectra and diffraction patterns, Geophysics, 36, 415-417, 1971.

Dohr, G.: Reflexionsseismische Messungen im Oberrheingraben mit digitaler Aufzeichnungstechnik und Bearbeitung. In: J.H. Illies and S. Mueller (eds.), Graben Problems, Schweizerbart'sche Verlagsbuchhandlung, Stuttgart, 207-218, 1970.

Edel, J.B., Fuchs, K., Gelbke, C., Prodehl, C.: Deep Structure of the Southern Rhinegraben Area from Seismic Refraction Investigations. J. Geophys., 41, 333-356, 1975.

Emmermann, R.: A petrogenetic model for the origin and evolution of the Hercynian granite series of the Schwarzwald. N. Jb. Miner., Abh. 128, 219-253, 1977.

Finckh, P., Frei, W., Fuller, B., Johnson, R., Mueller, St., Smithson, S. and Sprecher, Chr.: Detailed Crustal Structure From a Seismic Reflection Survey in Northern Switzerland, in: Reflection Seismology: A Global Perspective, (eds.: M. Barazangi, L. Brown), Geodynamics Series Volume 13, American Geophysical Union, Washington, D.C., 43-54, 1986.

Fountain, D.M.: The Ivrea-Verbano and Strona-Ceneri Zones, northern Italy: a cross-section of the continental crust - New evidence from seismic velocities of rock samples. Tectonophysics, 33, 145-165, 1976.

Fountain, D.M. and Salisbury, M.H.: Exposed cross-sections through the continental crust: implications for crustal structure, petrology, and evolution. Earth and Planetary Science Letters, 56, 263-277, 1981.

Fromm, G.: Ermittlung statischer Grundkorrekturen, in: Erfassung seismischer Daten, 4. Mintrop-Seminar 1984, (eds.: L. Dresen, J. Fertig, H. Rüter, W. Budach), 253-313, 1984.

Fuchs, K.: On the properties of deep crustal reflectors. Zeitschrift für Geophysik, 35, 133-149, 1969.

Fuchs, K. and Mueller, G.: Computation of synthetic seismograms with the reflectivity method and comparison with observations, Geophys. J.R. astr. Soc., 23, 417-433, 1971.

Fuchs, K., Bonjer, K.-P., Gajewski, D., Lüschen, E., Prodehl, C., Sandmeier, K.-J., Wenzel, F. and Wilhelm, H.: Crustal Evolution of the Rhinegraben Area. I. Exploring the Lower Crust in the Rhinegraben Rift by Unified Geophysical Experiments, Tectonophysics, 141, 261-275, 1987.

Fyfe, W.S. and Kerrich, R.: Fluids and thrusting. Chem. Geol., 49, 353-362, 1985.

Gajewski, D. and Prodehl, C.: Crustal evolution of the Rhinegraben area. II. Seismic refraction investigation of the Black Forest, Tectonophysics, 142, 27-48, 1987.

Gajewski, D., Holbrook W.S., and Prodehl C.: Three-dimensional crustal structure of Southwest-Germany, derived from seismic-refraction data, Tectonophysics, 142, 49-70, 1987.

Gazdag, J.: Modeling of the acoustic wave equation with transform methods, Geophysics, 46, 854-859, 1981.

Gebrande, H.: CMP-Refraktionsseismik, in: Seismik auf neuen Wegen, 6. Mintrop-Seminar 1986, (eds.: L. Dresen, J. Fertig, H. Rüter, W. Budach), 191-205, 1986.

Gerke, K.: Die Karte der Bouguer-Isanomalien 1:1.000.000 von Westdeutschland. Frankfurt/Main: Institut für Angewandte Geodäsie, 1957.

Giese, P.: Depth Calculation, in: P. Giese, C. Prodehl, A. Stein (eds.), Explosion Seismology in Central Europe, 146-161, 1976.

Gordon, R., Bender R., Herman G.T.: Algebraic reconstruction techniques (ART) for three-dimensional electron microscopy and x-Ray photography. J. theor. Biol. 29, 471-481, 1970.

Götze, H.-J., Schmalfeldt, J. and Wagener, M.: Gravimetrische und magnetische Vorerkundung im Schwarzwald, internal KTB communication, reported in v. Gehlen et al. (eds.), Kontinentales Tiefbohrprogramm der Bundesrepublik Deutschland KTB, Ergebnisse der Vorerkundungsarbeiten Lokation Schwarzwald, 2. KTB-Kolloquium Seeheim/Odenwald, 19.9.-21.9.1986, p. 29-30, 1986.

Hacker, W. and Hirschmann, G.: Ältere und jüngere Thrust-Tektonik im Zentralschwarzwälder Gneiskomplex, internal KTB communication, reported in v. Gehlen et al. (eds.), Kontinentales Tiefbohrprogramm der Bundesrepublik Deutschland KTB, Ergebnisse der Vorerkundungsarbeiten Lokation Schwarzwald, 2. KTB-Kolloquium Seeheim/Odenwald, 19.9.-21.9.1986, p. 76 ff., 1986.

Hagedoorn, J.G.: The Plus-Minus Method of Interpreting Seismic Refraction Sections. Geophysical Prospecting, vol. 7, 158-182, 1959.

Hajnal, Z. and Sereda, I.T.: Maximum uncertainty of interval velocity estimates, Geophysics, 46, 1543-1547, 1981.

Hale, L.D. and Thompson, G.A.: The Seismic Reflection Character of the Continental Mohorovicic Discontinuity. J. Geophys. Res., 87, 4625-4635, 1982.

Illies, J.H.: Mechanism of Graben Formation, Tectonophysics, 73, 249-266, 1981.

Illies, J.H. and Fuchs, K. (eds.): Approaches to Taphrogenesis, Schweizerbart'sche Verlagsbuchhandlung, Stuttgart, 460 p., 1974.

Illies, J.H. and Mueller, S. (eds.): Graben Problems. Schweizerbart'sche Verlagsbuchhandlung, Stuttgart, 316 p., 1970.

Jones, T.D. and Nur, A.: The Nature of Seismic Reflections From Deep Crustal Fault Zones. J. Geophys. Res., 89, 3153-3171, 1984.

Kern, H.: P- and S-Wave Velocities in Crustal and Mantle Rocks under the Simultaneous Action of High Confining Pressure and High Temperature and the Effect of the Rock Microstructure. In: High-Pressure Researches in Geoscience (ed.: W. Schreyer), E. Schweizerbart'sche Verlagsbuchhandlung, Stuttgart, 15-45, 1982.

Kern, H. and Schenk, V.: Elastic wave velocities in rocks from a lower crustal section in southern Calabria (Italy). Physics of the Earth and Planetary Interiors, 40, 147-160, 1985.

Ketelsen, K., Marschall, R., Fromm, G.: Automatic Picking of First Arrivals with two-sided signals (Vibroseis), 45th E.A.E.G.-Meeting, Oslo, Norway, 49 p, 1983.

Klein, H. and Wimmenauer, W.: Eclogites and their retrograde transformation in the Schwarzwald. N. Jb. Miner., Mh. 1984, 25-38, 1984.

Klemperer, S.L., Hauge, T.A., Hauser, E.C., Oliver, J.E. and Potter, C.J.: The Moho in the northern Basin and Range province, Nevada, along the COCORP 40°N seismic-reflection transect. Geological Society of American Bulletin, 97, 603-618, 1986.

Kosloff, D.D. and Baysal, E.: Forward modeling by a Fourier method. Geophysics, 47, 1402-1412, 1982.

Kremenetsky, A.A. and Ovchinnikov, L.N.: The Precambrian continental crust: its structure, composition and evolution as revealed by deep drilling in the U.S.S.R. Precambrian Res., 33,11-43,1987.

Lerwill, W.E.: The Amplitude and Phase Response of a Seismic Vibrator. Geophysical Prospecting, 29, 503-528, 1981.

Lippolt, H.J., Schleicher, H. and Raczek, I.: Rb-Sr systematics of Permian volcanics in the Schwarzwald (SW Germany). Contr. Min. Petr., 84, 272-280, 1983.

Lorenz, V. and Nicholls, I.A.: Plate and intraplate processes of Hercynian Europe during the Late Paleozoic. Tectonophysics, 107, 25-56, 1984.

LOTEM working group: Schwarzwald, LOTEM-Tiefensondierung, Abstract of Poster Program, 2. KTB-Kolloquium, 19.9.-21.9.1986, Seeheim, p. 48, 1986.

Matte, Ph.: Tectonics and plate tectonics model for the Variscan belt of Europe, Tectonophysics, 126, 329-374, 1986.

Matthews, D.H. and Cheadle, M.J.: Deep Reflections from the Caledonides and Variscides West of Britain and Comparison with the Himalayas, in: Reflection Seismology: A Global Perspective, (eds.: M. Barazangi, L. Brown), Geodynamics Series Volume 13, American Geophysical Union, Washington, D.C., p. 5-19, 1986.

Meissner, R. and Strehlau, J.: Limits of stresses in continental crusts and their relation to the depth-frequency distribution of shallow earthquakes, Tectonics, 1, 73-89, 1982.

Meissner, R., Lüschen, E. and Flüh, E.R.: Studies of the continental crust by near-vertical reflection methods: A review, Physics of the Earth and Planetary Interiors, 31, 363-376, 1983.

Meissner, R. and Wever, Th.: Nature and Development of the Crust According to Deep Reflection Data from the German Variscides, in: Reflection Seismology: A Global Perspective, (eds.: M. Barazangi, L. Brown), Geodynamics Series Volume 13, American Geophysical Union, Washington, D.C., P. 31-42, 1986.

Miller, M.K.: Stacking of Reflections From Complex Structures, Geophysics, 39, 427-440, 1974.

Morton, J.L. and Sleep, N.H.: Seismic Reflections From a Lau Basin Magma Chamber. In: Geology and offshore resources of Pacific island arcs - Tonga region, Circum-Pacific Council for Energy and Mineral Resources Earth Science Series, (eds.: D.W. Scholl and T.L. Vallier), vol. 2, Houston, 441-453, 1985.

Mueller, S.: A new model of the continental crust, in: The Earth's Crust, Geophys. Monogr. Series, 20, American Geophysical Union, Washington, D.C., 289-317, 1977.

Mueller, St., Peterschmitt, E., Fuchs, K. and Ansorge, J.: Crustal Structure Beneath the Rhinegraben From Seismic Refraction and Reflection Measurements. Tectonophysics, 8, 529-542, 1969.

NAGRA: Technischer Bericht 85-01, Textband und Beilagenbände A und B, Nationale Genossenschaft für die Lagerung radioaktiver Abfälle, Baden, 1985.

Oliver, J.E.: Changes at the crust-mantle boundary, Nature, 299, 398-399, 1982.

Peddy, C., Brown, L.D. and Klemperer, S.L.: Interpreting the Deep Structure of Rifts with Synthetic Seismic Sections. In: Reflection Seismology: A Global Perspective (eds.: M. Barazangi and L. Brown), Geodynamic Series Volume 13, American Geophysical Union, Washington, D.C., 301-311, 1986.

Plaumann, S., Groschopf, R., and Schädel, K.: Kompilation einer Schwerekarte und einer geologischen Karte für den mittleren und nördlichen Schwarzwald mit einer Interpretation gravimetrischer Detailvermessungen. Geol. Jb., E 33, 15-30, 1986.

Prodehl, C., Ansorge, J., Edel, J.B., Emter, D., Fuchs, K., Mueller S. and Peterschmitt, E.: Explosion Seismology Research in the Central and Southern Rhine Graben - A Case History, in: Explosion Seismology in Central Europe (eds.: Giese, P., Prodehl, C. and Stein, A.), Springer Verlag, Berlin, 313-328, 1976.

Rühl, Th.: Tauchwellen-Tomographie für Stripping-Korrekturen und geologische Interpretationen bei den reflexionsseismischen KTB-Profilen im Schwarzwald. Diploma Thesis, Geophysical Institute of the University of Karlsruhe, 1987.

Sandmeier, K.-J. and Wenzel, F.: Synthetic Seismograms for a Complex Crustal Model, Geophysical Research Letters, 13, 22-25, 1986.

Schenk, V.: Petrology of felsic granulites, metapelites, metabasics, ultramafics, and metacarbonates from Southern Calabria (Italy): propaganda metamorphism, uplift and cooling of a former lower crust. J. Petr., 25, 255-298, 1984.

Schmucker, U. and Tezkan, B., Zur Deutung regional einheitlicher Richtungsabhängigkeiten tellurischer Variationen, 47th Ann. Meeting of Dtsch. Geophys. Gesell., Clausthal-Zellerfeld, 31.3.-4.4.1987.

Schnell, H.P.: Untersuchung des Einflusses oberflächennaher Schichten auf seismische Signale und Beiträge zur Optimierung des Vibroseis-Verfahrens. Diploma Thesis, Geophysical Institute of the University of Karlsruhe, 1987.

Sheriff, R.E. and Geldart, L.P.: Exploration Seismology, Volume 2, Data-processing and interpretation, Cambridge University Press, Cambridge, 221 p, 1983.

Sittig, E.: Zur geologischen Charakterisierung des Moldanubikums am Oberrhein (Schwarzwald). Oberrh. Geol. Abh., 18, 119-161, 1969.

Slichter, L.B.. The Theory of the Interpretation of Seismic Travel Time Curves in horizontal Structures, Physics, 3, 273-295, 1932.

Smithson, S.B., Johnson R.A. and Hurich, C.A.: Crustal Reflections and Crustal Structure, in: Reflection Seismology: The Continental Crust, (eds.: M. Barazangi, L. Brown), Geodynamics Series Volume 14, American Geophysical Union, Washington, D.C., 21-32, 1986.

Stenger, R.: Petrology and Geochemistry of the Basement Rocks of the Research Drilling Project Urach 3., in: R. Haenel (ed.), The Urach Geothermal Project. Schweizerbart'sche Verlagsbuchhandlung, Stuttgart, 41-47, 1982.

Stiefel, A., Wilhelm, H., Finkbeiner, K., Haack, U.: Geothermal Studies, Schwarzwald, Abstract of poster, 2nd International Symposium on

Observation of the Continental Crust through Drilling, Alfred-Wegener-Stiftung, October 4-6, 1985, Seeheim, p. 72, 1985.

Vogelsang, D.: Geothermische und geologische Vorerkundung der Tiefbohrlokation Oberpfalz und Schwarzwald. Teilgebiet Bohrlochgeophysik Schwarzwald. Bericht NLfB Hannover, Archiv-Nr. 99126, 1986.

Warner, M.: Migration - why doesn't it work for deep continental data, Geophys. J.R. astr. Soc., 89, 21-26, 1987.

Walther, Ch., Trappe, H. and Meissner, R.: The detailed velocity structure of the Urach geothermal anomaly. J. Geophys. 59, 1-10, 1986.

Wenzel, F., Sandmeier, K.-J., and Wälde, W.: Properties of the lower crust from modelling refraction and reflection data. - submitted to J. Geophys. Res., 1987.

Wimmenauer, W. and Adam, A.: High-pressure phenomena in acid rocks accompanying eclogites in the Schwarzwald (SW Germany). Terra cognita, 5, p. 426, 1985.

Winterstein, D.F.: Anisotropy effects in P-wave and SH-wave stacking velocities contain information on lithology. Geophysics, 51, 661-672, 1986.

Zucca, J.J.: The crustal structure of the southern Rhinegraben from re-interpretation of seismic refraction data. J. Geophys., 55, 13-22, 1984.

Compressional and Shear-Wave Velocity Models of the Schwarzwald Derived from Seismic Refraction Data

Dirk Gajewski

Geophysical Institute, Hertzstr. 16, 7500 Karlsruhe, FRG
Present address: Geophysics Department, Stanford University, Stanford, California 94305, USA

ABSTRACT

To investigate the structure, physical properties and the composition of the crust beneath the Black Forest, southwest Germany, deep seismic sounding experiments were carried out in 1984. They comprised seismic refraction as well as seismic reflection measurements. This paper presents a summary of investigations concerning the Black Forest seismic refraction P– and S–wave three–component data, which were recoreded along a 240–km–long line, following the morphological axis of the Black Forest in a N–S direction. The data underscore the importance of joint P– and S–wave interpretations. Contrary to the common assumption, the reflectivity of the crust for P– and S–waves may differ significantly. The most surprising observations in the Black Forest data come from the middle and lower crust. Whereas the P–velocity model shows a distinct decrease in the middle crust (velocities of about 5.4 km/s are observed), the corresponding S–velocities show no decrease. The result is a low Poisson's ratio of 0.15 for the middle crust. It is interpreted as an area of fractured rocks with either dry cracks or fluid filled cracks at low pore pressure or both together. The reflected P–wavefield from the lower crust is characterized by long reverberating wavetrains followed by the reflection from the crust–mantle boundary (Moho). The Moho–reflection is also clearly visible in the S–wave seismograms; however, there does not exist a shear–wave response from the lower crust, which is comparable to that for the P–waves. The reflectivity of the lower crust is different for seismic P– and S–waves: the S–wave velocity distribution varies smoothly with depth, whereas the P–wave velocity distribution is laminated (as is, therefore, Poisson's ratio). The high P–wave velocity layers are of mafic composition, while the low P–wave velocity layers may have a higher quartz content. This supports the interpretation of the lower–crustal reflections as intrusions (underplating) in a previously more homogeneous, less mafic crust. A clear converted wave from the crust–mantle boundary is observed, indicating that the Moho along the eastern shoulder of the Rhinegraben rift is, at least locally, a first order discontinuity.

1 INTRODUCTION

To find a suitable site for a superdeep borehole, it is important to have some prior knowledge of the crustal structure, the physical properties and the composition of the crustal rocks. Therefore, a detailed seismic refraction survey was carried out in the Black Forest, southwest Germany, in August 1984 (Gajewski and Prodehl 1987). Along approximately the same line (Fig. 1), a reflection seismic survey was performed (Lüschen et al. 1987). While seismic reflection investigations reveal the structural details of the crust, the refraction method determines the gross velocity–depth structure, which is essential to migrate seismic reflection data and to estimate the composition of the deep crust.

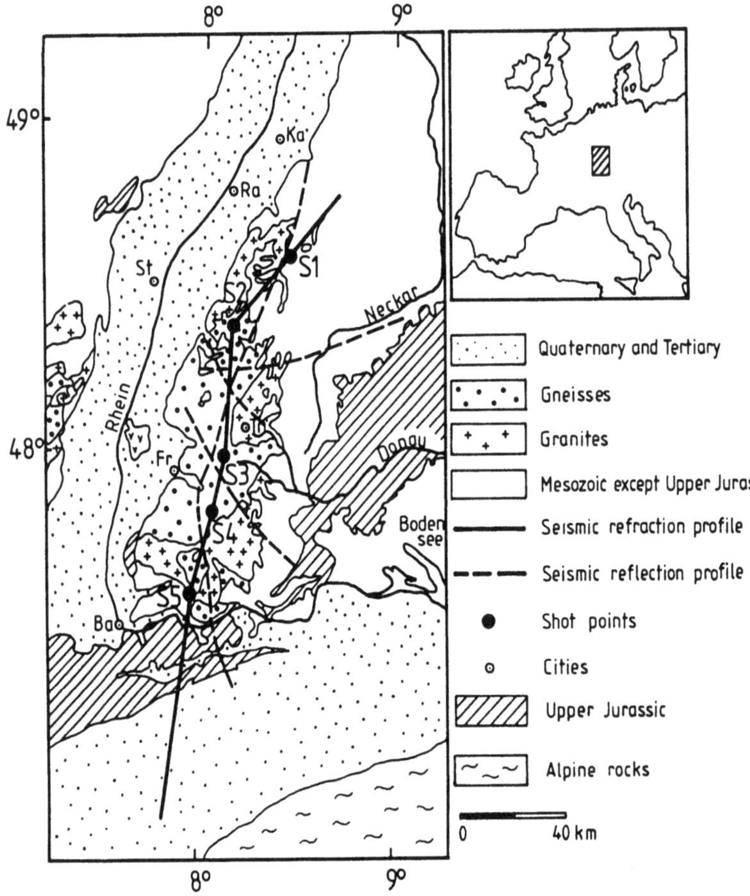

Fig. 1. Geologic sketch map of the Black Forest and adjacent areas, showing the position of the seismic surveys. Cities: *Ba* Basel; *Fr* Freiburg; *Ha* Haslach; *Ka* Karlsruhe; *Ra* Rastatt; *St* Strasbourg; *Tri* Triberg

Before 1984 a few regional seismic investigations in and around the Black Forest were performed to investigate the Rhinegraben rift zone (Emter 1971; Mueller et al. 1969, 1973; Edel et al. 1975; Prodehl et al. 1976; Zucca 1984). More recently some results from areas to the east and south became available (Egloff 1979; Deichmann and Ansorge 1983; Sierro et al. 1983; Finckh et al. 1984, 1986; Gajewski and Prodehl 1985). However, the existing data covered the Black Forest only sparsely. No multi-coverage seismic refraction data were recorded along the Black Forest before 1984.

The main purpose of the refraction survey in the Black Forest was to fill this data gap and to obtain a good picture of the velocity structure of P- and S-waves for the deep crust and in particular for the upper crust. It is well known that better constraints on the composition and the physical properties of the crust are possible if P- and S-wave velocities are jointly interpreted.

So far shear arrivals are not routinely being considered, although they are recorded in most seismic refraction data, especially if explosions are used as sources. The number of interpretations of seismic refraction S-wave data in the literature is nearly negligible in comparison with the number of P-wave interpretations (see, e.g., Braile et al. 1974; Keller et al. 1975; Zschau and Koschyk 1976; Assumpção and Bamford 1978; Banda et al. 1981; Gajewski 1981; Tarkov et al. 1981; Deichmann 1984; Brooks 1985; Gajewski and Prodehl 1985; Hall and Ali 1985; Kullinger and Lund 1986; Luosto and Korhonen 1986; Ruthardt 1986; Holbrook et al. 1987, 1988; El-Isa et al. 1987; Min and Wu 1987; Krammer 1988).

The interpretation of S–waves is most valuable if three–component recordings are available. However, they are also observed on vertical component recordings (e.g., Gajewski 1981; Gajewski and Prodehl 1985; Kullinger and Lund 1986; Ruthardt 1986), especially where no low–velocity surface layer is present. A low–velocity layer at the earth's surface, or more precisely, a strong velocity change close to the surface (like strongly weathered rocks or sediments overlying basement rocks), steepens the ray path and reduces the chances of observing shear waves, if only vertical components are available.

To a certain extent, this chapter is a review of the seismic refraction results obtained for the Black Forest. It takes into account results of several other investigations, derived from the 1984 Black Forest seismic refraction data (see Fuchs et al. 1987; Gajewski 1987; Gajewski and Prodehl 1987; Gajewski et al. 1987a, b, 1988a, b; Holbrook 1988; Holbrook et al. 1987, 1988; Krammer 1988; Lüschen et al. 1987; Lüschen et al. this Vol.). The author is indebted to the entire interdisciplinary Black Forest Working Group for many fruitful and encouraging discussions, which helped these investigation so much.

In this paper two–dimensional compressional and shear–wave velocity models will be presented, which were derived from the three component recordings obtained along the Black Forest profile. The unusually high signal–to–noise ratio and dense station spacing (about 2 km) of the data enabled the use of two–dimensional (2–D) ray–tracing (Červený and Pšenčik 1984) to model lateral variations in compressional (v_p) as well as in shear velocities (v_s). The derived v_p and v_s distributions were used to calculate the Poisson's ratio structure within the crust of the Black Forest. The results were compared with published P– and S–velocities and Poisson's ratios of crustal rocks to make inferences on the petrologic composition of the crust beneath the Black Forest.

2 PRESENTATION OF THE DATA

For details of the field experiment and the geolocigal and tectonic situation in the area of investigation the reader is referred to Gajewski and Prodehl (1987), Gajewski et al. (1988a), and other chapters of this Volume. The following section will first describe the data.

P– and S–wave data are presented with different reduction velocities: the P–wave data were reduced with 6.0 km/s and the S–wave data with 3.46 km/s, corresponding to a v_p/v_s–ratio of $\sqrt{3}$ or a Poisson's ratio of 0.25. The time scale for the S–wave record sections was compressed by a factor of $\sqrt{3}$ compared to the time scale of the P–wave record sections. This format allows easy comparison of P–wave and S–wave record sections: for a Poisson's ratio of 0.25 throughout the crust and upper mantle of the Black Forest, arrival–times of P– and S–waves would perfectly overlay. Deviations of the arrival–times thus indicate deviations from the crustal Poisson's ratio 0.25. The amplitudes of each trace of the record sections shown in the following sections are normalized, i.e., the amplitude of each trace is scaled to a fixed width. The differences in the frequency characteristics of P– and S–waves (Krammer 1988) allow the application of individual bandpass filters to the P–wave and S–wave record sections to optimize the signal–to–noise ratio.

For the presentation of S–waves, record sections constructed from transverse components were used, as they had a better signal–to–noise ratio than the radial components, which is probably an effect of the free surface (Gajewski et al. 1988b). During the Black Forest campaign, borehole explosions were used. Charges between 150 kg to 600 kg in one to three 50 m–deep drillholes were fired instantaneously.

The data for all shots show clear shear arrivals on all components (vertical, radial, i.e., in–line, transverse, i.e., cross–line, see Fig. 3 a–c), even for the single drillhole shots. The model of the explosive source (e.g., Aki and Richards 1980) cannot be applied to the explosions used. Strong arrivals are visible even for the P–waves on the transverse component (e.g., Fig. 3a, between –40 km and 40 km epicentral distances and 0 to –5 s reduced traveltime). Although the orientation of some seismometers may differ by a few degrees from the proper transverse direction, this observation is somewhat unexpected, as the amplitudes are strong compared to other arrivals. The wave propagation outside the profile plane is the most likely explanation, and might be influenced by three-dimensional velocity inhomogeneities or anisotropy. However, as no shear-wave splitting (Crampin 1981) is observed in the Black Forest (Gajewski 1987), three-dimensional effects are probably the preferred explanation.

The generation of S-waves from borehole explosions has been calculated by Heelan (1953) and was experimentally proved by White and Sengbush (1963) and Lash (1980). Fertig (1984) showed that S-wave energy can also be produced by P-wave surface conversions. Both explanations have in common that they do not predict a displacement vector with components in the transverse direction, which is, however, observed in the Black Forest data. Three-dimensional (3-D) effects, like strong topography of the earth surface at the source and/or the receiver (as along the Black Forest) or 3-D velocity variations, might be used to explain this feature. Normalized record sections covering both P- and S-waves show that the amplitudes of the P- and S-phases are about the same for all offsets, suggesting that P- and S-waves were generated with about the same strength.

After these introductory remarks on the Black Forest seismic refraction data, the P- and S-wave observations will be discussed in more detail.

3 P-WAVE OBSERVATIONS

The whole set of P-wave data recorded in the Black Forest is presented by Gajewski et al. (1988a) and discussed in detail by Gajewski and Prodehl (1987). We will concentrate here on the key observations, which mostly influence the resulting crustal P-velocity model. They can be best discussed with the recordings from shotpoints S3, S4 and S5 which are displayed in Fig. 2a, b and c.

The signal-to-noise ratio is mostly about 5 or greater for the recorders located in the Black Forest and is somewhat lower for the seismograms in the Swiss Molasse Basin. Viewing the data qualita-

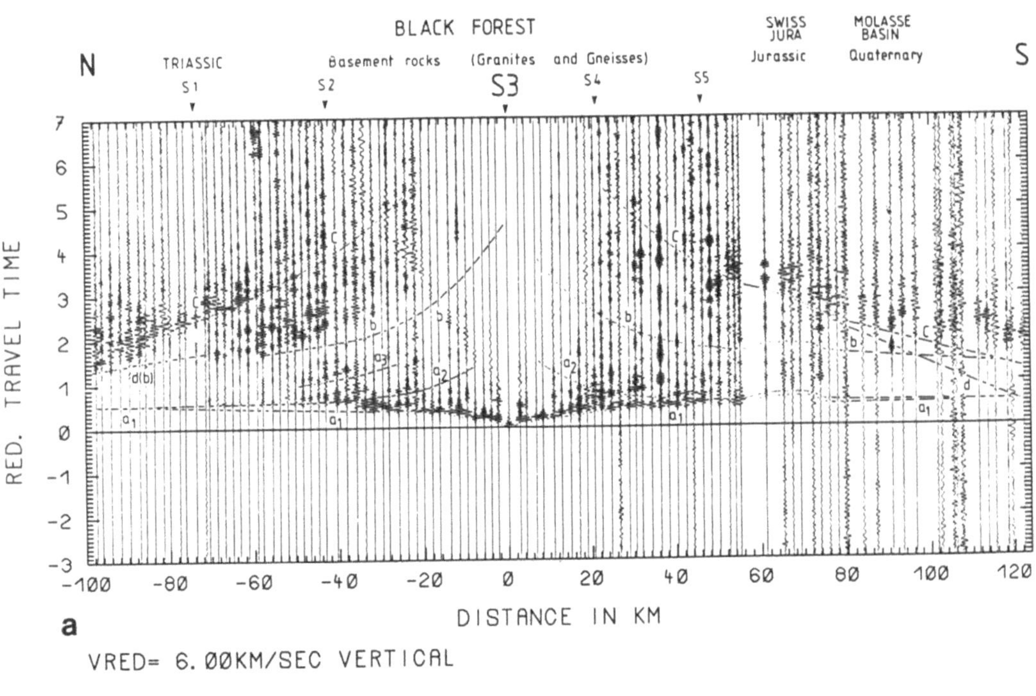

a

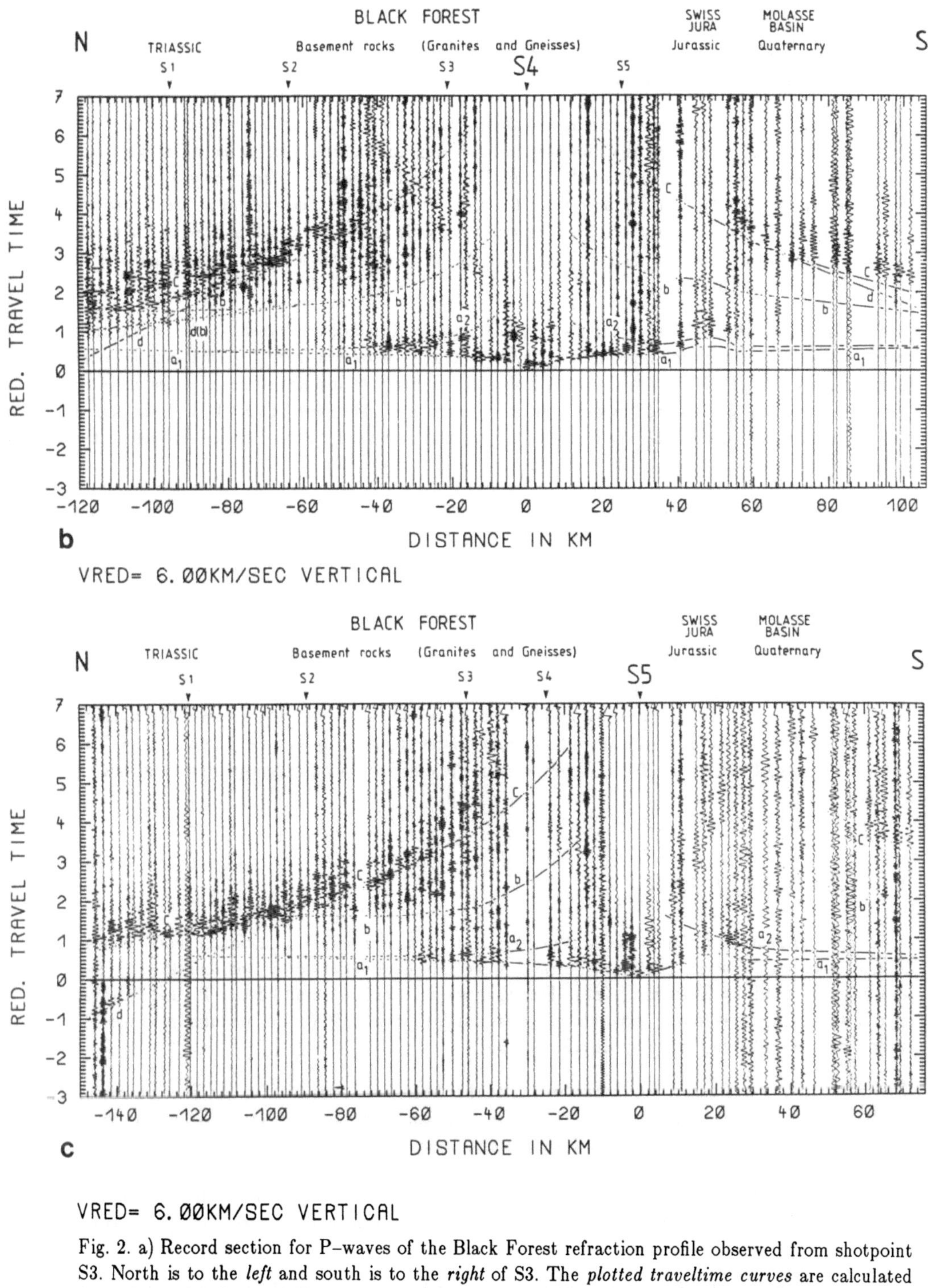

Fig. 2. a) Record section for P–waves of the Black Forest refraction profile observed from shotpoint S3. North is to the *left* and south is to the *right* of S3. The *plotted traveltime curves* are calculated from the model shown in Fig. 4a. Major geological units are indicated on the *top*. The positions of shotpoints are marked by *arrows*. b) Record section for P–waves of the Black Forest refraction profile observed from shotpoint S4. For further explanation see a. c) Record section for P–waves of the Black Forest refraction profile observed from shotpoint S5. For further explanation see a

tively, a strong lateral change of the crustal P-wave response moving from the Black Forest to the south into the Molasse Basin is visible. The phases labeled with a_1, a_2 and a_3 correspond to waves propagating in the upper crust. Phase a_1, also named P_g, occurs as a first arrival and shows a decrease in amplitude at distances greater than 40 km in the central Black Forest. The apparent velocity of this phase is about 6.0 km/s at 60 km.

A significant time gap between phase a_1 and the following major wave group labeled b, which will be discussed below, as well as the above-mentioned amplitude decrease of phase a_1, suggests the existence of a low-velocity zone beneath the Black Forest. Between 20–60 km, phase a_1 is followed by phase a_2, which is not clearly observed. It is interpreted as the reflection from the top of the low-velocity zone. Phase a_3 was observed for two shots only (S2 to the south and S3 to the north, shown in the left part of Fig. 2a), indicating the existence of a local reflector in the middle crust within the low-velocity zone.

For the response of the lower crust no "classical" phase correlation is possible. The whole time range between the traveltime branches of "phase" b and the reflection from the crust-mantle boundary (phase c or P_mP) is filled with oscillations of long, reverberating wavetrains. This is observed for seismograms in the central Black Forest only, as south of shotpoint S4 (marked on the top of the record sections) the reflectivity of the lower crust is decreased. "Phase" b corresponds to the reflection from the top of the lower crust (Conrad discontinuity).

The prograde traveltime branch of the refracted P-wave in the lower crust [labeled d(b)] is best visible for shotpoints S4 and S5 to the north (see Fig. 2b and c). The apparent velocity is about 6.6–6.7 km/s. For distances greater than 60 km phase c (or P_mP) is prominent in all record sections. It can be correlated clearly even to shorter distances, i.e., into the subcritical distance range, indicating an abrupt velocity change at the crust-mantle boundary (i.e., a first order discontinuity).

For the shots with the longest epicentral observation distances (shots S1 and S5, see Fig. 2c), a refracted wave from the upper mantle with apparent velocities of about 8.2 km/s is observed as a first arrival (Fig. 2c). Except phase a_3, all phases discussed in this section are observed for all shots in the Black Forest; however, the traveltime curves differ significantly for the different shots. The existence of lateral variations of the P-velocities in the crust is obvious from qualitatively viewing the data.

4 S-WAVE OBSERVATIONS

A complete presentation of all shear-wave observations (transverse and radial components) is given by Gajewski et al. (1988a). The Black Forest S-wave data were discussed by Holbrook et al. (1988), and in more detail by Krammer (1988). Discrepancies between the amplitudes of P- and S-waves in SW Germany were found earlier by Gajewski and Prodehl (1985) and are also reported for seismic data recorded in France by Hirn et al. (1987).

The most striking and surprising characteristics of the shear-wave sections (Fig. 3a, b) compared to the P-wave sections (Fig. 2a–c) is the lack of reverberations from the lower crust (Krammer 1988). This is best illustrated on shot S3. Although strong lower crustal reflections, enveloped by traveltime curves b and c, are present in the P-wave section (Fig. 2a) no corresponding S-wave reflections are visible (Fig. 3a). Phase b^s drawn in Fig. 3a, b and c is the theoretical traveltime determined from the derived S-wave model (see Fig. 4b, will be discussed below). The corresponding calculated amplitudes are negligible and agree with the observations.

This lack of lower-crustal S-wave reverberations cannot be attributed to the longer wavelength of S-waves relative to P-waves, as the frequency analysis of Krammer (1988) has shown. Both velocity and frequency for the S-waves are about a factor of $\sqrt{3}$ smaller than for P-waves, thus the resolution should be about the same for P–waves and S–waves. For shot S1 the spectra show a similar frequency content. The resolution of S–waves should be even higher than for P–waves here. An enhanced S–wave attenuation in the lower crust is also not a possible explanation, as strong S–wave reflections from the crust–mantle boundary (phase c^s) are observed. Nodal points in the radiation characteristics of the source can be also excluded. The S–wave reflection from the crust–mantle boundary (S_mS), which covers the same range of incidence angles as the possible lower crustal S–wave reflections, is clearly visible. The lack of the S–wave response from the lower crust is thus related to the physical properties there.

Another fundamental difference between the P– and S–wave data concerns the evidence for the mid–crustal low velocity zone (LVZ). Whereas for the P–wave data a large time gap is observed between traveltime curves a_1 and b (see Fig. 2a–c), it is almost possible to construct a closed traveltime curve system for the corresponding S–waves (phase a_1^s and b^s, best visible in Fig. 3c, because of large observation distances). This shows that there is no or only a small decrease of the S–wave velocity within the middle crust.

These observations suggest that the middle and lower crust is highly differentiated and inhomogeneous for P–waves, whereas for S–waves no significant velocity contrasts exist. Only two clear S–wave groups can be correlated: the refracted S–waves from the upper crust and the reflection from the crust–mantle boundary. Thus for shear waves the Black Forest crust seems to be homogeneous and undifferentiated.

A clear conversion from the crust–mantle boundary (phase cc^s, Fig. 3a) displays another extraordinary feature (Holbrook et al. 1988; Krammer 1988). This is probably the first time that a converted wave from the crust–mantle boundary with such strong amplitudes has been observed. Often it is much weaker in amplitude or even not visible at all (see, e.g., Jacob and Booth 1977; Gajewski 1981; Ruthardt 1986; Luetgert et al. 1987).

In the next section the modeling of the data will be described resulting in 2D–velocity distributions for P– (derived by Gajewski and Prodehl 1987) and S–waves (derived by Krammer 1988). They were combined by Holbrook et al. (1988), to calculate the Poisson's ratio structure of the Black Forest. The characteristics of the response of the lower crust will be discussed in detail.

5 P– AND S–WAVE VELOCITY DISTRIBUTIONS AND POISSON'S RATIO STRUCTURE

To model the observed P– and S–wavefields Gajewski and Prodehl (1987) and Krammer (1988) used the ray method (Červený et al. 1977). Traveltimes and amplitudes of P– and S–waves were calculated with the program package SEIS83 (Červený and Pšenčik 1984). It was assumed that the structural boundaries were the same for P– and S–waves. The interfaces found from the interpretation of the P–wave data were kept fixed when modeling the S–waves. It was further assumed that the source generates P– and S–waves with a homogeneous radiation pattern. Amplitude ratios between different phases for each trace were used to match qualitatively the observed amplitudes. To construct the models, a procedure described by Gajewski (1984) was applied. The velocity distributions were derived to explain the above mentioned key observations of P– and S–waves. The results will be discussed with special regard to the steep–angle reflection data recorded approximately along the same line (see Fig. 1 and Lüschen et al. 1987 as well as Lüschen et al. in this Vol.). Some of the conclusions would not be possible from the refraction data alone.

5.1 P– and S–Wave Velocity Models

The upper crust, except the near–surface structure, is laterally homogeneous for P– and S–waves (Fig. 4a and b). Moving down to the middle crust the main feature is a low–velocity zone for P–waves (Fig. 4a), which was postulated earlier by Mueller et al. (1969, 1973). The P–velocity

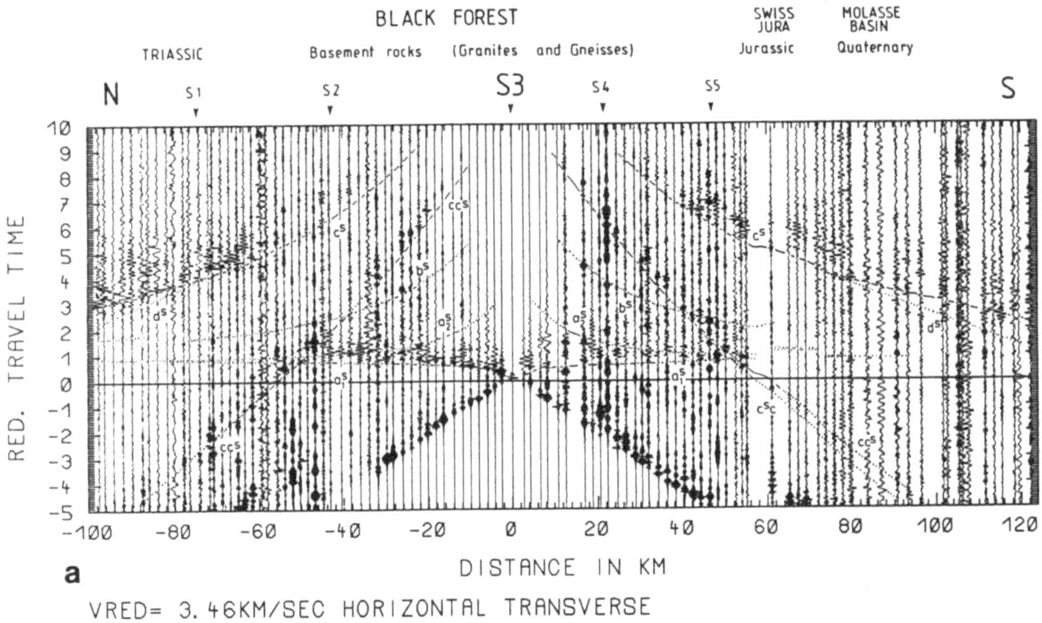

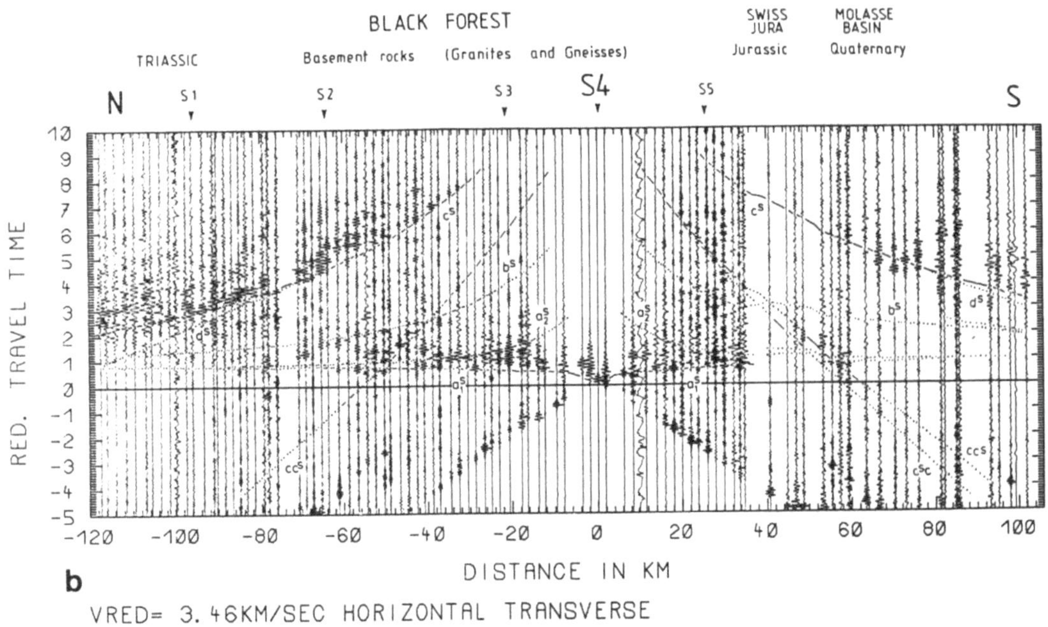

Fig. 3. a) Record section for S–waves of the Black Forest refraction profile observed from shotpoint S3. North is to the *left* and south is to the *right* of S3. The *plotted traveltime curves* are calculated from the model shown in Fig. 4b. Traveltime curves are *dashed* where ray theory predicts strong amplitudes, *dotted* where weak amplitudes are predicted. Major geological units are indicated on the *top*. The positions of shotpoints are marked by *arrows*. b) Record section for S–waves of the Black Forest refraction profile observed from shotpoint S4. For further explanation see a. c) Record section for S–waves of the Black Forest refraction profile observed from shotpoint S5. For further explanation see a

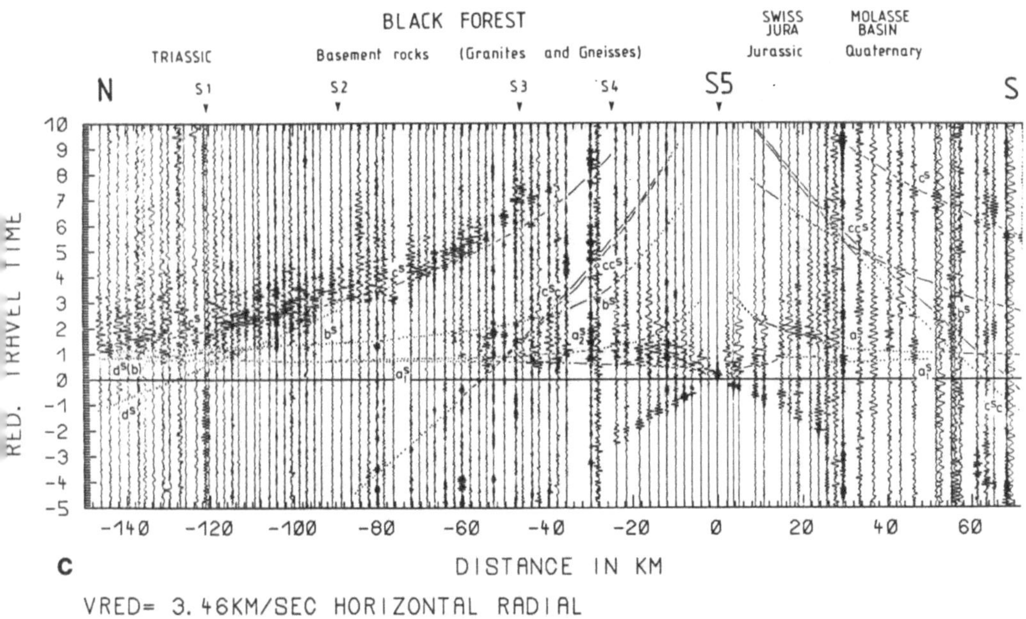

C

within this low–velocity zone decreases to 5.4 km/s in the northern and central Black Forest and increases gradually to the south, until a velocity inversion is no longer present in northern Switzerland.

Owing to the uncertainties related to phase a_2 (see Fig. 2a–c) and the difficulties in determining a critical distance for "phase" b, the evaluation of the P–velocity within the low–velocity zone is problematic. Raising the upper boundary of the low–velocity zone to a lower depth of 4 km would increase the P–velocity to 5.57 km/s. A low–velocity zone beginning at this level, however, is in contradiction with the observations of phase a_1, which is observed up to distances of 100–120 km (see Fig. 2b). The vertices of the corresponding rays are penetrating down to a depth of about 7 km (see Fig. 3b of Gajewski and Prodehl 1987). As the curvature of the rays and thus the deepest points of penetration of the rays are influenced by the magnitude of the v_p–gradient (which also influences the P_g–amplitudes giving the constraint for the intensity of the gradient), an additional convincing proof for the P–velocity within the low–velocity zone was obtained from an expanding–spread profile (Lüschen et al. 1987). Here the average velocity down to the top of the lower crust (the bottom of the low–velocity zone, see Fig. 4a) was determined to 5.62 km/s with error bounds of 2%. Since the average velocity for the upper crust down to the top of the low–velocity zone is 5.8 km/s (based on the model obtained for the P_g–phase), an average velocity of about 5.4 km/s results for the low–velocity zone.

The middle crust of the S–wave velocity model shows neither a decrease in v_s nor a clear distinction between the middle and lower crust (Fig. 4b). Although the Conrad discontinuity (as obtained from the P–wave data) is shown as a broken line in Fig. 4b, it does not appear as a change in the S–wave velocity. The Conrad discontinuity can be considered as a second–order boundary for S–waves, where only the first derivatives of the S–velocity change, in agreement with the missing shear–wave response from the lower crust (i.e., the small b^s amplitudes).

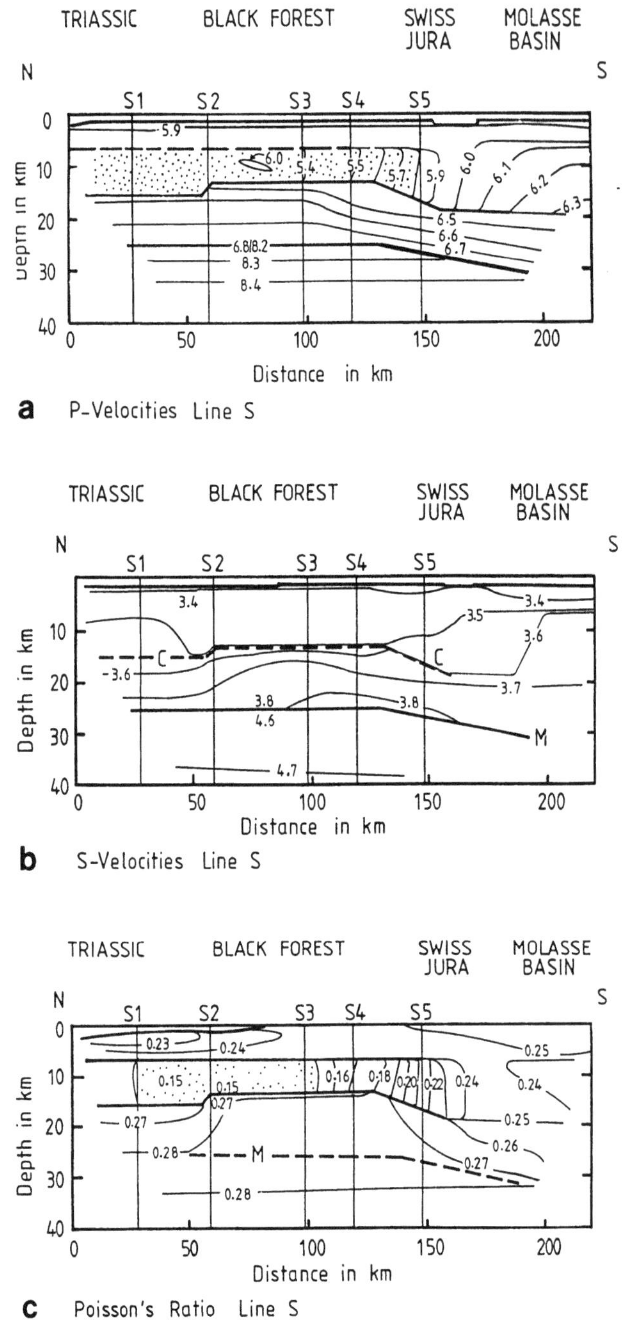

Fig. 4. a) P–wave velocity model for the Black Forest, isovelocity lines are contoured at 0.1 km/s interval. *Thick contour lines* correspond to velocity discontinuities. Velocities below 5.7 km/s are not shown. Low–velocity zones are *stippled*. Vertical exaggeration is 2:1. b) S–wave velocity model for the Black Forest, isovelocity lines are contoured at 0.1 km/s interval. *Thick contour lines* correspond to velocity discontinuities. Velocities below 3.3 km/s are not shown. The position of the Conrad discontinuity, as obtained from a, is indicated by a *dashed line*. Vertical exaggeration is 2:1. c) Poisson's ratio structure for the Black Forest, contoured at 0.01 interval. *Thick contour lines* correspond to discontinuous changes of the Poisson's ratio. The position of the Moho, as obtained from a, is indicated by a *dashed line*. Vertical exaggeration is 2:1

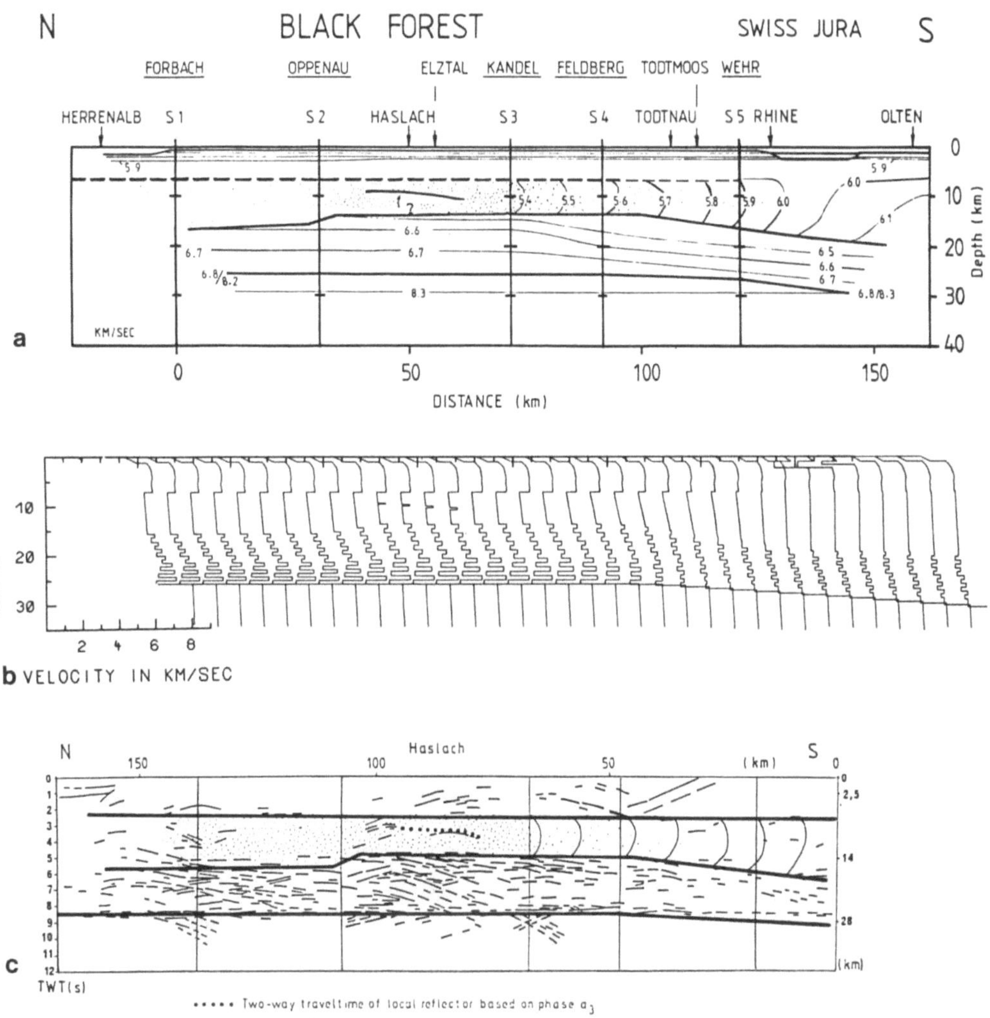

Fig. 5. a) P-wave velocity model similar to Fig. 4a, but without vertical exaggeration. b) Velocity depth sections along the Black Forest showing a model concept of a laminated and laterally varying lower crust. c) Line drawing of the reflection section along the Black Forest (simplified, after Lüschen et al., 1987). Two-way traveltimes as calculated for the model shown in a are plotted by *thick lines*. *Dotted line:* two-way traveltime of a local reflector based on phase a_3

The P-waves, on the other hand, show strong reflected energy from the lower crust. In the resulting P-wave model (Figs. 4a and 5a, showing the same model as Fig. 4a, but without exaggeration of the depth axis) the Conrad discontinuity is thus represented by a first-order change of the P-velocity. The structure for the lower crust shown in Fig. 5a presents average properties, as the model predicts a single reflected arrival only. But the data show a much more complicated pattern, as discussed above. Viewing the line drawing of the reflection data (Fig. 5c), the complicated structure of the lower crust is obvious. Numerous horizontal and dipping reflectors are present. Taking this observation into account, Sandmeier and Wenzel (1986) derived a velocity-depth function (v_p-model) for the data of shot S3 to explain not only single arrivals, but the whole wavefield. The result is a laminated lower crust containing a sequence of high and low P-velocity layers (Sandmeier and Wenzel 1986; Wenzel et al. 1987). The average thickness of the layers is about 120 m and the average velocity was determined to about 6.0 km/s at the top of the lower crust and 6.8 km/s close to the crust-mantle boundary.

Sandmeier and Wenzel (1986) used the reflectivity method (Fuchs and Müller 1971) to model the data; consequently, they assumed lateral homogeneity. This assumption may be valid in the northern Black Forest, but cannot be assumed for the southern part. Therefore Gajewski and Prodehl (1987) tried to combine both lateral variations and lamination within the lower crust. In order to minimize the input requirements and the computational efforts (using the ray mehod), they restricted the lower crust to 14 layers (resulting in an average thickness of about 700 m). It turns out that the lower crustal response is best modeled when the v_p contrast between the high- and low-velocity layers decreases to the south. For the resulting model shown in Fig. 5b, ray theoretical seismograms were calculated. Although the model of Fig. 5b can be considered only as a concept of the lower crust, the general amplitude characteristics calculated for shotpoint S4 match the observations well (compare Figs. 6 and 2b). A more detailed discussion and synthetic seismograms for all shots are given by Gajewski and Prodehl (1987).

As both P_mP and S_mS are the dominant phases in P-wave and S-wave record sections and can be observed to short distances from the shot, i.e., at the subcritical distance range, the transition from the crust to the upper mantle is being modeled as a first-order discontinuity in the v_p- and v_s-model. The assumption of a first-order discontinuity is additionally supported by the observation of a phase converted at the crust-mantle boundary. Synthetic seismograms of P-to-S conversions (Fuchs 1975) have shown that the amplitudes of this phase are sensitive to the character of the crust-mantle transition and that it results in strong reflections only, if a first-order discon-

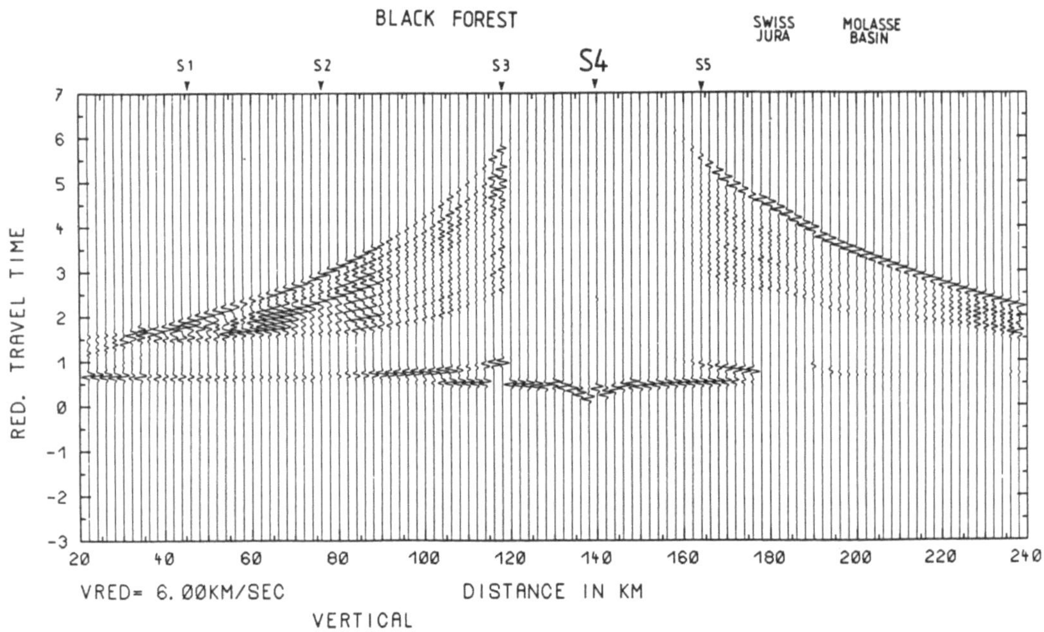

Fig. 6. Ray theoretical seismogram of P-waves for the model in Fig. 5b with a laminated laterally varying lower crust, calculated for shotpoint S4. Occasional gaps in some phases are due to numerical problems in the calculation of the synthetic seismograms

tinuity or a very thin transition zone (thickness less than the wavelength) is present. Along the Black Forest, as generally in SW Germany (Gajewski et al. 1987a), the Moho is a flat boundary at a shallow level (about 25.5 km), starting to descend at the southern end of the Black Forest underneath the Swiss Jura and the Molasse Basin towards the mountain root of the Alps.

The upper mantle velocity distribution of P– and S–waves is characterized by a gradient zone, where the gradient of the S–waves is weaker than the v_p–gradient. The phenomenon of a very strong P_n–phase (P_n/P_mP–ratios of about 1) combined with a weak and sometimes not visible S_n–phase is discussed in more detail by Gajewski et al. (1988b).

5.2 Poisson's Ratio Structure

The v_p– and v_s–velocity distributions by Gajewski and Prodehl (1987) and Krammer (1988) were used by Holbrook et al. (1988) to calculate the Poisson's ratio structure of the Black Forest. Although the results in Fig. 4c are already smoothed, one must take into account that the errors contained in the v_p– and v_s–velocity distributions are combined and squared here. Care is necessary to avoid over–interpreting the calculated Poisson's ratio structure, thus only the general features of Fig. 4c will be considered.

The upper crust shows a Poisson's ratio of 0.25 in the central and southern Black Forest, which is mostly considered as 'normal'. In the northern Black Forest, Poisson's ratio decreases to values of 0.23. A strong anomaly of the Poisson's ratio is observed in the middle crust. The strong v_p–velocity inversion with no corresponding v_s–velocity inversion results in a low Poisson's ratio of 0.15 (Fig. 4c), where a lateral change to the south is also observed. The low Poisson's ratio has been corroborated by X^2–T^2 analysis of the expanding spread data by Lüschen et al. (1987). Their Poisson's ratio of about 0.19 is somewhat higher than the 0.15 of our model. However, the shear–wave arrivals that Lüschen et al. (1987, see their Figs. 23 and 25) have interpreted to determine the Poisson's ratio within the v_p low–velocity zone are much less developed than the corresponding P–reflections. The correlation and determination of the traveltimes of these weak arrivals is uncertain, as the phase is hidden in scattered energy.

The lower crust and the upper mantle are characterized by Poisson's ratios of 0.27–0.28 in the central Black Forest. The Poisson's ratio for the lower crust reflects average properties only, as the P–velocity shows the lamination discussed above and the S–velocity is smooth. The Poisson's ratio is thus also laminated such that in the high v_p laminae a high Poisson's ratio is present, and in the low v_p laminae a low Poisson's ratio occurs. In the middle and lower crust, lateral changes of the Poisson's ratio are observed. In the middle crust the Poisson's ratio increases to the south, whereas the Poisson's ratio in the lower crust decreases in the same direction. The average Poisson's ratio for the whole crust does not change along the entire Black Forest and is about 0.25, which is in good agreement with the values derived by Lüschen et al. (1987). As v_p–, v_s– and Poisson's ratio distributions are now available, considerations regarding the composition of the crustal rocks in the Black Forest are possible and will be discussed in the next section.

6 PETROLOGICAL CONSIDERATIONS

A detailed quantitative investigation of possible compositions of the crustal rocks beneath the Black Forest based on the presented v_p– and v_s–distributions is given by Holbrook (1988) and Holbrook et al. (1988). The comparison of the velocities and Poisson's ratios is based on laboratory measured velocities, where the v_p– and v_s–values as a function of pressure and temperature of over 400 samples from about 50 published reports were used (for details see Holbrook 1988; Holbrook et al. 1988).

The upper crust is composed of granites or granitic gneisses with a quartz content between 15 and 40%. Due to the low Poisson's ratio in the mid-crustal v_p low-velocity zone, highly quartzitic rocks (like quartzite) are required. The quartz content decreases to the south corresponding to the increase of the Poisson's ratio. It is interesting to note that the obtained P- and S-velocity can be explained by pure compositional reasons. An alternative explanation for the derived v_p- and v_s-values is the presence of fluids at low pore pressure (Spencer and Nur 1976; Holbrook 1988). Fluids at this depth are also consistent with the high conductive layer near the base of the Black Forest v_p low-velocity zone (Berktold et al. 1985; Berktold this Vol.; Tezkan 1988).

For the upper and middle crust, the possible variation in compositions is small compared to the much broader range of possible rocks for the lower crust. The reasons are first the lateral velocity variations in the lower crust and, second, the considerable overlap between the v_p- and v_s-values of many lower crustal rocks (Holbrook 1988; Holbrook et al. 1988). Granulites of various compositions provided the best match to the observed P- and S-velocities of the lower crust. The lateral variations of the Poisson's ratio can be again explained by lateral changes of the quartz content. The velocities for the lower crust of the central Black Forest fit best with quartz-poor mafic rocks, like anorthosite, pyroxene granulite, and hornblende granulite. For the lower crust in the southern Black Forest and in northern Switzerland, the best match was obtained with quartz-rich rocks, like granodiorites, quartz diorites, and quartz monzonites. A compositional layering model, based on the lamination of the P-velocities and the Poisson's ratio in the lower crust, is given by Holbrook (1988). In this model the low v_p, low Poisson's ratio layers consist of quartz-rich rocks, while the high v_p, high Poisson's ratio lamellae have a more mafic composition.

As no hints on S-wave splitting were found neither for the S_g-phase nor for the S_mS-phase, crustal rocks can be considered as isotropic (or more precise: the anisotropy is less than the resolution of the data). In case of S-wave splitting a difference of the two S-wave velocities of about 2% for a transverse isotropic lower crust would result in a traveltime split of 0.5 s at 140 km epicentral distance for the two S-wave reflections from the crust-mantle boundary (Gajewski 1987). Such a split should be visible in the traveltime data. If anisotropy is present, it can only be less than 2% averaged along the whole ray path of the S_mS. This conclusion is valid for the intrinsic anisotropy as well as for the effective anisotropy (induced by oriented cracks and/or layering).

7 DISSCUSION

The seismic refraction and reflection data along the Black Forest have filled the gap between the detailed investigations performed earlier in the Rhinegraben (for a review see Prodehl et al. 1976) and along the Swabian Jura, located to the east of the Black Forest (Emter 1971; Jentsch et al. 1982; Prodehl et al. 1982; Bartelsen et al. 1982, Meissner et al. 1982; Deichmann and Ansorge 1983; Trappe 1983; Gajewski and Prodehl 1985; Walther et al. 1986). The derivation of v_p- and v_s-models allows conclusions regarding the causes of the v_p low-velocity zone, the lamination of the lower crust, and the crust-mantle transition zone.

7.1 Mid-Crustal P-Wave Low-Velocity Zone

The existence of a pronounced low-velocity zone beneath the Black Forest was already suggested in the late 1960's by Mueller et al. (1969, 1973). In the Rhinegraben proper, Edel et al. (1975) and Zucca (1984) did not model a v_p low-velocity zone within the middle crust. This is somewhat surprising, since the Black Forest (the eastern shoulder of the Rhinegraben) is located only 30 km to the east of the graben proper and thus strong lateral changes can be expected. This is also supported by reflection seismic data (Dohr 1970; Damotte et al. 1987; Lüschen et al. 1987; Fuchs et al. 1987), which show a remarkable difference in the lower crustal thickness of the Rhinegraben proper and the Black Forest. The duration of the lower crustal reverberations is about 1 s in the Rhinegraben proper and about 2 s beneath the Black Forest, indicating a thickness ratio of about 1:2 (Lüschen et al. 1987). Different thicknesses of upper and middle crust for both regions can be expected.

The shear-wave data offer an important information on the possible origin of the Black Forest v_p low-velocity zone, as the Poisson's ratio could be determined for this region. Since the temperature in the middle crust is too low for partial melts (Stiefel and Wilhelm 1986; Wilhelm et al. this Vol.), composition and fluids at low pore pressure might be considered as possible causes (Holbrook et al. 1988). As it is unlikely to have a quarzite in the middle crust to match the low observed Poisson's ratio, pore fluids at low pore pressure are probably the best explanation, especially taking into account the high electric conductivity in this region (Berktold this Vol.; Tezkan 1988). A low Poisson's ratio can be also achieved with dry cracks (O'Connel and Budiansky 1977). The formation of the Rhinegraben could produce the mechanism for both the above explanations: the uplift of the Black Forest during the Rhinegraben rifting combined with the erosion is responsible for a decompression, which results in an opening of new cracks as well as in a growing of the existing cracks. The increased temperatures at higher crustal levels combined with decreased pressure can support the increase of the crack density because of thermal crack growth.

Comparing the results of the refraction survey with a line drawing of the reflection data (Fig. 5a, b) two interesting aspects are obvious. First, the v_p low-velocity zone is connected with an area of low reflectivity and second, the lower crust is characterized by many reflecting elements. The low-velocity zone is located directly on the top of the lower crust, which seems to be exceptional compared to other combined seismic refraction-reflection experiments (for a review see Mooney and Brocher 1987).

7.2 The Lower Crust

As in many other areas (Mooney and Brocher 1987) the refraction Moho of the Black Forest coincides with the lower boundary of the reflective zone in the reflection data. On most reflection profiles in northern Switzerland and SW Germany, the Variscan crust is characterized by a reflective lower crust between a transparent middle crust and upper mantle (Dohr and Fuchs 1967; Bartelsen et al.; 1982; Meissner et al. 1982; Finckh et al. 1984, 1986; Bortfeld et al. 1985; Meissner and Wever 1986; Lüschen et al. 1987, see also Lüschen et al. this Vol.). Similar observations are also known from the BIRPS data collected around Britain (e.g., McGeary and Warner 1985; Matthews and Cheadle 1986), the ECORS project in France (Bois et al. 1986) and from the Basin and Range province of western North America (Allmendinger et al. 1986). The lamination of the lower crust thus cannot be caused by the graben formation alone.

Already in the 1960's, initial ideas on a lamination of the crust were proposed by Meissner (1967, 1973) and Mueller et al. (1969) and synthetic seismograms showing the effects of thin layers with alternating high and low velocities were calculated (Fuchs 1970). Precursors of the P_mP observed on a profile along the eastern border of the Black Forest led Deichmann and Ansorge (1983) to model the lower crust as a series of low- and high-velocity layers. Owing to the recently collected high-quality reflection and refraction data these ideas and observations were convincingly confirmed.

The P-velocities (about 6.8 km/s) and the Poisson's ratio (about 0.28) in the lower crust of the central Black Forest point to a mafic composition of the corresponding rocks. As it was mentioned earlier there exists also a lamination of the Poisson's ratio in the lower crust. The high-v_p laminae (with high Poisson's ratio) consist of mafic rocks while the low-v_p laminae (with low Poisson's ratio) show a higher quartz content. This correlation led Holbrook (1988) to interpret the lower-crustal reflections as mafic intrusions into an originally more homogeneous, less mafic lower crust. He suggests underplating and intrusion during the Variscan orogeny (in the way proposed by Furlong and Fountain 1986) as a likely scenario for the genesis of the laminated lower crust.

7.3 Lateral Variations and Crustal Types

Lateral variations were found in the seismic P– and S–wave data. From the interpretation of five seismic refraction profiles, and taking into account the results of reflection surveys, Gajewski et al. (1987a) defined two characteristic crustal types for SW Germany. Type I consists of a mid-crustal v_p low-velocity zone (v_p=5.4–5.8 km/s) and a thick (>10 km), high average velocity (v_p=6.6–6.8 km/s), high reflective lower crust. Type II has no prominent mid-crustal v_p-inversion, and a thin (< 10 km), low average velocity (v_p=6.3–6.5 km/s) lower crust with a decreased reflectivity compared to Type I. The central Black Forest can be considered as Type I, whereas the southernmost Black Forest and northern Switzerland match with the features of Type II. Although the v_p-average velocity of the lower crust does not change much along the Black Forest (Fig. 5a), the disappearance of the v_p-inversion and the decrease of the reflectivity of the lower crust (Fig. 5c) in the southern Black Forest and northern Switzerland suggest the Type II crust. Using the S-wave results of (Holbrook et al. 1987, 1988) and the model drawn in Fig. 4b, the crustal types can be characterized by an additional physical parameter: the Poisson's ratio. The Type I lower crust has an increased Poisson's ratio (values of about 0.27–0.28), whereas the Type II lower crust has Poisson's ratios of 0.22–0.25, suggesting a variation of the quartz content of the corresponding rocks.

Several other geophysical data show a similar lateral change along the Black Forest (see for instance the compilation of this data in Lüschen et al. 1987 or the corresponding papers in this Vol.). The heat flow is higher in the central Black Forest than in the southern part. The distribution of hypocenters of earthquakes shows an increased seismicity and a greater depth of earthquakes (down to 20 km) in the southern Black Forest (the northern part is almost aseismic). The Bouguer gravity is characterized by two anomalies located in the northern and southern Black Forest. This lateral change of geophysical data may reflect the existence of distinct blocks in the Variscan crust, which is supported by the existence of fault zones at the boundaries where the lateral variations of the geophysical data occur. In the southern Black Forest the Badenweiler–Lenzkirch fault zone (see Lüschen et al. 1987) separates the Type I crust from the Type II crust. It can be assumed that there is an other transition from Type I to Type II in the northern Black Forest related to the Baden–Baden suture zone (see Lüschen et al. 1987). A decrease of the thickness of the lower crust as well as a decreased reflectivity (Fig. 5c) in the northernmost Black Forest can be deduced from the existing data; however, conclusions on the continuation of the Black Forest v_p low-velocity zone to the north are not possible due to the lack of data.

7.4 The Crust-Mantle Boundary

The observation of a converted wave from the crust–mantle boundary is important for conclusions related to the the crust–mantle transition. Since the amplitudes of the corresponding phase (cc^s or c^sc see Fig. 3a–c) differ for the different shots, the sharpness of the crust–mantle transition varies laterally. At least locally the crust–mantle boundary is a first order discontinuity beneath the Black Forest, or more precise, the thickness of the transition zone is less than a wavelength (less than 600 m). For the Rhinegraben proper broader transition zones were derived. Edel et al. (1975) obtained a transition zone of 4 km thickness, and Zucca (1984), reinterpreting the data of Edel et al., derived a thickness of 1.5 km for the transition zone. Although opinions differ for the thickness of the transition, it is obvious from a qualitative comparison of amplitudes and critical distances of the P_mP-reflections recorded in the Black Forest (Fig. 2a–c) and in the Rhinegraben proper (see, e.g., Edel et al. 1975) that the crust–mantle transition varies for both locations. The depth of the Moho, however, is about the same for the shoulder and the graben proper and of about the same magnitude in whole SW Germany (Gajewski et al. 1987a). No mantle arch, as contoured by Edel et al. (1975), was found in the southern Black Forest.

It is interesting to note that the conversions are even visible on the transverse components. This is not expected from a 2-D structure and a flat crust–mantle boundary. If this observation is not influenced by the topography of the free surface along the Black Forest (differences of several hundreds of meters are present), 3–D inhomogeneities along the ray path have to be considered. This may result in uncertainties for the velocity distributions shown, as they were derived under the assumption of 2–D velocity variations.

Owing to the restricted recording distances, mantle phases were observed only for shotpoints S1 and S5 (see Fig. 1 and Fig. 2c and 3c). A strong P_n–phase was observed, whereas the corresponding S_n–phase is hardly visible. A stronger v_p–gradient and a weaker v_s–gradient was derived leading to an increase of the Poisson's ratio with depth. A similar behavior of P– and S–waves was found elsewhere in southwest Germany (Gajewski and Prodehl 1985; Holbrook et al. 1987, 1988), France, northeast America and Scandinavia (Gajewski et al. 1988b) and is interpreted as a depletion of the upper mantle rocks in basalts.

8 CONCLUSIONS

The striking contrasts between the P– and S–wave fields of the middle and lower crust along the Black Forest are surprising observations with important constraints on interpretations of the petrology and the physical properties there. The three–component data presented in this paper prove the advantage of combining v_p and v_s information in seismic refraction studies. The observations suggest that the mid–crust consists of cracked rocks with either dry cracks or fluid filled cracks at low pore pressure or a combination of both. The laminated lower crust consists of layers with more mafic and more quartzitic composition supporting an underplating model for the lower–crustal laminations. Strong lateral variations in the middle and lower crust can be expected on a laterally small scale (about 30 km), when moving from the Rhinegraben proper to the eastern shoulder: the Black Forest. Lateral variations of several geophysical data suggest the existence of different crustal blocks within the Variscan crust. The block boundaries, where the lateral changes occur, coincide with major fault zones.

Another significant point for the success of this investigation was the combination of seismic reflection and refraction data along the same line. The joint application of both methods results in considerably more detailed information than the use of only one method on its own. The existence of a v_p low–velocity zone is apparent as an almost transparent region in the reflection data and the fine structure of the lower crust, as seen in the reflection data, is represented by long reverberating wavetrains in the refraction data.

To study whether the presented results are general or more exceptional, the performance of combined P–wave, S–wave three–component seismic refraction and reflection surveys in different crustal regions is suggested. Regarding the interpretation of shear–wave amplitudes a controlled *shear– wave* seismology is recommended. The directivity of the source as well as the orientation of the displacement vector at the source influence the data. The comparison of observed and theoretical seismograms can be only properly achieved, if the source parameters are known.

Acknowledgments. The investigation was sponsored by the German Research Society (DFG) within the Continental Deep Drilling Project (KTB) of the Federal Republic of Germany. The interpretation work was additionally supported by the special research program, "Stress and Stress Release in the Lithosphere", of the University of Karlsruhe, also sponsored by the DFG. This paper would not have been possible without the continuous help of C. Prodehl, W. S. Holbrook, and A. Krammer. We are grateful to the Black Forest Working Group for encouraging discussions, especially to G. Eisbacher, K. Fuchs, E. Lüschen, F. Wenzel, K.-J. Sandmeier, H. Wilhelm and W. Wimmenauer. Special thanks are to the entire field crew of the experiment. Their engagement and effort was the basis for the high quality of the data which contributed so much to the success of this research. The following institutions participated within Germany: Universities of Berlin, Clausthal, Frankfurt, Göttingen, Karlsruhe, Kiel, München, Münster, Paris and Stuttgart and the Niedersächsisches

Landesamt für Bodenforschung, Hannover. The Eidgenössische Technische Hochschule, Zürich extended the measurements into Switzerland. The shots were carried out under the supervision of Dr. J. Wieck, Neckartailfingen. For the organization of the shots, the help of Mr. C. Behnke of Niedersächsisches Landesamt für Bodenforschung, and of Messrs. Kleinmann and Dudenhöfer of Landesbergamt Baden-Würtemberg in Freiburg are gratefully acknowledged. Without the support of W. Kaminski, U. Kästner, St. Zeiss and H. Fischer the digital data preparation and processing on the Raytheon 500 computer of the Geophysical Institute of Karlsruhe would not have been finished. Calculations were performed on the SIEMENS 7881 at the Computer Center of the University of Karlsruhe. Constructive reviews by W.S. Holbrook and C. Prodehl improved the content and readability of this manuscript.

REFERENCES

Aki K, Richards PG (1980) Quantitative seismology, theory and methods, Freeman, San Francisco

Allmendinger RW, Farmer H, Hauser E, Sharp J, Von Tisch D, Oliver J, Kaufman S (1986) Phanerozoic tectonics of the Basin and Range-Colorado Plateau transition from COCORP data and geological data. In: Barazangi M, Brown L (eds) Reflection seismology: A global perspective. Geodynamics Ser. Vol. 13, AGU, Washington, D.C., p 257

Assumpção M, Bamford D (1978) LISPB - V. Studies of crustal shear waves. Geophys. J. R. Astron. Soc. 54: 61-73

Banda E, Surinach E, Aparicio A, Sierra J, Ruiz de la Parte E (1981) Crust and upper mantle structure of the central Iberian Meseta (Spain). Geophys. J. R. Astron. Soc. 67: 779-789

Bartelsen H, Lüschen E, Krey Th, Meissner R, Schmoll H, Walter Ch (1982) The combined reflection-refraction investigation of the Urach geothermal anomaly. In: Haenel R (ed) The Urach geothermal project. Schweizerbart, Stuttgart, p 247

Berktold A, Musmann G, Tezkan B, Wohlenberg J (1985) Electrical conductivity studies, Schwarzwald Working Group (abs). Second International Symposium on Observation of the Continental Crust through Drilling, Alfred-Wegener-Stiftung, Bonn, 70 p

Bois C, Cazes M, Damotte B, Galdéano A, Hirn A, Mascle A, Matte P, Raoult JF, Torreilles G (1986) Deep seismic profiling in the crust of northern France: the ECORS project. In: Barazangi M, Brown L (eds) Reflection seismology: A global perspective. Geodynamics Ser. Vol. 13, AGU, Washington, D.C., p 21

Bortfeld RK, Gowin J, Stiller M, Baier B, Behr HJ, Heinrichs T, Dürbaum HJ, Hahn A, Reichert Ch, Schmoll J, Dohr G, Meissner R, Bittner R, Milkereit B, Gebrande H (1985) First results and preliminary interpretation of deep reflection seismic recordings along profile DEKORP 2-South. J. Geophys. 57: 137-163

Braile LW, Smith RB, Keller GR, Welch RM, Meyer RP (1974) Crustal structure across the Wasatch Front from detailed seismic refraction studies. J. Geophys. Res. 79: 2669-2677

Brooks SG (1985) Seismic velocities from crustal sections in northern Scandinavia, Ph.D. thesis, University of East Anglia.

Červený V, Molotkov IA, Pšenčik I (1977) Ray method in seismology, Univ. of Karlova, Prague

Červený V, Pšenčik I (1984) SEIS83- Numerical modelling of seismic wave fields in 2-D laterally varying layered structures by the ray method. In: Engdahl ER (ed) Documentation of earthquake algorithms, Rep. SE-35, World Data Center (A) for Solid Earth Geophysics, Boulder, Colorado, p 36

Crampin S (1981) A review of wave motion in anisotropic cracked elastic media. Wave Motion 3: 343-391

Damotte B, Fuchs K, Lüschen E, Wenzel F, Schlich R, Torreilles G (1987) Wide angle Vibroseis test across the Rhinegraben. Geophys. J. R. Astron. Soc. 89: 313-318

Deichmann N (1984) Combined travel-time and amplitude interpretation of two seismic refraction studies in Europe, Ph.D. thesis, Swiss Federal Institute of Technology, Zürich

Deichmann N, Ansorge J (1983) Evidence for lamination in the lower continental crust beneath the Black Forest (Southwestern Germany). J. Geophys. 52: 109-118

Dohr G (1970) Seismic reflection measurments in the Upper Rhinegraben with digital recording technique and processing (in German). In: Illies JH and Mueller St (eds) Graben problems. Schweizerbart, Stuttgart, p 207

Dohr G, Fuchs K (1967) Statistical evaluation of deep crustal reflections in Germany. Geophysics 32: 951-967

Edel JB, Fuchs K, Gelbke C, Prodehl C (1975) Deep structure of the Southern Rhinegraben area from seismic refraction investigations. J. Geophys. 41: 333-356

Egloff R (1979) Explosion seismic investigations of the earth's crust in Switzerland, Ph.D. thesis, Swiss Federal Institute of Technology, Zürich

El-Isa Z, Mechie J, Prodehl C (1987) Shear velocity structure of Jordan from explosion seismic data. Geophys. J. R. Astron. Soc. 90: 265-281

Emter D (1971) Results of seismic investigations of the earth's crust and upper mantle in Southwest Germany (in German), Ph.D. thesis, University of Stuttgart, Stuttgart

Fertig J (1984) Shear waves by an explosive point-source: the earth surface as a generator of converted P-S waves. Geophysical Prospecting 32: 1-17

Finckh P, Ansorge J, Mueller St, Sprecher Ch (1984) Deep crustal reflections from a Vibroseis survey in northern Switzerland. Tectonophysics 109: 1-14

Finckh P, Frei W, Fuller B, Johnson R, Mueller St, Smithson S, Sprecher Ch (1986) Detailed crustal structure from a seismic reflection survey in northern Switzerland. In: Barazangi M, Brown L (eds) Reflection seismology: A global perspective. Geodynamics Ser. Vol. 13, AGU, Washington, D.C., p 43

Fuchs K (1970) On the determination of velocity-depth distributions of elastic waves from the dynamic characteristics of the reflected wave field. Z. Geophys. 36: 531-548

Fuchs K (1975) Synthetic seismograms of PS-reflections from transition zones computed with the reflectivity method. J. Geophys. 41: 445-462

Fuchs K, Müller G (1971) Computation of synthetic seismograms with the reflectivity method and comparison with observations. Geophys. J. R. Astron. Soc. 23: 417-423

Fuchs K, Bonjer K-P, Gajewski D, Lüschen E, Prodehl C, Sandmeier K-J, Wenzel F, Wilhelm H (1987) Crustal evolution of the Rhinegraben area: I. Exploring the lower crust in the Rhinegraben rift by unified geophysical experiments. Tectonophysics 141: 261-275

Furlong KP, Fountain DM (1986) Continental crustal underplating: thermal considerations and seismic-petrologic consequences. J. Geophys. Res. 91: 8285-8294

Gajewski D (1981) Shear and converted waves in the interpretation of deep seismic measurements in the Kalahari craton, Namibia (in German), Diploma thesis, University of Clausthal, Clausthal-Zellerfeld

Gajewski D (1984) Interpretation of synthetic record sections for a 2-D laterally inhomogeneous structure. In: Ansorge J and Finlayson DM (eds) Workshop proceedings: Interpretation of synthetic record sections for a 2-D laterally inhomogeneous structure. Rep. 258, Bur. Mineral. Resour. Aust., Canberra, p 139

Gajewski D (1987) Ray-theoretical seismograms for arbitrarily heterogeneous, anisotropic media, with applications to the South German lithosphere (in German), Ph.D. thesis, Univ. of Karlsruhe, Karlsruhe

Gajewski D, Prodehl C (1985) Crustal structure beneath the Swabian Jura, SW Germany, from seismic refraction investigations. J.Geophys. 56: 69-80

Gajewski D, Prodehl C (1987) Seismic refraction investigation of the Black Forest. Tectonophysics 142: 27-48

Gajewski D, Holbrook WS, Prodehl C (1987a) A three-dimensional crustal model of Southwest Germany derived from seismic refraction data. Tectonophysics 142: 49-70

Gajewski D, Holbrook WS, Prodehl C (1987b) Combined seismic reflection and refraction profiling in southwest Germany - detailed velocity mapping by the refraction survey, Geophys. J. R. Astron. Soc. 89: 333-338

Gajewski D, Prodehl C, Ritter J, Feddersen J (1988a) A compilation of data from the 1984 seismic-refraction experiment in SW-Germany, Open File Report 88-1, Geophysical Institute, University of Karlsruhe, Karlsruhe

Gajewski D, Fuchs K, Sandmeier K-J, Stangl R (1988b) Constraints on composition of the topmost continental mantle from widespread observations of weak Sn amplitudes. submitted to Geophys. J. R. Astron. Soc.

Hall J, Ali M (1985) Shear waves in a seismic survey of Lewisian basement: an extra control on lithological variation and porosity. J. Geol. Soc. London 142: 677-688

Heelan PA (1953) Radiation from a cylindrical source of finite length. Geophysics 18: 685-696

Hirn A, Damotte B, Torreilles G, ECORS scientific party (1987) Crustal reflection seismics: the contribution of oblique, low frequency and shear-wave illuminations. Geophys. J. R. Astron. Soc. 89: 287-296

Holbrook WS, Gajewski D, Prodehl C (1987) Shear-wave velocity and Poisson's ratio structure of the upper lithosphere in Southwest Germany. Geophys. Res. Lett. 14: 231-234

Holbrook WS, Gajewski D, Krammer A, Prodehl C (in press, 1988) An interpretation of wide-angle compressional and shear wave data in Southwest Germany: Poisson's ratio and petrological implications. J. Geophys. Res.

Holbrook WS (in press 1988) A petrological model of the laminated lower crust in southwest Germany based on wide-angle P- and S-wave seismic data. AGU Geophysical Monograph of the IUGG 1987, Volume on the lower crust

Jacob AWB, Booth DC (1977) Observation of PS reflections from the Moho. J. Geophys. 43: 687-692

Jentsch M, Bamford D, Emter D, Prodehl C (1982) A seismic refraction investigation of the basement structure in the Urach geothermal anomaly, southern Germany. In: Haenel R (ed) The Urach geothermal project. Schweizerbart, Stuttgart, p 247

Keller GR, Smith RB, Braile LW (1975) Crustal structure along the Great Basin - Colorado Plateau transition from seismic refraction studies. J. Geophys. Res. 80: 1093-1098

Krammer A (1988) Evaluation and interpretation of seismic shear- and converted-waves for a refraction profile in the Black Forest (in German), Diploma thesis, University of Karlsruhe, Karlsruhe

Kullinger B, Lund C-E (1986) A preliminary interpretation of S-wave traveltimes from Fennolora data. Tectonophysics 126: 375-388

Lash CC (1980) Shear waves, multiple reflections, and converted waves found by a deep vertical wave test (vertical seismic profiling). Geophysics 45: 1373-1411

Lüschen E., Wenzel F, Sandmeier K-J, Menges D, Rühl Th, Stiller M, Janoth W, Keller F, Söllner W, Thomas R, Krohe A, Stenger R, Fuchs K, Wilhelm H, Eisbacher G (1987) Near-vertical and wide-angle seismic surveys in the Black Forest. J. Geophys. 62: 1-30

Luetgert JH, Mann CE, Klemperer SL (1987) Wide-angle deep crustal reflections in the northern Appalachians, Geophys. J. R. Astron. Soc. 89: 183-188

Luosto O, Korhonen H (1986) Crustal structure of the Baltic Shield based on off-Fennolora refraction data. Tectonophysics 128: 183-208

Matthews DH, Cheadle MJ (1986) Deep reflections from the Caledonides and Variscides west of Britain and comparison with the Himalayas. In: Barazangi M, Brown L (eds) Reflection seismology: A global perspective. Geodynamics Ser. Vol. 13, AGU, Washington, D.C., p 5

McGeary S, Warner MR (1985) Seismic profiling of the continental lithosphere. Nature 317: 795-797

Meissner R (1967) On the structure of the earth's crust (in German). Gerlands Beitr. Geophys. 76: 211-254

Meissner R (1973) The 'Moho' as a transition zone. Geophys. Surveys 1: 195-216

Meissner R, Bartelsen H, Krey Th, Schmoll J (1982) Detecting velocity anomalies in the region of the Urach geothermal anomaly by means of new seismic field arrangements. In: Čermák V, Haenel R (eds) Geothermics and geothermal energy. Schweizerbart, Stuttgart, p 285

Meissner R, Wever T (1986) Nature and development of the crust according to deep reflection data from the German Variscides. In: Barazangi M, Brown L (eds) Reflection seismology: The continental crust. Geodynamics Ser. Vol. 13, AGU, Washington, D.C., p 31

Min Z, Wu FT (1987) Nature of the upper crust beneath Central Tibet. Earth Planet. Sci. Lett. 84: 204-210

Mooney WD, Brocher TM (1987) Coincident seismic reflection/refraction studies of the continental lithosphere: a global review. Rev. of Geoph. 25: 723-742

Mueller St, Peterschmitt E, Fuchs K, Ansorge J (1969) Crustal structure beneath the Rhinegraben from seismic refraction and reflection measurements. Tectonophysics 8: 529-542

Mueller St, Peterschmitt E, Fuchs K, Emter D, Ansorge J (1973) Crustal structure of the Rhinegraben area. Tectonophysics 20: 381-391

O'Connel RJ, Budiansky B (1977) Viscoelastic properties of fluid-saturated cracked solids. J. Geophys. Res. 82: 5719-5735

Prodehl C, Ansorge J, Edel JB, Emter D, Fuchs K, Mueller St, Peterschmitt E (1976) Explosion seismology research of the central and southern Rhinegraben - a case history. In: Giese P, Prodehl C, Stein A (eds) Explosion seismology in central Europe - data and results. Springer, Berlin, Heidelberg, New York, p 313

Prodehl C, Emter D, Jentsch M (1982) Seismic refraction studies of the geothermal area of Urach, southwest Germany. In: Haenel R (ed) The Urach geothermal project. Schweizerbart, Stuttgart, p 277

Ruthardt A (1986) Interpretation of deep seismic measurements in the Kalahari craton, Namibia, with special consideration of shear and converted waves (in German), Diploma thesis, University of Karlsruhe, Karlsruhe

Sandmeier K-J, Wenzel F (1986) Synthetic seismograms for a complex crustal model. Geophys. Res. Lett. 13: 22-25

Sierro N, Bindschaedler A, Ansorge J, Mueller St (1983) A geophysical investigation of northern Switzerland: Regional seismic-refraction measurments (in German). Tech. Rep. 21-83, NAGRA (National association for the deposit of nuclear waste), Baden

Spencer JW, Nur A (1976) The effect of pressure, temperature and pore water on velocities in Westerly granite. J. Geophys. Res. 81: 899-904

Stiefel A, Wilhelm H (1986) Geothermal investigations in the Black Forest (in German). Berichtsband 1984-1986 des Sonderforschungsbereichs 108 der Uni. Karlsruhe: Spannung und Spannungsumwandlung in der Lithosphäre, p 213

Tarkov AP, Basula IP, Generalov VP, Dubyansky AI, Chernykh VV (1981) Composite travel times of seismic waves and general velocity models of the Voronezh Shield crust and upper mantle. Geophys. J. R. Astron. Soc. 67: 137-143

Tezkan B (1988) Electromagnetic sounding experiments in the Schwarzwald central gneiss massif. J.Geophys 62: 109-118

Trappe H (1983) An interpretation of wide-angle reflections along the Urach profile and a correlation to the results of the steep-angle seismic reflection experiment (in German), Diploma thesis, University of Kiel, Kiel

Walther Ch, Trappe H, Meissner R (1986) The detailed velocity structure of the Urach geothermal anomaly. J. Geophys. 59: 1-10

Wenzel F, Sandmeier K-J, Wälde W (1987) Properties of the lower crust from modeling refraction and reflection data. J. Geophys. Res. 92: 11575-11583

White JE, Sengbush RL (1963) Shear waves from explosive sources. Geophysics 28: 1001-1019

Zschau J, Koschyk K (1976) Results of a combined evaluation of longitudinal and transverse waves on a seismic profile along the northern margin of the Alps. In: Giese P, Prodehl C, Stein A (eds) Explosion seismology in central Europe. Springer, Berlin, Heidelberg, New York, p 332

Zucca JJ (1984) The crustal structure of the southern Rhinegraben from re-interpretation of seismic refraction data. J. Geophys. 55: 13-22

Electrical Resistivity Studies at the KTB Location Schwarzwald

Alfred Berktold

Institut für Allgemeine und Angewandte Geophysik der Universität München, Theresienstr. 41/IV, 8000 München 2, FRG

Pre-site studies in the Black Forest for the Continental Deep Drilling Program using electromagnetic methods were performed by the following working groups:

TU Aachen:	Lehr- und Forschungsgebiet für Angewandte Geophysik, J. Wohlenberg, G. Voigt, C. Wendel
TU Braunschweig:	Institut für Geophysik und Meteorologie, G. Musmann, Ch. Drews, H.-M. Maurer
U Köln:	Institut für Geophysik und Meteorologie, K.-M. Strack, H.N. Eilenz, T. Hanstein, C.H. Kalle, A.W. Kötz
U München:	Institut für Allgemeine und Angewandte Geophysik, A. Berktold, H. Dittus, U. Teufel
U Göttingen:	Institut für Geophysik, U. Schmucker, B. Tezkan

This report is given by the speaker of the five working groups participating in the Black Forest program.

(See also in this Volume the chapter of V. Haak: Electrical resistivity studies in the vicinity of the deep drilling site Oberpfalz)

Introduction

Vertical electric soundings, controlled source frequency domain electromagnetic soundings (active audiomagnetotellurics), controlled source time domain electromagnetic soundings (long offset transient electromagnetic soundings), passive audiomagnetotellurics, magnetotellurics and magnetovariational studies were carried out in the Haslach-Hausach area of the Kinzigtal (Central Black Forest/SW Germany), one of the two envisaged Continental Deep Drilling (KTB) sites (see Fig. 1a,b and 8a,b). Magnetotelluric and magnetovariational studies were undertaken additionally in the whole Black Forest for an optimal understanding of the natural induced current system. The aim of the field projects in the proposed deep drilling area was to determine the conductivity-depth distribution in the uppermost ca. 15 km. Crystalline rocks (gneisses and granites), generally poorly conducting, occur in the Black Forest crust. Highly conductive structures, probably existing within the poorly conducting crystalline crust, may be either caused by increased electrolytic conductivity (increased porosity, permeability, ion concentration of the fluids, etc.) or by electronic conductors (sulfides or graphite). The field measurements do not indicate the cause of such a possible high conductivity. Therefore additional information is necessary. Such information may be the probable correlation of the conductive structures with the distribution of other physical parameters and/or with the known geological structures.

Field Techniques

Vertical Electric Sounding Method (VES)

In this well known method, an outer pair of transmitter and an inner pair of receiver electrodes are set up along a line, the electrode locations being symmetric to the center of the configuration. A direct (or quasi-direct) current is injected into the earth through the transmitter electrodes. A potential difference correlated with the injected current flow is measured through the receiver electrodes. The conductivity-depth distribution can be determined by successively increasing the separation of the transmitter electrodes.

Active Audiomagnetotelluric Method CSAMT (Controlled Source Frequency Domain Electromagnetic Method)

A pair of electrodes (a grounded wire source) is used as transmitter. A sinusoidal current is injected into the earth successively from 0.001 s to 10 s period. To optimize the signal/noise ratio the transmitter signals are stacked. A secondary current system is induced within the earth, being a function of dipole length, frequency and conductivity distribution. As in magnetotellurics, two pairs of orthogonal electric and magnetic field components are measured at each receiver site. These field components are transformed into two apparent resistivity and two phase curves. One pair of resistivity-/phase curves is calculated from the radial electric field component E and the magnetic field component perpendicular to E (R-Psi curves), and the second pair is calculated vice versa (Psi-R curves).

With controlled source frequency domain methods the signal is directly coupled to resistivity and it is coupled to the earth more strongly than in magnetotellurics. Contrary to the time domain methods such as LOTEM (see below), the secondary as well as the primary field is measured. At low frequencies the primary field dominates the signal. This fundamentally limits the depth of investigation to about 3 km, if impractically large source moments are not to be used. Regardless of frequency, the primary field is strongest near the source. Therefore any lateral distortion caused by a resistivity change near the source will show up at all frequencies. The measurements are sensitive also to lateral changes between the source and receiver.

In active audiomagnetotellurics the receiver signals of only the "far field" and the "transition field" distance range is interpreted: The source-receiver separation must be several times the skin depth at the period used, the primary field then being approximately a plane wave at the receiver sites. The "near field" distance range, independent of the conductivity-depth distribution, is indicated by a 45° increase of the apparent resistivity curves and by zeros in the phase curves (Fig. 3). The depth of investigation is limited by the skin depth at the period used as well as by the transmitter-receiver separation.

Long Offset Transient Electromagnetic Method LOTEM (Controlled Source Time Domain Electromagnetic Method)

The LOTEM sounding system consists of a grounded wire transmitter and a horizontal induction loop receiver. The transmitter is kept fixed and the receiver is moved about the survey area. A square wave current step is injected into the ground through the transmitter wire. This current step induces eddy currents in the earth which diffuse deeper with time. At the earth's surface the vertical magnetic field component of the eddy currents is measured. Contrary to the frequency domain methods (such as CSAMT) the induced secondary field is measured in the absence of a primary field, the masking effect of the primary field being absent. The vertical magnetic field component, induced by the current step, is a decay function. These decay functions are transformed into two apparent resistivity curves as a function of the decay time (the early-time and the late-time apparent resistivity curve) for data display and comparison between stations.

The late-time behavior of the induced field is the key difference between the time domain and the frequency domain methods. In the late-time range the signal shows the strongest coupling to the earth of any electrical method. The late-time field achieves a reduced sensitivity to lateral distortions, and therefore improved lateral resolution. Lateral boundaries may be indicated by the existence of distorted transients.

To optimize the signal/noise ratio, the digitized transients are selectively stacked after digital filtering and the optimal transients are deconvolved from the transfer function of the equipment. Early-time and late-time apparent resistivity curves are calculated from the optimal transients. These curves are interpreted by 1-D model calculations and by a 1-D inversion program.

(Passive) Audiomagnetotelluric, Magnetotelluric and Magnetovariational Methods

While the methods discussed up to now use artificial current sources, the magnetotelluric and magnetovariational methods use natural sources with a time-varying signal. Two orthogonal horizontal electric and magnetic field components, and the vertical magnetic field component as well, are measured at the earth's surface in a magnetic NS-EW coordinate system. These electric and magnetic field variations are measured as a function of time (in the time domain) and they are transformed into the frequency domain calculating complex Fourier coefficients as a function of the period of the field variations. Transfer functions are calculated between the Fourier coefficients of distinct horizontal electric and magnetic field components. These transfer functions of magnetotellurics can be presented as the elements of an impedance tensor. For an optimal interpretation of the analyzed data, the coordinate system of the impedance tensor often has to be rotated, according to the lateral variation of the integrated conductivity. Data interpretation may then be done separately for the current system parallel to the structure's striking direction (TE-mode of the induced current system) and for the current system perpendicular to the striking direction (TM-mode of the induced current system). The magnetotelluric transfer functions may be transformed into apparent resistivity and phase curves as a function of the field variation's period or into the $\rho^*(z^*)$ representation of the magnetotelluric field data.
For magnetovariational studies, transfer functions are calculated between the vertical magnetic field component at a field site and the optimal coherent horizontal magnetic field component at the same site or alternatively at a reference site. These complex transfer functions can be plotted graphically by real and imaginary induction vectors as a function of the period T.

Field Measurements and Their Interpretation

Vertical Electric Soundings

These soundings were performed by the working group at the RWTH Aachen (Wohlenberg, Wendel, Voigt 1985). Vertical electric soundings using the Schlumberger configuration were carried out at 16 sites near the proposed drilling area (Fig. 1a). The electrode spacing of the current electrodes MN was 1000 m in general, and up to 2000 m at few sites. At some of the sites cross-soundings were carried out. The field measurements were interpreted approximately by plane layer models.

The resistivities of the loose sediments (gravel, sand, clay) within the uppermost 8-15 m are 60-4000 Ωm and of the weathered crystalline rocks 60-250 Ωm. The surface of the crystalline basement is located at a depth of 8-15 m. Within the uppermost ca. 200 m of the crystalline basement the resistivities vary in a wide range of 400-9000 Ωm. In a depth interval of 40-100 m, zones of lower resistivities (< 400 Ωm) were found. These low resistivities may be correlated with more or less moistured shear zones. The different types of gneisses occurring in the area cannot be distinguished by their resistivities.

Active Audiomagnetotelluric Measurements

These measurements were carried out by the working group at the University of Braunschweig (Maurer 1986). Active audiomagnetotelluric measurements were undertaken at 12 sites within the proposed drilling area (Fig 1b). Transmitter dipoles were set up at two different locations (NI and FLA, see Fig. 1b). Two roughly orthogonal dipoles were used at each transmitter location. The transmitter-receiver separations varied between about 3.5 and 9 km. Injected current intensity was up to 9 A. In the side valleys of the Kinzigtal, where artificial current intensity is lower, the data quality was sufficiently good. Within the Kinzig valley proper the data quality was also sufficient, despite the strong artificial currents, if the distance between the transmitter dipole and the receiver equipment was not too great.

Two apparent resistivity and two phase curves were calculated for each receiver site (Fig. 3), one pair of resistivity/phase-curves from the radial electric field E and the magnetic component orthogonal to E (R-Psi-curves) and the second one vice versa (Psi-R-curves). These curves were interpreted approximately by plane layer models using Marquart's inversion scheme (Fig. 2). Depending on the site, two-layer or three-layer model curves fitted the measurements sufficiently

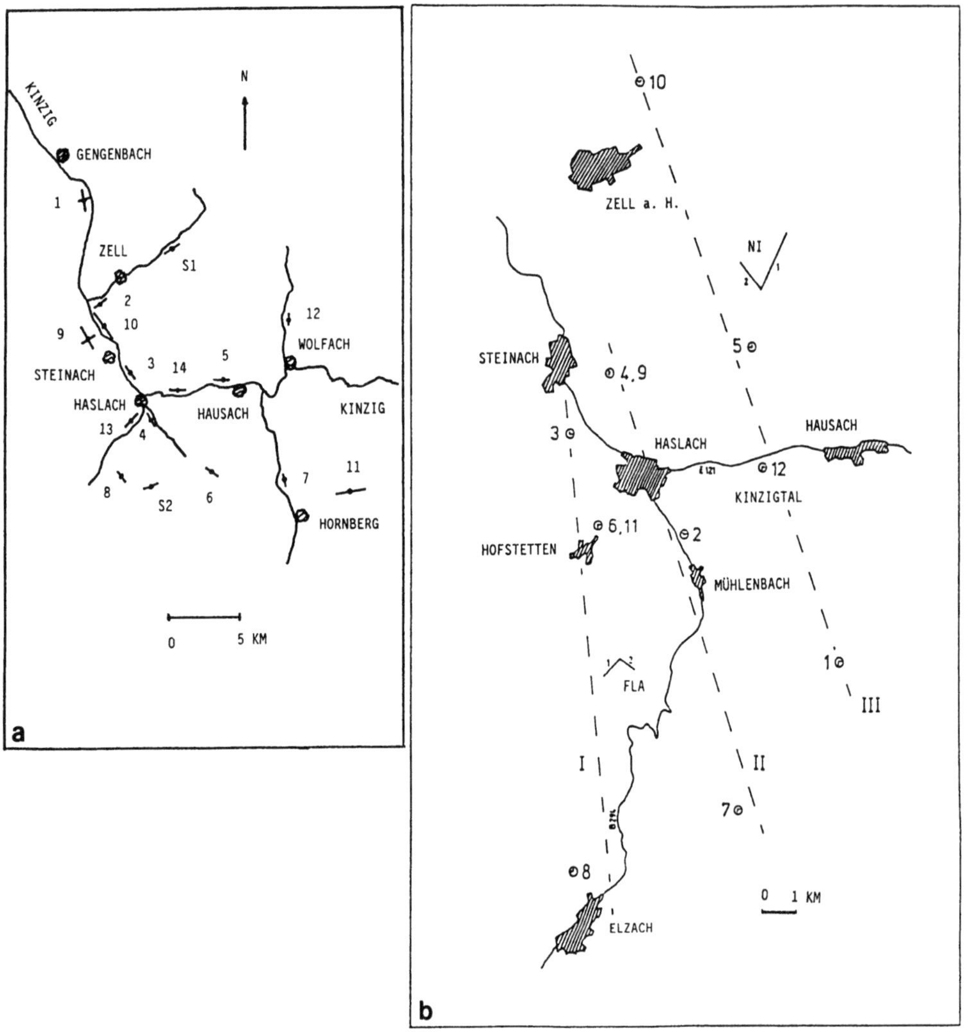

Fig. 1. a Location map of the 16 vertical electric soundings. The deep drilling site was envisaged in the area of Haslach-Hausach (see the location of the city of Haslach in Fig. 8a), b Location map of the 12 active audiomagnetotelluric soundings (see the location of the city of Haslach in Fig. 8a)

well. At some of the sites the model found for the R-Psi curve differed from that found for the Psi-R curve, indicating the influence of 3-D effects on the curves.

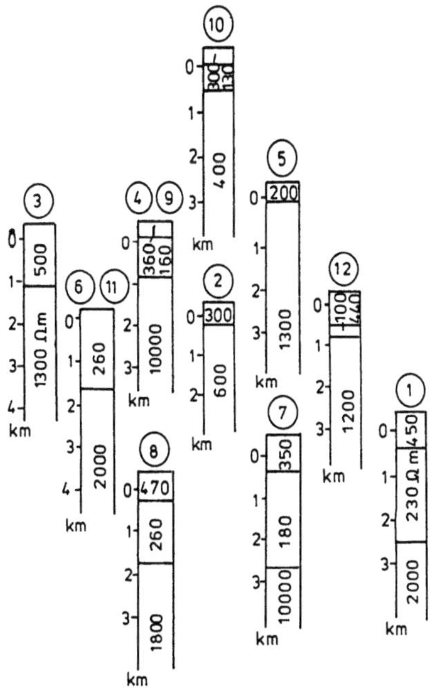

Fig. 2. Resistivity-depth variation at the 12 audiomagnetotelluric sites, calculated for plane layer models

Resistivities of 200-500 Ωm were observed down to 1000-2500 m (Fig. 2). Such values are low for gneisses. Resistivity increases with depth to > 1000 Ωm. For depths below about 4000 m, true resistivities could not be resolved, but a resistivity increase with depth is indicated by the data. No high conductivity zone was observed within the uppermost 4000 m.

Long Offset Transient Electromagnetic Measurements

These measurements were performed by the working group at the University of Köln (LOTEM-group, 1986). Transmitter dipoles were set up at two different locations (Fig. 4). The transmitter locations are marked in Fig. 4 by "transmitter A" and "transmitter B". The dipole length was about 2 km. Current intensity was up to 60 A. The average transmitter-receiver separations were 12 km for transmitter A and 10 km for transmitter B. The receiver equipment was set up at 57 sites. In total, 13,100 transients were recorded, on the average 229 transients per measuring site. Although strong cultural noise of 16 2/3 Hz, 50 Hz, and their higher harmonics occurs in the measuring area, the transients became sufficiently good after digital filtering and stacking (Fig. 5). The transients proved to be twice as long as those near the alternative proposed drilling site in the Oberpfalz

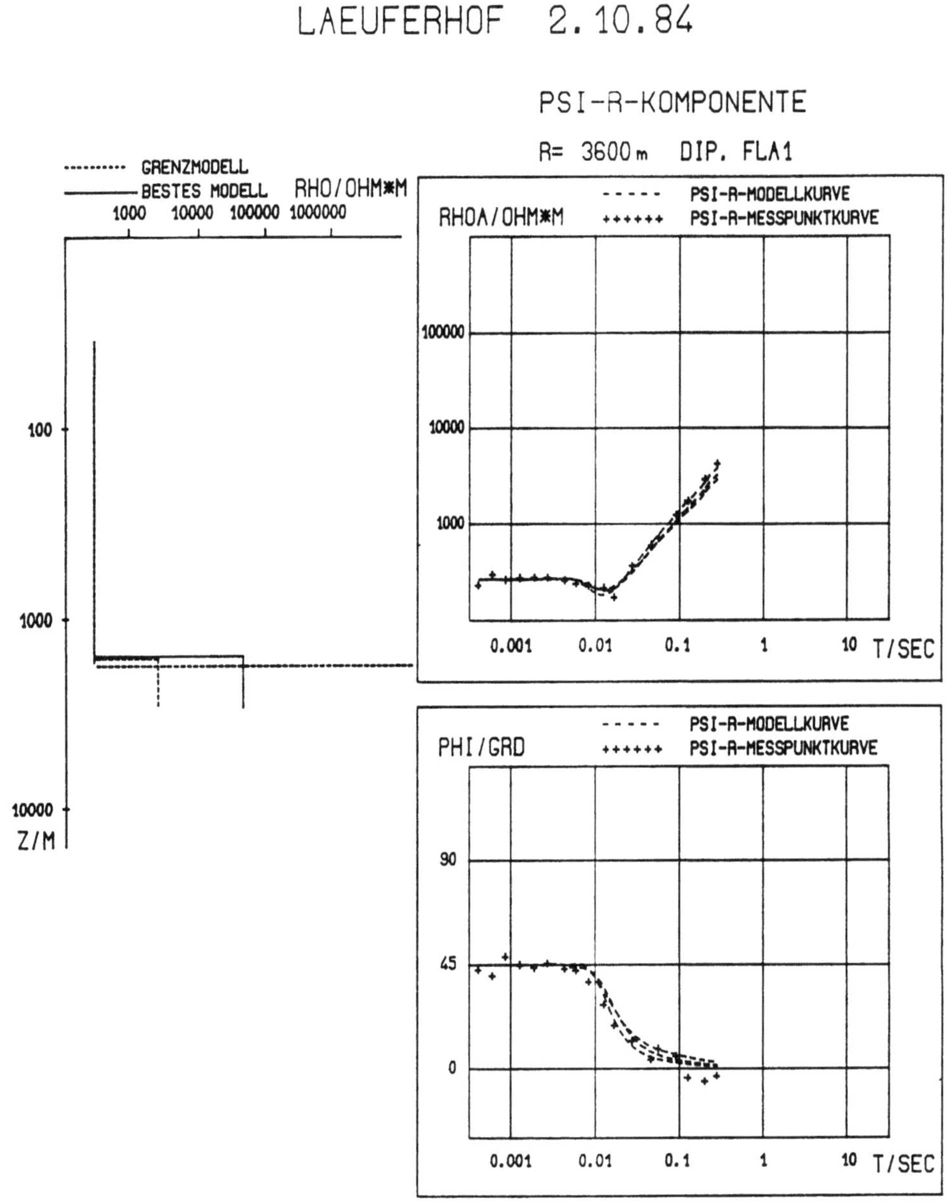

RHO/OHM*M	BESTES MODELL :	2.69E+02	4.31E+04
	GRENZMODELL :	2.69E+02	4.70E+07
	GRENZMODELL :	2.69E+02	2.27E+03
D/M	BESTES MODELL :	1.61E+03	
	GRENZMODELL :	1.79E+03	
	GRENZMODELL :	1.56E+03	

Fig. 3. Apparent resistivity and phase curves (Psi-R component) of active audiomagnetotellurics at site 6 (Laeuferhof)

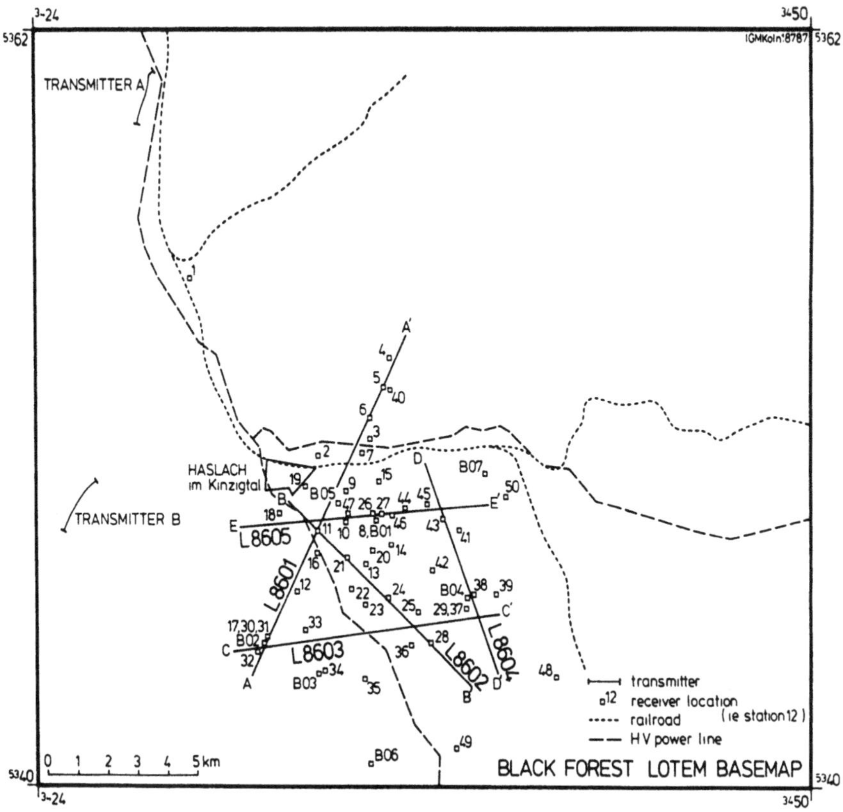

Fig. 4. Basemap of the 57 LOTEM field sites south of the Haslach-Hausach area in the Kinzig valley (see the location of the city of Haslach in Fig. 8a)

area. As a consequence the deconvolution of the system response of the data brought better results. At five sites reversals of the transients were observed, which may be partly correlated with lateral resistivity variations. For interpretation, a number of sites were connected to a profile. Five profiles were used for interpretation (Fig. 4).

The early-time asymptote and the late-time asymptote were transformed into apparent resistivity curves (Fig. 6a,b). For interpretation, 1-D model calculations and 1-D inversions were carried out.

Two resistivity contrasts (resistivity-depth variations) were observed within the depth of investigation (Fig. 7). One resistivity contrast was observed at about 2 km depth from some few hundreds of Ωm above to about 1000 Ωm below. This resistivity contrast may be correlated with the increase of resistivity with depth found by active audiomagnetotellurics in about the same depth range. A

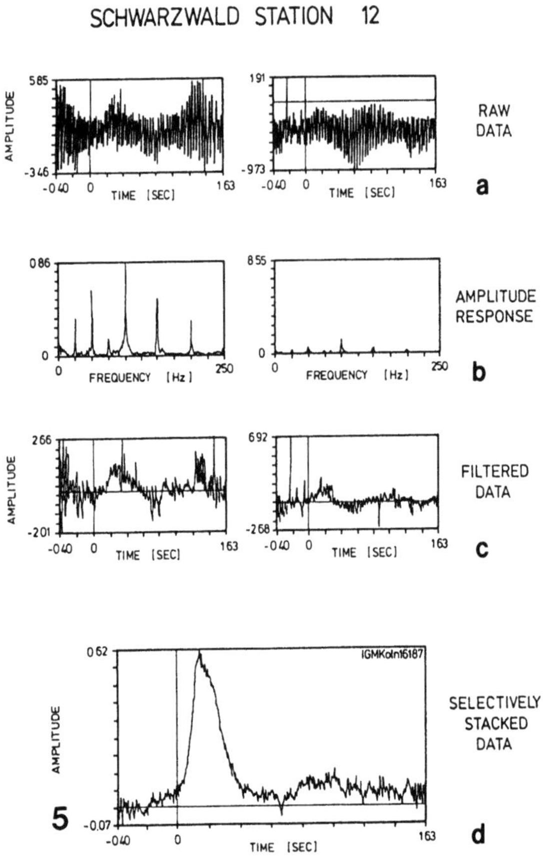

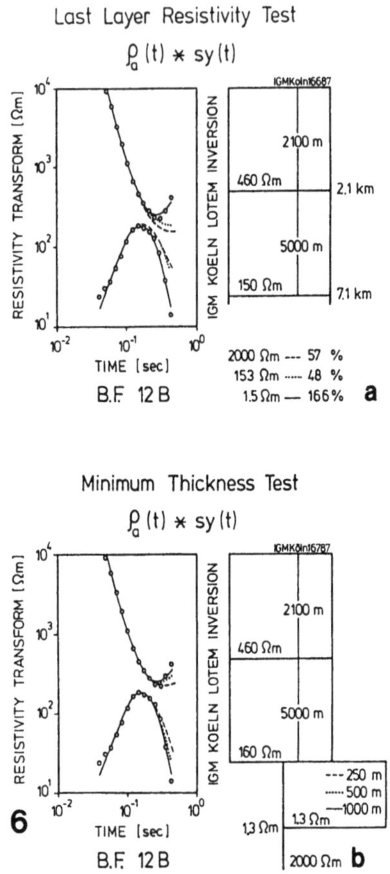

Fig. 5. Steps of data analysis in LOTEM. a raw data display of two transients measured at 12 s distance; b amplitude response functions with partly strong amplitudes of the industrial noise harmonics; c filtered data, filtered with a true amplitude digital recursive sperr filter; d selectively stacked transient (about 100 filtered stacks)

Fig. 6 Early-time and late-time apparent resistivity curves at site 12; a last layer resistivity test; b minimum thickness test

Fig. 7. Resistivity-depth variation along the EW-profile 8605 and the profile 8602 (see Fig. 4): The models are combined from 1-D inversion results at each field site, the hatched depth intervals marking the area of increased conductivity. Thickness and integrated conductivity of the zones could not be determined

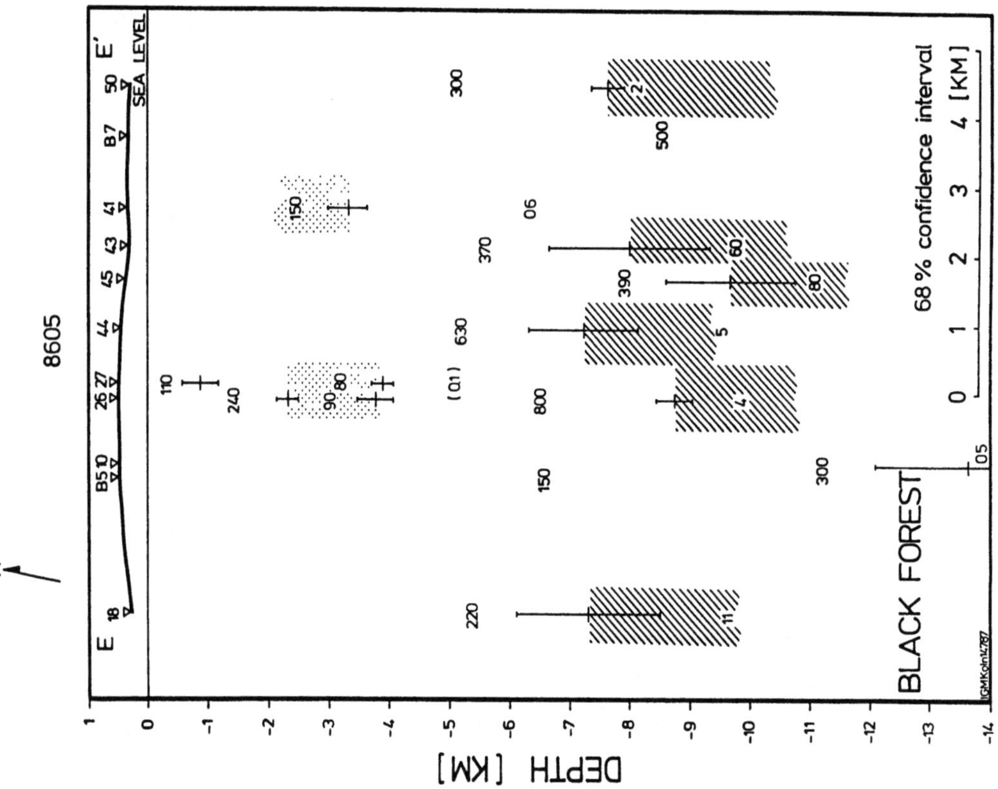
8605
E
E'
SEA LEVEL
BLACK FOREST
68 % confidence interval
DEPTH [KM]
0 1 2 3 4 [KM]

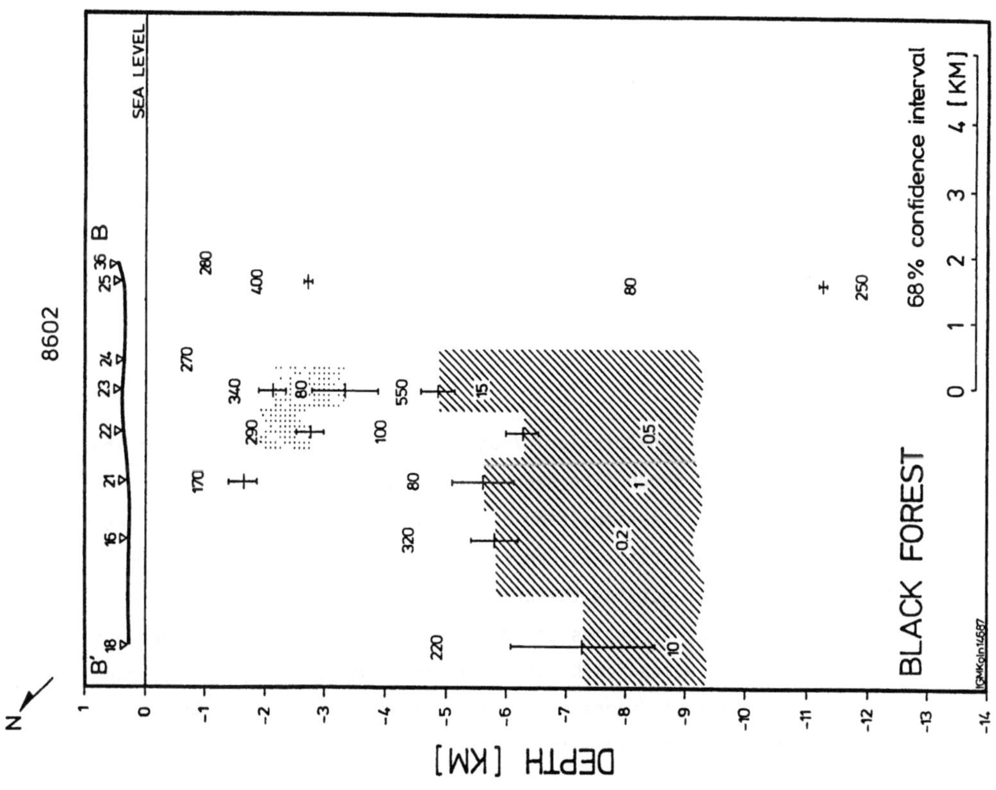
8602
B'
B
SEA LEVEL
BLACK FOREST
68 % confidence interval
DEPTH [KM]
0 1 2 3 4 [KM]

second resistivity contrast was found in about 6-8 km with a distinct decrease of resistivity to as low as 1-10 Ωm. The deeper resistivity contrast is more pronounced than the shallower one. On the average, the depth to the resistivity contrast in 6-8 km increases to NW. The thickness of this conducting layer could not be reliably interpreted because of the increasing cultural noise at later times. (The minimum thickness test (Fig. 6b) suggests a layer thickness of at least 500 m at a resistivity of 1.3 Ωm). The conductivity-depth distribution beneath 6-8 km depth could not be resolved by LOTEM.

(Passive) Audiomagnetotelluric Measurements

Audiomagnetotelluric measurements using natural sources were performed by the working group in Munich at 35 sites in the Black Forest (Dittus 1986; Teufel 1986; Teufel et al. 1986). The two electric and three magnetic field components were recorded in the period range of 0.001 to 10 s. As is observed in all electromagnetic methods, the industrial noise was found to be rather strong, mainly in the northern part of the Black Forest. This limits the information available with audiomagnetotellurics. The telluric fields in the audiofrequency range were found to be only weakly polarized, in contrary to the strong polarization in the period range of magnetotellurics. Apparent resistivities are in the range of 500 to 2000 Ωm. Only a weak dependence of the period is observed. The results were similar at the different sites. No indications were found for an increased conductivity in the depth range of 6-8 km (as is observed in LOTEM). Such indications should have been observed in the period range of > 0.1 s by passive audiomagnetotellurics. The most probable explanation for the missing effect of the well-conducting layer in audiomagnetotellurics is a much too short duration of the field measurements, taking the strong industrial noise into account. New field measurements of much longer duration hopefullly will show the expected effect of the well-conducting zone.

Fig. 8. a Location of 12 magnetotelluric/magnetovariational field sites of the Göttingen working group (The deep drilling site was envisaged near the city of Haslach) in a simplified geological map of Schwarzwald, Vogesen and the Rhinegraben rift. 1: Cenozoic and Mesozoic graben sediments; 2: Mesozoic sediments outside the graben; 3: Permian sediments; 4: Pre-Permian sediments; 5: gneiss and migmatites; 6: granite; 7: granodiorites; 8: tertiary volcanics/Kaiserstuhl; (see Tezkan 1988), b. Basemap of 24 magnetotelluric/magnetovariational field sites of the Munich working group (see the location of the city of Haslach in Fig. a)

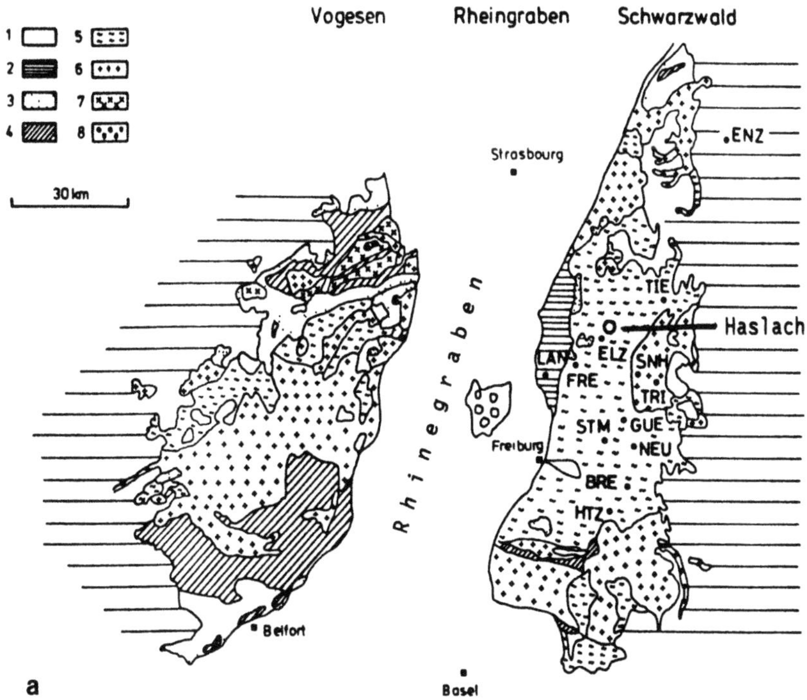
Vogesen
Rheingraben
Schwarzwald
1
2
3
4
5
6
7
8
30 km
Strasbourg
Rheingraben
Freiburg
Belfort
Basel
ENZ
TIE
Haslach
LAN
ELZ
SNH
FRE
TRI
STM
GUE
NEU
BRE
HTZ

a

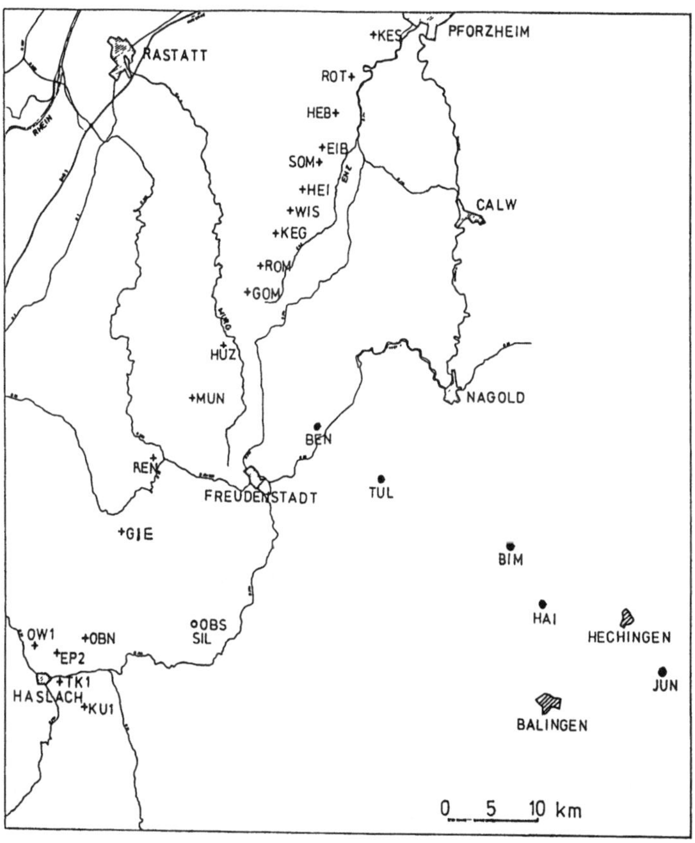
PFORZHEIM
+KES
RASTATT
ROT+
HEB+
+EIB
SOM+
+HEI
+WIS
+KEG
+ROM
+GOM
CALW
HÜZ
+MUN
NAGOLD
BEN
REN
FREUDENSTADT
TUL
+GIE
BIM
HAI
HECHINGEN
OW1
+OBN
OOBS
SIL
EP2
+TK1
HASLACH
+KU1
JUN
BALINGEN
0 5 10 km

b

Magnetotelluric and Geomagnetic Depth Sounding Measurements

Magnetotelluric and geomagnetic depth sounding measurements were performed in 1984 and 1985 by the working groups in Göttingen (Schmucker et al. 1985; Tezkan 1986) and in Munich (Berktold 1988; Dittus 1986; Teufel 1986). The approximately 33 measuring sites covered, as can be seen from Fig. 8a and b, a large area of the Black Forest. At most of the sites induction coil magnetometers ($T \approx 6$-1000 s) and at some sites fluxgate magnetometers ($T \approx 50$-10000 s) were used for the magnetovariational field measurement.

Strong industrial currents, that superimpose and therefore distort the natural current systems, occur in the northern Black Forest north of the Kinzig valley (north of Haslach in Fig. 8a). The industrial current intensity increases from the Kinzig valley to the north. The field data from the most northern Black Forest could therefore not be used for interpretation.

The magnetovariational studies:
For the magnetovariational data analysis complex transfer functions were calculated between the vertical magnetic field component Z and the horizontal field component correlated maximal with Z. The real part of these complex transfer functions is presented in Fig. 9 graphically by "real" induction vectors. The induction vectors are plotted in the figures as a function of the period T.

Supposing the Rhinegraben to be the dominating inductive structure near the Black Forest, a direction of the real induction vectors perpendicular to the average Rhinegraben strike will occur in the Black Forest. This, however, is not observed. Rather, in the Black Forest, the local Z-component correlated with the Rhinegraben (maximum border Z at about 30 s) is superimposed (nonadditive) by a "regional" Z-component, which is approximately constant in amplitude and direction all over the Black Forest (maximum Z at about 500 s). This regional Z-component is known to occur in the whole of Southern Germany as well as in the surrounding areas, at least from the western part of the Vosges to the central part of the Bohemian Massif. In the Black Forest the real induction vectors of this regional Z-component are about 0.30 in length and about E 60° S in direction. The regional Z-component seems to be caused by an elongated regional inductive structure approximately in the middle of the crust striking south of the Moldanubikum/Saxothuringikum border at least from west of the Vosges to the

→

Fig. 9. Real induction vectors as a function of the period of the magnetic variations, along the Rhinegraben parallel profile shown in Fig. 8b

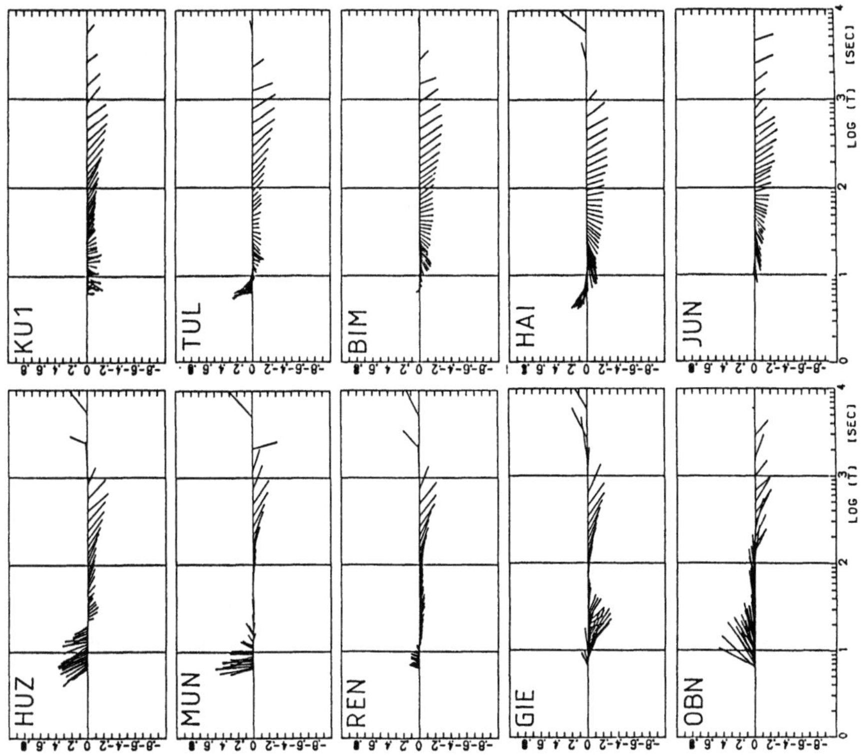

REALPFEILE

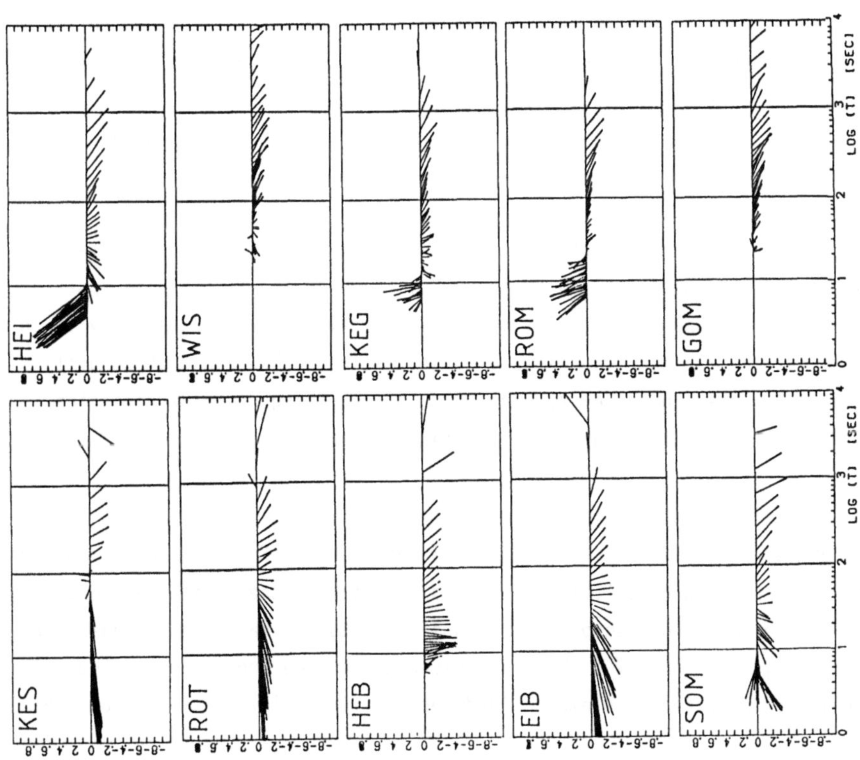

REALPFEILE

central Bohemian Massif. The superimposition of the Z-components correlated with both structures, the Rhinegraben and the regional inductive structure, causes a clockwise rotation of the real induction vectors' direction, depending on the period, from about E20°S near the eastern Rhinegraben border to about E60°S east of the Black Forest. The magnetovariational results presented here will help to understand the telluric results discussed below.

The induced electric (telluric) field:
Supposing the Rhinegraben to be the dominating inductive structure near the Black Forest, the telluric field will be directed perpendicular to the Rhinegraben strike in the Black Forest. A telluric anisotropy will exist, being a function of the period. The telluric anisotropy will decay with increasing distance from the Rhinegraben. All these effects are not observed in the Black Forest:

1. At most sites in the Black Forest the telluric field is directed not perpendicular to the Rhinegraben strike of ca. N20°E/S20°W, but approximately in NW/SE. This means that it is rotated clockwise about 25° from the expected telluric direction in the Black Forest.

2. The telluric anisotropy does not decay with increasing distance from the Rhinegraben; the telluric field is highly anisotropic all over the Black Forest. The anisotropy is about 5 and more and is a function of the period. As the strong anisotropy is more or less constant in the Black Forest, it cannot be caused solely by the inhomogeneous conductivity distribution near the measuring sites (direct current distortion). Both effects, the uniform NW/SE direction as well as the strong anisotropy of the telluric field, must be caused mainly by approximately 2-D conductive structures in the Black Forest.

Both effects can be interpreted sufficiently well by a 2-D model suggested by Schmucker (1987). The intention of the model is to create a horizontal pseudo-anisotropy with higher resistivity in NW/SE direction and lower resistivity in NE/SW direction. The model consists of a succession of low and high conductive 2-D vertical plates, parallel to each other and striking in NE/SW direction. The vertical plates may occur only in a limited depth range.

The magnetotelluric studies:
In magnetotellurics impedances were calculated from the Fourier coefficients of the telluric and magnetic field components. At most of the sites the coordinate system of the impedance tensor being transformed to its main axes is near the coordinate system of NE/SW - NW/SE discussed above. The properties of the off diagonal impedance tensor elements (and the additional information about the

striking direction of the Rhinegraben and of the superimposed regional current system) suggest the NE/SW-directed current flow to be approximately the TE-mode and the NW/SE-directed current flow to be approximately the TM-mode of the induced current system in the Black Forest.

The off diagonal impedance tensor elements (in the coordinate system of about NE/SW - NW/SE) were transformed at each site into apparent resistivities (ρ_a) and phases as a function of the period T. Such curves are shown in Fig. 10a and b (Tezkan 1986). According to the assumptions discussed above, the lower value ρ_a-curves and the corresponding phase curves are supposed to represent approximately the TE-mode and the higher value ρ_a-curves with their phases the TM-mode of the induced current system. The period dependencies of the TE- and TM-modes differ, as expected near 2-D structures. There exists a static shift (direct current distortion and anisotropy effect) of the TE- and TM-mode ρ_a-curves at each site, as discussed above.

The phases of the TE-mode are about 80° at 10 s period and they decrease with increasing period to nearly 45°. Corresponding to the period dependence of the phases, the ρ_a-curves of the TE-mode decrease with increasing period. Both results indicate the existence of a well-conducting zone beneath the Black Forest. The ρ_a- and phase curves of the TE-mode were used to determine the conductivity-depth distribution at each site by 1-D model calculations. The integrated conductivity of the well-conducting zone was found with Schmucker's Psi-algorithm to be about 650 Siemens. Because of the ρ_a-curves' static shift, the depth down to the top of the well-conducting zone could be determined only approximately with 1-D model calculations. Depending on the site, this depth was found to be between 12 and 25 km. Magnetovariational studies indicate no lateral variations of the well conducting zone's depth. Therefore the lateral depth variation between 12 km and 25 km found by magnetotellurics is only apparent and is most probably caused by the static shift effect.

Fortunately the existence of the Rhinegraben enabled us to confine the values found by 1-D model calculations for the depth of the well conducting layer.

A number of 2-D models were calculated with the Rhinegraben as a 2-D inductive structure (Tezkan 1986; see Fig. 11). According to borehole measurements and magnetotelluric results, the conductance of the Rhinegraben sedimentary filling was taken to 1500 Siemens. Beneath the Black Forest, a well-conducting zone was introduced roughly in the middle of the crust. The depth down to this well-conducting zone was modified with increasing model number until the fit of field and model data was optimal. The conductance of the well-conducting zone was

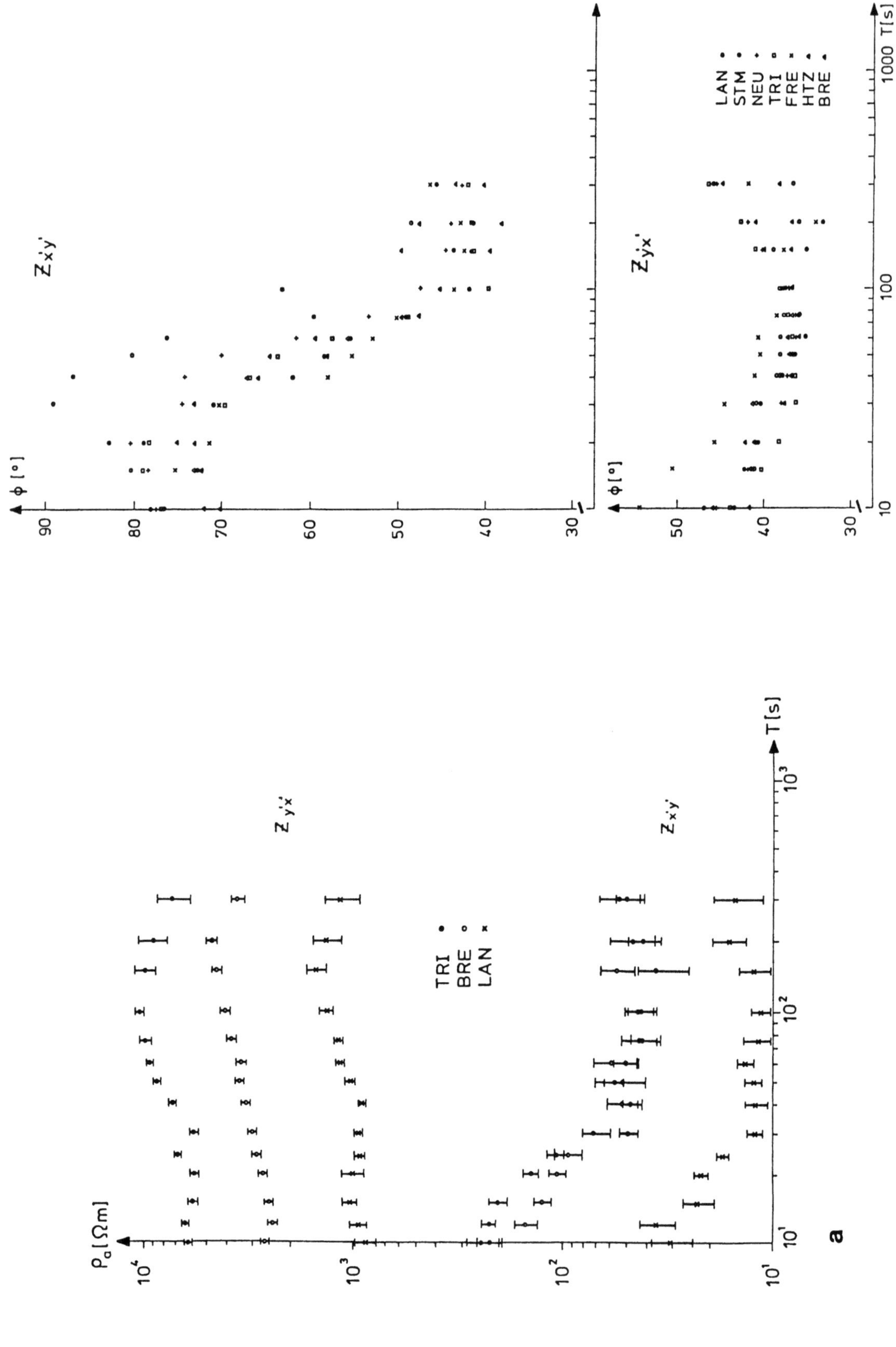
Z'x'y'
Φ [°]
90
80
70
60
50
40
30
Z'y'x'
Φ [°]
50
40
30
LAN
STM
NEU
TRI
FRE
HTZ
BRE
10
100
1000
T[s]
Z'y'x'
Z'x'y'
ρa [Ωm]
10⁴
10³
10²
10¹
TRI
BRE
LAN
T[s]
a

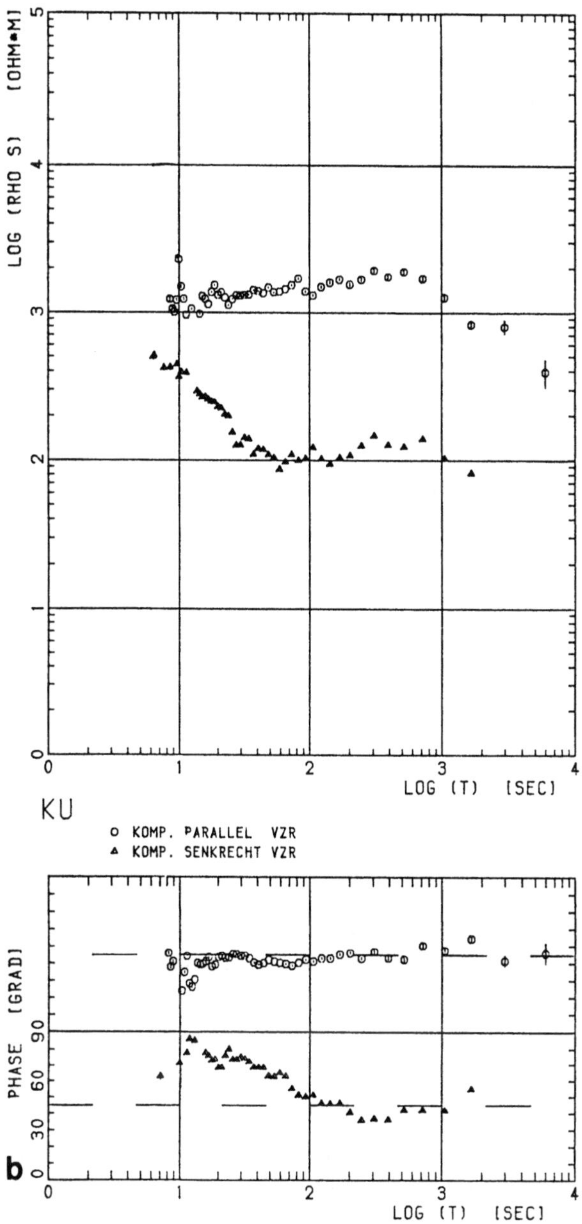

Fig. 10. a Apparent resistivity (ρ_a) and phase curves (Φ) derived by the Göttingen group for selected field sites (e.g. TRI, BRE, LAN) in a rotated coordinate system x'-y' (x' in N45°E/S45°W direction) as shown in Fig. 11, b A representative couple of apparent resistivity and phase curves for site KU1 near the Kinzigtal derived by the Munich working group in the rotated coordinate system x'-y' as discussed above

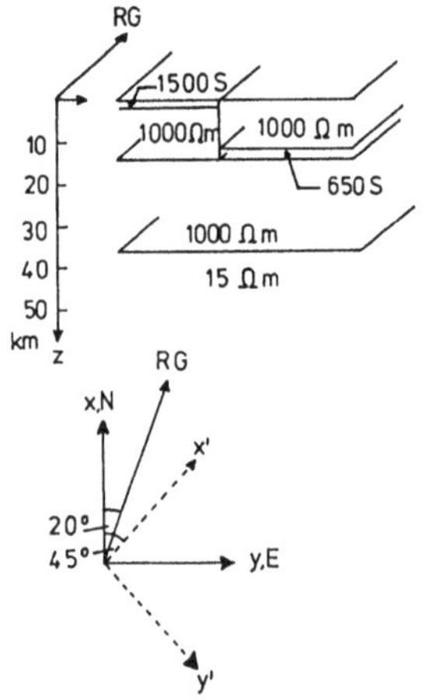

Fig. 11. 2-D model of conductivity distribution beneath the southern Rhinegraben and Black Forest for interpretation of the magnetotelluric and magnetovariational field measurements

taken from 1-D model calculations to 650 Siemens. The model data were calculated for a profile perpendicular to the Rhinegraben strike.

One item of field data being compared with the model data was the vertical component transfer function - more exactly, its projection to the profile direction. The coincidence of the field and model data was found to be optimal near the Rhinegraben border for the short periods of $T < 100$ s (as the real induction vectors are perpendicular to the Rhinegraben strike there).

The other field data being compared with the model data were the phases of the off diagonal impedance tensor elements. Because of the static shift problem, the ρ_a-curves could not be used for the comparison. As was described above, the coordinate system of the observed impedances is rotated clockwise about 25° from the coordinate system of the Rhinegraben to the x'-y' coordinate system of Fig. 11. As a consequence, the observed impedances and the model impedances compared to the observed ones do not refer to the same coordinate system.

The 2-D model presented in Fig. 11 explains the Z-component (best near the Rhinegraben for short periods) and the magnetotelluric phases, their period dependence, and the large offset between the phase curves of TE- and TM-mode. The results of the 2-D model calculations can be summarized as follows:

Magnetotelluric and magnetovariational studies using periods $T > 10$ s indicate a high resistivity in the uppermost ca. 12 km ($\varrho > 200\ \Omega m$). This result does not necessarily contradict the LOTEM results: For periods $T > 10$ s, the conductivity-depth distribution in the uppermost 12 km cannot be resolved by magnetotellurics. The highly conducting zone indicated by LOTEM at about 6-8 km depth can therefore not be detected by magnetotellurics if the conductance of the zone is not too high. No quantitative information exists about the depth range between 6-8 km and about 12 km.

Below about 12 km (at most 18 km) depth, a well-conducting zone occurs with a conductance of 650 S. $\varrho^*(z^*)$-calculations suggest the resistivity of the conductor to be below 10 Ωm. The depth down to the well-conducting zone was found to be roughly the same calculated by 1-D or by 2-D models. This indicates, that the TE-mode telluric field component is not distorted significantly by static shift or other effects. To fit the longer period magnetotellurics a high resistivity (1000 Ωm, Fig. 11) had to be introduced into the model down to 40 km depth followed by a well conducting halfspace of 15 Ωm. The well conducting zone in a depth of about 12 km was found beneath all measuring sites in the Black Forest. This well conducting zone therefore seems to be a property of the whole Black Forest crust.

Not much is known about the lateral variation of the conductance in the Black Forest crust. Magnetovariational studies available up to now give no indications of strong lateral variations. The magnetotelluric studies indicate smooth lateral variations mainly in the central part of the Black Forest. On the other hand, structural differences are apparent in the seismic records and in the heat flow measurements. They are especially strong in the seismicity distribution with depth and lateral distance.

Summary and Conclusions

In the Haslach-Hausach area of the Kinzig valley (Central Black Forest/SW Germany), one of the two envisaged Continental Deep Drilling (KTB) sites, geoelectric and electromagnetic methods were carried out to determine the conductivity-depth distribution within the uppermost ca. 15 km of the crust. Crystalline rocks (gneisses and granites), generally poorly conducting, occur in the Black Forest crust. One problem of interest was the probable existence of highly conductive structures within the poorly conducting crystalline crust. Resistivities of 200-500 Ωm occur down to 1000-2500 m. These values are low for gneisses. Resistivities increase to $> 1000\ \Omega m$ for depths less than 6000 m. No highly conductive structures were found in the uppermost ca. 6000 m of the crystalline crust. A highly conductive zone is indicated by LOTEM measurements (but not by passive audiomagnetotellurics) at a depth of 6-8 km, with a resistivity of about 1-10 Ωm. No information exists about the integrated conductivity and the thickness of this zone. The depth down to this well-conducting zone seems to vary in lateral direction. A pronounced conductive zone was found by magnetotelluric/magnetovariational studies at a depth of about 12 km, having a conductance of about 650 Siemens. According to magnetotelluric/magnetovariational measurements, this zone seems to exist beneath the whole Black Forest. The conductivity distribution in the depth range of about 6-12 km could not yet be determined quantitatively. A smoothly increased conductivity within this depth range is, however, not ruled out by the existing data. Nothing is known about the cause of the increased conductivity in 6-8 and 12-14 km depth respectively. It may be a relic of overthrusted, graphite-containing metasediments recently found among surface rocks. It may, however, also be caused by an increased fluid content. Magnetotelluric studies in and around the Black Forest indicate the conductance of the highly conducting zone to be higher in the Black Forest crust than in the crust of the surrounding area. It may be a property of the Black Forest crust.

Seismic measurements indicate a pronounced low velocity layer between 7 and 14 km depth. Consequently the question arises whether there is a common cause for both the enhanced electrical conductivity and the reduced seismic velocity. Until now this is an unsolved problem.

References

Berktold, A.: Aktive und passive Audiomagnetotellurik, Magnetotellurik und erdmagnetische Tiefensondierung sowie Gleichstrom-Geoelektrik im Zielgebiet Kinzigtal des Kontinentalen Tiefbohrprogrammes. Abschlußbericht zum Forschungsvorhaben an die Deutsche Forschungsgemeinschaft, 1988

Dittus, H.-J.: Methoden der Registrierung und Datenverarbeitung für das Prospektions-Verfahren "Passive Audio-Magnetotellurik", Dissertation, Institut für Allgemeine und Angewandte Geophysik, Universität München, 1986

Fuchs, K., Bonjer, K.-B., Gajewski, D., Lüschen, E., Prodehl, C., Sandmeier, K.-J., Wenzel, F., Wilhelm, H.: Crustal evolution of the Rhinegraben area. 1. Exploring the lower crust in the Rhinegraben rift by unified geophysical experiments, Tectonophysics, 141, pp. 261-275, 1987

LOTEM-group: Long Offset Transient Electromagnetic (LOTEM) - Tiefensondierungen im Schwarzwald, Arbeitsbericht, Institut für Geophysik und Meteorologie der Universität Köln, 1986

Maurer, H.-M.: Ermittlung der Tiefenverteilung der elektrischen Leitfähigkeit im mittleren Kinzigtal mit Hilfe der Aktiven Audiomagnetotellurik, Diplomarbeit, Institut für Geophysik und Meteorologie der TU Braunschweig, 1986

Schmucker, U., Tezkan, B.: Erdmagnetische und tellurische Untersuchungen im Schwarzwald, Bericht zum Unterantrag der Univ. Karlsruhe zum BMFT Förderungsvorhaben RG83146, Inst. f. Geophysik d. Univ. Göttingen, 1985

Schmucker, U., Tezkan, B.: Zur Deutung regional einheitlicher Richtungsabhängigkeiten tellurischer Variationen, Tagungsheft zur 47. Jahrestagung der Deutschen Geophysikalischen Gesellschaft in Clausthal-Zellerfeld, 31.03.-04.04. 1987, S. 195

Teufel, U.: Die Verteilung der elektrischen Leitfähigkeit in der Erdkruste unter dem Schwarzwald, ein Beispiel für Möglichkeiten und Grenzen der Interpretation von Audio-Magnetotellurik, Magnetotellurik, Erdmagnetischer Tiefensondierung, Dissertation, Institut für Allgemeine und Angewandte Geophysik der Universität München, 1986

Teufel, U., H.-J. Dittus, A. Berktold: Audiomagnetotellurik, Magnetotellurik und erdmagnetische Tiefensondierung im mittleren und nördlichen Schwarzwald, Protokoll Koll. Elektromagn. Tiefenforschung, Lerbach bei Köln, editors: V. Haak and J. Homilius, Univ. Frankfurt, S. 137-150, 1986

Tezkan, B.: Erdmagnetische und magnetotellurische Untersuchungen auf hochohmigen Kristallinstrukturen des Hochschwarzwaldes und des Bayerischen Waldes bei Passau, Dissertation, Institut für Geophysik der Universität Göttingen, 1986

Tezkan, B.: Electromagnetic sounding experiments in the Schwarzwald central gneiss massif, Journal of Geophysics, v. 62, no. 2, pp 109-118, 1988

Untiedt, J.: Die elektrische Leitfähigkeit in der Erdkruste in den Zielgebieten Oberpfalz und Schwarzwald - Zusammenfassung der Ergebnisse der elektromagnetischen Voruntersuchungen, Protokoll Koll. Elektromagn. Tiefenforschung, Lerbach bei Köln, editors: V. Haak and J. Homilius, Univ. Frankfurt, S. 177-204, 1986

Wilhelm, H., Berktold, A., Bonjer, K.-P., Jäger, K., Stiefel, A., Strack, K.-M.: Heat flow, electrical conductivity and seismicity in the Black Forest crust, SW Germany, in press

Wohlenberg, J., C. Wendel, G. Voigt: Bericht zu den Voruntersuchungen des KTB: Geoelektrische Tiefensondierung im Raum Haslach i.K., Schwarzwald, Lehr- und Forschungsgebiet für Angewandte Geophysik der RWTH Aachen, 1985
2. KTB-Kolloquium, Seeheim/Odenwald: Magnetotellurische, erdmagnetische und elektromagnetische Tiefensondierung, S. 30-33, Ergebnisse der Vorerkundungsarbeiten Lokation Schwarzwald, 1986

Gravity and Magnetic Investigations at the KTB Locations Schwarzwald and Oberpfalz

H. C. Soffel[1], S. Plaumann[2], R. Pucher[2], C. Bücker[1], H.-J. Götze[3], M. Wagener[4] and V. Haak[5]

[1] Institut für Allgemeine und Angewandte Geophysik, Universität München, Theresienstr. 41, D-8000 München 2, FRG
[2] Niedersächsisches Landesamt für Bodenforschung (NLFB), Stilleweg 2, D-3000 Hannover 51, FRG
[3] Institut für Geophysikalische Wissenschaften, Freie Universität Berlin, Malteserstr. 74–100, D-1000 Berlin 46, FRG
[4] Institut für Geophysik, Technische Universität Clausthal, Arnold Sommerfeld Str. 1, D-3392 Clausthal-Zellerfeld, FRG
[5] Institut für Geophysik und Meteorologie, Universität Frankfurt, Feldbergstr. 45, D-6000 Frankfurt, FRG

Abstract

Gravimetric and magnetic measurements were made in the areas of the two proposed sites for the ultra deep drill hole (KTB) in order to evaluate the lateral and vertical distribution of the density and of the magnetic properties of the earth's crust. The field data (anomaly pattern) and the laboratory data obtained from drill cores taken in the vicinity of both proposed drill sites indicate that the Black Forest site is characterized by a more or less homogeneous upper crust. The drill site in the Upper Palatinate, on the other hand, is situated near the center of a strong positive magnetic and gravimetric anomaly. These anomalies are associated with outcropping amphibolite bodies, which extend to a depth of about 5 to 6 km. The upper crust near the Upper Palatinate drill site is therefore more inhomogeneous than at the Black Forest locality. From the gravimetric and magnetic point of view, the area of the Upper Palatinate is therefore the more interesting drill site because of its variable physical parameters with depth due to a more variable lithology.

1 Gravimetric Investigations

1.1 Introduction

The Bouguer gravity field in southern Germany (Gerke 1957) is characterized by the marked regional minimum of the Alps and of the Rhine valley graben structure. A series of gravity maxima crosses southern Germany from WSW to ENE. Strong positive anomalies are also encountered in the Rheinische Schiefergebirge (Rhenish Massif) and in the Pfalz (Palatinate). Weak relative maxima are found in the areas of the outcropping crystalline basement rocks of the Schwarzwald (Black Forest) and of the Böhmische Masse (Bohemian Massif).

In the vicinity of both proposed drill sites (near Haslach in the Black Forest and near Erbendorf in the Upper Palatinate) detailed areal and profile measurements of the gravity fields were made in order to improve the knowledge of local geological structures and to provide arguments for the decision with regard to the most appropriate drill site of the KTB project from the gravimetric and magnetic point of view. The following questions were of major importance:

1) Are there any larger geological bodies with different densities at the two drill sites of a geometry and extension which could eventually influence the choice of the drilling area and the position of the drill site?
2) What is the lateral and vertical density distribution at both drill sites? Is the density field at greater depth homogeneous or inhomogeneous?
3) Is it possible to locate deep-reaching faults with gravimetric methods?
4) Can gravity data help to improve the knowledge of the structures in greater depth and of the general geological situation at both drill sites?

1.2 Gravimetric Field Measurements

Gravimetric measurements in the Schwarzwald (Black Forest) and in the Oberpfalz (Upper Palatinate) were made first by the Niedersächsische Landesamt für Bodenforschung (NLFB). Earlier regional and local measurements (for example near Erbendorf) were updated and complimented by more recent ground surveys, which ended in 1985. In connection with the pre-site investigations in the Black Forest and in the Upper Palatinate other groups were also involved with additional regional measurements and measurements along profiles between 1984 and 1986. More details can also be taken from Table 1.

The NLFB work in the Black Forest and in the Upper Palatinate consisted in areal surveys covering about 20 to 40 gravity stations per 100 km^2 and in profiles with much closer spacing (varying from less than 100 m to several 100 m) across special structures near Erbendorf, Baden-Baden and Haslach. The group from Clausthal and Berlin also made areal measurements in a smaller region near the drill site Haslach in the Black Forest with about two gravity stations per km^2. They also measured a north-south profile across the Black Forest with a length of about 160 km and an average distance of about 600 m between the stations. The group from Munich made profile measurements in the Upper Palatinate region with average distances of about 100 m between stations. The data of the two areas are discussed separately in the following sections.

Table 1. Gravimetric measurements in the Black Forest and in the Upper Palatinate

Locality: Upper Palatinate	Locality: Black Forest	Institution	Scientists
XXXXXXXXXXX	XXXXXXXXXXXX	NLFB Hannover	Plaumann
	XXXXXXXXXXXX	TU Clausthal	Götze, Wagener
XXXXXXXXXXX		Univ. München	Bücker, Soffel

1.3 Gravimetric Measurements in the Black Forest

Plaumann et al. (1986) compiled gravity measurements made specially for the KTB project, earlier measurements of the NLFB in the Black Forest and older gravity data from regional surveys made in the 1930s and 1940s (Geophysikalische Reichsaufnahme) and combined them into a Bouguer gravity anomaly map of the northern and central Black Forest. Figure 1 shows part of this map and the Bouguer gravity field in the vicinity of the Black Forest drill site near Haslach. Figure 2 is the Bouguer map of essentially the same area around Haslach resulting from the gravity survey made by the group Berlin/Clausthal (Götze 1986), which contains a higher density of gravity stations than the map of Plaumann et al. (1986). Both maps show more or less the same pattern of anomalies. The Bouguer gravity field in the area southeast of

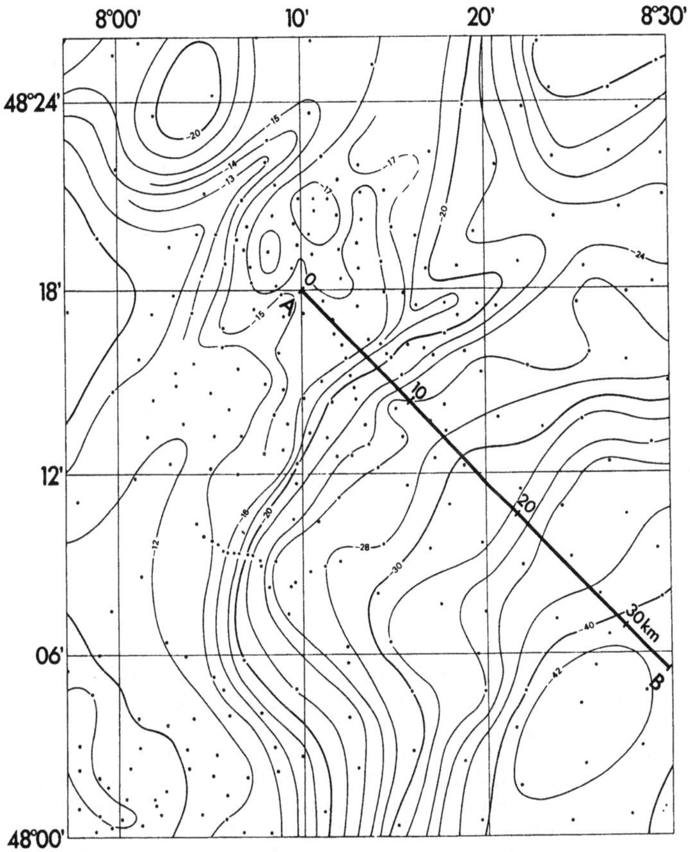

Fig. 1. Bouguer gravity map of the region near the proposed drill site in the Black Forest as part of the larger map compiled by Plaumann et al. (1986). Isoline spacing: 2 mgal. A-B Profile used for the two-dimensional modelling of Fig. 3. Dots: gravity stations. The position of the proposed drill site was near Haslach (HAS) (8°06'E, 48°17'N)

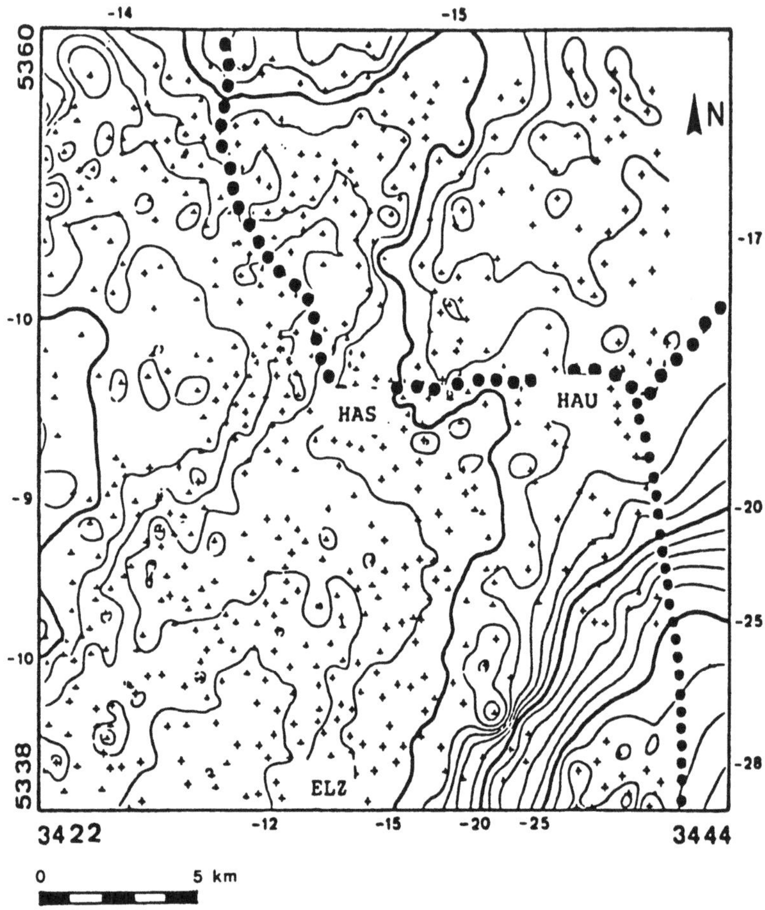

Fig. 2. Bouguer gravity map in the vicinity of the proposed drill site in the Black Forest (after Götze 1986, Wagener and Götze 1987, 1988). Small crosses: gravity stations. The position of the proposed drill site was near Haslach (HAS). Dots: main roads. HAU Hausach; ELZ Elzach; boundaries given in Gauss-Krüger coordinates.

Haslach is characterized by the strong gravity minimum associated with the Triberg granite body, which is better seen on the larger map shown in Fig. 1. Figure 3 shows a gravity profile following roughly line AB in Fig. 1, for which a two-dimensional interpretation was made. According to this model, the Triberg granite reaches a depth of about 8 km. Plaumann et al. (1986) critically discuss the influence of the density distribution in the area on this result. Whatever the thickness of the Triberg granite may be, the most important result for the drill site Haslach in the Black Forest was that no larger granite bodies can be expected immediately near or below Haslach. The contact between the Haslach gneisses and the Triberg granite is about 10 km southeast of Haslach (Fig. 3) and dips towards SE, that means even away from the planned drill site.

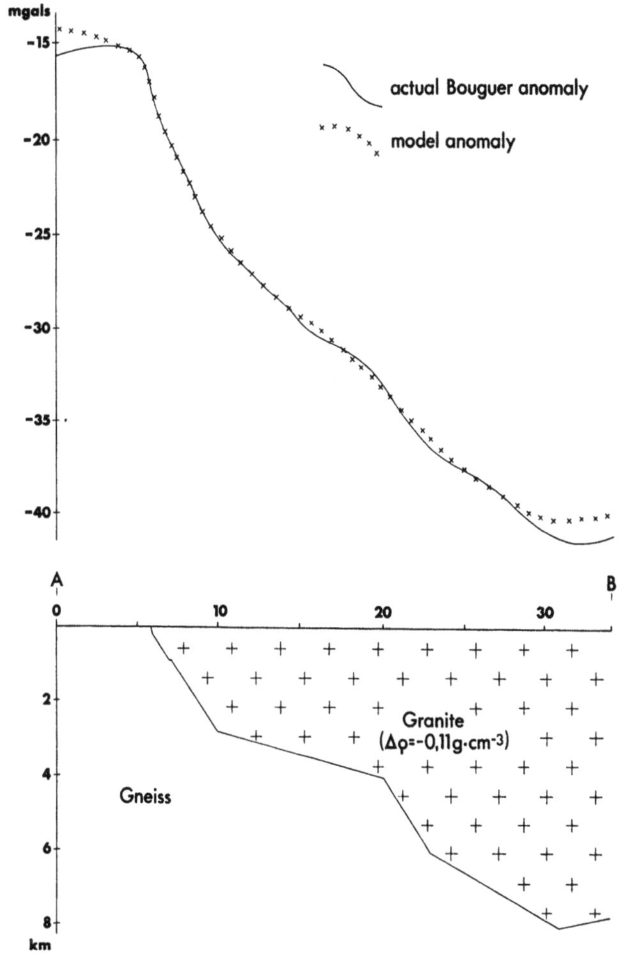

Fig. 3. Gravity profile A-B from Fig. 1 with a two-dimensional interpretative model of the Triberg granite (after Plaumann et al. 1986)

By the measurement of a long profile from the northern to the southern part of the Black Forest, Götze (1986) and Wagener and Götze (1987, 1988) intended to combine gravity data with the results of reflection and refraction seismic profiles. The gravity profile follows the seismic KTB-profile No. 8401. Seismic velocities were transformed into density values using the velocity/density relations by Woolard (1975). Additionally, density values were attributed to a tectonic model by Eisbacher and Krohe (1985). Figure 4a shows the stations of the measured gravity profile. Figure 4b is a plot of the measured Bouguer anomaly (dashed line) and the modelled gravity of the "seismic" model (dotted line) and of the "tectonic" model (full line). The "seismic" model and the "tectonic" model can be seen in Fig. 4c and d respectively. Both models assume density inversions in the area of the Black Forest drill site in the depth range 8 to 15 km.

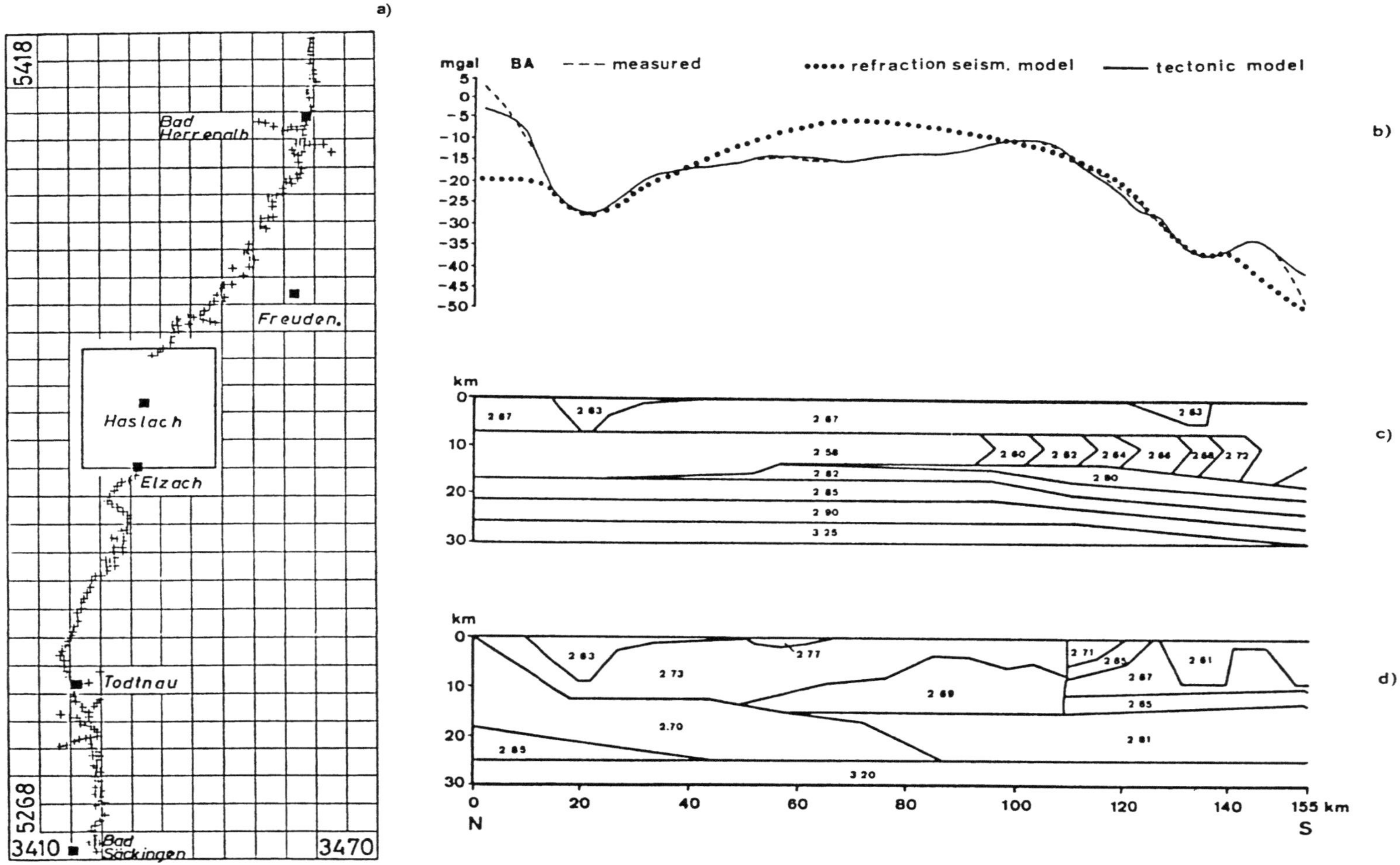
a)
5418
Bad Herrenalb
Freuden.
Haslach
Elzach
Todtnau
Bad Säckingen
5268
3410
3470
b)
mgal
BA
measured
refraction seism. model
tectonic model
c)
km
d)
155 km
N
S

Fig. 4. Two-dimensional interpretation of the Bouguer gravity anomaly along a north-south profile in the Black Forest (after Götze 1986, Wagener und Götze 1987, 1988). The position of the planned drill site was near Haslach.
a Location map of the measured profile. b Measured Bouguer anomaly (dashed line), gravity anomaly of the refraction seismic model (dotted line), gravity anomaly of the tectonic model (full line). c Density model derived from the refraction seismic model (after Gajewski and Prodehl 1985) with density/P-wave velocity correlation (after Woolard 1975). d Density model derived from a tectonic model proposed by Eisbacher and Krohe (1985)

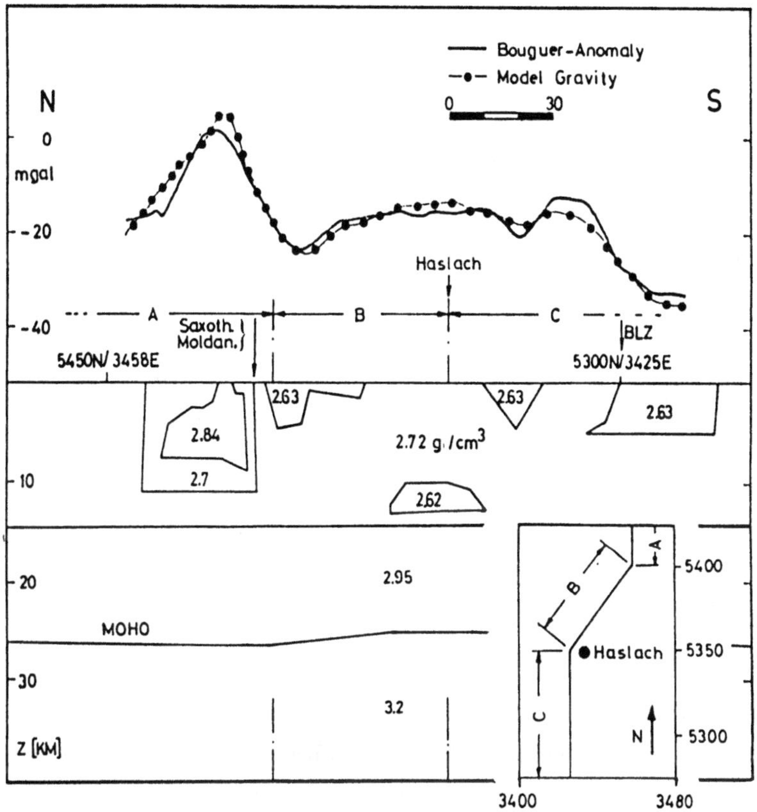

Fig. 5. Vertical section through a three-dimensional density model computed for the profile A-B-C (see insert) above the zone of increased reflectivity (bright spot) underneath the proposed drill site in the Black Forest near Haslach (Wagener and Götze 1987, 1988). The granitic cores (2.63 g cm^{-3}) causes the modelled gravity as well as the structure with basic rock behaviour beneath the prominent high in the north. BLZ Zone of Badenweiler-Lenzkirch

However, it must be pointed out that the conditions for a two-dimensional model calculation are not completely fulfilled for this long profile. In parts it even follows the Bouguer isolines instead of being normal to them (see Gerke 1957).

The dense areal measurements by Götze (1986) in the region near Haslach allowed a three-dimensional model calculation for the investigation of the zone of high reflectivity (bright spot, Lüschen et al. 1985) below Haslach at a depth of about 10 km. These model calculations (see Fig. 5) allow the assumption of a body of reduced density at that depth (Wagener and Götze 1988).

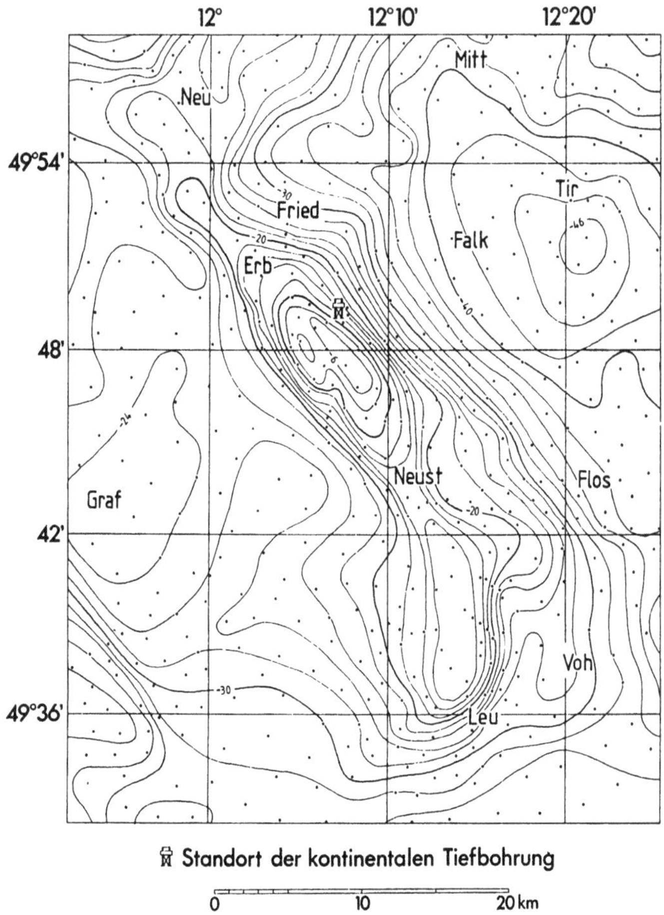

Fig. 6. Bouguer gravity map in the vicinity of the proposed drill site in the Upper Palatinate. Part of a larger map compiled by Plaumann (1986). Isoline spacing: 2 mgal. Circle: position of the drill site. Dots: gravity stations; not.: between Erb and Neust nearly 200 stations have not been plotted (mostly along profiles) in order to obtain a clearer presentation. Erb Erbendorf; Falk Falkenstein; Flos Flossenbürg; Fried Friedenfels; Graf Grafenwöhr; Leu Leuchtenberg; Neu Neusorg; Neust Neustadt a.d.W.; Tir Tirschenreuth; Voh Vohenstrauß; Mitt Mitterteich

1.4 Gravimetric Measurements in the Upper Palatinate

In the same way as in the Black Forest area, previous regional surveys and detailed local investigations (along profiles) in the Erbendorf area were combined by Plaumann (1986) with newly acquired areal data of the years 1984 and 1985 to produce a Bouguer gravity map of the Upper Palatinate region. Part of the map in the area around the drill site is presented in Fig. 6. Compared with the map of the Black Forest drill site (see Fig. 1) the Bouguer map of the Upper Palatinate is characterized by many local anomalies indicating larger variations of the upper crustal density. The most prominent anomaly is the Bouguer gravity maximum southeast of Erbendorf, which is obviously closely related to the structural unit of the Zone Erbendorf-Vohenstrauß (ZEV). Bouguer gravity minima occur mainly in the areas of larger granite bodies (Fichtelgebirge, Steinwald, Falkenberg, Leuchtenberg, Flossenbürg).

In addition to the areal and profile measurements made by Plaumann (1986) and his discussion of the relationships between the gravity anomalies and the geotectonic units of the region, further gravity data along densely measured profiles (distance between stations: about 100 m) were carried out by Bücker (1986). Figure 7 shows the position of the profiles in relation to the structural units of the area. The most informative profile, No. 1, extending from the sediments in the west across the ZEV to the eastern border of the Falkenberg granite body, is shown in Fig. 8. The position of this profile is close to that of the seismic profile KTB No. 8502 and to another gravity profile measured and interpreted by Plaumann (1976). The measurements by Plaumann (1976, 1986) as well as the data of Bücker (1986) indicate that the ZEV seems to extend towards the west beyond the Fränkische Linie (Franconian Line) and below the sediments of the foreland. For the ZEV as well as for the granite bodies of the Upper Palatinate a thickness of up to 6 km can be assumed. This result is in good agreement with the models deduced from reflection and refraction seismic data (see corresponding works, this Vol.).

The Erbendorfer Linie (EL, Erbendorf Line), which is believed to be the boundary between the Saxothuringian zone in the north and the Moldanubian zone in the south, is not clearly manifested in the Bouguer gravity data. It is possible that the intrusion of the Steinwald granite body into this border zone with its gravity minimum masks any other gravity effects associated with the Erbendorf Line.

1.5 Comparison of the Bouguer Gravity Anomalies in Both Areas and Conclusions

Regional trends of the Bouguer gravity fields further away from the two drill sites shall not be discussed here in detail. For this we refer to the works of Plaumann (1986) for the Upper Palatinate as well as Plaumann et al. (1986) and Götze (1986) for the Black Forest. As far as the innermost zone with a radius of 20 km around both drill sites is concerned, the Black Forest locality shows a significantly quieter Bouguer gravity field than the Upper Palatinate site. At the site near Haslach much lower lateral and presumably also vertical density variations can be assumed for the uppermost 10 km of the crust than at the Erbendorf site. At the Upper Palatinate site drilling begins nearly in the maximum of a Bouguer gravity anomaly in the northern part of the ZEV, where also the maximum of a total magnetic

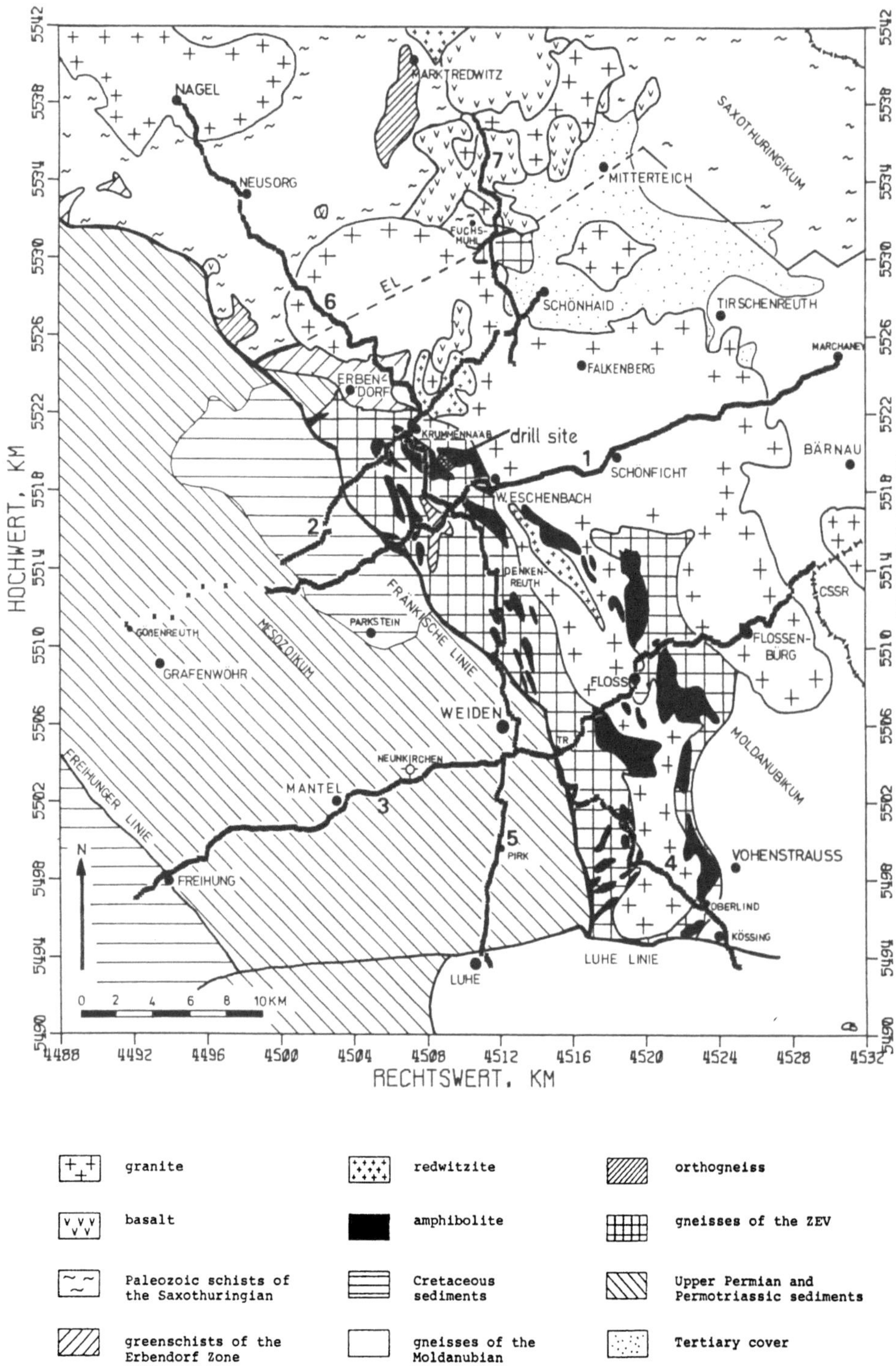

<u>Fig. 7.</u> Schematic geology of the Upper Palatinate drill site region and position of the gravity profiles measured by Bücker (1986). Gauss-Krüger coordinate system. <u>EL</u> Erbendorf line. <u>1</u> to <u>6</u> Number of profiles. The drill site is located in an amphibolite body half way between Windischeschenbach and Krummennaab

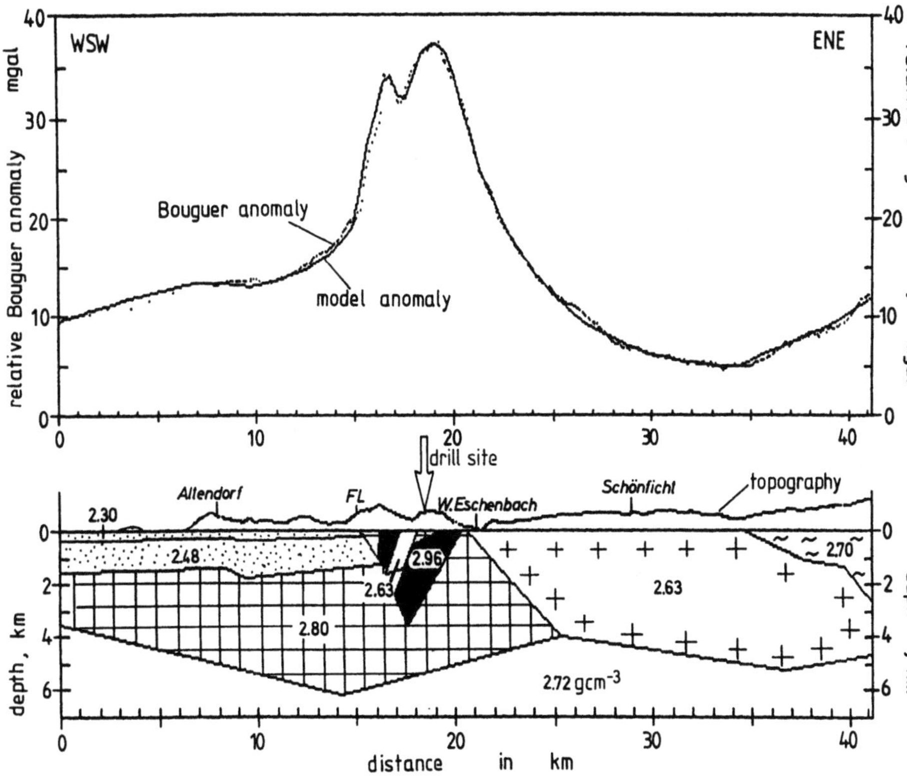

Fig. 8. Bouguer gravity profile number 1 (see Fig. 7) and two-dimensional density model after Bücker (1986). Vertical arrow: position of the drill site; FL Franconian line

intensity anomaly (see sect. 2) associated with a complex of amphibolite bodies occurs. Such amphibolite bodies of various sizes are typical for the structural unit of the ZEV and are presumably responsible for their higher block density of about 2.8 g cm^{-3} on the average. Gravity data show that both the lateral and the vertical density variations are very likely at the Upper Palatinate drill site. On the other hand, it could be expected for the Black Forest crust that drilling would start in gneisses and would remain in these rock units of presumably constant density over long distances during drilling. The relatively quiet Bouguer gravity field near the Black Forest drill site, which is only peripherally influenced by the Triberg granite body, favours the model of a more homogeneous basement with regard to density.

For more details of the gravity field and associated model computations we refer to Plaumann (1986), Plaumann et al. (1986), Götze (1986), Wagener and Götze (1987, 1988), Bücker (1986) as well as to the abstracts and reports of the Seeheim Conference in 1985 and 1986. Plaumann et al. (1986) are also treating the Baden-Baden/Gaggenau zone, which was discussed temporarily as a drill site in the Black Forest.

2. Magnetic Investigations

2.1 Introduction

The earth's magnetic field of the Federal Republic of Germany has been measured in the form of a total intensity aeromagnetic survey between 1965 and 1971 by the Bundesanstalt für Geowissenschaften und Rohstoffe (Bundesanstalt für Geowissenschaften und Rohstoffe 1976). The areas of outcropping Tertiary volcanic rocks (e.g. Vogelsberg, Rhön, Oberpfalz, Kaiserstuhl, Hegau, etc.) and the Ries meteoritic impact crater are characterized by strong disturbances of the local geomagnetic field. In southern Germany, a line of positive anomalies is found in the region of the Swabian-Franconian Jura. These anomalies are associated with amphibolite bodies in the crystalline basement below the Mesozoic sediments. Another trend of WSW-ENE striking positive anomalies is found along a line reaching from the Palatinate west of the river Rhine into Thuringia. These anomalies belong to the Saxothuringian zone.

The pattern of anomalies of the total intensity in the vicinity of the two proposed drill sites are quite different. At the Black Forest drill site only weak anomalies are found (Fig. 9).

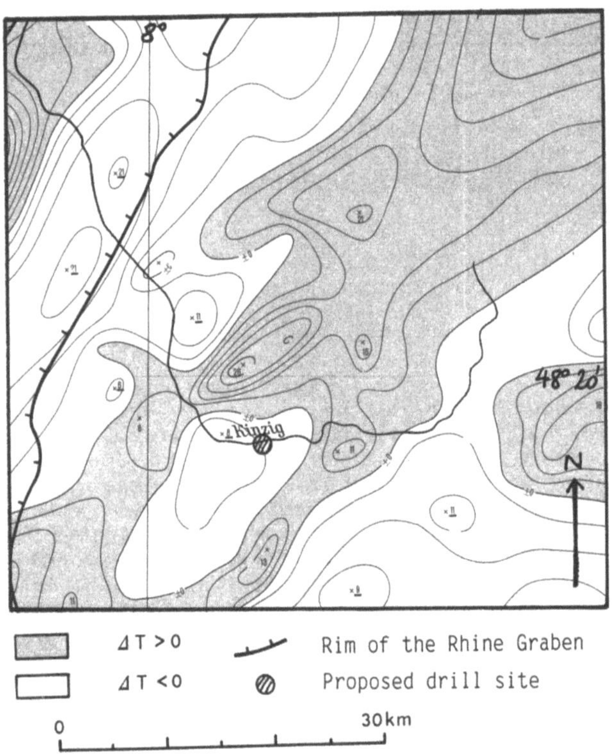

Fig. 9. Anomalies of the total intensity of the earth's magnetic field 1500 m above mean sea level in the area of the drill site Black Forest (Bundesanstalt für Geowissenschaften und Rohstoffe 1976). Hachured circle: position of the proposed drill site

The obtained "regional" trend of NE-SW-striking T-field anomalies is also found in the gravity field (see Fig. 2). The structure and the depth of the magnetized source bodies could not be determined from the anomalies in the flight level 1500 m above sea level without further information about the rocks near the surface and their magnetic properties. In the region of the Upper Palatinate drill site near Erbendorf, however, a strong anomaly of the total intensity has been found in the aeromagnetic survey (Fig. 10) which correlates with the strong positive Bouguer gravity anomaly (Plaumann 1986) in section 1.3.

Similar to the gravimetric investigations discussed in section 1, the magnetic measurements in the areas of both proposed drill sites should provide answers to the following questions:

1. Are there any magnetized geological bodies near the drill sites?
2. What is the relationship between such magnetized bodies and the local geological structure?
3. What is the lateral and vertical variation of the magnetic properties of the rocks at both drill sites?

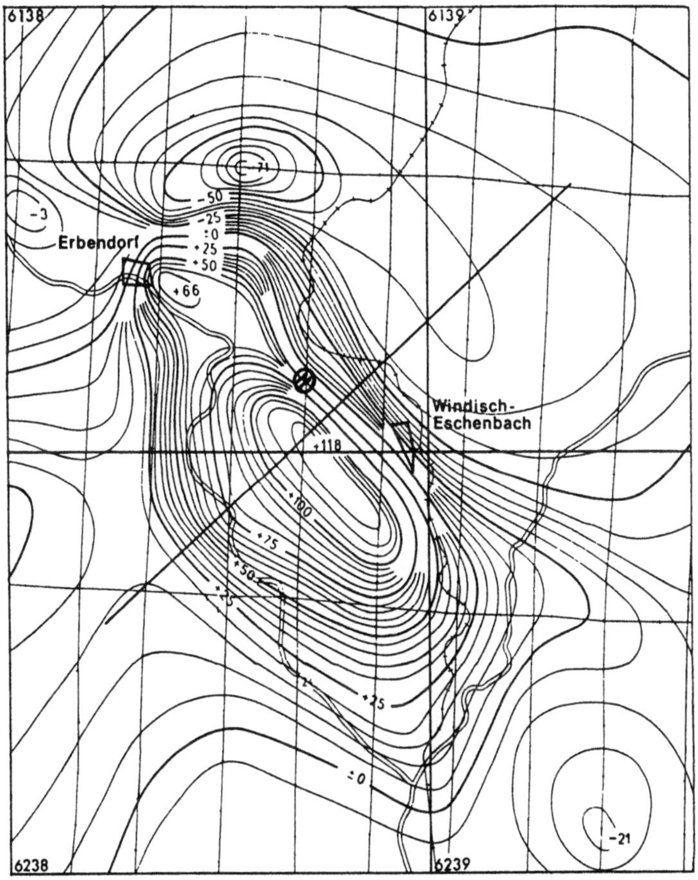

Fig. 10. Anomaly of the total intensity of the earth's magnetic field 1500 m above mean sea level in the area of the drill site Upper Palatinate (Bundesanstalt für Geowissenschaften und Rohstoffe 1986). Hachured circle: position of the drill site

2.2 Magnetic Field Measurements

In addition to the aeromagnetic survey, groundborne magnetic measurements of the total field and partly of the vertical intensity were made at both proposed drill sites in the Black Forest and in the Upper Palatinate in larger areas around the sites and along densely measured profiles. They were made to determine the detailed anomaly pattern and to correlate it with outcropping rock series. Similarly to the gravity surveys described in section 1.2, different groups were active in both regions.

2.3 Magnetic Measurements in the Black Forest

Together with the gravity survey described in section 1.3, Götze and co-workers (Götze 1986) made areal measurements of the anomalies of the vertical component of the earth's magnetic field in the vicinity of the proposed drill site in the Black Forest, which are shown in Fig. 11. The weak anomalies of the large-scale aeromagnetic total field intensities, as shown in Fig. 9, appear in the ground magnetic survey in the form of many locally limited small anomalies, which cannot be associated with larger geologi-

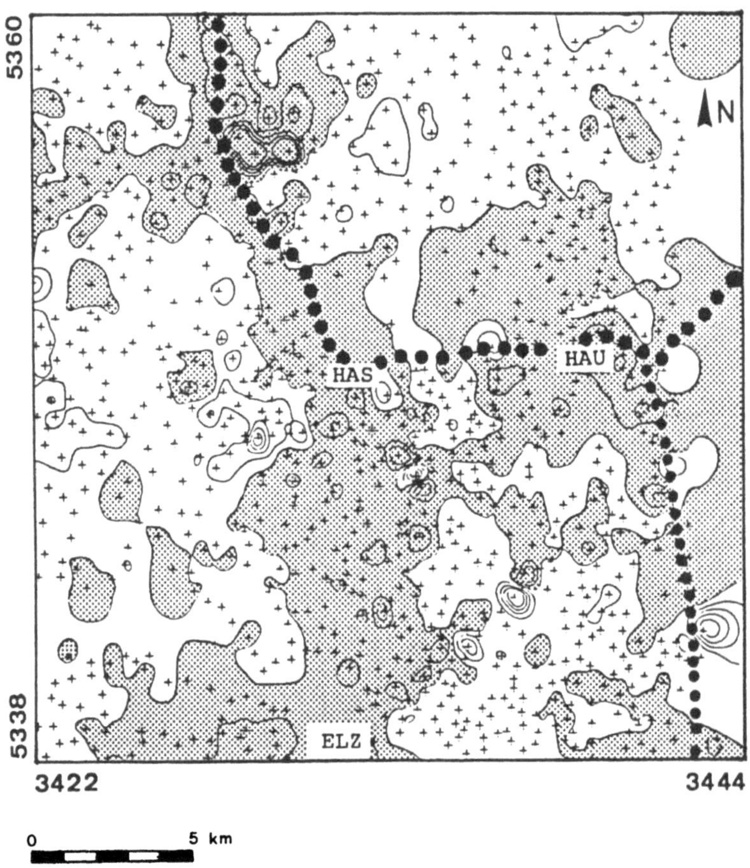

Fig. 11. Ground anomalies of the vertical component of the earth's magnetic field in the area of the drill site Black Forest compiled by G. Eilers (in Götze 1986). HAS (Haslach): position of the proposed drill site. See also legend of Fig. 2

Table 2. Magnetic measurements in the Black Forest and in the Upper Palatinate

Locality Upper Palatinate	Locality Black Forest	Institute	Scientists
	XXXXXXXXXXXX	TU Clausthal	G. Eilers
	XXXXXXXXXXXX	FU Berlin	H.J. Götze
XXXXXXXXXXXX	XXXXXXXXXXXX	NLFB Hannover	R. Pucher
XXXXXXXXXXXX		U. München	H. Soffel, C. Bücker, R. Waitz
XXXXXXXXXXXX		U. Frankfurt	V. Haak

cal bodies. Local variations in the content of magnetite or pyrrhotite of the outcropping gneisses can be regarded as the most probable reason for these anomalies. This is also confirmed by densely measured magnetic profiles crossing petrologically well-mapped areas near the drill site (Pucher 1985). Rock magnetic measurements carried out on cores from the geothermic drill holes in the Black Forest also yielded only quite low values of mostly less than 30 mA m^{-1} for the magnetization (Pucher 1986c).

Thus the areal as well as the additional profile magnetic measurements, in combination with the rock magnetic investigations, confirm the relatively undisturbed geomagnetic field near the Black Forest drill site. Larger magnetic anomalies in other parts of the Black Forest are only encountered in association with the Tertiary volcanic rocks of the Kaiserstuhl and Hegau. There is also a remarkable anomaly in the northern part of the Black Forest near Baden-Baden, which is connected with a large positive gravity anomaly. This area is associated with the border between the geological units of the Saxothuringian in the north and the Moldanubian in the south. Both the magnetic (Pucher 1986a) and the gravimetric anomalies (Plaumann et al. 1986) have been studied in detail, because the anomaly region in the northern Black Forest was also considered as a possible site for the KTB drill hole at an earlier stage of the site selection process.

2.4 Magnetic Measurements in the Upper Palatinate

The geomagnetic field near this drill site is characterized by a strong positive magnetic anomaly in the area of Erbendorf/-Windischeschenbach (Fig. 10). This anomaly was interpreted by the scientists of the NLFB immediately after the aeromagnetic survey of the Federal Republic of Germany was finished in the mid-1970s, also in combination with a ground magnetic survey. Figure 12 shows a plan of the ground magnetic profiles measured by Pucher (1986b) and a map of the isolines of the widespread total intensity anomaly. The parts of the profiles with heavy lines represent zones of strong local anomalies due to source bodies near the earth's surface. They were interpreted by relatively small bodies, because their anomalies were only rarely found on nearby positioned profiles. The ground magnetic data have of course a higher degree of resolution than the aeromagnetic data at a flight level of about 1000 m above the earth's surface. The use of the data of both measuring levels reduces the ambiguity of the

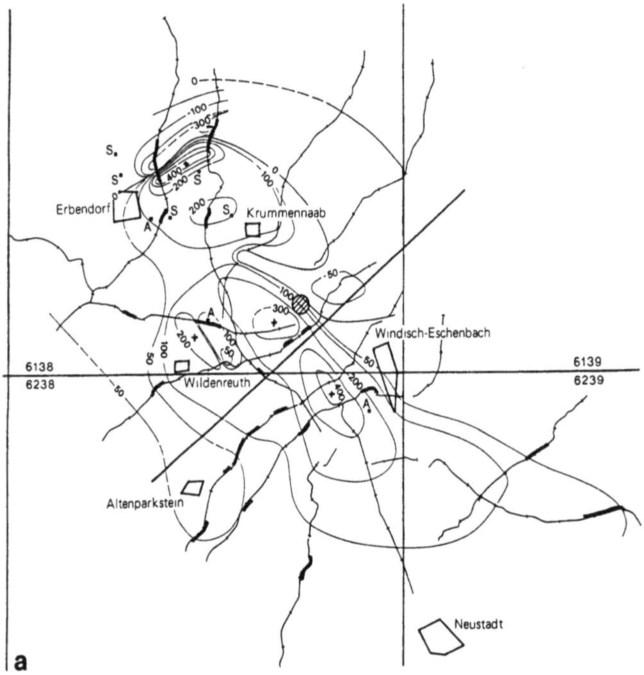

Erbendorf
Krummennaab
Windisch-Eschenbach
Wildenreuth
Altenparkstein
Neustadt
6138
6238
6139
6239
a

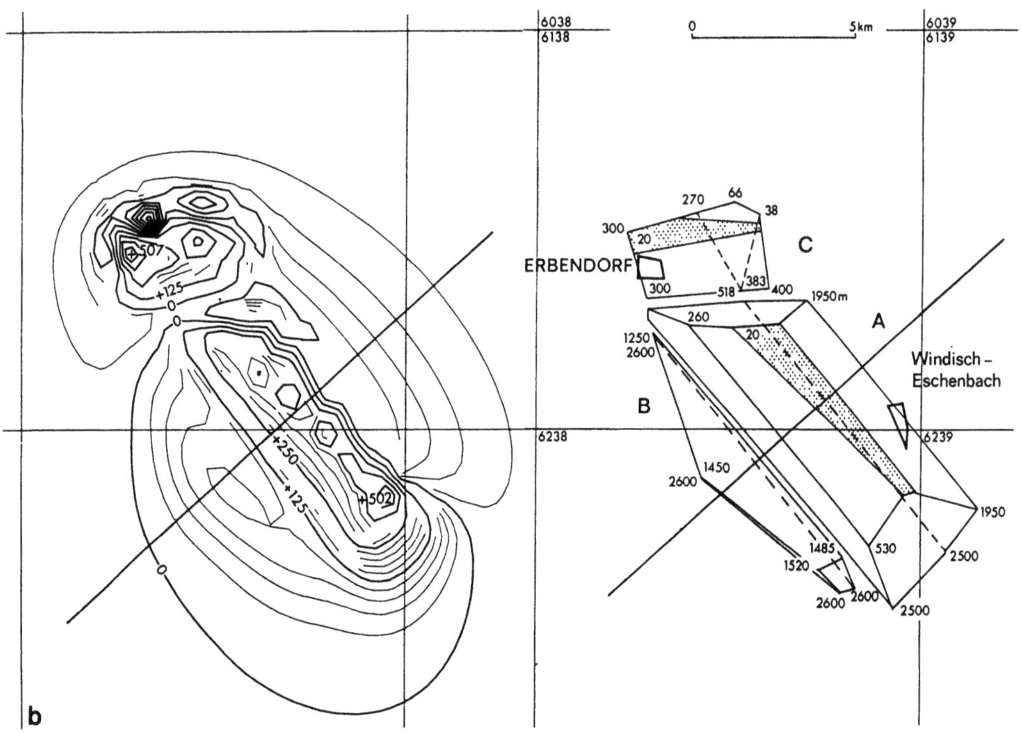

6038
6138
0
5km
6039
6139
6238
6239
+507
+125
+250
+502
ERBENDORF
C
A
B
Windisch-Eschenbach
1950m
b

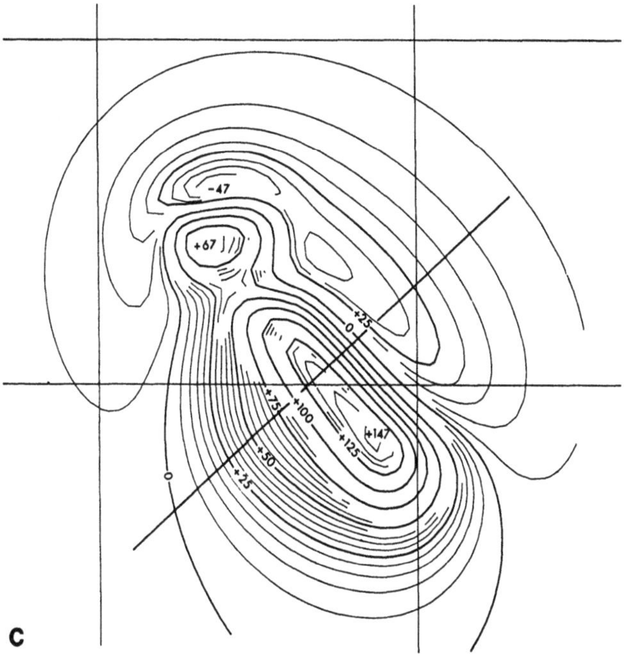

<u>Fig. 12a.</u> Sketch map of the profiles of a ground magnetic survey in the area of Erbendorf/Windischeschenbach with the isolines of the total intensity of the earth's magnetic field. Profile parts with heavy lines are characterized by strong local anomalies. These local anomalies have been subtracted for the construction of the main anomaly (Pucher 1986b). <u>Circle:</u> proposed drill site. <u>b</u> Sketch map of the three-dimensional model body and its total magnetic field computed for the level of the earth's surface (cf. <u>a</u>). Parameters of magnetization: <u>Body A:</u> M = 0.65 A/m, D = 350°, I = 73°. <u>Body B:</u> M = 1.1 A/m, D = 350°, I = 73°. <u>Body C:</u> M = 1.7 A/m, D = 350°, I = 66°.
<u>c</u> Total intensity of the magnetic field of the same model body for the flight level 1500 m above mean sea level (cf. Fig. 10)

magnetic models considerably. Both two- and three-dimensional model calculations have been made. Figure 12 shows also that the conditions for a two-dimensional interpretation, namely an infinite extension of the models normal to the interpretation plane, is not fulfilled even approximately. Therefore the two-dimensional interpretations have to be regarded as a more or less satisfactory first approximation. Figure 12 shows a top view of the three-dimensional source body and the model anomaly of the total intensity computed for the level of the earth's surface (Pucher 1986b). The aeromagnetic anomaly of the same body, computed for a flight level of about 1000 m above ground, is shown in Fig. 12c. A comparison with the actually measured aeromagnetic anomaly (Fig. 10) manifests the good coincidence between measured and modelled anomaly. The anomalies along a section, which is shown in Fig. 12b, displays the cross-section and the vertical extension of the model bodies. For the determination of these models, Pucher (1986c) also included results of detailed rock magnetic measurements.

As mentioned above, Bücker (1986) made total magnetic field measurements along his gravimetric profile No. 1 in Fig. 8 (see sect. 1.4). Figure 13 displays the measured anomaly, the two-dimensional model anomaly and the cross-section of the source body. Haak (1966) comes to a similar result for his profile (see Fig. 13c), which quite satisfactorily explains both the densely measured far-field of the ground survey and the aeromagnetic anomaly. The investigations of all three groups yielded the concurring result that the magnetic anomalies of Erbendorf/Windisch-eschenbach are caused by several massive bodies (presumably amphibolites), which seem to dip towards the west according to surface geological observations and model calculations on the basis of a predominantly induced magnetization. They reach presumably to a depth of about 3 to 5 km.

a

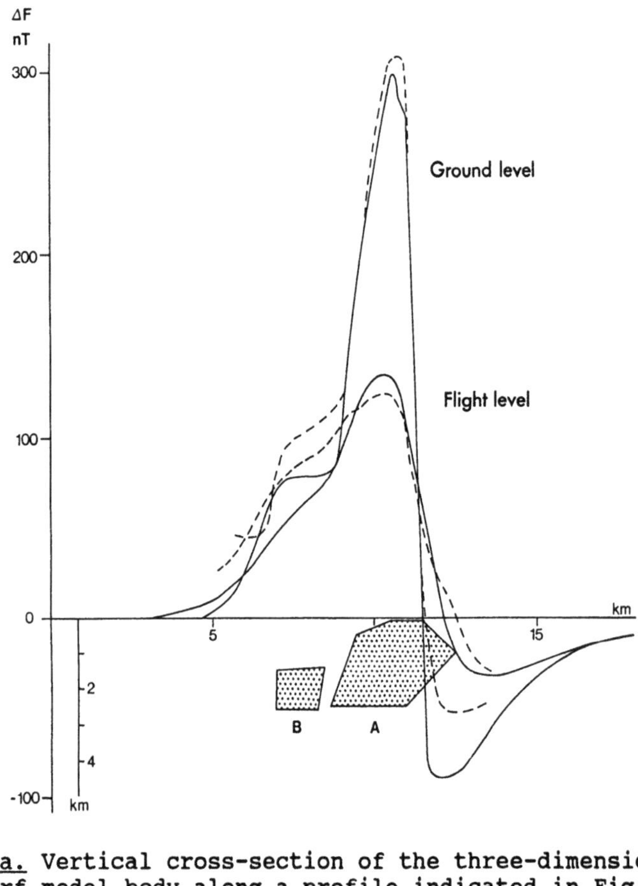

Fig. 13a. Vertical cross-section of the three-dimensional Erbendorf model body along a profile indicated in Fig. 12 and comparison of the computed and measured field for the ground and flight level (Pucher 1986b). b Ground magnetic profile of the total intensity of the earth's magnetic field crossing the Erbendorf anomaly and its two-dimensional interpretation by Bücker (1986); the profile is the same as the gravity profile number 1 (see Fig. 8). c Ground magnetic profile of the total intensity of the earth's magnetic field crossing the Erbendorf magnetic anomaly and its two-dimensional interpretation by Haak (1986)

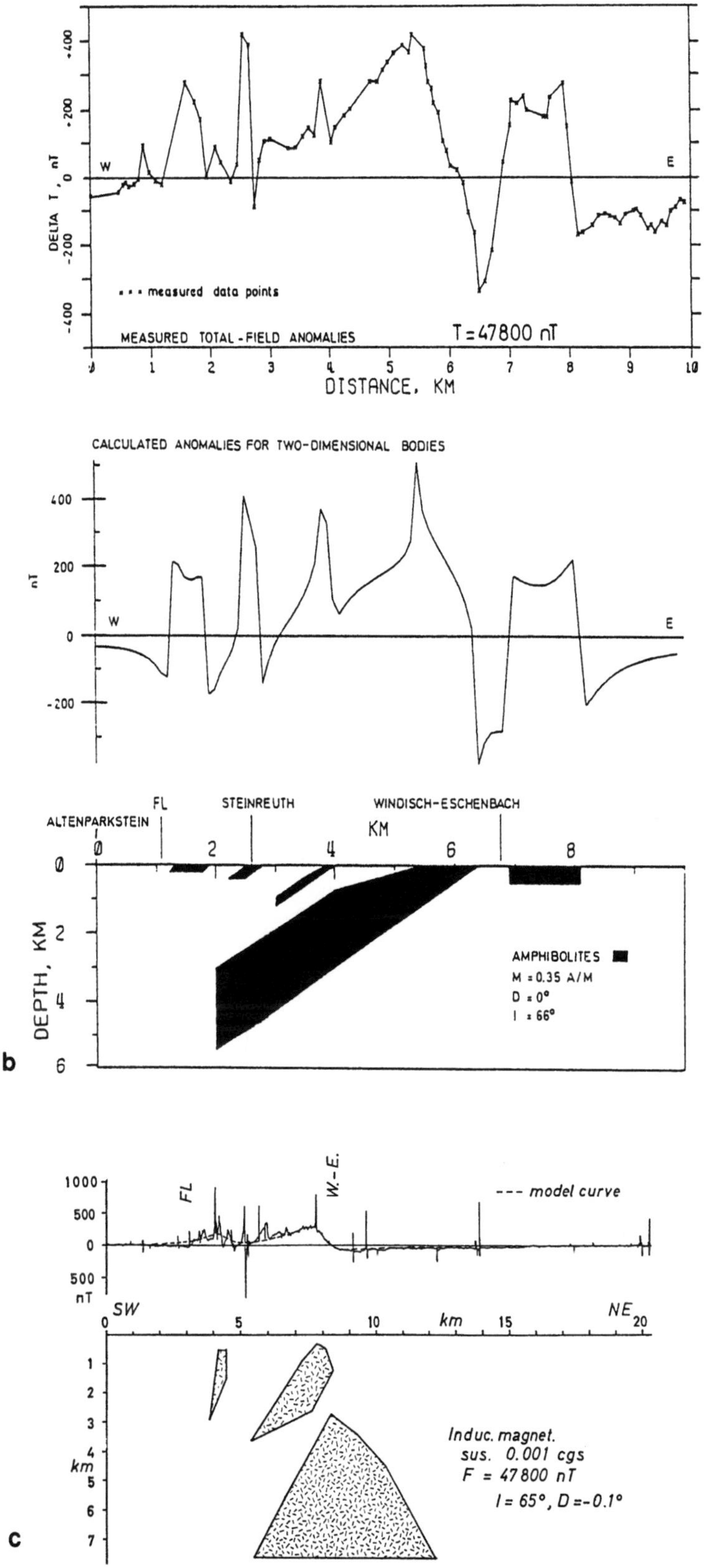

DELTA T, nT
W
E
measured data points
MEASURED TOTAL-FIELD ANOMALIES
T=47800 nT
DISTANCE, KM
CALCULATED ANOMALIES FOR TWO-DIMENSIONAL BODIES
nT
ALTENPARKSTEIN
FL
STEINREUTH
WINDISCH-ESCHENBACH
KM
DEPTH, KM
AMPHIBOLITES
M = 0.35 A/M
D = 0°
I = 66°
b
--- model curve
FL
W.-E.
nT
SW
NE
km
Induc. magnet.
sus. 0.001 cgs
F = 47 800 nT
I = 65°, D = -0.1°
c

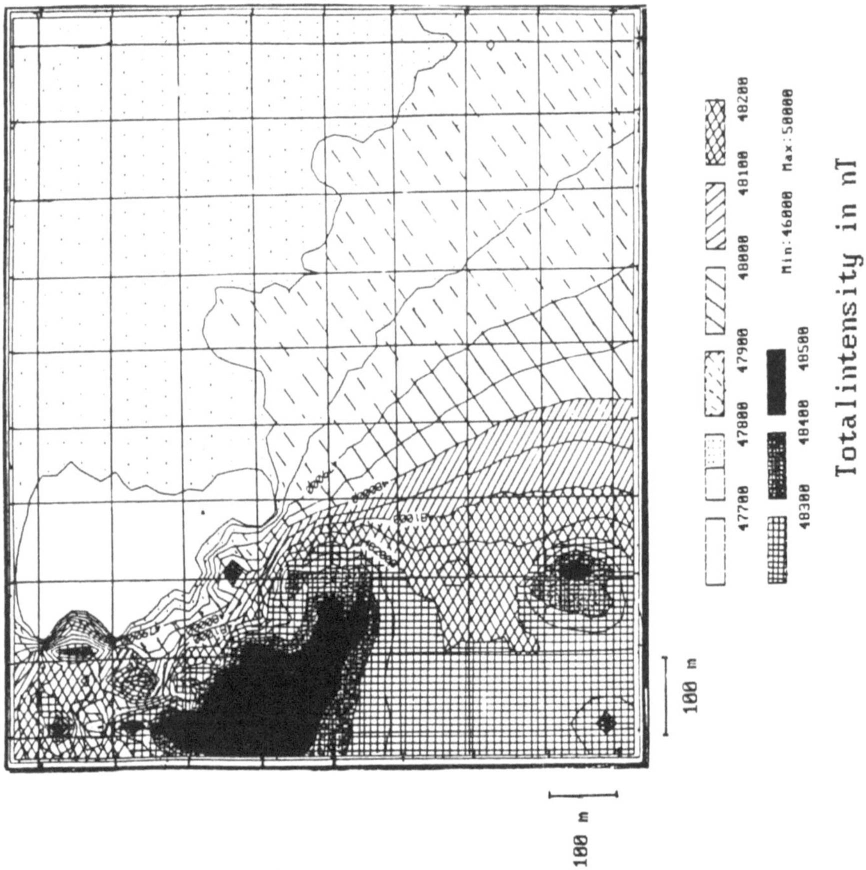
100 m
100 m
47700
47800
47900
48000
48100
48200
48300
48400
48500
Min:46000 Max:50000
Totalintensity in nT

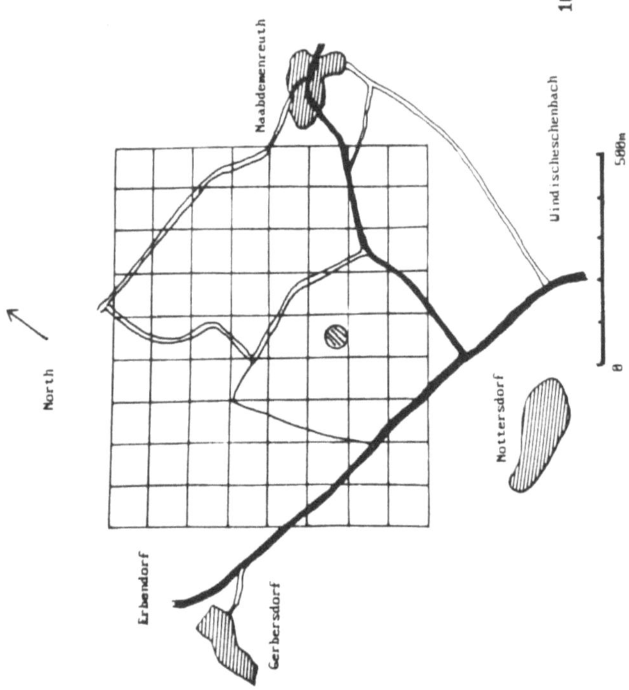
North
Naabdemenreuth
Windischeschenbach
Mottersdorf
Erbendorf
Gerbersdorf
0
500m

In the immediate vicinity of the drill site in the Upper Palatinate south of the village of Naabdehmenreuth, on an area of 1 x 1 km, a dense total magnetic intensity survey was made by Soffel et al. (1987) along profiles 50 m apart from each other with a distance of 10 m between stations. The map with the isolines is shown in Fig. 14. The positive anomalies in the northern part of the map are associated with outcropping amphibolites.

Other magnetic anomalies are also found further away from the drill site. North of Erbendorf there is a large positive anomaly caused by a serpentinite body. This anomaly was investigated by Hess (1964) with a ground magnetic survey of the vertical intensity. A strong positive aeromagnetic total intensity anomaly is found in the southern part of the ZEV near Vohenstrauß, which was modeled and interpreted by Eberle et al. (1977) with the help of additional ground magnetic profile measurements. This anomaly is thought to be caused by amphibolite bodies as well. This interpretation is supported by rock magnetic measurements made by Eberle et al. (1977) on outcropping amphibolites. In the eastern part of the Upper Palatinate strong local anomalies of different polarities point to Tertiary basalts. Smaller anomalies in the entire region are either caused by serpentinites or amphibolites, some of which have been investigated by Haak (1986). A review of the results of older field magnetic measurements in five regions in the western part of the Bohemian Massif has been published by Angenheister (1973). A helicopter aeromagnetic survey has been made in the northeastern part of the Upper Palatinate by the NLFB in the 1970s, however, this survey unfortunately does not reach the area of the drill site.

2.5 Comparison of the Magnetic Measurements in Both Areas and Conclusions

In the following we will only discuss and compare the anomalies of the earth's magnetic field near both drill locations.

The site in the Black Forest is characterized by a quite smooth magnetic field, which is rather exceptional for an area with outcropping crystalline basement rocks. A larger magnetized body could not be detected near the surface or at greater depth, at least not within the first 5 to 6 km. The observed small and only locally occurring variations of the geomagnetic field can presumably be associated with likewise small local variations of the magnetite or pyrrhotite content. This is confirmed by rock magnetic measurements on drill cores from the Black Forest. In lateral as well as vertical extension the magnetic properties seem to be uniform or changing only slightly.

In contrast, the drill site in the Upper Palatinate is situated within the maximum of a strong aeromagnetic and ground magnetic anomaly, which is associated with outcropping and deep-reaching amphibolite bodies or possibly also strongly magnetized contacts between the amphibolites and their adjacent rocks (gneisses). Strong lateral and vertical variations of the magnetic properties seem to be characteristic for this drill site, at least within the first 5 to 6 km. No conclusions are possible for

Fig. 14. Isolines of the total intensity of the earth's magnetic field of a special ground survey around the drill site in the Upper Palatinate (Soffel et al. 1987). For technical details see text. Hachured circle: drill site

greater depths at either drill sites. The Upper Palatinate locality therefore seems to be more attractive for magnetic investigations on cores and for magnetic borehole measurements, because of the variations of several magnetic parameters with depth.

The group of scientists responsible for the choice of the drill site therefore favoured, besides many other arguments, the Upper Palatinate drill site because of its more inhomogeneous uppermost crust compared with the more homogeneous upper crust at the Black Forest site.

Acknowledgements The measurements in the field and the interpretation of the data were possible only with the help of many colleagues and students from the institutions of the various groups. We greatly appreciate their support, without which a large part of the data could not have been acquired in a relatively short time. Thanks are also due to the Deutsche Forschungsgemeinschaft and the corresponding surveys (NLFB Hannover, Bayerisches Geologisches Landesamt in München, Landesvermessungsämter in Stuttgart, Karlsruhe and München) for their financial and logistic support.

References

Angenheister G (1973) Die Interpretation der magnetischen Störfelder (Anomalien) von mehreren Serpentinit-Körpern in fünf Arealen im Westen der Böhmischen Masse. Geol Bavarica 67: 35-63

Bücker C (1986) Die Anomalien der Schwere im Bereich der Kontinentalen Tiefbohrlokation (KTB) Oberpfalz und ihre Interpretation. Diss, Fak Geowiss, Univ München

Bundesanstalt für Geowissenschaften und Rohstoffe, BGR (ed) (1976) Karte der Anomalien der Totalintensität des erdmagnetischen Feldes in der Bundesrepublik Deutschland, 1:500 000. BGR, Hannover

Eberle D, Schmidt H, Stettner G (1977) Bericht über die aeromagnetische Anomalie Vohenstrauß (Opf.) in Verbindung mit magnetischen Untersuchungen am Erdboden. Rep Niedersächs Landesamt Bodenforsch, Hannover, Arch No 68270

Eisbacher G, Krohe A (1985) The tectonic setting of the Variscan Schwarzwald - a brief synopsis. 2nd Int Symp Observ Cont Crust through Drilling, Seeheim, Excursion Guide Schwarzwald

Gajewski D, Prodehl C (1985) Refraction seismic investigation, Schwarzwald. 2nd Int Symp Observ Cont Crust through Drilling, Seeheim, Deutsche Forschungsgemeinschaft

Gerke K (1957) Die Karte der Bouguer-Isanomalen 1:1 000 000 von Westdeutschland. Dtsche Geodät Komm, B 46, I

Götze H J (1986) Schwarzwald Gravimetrie/Magnetik. Deutsche Forschungsgemeinschaft (unpublished)

Haak V (1986) Detail-Magnetik in der Oberpfalz (Erbendorfer Linie und ZEV). Deutsche Forschungsgemeinschaft, unpublished

Hess Chr (1964) Die Delta-Z-Anomalien im Gebiet der Serpentinitvorkommen bei Erbendorf/Oberpfalz und ihre Interpretation. Dipl Arb Inst Angew Geophys, Univ München

Lüschen E, Menges D, Rühl T H, Sandmeier K J, Wenzel F, Gowin J, Janoth W, Keller F, Stiller M, Söllner W, Trappe H (1985) Presite seismic reflection survey. Abstr 2nd Int Symp Observ Cont Crust through Drilling, Seeheim

Plaumann S (1976) Schweremessungen im Bereich Erbendorf. Rep Niedersächs Landesamt Bodenforsch, Arch No 75564, Hannover (unpublished)

Plaumann S (1986) Die Schwerekarte der Oberpfalz und ihre Bezüge zu Strukturen der oberen Erdkruste. Geol Jahrb E 33: 5-13

Plaumann S, Groschopf R, Schädel K (1986) Kompilation einer Schwerekarte und einer geologischen Karte für den mittleren und nördlichen Schwarzwald mit einer Interpretation gravimetrischer Detailvermessungen. Geol Jahrb E 33: 15-30

Pucher R (1985) Magnetfeldmessungen im Gebiet Haslach/Kinzig im Nordschwarzwald zur Vorbereitung der Kontinentalen Tiefbohrung. Rep Niedersächs Landesamt Bodenforsch, Hannover, Arch No 97033, (unpublished)

Pucher R (1986a) Interpretation der magnetischen Anomalie bei Baden-Baden (Schwarzwald). Geol Jahrb E 33; 53-64

Pucher R (1986b) Interpretation der magnetischen Anomalie von Erbendorf (Oberpfalz) und dazugehörige gesteinsmagnetische Untersuchungen. Geol Jahrb E 33; 31-52

Pucher R (1986c) Magnetische Übersichtsmessungen der Kerne der geothermischen Erkundungsbohrungen in der Umgebung von Haslach (Schwarzwald). Rep Niedersächs Landesamt Bodenforsch, Hannover, Arch No 99995

Seeheim Conference (1985) 2nd Int Symp Observ Cont Crust through Drilling, Seeheim. Deutsche Forschungsgemeinschaft

Seeheim Conference (1986) 2nd KTB-Kolloq. Ergebnisse der Vorerkundungsarbeiten Lokation Schwarzwald und Lokation Oberpfalz. Deutsche Forschungsgemeinschaft

Soffel H, Schult A, Bücker C, Waitz R (1987) Feldmagnetische und radiometrische Messungen am Bohrplatz KTB Oberpfalz. Deutsche Forschungsgemeinschaft (unpublished)

Wagener M, Götze H J (1987) Gravity anomalies of the Black Forest related to lithospheric structures. (Abstract). Terra Cognita 7 (2/3): 167

Wagener M, Götze H J (1988) Regional three-dimensional gravity investigations in the Black Forest, West Germany. Tectonophysics (in press)

Woolard G P (1975) The regional changes in gravity and their relation to crustal parameter. Bull Int Bur Gravimet 36: 106-110

Geothermal Investigations in the KTB Locations Oberpfalz and Schwarzwald

H. Burkhardt[1], U. Haack[2], A. Hahn[3], H. Honarmand[1], K. Jäger[4], A. Stiefel[4], P. Wägerle[1] and H. Wilhelm[4]

[1] Institut für Angewandte Geophysik, Technische Universität Berlin, Ackerstr. 71-76, 1000 Berlin 65, FRG
[2] Geochemisches Institut, Universität Göttingen, Goldschmidtstr. 1, 3400 Göttingen, FRG
[3] Niedersächsisches Landesamt für Bodenforschung, Stilleweg 2, 3000 Hannover 51, FRG
[4] Geophysikalisches Institut, Universität Karlsruhe, Hertzstr. 16, 7500 Karlsruhe 21, FRG

1 INTRODUCTION

Temperature as a thermodynamic variable is of fundamental importance for the physical and chemical properties of rocks at depth as well as for temperature-controlled processes within the earth such as, e.g., deformation and metamorphic reactions. Most of these processes have certain threshold temperatures beyond which the reaction rate increases significantly. On the other hand, the temperature is a stringent limiting factor for the use of drilling tools and borehole logging instruments. Therefore the investigation of the temperature field within both KTB target areas was an important task of the pre-drilling site investigations, in order to provide a reliable basis for predicting the temperature at the projected final depth of the borehole. For this purpose it is necessary to obtain as precisely as possible reliable values of the undisturbed near-surface heat flow for the area under investigation.

The most serious factor disturbing the determination of heat flow is the influence of convection on the measured temperature gradient. It has been shown (Drury 1984) that the influence of convective heat transport beneath the bottom of a borehole may not be detected in the measured temperature gradient. Therefore as long as convection and lateral temperature variations cannot be ruled out, no single borehole, irrespective of its depth, can give truly reliable information on the required regional heat flow. This can, at least, only be provided by measurements in a regional array of boreholes. Prior to the geothermal KTB-investigations no such comprehensive information was available in both areas.

Consequently, to detect local perturbations of the temperature gradient by water movements in the vicinity of a borehole, regional borehole arrays were designed and drilled, in the central Schwarzwald and in the Oberpfalz. The array patterns were selected by the requirement not only to cover the whole region under investigation, but also to comprehend all major geological-petrological units, to obtain valuable information for structural geology and petrology, too. To avoid near-surface effects such as lateral heterogeneities and climatic influences, it was decided that the depth of the holes should be at least 150 m and that the maximum depth be less than 300 m because of limited funds.

In addition, other available boreholes drilled for commercial or public purposes in these regions were utilized for geothermal measurements.

The geologic and tectonic evolution of the Oberpfalz and the Schwarzwald crust is of fundamental importance for its structural and compositional properties, which are intimately related to the crustal temperature field and heat flux. Also its hydrological properties, which can substantially influence the mechanism of heat transfer, are essentially based on the tectonic style which has evolved from the Variscan orogeny and, in addition, in the Schwarzwald from the Rhinegraben rift process.

The present-day thermal surface heat flux may still contain signatures of the past. In the Schwarzwald, hydrothermal alterations and thrust tectonics may have changed substantially the thermal and seismic properties of the rocks at depth. Granitization and granulitization may have caused enhanced or reduced concentrations of radiogenic elements, respectively, and a redistribution mainly of quartz with strong consequences for the thermal conductivity of the corresponding rocks.

2 GEOTHERMAL INVESTIGATION IN THE OBERPFALZ AREA

2.1 Geothermal Drill Sites

The crustal structure of the Oberpfalz is characterized by intensive Variscan overthrust tectonics, resulting in an exposure of three main crustal zones, which can be clearly delineated according to their different tectonic development: Saxothuringian, Moldanubian and several metamorphic nappe complexes. The target area is situated on the western boundary of the Bohemian Massif in the southwestern continuation of the tertiary-Pleistocene Eger-rift.

This complex geologic and tectonic situation, with its implications for the geothermal field, has been the starting point for the selection of the drill sites for the geothermal investigations. Prior to these investigations only data from three holes were available with different levels of accuracy. As far as possible, these data were re-evaluated and supplemented, but even so could not suffice for a reliable regional survey.

Therefore, following the criteria as outlined in the introduction, nine new drill sites were selected, covering all main geologic-tectonic provinces. This not only ensured representative regional heat flow values, but also the necessary information about the physical and petrological parameters of the main rocks, which were supposed to be constituent parts of the crust in this region (see Fig. 1). Two drill holes were positioned in the ZEV-complex, one of them (PÜ) near a prospective site for the ultradeep well, the other (PE) in amphibolitic rock. For the majority (five) of drillhole sites in the Saxothuringian were chosen, covering different rock units: predominantly granites (WE), micaschists (WÜ and NA), metavolcanics and phyllites (NE), and redwitzits (RÖ). One site (GR) belongs to the Moldanubian (gneisses) and one hole (PO) was drilled within the boundary zone between Saxothuringian and Moldanubian. The already existing holes, which could be utilized, were all situated in the Saxothuringian, comprising Münchberger gneisses (WS), Falkenberg-granites (FA), phyllites and chlorites (AL) and quartzitic schists (HA).

The KTB-holes have depths between 140 m and 300 m and were completely cored for subsequent geophysical, geological and petrological investigations. Data for the other holes were taken from publications (Hänel 1971, 1974; Hänel and Schellschmidt 1984; Bader 1971); in one case (HA) the missing laboratory measurements from thermal conductivity on rock samples had to be supplemented.

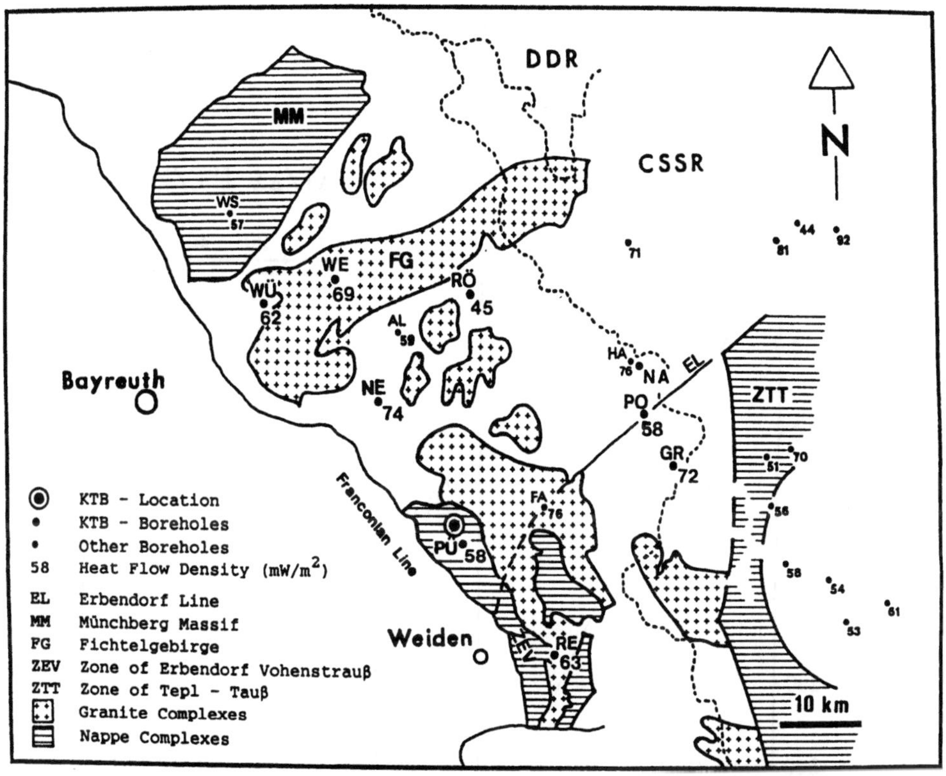

Fig. 1. Heat flow values in mW/m^2 of the Oberpfalz (this investigation) and the Western Bohemian Massif. (Cermak 1975)

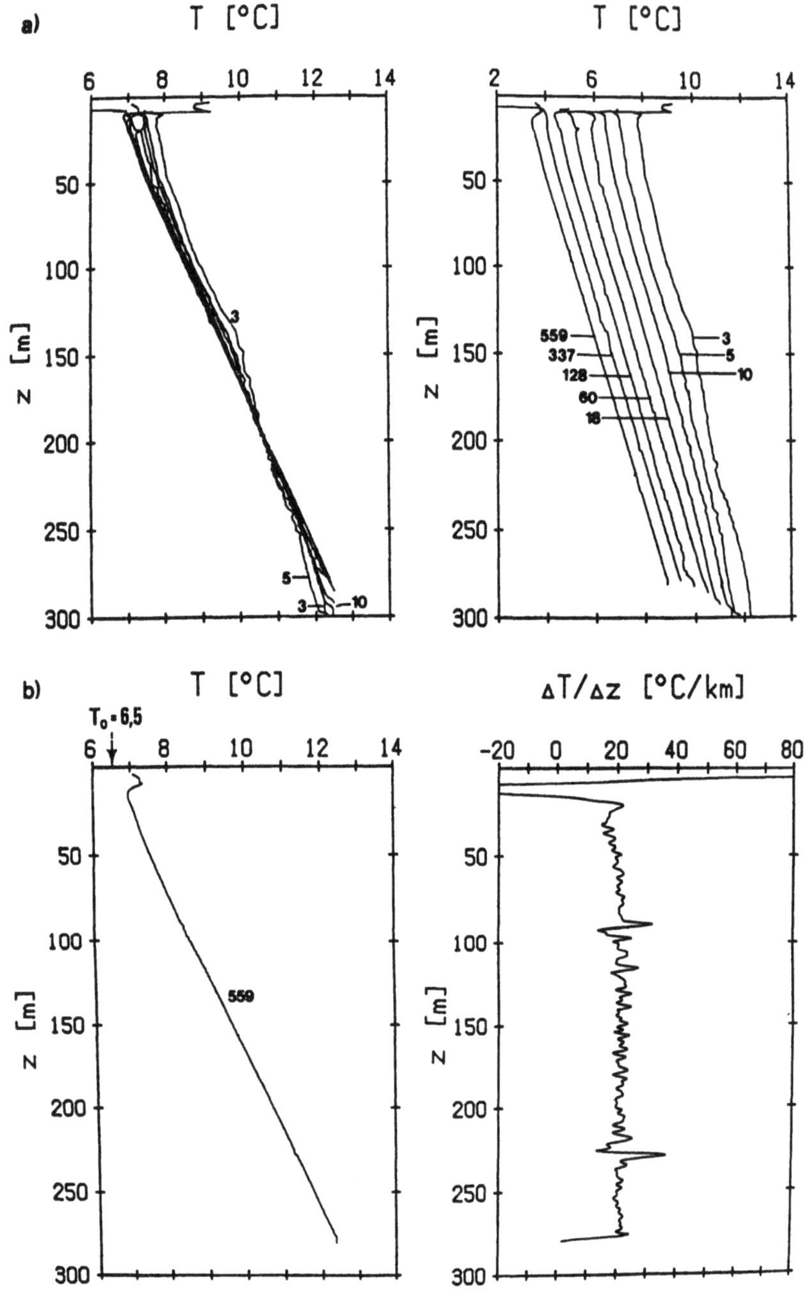

Fig. 2a. Temperature logs of the Griesbach-well. Label: shut-in-time in days. Left: True temperature logs. Right: The log with shut in time 3 corresponds to the true temperature; the other logs are shifted by 0.5°C respectively

Fig. 2b. Temperature and temperature gradient records for the Griesbach-well. Label: Shut-in-time in days

2.2 Temperature Measurements

In all KTB-wells continuous high resolution and bottom hole temperature measurements were carried out during drilling stops to observe the influence of possible water movements in the vicinity of the boreholes. After completion of the drilling, repeated runs of continuous temperature logs were made to determine the equilibrium temperature field prior to drilling. This adjustment process can be seen in Fig. 2, where temperature logs at different times after drilling are shown for the Griesbach well. After 559 days there are no significant disturbances left, indicating equilibrium conditions. Except for the Neualbenreuth well, all other KTB-boreholes yielded similar undisturbed temperature profiles after appropriate waiting times, as shown in the compilation of Fig. 3.

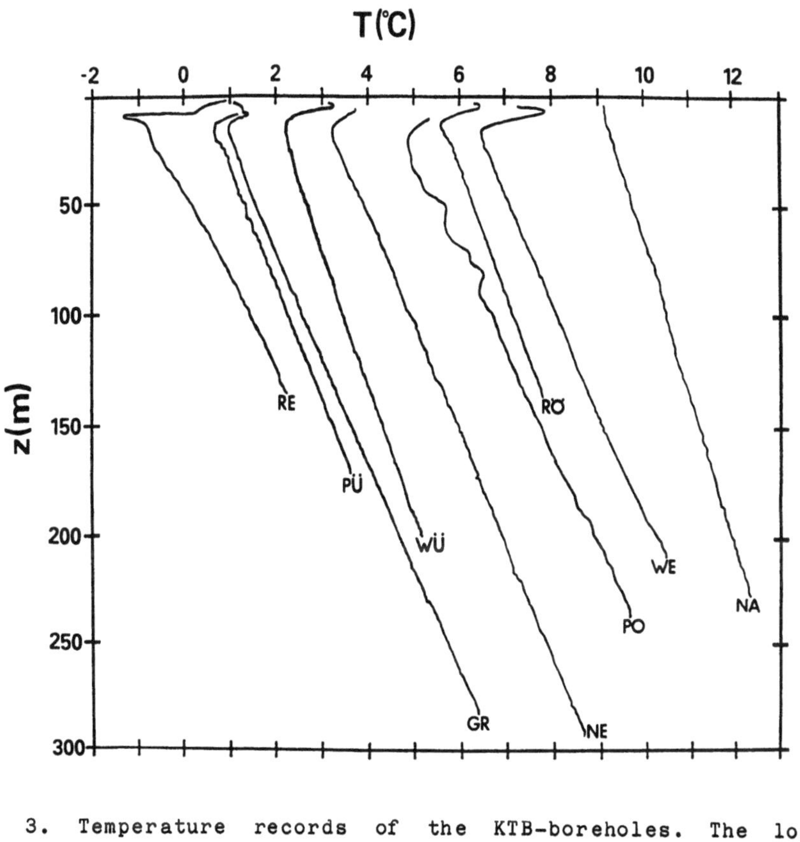

Fig. 3. Temperature records of the KTB-boreholes. The log WE corresponds to the true temperature; the other logs are shifted by 0.5°C respectively

2.3 Corrections

The measured near-surface temperature field is influenced by several factors. The most relevant are topography in the vicinity of the well, climatic history and erosion, which all have to be taken into account and corrected for. Topographic corrections for the temperature gradient were calculated by means of a 3-D finite-difference program. The results indicate that due to only little topographic relief in the vicinity of the drill sites the topographic corrections are relatively small, with values ranging between 0.01 to 0.6 K/km.

Various geological processes can be reflected in the subsurface temperature field. For these investigations erosion, connected with tectonic uplift and with glacial processes may be of importance. Estimations on the basis of published uplift rates (Wagner et al., this volume), however, led to the conclusion that erosion does not play a significant role and that the corresponding correction can be neglected for the Oberpfalz region.

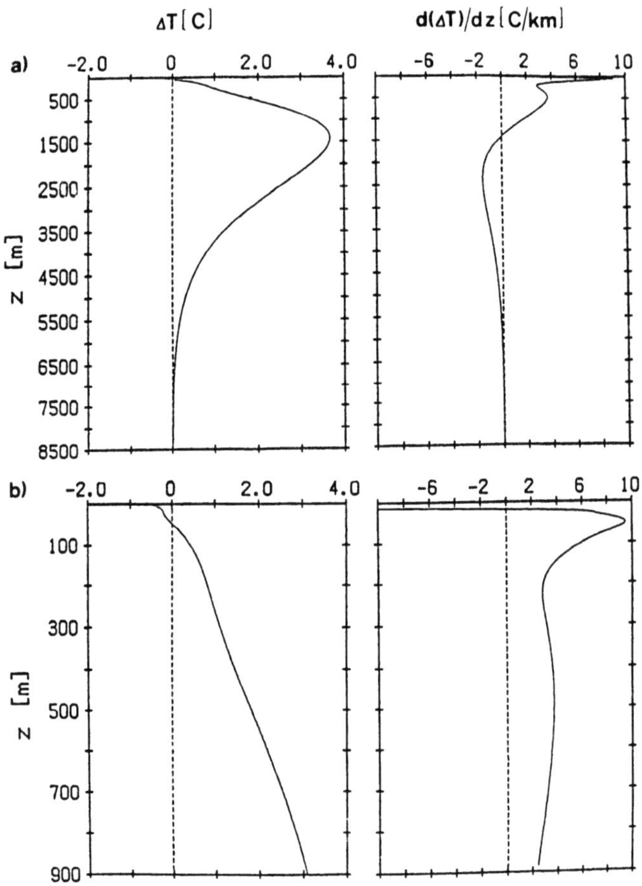

Fig. 4. The influence of paleotemperatures down to a: 8500 m; b: 900 m. Assumed thermal diffusivity: 1.26 10^{-6} m^2/s . Corrections for temperature (left) and temperature gradient (right) must be added to actual values

The influence of climatic history on the temperature gradient has long been a matter of debate with regard to the way of taking it into account. Although there are some recent investigations demonstrating a successful application of paleoclimatic corrections, the convincing proof that these corrections are necessary in all cases is still missing. To estimate the magnitude of the paleoclimatic influence temperature gradient corrections were calculated for a homogeneous half-space and with a surface temperature time series consisting of temperatures from Bayreuth for the last 130 years of observation in combination with data up to 100.000 years backward after Hänel (1983a) (Fig. 4). Temperature correction is maximum at 1500 m depth; the gradient correction shows the largest variations in the upper 150 m, then remaining relatively constant down to 900 m depth. From this follows that with regard to the uncertainty of paleoclimatic corrections the minimum depth of boreholes for heat-flow determinations should be 150 m approximately.

In the considered depth range the correction will always tend to increase the measured temperature gradient, resulting in higher heat flow values. In the present case this increase is in the order of 15% to 25% for the different drill holes.

A comparison of measured and corrected temperature profiles for the KTB-wells did not show any decisive hint towards the necessity of applying paleoclimatic corrections (see Fig. 6). So it was decided to use the data without this correction, but to include it in the error estimation (see also Figs. 8, 9).

2.4 Thermal Conductivity

Thermal conductivity was measured on water-saturated core samples with three different methods (modified divided bar, half space line source, quick thermal conductivity-meter QTM) at ambient pressure and temperature. Data were consistent within 5%. Most of the data were measured in the direction parallel to the borehole axis, but in some cases it was necessary to consider anisotropy, too. Fig. 5 shows the results for the Griesbach and the Neusorg wells. Whereas the Griesbach data indicate constant values for the entire depth range there is a distinct change of k-values in the Neusorg data corresponding to the petrological profile with a change from intermediate volcanics to phyllites in 180 m depth.

Radiogenic heat production was determined (see Section 4) for a large number of representative core samples.

2.5 Surface Heat Flow

Assuming pure heat conduction heat flow is determined as the product of thermal conductivity and temperature gradient. A modification of this straightforward procedure, the thermal resistance method of Bullard (1939) was used which can give more information about the depth dependence and about possible disturbances of the heat flow. In this method the thermal resistance,

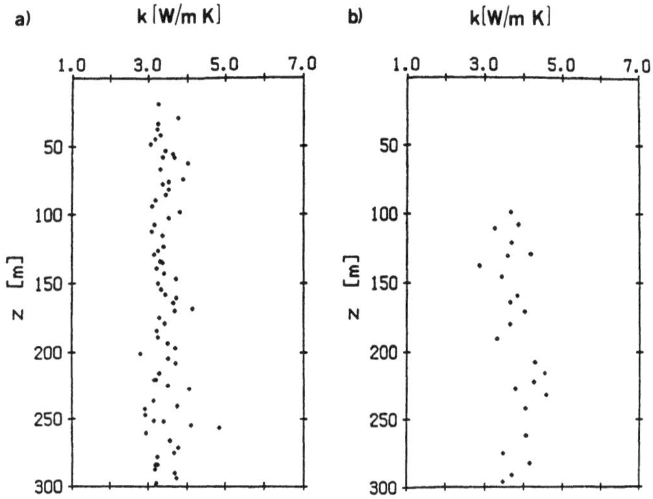

Fig. 5. Depth profiles of the thermal conductivity k for the boreholes Griesbach (a) and Neusorg (b)

defined as the ratio of layer thickness and of the corresponding thermal conductivity, and summed up to a specific depth, is plotted against temperature at this depth. The slope of this curve gives the heat flow; deviations from a linear dependence are indications of additional convective heat transport and/or of variations in the radiogenic heat production in that depth interval.

Figure 6 shows this so-called Bullard plot for the Griesbach well with the original data and with data corrected for the paleoclimatic influence. In the first case there is, in good approximation, a straight line with a slope, yielding a reliable surface heat flow density value of 72 mW/m^2, whereas a non-linear relationship results with the corrected values. Since neither contrasts in the radiogenic heat production nor convective influences in the temperature gradients were observed, it follows that in this case the paleoclimatic correction is not appropriate. The same evidence is coming from the results of all other logs, strengthening the conclusion, mentioned above, not to apply the paleoclimatic correction.

Table 1 gives heat flow data of this region, with topographic correction but without paleoclimatic correction. In some cases, where the thermal conductivity changes with depth mean values are listed, too.

The regional distribution of these data is shown in Fig. 1 together with data for the Bohemian Massif. The new values fit very well in the regional distribution which is typical for tectonically stable continental regions (Cermak and Bodri 1986).

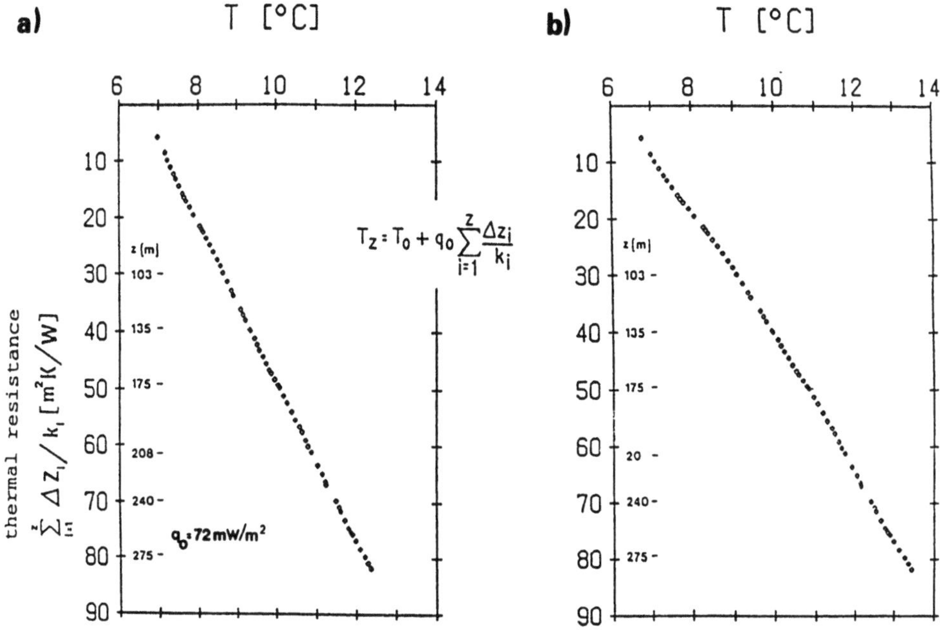

Fig. 6. Surface heat flow determined by thermal resistance method (Bullard). a) with topographic correction b) with additional climatic correction

The empirical discovery of Birch et al. (1968) and Roy et al. (1968) that for certain 'heat flow provinces' there is a linear relationship between surface heat flow and surface heat production opened the possibility to get some insight into the thermal structure of continents. Although the full implications of this relationship are not yet clear, it allows the observed surface heat flow to be separated into two components, heat generated within an upper crustal zone enriched in radiogenic isotopes and heat from below this zone, the so-called reduced or basal heat flow. Recent work however shows that interpretations of the actual distribution of heat sources in the crust are not possible unless quite detailed information about crustal structure and thermal properties are available (Morgan et al. 1987; Crowley 1987).

For the Oberpfalz region no such linear relationship exists (Fig. 7). Even if the values for the Röthenbach well (RÖ), where there are indications for convection-induced disturbances of the measured temperature gradient, and for the Weissenstadt well (WE), where the measured very high heat production rates may not be typical for the region under investigation, are excluded, the correlation is still very poor. This is also the case when only data for the same geological unit are considered. A possible reason for this behaviour is thought to be the complex geological structure of the Oberpfalz region with its complicated overthrust tectonics. Another possibility could be the additional effect of heat advection. But with regard to the regional distribution of heat flow values, this

Table 1. Surface heat flow values (topographically corrected) for the Oberpfalz region

Borehole	Depth (m)	$\Delta T/\Delta z$[a] (K/km) top.corr.	k[b] (W/(m K))	A[c] ($\mu W/m^3$)	q[d] (mW/m^2) calculated	q[d] (mW/m^2) means
Griesbach GR	300	21.1	3.43	2.10	72	72
Püllersreuth PÜ	300	19.3	3.02	1.84	58	58
Neusorg						
NE <170m	300	20.6	3.59	2.81	74	74
NE >170m	300	18.8	3.94	2.63	74	
Röthenbach* RÖ	140	16.9	2.64	3.45	45	45
Remmelberg RE	140	22.9	2.73	0.19	63	63
Weissenstadt						
WE <160m	220	19.4	3.46	9.00	67	69
WE >160m	220	21.3	3.37	10.60	72	
Wülfersreuth WÜ	220	17.1	3.61	2.33	62	62
Poppenreuth						
PO <160m	240	?	3.56	2.40	?	58
PO >160m	240	22.8	2.54	1.76	58	
Neualbenreuth A NA	250	-	3.42	1.30	-	-
Hardeck						
HA <210m	502	20.4	3.62	2.56	74	76
HA >210m	502	16.8	4.64	-	78	
Falkenberg FA	500	22.0	3.45	6.74	76	76
Alexanderbad AL	115	17.3	3.43	-	59	59
Weissenstein WS	231	20.0	2.84	0.22	57	57

a: temperature gradient; b: thermal conductivity; c: heat production; d: surface heat flow; * temperature gradient probably disturbed

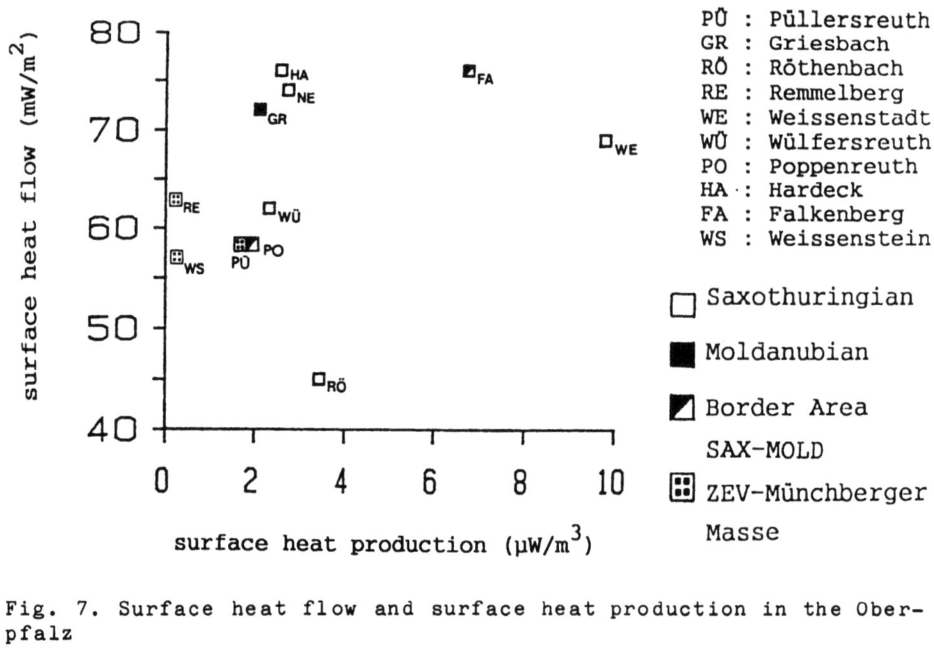

Fig. 7. Surface heat flow and surface heat production in the Oberpfalz

would have to be a very complicated convection system. An important contribution towards the solution of this problem can be expected from the ultra deep borehole.

2.6 Geothermal Models

The principle goal of the geothermal investigation was the prediction of temperature at depth. To infer these temperatures from surface heat flow, data model calculations are necessary. The type of models can only be assumed on the basis of the empirical evidence. This evidence includes the regional distribution of heat flow data and lithological models of the crust.

From the rather heterogeneous distribution of heat flow values in the Oberpfalz region and in the adjacent part of the Bohemian Massif, there is no indication of regional trends. Therefore it can be concluded that there is no stringent evidence for large-scale convective type of heat transport and within a sufficient degree of approximation it was possible to restrict the calculations to purely conductive mechanisms of heat transport and to neglect heat advection.

As a first step, one-dimensional basic models were constructed for several regions, derived from seismic, gravimetric, geological, and petrological data (Gebrande, Schmoll, Bucker, Weber, Vollbrecht, this Vol.). Necessary model parameters are thermal conductivity and heat production. They were evaluated on the basis of the measurements on characteristic Oberpfalz-core samples, supplemented by data from literature (Haack 1982; Cermak and Rybach 1982; Touloukian 1981). Thermal conductivities were corrected for in situ pressure and temperature (Schön 1983; Cermak 1982b). For deeper parts of the models thermal conductivities were also estimated from seismic data by means of thermal conductivity - compressional wave velocity correlations (Karl et al. 1965). A linear decrease of heat production with depth has been assumed.

The results of model calculations for two specific regions are shown in Figs. 8 and 9. The region ZEV includes the actual drill site of the KTB ultra deep hole; the boundary values are based on data from the Püllersreuth drillhole (PÜ). Calculations were made with surface heat flow values corrected for topography (q0).

For all corresponding basic models with the boundary value q0 temperatures at 15 km depth are quite similar. The same is true for the basal heat flows at 30 km depth (mantle heat flow) with values corresponding to results for the Western Bohemian Massif (Cermak and Bodri 1986). So it can be concluded that the relatively small regional differences in surface heat flow are most probably due to sources within the upper crust.

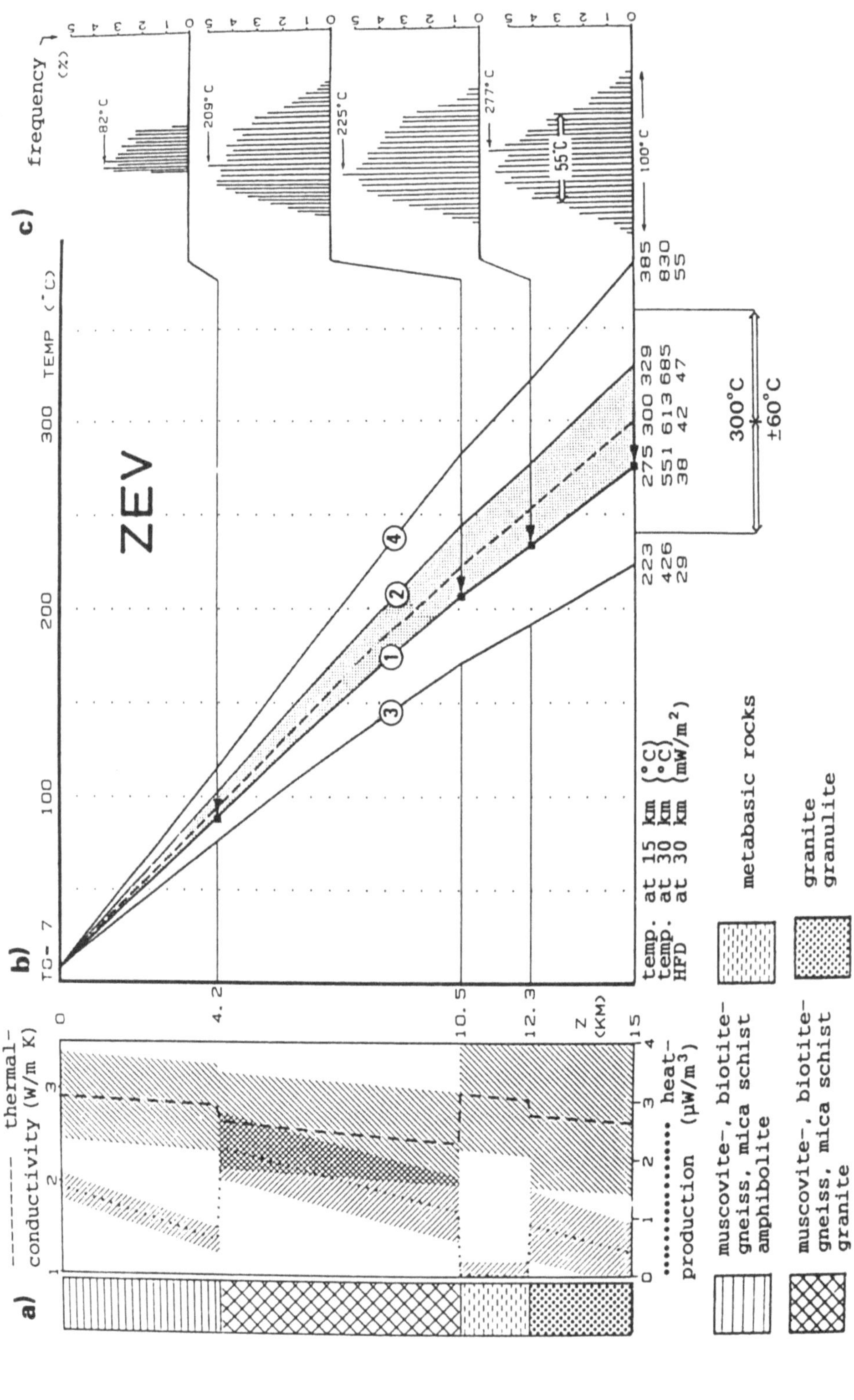
a)
b)
c)
ZEV
thermal-
conductivity (W/m K)
heat-
production (µW/m³)
Z
(KM)
4.2
10.5
12.3
15
TG- 7
100
200
300
TEMP (°C)
frequency
(%)
82°C
209°C
225°C
277°C
55°C
100°C
temp. at 15 km (°C)
temp. at 30 km (°C)
HFD at 30 km (mW/m²)
223 426 29
275 551 38
300 613 42
329 685 47
385 830 55
300°C
±60°C
muscovite-, biotite-gneiss, mica schist amphibolite
muscovite-, biotite-gneiss, mica schist granite
metabasic rocks
granite granulite

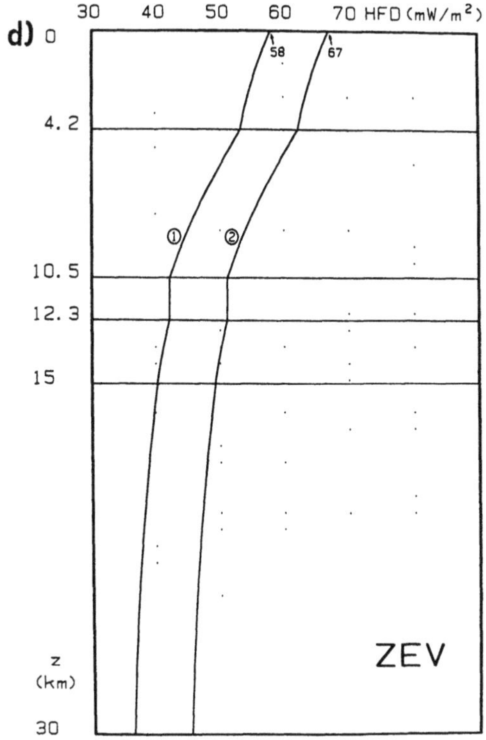

Fig. 8. One-dimensional calculations for the ZEV region

Fig. 8a. Left: Geological model, derived from seismic and gravity data. Right: Assumed heat production and thermal conductivity (temperature and pressure dependent). Hatched areas show the range of parameters used by Monte-Carlo method in part c

Fig. 8b. Calculated temperature-depth profiles assuming a variation of surface heat flow

Curve 1: q0 = 58.0 mW/m^2 Surface heat flow topographically corrected value

Curve 2: q1 = 66.7 mW/m^2 (q0+15%) Surface heat flow with maximum paleoclimatic correction

Curve 3: q3 = 49.3 mW/m^2 (q0-15%) Surface heat flow with lower bound of estimated correction for influence of convection and measuring errors in the determination of q0

Curve 4: q2 = 75.4 mW/m^2 (q0+30%) Surface heat flow with upper bound of estimated correction including maximum paleoclimatic influence

Below 15 km: In situ thermal conductivity corresponding to 2.8 W/(m K) at 25°C
Heat production 0.4-0.05 μW/m^3 (linear decrease)

Fig. 8c. Frequency distribution of temperatures at different depths assuming a variation of parameter values with constant surface heat flow 58 mW/m^2 (Monte Carlo method with 10.000 model calculations and assumed uniform distribution of parameter values within the hatched areas in a)

Fig. 8d. Calculated heat flow versus depth distribution for the model ZEV with assumed surface values 58 mW/m^2 and 67 mW/m^2 as indicated

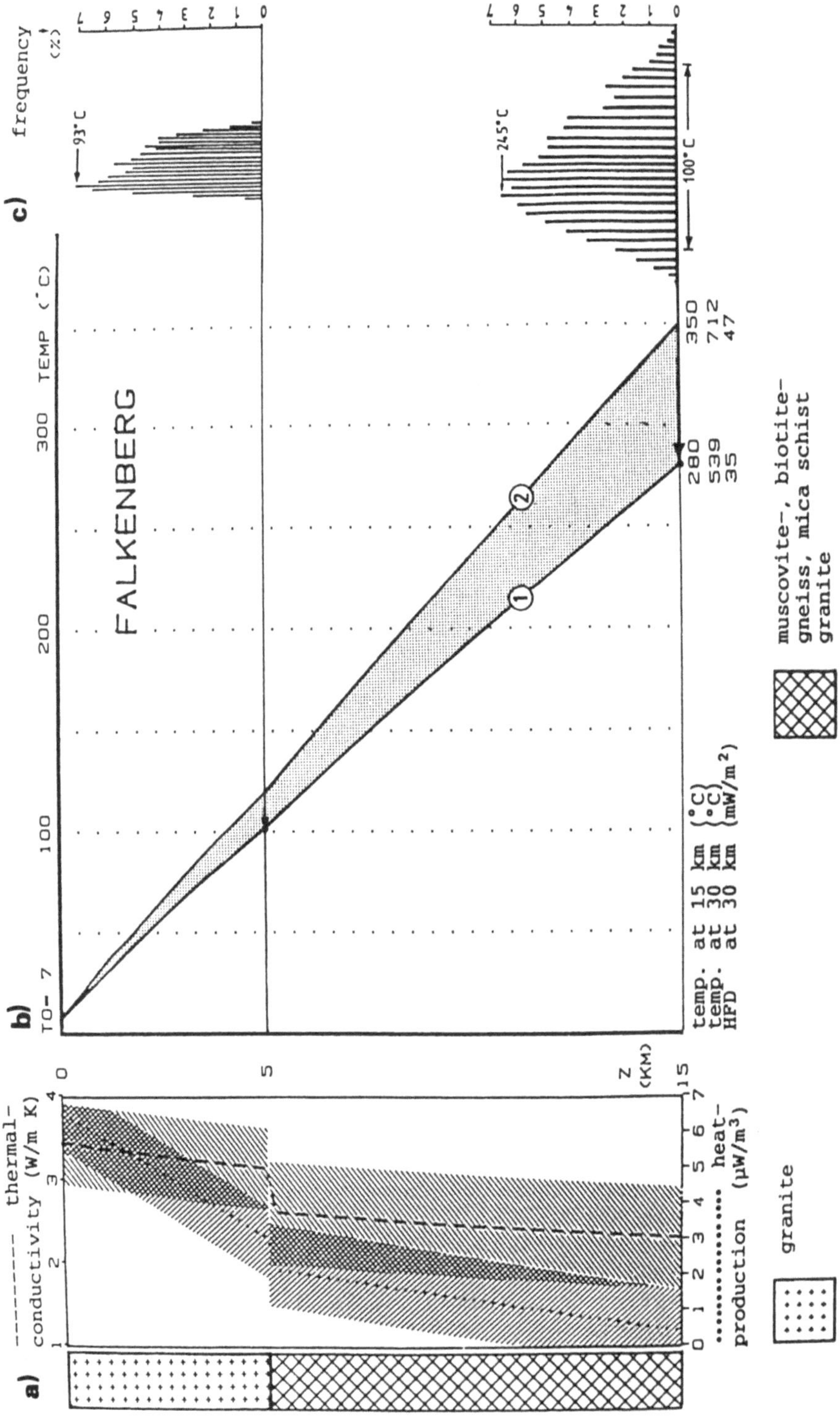
a)
thermal-conductivity (W/m K)
heat-production (μW/m³)
granite
muscovite-, biotite-gneiss, mica schist granite
b)
FALKENBERG
T0- 7
100
200
300
TEMP (°C)
Z (KM)
temp. at 15 km (°C)
temp. at 30 km (°C)
HFD at 30 km (mW/m²)
280
539
35
350
712
47
c)
frequency (%)
93°C
245°C
100°C

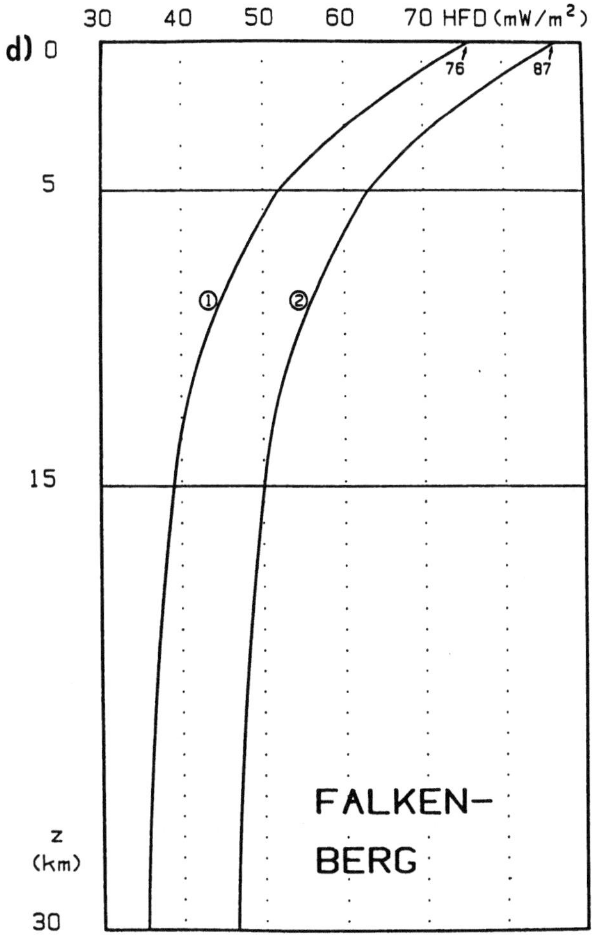

Fig. 9. One-dimensional calculations for the Falkenberg region

Fig. 9a. Left: Geological model, derived from seismic and gravity data. Right: Assumed heat production and thermal conductivity (temperature and pressure dependent). Hatched areas show the range of parameter values used by Monte-Carlo method on part c

Fig. 9b. Calculated temperature-depth profiles assuming a variation of surface heat flow

Curve 1: q0 = 76.0 mW/m^2 Surface heat flow topographically corrected value

Curve 2: q1 = 87.0 mW/m^2 (q0+15%) Surface heat flow with maximum paleoclimatic correction

Below 15 km: In situ thermal conductivity corresponding to 2.8 W/(m K) at 25°C
Heat production 0.4-0.05 µW/m^3 (linear decrease)

Fig. 9c. Frequency distribution of temperatures at different depths assuming a variation of parameter values with constant surface heat flow 76 mW/m^2 (Monte-Carlo method with 10.000 model calculations and assumed uniform distribution of parameter values within the hatched areas in a)

Fig. 9d. Calculated heat flow versus depth distribution for the model Falkenberg with assumed surface values 76 mW/m^2 and 87 mW/m^2 as indicated

To estimate the influence of lateral inhomogeneities on the temperature field in the target area a two-dimensional model was derived for a profile crossing the Frankonian Line (FL) and the ZEV in east-west direction. Calculations were made with a finite-difference program (Clauser 1988) for purely conductive heat transport. The structural model is again based on geophysical and geological evidence, with the additional sedimentary layer and correspondingly displaced sublayers as major features for the region west of the Frankonian line (Fig. 10). A constant basal heat flux of 37 mW/m^2, as deduced from the one-dimensional model, was assumed. As shown in Fig. 10 there are only minor differences compared to the 1D model for the temperature at 15 km depth, demonstrating the small effect of this prominent tectonic feature.

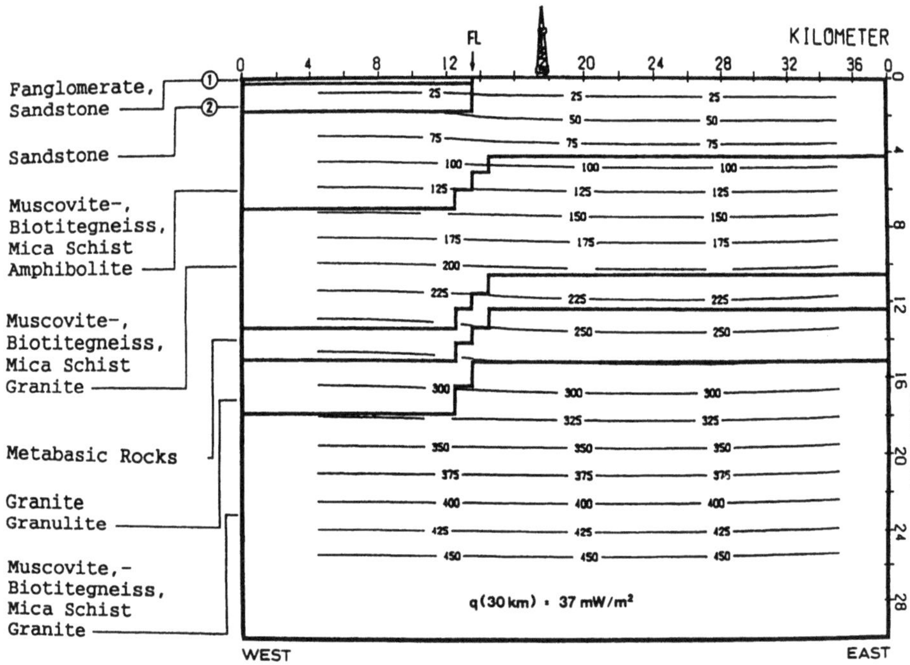

Fig. 10. Calculated 2D temperature distribution (°C) along EW-profile crossing the Franconian Line (FL) with model layer boundaries superimposed

	Thermal conductivity	Heat production
1	1.8 W/(m K)	0.7 $\mu W/m^3$
2	2.2 W/(m K)	0.7 $\mu W/m^3$

2.7 Error Estimation for Temperature Predictions

The reliability of temperatures at depth predicted from surface measurements is heavily restricted by a number of possible sources of error, the most important ones being firstly the error in the surface heat flow, and secondly the uncertainty of the model and its parameters.

The following errors in the deduced surface heat flow values have been considered:

- measuring errors for both temperature gradient and thermal conductivity depend on a large number of sources, e.g., number of measurements, measuring point density, disturbances of the temperature gradient, transferability of laboratory data to in-situ conditions. For the present investigations, which are rather carefully controlled, a maximum measuring error for the heat flow value was estimated to ± 7%.

- another serious error can occur due to fluid movements beneath the borehole, which do not show up in the measured temperature gradient. According to Drury (1984), this influence can lead to errors up to ± 8%, approximately.

- for our basic models topographically corrected heat flow values (q0) were used without paleoclimatic correction. Although no conclusive evidence for an influence of the paleoclimate on the near-surface temperature field could be observed, it nevertheless cannot be ruled out completely. According to the assumed climatic history a maximum correction of +15% was deduced for a 300-m drillhole.

Certainly the decisive source of errors is due to insufficient information about the structure and thermal properties of the crust in this region. On the basis of the one-dimensional purely conductive model for the ZEV region, the influence of the model parameters on the predicted temperature at 15 km depth has been investigated. The results (Fig. 11) indicate that in general, changes of the thermal conductivity have a stronger influence on the temperature than have changes of the heat production.

Since it is not very probable that all the considered errors of the thermal parameters will add successively in the same direction, a statistical Monte-Carlo-method was used. For all models, the thermal parameters conductivity and heat production were randomly assumed with uniform probability within the indicated limits (see hatched areas in Figs. 8a and 9a). Although, of course, nothing is known about the actual statistical properties of the parameters, this assumption gives a conservative estimation of the effect. The results of 10,000 model calculations are shown in the right part (c) of Figs. 8 and 9, depicting the frequency distribution of temperatures in different depths referred to the basic models (q0).

The half-width of this distribution, e.g., ± 30°C at 15 km depth is considered as a plausible estimation of the error in the temperature at depth due to the uncertainties of model parameters.

To arrive at maximum error bounds for the temperature prediction, the surface heat flow with maximum errors was taken as boundary value, yielding the temperature curves 3 (q3 = q0-15%) and 4 (q2 = q0+30%). (Paleoclimatic correction (+15%) only positive.) This maximum estimation leads to a possible temperature variation between 220 and 380°C.

A more conservative and probable error estimation will lead to temperatures between the curves 1 (q0) and 2 (q1 = q0+15%) (paleoclimatic influence). This gives 300 ± 30°C for 15 km depth. In addition there is the error due to the model parameters, which amounts to another ±30°C. So in total the predicted temperature at

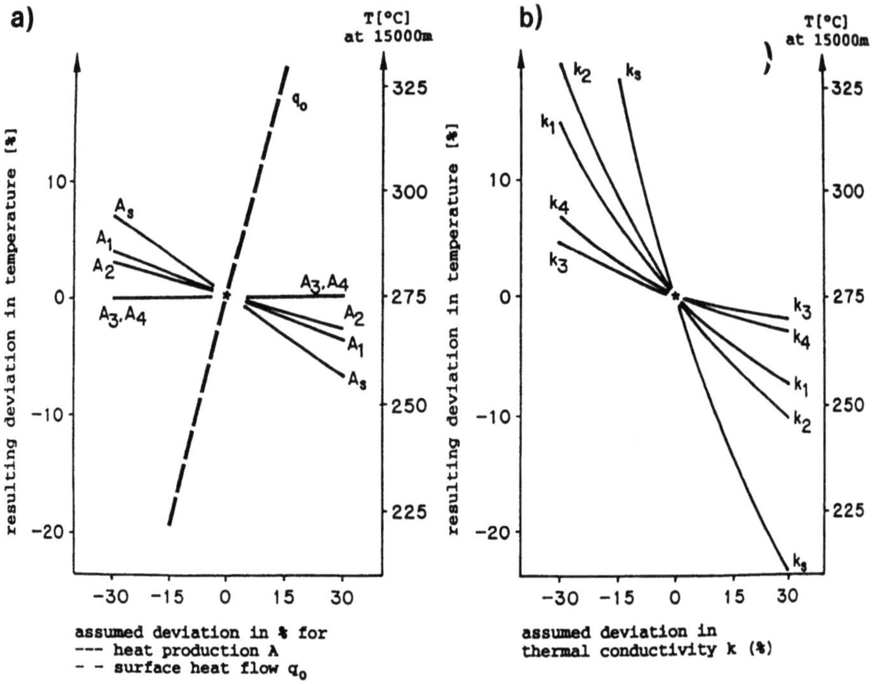

Fig. 11. Influence of the model parameter deviations (see Fig. 8, ZEV region) on the predicted temperature. A heat production, k thermal conductivity, q_o surface heat flow, 1-4 number of layer, s deviation for all layers, * temperature at 15 km depth for the basic model

Fig. 11a. Calculated temperature deviation from temperature at 15 km depth is derived assuming errors in heat production and surface heat flow (in %).

Fig. 11b. Calculated temperature deviation assuming errors in thermal conductivity (in %) for temperature at 15 km depth

15 km depth is estimated to be 300°C$\pm$60°C for the target area of the ultra deep KTB-drillhole.

Possible effects of heat convection have not been included in this temperature prediction since there is virtually no evidence for this type of heat transport. If, despite this lacking evidence, there were still such convection processes active, their influence would tend to lower the predicted temperatures.

2.8 Summary and Conclusions

In the course of the KTB geothermal survey of the Oberpfalz nine new surface heat flow values in special scientific boreholes have been determined. The drill sites are covering all main geological provinces of that region to get representative regional heat flow values. The distribution fits very well into already existing data for the western part of the Bohemian Massif, with average values which are typical for stable continental regions. An influence of paleoclimatic effects on the measured temperature gradients could not be observed. There is also no linear relationship between surface heat flow and heat production for the Oberpfalz region. Since no evidence for heat convection could be observed, the temperature prediction for the KTB ultra deep drillhole was done by means of models with heat conduction only. Calculations for one- and two-dimensional models yield similar results, with regionally rather constant values of mantle heat flow, in accordance with other investigations. Differences in surface heat flow values seem to have their sources within the upper crust. From a discussion and estimation of possible errors for the temperature prediction in 15 km depth in the KTB-target area an expected value of 300 ± 60°C is derived.

3 GEOTHERMAL INVESTIGATION IN THE SCHWARZWALD

3.1 Geothermal Drill Sites

Prior to the geothermal KTB investigations almost nothing was known about surface heat flow in the Schwarzwald. There was only one value, i.e., 68.8 mW/m^2, observed in the sediments of the Feldsee (Hänel 1983b). Temperature extrapolations to 10 km depth were therefore largely influenced by surface heat flow determinations in the adjacent Rhinegraben and Swabian Jura.

Table 2. Surface Heat Flow in the Schwarzwald

Hole Number	Hole Name	Depth (m)	Heat Flux (mW/m^2)	± σ	Heat Production ($\mu W/m^3$)
1	Hechtsberg	300	89	1.8	1.72
2	Schönmatt	272	81	1.2	2.36
3	Moosengrund	265	88	2.6	4.51
4	Kunklerwald	300	71	1.5	1.39
5	Ettersbach	182	92	5.2	1.66
6	Geschahse	190	77	3.2	3.81
7	Vulkan	200	82		
8	Berghaupten	762	89		
9	Griesbach	283	96		
10	Silberberg	331	103		
11	Waldkirch	662	72		
12	Herrenalb	488	94		
13	Wildbad	79	88		
14	Peterstal	200	105		
15	Kirchzarten	471	72		

Nos. 1-6: Heat flux values computed from thermal resistivity versus temperature plot (Bullard 1939) with 2D-topography corrected temperatures

Nos. 7-15: To obtain the heat flux the measured temperature gradient was multiplied by an average value of thermal conductivity for comparable rocks in the Schwarzwald

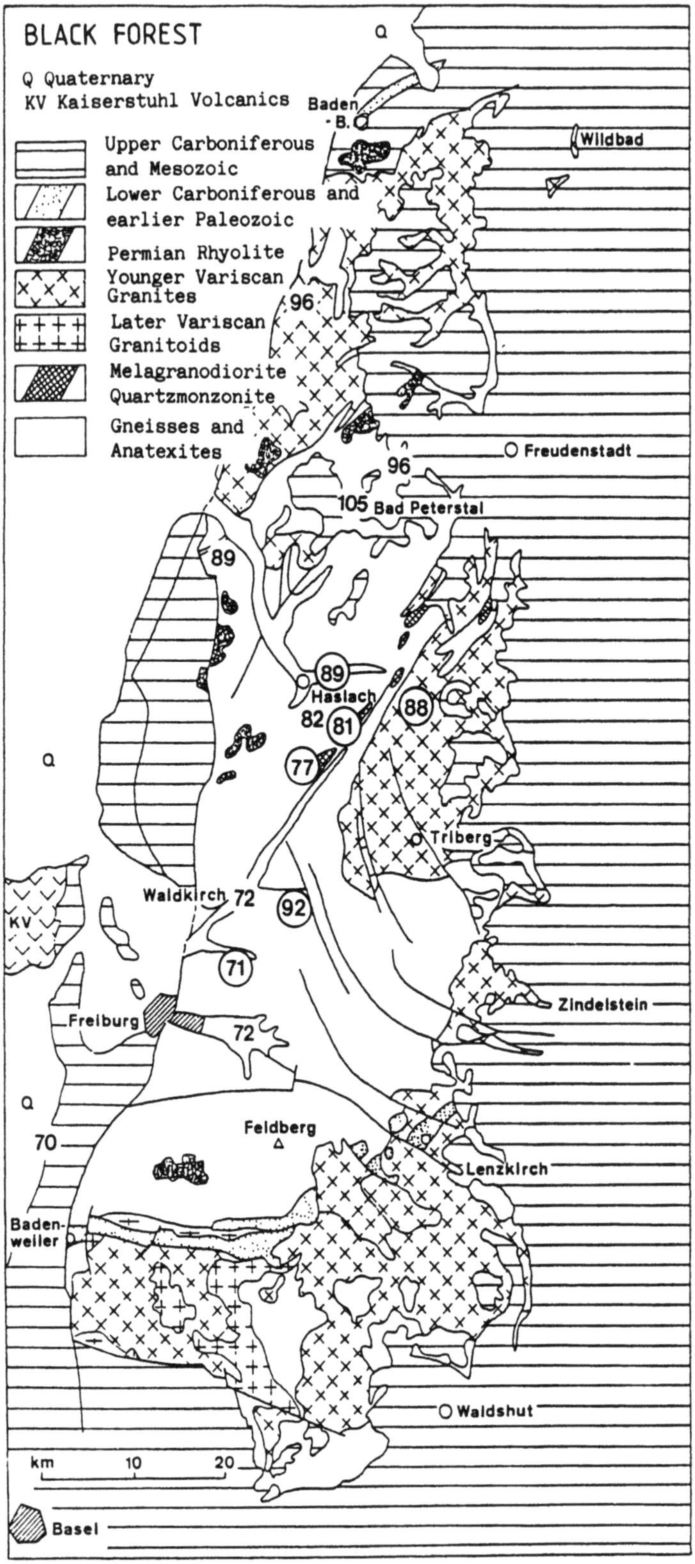
BLACK FOREST
Q Quaternary
KV Kaiserstuhl Volcanics
Upper Carboniferous and Mesozoic
Lower Carboniferous and earlier Paleozoic
Permian Rhyolite
Younger Variscan Granites
Later Variscan Granitoids
Melagranodiorite Quartzmonzonite
Gneisses and Anatexites
Q
Baden-B.
Wildbad
96
Freudenstadt
96
105 Bad Peterstal
89
89
Haslach
88
82
81
77
Q
Triberg
KV
Waldkirch 72
92
71
Zindelstein
Freiburg
72
Q
Feldberg
70
Lenzkirch
Baden-weiler
Waldshut
km 10 20
Basel

The Schwarzwald crust at the surface is mainly composed of different types of gneisses and granites which have been investigated with respect to their thermal properties during the KTB survey by probing surface rocks and core samples of six drill holes which were drilled to a depth between 180 m to 300 m into the crystalline basement of the central Schwarzwald gneiss complex and into the Triberg granite (Jenkner et al. 1986). These holes were completely cored for geological, mineralogical, petrological and geophysical laboratory investigations. Holes No. 1 (Hechtsberg) and No. 2 (Schönmatt) were prepared in the vicinity of the proposed deep drilling site near Haslach. Hole No. 3 (Moosengrund) was positioned to yield compositional and structural information about the properties of granites in the surroundings of Haslach and hole No. 4 (Kunklerwald) was chosen to explore the properties of a special granulite gneiss facies which is supposed to represent the main rock type in the lower crust. Holes Nos. 5 (Ettersbach) and 6 (Geschahse) were drilled into other specified types of gneiss (cf. Jenkner et al. 1986). All wells except No. 3 are approximately situated on a continuous NS geothermal profile. In Fig. 12 these holes can be identified by their corresponding encircled heat flow values listed in Table 2.

3.2 Temperature Measurements

In these holes high resolution temperatures (HRT) at 10 cm intervals and bottom hole temperatures (BHT) were measured already during pauses in drilling at weekends (Stiefel 1986). Thus regular temperature profiles could be obtained even for depth regions where, by subsequent drilling, the temperature logs later got disturbed when circulating water infiltrated the hole, see Fig. 13 for an undisturbed and Fig. 14 for a disturbed temperature profile.

From other already existing holes temperature profiles could also be obtained, mainly from mineral and thermal water wells. In these commercial wells temperature was measured if possible, otherwise the data had to be accepted as published, for example from Zoth (1985) for the Kirchzarten well and from Kiderlen (1977, 1981) for thermal wells in the northern Schwarzwald.

3.3 Corrections

Topographic corrections had to be applied to the temperature gradients obtained in the KTB drill holes. The analytical 1D-Bullard method (Bullard 1940) and a numerical 2D-finite difference correction were used separately for verification of the results. A comparison shows that the topographic effects are mostly overcorrected by the Bullard method because of the assumed radial symmetry. Below 1600 m depth the topographic correction is smaller than 0.5 K/km everywhere in the central Schwarzwald. The importance of the topographic correction for temperature estimates in the Schwarzwald subsurface region is obvious from Fig. 15.

Climatic corrections were estimated from temperature time series measured at Gengenbach, Strassburg, Freiburg and Basel. The effect of the variation of the mean annual temperature between 1755 and 1985 has been calculated to be smaller than 0.5 K/km below 250 m.

Fig. 12. Surface heat flow in mW/m^2 in the Schwarzwald (Black Forest); encircled values from KTB drill holes, listed in Table 2

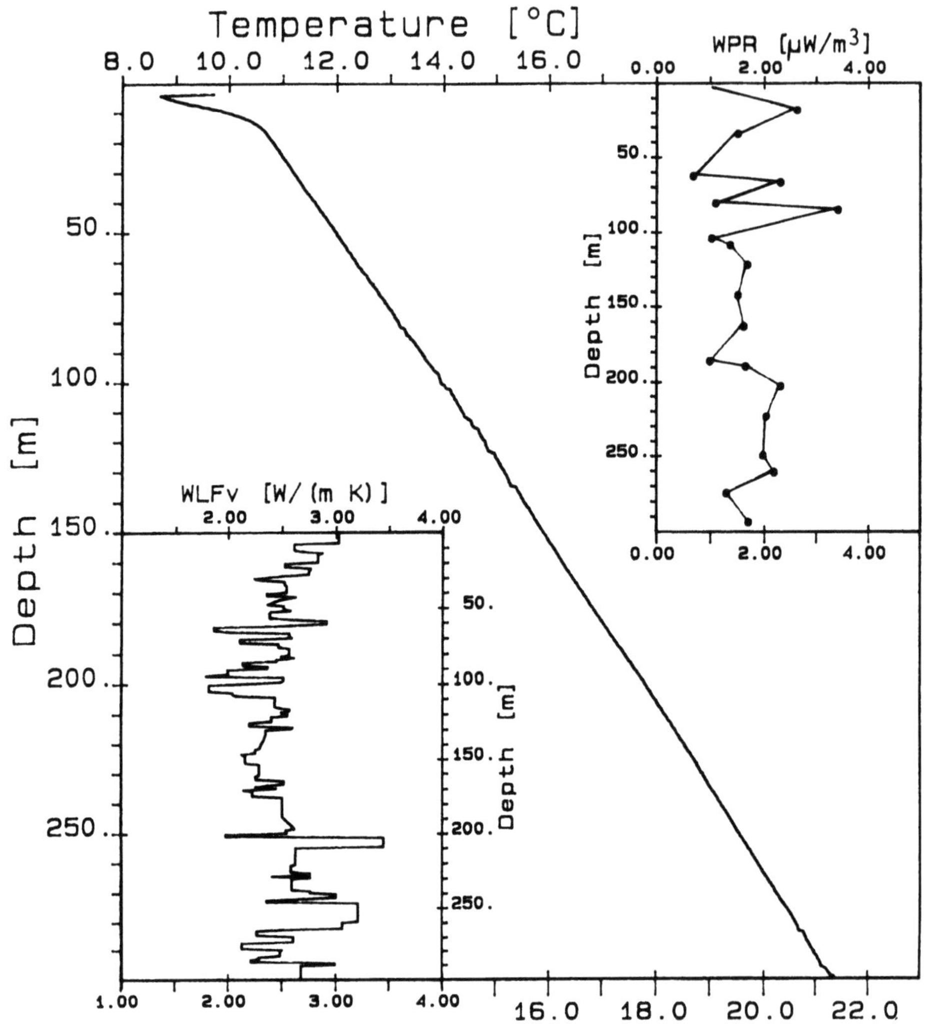

Fig. 13. Geothermal measurements in hole No. 1 (Hechtsberg). High resolution temperature profile, measured 426 days after drilling finished. Heat production WPR in $\mu W/m^3$. Thermal conductivity WLFV in W/m/K, measured parallel and perpendicular to foliation and projected to the vertical direction. Surface heat flow q_o = 89.0 mW/m^2

Paleoclimatic effects were determined from a temperature time series derived from different sources for the time span from 1700 to 50 My before present. The correction is strongest above 1500 m depth, reaching 7.8 K/km near the surface for a temperature diffusivity of 1.2×10^{-6} m^2/s (Stiefel et al. 1987). However, the expected increase of the temperature gradient with depth near the surface did not appear in the borehole temperature, so the paleoclimatic corrections were only tentatively calculated but were actually not applied to the data.

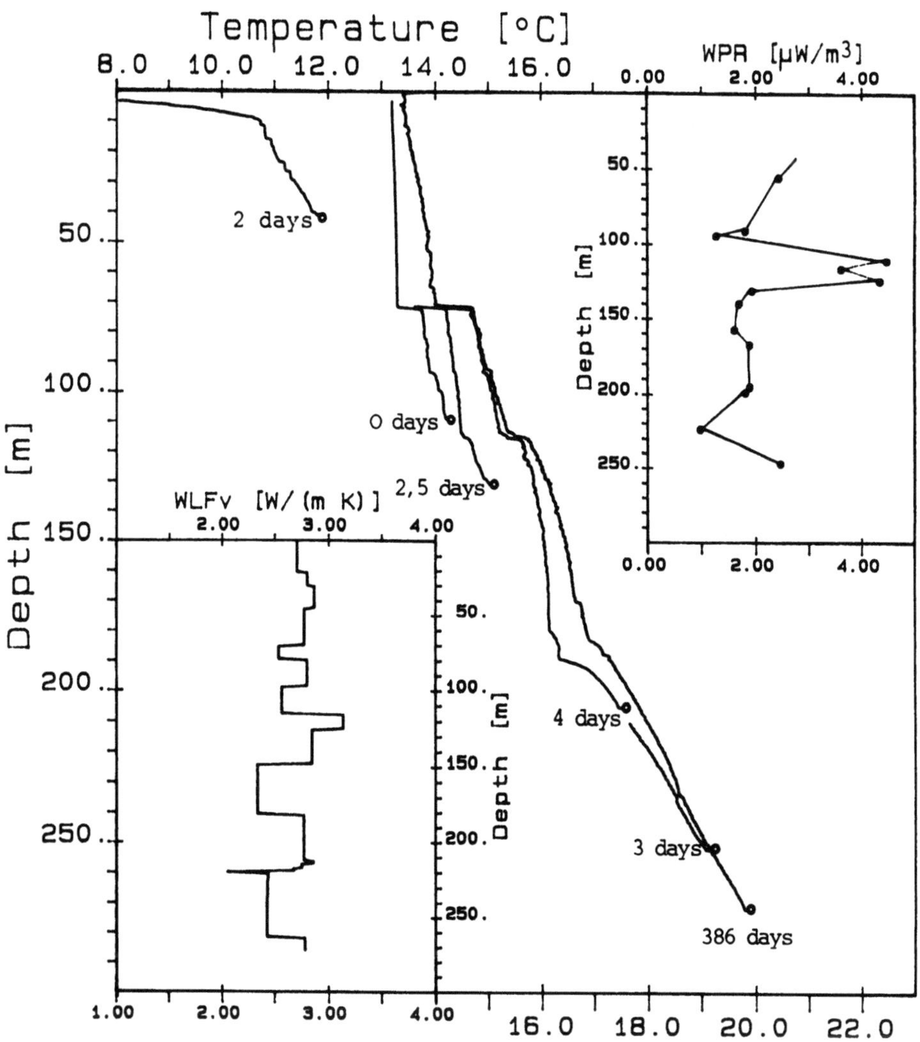

Fig. 14. Geothermal measurements in hole No. 2 (Schönmatt). High resolution temperature profile, measured 386 days after drilling stopped, temperature profiles and bottom hole temperatures taken in intermissions during the drilling phase. Heat production WPR in $\mu W/m^3$. Thermal conductivity WLFV in W/(m K), measured parallel and perpendicular to foliation and projected to the vertical direction. Surface heat flow q_o = 81.0 mW/m^2

In addition to topographic effects, erosion may have changed significantly the subsurface temperature field in some of the borehole sites situated in deep valleys. If these valleys were formed in glacial or post-glacial time erosion rates of 0. 5 mm/a for broader valleys and 1.5 to 4.5 mm/y for narrow valleys are conceivable and corresponding corrections are significant for holes No. 1 and No. 5. In Table 3 the temperature gradient values corrected for the individual effects are given.

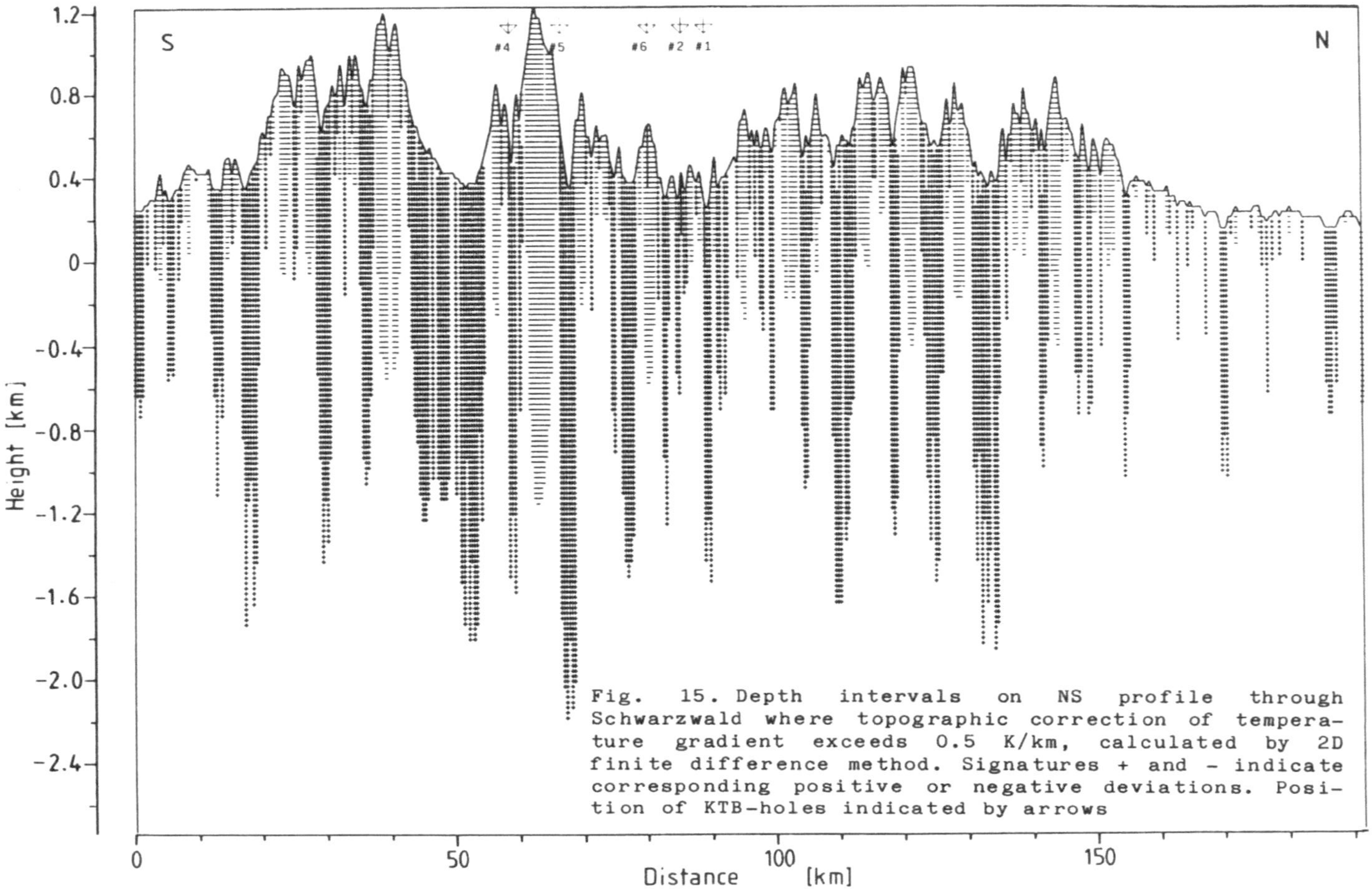

Fig. 15. Depth intervals on NS profile through Schwarzwald where topographic correction of temperature gradient exceeds 0.5 K/km, calculated by 2D finite difference method. Signatures + and - indicate corresponding positive or negative deviations. Position of KTB-holes indicated by arrows

Table 3. Thermal Gradient and Corrections

Hole Number	Thermal Conductivity n^a	$(W/m/k)^b$	$\parallel/\perp^c$	Thermal Gradient[d]	Corrected Thermal Gradient Bul[e]	2D[f]	Erosion[g]
1	55	2.52	1.26	36.0	32.6	33.7	25.7
2	17	2.66	1.08	34.1	29.9	30.0	25.0
3	31	2.93	-	32.0	30.2	30.1	27.6
4	39	2.81	1.13	25.5	23.7	24.4	23.6
5	9	2.50	1.29	43.6	24.6	35.3	24.5
6	11	2.87	-	27.0	27.8	26.9	25.7

[a] Number of core samples, [b] vertical component of thermal conductivity, [c] anisotropy factor (parallel to perpendicular to foliation), [d] unit: K/km, [e] Bullard (1940) correction applied, unit: K/km, [f] 2D numerically computed correction applied, unit: K/km, [g] 1D erosion correction applied, unit: K/km.

3.4 Thermal Conductivity

From surface rocks and from cores of the scientific holes water-saturated samples were taken to measure thermal conductivity at atmospheric pressure and room temperature using a needle probe (Sattel 1979) and a QTM probe, which was kindly supplied by Dr. Rybach, Zürich University (ETH). No statistically significant deviations between the results of both kinds of thermal conductivity measurements could be detected. Measurements were performed parallel and normal to the texture of the rocks. Anisotropy is significant in the holes Nos. 1, 4 and 5 with a corresponding anisotropy factor between 1.13 and 1.29 derived from conductivity measurements parallel and perpendicular to the texture.

In addition, an in-situ measurement of thermal conductivity was performed in hole No. 1 applying a temperature relaxation method TAV. In this method a known heat-flow disturbance is injected from the borehole to the surrounding rock and the resulting temperature disturbance depending on the thermal conductivity of the rock is measured as a function of time (Dornstädter and Sattel 1985; Dornstädter 1987). Fig. 16, taken from this thesis, shows the in-situ thermal conductivity determined by the TAV-method in the Hechtsberg borehole in comparison with laboratory thermal conductivity measurements on core samples from the same hole. Irregularities of the in-situ profile are mainly due to convective disturbances in the temperature profile.

Radiogenic heat production was measured at surface rocks and core samples. The results are reported separately in Section 4. Figs. 13 and 14 show the temperature profile, the thermal conductivity and heat production for hole No. 1 with a rather undisturbed temperature profile and for hole No. 2 with a heavily disturbed profile where it was only possible to determine a reliable value for the temperature gradient by evaluating the bottom hole temperatures taken during drilling stops.

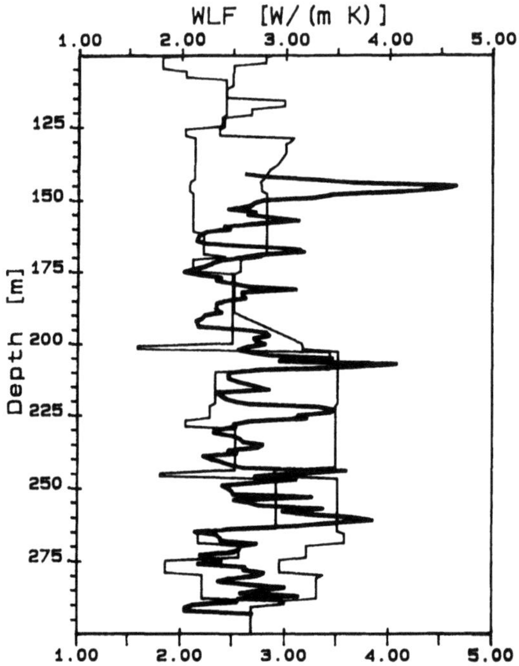

Fig. 16. Thermal conductivity WLF in borehole Hechtsberg, in-situ conductivity (thick solid line), determined by TAV-method (Dornstädter 1987), and corresponding laboratory measurements on core samples (left thin solid line for values in direction perpendicular, right thin solid line for direction parallel to foliation)

3.5 Surface Heat Flow

The overall distribution of surface heat flow in the northern and central Schwarzwald is shown in Fig. 12. The temperature gradients were only corrected for topographic effects. In the northern Schwarzwald down to the Kinzig valley passing by Haslach the heat flow is at a level of 90 mW/m^2 partly exceeding this value whereas south of the Kinzig valley the heat flow decreases to about 70 mW/m^2 across a distance of approximately 30 km. This strong variation turns out to be a severe constraint to geothermal models. With exception of hole No. 5 where topographic and erosion effects cannot be exactly determined the mean errors are smaller than ± 3 mW/m^2 determined from the m.s.e. of the slope of the corresponding Bullard plots. From the regional variation of the surface heat flow of about 20 mW/m^2 across a horizontal distance of 30 km the conclusion is almost inevitable that there must be additional heat sources in the upper crust which cannot be resolved from the surface or that heat must be advected, perhaps at midcrustal level, to supply the observed surface heat flux.

Usually a linear relationship is assumed to be valid between surface heat flow and surface heat production for sites with a common basement and with the same tectonic and thermal history (Birch et al. 1968; Roy et al. 1968) although this is now questioned for larger regions (Drury 1987). This linear relation simply means that there is a common basal heat flow q* from the lower crust given by the intersection of the line with the ordinate and that the vertically integrated heat production divided by its surface value A_o is the same for the whole region under consideration. In Fig. 17, the surface heat flow corrected for topography is plotted against surface heat production at the six scientific holes (black points). Because of the large scatter no reliable linear relationship can be deduced. A better fit is obtained by applying a correction for erosion to the data. The resulting (open) points are highly correlated. The corresponding linear relationship yields a basal heat flow as high as 61 mW/m^2 and a characteristic depth of only 4 km. If conduction and radiogenic heat production would determine the surface heat flow alone this result would mean that the contribution of radiogenic heat production would be rather small and that the basal heat flux would be as high as for the Basin and Range province for which the linear relationship is displayed in the upper part of Fig. 17 (Lachenbruch et al. 1985). The upper Rhinegraben is, however, much less tectonically active than the Basin and Range province so that the reduced heat flow should be smaller. From a regional point of view heat flow values in the Schwarzwald region would rather be expected to correspond to the relation obtained by Cermak (1975, 1982a) for the Bohemian Massif (Fig. 17), perhaps with a somewhat higher basal heat flow between 40 and 50 mW/m^2 in order to account for the present weak rift situation of the Rhinegraben. Another explanation for the apparently augmented value of the basal heat flow may be that it is

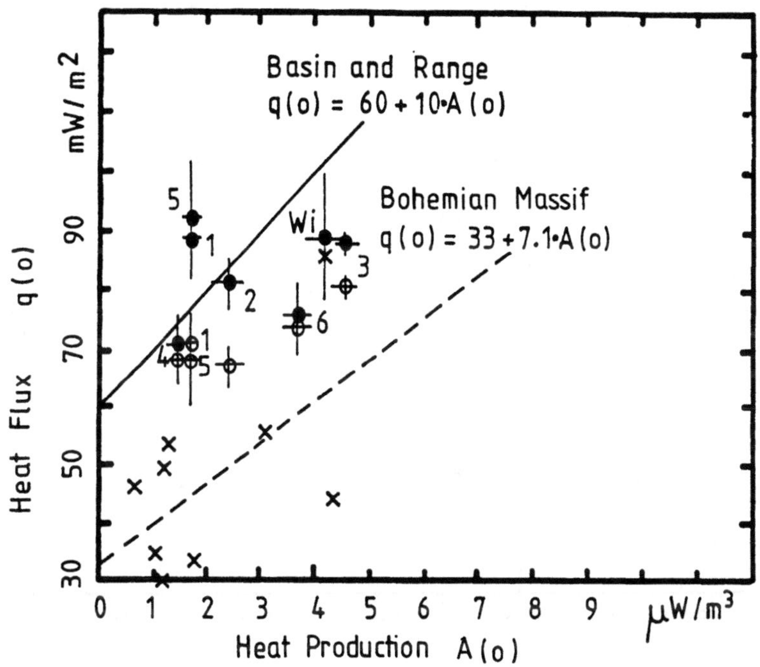

Fig. 17. Surface heat flow and heat production measured in the KTB-holes and on corresponding cores. Black circles: heat flow corrected for topographic effects; open circles: correction for erosion effects applied in addition; crosses: values for Bohemian Massif from Cermak (1982a). For hole numbers refer to Table 2, Wi is hole No. 13

caused by lateral heterogeneities, as Furlong and Chapman (1987) have shown by 2D-model simulations.

3.6 Geothermal Models

There are two aspects constraining geothermal models for the central Schwarzwald crust, i.e. the lateral variation of 20 mW/m^2 on a horizontal distance of 30 km and a high basal heat flow which presumably is, perhaps at least partly, a result of heat advection from the surrounding. There is no possibility to model the regional surface heat flow variation by lateral variations in the mantle heat flux. The cause of this change has to be sought at a shallower depth. Possible explanations are variations in the heat production rate or convective heat transport at high or midcrustal level.

At a first stage one-dimensional conductive models were derived for the five borehole sites, with measured surface heat flow and 10°C surface temperature as boundary conditions. Also, the measured thermal conductivity and heat production were included as surface values. The thermal conductivity is assumed to be temperature and pressure dependent and the heat production to be exponentially varying in the upper crust with a characteristic depth of 10 km. In Fig. 18 the assumed heat production rate, the calculated tempera-

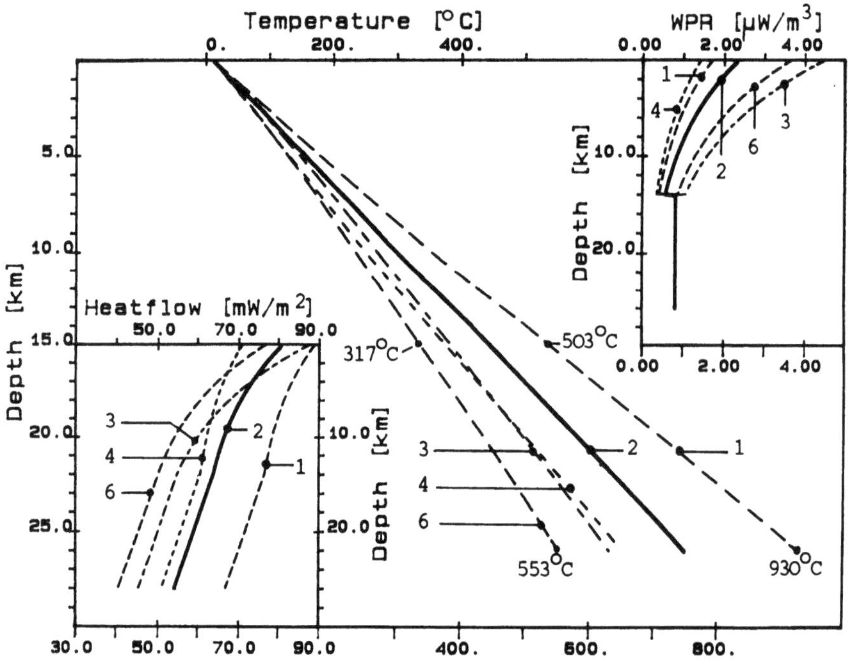

Fig. 18. 1D conductive temperature models for KTB-holes Nos. 1-6 except No. 5. Assumed conditions: surface temperature 10°C, surface heat flow, temperature gradient and heat production from Table 2 and Table 3, heat production WPR exponentially decreasing with depth as exp (-z/D) with D = 10 km in the upper crust (0-14 km), lower crust (14-26 km) with heat production 0.8 µW/m^3 assumed constant

ture distribution and the resulting heat flow for five holes are shown. At 14 km depth the temperature varies between approximately 300 and 500 °C and the heat flux between 50 and 75 mW/m^2. These lateral differences will lead to horizontal heat flow and the one-dimensional approach is therefore insufficient.

Two-dimensional simulations were performed with the finite element (FE) code ADINAT (Bathe 1978). Structural information was drawn from reflection and refraction seismic observations (Lüschen et al. 1987; Gajewski and Prodehl 1987). Thermal conductivity and heat production were assigned to the blocks according to different assumptions about their lateral and vertical distribution. Surface temperature (10°C), basal heat flow (50 mW/m^2) in 26 km depth and insulating vertical boundaries at the sides were assigned as boundary conditions. Fig. 19 shows an example where at midcrustal level in the low velocity zone a strong variation in the heat production was assumed. This distribution of heat production is fictitiously assumed, as it would imply the presence of a large granitic batholith in the midcrustal low velocity zone determined from seismic refraction observations (Gajewski and Prodehl 1987) for which, until now, there are no plausible model interpretations from gravity observations (Plaumann et al. 1986). On the other hand, Müller (1978) has postulated a granitic midcrustal layer causing the frequently observed reduction of P-wave velocity at the corresponding depth. On top (a) of Fig. 19 the assumed conductivity and heat production distribution is shown. In the centre (b) the calculated vertical profile of temperature and temperature gradient and in addition of thermal conductivity, heat production and of the resulting heat flux are displayed for the proposed deep drilling site Haslach. Below in (c), the calculated surface heat flux together with the observed values only corrected for topography and in (d), the thermal field in the crust is shown. The trend of the observed surface values has been accurately modeled assuming rather high values of heat production in the crust but even higher values would have to be assumed in the northern (left-hand) part of the model to produce a much better fit to the observations. Although such a model is not completely unreasonable, there is some doubt concerning the positional stability of such a large extended granitic batholith within the crust.

Another possibility to produce the observed variation in the surface heat flux is by taking convection into account. To estimate the first order effect of convective heat transport, a one-dimensional model was considered with purely vertical flow through a permeable crust, for which an analytical solution exists (Mansure and Reiter 1979). Inflow to and outflow from the model are disregarded and sources or sinks of the flow have to be assumed for the 1D-model at the upper crust-lower crust boundary at 14 km depth in order to allow for different velocities in both crustal parts. These velocities were determined from the difference between the conductive heat flow and the assumed basal heat flow, both at 14 km depth. The results for the temperature-depth extrapolation are displayed in Fig. 20. A vertical flow was assumed which produces a convective heat flux $\rho c v_d (T-T_o)$, where $\rho c = 4.18 \cdot 10^6$ Ws/(K m^3) is the volumetric heat capacity of water, v_d the Darcy velocity and T_o the surface temperature. This flow reduces the conductive heat flux and hence the temperature gradient. As the corresponding conductive models in Fig. 18 are calculated with the same parameters the difference in conductive heat flux between Figs. 18 and 20 yields the convective heat flux at the corresponding depth. The Darcy

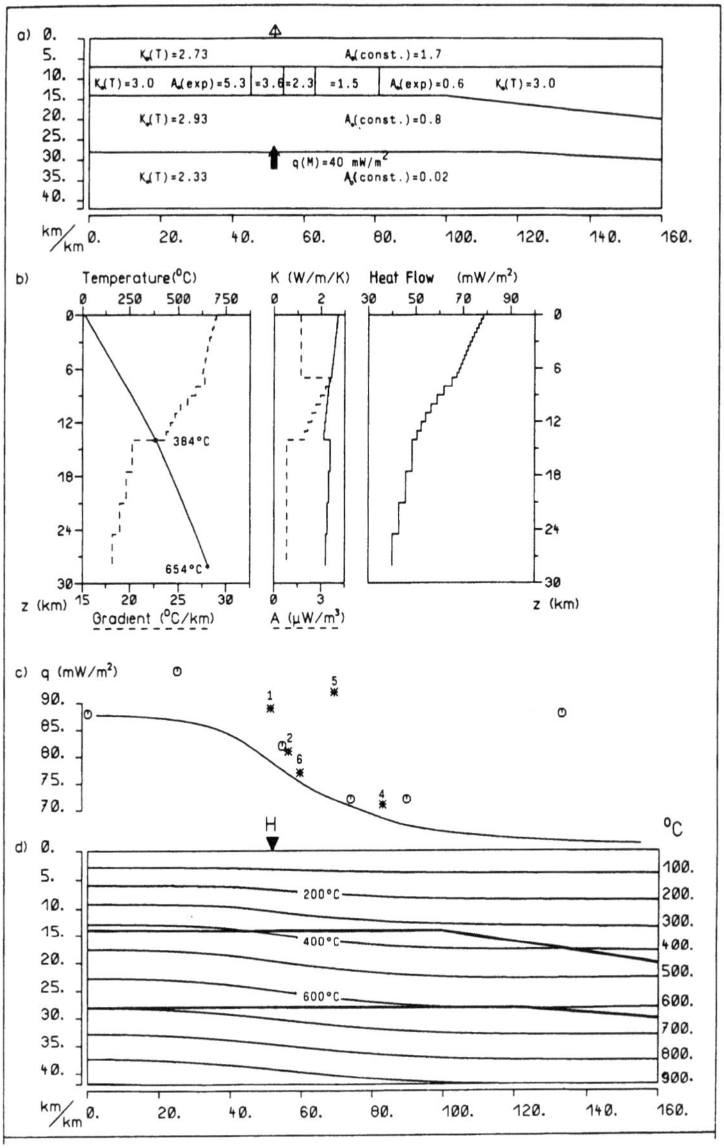

Fig. 19. Conductive geothermal 2D-model, NS profile

Fig. 19a. Assumed distribution of geothermal parameters. K thermal conductivity in W/(m K), temperature dependent; A heat production in $\mu W/m^3$ constant or exponentially varying as exp (-z/D) with D = 10 km, basal heat flow 40 mW/m^2

Fig. 19b. Depth distribution of temperature, temperature gradient, thermal conductivity K, heat production A and heat flow at assumed deep drilling site, indicated by H in Fig. 19d

Fig. 19c. Surface heat flow; values measured in KTB holes (stars) and in commercial holes (circles), calculated heat flow (solid line)

Fig. 19d. Temperature isolines and structural crustal model

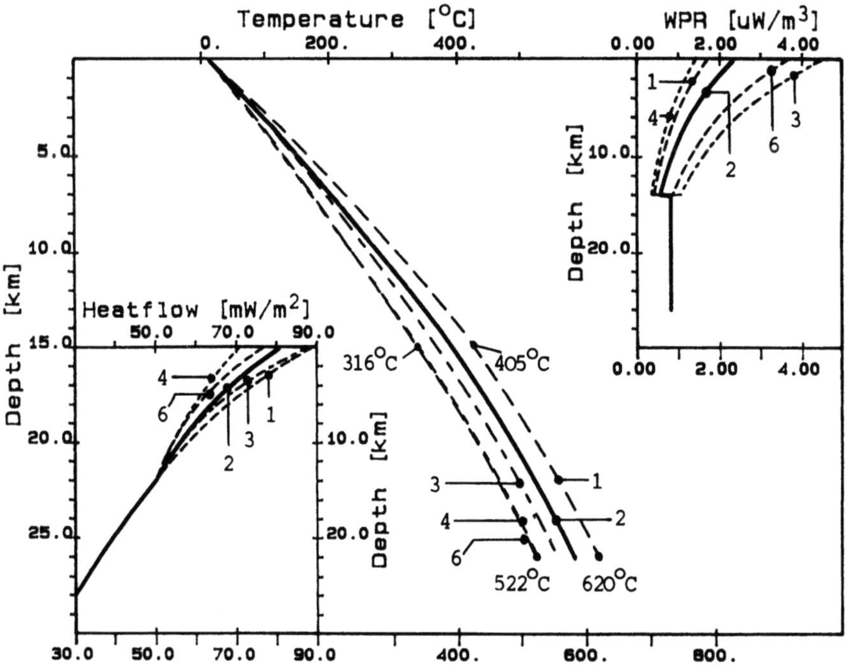

Fig. 20. 1D convective temperature models for KTB-holes Nos.1-6; except No. 5. Assumed conditions: surface temperature 10°C, surface heat flow and heat production from Table 2 and Table 3, heat production exponentially decreasing in the upper crust (0-14 km) as in Fig. 18, lower crust (14-26 km) with constant heat production 0.8 $\mu W/m^3$. Assumed heat flow from the mantle 30 mW/m^2 and 20 W/m^2 in addition from the lower crust. Corresponding Darcy velocities v_D are 1.8 10^{-12} m/s in the lower crust and varying between 1.0 10^{-12} m/s (hole No. 1) and 2.5 10^{-12} (hole No. 6) in the upper crust

velocity v_d = P/μ grad(p- ρgz) which can be determined if surface temperature T_o, surface heat flux q_o and the basal heat flux q* are given, is rather small, i.e. O(0.1 mm/a). The viscosity μ of water is expected to be about 10^{-3} Pa s for near surface conditions and 10^{-4} Pa s at a temperature of some hundred degrees. The permeability P for granitic rocks is between 10^{-18} m^2 to 10^{-20} m^2, the pressure p exerted on the fluid has to exceed the hydrostatic pressure ρgz by a factor 10^{-1} to 10 to yield the calculated Darcy velocity. On the other hand, this Darcy velocity is of the order of the observed vertical uplift in the Schwarzwald (Mälzer and Zippelt 1987). It is therefore difficult to separate both effects especially since the standard deviation of the uplift rate is also 0.1 mm/a.

As in the case of purely conductive heat transport the one-dimensional model can only give an indication of the expected effects. In order to get quantitative results two-dimensional FE-calculations for forced convection were performed with different assumptions about the permeability distribution and the forcing hydraulic potential H = p/(ρg)-z. Fig. 21 shows the results for an example where the hydraulic potential was assumed to vary linearly

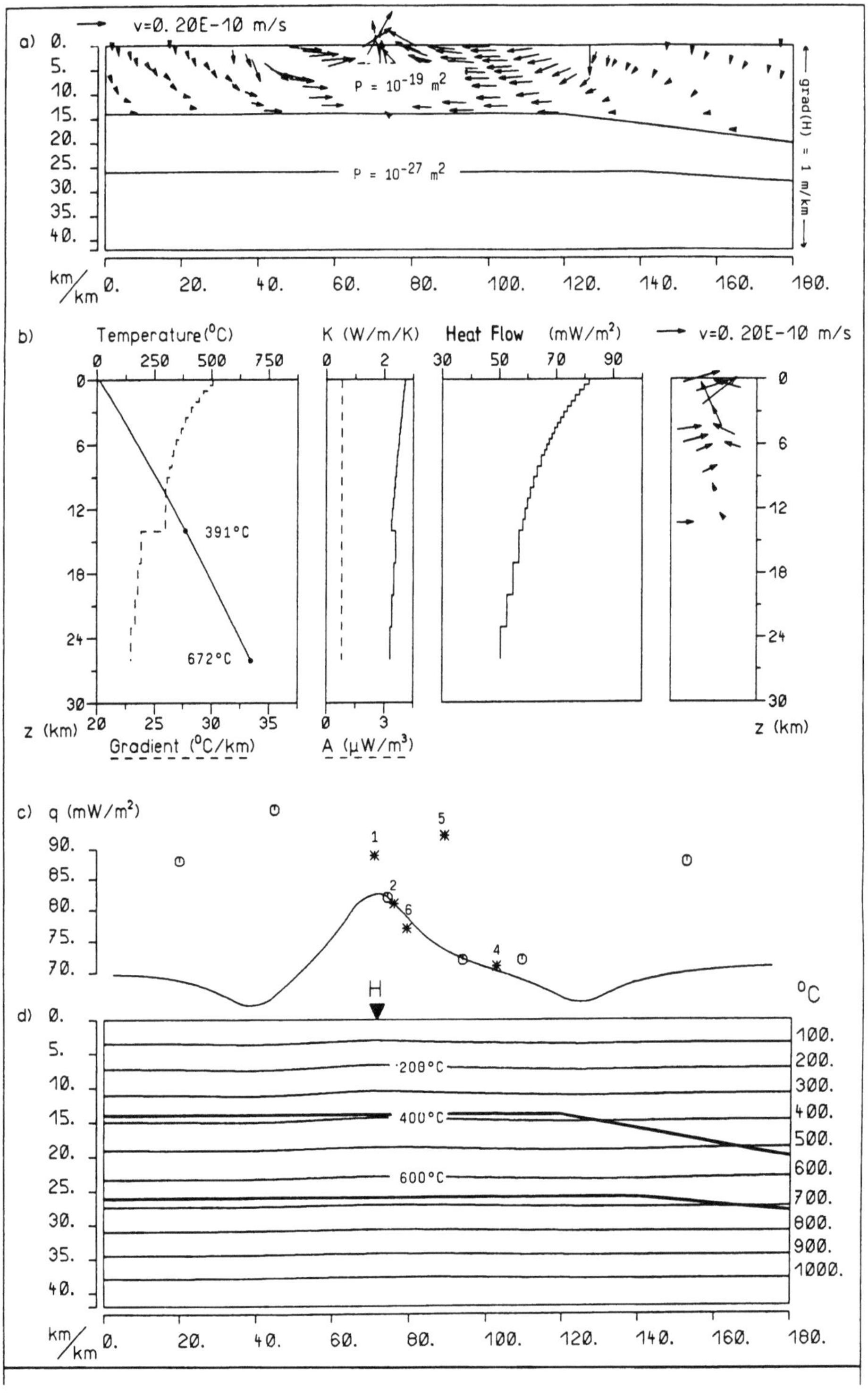

a)
v=0.20E-10 m/s
P = 10^{-19} m²
P = 10^{-27} m²
grad(H) = 1 m/km
km/km
b)
Temperature (°C)
K (W/m/K)
Heat Flow (mW/m²)
v=0.20E-10 m/s
391°C
672°C
z (km)
Gradient (°C/km)
A (µW/m³)
c) q (mW/m²)
H
°C
d)
200°C
400°C
600°C
km/km

corresponding to the regional surface topography which is lowest in the central Schwarzwald. Surface temperature and mantle heat flux were prescribed as upper and lower boundary conditions, the vertical sides of the model were assumed to be impermeable and isolating. To the upper crust the permeability of granitic rocks (10^{-19} m^2) was assigned, the lower crust is practically impermeable (10^{-27} m^2). The resulting velocity field (a) shows recharge and downward flow induced by the high topography in the N and S and discharge and upward flow in the central Schwarzwald.

The resulting surface heat flow (c) is in correspondence with the observed values in the central Schwarzwald, if erosion corrections are applied to the flux at the holes No.1 and No. 5. The large discrepancies in the N and S can be eliminated by accounting for the observed heat production. Obviously the calculated flow field is based on simplifying assumptions, for example neglecting the Rhinegraben influence, however, the model shows that the observed regional variation of surface heat flux can be produced by fluid flow in the crust. Also it has to be stated, that the surface heat flow critically depends on the value of the chosen permeability. This has already been verified in one-dimensional simulations, as described in Mansure and Reiter (1979).

3.7 Error Estimation for Temperature Predictions

The geothermal KTB investigations in the Schwarzwald have shown that in the northern and central part the surface and the basal heat flux are greater than in the Bohemian Massif and that there is a considerable lateral variation in the surface heat flow which can be modelled either by an extended midcrustal granitic layer, for which, however, there are no indications from gravity measurements, or by forced convective heat transport. Both conductive and convective model calculations have shown that there are lateral components up to 10% of the temperature gradient and of the heat flux.

Corrections for topography and in some places for erosion have to be applied to the observed surface temperature gradients in order to get reliable starting values for temperature and heat flow extrapolations. Temperature extrapolations from conductive modelling with temperature dependent thermal conductivity yield a value

---

Fig. 21. Convective geothermal 2D-model, convection forced by linearly varying hydraulic potential according to a corresponding topography

Fig. 21a. Resulting velocity field for a crustal model with permeabilities 10^{-19} m^2 and 10^{-27} m^2 for upper crust and space beneath respectively; heat production 0.8 $\mu W/m^3$ assumed constant, thermal conductivity temperature dependent, basal heat flux 50 mW/m^2 at 26 km depth.

Fig. 21b. Depth distribution of temperature, temperature gradient, thermal conductivity K, heat production A, heat flow and velocity field at assumed deep drilling site, indicated by H in Fig. 21d

Fig. 21c: Surface heat flow, measured values in KTB holes (stars) and in commercial holes (circles), calculated heat flow (solid line)

Fig. 21d. Temperature isolines and structural crustal model

of 400°C at 14 km depth with estimated bounds of ± 80°C. This spread mainly reflects the uncertainties in the assumed thermal conductivity profile and in the surface heat flux.

From purely conductive models it appears that variations in temperature down to a depth of 14 km are mainly determined by variations in thermal conductivity and surface heat flux, whereas variations in the heat production in the upper crust do not affect the temperature as much. Deviations of ± 20% in the assumed thermal conductivity distribution are conceivable as a consequence of differences in type and composition of rocks, which may have changed, e.g., by hydrothermal mineralization, granitization and mylonitization. At 14 km depth this could result in a temperature variation of ± 80°C, and at 10 km depth of ± 60°C.

3.8 Summary and Conclusions

The KTB geothermal survey of the central Schwarzwald has established very well determined values of near-surface temperature gradients in the KTB holes. This could be achieved by recording high resolution and bottom hole temperatures as early as possible during weekend drilling stops and later on in a time span of two years after finishing the six scientific holes especially prepared for this project. Corrections had to be applied to account for topographic effects whereas paleoclimatic effects could not clearly be identified.

Thermal conductivity and heat production rate were measured in the laboratory at core samples and surface rocks. No surprising deviations from expected values, known from the literature, could be detected. The resulting surface heat flow is comparatively high in the northern part of the central Schwarzwald and is systematicly decreasing to the south. From the magnitude of the surface heat flux a relatively high basal heat flow of at least 50 mW/m^2 must be inferred, whereas the regional variation can only be modelled by assuming enhanced (radiogenic or chemical) heat production or additional convective heat transport in the upper crust.

4 RADIOGENIC HEAT PRODUCTION IN BOTH AREAS

4.1 Heat Flow and Heat Production

Between 1/3 and 2/3 of the surface heat flow of the continents is produced by the radiogenic decay of U, Th, and K in the crust itself. Therefore, the concentration-depth distributions of these elements are important for models of the temperature depth variation in the crust. Without the deep drill hole we can only estimate the vertical distribution of these three elements assuming certain crustal models with rock units characterized by typical heat production rates. Combined with heat flow measurements the observed heat production of surface rock and core samples were used to model temperature-depth profiles at the proposed drilling sites as desribed in Sections 2 and 3.

Details of the analytical procedure, as well as the individual results and the errors for 482 rock samples will be given in a separate KTB Report (Haack et al. 1988). The drill hole samples were taken from the same cores which were used for the measurements of the heat conductivities. For all samples radioactive equilibrium was assumed. The heat production A was calculated according to the formula (Rybach 1976):

$$A = 10^{-2}\ \rho\ (9.52\ U + 2.56\ Th + 3.48\ K)$$

Here, ρ is the rock density based on data of Plaumann (1985), Plaumann and Wohlenberg (1982), Plaumann et al. (1985), Wohlenberg (1982). U and Th are the concentrations of these elements in ppm, K is the concentration of potassium in %.

4.2 Heat Production in the Schwarzwald

Fig. 22 shows the depth distribution of the measured heat production for the KTB holes listed in Table 4. Fig. 23 shows the corresponding histograms for gneisses and granites separately. Surface samples yield approximately the same mean Th/U ratios and heat production rates as drill core samples. Examples are the Triberger granite near the drill hole Moosengrund as well as the gneisses near the drill holes Hechtsberg and Schönmatt. The results are listed in Table 4, where the heat flow is corrected for topography only, see Section 3.

Table 4. Heat production A (arithmetic means) and heat flow Q according to Section 3 (n frequency, σ standard deviation)

Drill Site	A ($\mu W/m^3$)	$\sigma/\sqrt{n}$ ($\mu W/m^3$)	n	Q (mW/m^2)
Hechtsberg	1.72	0.15	20	89
Schönmatt	2.36	0.28	15	81
Moosengrund	4.51	0.23	15	88
Kunklerwald	1.39	0.16	13	71
Ettersbach	1.66	0.14	11	92
Geschahse	3.81	0.29	10	77

A comparison of the heat production rates of the gneisses with those of the Oberpfalz (compare Figs. 23 and 25) shows no remarkable differences. The Oberpfalz granites, however, are much more radioactive than the Schwarzwald granites. The higher geothermal gradients and heat flow values of the Schwarzwald discussed in Section 3, therefore, cannot be explained by radioactivity of the uppermost crust.

One aim of the project was to find out whether the Schwarzwald is a heat flow province in the sense of Lachenbruch (1968) and Roy et al. (1968). As Fig. 17 shows, such a correlation is not evident as long as the values of the surface heat flow remain corrected only for topographic effects.

It is not possible to give a unique interpretation of the Schwarzwald data. The radioactivity of the surface rocks is insufficient to explain the high average heat flow of about 80 mWm^{-2}. Taking the seismic results of Lüschen et al. (1987) into account and assuming

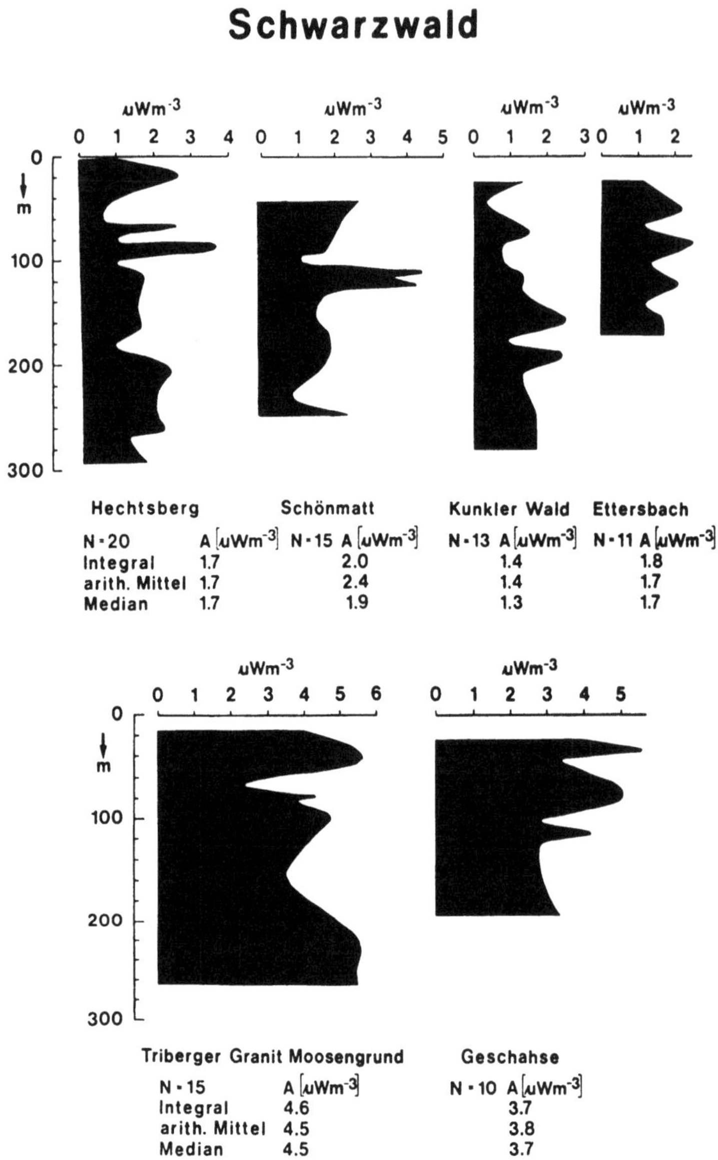

Fig. 22. Depth distribution of heat production of core samples from boreholes listed in Table 4

an upper crustal layer of 6 km thickness and a constant heat production the heat flow from this layer contributes 18 mWm^{-2}, i.e. less than 1/4 of the total. This means that either much heat is carried upwards by hydrothermal convection, or that the mantle contributes an unusually high proportion of heat, or that the lower and/or middle crust is not, or only slightly depleted of radioactive elements. All three explanations and combinations of them are possible. If, alternatively, one assumes that the upper crust which according to Lüschen et al. (1987) is 14 km thick would consist of the same rocks as occur at the surface then the upper crust could contribute about 50% of the observed surface heat flow. The remaining 12 km of the crust would then be composed of rocks depleted in U, Th, and K, e.g. granulites. The measurements do not permit to distinguish between these possibilities. Therefore the calculation of the temperature depth profile strongly depends on the geological model.

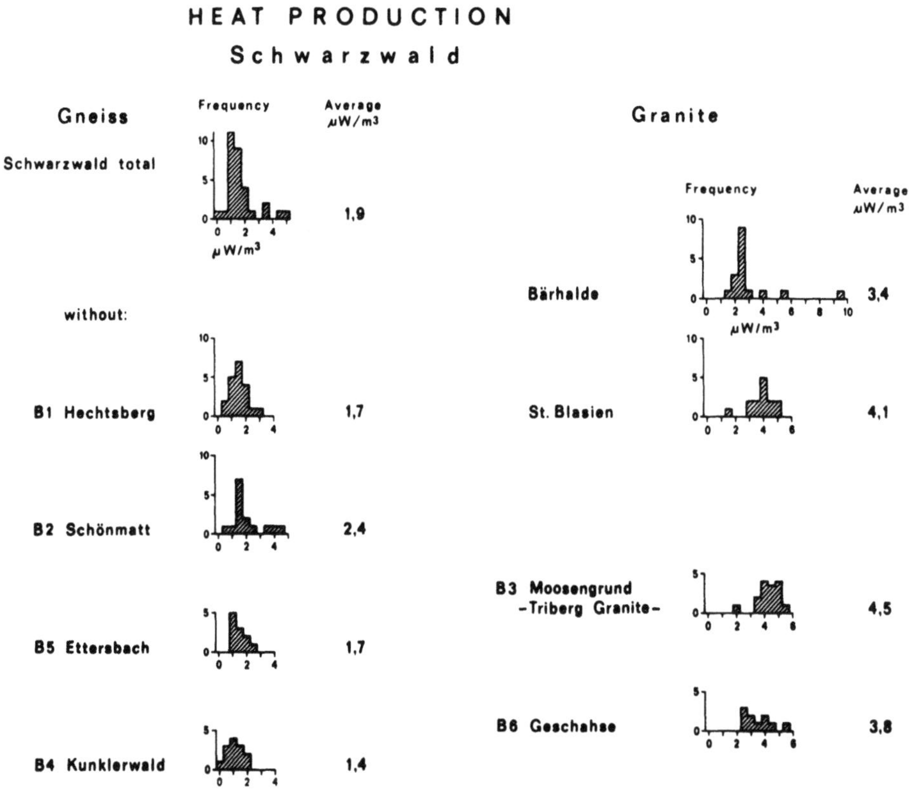

Fig. 23. Histograms of heat generation of gneisses and granites from the Schwarzwald

4.3 Heat Production in the Oberpfalz

Fig. 24 shows the depth distribution of the measured heat production for the drill sites listed in Table 5. Fig. 25 shows the histograms for gneisses and granites separately. The Oberpfalz data for the drill holes are displayed in Table 5 where the heat flow values are corrected for topography only, see Section 2.

The data in Fig. 26 show clearly that the heat flow in the Oberpfalz is considerably lower than in the Schwarzwald although the heat production of the gneisses are similar and those of the Oberpfalz granites are higher. For both areas there is no statistically significant linear correlation of topography corrected heat flow with heat production.

The Oberpfalz results can be discussed in terms of models in which a slice of Moldanubian crust is superimposed on Saxothuringian crust. All these models are based on the assumption of equilibrium heat flow and the absence of convective heat transport. Depending

Oberpfalz

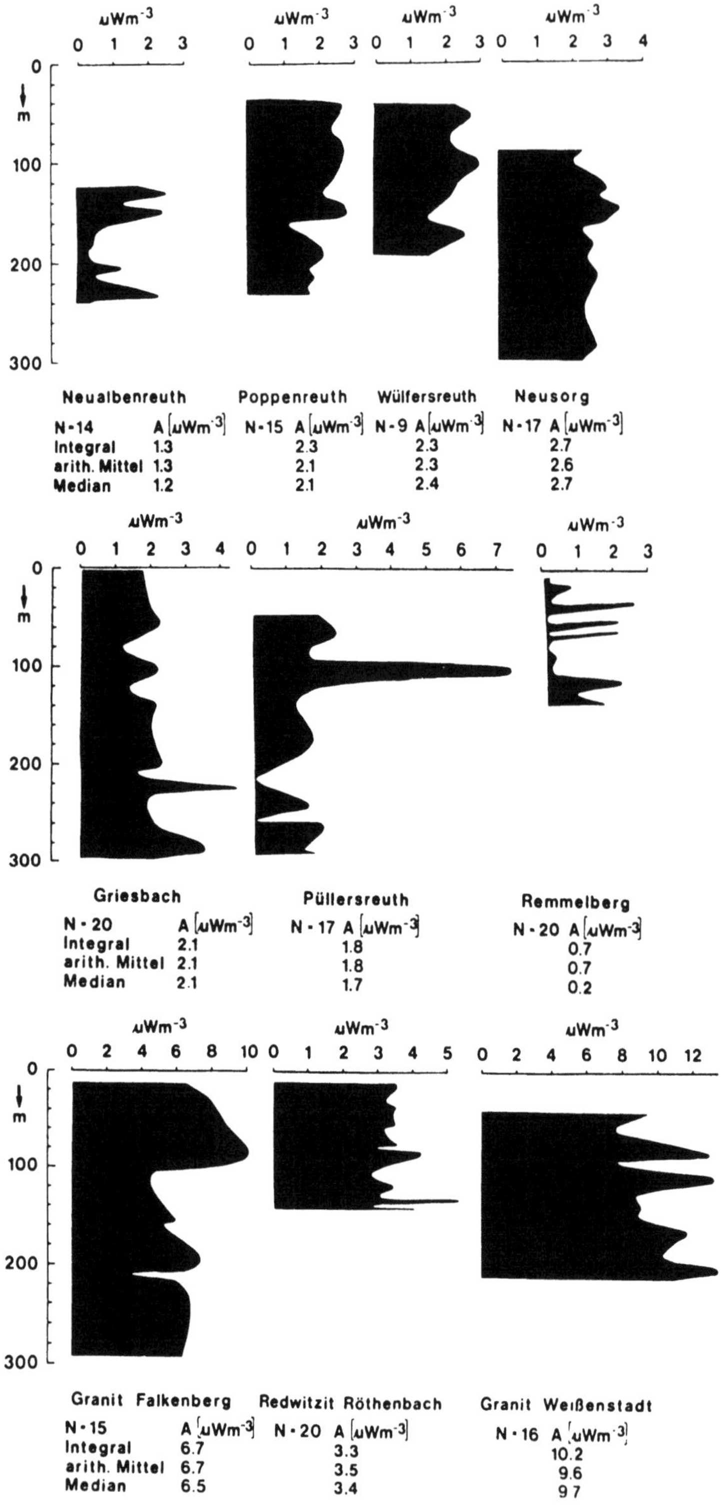
µWm⁻³
0 1 2 3
0 1 2 3 4
0 1 2 3 4 5 6 7
0 2 4 6 8 10
0 1 2 3 4 5
0 2 4 6 8 10 12
0
m
100
200
300
Neualbenreuth
N = 14 A [µWm⁻³]
Integral 1.3
arith. Mittel 1.3
Median 1.2
Poppenreuth
N = 15 A [µWm⁻³]
2.3
2.1
2.1
Wülfersreuth
N = 9 A [µWm⁻³]
2.3
2.3
2.4
Neusorg
N = 17 A [µWm⁻³]
2.7
2.6
2.7
Griesbach
N = 20 A [µWm⁻³]
Integral 2.1
arith. Mittel 2.1
Median 2.1
Püllersreuth
N = 17 A [µWm⁻³]
1.8
1.8
1.7
Remmelberg
N = 20 A [µWm⁻³]
0.7
0.7
0.2
Granit Falkenberg
N = 15 A [µWm⁻³]
Integral 6.7
arith. Mittel 6.7
Median 6.5
Redwitzit Röthenbach
N = 20 A [µWm⁻³]
3.3
3.5
3.4
Granit Weißenstadt
N = 16 A [µWm⁻³]
10.2
9.6
9.7

Table 5. Heat production A (arithmetic means) and heat flow Q according to Section 2 (n frequency, σ standard deviation)

Drill Site	A ($\mu W/m^3$)	$\sigma/\sqrt{n}$ ($\mu W/m^3$)	n	Q (mW/m^2)
Griesbach	2.10	0.17	20	72
Püllersreuth	1.84	0.38	17	58
Neusorg	2.72	0.08	17	74
Röthenbach	3.45	0.12	20	45
Remmelberg	0.71	0.18	20	63
Weißenstadt	9.60	0.47	16	69
Wülfersreuth	2.33	0.16	9	62
Poppenreuth	2.10	0.14	16	58
Neualbenreuth	1.30	0.22	14	-
Hardeck	2.56	0.49	5	76
Falkenberg	6.74	0.43	15	76
Weißenstein	0.22	-	1	57

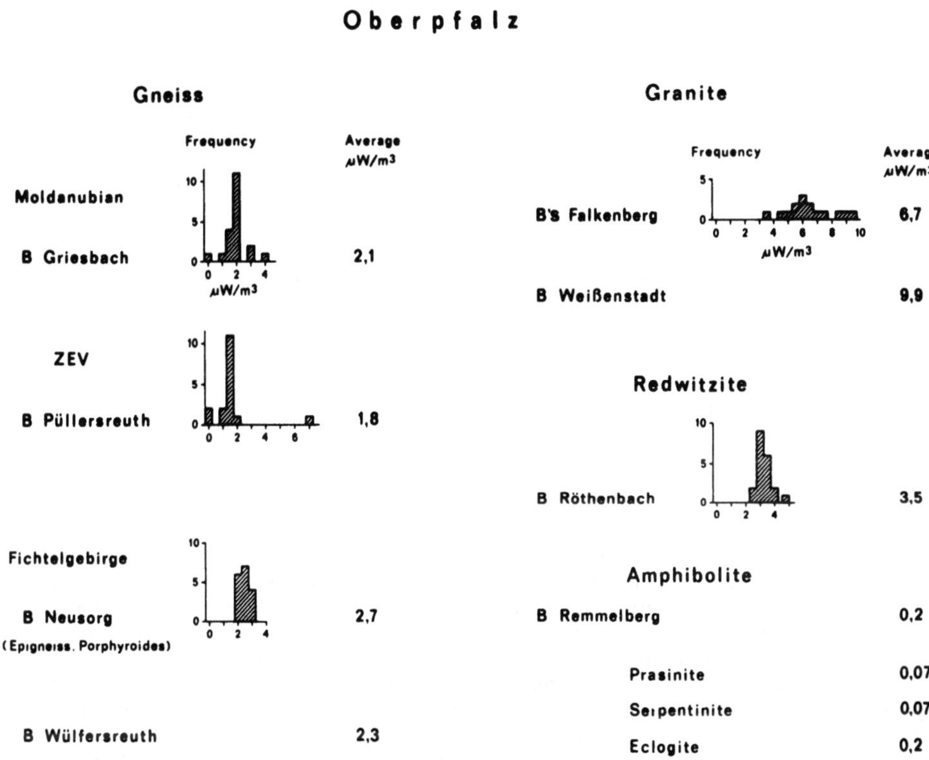

Fig. 25. Histograms of heat generation of gneisses, granites and less common rocks from the Oberpfalz

Fig. 24. Depth distribution of heat production of core samples from boreholes listed in Table 5

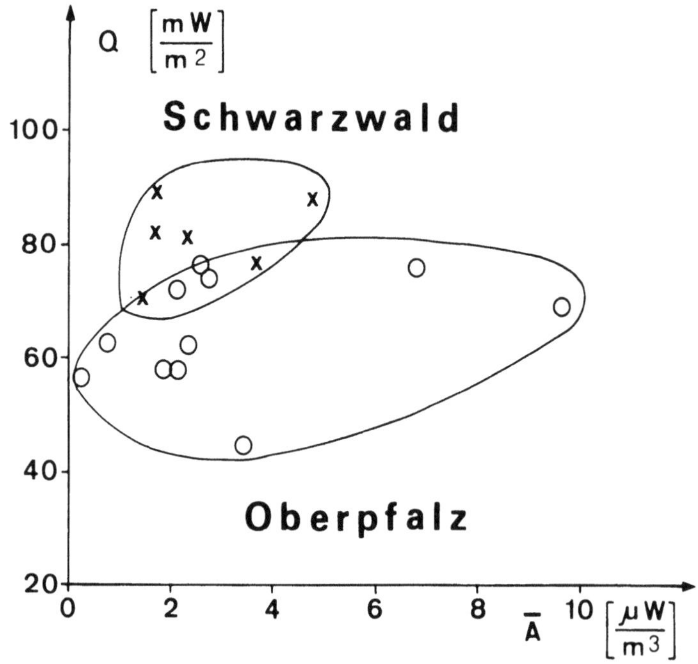

Fig. 26. Surface heat flow (topography corrected for the Schwarzwald) and surface heat production

on the choice of a model, 20-24 mW/m^2 of the heat flow have to be ascribed to the upper 12.5 km of the crust. The remainder is due to the lower crust and mantle. If the latter contributes only the low value of 20 mW/m^2 (arbitrary assumption), the lower crust produces 13 mW/m^2. Because it is 20 km thick the lower crust can have a mean heat production of only 0.65 $\mu W/m^3$ which is less than the world average for granulites (Haack 1982). If the mantle yields a higher heat flow, the lower crust must be even more U-, Th-, and K-depleted.

These considerations practically exclude the possibility that granites occur below the chosen drill site. Equally excluded are all models envisaging large parts of unchanged Saxothuringian crust below 12 km depth. The resulting heat flow would be much too high. On the other hand, such models are deduced from the seismic results and are, therefore, generally accepted. They can be only reconciled with the heat flow/heat production constraints if the Saxothuringian portion of the crust below the drill site has lost the major portion of its heat producing elements thus leaving a crust with a U, Th, and K content comparable to granulites. One possibility how this could have been accomplished is that this portion of the crust has produced the granites. The deep drill hole is expected to reach this interesting zone.

5 COMPARISON OF THE GEOTHERMAL ASPECTS OF THE KTB SITES OBERPFALZ AND SCHWARZWALD

One of the most important aspects upon which the site decision of the Continental Deep Borehole (KTB) had to be based is an estimate of the expected temperature depth profile. On one hand the temperature is a fundamental parameter with respect to the physical and chemical conditions and processes in the rocks to be studied, and on the other hand it is the main factor determining the maximum depth at which drilling tools and downhole measuring equipment can be safely used.

In order to provide a reliable basis for predicting the temperatures at the two possible KTB drillhole sites, it was necessary to obtain near surface heat-flow values that are really representative for the areas under investigation. Therefore, fifteen shallow boreholes (~ 300 m) were drilled, six in the Schwarzwald and nine in the Oberpfalz. The drill sites were chosen to obtain representative values of the geothermal gradient for an area of about 10 km radius around the expected KTB borehole. A second aspect was that in both regions different petrological units crop out which possibly exhibit different heat flow values. Consequently, a borehole was positioned roughly in the centre of each main unit. Last but not least, fault zones containing groundwater had to be avoided as far as possible.

The results worked out in Sections 2 to 4 are summarized here in order to compare the two possible KTB sites in the Schwarzwald and in the Oberpfalz.

Fig. 27 gives an impression of the spatial variability of the predicted temperature distributions. As can be seen, the difference between Schwarzwald and Oberpfalz sites is clearly greater than the spatial variability at either of them.

In Fig. 28 the estimated variations in the temperature predictions at 15 km depth are shown.

The error in the heat flow density at the surface is obtained from the errors in the measured values of the temperature gradient and the heat conductivity. All other contributions to the total variation are estimated. The contributions from the heat conductivity and heat production between the surface and 15 km depth are roughly estimated from the uncertainty of the assumptions about those two parameters, which themselves are dependent on the rock types expected at this depth interval of the crust.

The errors in determining the effect of paleoclimatic changes and erosion are estimated from model considerations. The influence of fluid convection was estimated for the Oberpfalz to be negligible. In the Schwarzwald several models were designed which show the effect of fluid convection on the distribution of heat flow density.

General results are presented in Fig. 29 in the form of two bands widening downwards reflecting the increase in the temperature variation with depth. The heat flow values at the surface and at a depth of 15 km are also shown, the latter being based on reasonable assumptions on the crustal heat production.

The heat flow as a function of depth within the crust is dependent on the heat flow entering at the Moho and on the heat production

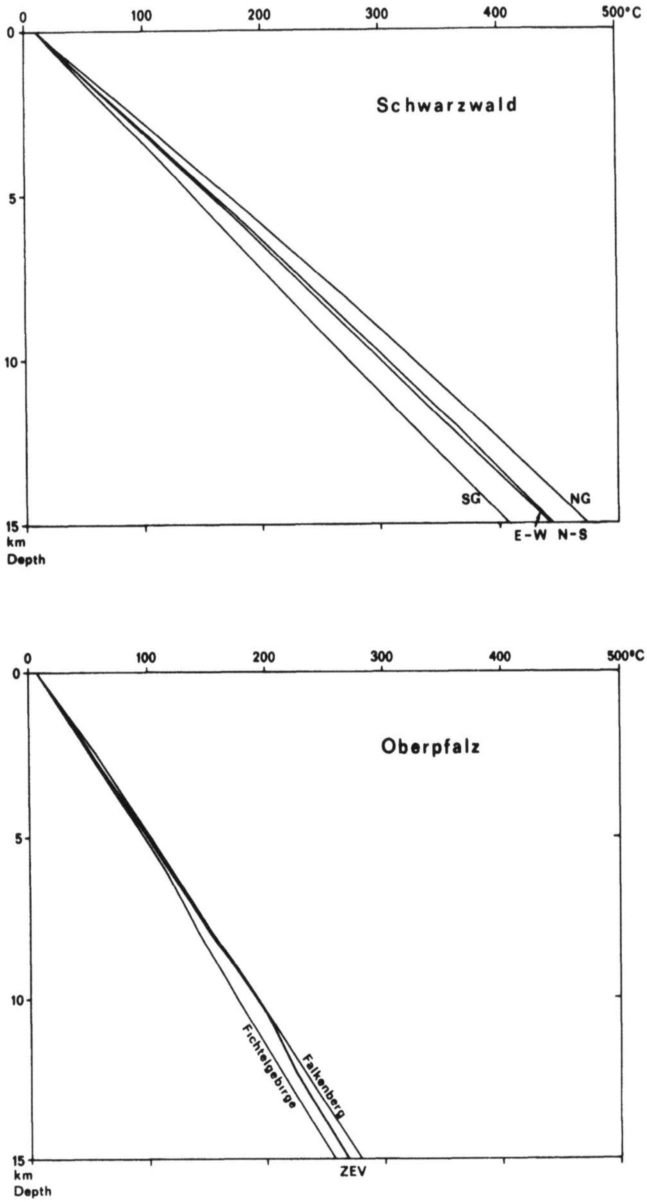

Fig. 27. Temperature profiles for different areas of the central Schwarzwald (top) and the Oberpfalz (bottom). The temperature-depth functions are calculated from near-surface heat flow values using distributions of thermal conductivity and heat generation which are estimated on the basis of seismic and petrological constraints. Top: Schwarzwald: E-W and N-S temperature profiles near Haslach, two-dimensional models represented by a WNW-ESE and a NNE-SSW section respectively. NG, SG temperature profiles in the northern and southern gneiss massif respectively, one-dimensional distribution of heat conductivity and heat generation. Bottom: Oberpfalz: Temperature profiles in the Fichtelgebirge, Erbendorf - Vohenstrauß Zone (ZEV), Falkenberg Granite Massif; one-dimensional models

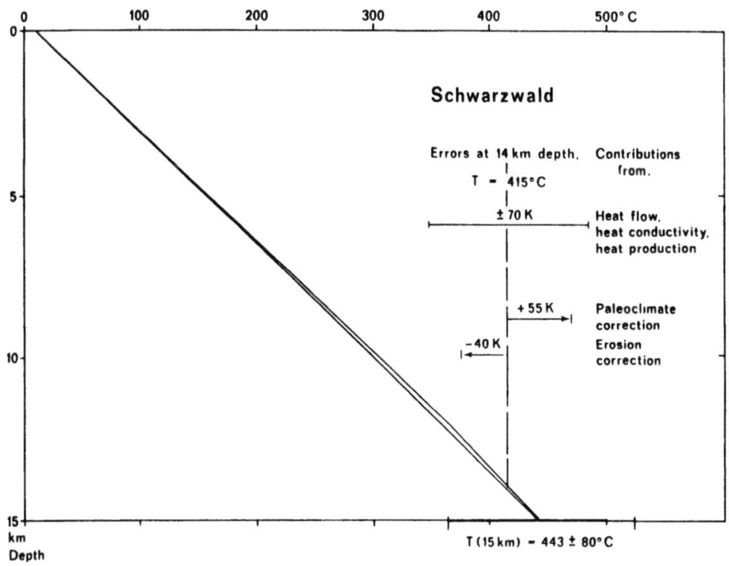

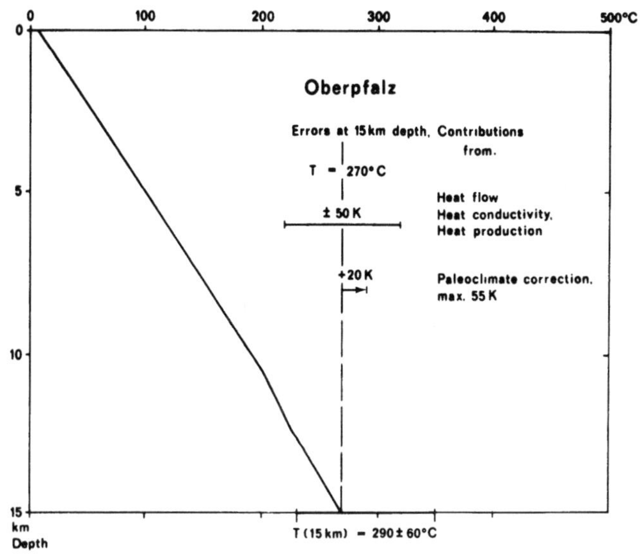

Fig. 28. Estimated temperature variations for conductive models. Resultant uncertainties are due to errors in the near surface heat flow, obtained from shallow boreholes; differing assumptions about the distribution of thermal conductivity and heat production between the surface and 15 km depth; maximum correction for paleoclimatic variations; maximum correction for effect of erosion (Schwarzwald only). Top: In the Schwarzwald the paleoclimatic and erosion corrections have the same order of magnitude but opposite directions; moreover, they are not independent of each other. Bottom: In the Oberpfalz the effect of erosion is estimated to be negligible. Therefore the sum of the paleoclimatic and erosion corrections is assumed to be about a third of the maximum +55 K, i.e. +20 K

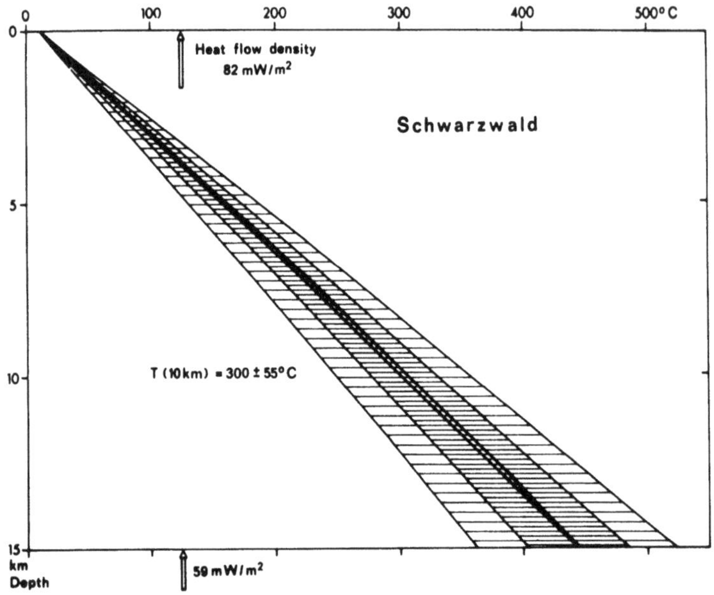

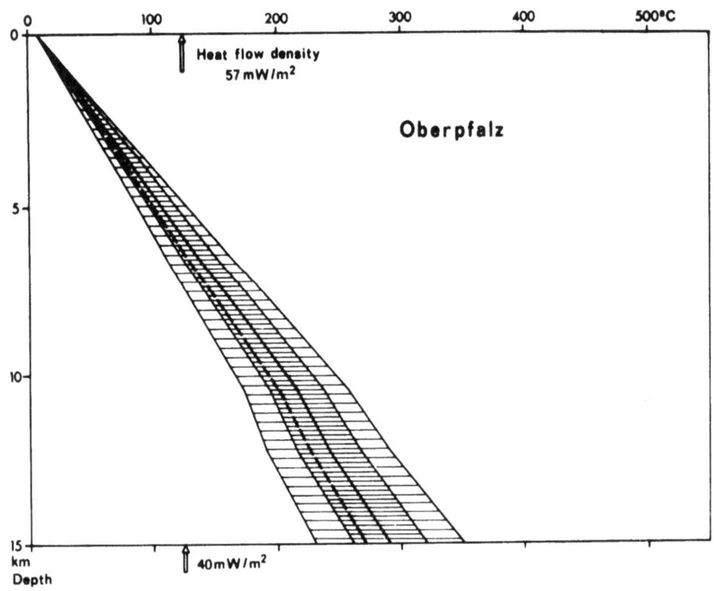

Fig. 29. General result of the geothermal studies. Shaded areas correspond to the variations as given in Fig. 28, the darker central area represents 50% of this error. The arrows indicate the heat flow at 0 and 15 km depth. Top: Schwarzwald: The double line in the middle refers to the Haslach area (Fig. 27 top). Bottom: Oberpfalz: The dashed line is the same as the line 'ZEV' in Fig. 27 bottom); the central continuous line has been shifted by +20 K at 15 km depth to take the effect of paleoclimatic changes into account

from crustal rocks. As can be seen from Figs. 23 and 25 the heat production of gneiss is more or less the same in both areas, whereas the Oberpfalz granites are about twice as radioactive as the Schwarzwald granites. The higher surface heat flow in the Schwarzwald therefore cannot be ascribed to a higher rate of heat production of the most frequent surface rock types in that area (cf. Sect. 4).

Thus, summarizing, we may state that the crust under the Schwarzwald is warmer than that under the Oberpfalz; in the Schwarzwald a temperature of 300°C will presumably be reached at about 10 km, and in the Oberpfalz at about 15 km.

Further, the higher temperature gradient in the Schwarzwald is probably a consequence of a higher heat flow from the upper mantle, about 40-60 mW/m^2, compared to about 20-30 mW/m^2 in the Oberpfalz.

Acknowledgements: The work described in this paper was funded by the Ministry of Research and Technology (BMFT) of the Federal Republic of Germany and by the German Research Society (DFG) within the framework of the Continental Deep Drilling Program (KTB). We are grateful to J. Behrens, G. Buntebarth, D.S. Chapman, C. Clauser, K. Finkbeiner, K. Fuchs, W. Grimmeisen, G. Gudden, R.Hänel, O. Kappelmayer, E. Lüschen, T. Massalsky, J. Meunier, W. Ott, L. Rybach, K. Schädel, G. Stettner, I. Stober, A. Vollbrecht and K. Weber for fruitful discussions. Numerous scientists were involved in the field measurements. Their personal engagement is gratefully acknowledged. We thank Gaby Bartman for typing the manuscript. We also thank for the support by federal, provincial, and local authorities whose cooperation was an indispensable aid for the field work and for the preparation of the boreholes.

REFERENCES

Bader K (1971) Bericht über Gamma-Log-Messungen in den Bohrungen H1 und T1, Neualbenreuth Projekt Sybillenbad. Bayer Geolog Landesamt Archiv-Nr 318e-II/5-742

Bathe KJ (1978) A finite element program for automatic dynamic incremental nonlinear analysis of temperatures. MIT Report 82448-5

Birch F, Roy RF, Decker ER (1968) Heat flow and thermal history in New York and New England. In: E-an Zen, White WS, Haddley JB, Thompson Jr JB (eds) Studies of Appalachian Geology: Northern and Maritime. Wiley, New York, pp 437-451

Bullard EC (1939) Heat flow in South Africa. Proc R Soc Lond A 173:474-502

Bullard EC (1940) The disturbance of the temperature gradient in the earth's crust by inequalities of height. Monthly Not R Astr Soc, Geophys Suppl 4:300-362

Cermak V (1975) Combined heat flow and heat generation measurements in the Bohemian Massif. Geothermics 4:19-26

Cermak V (1982a) Crustal temperature and mantle heat flow in Europe. Tectonophysics 83:123-142

Cermak V (1982b) Regional pattern of the lithospheric thickness in Europe. In: Cermak V, Hänel E (eds) Geothermics and geothermal energy. Schweizerbartsche Verlagsbuchhandlung, Stuttgart, pp 1-10

Cermak V, Bodri L (1986) Two-dimensional temperature modelling along five East-European geotraverses. J Geodynamics 5:133-163
Cermak V, Rybach L (1982) Thermische Eigenschaften. Kapitel 4 In: Angenheister G (ed) Landoldt Börnstein, Zahlenwerte und Funktionen aus Naturwissenschaften und Technik. Neue Serie, Gruppe V, Band 1a, Physikalische Eigenschaften der Gesteine, Springer, Berlin, Heidelberg, New York, Tokyo, pp 305-371
Clauser C (1988) Untersuchungen zur Trennung der konduktiven und konvektiven Anteile im Wärmetransport in einem Sedimentbecken am Beispiel des Oberrheingrabens. Ph D Thesis TU Berlin
Crowley KD (1987) Distribution of readioelements and heat production in continental crust: Testing hypotheses with the drill. EOS Transactions, AGU Vol 68 22:553-558
Dornstädter J (1987) Das Temperaturangleich-Verfahren. Diploma Thesis Geophysik Univ Karlsruhe
Dornstädter J, Sattel G (1985) Thermal measurements in the Rhenish Borehole Konzen, Hohes Venn (West Germany). N Jb Geol Paläont Abh 171:117-130
Drury MJ (1984) Borehole temperature logging for the detection of water flow. Geoexploration 22:231-243
Drury MJ (1987) Heat flow provinces reconsidered. Phys Earth Planet Inter 49:78-96
Furlong KP, Chapman DS (1982) Thermal modeling of the geometry of subduction with implications for the tectonics of the overriding plate. JGR Vol 87 B3:1786-1802
Gajewski D, Prodehl C (1987) Crustal evolution of the Rhinegraben area II. Seismic refraction investigation of the Black Forest. Tectonophysics 142:27-48
Haack U (1982) Radioaktivität der Gesteine. Kapitel 7. In: Angenheister G (ed) Landolt-Börnstein, Zahlenwerte und Funktionen aus Naturwissenschaften und Technik. Neue Serie, Gruppe V, Band 1b, Physikalische Eigenschaften der Gesteine. Springer, Berlin, Heidelberg, New York, Tokyo, pp 433-481
Haack U, Gohn E, Laschtowitz K, Becker KH (1988) Uran, Thorium und Kalium in Gesteinen des Schwarzwaldes und der Oberpfalz. KTB-Report (in prep.)
Hänel R (1970) Bestimmung der terrestrischen Wärmestromdichte in Deutschland. Z Geophys 37, Physika Verlag, Würzburg Archiv-Nr 5836:119-134
Hänel R (1971) Bericht über geothermische Messungen in der Bohrung Weißenstein. Niedersächsisches Landesamt f Bodenforschung Hannover Archiv-Nr T147
Hänel R (1974) Bericht zur Wärmestromdichtebestimmung der Bohrung Alexanderbad. Niedersächsisches Landesamt f Bodenforschung Hannover Archiv-Nr. 30173
Hänel R (1983a) Schlußbericht des Forschungsvorhabens: Vertikalbewegung und ihre Ursache am Beispiel des Rheinischen Schildes. Niedersächsisches Landesamt f Bodenforschung Hannover Archiv-Nr 93273
Hänel R (1983b) Geothermal investigations in the Rhenish Massif. In: Fuchs K et al. (eds) Plateau uplift. Springer, Berlin, Heidelberg, New York, Tokyo, pp 228-246
Hänel R, Schellschmidt R (1984) Bestimmung der Zufluß- und Abflußgebiete sowie der quantitativen Zufluß- und Abflußmengen mit Hilfe von Temperaturmessungen in den Bohrungen des HDR-Experimentierfeldes Falkenberg. Niedersächsisches Landesamt f Bodenforschung Hannover Archiv-Nr 96987
Jenkner B, Jenkner I, Klein H, Schädel K, Stober I (1986) Geothermievorbohrungen im mittleren Schwarzwald. KTB-Abschlußbericht Geol Landesamt Baden-Württemberg
Karl R, Manthey W, Schuster K (1965) Gesteinsphysikalische Parameter. Freiberger Forschungshefte Leipzig C 197

Kiderlen H (1977) Die Thermalquellen von Wildbad (Schwarzwald), ihre Mechanik und Genese. Jb Geol Landesamt Baden-Württemberg 19:165-217

Kiderlen H (1981) Die thermalen Mineralquellen von Bad Liebenzell (Schwarzwald). Jb Geol Landesamt Baden-Württemberg 22:7-34

Lachenbruch AH (1968) Preliminary geothermal model of the Sierra Nevada. J Geophys Res 73:6977-6989

Lachenbruch AH, Sass JH, Galanis, SP Jr (1985) Heat flow in southernmost California and the origin of the Salton Trough. J Geophys Res 90:6709-6736

Lüschen E, Wenzel F, Sandmeier K-J, Menges D, Rühl T, Stiller M, Janoth W, Keller F, Söllner W, Thomas R, Krohe A, Stenger R, Fuchs K, Wilhelm H, Eisbacher G (1987) Near-vertical and wide-angle seismic surveys in the Black Forest, SW Germany. J Geophys 62:1-30

Mansure AJ, Reiter M (1979) A vertical groundwater movement correction for heat flow. J Geophys Res 84:3490-3496

Mälzer H, Zippelt K (1987) Kriechende Spannungsumwandlungen: Rezente vertikale und horizontale Bewegungen. SFB 108 Rep 1984-1986 Teil A Univ Karlsruhe:47-97

Morgan P, Sawka WN, Furlong KP (1987) Introduction: Background and implications of linear heat flow-heat production relationship. Geophys Res Lett Vol 14 3:248-251

Müller S (1978) Evolution of the earth's crust. In: Spilhaus AF (ed) The Earth's crust. Geophys Monograph Ser 20, Am Geophys Union Washington DC, pp 11-28

Plaumann S (1985) Die Schwereanomalie im Bereich der Münchberger Gneismasse und ihre Interpretation. Z Dt Geol Ges 133:649-665

Plaumann S, Wohlenberg S (1982) Determination of rock densities in the Urach borehole. In: Hänel R (ed) The Urach Geothermal Project. Schweizerbartsche Verlagsbuchhandlung, Stuttgart, pp 101-105

Plaumann S, Groschopf R, Schädel K (1986) Kompilation einer geologischen Karte für den mittleren und nördlichen Schwarzwald mit einer Interpretation gravimetrischer Detailvermessungen. Geol Jb E33:15-30

Roy RF, Blackwell DD, Birch F (1968) Heat generation of plutonic rocks and continental heat flow provinces. Earth Planet Sci Lett 5:1-12

Roy RF, Blackwell DD, Decker ED (1971) Continental heat flow. In: Robertson EC (ed) The nature of the solid earth. McGraw-Hill, New York, pp 506-543

Rybach L (1976) Radioactive heat production. A physical property determined by the density of rocks. In: Strens RGJ (ed) The physics and chemistry of minerals and rocks. Wiley, London, pp 309-318

Sattel G (1979) Aufbau einer Meßapparatur zur Bestimmung thermischer Parameter von Gesteinen und Messungen an tertiären Sedimenten. Diploma Thesis Geophysik Univ Karlsruhe

Schön J (1983) Petrophysik, Physikalische Eigenschaften von Gesteinen und Mineralen. Enke, Stuttgart

Stiefel A (1986) Geothermische Untersuchungen im Mittleren Schwarzwald. KTB Rep 1984/1985 Geophysik Universität Karlsruhe

Stiefel A, Jäger K (1987) Black Forest: New geothermal aspects for SW Germany. XII Gen Ass Europ Geophys Soc Strasbourg, Terra Cognita 7:479

Stiefel A, Jäger K, Bried M, Reifenstahl F (1987) Geothermische Untersuchungen im Mittleren Schwarzwald. BMFT RG84012 KTB Rep 1986 Geophysik Univ Karlsruhe

Touloukian YS, Judd W, Roy RF (1981) Physical properties of rocks and minerals. McGraw-Hill, London

Wohlenberg J (1982) Dichte der Gesteine, Tab. 1.2.1, 1.2.4, 1.2.5. In: Angenheister G (ed) Landolt-Börnstein, Zahlenwerte und Funktionen aus Naturwissenschaften und Technik. Neue Serie, Gruppe V, Band 1a, Physikalische Eigenschaften der Gesteine. Springer, Berlin, Heidelberg, New York, Tokyo, pp 114-118
Zoth G (1985) Untersuchungen zum Temperaturfeld in Raum Kirchzarten. Geol Jb E28:175-190

Apatite Fission Track Dating of the Central European Basement. Postvariscan Thermo-Tectonic Evolution

G. A. Wagner, I. Michalski and P. Zaun*

Max-Planck-Institut für Kernphysik, Postfach 103 980, 6900 Heidelberg, FRG
*Duvenwischen 25, 2000 Hamburg 67, FRG

ABSTRACT

Apatite fission track (FT) dating reveals the thermotectonic evolution of rocks below 140°C. It is applied to 75 samples from the Central European crystalline basement in the Oberpfalz, Schwarzwald, and Vosges regions. Most intensively studied are the areas around the potential target locations for the deep continental drilling project KTB near Erbendorf/Oberpfalz and Haslach/Schwarzwald. The measured apatite FT ages range from 30 to 200 Ma. For nearly all samples, except for a few samples in the Oberpfalz, confined track length measurements indicate steady cooling below 140°C. The measured apatite FT ages date the cooling to ca. 100°C. In addition, projected track length measurements allow one to calculate the cooling age to ca. 60°C. Consequently, for each sample the temperature vs time path can be reconstructed. Since for most of the areas the cooling is due to tectonic uplift, uplift rates between 0.02 and 0.08 km Ma^{-1} can be inferred. Block faulting prevails in the Schwarzwald. In some cases elevated palaeo-thermogradients must be assumed in order to reconcile the apatite FT data. Notably, the Steinwald Höhenzug, in the southwestern extension of the Eger-Graben, and the Haslach area were likely geothermally active during the Tertiary. The selected KTB location near Windisch-Eschenbach/ Oberpfalz seems to have experienced uniform uplift (0.05 km Ma^{-1}) without significant thermal disturbance during the last 60 Ma.

1. INTRODUCTION

Since the early days of fission track (FT) dating in the Odenwald (Wagner 1968) there has been little application of this technique to the Central European crystalline basement. In that work it was originally established that the apatite (FT) clock gives the age when the basement cooled to ca. 100°C during slow uplift. It also initiated the development of the apatite FT system as a sensitive and versatile palaeo-geochronometer. The technique is now routinely used for studying the uplift behaviour of orogenic belts and the thermotectonic history of sedimentary basins.

Due to the growing interest in the Central European basement in connection with the deep continental drilling project KTB, a more extensive application of the FT method to the outcropping basement rocks, especially in the two KTB target areas, Schwarzwald and Oberpfalz, became desirable. Since fission tracks record the low temperature tail of the thermal evolution, it was hoped to gather basic data on the Post-Variscan thermotectonic history of these areas and, thus to contribute towards the decision on the final drilling site.

With regard to the short time available within the preparation phase of the KTB project, we concentrated on the mineral apatite as the most temperature-sensitive FT system. It is now well established that apatite records palaeo-temperatures around 100°C, whereas FT in zircon and titanite are less sensitive, probably in the 200° and 250°C temperature zones, respectively.

There are three different sources of information by which the apatite FT system reveals the thermotectonic evolution of rocks, that is the measured age, the age vs altitude/depth profile, and the fission track length. The *measured fission track age* may simply date the cooling to 100°C during uplift or, depending on the geological history, give the age and intensity of a weak thermal overprint. The regional distribution of such ages indicates areas of fast uplift and/or high palaeo-thermogradients by low age values. The *age vs altitude/depth profile* of a rock column contains the record of vertical movements provided the geothermal gradient remained constant. Uplift rates can be directly read from such profiles. *Fission track lengths* facilitate the diagnosis of the type of thermal history such as fast or slow steady cooling or even thermal overprinting. Fission tracks exposed to temperatures in the partial stability zone, i.e. between ca. 140 and ca. 60°C, gradually shorten. For instance, fission tracks with full length indicate fast, "volcanic-type" cooling. The longer the time the rock sample spent in the partial stability zone, the larger is the fraction of shortened fission tracks. The fission track technique and its potential for thermotectonic investigations has recently been reviewed by Wagner and Zaun (1985).

2. SAMPLES, EXPERIMENTAL PROCEDURES, AND AGE RESULTS

Within the present project, 75 rock samples from the Oberpfalz, Schwarzwald, and Vosges regions were studied. They were collected mostly from outcrops and a few from drill cores in the crystalline basement. The samples were selected according to the following criteria:

1. Fresh rocks containing apatite as accessory mineral;
2. Various rock types and rock units;
3. Different altitudes and bore hole depths;
4. The same rock samples were also preferably studied with other isotope-geological techniques.

Information on sampling locality, altitude, and the rock type is given in Table 1a for the Oberpfalz, Table 1b for the Schwarzwald, and Table 1c for the Vosges samples. The Schwarzwald and Vosges samples are described in more detail by Michalski (1987).

After crushing the rock samples (1-3 kg), at least 10 mg apatite concentrate was separated from the heavy mineral 100-200 μm size fraction. The apatites were prepared according to the grain population technique, i.e. one subpopulation of each concentrate was

Table 1a. Sample localities in the Oberpfalz

HDSP number	Other number	Locality	Rock type[a]	Elevation (m above N.N.)	Coordinates R	H
70	TB 47[b] (170 m)	Mähring-Wäldel	gr	515	45 37 513	55 32 075
71	TB 103[b]	Mähring	gn	652	45 37 800	55 30 575
72	TB 105[b]	Weiße Marter	gn	670	45 35 363	55 23 513
73	TB 104[b]	Rosall	sc	585	45 27 275	55 32 850
74	TB 101[b]	Theisseil	gn	503	45 15 850	55 05 075
75	TB 120[b]	Oedenthal	gn	552	45 17 113	55 00 675
76	TB 9/5[b]	Michldorf	gn	448	45 17 675	54 96 800
77	BOG 30[c] (30 m)	Griesbach	gn	690	45 35 750	55 25 350
78	BOG 296[c] (296 m)	Griesbach	gn	424	45 35 750	55 25 350
79	BOG 199[c] (199 m)	Griesbach	gn	521	45 35 750	55 25 350
80	BOP 36[c] (36 m)	Püllersreuth	gn	512	45 09 675	55 16 325
81	BOP 295[c] (295 m)	Püllersreuth	gn	253	45 09 675	55 16 325
82	BON 100[c] (100 m)	Neusorg	gn	536	44 99 800	56 35 575
84		Napfberg	gr	700	45 03 700	55 28 550
85		Phaben-Brand	gr	785	45 02 400	55 27 850
88		Reuth-Drahthammer	di	470	45 08 700	55 23 450
89		Bernstein	gr	480	45 12 000	55 20 950
90		Windischeschenbach	am	440	45 11 075	55 19 500
91		Beidl Teich	gr	500	45 19 100	55 19 250
96		Iglersreuth	gr	540	45 27 500	55 19 200
98		Silberhütte	gr	840	45 28 150	55 14 000
99		Flossenbürg	gr	670	45 25 500	55 11 475
101		Kainzelmühle	gn	460	45 21 400	54 92 950
	BOH,DF-3077[d]	Böhmisch-Bruck		520	45 25 750	54 92 500
102		Leuchtenberg	gr	500	45 18 375	54 95 625
107		Tröstau	gr	580	44 95 950	55 41 075
108		Weißenstein	ek	668	44 78 000	55 54 900
109	68726[c,e] (293 m)	Falkenberg	gr	214	45 14 400	55 24 700
110	68729[c,e] (54 m)	Falkenberg	gr	453	45 14 400	55 24 700

b. Sample localities in the Schwarzwald

HDSP number	Locality	Rock type[a]	Elevation (m above N.N.)	Coordinates R	H
2	Herrenberg	gn	230	34 32 95	53 49 85
4	Steinebühl	gn	580	34 34 67	53 56 55
5	Riersbach	gn	390	34 37 20	53 60 73
6	Löcherbergwasen	gn	395	34 38 30	53 67 90
9	Vorbäch	gn	290	34 34 25	53 45 65
10	Mattenhof	gn	470	34 32 98	53 42 58
11	Oberwinden	gn	350	34 32 98	53 42 58
12	Gschasifelsen	gn	1014	34 35 10	53 37 95
13	Wolfsgrube	gn	925	34 35 03	53 37 80
14	Friedhofkehre	gn	600	34 33 83	53 37 83
15	Hechtsberg	gn	230	34 35 78	53 49 60
16	Artenberg	gn	210	34 30 45	53 51 00
18	Wolfsbrunnen	gr	690	34 41 38	53 83 50
19	Hohfelsen	gr	900	34 42 08	53 83 20
22	Steinbühl	gn	640	34 30 68	53 34 40
24	Längenberg	gn	500	34 42 35	53 93 15
25	Roßdobel	gn	700	34 29 75	53 32 55
27	Hausach	gn	270	34 41 50	53 50 10
33	Rotwasser	gn	940	34 30 55	53 04 28
34	Eckkopf	gn	915.2	34 24 73	52 97 05
36	Todtnau	gn	730	34 20 53	53 00 55
39	Schweizerhof	gn	700	34 26 03	53 08 08
40	Kaibenbühl	mi	630	34 28 28	53 14 83
41[g]	Siensbachereck	gn	700	34 25 50	53 27 90
43[g]	Kandelrücken	gn	1030	34 26 13	53 26 60
44[g]	Kandelgipfel	gn	1230	34 26 45	53 25 50
45	Unterer Felsen	gn	370	34 30 25	53 29 05

b. Sample localities in the Schwarzwald

HDSP number	Locality	Rock type[a]	Elevation (m above N.N.)	Coordinates R	H
46[g]	Hornhausle	am	720	34 30 73	53 27 45
47[c] A	Bohrung Kirchzarten	mi	- 64	34 23 10	53 16 34
47[c] B	Bohrung Kirchzarten	mi	-618	34 23 10	53 16 34
49[c] A	Bohrung Schonmatt	gn	214	34 35 58	53 45 88
49[c] B	Bohrung Schönmatt	gn	- 31	34 35 58	53 45 88
50[f]	Waldshut	mi	440	34 43 59	52 79 75
51[f]	Heimbach	gn	600	34 20 80	52 71 50
52[f]	Staubecken	gn	960	34 21 18	52 80 50
53[f]	Tiefenstein	gr	530	34 32 00	52 77 30
54[f]	Weiherkopf	gr	980	34 09 00	52 97 15
55[f]	Elbenschwand	mi	600	34 12 20	52 91 28
56[f]	Pfeifershutte	mi	660	34 20 65	52 82 35

c. Sample localities in the Vosges

HDSP number	Locality	Rock type[a]	Elevation (m above N.N.)	Coordinates R	H
BV 7	Dusenbach	gn	340	33 73 75	53 40 25
BV 11	Tete des Immerlins	di	1120	33 58 40	53 32 73
BV 12	Lac Blanc	di	1050	33 58 90	53 31 85
BV 16	Col de Bussang	vo	730	33 43 63	53 06 63
BV 17	Ballon d'Alsace	gr	800	33 39 78	53 00 05
BV 19	Sewen	di	500	33 44 200	52 97 45
BV 20	Lauw	vo	400	33 50 95	52 91 25

[a] Abbreviations: am=amphibolite; di=diorite; ek=eclogite; gn=gneiss; gr=granite; mi=migmatite; N.N.=Normal Null= Sea Level; sc=schist; vo=volcanic rock
[b] Samples collected by St. Teufel, Universitat Gottingen
[c] Samples from drill cores, depth is given in metres
[d] Sample collected by D. Gebauer, ETH Zürich
[e] Samples from I. Wendt, BGR Hannover
[f] Samples collected by F. Taborszky, Universitat Freiburg
[g] Samples collected by M. Wagner, MPI Heidelberg

heated (450°C, 10 h) in order to anneal the fossil tracks and then irradiated with thermal neutrons.

The irradiations were performed in channel 8 (nominal neutron flux 1.6×10^{11} n cm^{-2}, cadmium ratio for gold=11, $\Phi_{epith}/\Phi_{th}=1/152$) of the Thetis reactor at Rijksuniversiteit Gent. The neutron dosimetry was carried out with Au-, Co-, and Cu-activation monitors as well as with uranium dosimeter glass, moldavite Trebice. The Cu-activation monitors were used for flux gradient correction only. The neutron dose determination of the Au, Co, and glass dosimeters agreed within 1%. The "Au" neutron dose values are given in Table 2.

After irradiation both subpopulations of each concentrate were polished and simultaneously etched for 50 s in 5% HNO_3 at 20°C. The tracks were counted in transmitted light (magnification $100 \times 1.25 \times 25$) under oil immersion. At least 500 tracks and 100 crystals were considered for determining both the spontaneous and induced fission track densities. As constants $\lambda_f=8.46 \times 10^{-17} a^{-1}$, $\sigma_f=580.2 \times 10^{-24}$ cm^2, U-235/U-238 = 7.25×10^{-3} were used for age calculation. Under these conditions an FT-age of 26.7 ± 1.1 Ma is found for the Fish Canyon apatite FCT3. The analytical data and the resulting ages are presented in Table 2a for the Oberpfalz, Table 2b for the Schwarzwald, and Table 2c for the Vosges apatites.

Table 2a. Fission track data for Oberpfalz apatites

HDSP number	ρ_s [a] $10^5/cm^2$	ρ_i [b] $10^5/cm^2$	ρ_s/ρ_i	Neutron fluence ϕ $10^{14}/cm^2$	Age t_m (Ma)	1σ [c] (Ma)	c_s/c_i	Age t_f [d] (Ma)	$\bar{l}_s^*$ [e] (μm)	1σ [e] (μm)
70	11.2 (3063)	8.8 (2400)	1.276	9.04	57.1	3.8	-	-	-	-
71	13.3 (1972)	8.1 (1198)	1.646	8.96	72.9	4.2	-	-	13.6	2.3
72	18.1 (4091)	10.9 (2451)	1.669	9.12	75.3	5.1	0.39	29	-	-
73	20.1 (3200)	13.4 (2133)	1.500	9.20	68.	4.4	0.43	29	-	-
74	18.2 (3339)	6.5 (1186)	2.816	8.99	124.7	8.4	0.33	41	13.2	2.2
75	20.3 (4835)	5.9 (1407)	3.436	9.78	165.0	11.7	-	-	11.5	3.0
76	26.4 (9540)	5.9 (2130)	4.479	8.90	195.3	12.5	0.23	44	11.1	2.7
77	22.4 (1845)	14.4 (1184)	1.558	8.55	65.9	5.2	-	-	-	-
78	10.9 (2939)	6.3 (1702)	1.727	8.46	72.2	3.1	-	-	-	-
79	17.6 (1500)	12.6 (1072)	1.398	8.54	59.1	3.9	-	-	-	-
80	11.4 (2126)	8.6 (1611)	1.320	8.86	57.9	2.7	0.60	35	-	-
81	27.9 (2304)	20.5 (1689)	1.364	8.63	58.3	2.6	-	-	-	-
82	16.1 (2515)	12.4 (1939)	1.297	9.13	58.6	2.9	-	-	-	-
84	30.9 (5409)	23.4 (4092)	1.322	8.59	56.1	2.4	0.46	26	-	-
85	17.8 (2529)	15.7 (2227)	1.136	8.77	49.3	3.2	0.37	18	-	-
88	6.2 (970)	4.4 (693)	1.400	9.26	64.1	4.6	0.61	39	-	-
89	13.2 (1776)	10.5 (1411)	1.259	9.35	58.3	3.0	-	-	-	-
90	5.0 (1120)	3.5 (785)	1.427	9.31	65.7	5.0	0.56	37	-	-
91	16.4 (3148)	12.7 (2429)	1.296	9.22	59.2	2.8	-	-	-	-
96	12.7 (1901)	8.9 (1335)	1.424	8.24	58.1	3.1	0.48	28	-	-
98	42.6 (5775)	31.1 (4240)	1.362	8.72	58.8	2.4	0.50	29	-	-
99	13.4 (1575)	9.5 (1120)	1.406	8.81	61.3	4.3	0.57	35	-	-
101	31.6 (5240)	8.9 (1483)	3.534	8.80	152.9	7.4	0.61	93	-	-
102	4.8 (4838)	1.6 (1599)	3.024	8.41	125.3	16.0	0.30	37	12.6	1.9
107	0.66 (834)	0.44 (550)	1.517	9.11	68.4	9.0	0.76	52	-	-
108	0.83 (971)	0.40 (472)	2.058	8.54	86.6	10.0	0.75	65	-	-
109	8.3 (3303)	8.12(3248)	1.017	8.28	41.7	3.8	-	-	-	-
110	14.2 (3554)	10.5 (2625)	1.354	8.59	57.6	3.0	-	-	12.8	2.0
BOH, DF-3077	7.14 (926)	14.23 (922)		53.4	159[f]	4.5	-	-	-	-

b. Fission track data for Schwarzwald apatites

HDSP number	ρ_s [a] $10^5/cm^2$	ρ_i [b] $10^5/cm^2$	ρ_s/ρ_i	Neutron fluence ϕ $10^{14}/cm^2$	Age t_m (Ma)	1σ [c] (Ma)	c_s/c_i	Age t_f [d] (Ma)	$\bar{l}_s^{*}$ [e] (μm)	1σ [e] (μm)
2	0.18(3189)	0.22(3842)	0.83	9.37	38.6	2.4	-	-	-	-
4	0.58(2953)	0.51(2590)	1.14	9.37	53.1	2.8	-	-	-	-
5	2.34(2521)	1.59(1714)	1.47	9.37	68.6	2.8	-	-	13.01	1.34
6	2.59(3519)	2.02(2749)	1.28	9.37	59.2	2.9	-	-	13.81	1.29
9	1.98(3463)	0.81(1408)	2.45	8.83	107.0	4.3	0.45	48	13.21	1.37
10	1.77(3913)	1.03(2275)	1.72	9.37	79.9	3.2	-	-	13.32	1.23
11	0.42(2879)	0.27(1869)	1.54	8.78	66.9	3.9	0.31	21	-	-
12	2.40(3257)	1.58(2143)	1.52	9.37	70.8	3.1	-	-	-	-
13	2.22(4119)	1.56(2900)	1.42	9.37	65.9	2.5	0.32	21	12.91	1.34
14	1.64(3011)	1.30(2389)	1.26	9.37	58.7	2.1	-	-	-	-
15	1.48(2957)	1.21(2424)	1.22	9.37	56.9	2.5	-	-	-	-
16	0.56(3017)	0.54(2900)	1.04	9.37	48.1	3.1	-	-	-	-
18	1.50(2071)	1.72(2380)	0.87	9.37	39.5	1.7	0.57	23	-	-
19	2.07(2927)	2.28(3216)	0.91	9.37	42.0	1.4	-		13.87	1.27
22	2.15(2871)	1.65(2208)	1.30	8.56	55.1	2.4	0.48	27	-	-
24	2.16(3065)	1.73(2476)	1.25	9.15	56.8	2.5	-	-	13.34	1.20
25	2.46(4725)	1.97(3780)	1.25	9.37	58.1	2.4	-	-	-	-
27	2.38(4256)	1.97(3517)	1.21	9.37	56.0	1.8	-	-	-	-
33	0.80(3215)	0.73(2950)	1.09	8.6	46.6	2.3	-	-	12.98	1.58
34	0.32(2356)	0.46(3414)	0.69	8.37	29.0	1.3	-	-	-	-
36	2.44(4237)	2.37(4116)	1.03	9.37	47.2	2.6	0.57	27	-	-
39	0.74(1293)	0.57(1003)	1.29	9.24	59.4	3.4	0.48	28	-	-
40	1.26(3219)	0.91(2316)	1.39	8.97	61.7	2.8	0.34	21	-	-
41	0.72(2215)	0.6 (1845)	1.20	8.88	52.8	3.2	0.23	12	-	-
43	0.89(3256)	0.65(2377)	1.37	9.1	61.9	2.8	0.25	15	-	-
44	1.59(2746)	1.01(1747)	1.57	9.06	70.0	3.1	-	-	13.44	1.48
45	1.24(2876)	1.17(2713)	1.06	9.37	49.4	2.2	0.36	18	-	-
46	1.51(2908)	1.20(2307)	1.26	8.97	55.9	2.6	-	-	13.22	1.55
47A	1.68(3628)	1.29(2790)	1.30	8.51	55.2	2.2	-	-	-	-
47B	1.14(2785)	1.58(3868)	0.72	9.06	32.5	1.4	-	-	-	-
49A	2.0 (3219)	1.09(1759)	1.83	8.85	75.2	3.7	-	-	-	-
49B	1.48(2879)	1.20(2340)	1.23	8.41	51.5	2.6	-	-	-	-
50	1.0 (1745)	0.72(1255)	1.39	8.69	59.6	3.0	0.39	23	-	-

51	1.01(2750)	0.63	1.61	8.83	70.3	3.5	0.31	22	-	-
52	1.1(2537)	0.73(1703)	1.49	8.46	62.7	3.2	0.37	23	-	-
53	0.97(3357)	0.7 (2433)	1.38	8.46	58.0	2.8	0.43	25	-	-
54	0.97(2895)	1.33(3965)	0.73	8.6	31.2	1.5	-	-	12.76	1.37
55	0.5 (3173)	0.57(3605)	0.88	8.74	38.2	2.1	0.49	19	-	-
56	1.26(2497)	1.27(2522)	0.99	8.74	43.0	2.1	-	-	-	-

c. Fission track data for Vosges apatites

BV number	p_s[a] $10^5/cm^2$	p_i[b] $10^5/cm^2$	p_s/p_i	Neutron fluence ϕ $10^{14}/cm^2$	Age t_m (Ma)	1σ[c] (Ma)	c_s/c_i	Age t_f[d] (Ma)	$\bar{l}_s^{*e}$ (μm)	1σ[e] (μm)
7	1.33(2374)	1.38(2472)	0.96	9.18	44.0	2.2	0.40	18	12.94	1.66
11	1.43(2482)	1.14(1969)	1.26	9.32	58.1	3.0	-	-	-	-
12	1.18(2046)	1.01(1748)	1.17	9.18	53.4	2.8	0.48	27	13.01	1.63
16	0.76(1326)	1.40(1339)	0.99	9.01	44.1	2.3	-	-	-	-
17	1.40(2434)	1.43(2483)	0.98	8.7	42.3	1.7	0.45	19	12.75	1.47
19	1.1(1897)	1.25(2180)	0.87	9.14	39.8	1.7	0.65	26	13.39	1.38
20	0.31(919)	0.34(1009)	0.91	8.83	39.9	2.6	-	-	-	-

[a] Spontaneous track density with number of track counts in parentheses.
[b] Induced track density with number of track counts in parentheses.
[c] The error of the age t_m takes into account the 1σ deviation for the track counts and 2.5% for neutron dosimetry.
[d] The 60°C cooling age t_f has a 1σ error of ca. ± 15%.
[e] The mean lengths of confined spontaneous fission tracks with 1σ deviation.
[f] Fission track age determined by C.W. Naeser, U.S.G.S. Denver (personal communication).

3. TRACK LENGTH MEASUREMENTS

As outlined above, the length distribution of fission tracks bears information on the type of thermal evolution. Lengths of fission tracks in apatite are measured either on tracks intersected by and projected on a polished internal face ("projected track length"), i.e. on the tracks which are also used for age counting, or on horizontal tracks confined within the crystal and intersected by host tracks or cleavage planes ("confined track length"). Clearly, projected track lengths do not represent the true lengths of the tracks. However, their advantage is common occurrence and fast and precise measurement with image analyzing systems in one and the same operation as track counting. On the other hand, although the confined track lengths represent much better the true length distribution, they are biased in favour of long tracks.

Using confined track lengths in apatite, Gleadow et al. (1986) distinguish several cases of thermal evolution such as rapid cooling, slow continuous cooling and complex thermal history. Table 3 lists these types of thermal history with the corresponding mean length and standard deviation of the spontaneous fission tracks. Recently, Wagner (1988) proposed an approach for deciphering the low temperature (i.e. below 140°C) path of steadily cooling rocks. This method was first corroborated in the context of the present project. It is based on projected length measurements of spontaneous and induced fission tracks in apatite. It enables one to distinguish different cooling patterns such as constant, decelerated and accelerated rates as well as to date the cooling age to ca. 60°C, in addition to the conventional ca. 100°C cooling age. The observation that all tracks with projected lengths longer than 10 µm were formed in the zone of full track stability (i.e. below ca. 60°C) only, underlies this approach. A ratio c_s/c_i is defined in which c_s and c_i are the fractions of spontaneous and induced tracks respectively, longer than 10 µm (Fig. 1). As shown by Wagner (1988) this ratio is practically equal to the fraction of tracks produced below 60°C compared to the total number of spontaneous fission tracks. In other words, from the total spontaneous fission track density ρ_s in apatites with a steady cooling history one calculates the 100°C cooling age t_m and from the fraction $c_s/c_i x \rho_s$ one calculates the 60°C cooling age t_f.

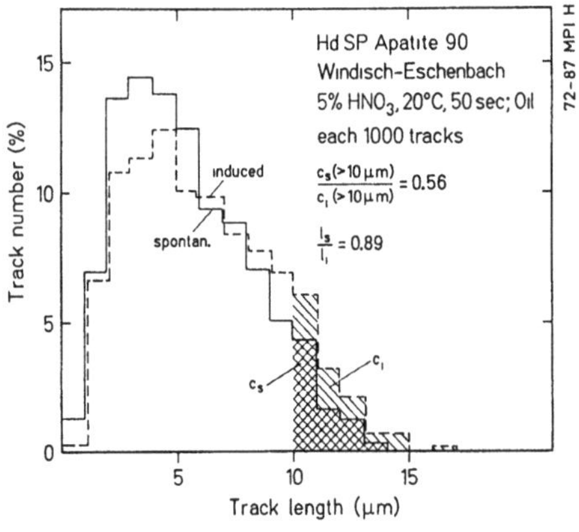

Fig. 1. Length histogram for projected fission tracks in apatite grains from sample HDSP 90

In the present study, the projected lengths were measured in most of the samples. This was done in the same operation as track counting, i.e. more than 1000 tracks were considered at times. An example of the distribution of projected spontaneous and induced fission tracks is given in Fig. 1. The projected length measurements are given in the form of the c_s/c_i ratio together with the resulting 60°C cooling ages t_f in Table 2 (a, b, c).

For most of the Schwarzwald, Vosges, and some Oberpfalz apatites the confined track lengths were also determined (magnification 63 × 1.25 × 25, in transmitted light, air). This requires stronger etching conditions (5 m HNO_3, 20 s, 20°C) after the tracks have been counted (for details see Michalski 1987). An example of the distribution of confined spontaneous fission tracks is presented in Fig.2. These measurements are also listed (as $\bar{l}_s^* \pm 1\sigma$) in Table 2 (a, b, c).

4. GEOLOGICAL INTERPRETATION

Under geological conditions fission tracks in apatite are completely erased at temperatures above 150°C but fully stable only below 50°C. In the intermediate temperature zone they are partially stable. Therefore, when interpreting FT ages, one has to distinguish two possible model cases: firstly, the rock cools steadily, i.e. the fission tracks become increasingly stable. The FT age is then a cooling age for a temperature at which the fission tracks become effectively stable, i.e. about halfway within the partial stability zone. Secondly, when reheating occurs, fission tracks become increasingly instable. The FT age is then a mixed age without any direct geological meaning. As discussed above, these cases can be recognized and distinguished by confined track length criteria. From the apatites checked (Table 2 a, b, c) in the present investigation only two samples, namely HDSP 75 and 76 from the Oberpfalz (southern zone of Erbendorf-Vohenstrauß), have mean confined lengths $\bar{l}_s^*$ below 12 μm with standard deviations above 2 μm, indicating a complex thermal evolution according to Table 3. The remaining apatite samples which were checked for confined lengths are all compatible

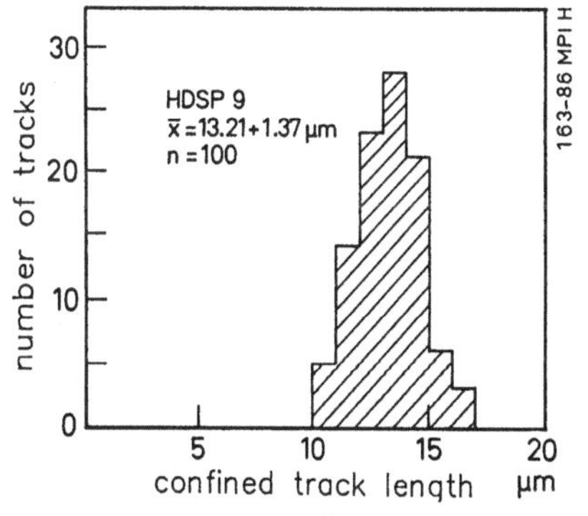

Fig. 2. Length histogram for confined spontaneous fission tracks in apatite grains from sample HDSP 9. x̄=mean length ± 1 SD of 100 tracks

Table 3. Confined lengths of fossil tracks in apatite and the type of thermal history (Gleadow et al. 1986)

Thermal history	Mean length (μm)	Standard deviation (μm)
Very rapid cooling	14.5-15	≈1.0
Slow continuous cooling	12 -13	1.2
Complex thermal cooling	≤11.5	≥2
For comparison: freshly induced tracks	16.3	≈0.9

with slow but steady cooling. The two "complex-cooling type" samples give also the highest measured apatite FT ages. The low c_s/c_i ratio = 0.23 for sample HDSP 76 indicates a relatively long residence time within the partial track stability zone. In short, for the majority of the Variscan basement samples so far studied, the apatite FT ages are cooling ages whereas some samples from the Oberpfalz seem to have experienced a more complex Post-Variscan thermal history.

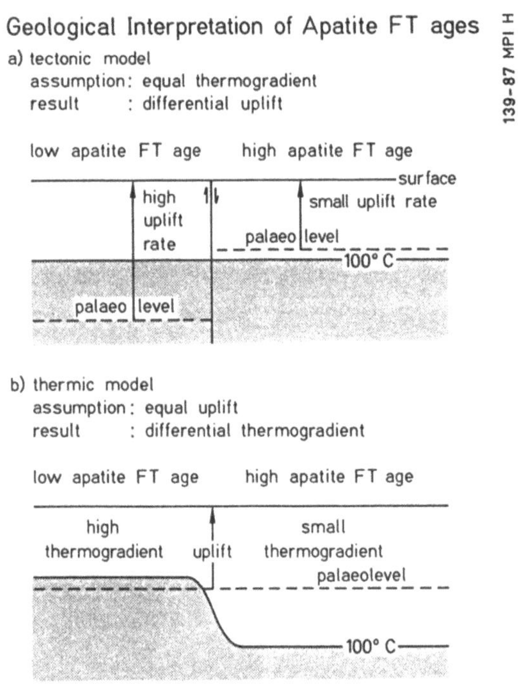

Fig. 3. Two principal models of interpreting fission track ages of apatites with steady cooling history. The low apatite FT age presents a "terminus post quem" for the last tectonic and/or thermic activity

Another important aspect for geological interpretation of apatite FT ages is the role of the geothermal gradient. Let us assume cooling ages with a significant regional variation, a pattern commonly observed in the present study. Such lateral age variation may principally be explained by *different uplift rates* in an environment of an equal geothermal gradient ("tectonic model") or by *different geothermal gradients* for an equal uplift rate ("thermal model") (see Fig. 3). Of course, all gradual transitions between these two extreme cases are also conceivable. A further complication is introduced if one takes into account the possibility of palaeo-gradients changing with time. For deciphering such complex cases one needs samples from three dimensions, especially deep drill holes, and independent geological information.

4.1 Oberpfalz

Figure 4 shows the sample localities in the Oberpfalz with the measured apatite FT ages. They range from the Münchberger Gneismasse in the NW, across the Fichtelgebirge/Steinwald, the zone Erbendorf/ Vohenstrauß (ZEV), the Luhe-Line into the Moldanubikum gneisses in the SE. The regional distribution of the ages consists of a broad, central zone with relatively low FT ages (50-70 Ma) extending from Erbendorf/Windisch-Eschenbach area eastward to Mähring near the Bohemian border. Much higher FT ages (120-200 Ma) with pronounced differences are found south of this zone in the area Leuchtenberg/Vohenstrauß/Böhmisch-Bruck. These dates are the highest apatite FT ages so far found in the Variscan basement. According to the confined track length criteria, discussed above, the apatite FT ages of the first group, i.e. below 100 Ma, are cooling ages to ca. 100°C. On the other hand, some apatite FT ages from the southern ZEV are probably mixed ages. For most of the apatites also the 60°C cooling ages, based on projected length measurements, were determined which are also shown in Fig. 4. In the following the samples from the central and northern region are discussed separately.

In the *central area*, which is outlined by the triangle Windisch-Eschenbach/Mähring/Flossenbürg, the FT ages t_m lie between 58 and 75 Ma and the FT ages t_f between 29 and 39 Ma. More to the north two Steinwald samples show as t_m 49-56 Ma and as t_f 18-26 Ma, whereas higher ages are found for Tröstau/Fichtelgebirge (t_m:68 Ma, t_f: 52 Ma) and Weißenstein/Münchberger Gneismasse (t_m: 87 Ma, t_f: 65 Ma). The apatite FT age pattern turns out to be characteristic for each region but differs from region to region. From the 100°C ($\hat{=}t_m$) and 60°C ($\hat{=}t_f$) cooling ages the time vs temperature path can be reconstructed for each region separately. This is illustrated in Fig. 5. The different cooling behaviour may be caused by different uplift rates and/or different palaeo-thermogradients as mentioned above. Along a profile from the Münchberger Gneismasse in the NW to the Entenbühl in the SE two possible interpretations, according to the tectonic and the thermic models are depicted in Fig. 6.

In Fig. 6a, the tectonic model, a constant geothermal gradient of 30° km^{-1} is assumed. The cooling is purely due to uplift. The resulting palaeo-positions of the present 500-m altitude level ($\hat{\approx}$ present surface) with the corresponding uplift rates is calculated for the moments 30, 60, and 90 Ma ago. A lower value for the thermogradient would increase the depths and uplift rates, but would not affect relative differences. From this model an early fast uplift of the Fichtelgebirge and the Münchberger Gneismasse 90 to 60 Ma ago with subsequent slowing down appears. The Steinwald shows increasing and fast uplift during the last 30 Ma. A uniform uplift history dur-

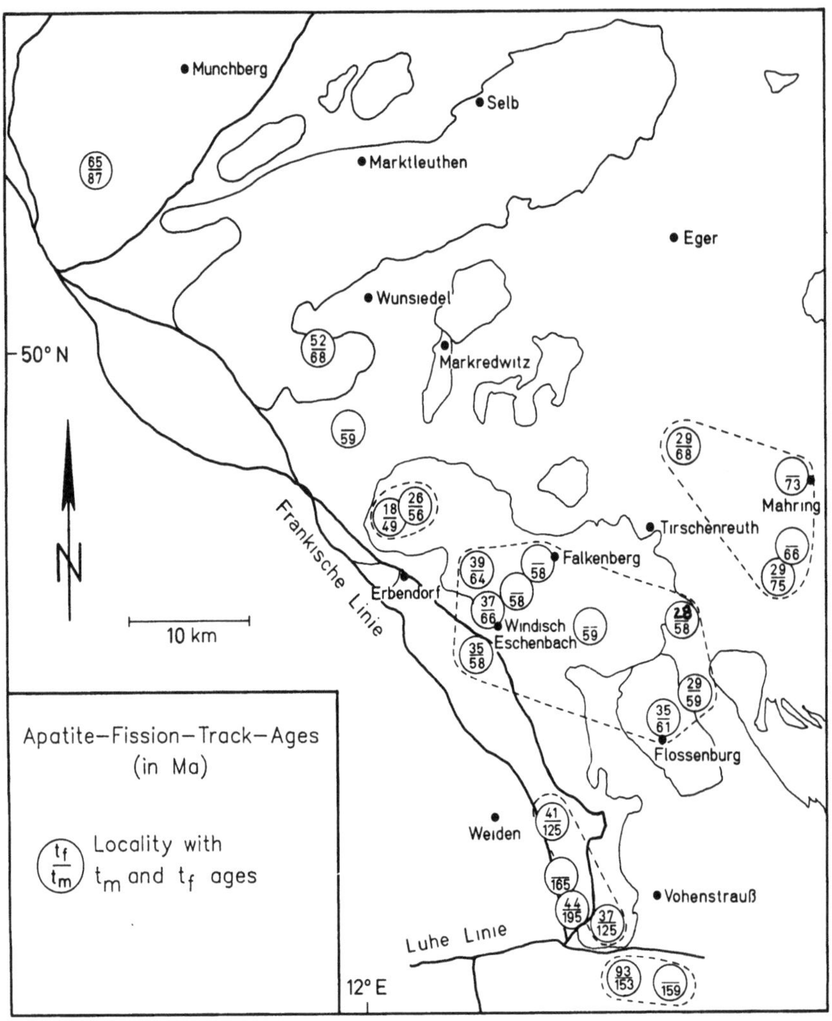

Fig. 4. Sample localities in the Oberpfalz region with measured apatite fission track ages t_m ($\hat{=}$ ca. 100°C) and the 60°C cooling ages t_f. The <u>dashed outlines</u> combine samples with similar age patterns

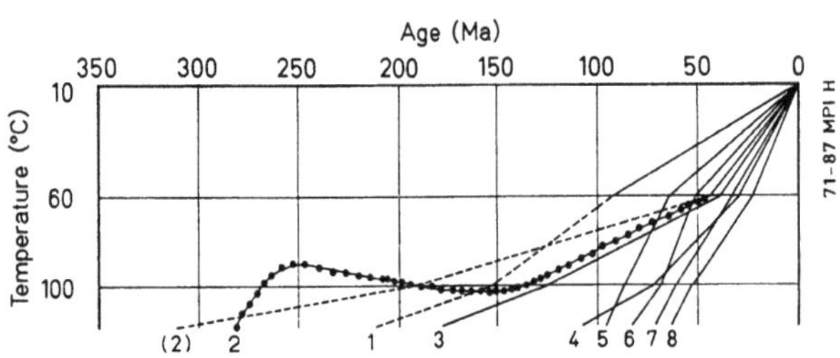

Fig. 5. Thermotectonic evolution of various regions of the Oberpfalz (cf. Fig. 4) based on FT analysis: <u>1</u> HDSP 76, <u>2</u> HDSP 101, <u>3</u> HDSP 74, 102, <u>4</u> HDSP 72, 73, <u>5</u> HDSP 108, <u>6</u> HDSP 107, <u>7</u> HDSP 80, 88, 90, 96, 98, 99, <u>8</u> HDSP 84, 85. The <u>dashed lines</u> assume steady cooling for regions <u>1</u> and <u>2</u>, however, more realistic is thermal overprint during Mesozoic subsidence (<u>dotted line</u>)

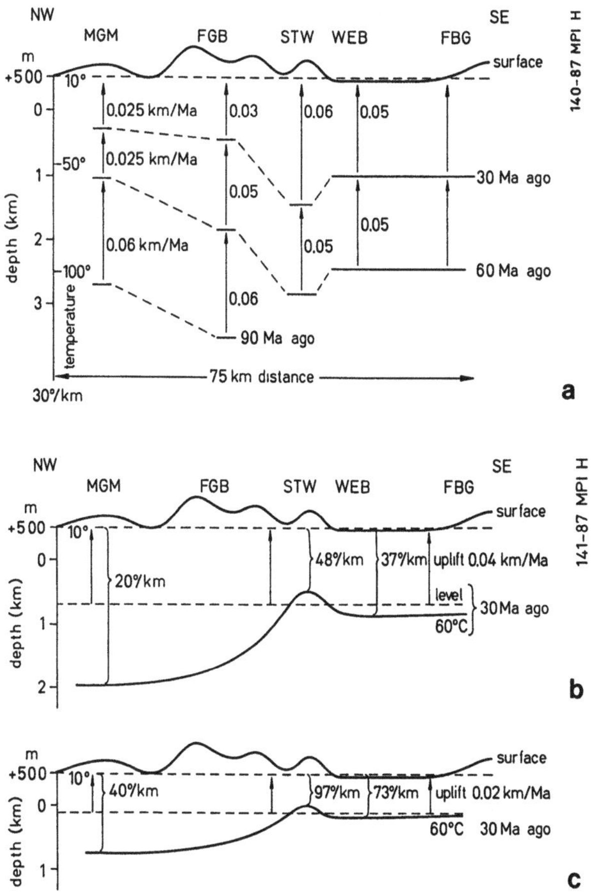

Fig. 6. Interpretation of the apatite FT ages along a NW-SE-profile through the Bohemian Massif from Münchberger Gneismasse (MGM), across the Fichtelgebirge (FGB), the Steinwald (STW), Windisch-Eschenbach (WEB) to Flossenbürg (FBG). The tectonic model (*a*) shows the palaeo-positions of the present +500 m level 30, 60, and 90 Ma ago with corresponding average uplift rates. The thermic model (*b*, *c*) shows the depths of the 60°C isotherm 30 Ma ago with the corresponding palaeo-gradients for assumed general uplift rates of 0.04 and 0.02 km Ma^{-1} respectively

ing the last 60 Ma is indicated for the Windisch-Eschenbach/Mähring/Flossenbürg area. This uplift picture seems to some degree to be supported by independent geological evidence. (1) The Fichtelgebirge is the most pronounced hydrographic centre of Central Europe. Therefore, in the past it was frequently argued (see for instance Louis 1984) that this centre represents an early uplift area at least since the Upper Cretaceous. The Münchberger Gneismasse forms the watershed between the Saale and Main which has its beginning also in the Late Mesozoic (Louis 1984). (2) The apatite FT dates indicate a strong differential uplift of the southwestern Steinwald by about 400 m relative to the bordering lands in the NW and SE since the Early Miocene. A relative young "post-basaltic" uplift by 200 m was recently postulated by Louis (1984) for the Steinwald-Höhenzug. Based on K-Ar ages (Todt and Lippolt 1975) for the basalt cover on the Steinwald, Louis (1984) places this uplift into the Upper Miocene. However, the total uplift of the Steinwald by 1.2 km during the last 20 Ma, predicted by the tectonic model, is probably somewhat too high since the Miocene basalt presently covers the Steinwald only up to 820 m NN (Plößberg) although it is unknown how much of this cover has already been lost by erosion. Nevertheless, higher geothermal gradients than the assumed 30° km^{-1} model gradient could resolve such difficulty.

In Fig. 6b and c, the thermic model with constant uplift rates of 0.04 and 0.02 km Ma^{-1}, respectively is assumed. The cooling is due to uplift in an inhomogeneous geothermal environment. The palaeo-position of the 60° isotherm and corresponding gradients are calculated for the moment 30 Ma ago. A model for the Miocene would look practically identical. As is evident from Fig. 6c, smaller uplift rates would increase the palaeo-gradients unrealistically. The thermic model predicts a relatively low palaeo-gradient for the northern region of the Fichtelgebirge, an intermediate palaeo-gradient for the area south of the Erbendorf-Linie and a high palaeo-gradient for the Steinwald. A geothermal palaeo-anomaly underneath the Steinwald has a certain appeal since the Steinwald Höhenzug is situated in the SW extension of the Eger Graben. Especially in the context of the Miocene volcanism at Grosser Teichelberg (K-Ar age 21.4 ± 1.1 Ma according to Todt and Lippolt 1975) elevated geothermal gradients are easily conceivable for the Steinwald Höhenzug. However, it is questionable whether such strong and steep geothermal gradients, as derived from the model, are realistic. If one introduces into the model preferential uplift for the Steinwald, the anomaly is reduced. It is likely that neither the thermic nor the tectonic model in their strict forms are valid for the Steinwald. Rather an intermediate model is preferred which, however, at the present stage of investigation cannot be stated more precisely. Notwithstanding, during the Miocene the Steinwald Höhenzug appears as a tectonically and/or thermally active area which is probably related to the formation of the Eger Graben.

In the *southern area*, which comprises the southern parts of the ZEV as well as of the Leuchtenberg Granite and which extends across the tectonic Luhe-Line, the FT ages t_m lie between 125 and 195 Ma. For the three samples north of the Luhe-Line the FT ages t_f cluster between 37 and 44 Ma, whereas the sample Kainzelmühle (HDSP 101) south of the Luhe-Line has the highest t_f age so far found in the Oberpfalz, namely 93 Ma. Of the five apatite samples available from the southern area four were suitable for confined track length measurements. For the samples Oedenthal (HDSP 75) and Michldorf (HDSP 76) the confined track length criteria (Table 3) indicate complex thermal evolution, whereas the samples Theisseil (HDSP 74) and Leuchtenberg (HDSP 102) seem to have experienced steady, slow cooling below $150^{o}C$. The cooling path of the latter two samples, constructed from their t_m and t_f ages, is practically identical to each other and is illustrated in Fig. 5. The apatite FT ages of these two samples are best interpreted according to the tectonic model. Their average uplift rate increased from 0.02 km Ma^{-1} between 125 and 40 Ma ago to 0.04 km Ma^{-1} within the last 40 Ma. Geologically, the $100^{o}C$ cooling ages of these two samples correspond to the Lower Cretaceous and the $60^{o}C$ cooling ages to the Palaeogene which both are long periods of uplift and erosion for the western margin of the Bohemian massif (Schröder 1976, personal communication). The $60^{o}C$ cooling age of sample HDSP 76 is very similar to those of samples 74 and 102 and thus experienced probably a similar uplift history during the last 40 Ma. However, its high age t_m and the confined track length are likely due to a complex thermal evolution. In his palaeo-geographic reconstruction of the western margin of the Bohemian massif, Schröder (1976, personal communication) assumes increasing sedimentation during the Triassic and the Jurassic which may have reached a total thickness up to ca. 1000 m immediately east of the Franconian Line. This would require subsidence and heating of the crystalline basement by about 30°C. This thermal overprint, which reached its maximum about 150 Ma ago, could be responsible for the mixed FT age of sample Michldorf (HDSP 76). The maximum temperature required for the observed track fading is about 105°C, which corresponds to 2-3 km depth. The temperature and depth vs time path of sample Michldorf is illustrated in Fig. 5. Consequently, the samples HDSP 74 and 102 experienced higher temperatures during the Mesozoic. Their different uplift behaviours between 150 and 40 Ma ago may indicate some block-like tectonics during that period for the southern ZEV.

For the apatite FT age of the single sample HDSP 101 south of the Luhe-Line it is not yet clear whether it represents steady uplift or thermal overprinting. As in case of the zone immediately north of this tectonic line more studies are necessary. However, the large 60°C cooling age of this sample might reveal that the Luhe-Line was still tectonically active during the last 40 Ma.

4.2 Schwarzwald

The sampling localities with the corresponding apatite FT ages t_m are shown in Fig. 7. The ages range from 29 to 107 Ma. Compared to the Oberpfalz, it strikes immediately that there is no clear regional distribution of these ages. Especially near Haslach, age values around 40 Ma are found next within a few kilometres distance, to 80-107 Ma FT ages. In contrast, the apatite FT ages t_f, so far determined, show a much smoother distribution (for comparison, see Table 2b). Confined track length measurements on ten selected samples indicate that all the FT apatite ages from the Schwarzwald represent steady cooling. Consequently, the ages t_m and t_f can be straightforwardly interpreted as cooling ages to 100° and 60°C respectively. When the temperature vs time paths of the single samples are constructed, it appears that they are rather similar below 60° but differ remarkably from each other above 60° cooling temperature (Fig. 8). In the following geological interpretation of this cooling behaviour, the tectonic and the thermic models are discussed separately. Since the details of the fission track study in the Schwarzwald will be presented by Michalski (1987), we concentrate here on some important aspects.

In Fig. 9a, the tectonic model, the same assumptions as for the Oberpfalz are presumed. According to Fig. 9a, which represents the palaeo-positions of the present +500 m level 20 and 50 Ma ago along a N-S profile through the Schwarzwald, young block faulting seems to be the dominant tectonic process. With the exception of the Haslach area, the tectonic blocks, derived from the fission track analysis, compare rather well with independent evidence. Based on geological, tectonic, and geomorphological observations, van Werwecke (1925), Deecke (1933), Paul (1955), and Kleinschmidt et al. (1986) subdivided the crystalline basement of the Schwarzwald into several "Schollen" (blocks). During the formation of the Rheingraben and the uplift of the Schwarzwald the blocks moved relative to each other along faults such as the Elztal-fault. Most notably are the Mooswald, the Kandel, and the Feldberg-Schollen as well as the Heidburg- and Dreisam-Graben which are also revealed by the tectonic model in Fig. 9a. Apart from the average uplift rates, even the amount of

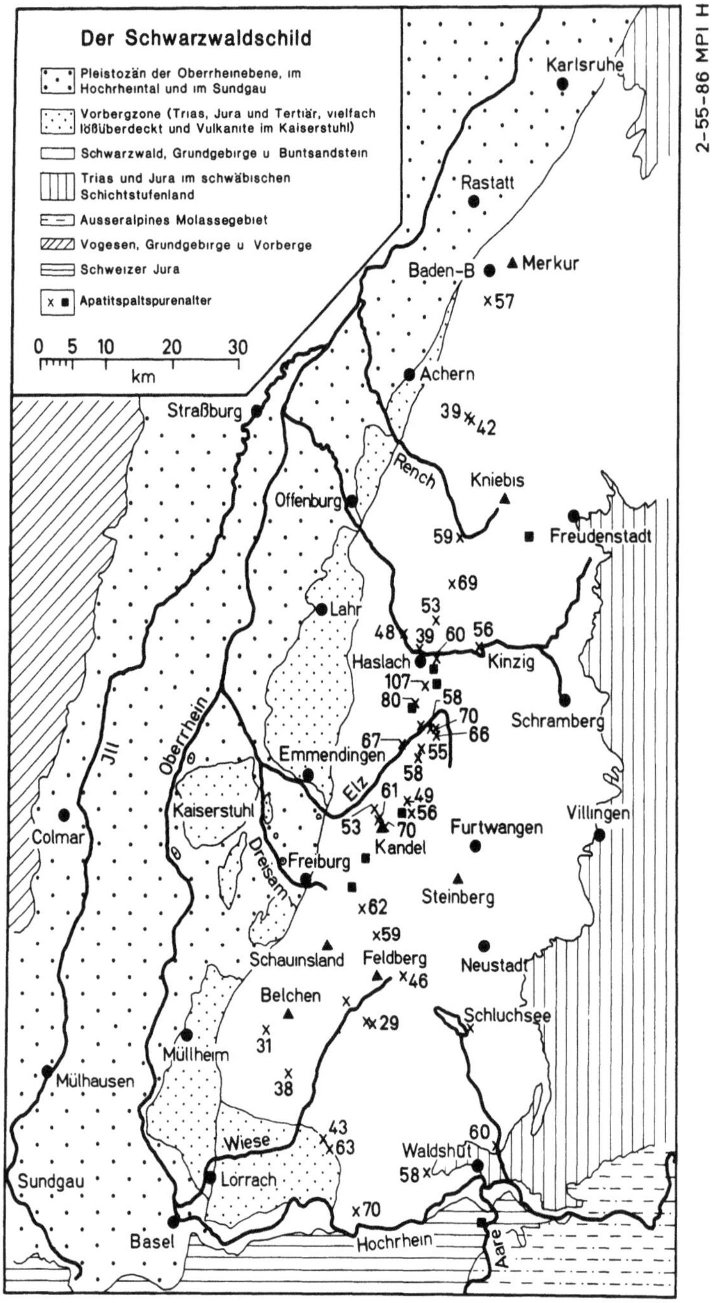

Fig. 7. Sample localities in the Schwarzwald region with measured apatite fission track ages t_m (from Michalski 1987)

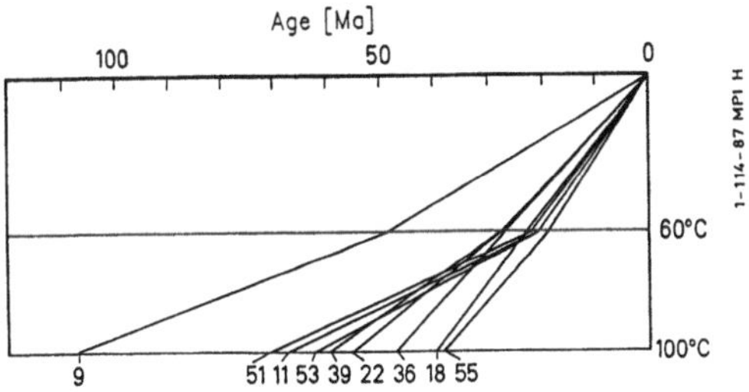

Fig. 8. Thermotectonic evolution of single samples from the Schwarzwald based on FT ages

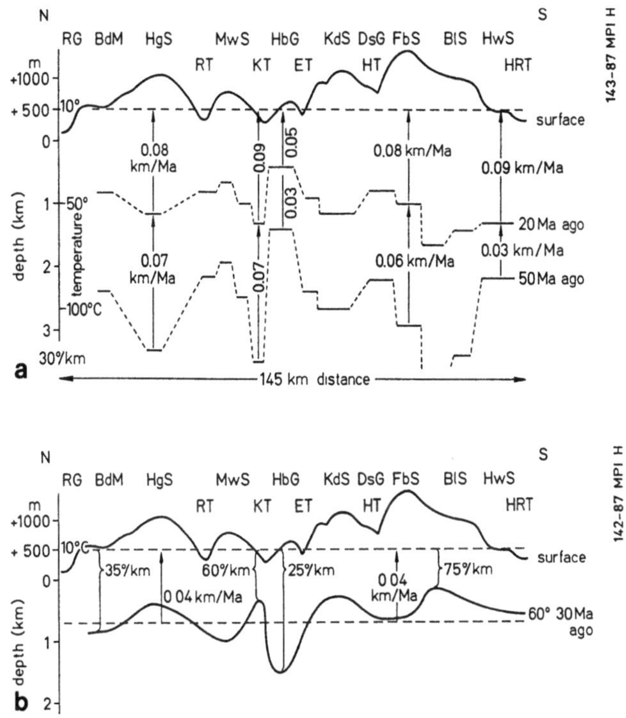

Fig. 9. Interpretation of the apatite FT ages along a N-S profile through the Schwarzwald from RG Rheingraben; BdM Badener Mulde; HgS Hornisgrinde Scholle; RT Renchtal; MwS Mooswald Scholle; KT Kinzigtal; HbG Heidburg Graben; ET Elztal; KdS Kandel Scholle; HT Höllental; DsG Dreisam Graben; FbS Feld berg Scholle; BlS Blauen Scholle; HwS Hotzenwald Scholle to HRT Hochrheintal. The tectonic model (*a*) shows the palaeo-positions of the present + 500 m level 20 and 50 Ma ago with corresponding uplift rates. The thermic model (*b*) shows the depth of the 60°C isotherm 30 Ma ago with the corresponding palaeo-thermogradients for an assumed general uplift of 0.04 km Ma^{-1}

relative movements can be derived from the model. For regions with high sampling density, including also the samples from different altitudes, the uplift behaviour of the blocks can be studied in more detail. This is illustrated in Figs. 10 and 11 for the Gschasi/Kandel region in the central Schwarzwald. The age vs altitude variation of the apatite FT ages reveals an uplift rate of 0.04 km Ma^{-1} for both the Gschasi and the Kandel-Scholle 70 to 50 Ma ago (Fig.10). Subsequently, the Kandel-Scholle experienced uplift by 200 m relative to the Gschasi-Scholle (Fig. 11) which confirms geomorphological evidence (Paul 1955). Although the tectonic model explains the data rather satisfactorily, it also presents some difficulties, especially in the Haslach area, originally selected as one of the KTB target sites. It would require young uplift, i.e. during the last 50 Ma, of the rocks immediately north of the Kinzig valley by about 2 km relative to the area south of the valley. Such a strong vertical displacement along the Kinzig-fault cannot be accepted on geological grounds. Therefore, due to the thermic model, one has also to consider changing thermogradients.

The thermic model along the same N-S profile through the Schwarzwald with the depth of the 60°C palaeo-isotherm 30 Ma ago is shown in Fig. 9b. To make it clear, this position of this 60°C isotherm model would be required from the apatite FT ages if the Schwarzwald moved up as a rigid block with 0.04 km Ma^{-1} without any differential uplift. Although, in general, this thermic model does not seem realistic for the Schwarzwald, some points are worthwhile discussing. Across the Kinzig valley, the thermogradient decreases from N to S. A similar trend appears in the recent heat flow data (Stiefel et al., 1986). Significant in the model is a narrow but strong heat anomaly underneath Haslach, which results from the low apatite FT age (39 Ma, the Schnellingen quarry sample). The shape of this anomaly points to convective rather than conductive heat transport in the underground 30 Ma ago. The high density of hydrothermal veins in this area (Bliedtner and Martin 1986) may be connected with such an anomaly. Therefore, the thermic model seems more realistic than the tectonic model in order to explain the low apatite FT ages in the immediate surroundings of Haslach.

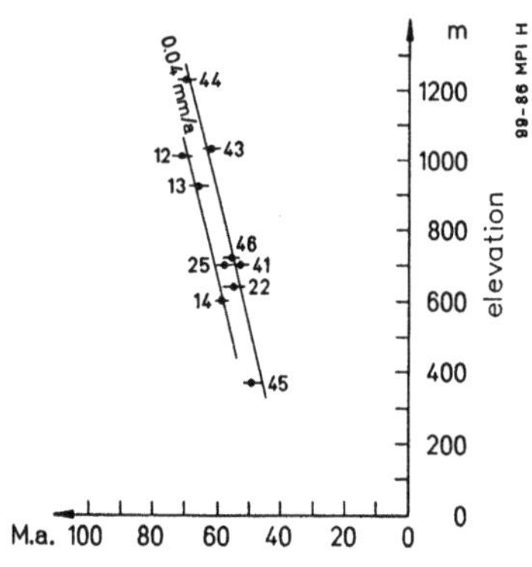

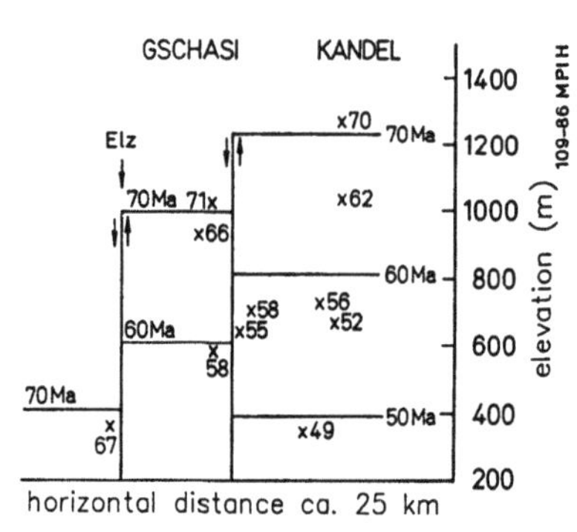

Fig. 10. The apatite FT ages vs altitude for the samples from the Kandel and the Gschasi Schollen reveal palaeo-uplift rates of 0.04 km Ma^{-1} for both tectonic blocks

Fig. 11. The apatite FT ages are plotted into the profile from the Heidburg Graben across the Elz fault, the Gschasi Scholle into the Kandel Scholle of the central Schwarzwald. The displacement of the FT isochrons indicate younger block faulting

4.3 Vosges

A few samples (Table 1c) from the outcropping crystalline Vosges basement have been included in the dating program for comparison with the Schwarzwald. The sampling density is too small for profound geological statements. However, the FT data are presented in Table 2c. The ages t_m range from 39 to 59 Ma and the ages of t_f from 18 to 27 Ma. Confined track length measurements indicate that all ages are cooling ages. The temperature vs time paths for a few samples are illustrated in Fig. 12. These samples experienced a rather even cooling behaviour similar to many samples from the Schwarzwald. Contrary to the Schwarzwald, apatite FT ages above 60 Ma and strong regional age variation are still missing in the Vosges. Whether this difference indicates different uplift/cooling behaviour on the opposite graben shoulders or is simply due to insufficient sampling, has to be clarified in the future. According to the geomorphological and tectonic studies by van Wervecke (1925) and Rolland (1975), young block faulting, with similar effects on the apatite FT ages as in the Schwarzwald, is likely in the Vosges basement.

5. CONCLUSION

Fission tracks in apatite contain a wealthy record on the Post-Variscan thermotectonic evolution of the Central European basement. Due to Mesozoic subsidence, most of the studied rock samples from the crystalline massifs bordering the Rheingraben and from the NE Bavarian part of the Bohemian massif suffered sufficient heating, i.e. $\gtrapprox 140^oC$, thus losing completely their previous track record. Only few samples from the southern "zone Erbendorf/ Vohenstrauß"/ Oberpfalz escaped this fate. Subsequently, the basement rocks underwent cooling caused by Cretaceous and Cainozoic uplift. The ages of cooling to 100° and 60^oC rock temperatures are preserved in the FT data. Average uplift rates ranging from 0.02 to 0.08 km Ma^{-1} are found.

For the Schwarzwald the FT data reveal pronounced block faulting, especially during the Palaeogene, a pattern which may be related to the formation of the Rheingraben. In the NE Bavarian basement, on the other hand, regions with specific uplift behaviour can be recognized. Thermotectonic activities in the SW extension of the Eger Graben appear in the apatite FT data of the Steinwald Höhenzug.

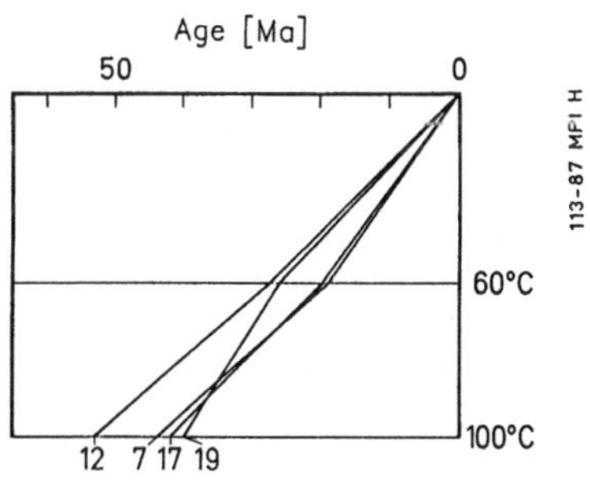

Fig. 12. Thermotectonic evolution of single samples from the Vosges based on FT analysis

The site selected for deep continental drilling near Windisch-Eschenbach/Oberpfalz apparently experienced rather uniform uplift by about 3 km with 0.05 km Ma^{-1} during the Cainozoic. For the alternative KTB location near Haslach in the Schwarzwald the apatite FT data seem to indicate a convective geothermal anomaly during the Tertiary which may be related to the hydrothermal mineral veins commonly observed in that area.

Acknowledgements. Samples were supplied by Dr. D. Gebauer, ETH Zürich, Prof. F. Taborszky, Universität Freiburg, Dr. S. Teufel, Universität Göttingen, Dr. M. Wagner, MPI Heidelberg, now TU Karlsruhe, and Prof. I. Wendt, BGR Hannover. Dr. C.W. Naeser, U.S.G.S. Denver contributed unpublished fission track data. The neutron irradiations were arranged by Dr. P. van den Haute, Rijksuniversiteit Gent. We appreciate the geological discussions with Prof. B. Schröder, Universität Bochum, Dr. A. Vollbrecht, Universität Göttingen, Prof. W. Wimmenauer, Universität Freiburg, Dr. R. Groschopf and Dr. K. Schädel, both GLA Freiburg. The project was financially supported by the KSB Stiftung and the Deutsche Forschungsgemeinschaft.

REFERENCES

Bliedtner, M, Martin M (1986) Erz- und Minerallagerstätten des mittleren Schwarzwaldes. Geol Landesamt Baden-Württemberg, Mineral Rohst 1

Deecke W (1933) Die Tektonik des Schwarzwälder Grundgebirges. Fortschr Geol Palaeont 10 (31): 141-346

Gleadow AJW, Duddy IR, Green PF, Lovering JF (1986) Confined fission track lengths in apatite: a diagnostic tool for thermal history analysis. Contrib Mineral Petrol 94: 405-415

Kleinschmidt G, Flöttmann T, Gallus B (1986) Mittlerer Schwarzwald, Strukturdaten und ihre räumliche Verteilung. 2. KTB-Kolloq, Seeheim, Odenwald, p 39

Louis H (1984) Zur Reliefentwicklung der Oberpfalz. Relief, Boden, Paläoklima 3: 1-66

Michalski I (1987) Apatit-Spaltspuren-Datierungen des Grundgebirges von Schwarzwald und Vogesen: Die postvariszische Entwicklung. Diss, Univ Heidelberg 125 Seiten

Paul W (1955) Zur Morphogenese des Schwarzwaldes (1). Jahresh Geol Landesamt Baden-Württemb 1: 395-427

Rolland A (1975) Fractionation et Géomorphologie dans le Socle Vosgien. Essai de comparaison avec la Forêt Noire I. Mosella, Tome V - No 2 and 3, Metz

Schröder B (1976) Saxonische Tektonik im Ostteil der Süddeutschen Scholle. Geol Rundsch 65 (1): 34-54

Stiefel A, Dornstädter J, Reifenstahl F, Bried M, Jäger K (1986) Geothermische KTB-Voruntersuchungen im Mittleren Schwarzwald. Sitzungsber FKPE 16: 88-109

Todt W, Lippolt H J (1975) K-Ar-Altersbestimmungen an Vulkaniten bekannter paläomagnetischer Feldrichtung. I Oberpfalz und Oberfranken. J Geophys 41: 43-61

Wagner G A (1968) Fission track dating of apatites. Earth Planet Sci Lett 4: 411-415

Wagner G A (1988) Apatite fission track geochrono-thermometer to 60°C: Projected length studies. Chem Geol (Isotope Geosci Sect) 72: 145-153

Wagner G A, Zaun P (1985) Spaltspuren - Uhr und Thermometer für Gesteine. Geowiss Unserer Zeit 4: 127-134

Werwecke van L (1925) Der Verlauf und das Alter der Hauptverwerfungen und der übrigen wichtigeren Störungen und Bewegungen im Gebiet des Mittelrheingrabens. Abh Heidelberger Akad Wiss Math - Nat Kl 21: 97 p

A Reinterpretation of the Gravity Field in the Surroundings of the KTB Drill Site – Implications for Granite Plutonism and Terrane Tectonics in the Variscan

H.-J. Behr, S. Große, T. Heinrichs and U. Wolf

Institut für Geologie und Dynamik der Lithosphäre, Universität Göttingen, Goldschmidtstr. 3, D-3400 Göttingen, FRG

ABSTRACT

To complement deep seismic sounding around the KTB, field transformations and 2-D as well as 3-D inversions of the gravity field have been undertaken. Whereas the ZEV is characterized by anomalously dense gneisses down to an unspecified depth, the remainder of the gravity field is controlled by large masses of very light rocks which in the Karlsbad area must reach an extreme depth of at least 18 km if granite is used to explain the mass deficiency. On combined gravimetric and geologic grounds the entire regional gravity low - the "Erzgebirge Minimum"- is interpreted as being due to a belt of granite batholiths which formed through hypercollision at the leading edge of a Bohemicum microplate. The KTB site is thought to be situated in the "Upper Palatinate transcurrent fault block" representing the western transformlike boundary of the "Bohemicum Terrane" with left-lateral, largely transpressive displacements.

1 INTRODUCTION

In the KTB pilot borehole (final depth of 4000 m) some differences became evident with respect to the the overall structure and what had been assumed in the predictive profile:

1. A steeply dipping foliation of paragneisses and amphibolites with a dip ranging between 60° and 90° was found rather than a relatively shallow nappe synform. Only few depth sections tend to show shallower dips of foliation. In contrast, an initial surficial dip of 60° was expected which becomes shallower with depth down to an assumed subhorizontal nappe base of the ZEV (MP metamorphites of the zone of Erbendorf-Vohenstrauß) where another tectonometamorphic unit was predicted at a depth between 3000 and 4000 m. However, as the final depth of the pilot borehole is still within the assumed transition zone the initial hypothesis cannot be falsified yet.
2. The rock has been strongly affected by faulting, especially in the intermediate and micro-scale.
3. The immediate and wider environment is cut by numerous fault systems, among which, in addition to the already known directions of the Franconian Line and the Fichtelnaab, the NNW and EW directions predominate to an unexpected extent.
4. The temperature rises faster than expected and at 4000 m reaches about +118 °C. It is not clear whether this is due to anisotropy of heat flow due to the steep foliation and/or to the neighbouring Eger rift. The need for a predictive profile for the main drilling of about 10 km depth therefore requires critical examination of the initial model both with respect to scientific aims and to the technical realization of the project. The borehole stability is affected by hydrologically induced reduction of rock strength due to cataclastic processes and by stress. The structure influences borehole deviations and a temperature of about +300 °C limits drilling and logging technology and thus depths to be reached.

In order to verify the geological model it has proven inadequate to restrict investigations to the immediate surroundings of the drilling site or even to the area of 16 x 16 km^2 planned for a 3-D reflexion-seismic experiment.

The ZEV considered a nappe unit (Weber and Vollbrecht 1986) as an integrated component of a crustal domain, the geodynamic structure of which can only be assessed and explained in relation to the neighbouring large-scale units. The fault patterns and most of all those related to a transcurrent fault block crossing the Variscan NE-SW direction have not yet been sufficiently considered. These structures have up to now been classified as late to post-Variscan and have been assumed to have no influence on the pre-granitic crustal dynamics. As the interpretation of the reflexion-seismic profile grid influenced the current concepts to a high degree further geophysical sources of information have now to be examined to supply checks and balances. Here, the gravity map lends itself to an in-depth study as the region is characterized by particularly striking gravity gradients. Particularly the most important gravity minimum of Central Europe, e.g. between the western Erzgebirge and the Iser- and Riesengebirge attracts attention.

The data and geological interpretations presented here have been provided by a DEKORP working group. The study was started after the original pre-site survey had been completed. We have therefore been able to incorporate findings from the pilot borehole. Results were first presented at the 2nd KTB colloquium in March 1989. The paper was accepted to be included in this volume in order to present an attempt of an alternative interpretation.

2 New Interpretations of the Gravity Field around the KTB Pilot Borehole

The gravity maps for the area investigated (Figs. 1 and 2) had to be compiled from various publications, which was complicated by the varying quality of the available data sets. There is, for example, only an incomplete gravity map of the SW part of the GDR by Siemens (1953). On the gravity maps of the CSSR there are neither coordinates nor, at least partly, isogam values indicated. So the only means to put the various gravity maps together at the right level is to use national boundaries and comparative observations. It is estimated that

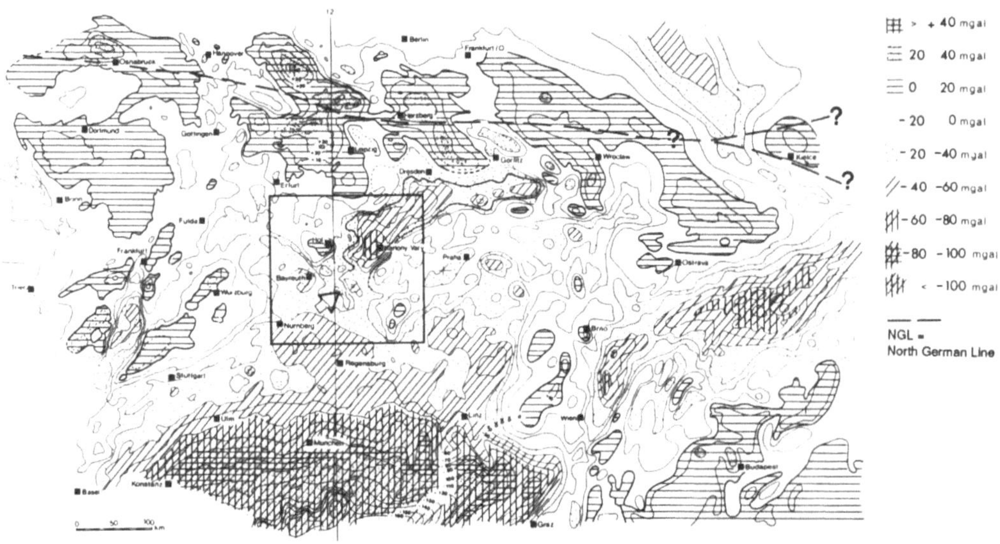

Fig. 1. Gravity map of South and Central Europe. Bouguer gravity anomaly on the level of the IGSN 71, compiled according to Plaumann (1987), Große et al. (1961a) and Ibrmajer (1981). Arrow KTB situation; box area of the following detailed analysis (cf. Figs. 2, 3)

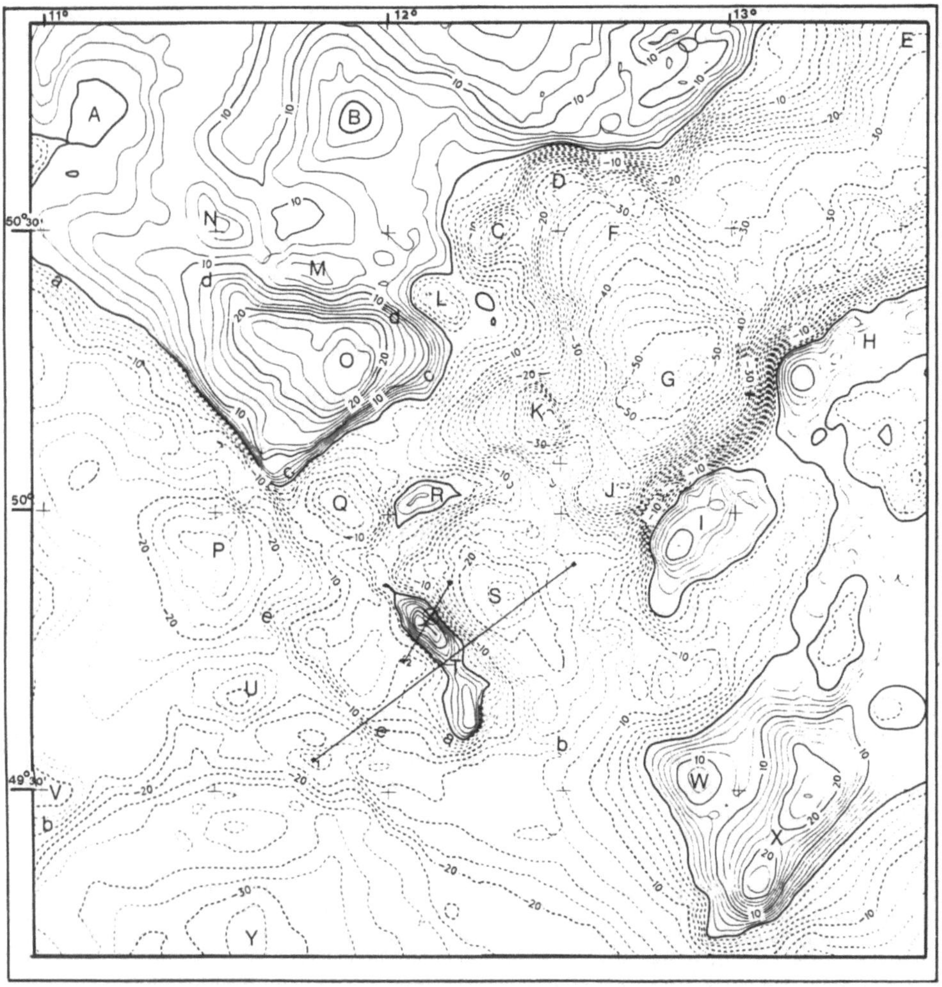

Fig. 2. Gravity map of the Erzgebirge minimum, western part. Bouguer gravity anomaly on the level of the IGSN 71 + 20 mgal, compiled according to Plaumann (1982, 1986, 1987), Siemens (1953), Große et al. (1961), Ibrmajer (1981) and Polansky (1975). A Gravity minus axis in the area of the Schwarzburg anticline; B gravity minimum of Auma; C gravity minimum of Bergen; D gravity minimum of Kirchberg; E gravity minimum of the east Erzgebirge (Freiberg); F gravity minimum of Eibenstock; G gravity minimum of Karlsbad; H gravity maximum of Doupov-Zatec; I gravity maximum of Tepla; J gravity minimum of Marienbad; K gravity minimum of Milhoslov (Mühlessen); L gravity minimum of Eichigt; M gravity minimum of Pottiga and Sparnberg; N gravity minimum of Henneberg; O gravity maximum of Hof; P gravity minimum of Bayreuth; Q gravity minimum of the Fichtelgebirge; R gravity maximum of Arzberg; S gravity minimum of Tirschenreuth; T gravity maximum of the ZEV (zone of Erbendorf-Vohenstrauß); U gravity maximum of Auerbach / Oberpfalz; V gravity maximum of Fürth; W gravity maximum of Taus; X gravity maximum of Neukirchen-Svihoy; Y gravity minimum of Neumarkt / Oberpfalz. a - a Gravity gradient of the Franconian Line; b - b concentrations of isogams in the area of the Luhe Line; c - c gravity gradient at the SE flank of the gravity maximum of Hof; d - d gravity gradient at the N flank of the gravity maximum of Hof; e - e gravity gradient at the SW flank of the Hahnbach dome; straight lines 1 and 2: situation of 2-D models of Figs. 8, 9.

except for a level variation of ±2 mgal this has been successful. In closing the isogams across national boundaries attention was paid not to produce any gradients or bends not evident in the original maps. The level of the gravity map of the drilling site surroundings (Fig. 2) was raised by a constant of +20 mgal to achieve a more legible distribution of plus and minus values in the investigated area. This does not have any effect on significance and interpretation.

2.1 Gravity Map of Europe

The ZEV and the drilling site are situated between a zone of positive anomalies in the north and the Alpine minimum in the south. The zone of positive anomalies stretches throughout Central Europe from South England to South Poland (Conrad 1980; Schubert and Conrad 1981). In addition to the Saxonian uplift of these regions the origin of these anomalies are, according to both authors, basic intrusions which are bound to a W-E trending transform fault zone which can be traced from the continent into the Atlantic Ocean. According to recent findings (Große et al. 1986; Bachmann and Große 1989) on the basis of the corrected gravity map of North Germany (Hermes 1986a, 1986b), these positive gravity anomalies accompany the so-called North German Line, a fundamental W-E extending density boundary in the crystalline crust with higher density in the north and lower density in the south. In Fig. 1 this density boundary runs along the southern margins of the anomalies of Bramsche, Magdeburg, Dessau and Hillmersadrof-Sonnewalde. Thus, it is not controlled by the Variscan.

South of our region the Alpine minimum can be clearly identified. Above all, it originated from the depression of the crust-mantle boundary and the significantly increased sediment thicknesses in the molasse trough (Müller 1988).

The ZEV is situated at the western boundary of a very strong negative anomaly, the so-called Erzgebirge minimum. In the east it stretches to the Isergebirge and the Riesengebirge. In the west it stretches to the Fichtelgebirge and ends with a second branch southwards into the Alpine minimum. West of the ZEV there is a SE-NW directed minus axis which develops from the Alpine minimum. According to Müller (1988) this is due to a depression of the basement by about 5 km and of the Moho by about 2-3 km.

What is the origin of the Erzgebirge minimum? Since 1959 a "granitic deep body" has been postulated as origin of this anomaly (Große 1959; Große et al. 1961a, 1961b; Kopf et al. 1961) since only granite as mass rock has the necessary low density in the basement. As early as 1954 on the basis of geological considerations Watznauer spoke of a "granite flood", which, however, was assumed to extend from the Lausitz granodiorites across the Erzgebirge into Thuringia.

But according to the gravity map the granitic deep body differs in several aspects from this early concept. It very clearly ends at the northwestern boundary of the granitic body of Kirchberg, Bergen and Eibenstock and it is not at all related to the granodiorites of the Lausitz block as the gravity minimum is clearly forced southwards along the Elbe Line. In the west it ends with the Fichtelgebirge granite and the Falkenberg granite at the ZEV and in the east with the granites of the Isergebirge and the Riesengebirge. Contrary to the northern flank the southern flank of the minimum is extraordinarily homogeneous and straight. This can actually be considered as evidence of a homogeneous and self-contained body. In this connection the term "granitic deep body" is to be defined. The high amplitude and the homogeneity of the Erzgebirge anomaly as a whole and particularly the consistency of the above-mentioned southern flank require the equally "homogeneous" existence of one or several rocks with very low density. In various places the minimum strikingly corresponds to outcropping or otherwise known granites. From the total anomaly outwards these granites are always connected with important gravity gradients (for example, in Fig. 2 the northern and western flank of the granites of Kirchberg D, Bergen C and Eibenstock F, the western flank of the Fichtelgebirge granite Q and the Falkenberg granite S, the southern flank of the Karlsbad granite G), whereas within the anomaly the gradient is always low as, for example, the eastern margin of the Falkenberg granite, which does not at all appear gravimetrically in the minimum of Tirschenreuth. Thus, the question arises: What produces the negative anomaly where no granites crop out? The missing or low gradient along the trend of the

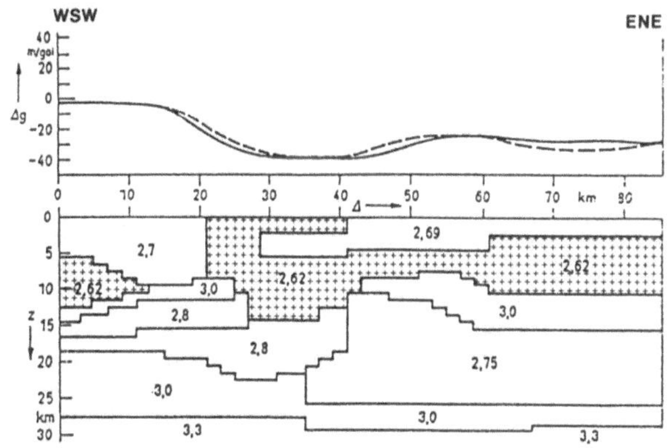

Fig. 3. Crustal density section derived from combined reflexion and refraction seismic deep sounding, also its inversion; continuous line observed Bouguer anomaly: Δg_0; dashed theoretical anomaly calculated from the model; from Apitz et al. 1987, orientation and signature of the granite added.

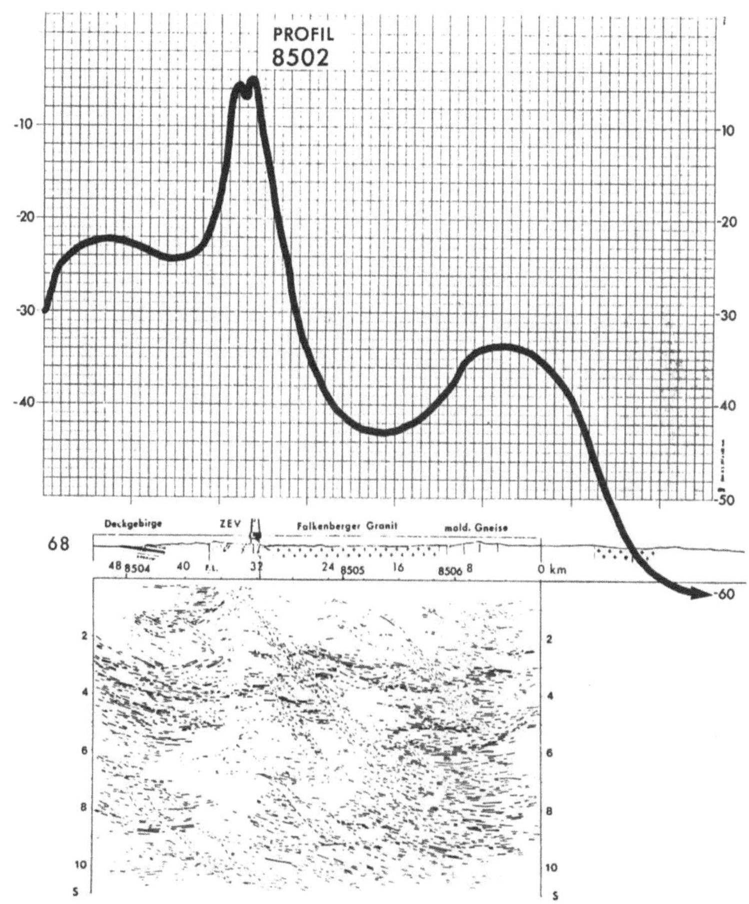

Fig. 4. Line drawing of KTB profile 8502 (Weber and Vollbrecht 1986) with the Bouguer gravity anomaly; note correlation of gravity lows with Falkenberg granite (crosses)

Erzgebirge anomaly requires a disturbing body of the same or similar mass deficit (volume and density) as the granites. It is extremely likely to be granite again with a respective cover as only granite has a sufficiently low density. As is shown by the Lausitz example even granodiorites with densities of 2.67 to 2.72 cannot produce negative gravity anomalies. Also, in the case of metamorphic rock no mass rock is known with a density below 2.72-2.75.

The horizontal and vertical structure of such a granitic deep body has to be determined in every individual case. Volume and difference in density necessary for producing the anomaly are essential. A good idea of what the granite body could be like is given by a profile published by Apitz et al. (1987) which presents densities inferred from seismic data in a "gravimetric control calculation" (Fig. 3). From the outcropping Eibenstock granite (profile km 20 to 40) a continuous layer of a density of 2.62 extends eastwards with a thickness of 8 km which can actually only be interpreted as a granitic layer in the sense of a postulated granitic deep body. The same impression is provided by the KTB profile 8502 (Weber and Vollbrecht 1986); (Fig. 4). The gravity minimum of Tirschenreuth (Falkenberg granite, minimum "S" of Fig. 2) is situated in the area of a zone nearly free of reflexions which can probably be considered the indirect image of the granite body.

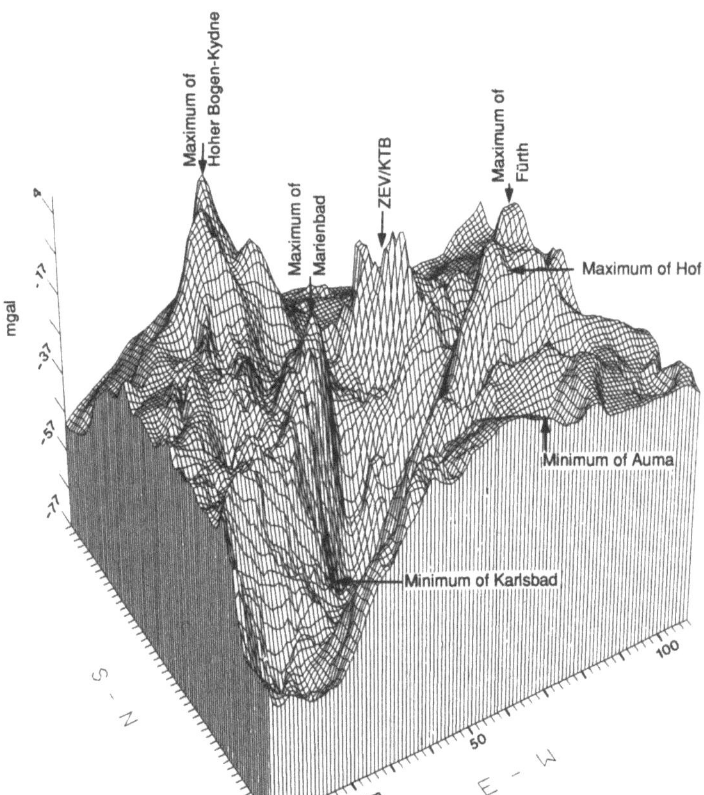

Fig. 5. 3-D presentation of the gravity map (Fig. 2). The observer is in the NE and looks in WSW direction onto the western Erzgebirge minimum and the gravity maximum of the ZEV.

What does this anomaly signify for the geological situation in the Oberpfalz? With respect to extension it is of regional significance and its amplitude is by far higher than all minima in the middle European region which do not originate from the depression of the Moho. The latter is ruled out for the Erzgebirge anomaly by the international profile No. VI (1972) and a publication on the position of the Moho in the CSR by Mayerova et al. (1985).

If the interpretation of the Erzgebirge anomaly as a granitic deep body is true, the structural development must at least be analyzed in view of this granite body and the gravimetric anomaly bound to it.

The Erzgebirge anomaly has two western branches: (1) in the direction towards the Fichtelgebirge and (2) in the direction towards Falkenberg - Tirschenreuth and farther to the south. The ZEV constitutes the western boundary of the large anomaly. Thus, it has to be determined how the deep body influenced the present structure (see Sect. 2.3). Figure 5 gives a three-dimensional view of the detailed gravity map that helps appreciate the importance of the gravity low of the western Erzgebirge for the drilling site interpretation.

2.2 The Gravity Map of the Investigated Area and Results of Two-Dimensional Inversions

The gravity map (Fig. 2) was compiled from very precise gravity maps at 2 mgal intervals by Plaumann (1982, 1986), Große et al. (1961a) and at 1-mgal intervals by Polansky (1978). Only at the margins was it complemented by maps of minor accuracy. The significant anomalies were marked with capital letters, important gradient lines with lower-case letters (Fig. 2). The individual Bouguer maps were redigitized on a regular grid with an interval of 2 km. The original maps of Große et al. (1961) were based on measurements with an average spacing of 1.2 km and an average accuracy of $\pm$0.06 mgal. The maps of Plaumann (1982, 1986) are based on an average point spacing of 2 km which locally increases to 0.1 km; the average accuracy is higher than 0.1 mgal. For the map by Polansky (1978) no information on spacing and accuracy of measurement were provided; however, from isoline intervals and mapped detail a corresponding accuracy of measurement can be inferred.

For a detailed qualitative analysis of gravity maps field transformations (field continuations, higher potential derivatives, wavelength filtering) have always proven useful. For the purpose of identifying local structures so-called second derivatives with various damping coefficients are particularly suitable for the anomalies of disturbing bodies near the surface as to some extent they enable a detailed presentation of the gravity field in a depth-related sequence. The transformation characteristics of all field transformations and filter operations can be well estimated with so-called depth characteristics (Große 1967). It has to be borne in mind, however, that the dependency on depth depends also on the form of the disturbing body which can even reach the complete independence of depth in the case of the Bouguer plate. The depth transformation characteristics indicate from which depth z a certain disturbing body contributes a maximum [$N(z) = 1$] or a minimum [$N(z) = 0$] to the anomalies in the transformed map. A set of such transformation formulae (Fig. 6) makes it possible to design a differentiated picture of the geological structures in their horizontal and vertical dimension and to interpret the original Bouguer map qualitatively on this basis. It is definitely not to be recommended to work quantitatively on transformed or even filtered maps as there is always uncontrolled splitting up of the anomalies into regional and local components which are inappropriate for quantitative analysis. Yet they are of high value for qualitative observations and analysis of the Bouguer gravity field.

To present several calculated field transformations would go beyond the scope of this contribution. Thus, only one field transformation is presented here (Fig. 7). It is used as an example and will be referred to in the brief analysis of the gravity field.

The zone of Erbendorf-Vohenstrauß (ZEV) appears as the local gravity maximum T with very steep flanks in the gravity map (Fig. 2). It is due to the high density of the outcropping gneisses drilled by KTB down to 4000 m with an average density of 2.75 - 2.78 as opposed to the granite in the east (2.62 - 2.64) and the younger sediments in the west. Added to this is the effect of the intercalation of amphibolites with very high densities (2.90 - 3.10). An

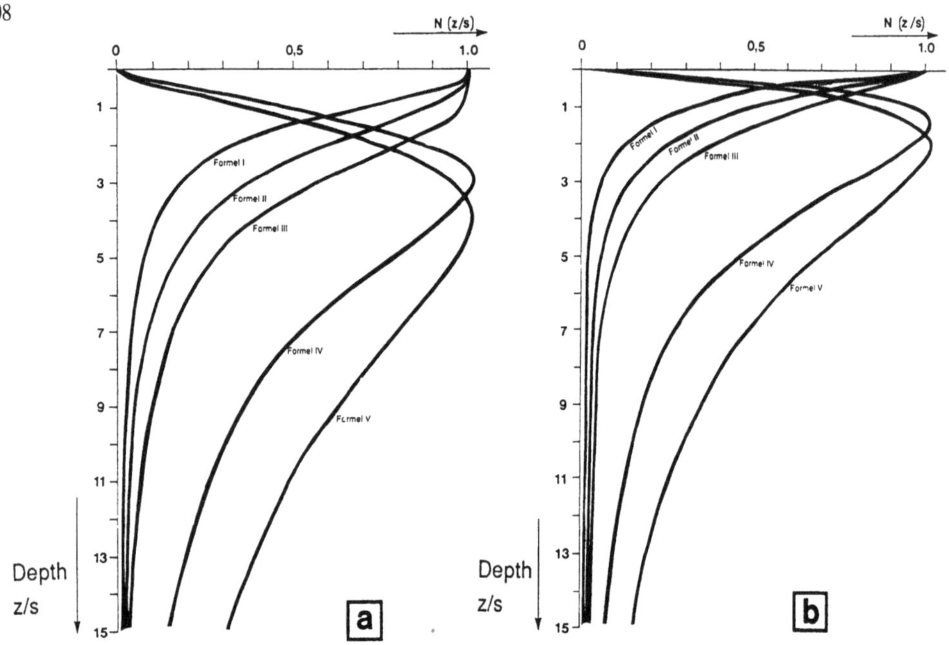

Fig. 6. Depth characteristics of five different transformation formulae (Große 1967) used to calculate the second gravity derivative as a function of the vertical with different damping of surficial anomalies. a Three-dimensional structures, disturbing body is a sphere; b two-dimensional structures, disturbing body is a horizontal cylinder; for further explanation see text.

independent density block as assumed by Bücker 1986 is not essential with respect to the gravity picture as the development of gravity can also be explained without postulating such a block (Fig. 8). Also the gravity plateau stretching westwards to the gradient line e - e is not a sound argument in favour of the existence of a particular density block extending further westwards across the Franconian Line (gradient line a - a), as together with the system of step faults in front of the Franconian Line this gravity plateau can by all means be considered complete (see profile calculations in Fig. 8 and map of the second derivative of gravity in Fig. 7).

In the east the strong gradient is produced by the steep flank of the granites of Falkenberg and Friedenfels which are clearly distinguished in the map of the second derivative (Fig. 7). The granite flank runs at a distance of about 700 and 1000 m of the drilling site and dips down by about 60° to 70° into depths of at least 4 to 6 km (see Fig. 8). How far the granite reaches down is yet uncertain as the depth can only be calculated using the difference in density at the base of the granite on which there are are no reliable data as yet. The situation is presented in Fig. 9 as a profile of the drilling site. The gravity maximum of the ZEV can be completely explained by means of the rocks found in the borehole, their densities and the steep granite margin in the northeast.

North of the ZEV there is the gravity minimum of the Fichtelgebirge granite Q and to the south, in front of it, there is the gravity maximum of Arzberg R. The latter is a region largely free of granites and consists of phyllites and mica schist which, according to both the gravity map (Fig. 2) and its derivative map (Fig. 7), shows tendencies of a continuation towards the ZEV. In the Bouguer map the Fichtelgebirge appears as a compact minimum Q, but with the branch of a minus axis towards the main Erzgebirge minimum. In the derivative map the minimum breaks up into the minima of the individual Fichtelgebirge granites and the minima that are assigned to the relatively small granite occurrences (Kirchenlamitz, Schönwald) at the southern margin of the anomaly of Hof O. To the east the minima of the Fichtelgebirge granite can be distinguished clearly although granites crop out. This is obviously due to the influence of the floor of the granite which seems to have its own roots in the region of the Fichtelgebirge (see Sect. 2.3).

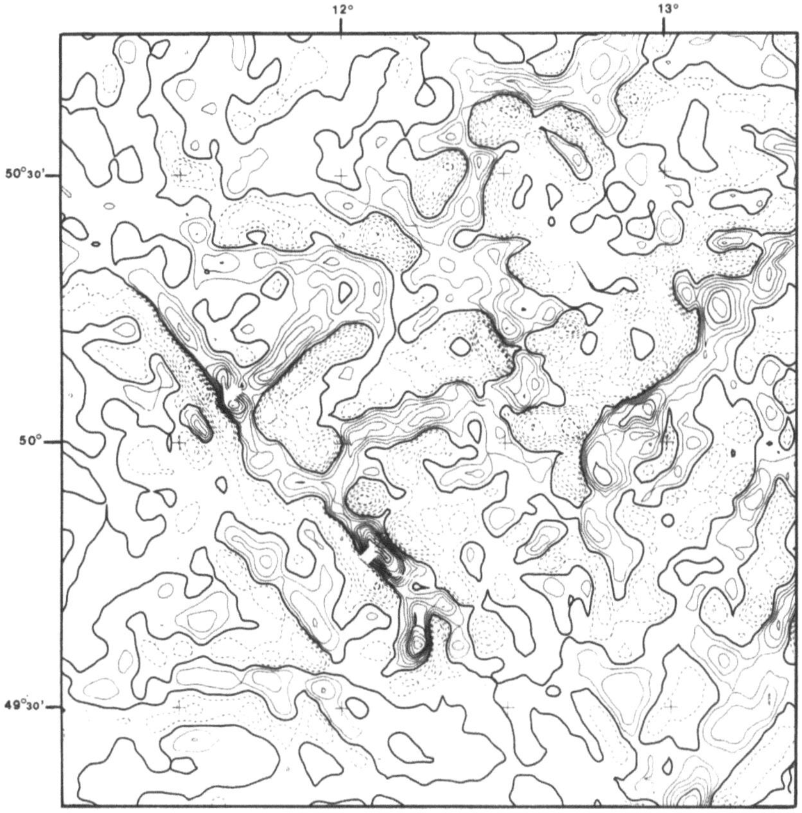

Fig. 7. Map of the second derivative of gravity as a function of the vertical, according to formula 3 (Große 1967)

The gravity minimum of Hof O is an extremely interesting and significant anomaly. This anomaly has been given a new name as the denotation as gravity maximum of the Münchberg massif is misleading, the rock of the Münchberg massif only contributing slightly to the gravity effect in the southern part of the total anomaly. This has already been shown in the three-dimensional calculation of Plaumann (1982). According to this calculation, the major effect of the positive gravity anomaly of Hof is produced by a plate-like body of increased density (2.85) from a depth of about 2.5 to 7 km.

If looking for criteria for genetic characterization of this area the immediate gravimetric environment has to be considered. Then it becomes evident that both the southern flank c - c and the northern flank d - d are accompanied by minima. In the south it is the already mentioned minima of the granites of Kirchenlamitz and Schönwald and in the north the minima of the granites of Eichigt L, Pottiga and Sparnberg M and Henneberg N. For the western boundary a participation of granites at present cannot be excluded either. Details can only be given when the improved gravity map along the Franconian Line is available.

The positive gravity anomaly of Hof, surrounded by relatively small granite occurrences, provides evidence of faults at the margin of the anomaly which traced the penetration of the granites. This precludes an interpretation of the deep-seated plate in terms of an overthrust nappe and thus a genetic relation between the lower plate and the clearly overthrust nappe of the Münchberg massif. Their occurrence may be coincidental unless the entire region of the anomaly is assumed to have been independently moved in a vertical direction. But there is no evidence of this in the mapping in the area of the northern flank d - d of the anomaly.

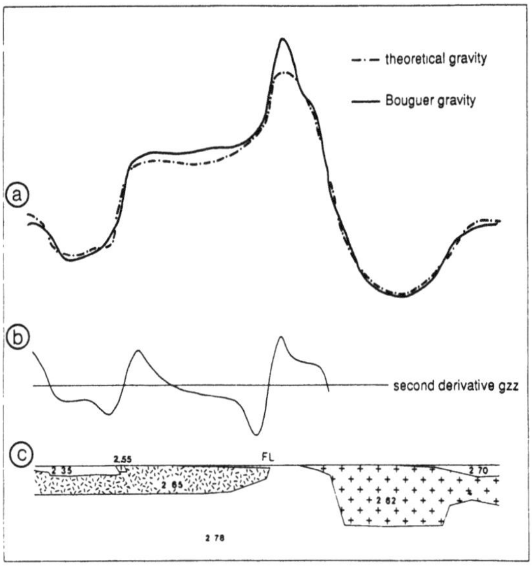

Fig. 8.
Gravimetric profile calculation profile 1. Course of profile see Fig. 2. a Continuous line observed; dash - dot calculated gravity profile. b Second derivative of observed gravity profile. c Geological model; crosses granite; FL Franconian Line; hatching Permo-Mesozoic sediments.

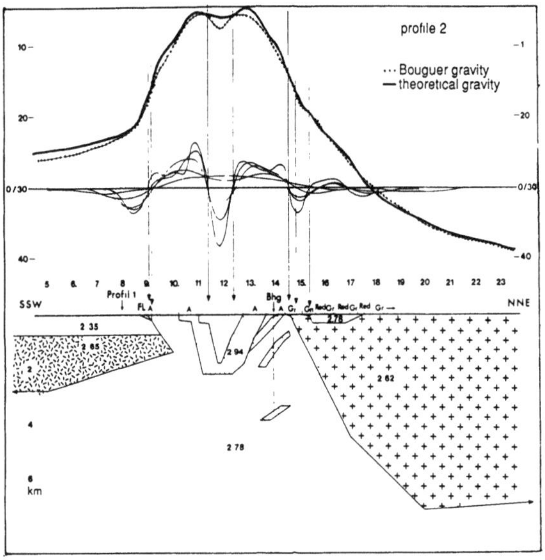

Fig. 9.
Gravimetric profile calculation profile 2. Course of profile see Fig. 2; top = dotted line; observed continuous line calculated gravity anomaly; middle second derivative of observed gravity anomaly; bottom geologic model; FL Franconian Line; Bhg KTB; A amphibolite (dots); Gr granite (crosses); Gn gneiss; Red redwitzite

West of the Franconian Line there are two more relatively large anomalies, the minimum of Bayreuth P and the gravity maximum of Auerbach U. Both anomalies do not appear in the second derivative (Fig. 7) and can thus be clearly assigned to a deeper level. They will be interpreted when an improved gravity map west of the Franconian Line is available.

South of the anomaly of Auerbach U some isogam bends and a gravity gradient b - b can be seen. In the map of the second derivative this breakup into a varied series of W-E directed plus and minus axes which obviously trace the Luhe Line and can be followed westwards to the east branches V of the Kraichgau anomaly. In the area of the point of intersection with the gravity flank e - e of the Hahnbach Anticline the anomalies of the second derivative correspond precisely to the mapped geologic contacts. Further to the west the anomalies weaken which is obviously due to the structures along the Luhe Line being covered by younger sediments.

2.3 3-D-Inversion of the Western Extremities of the Erzgebirge Gravity Low

In Figs. 10 to 13 a model of the western Erzgebirge minimum is presented as a three-dimensionally calculated granitic deep body.

The model is based on the gravity map presented in Fig. 2. The level of the granite-induced anomalies was fixed at -15 mgal (in the IGSN 71) in the model on the understanding that larger gravity values are due to structures with larger density. Examples are the anomaly of the ZEV T, the anomaly of Tepla I and the gravity maximum of Hof O (see Fig. 2). The modelling was carried out according to Cordell and Henderson (1968).

The calculation is based on a regular, square grid with a spacing of 2 km. To each point in the grid a vertical, square prism of edge length 2 km is assigned. The shape of the sought disturbing body, which is composed of the sum of all prisms, is modelled with the height of the prism. The model is iteratively improved by comparison of the calculated gravity with the given gravity for each point. In the present case the matrix consisted of 74 x 68 values.

As a reference plane for the upper margin of the prisms the top of the granite presented in Fig. 10 was used, which was designed on the basis of investigations by Tischendorf et al. (1965) and Forman and Obr (1974) as well as the geological map Bayreuth 1:200000 (BGR 1981). Granites cropping out at the earth's surface are marked by hatching.

The isoline diagram of the lower granite margin presented in Fig. 11 has been obtained after modelling in ten iterative steps. The isolines refer to a datum of 500 m above sea level. The interval between the isolines is 1 km. The density difference used is 0.18 g/cm^3 so that a density of 2.80 g/cm^3 is obtained for the rock surrounding the granite with a density of 2.62-2.63 g/cm^3. This is regarded as the average value of all rocks beside and below the granites and makes it possible to calculate the minimal granite thicknesses.

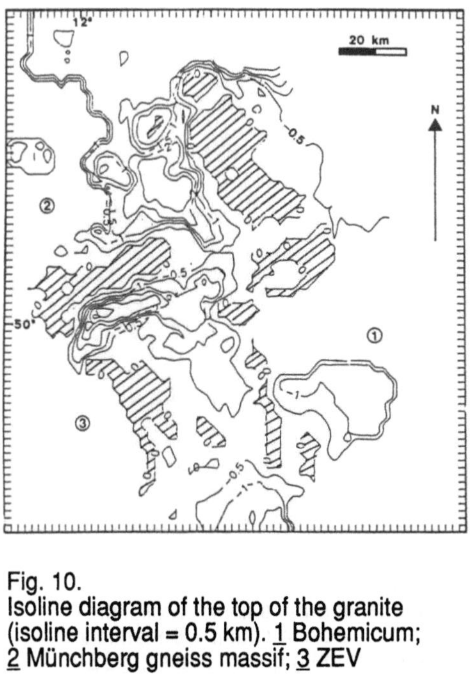

Fig. 10.
Isoline diagram of the top of the granite (isoline interval = 0.5 km). 1 Bohemicum; 2 Münchberg gneiss massif; 3 ZEV

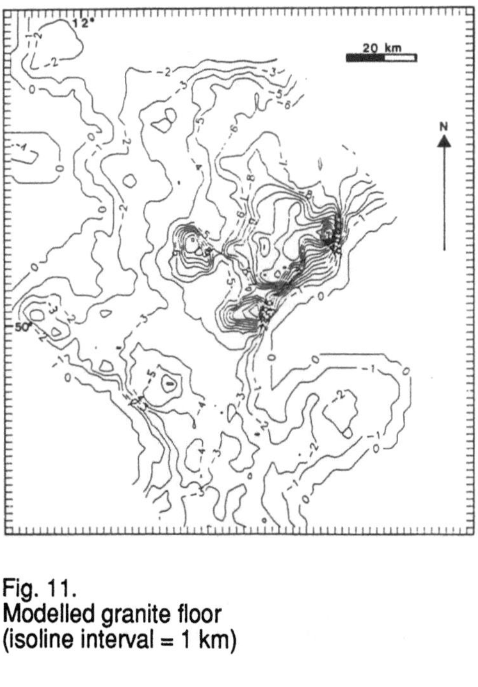

Fig. 11.
Modelled granite floor (isoline interval = 1 km)

The Karlsbad granite is the thickest granite body of the investigated area. Its base is modelled at a depth of 12-14 km. It has a steep SW-NE striking southern flank which dips nothwestwards. At its NE end there is a deep depression at its bottom extending down to 18 km. At the northern flank of the pluton the floor gradually rises to 8-9 km depth. Here the transition to the granite of Eibenstock begins, its floor equally rising to a higher level from south to north. By comparison, below the Eibenstock granite massif, the published 2-D gravity model by Apitz et al. 1987 (Fig. 3) shows a lower granite contact at a depth of 14 km, whereas our 3-D model gives only 6 km. The difference may be due to the fact that they assume lower densities for the country rock, namely 2.7 and 2.69 g/cm^3, which seem to be derived from conversions of seismic velocities. In addition, they insert a 4-km-thick phyllite inlier beneath the northeastern lobe of the Eibenstock granite which pushes down the lower contact of the granite by a roughly similar amount. Northwest of the Eibenstock granite the granites of Kirchberg, Bergen and Eichigt follow. The granite of Eichigt is not cut by the upper field margin, but on the basis of results obtained in drilling its upper margin could be determined at a level of 200-300 m below sea level. The granite of Kirchberg is a 4-5-km-thick granite pluton with a steep northern flank which begins directly at its mapped northern margin. The granite of Bergen was calculated with a thickness of 3-4 km.

The granite of Marienbad J is a stocklike granite body with a funnel-shaped root zone with a basis at a depth of 11-12 km. A similar structure is inferred for the granite of Selb, which in the northern section of the Tertiary basin of Eger (Cheb) where granite was detected by numerous drillings, also has a steep root zone (anomaly K). To the west the lower margin rises to a shallower level. The granite of Weißenstadt-Marktleuthen was modelled as a flat granite pluton of 1-2 km thickness.

In the area of the Fichtelgebirge granites, the floor again lowers to 3-4 km depth. Thus, in comparison with the older granites of Weißenstadt-Marktleuthen and Selb the Fichtelgebirge granites have an autonomous character at the least with respect to their area of emplacement.

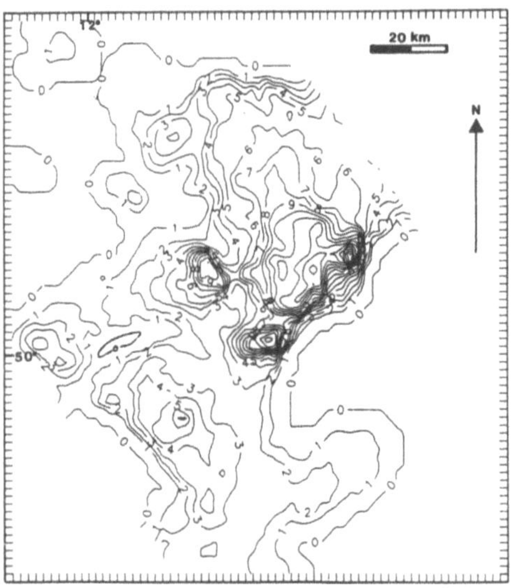

Fig. 12.
Granite thickness calculated from Figs. 10, 11. Isoline interval = 1 km

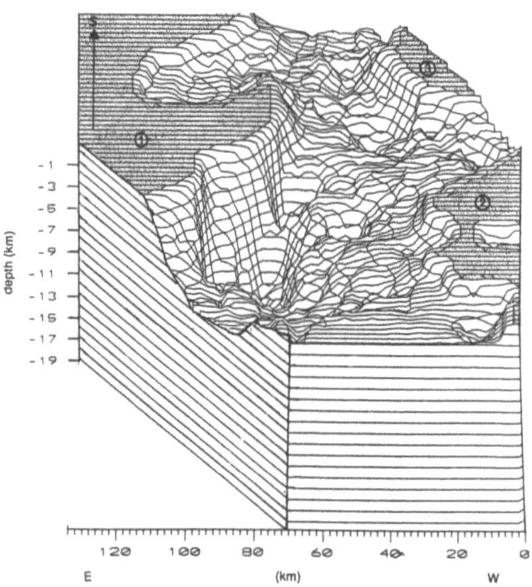

Fig. 13.
Model of granite floor; parallel projection looking south. 1 Bohemicum; 2 Münchberg gneiss massif; 3 ZEV

The Oberpfalz granites show a steep SW boundary which dips northeastwards. The Leuchtenberg granite was modelled as a flat, 1-2-km-thick granite plate. At the eastern margin of the Falkenberg granite (anomaly S) the granites are thickest where the floor sinks to a level of 6-7 km depth. The root of the Oberpfalz granite is therefore presumably situated northeastwards of the present outcrop. The depth extent of the granites of Bor and Kladruby cannot be verified by means of the present model. In contrast to the other granites of the investigated area, the two of them only cause relatively minor gravity minima. Correspondingly, the floor was modelled here at a shallow level of 1-2 km depth. According to Neuzilova et al. (1966) and Vejnar et al. (1969), both plutons largely consist of granodiorites and diorites with a density of 2.68-2.74 g/cm^3, according to Elias and Uhmann (1968). Thus, modelling with the density of 2.62-2.63 results in depths too low for the lower margins.

In fixing the anomalies due to granites at a level of <-15 mgal a region remains east of the pluton of Bor which continues to be characterized by negative gravity values and which is therefore included in modelling the lower granite margin. It is still a moot question whether there are granites in this region as here no granite body is exposed and the upper granite margin had to be fixed arbitrarily due to the lack of drilling data.

Figure 12 shows an isoline pattern of the granite thickness obtained by deducting the top of the granite from the calculated floor of the granite. It shows a largely compact granitic deep body. Only the granites of Auma and Pottiga/Sparnberg (GDR) in the NW of the investigated area are clearly isolated granite intrusions. In addition, the area north of the Oberpfalz granites (area of Ebnath-Arzberg) was modelled as a NE striking narrow stretch free of granites.

3 Interrelation of Gravity Low, Granites and Permo-Carboniferous Basins: The Erzgebirge Granite Belt - a Product of Carboniferous Hypercollision at the Leading Edge of a Bohemian Microplate?

Gravimetric arguments and the inversion of the western part of the gravimetric anomaly presented in the previous section suggest large masses of light granitoid rocks extending throughout the entire anomaly and reaching between 10 and 15 km depth. As pointed out above, within the Upper Palatinate and the western Erzgebirge there is an obvious coincidence of the anomaly with large tracts of exposed Upper Carboniferous granites (Fig. 14). In the eastern Erzgebirge there are only small areas with granite outcropping. However, more granite is known to exist from contact aureoles as well as from drilling and mining. Lastly, the vast number of vein deposits, if seen in context, point to large granites at depth, the roof region of which has just been exposed locally (Tischendorf 1986; Pietsch 1962).

Only about 20% of the granite crop out at the surface, 80% are covered. The erosional level is 5-6 km (Lange et al. 1972). The Erzgebirge part of the granite body consists of three large NW-SE extending partial plutons and smaller NE-SW bodies at the surface. In the eastern and middle Erzgebirge the partial plutons are in contact with HT metamorphites. In the eastern Erzgebirge the largest partial pluton has risen to the phyllite level.

These buried granite massifs of Annaberg, Marienberg and Freiberg seem to extend NW-SE somewhat similarly to the exposed Karlsbad-Eibenstock-Kirchberg granite complex in the west. Some of the buried massifs are flanked by mineralized faults with the same strike (Watznauer 1954). Such a NW-SE segmentation may also be gleaned from the shape of the northern edge of the gravity anomaly and it is also obvious from the right-lateral swing at the Elbe Line beyond which the anomaly again is clearly superposed on exposed granites of the Riesengebirge extending E-W. Though the segmentation may reflect certain subtle controls on the ascent, emplacement or deformation of the granites, the overall extension and the steep southern flank of the gravity low as a whole, following smoothly the main Variscan trend, point to a major influence of Variscan compressive deformation on the origin of the inferred granite belt.

The gravity low founders in the W where a number of NW-SE aligned basins like those of Weiden, Erbendorf, Stockheim mark the western border of the Bohemian Massif (Fig. 14).

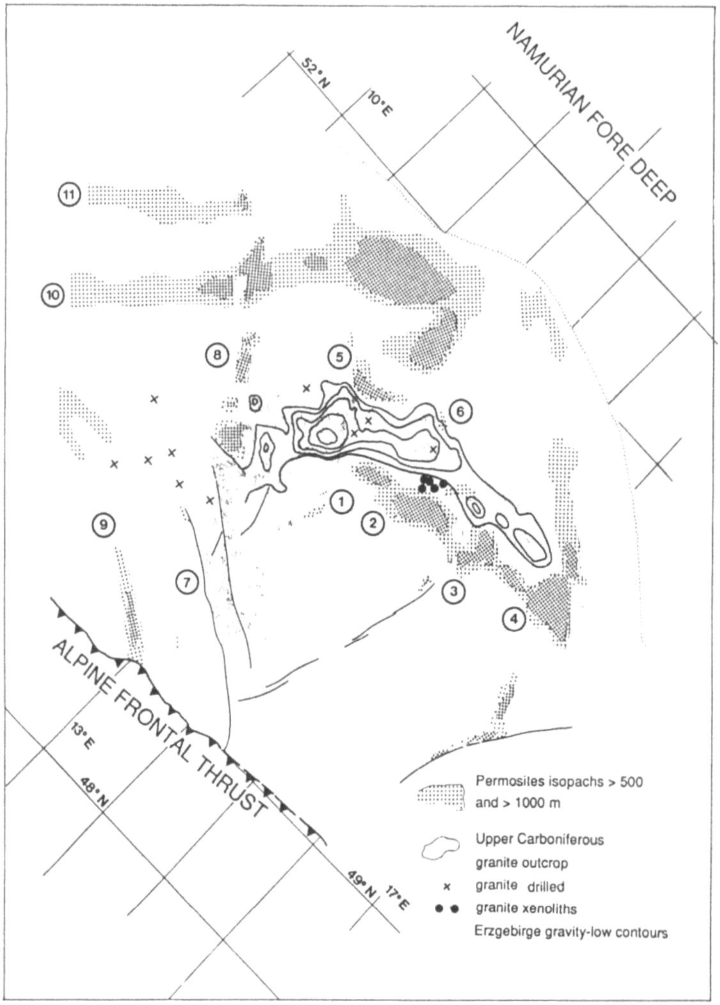

Fig. 14. Spatial relation between Permosilesian molasse basins and the regional gravity low. 1 Plzn basin; 2 Central Bohemian basins; 3 Mseno basin; 4 Intra-Sudetic basin; 5 Erzgebirge basin; 6 Döhlen basin; 7 Donau and Pfahl fault zones; 8 Weiden / Erbendorf / Stockheim basins; 9 Ries-Salzach basin; 10 Kraichgau-Saale trough; 11 Hessian basin; mainly compiled from Andreas et al. (1974), Bojkowski and Porzycki (1983), Dill (1986), Holub et al. (1975), Katzung (1982), Kulick et al. (1984), Lemcke (1988), Malkowski (1987), Nachtmann-Wagner (1987), Schröder (1988), Sokolowski et al. (1970), Trusheim (1964)

Similarly the gravity low ends in the E where the North-Sudetic Basin also follows the NW-SE trending Marginal Sudetic Fault. Most of these basins must probably be seen in the context of right-lateral transcurrent fault zones superposed on and interfering with Variscan structure during Late Carboniferous to Early Triassic times (Arthaud and Matte 1977).

The southern flank of the gravity anomaly is accompanied by a continuous string of internal molasse basins from the only partially preserved Plzn Basin in the west to the Intra-Sudetic Basin in the east (Fig. 14). Deposition in these basins started mainly during the Westphalian, partly even Namurian, thus slightly lagging behind the early granite intrusions (Holub et al.

1975; Tischendorf 1986). Deposition continued through the Autunian, locally even Saxonian, and so outlasted intrusion in the granite belt by some 10 m.y. Uplift of the granite belt and sedimentation in the sinks on its southern fringe have also been linked by sedimentologic studies stressing the importance of clastic influx of granite and gneiss detritus from the north since Westphalian B compared to only a minor supply of a southern provenance (Holub et al. 1975). In contrast to the Central Bohemian Basins south of the granite belt, the Permo-Carboniferous basins to the north are discontinuous. Also, depositional axes of the Döhlen and of part of the Erzgebirge Basins cross the Variscan trend, mirroring the NW-SE segmentation. In the south a corresponding situation is observed only where the Elbe Line crosses the trend, e.g. Mseno Basin etc., whereas the Plzn, Rakovnik, Klodno, Roudnice, Krkonose-Piedmont basin axes all follow more or less the southern flank of the gravity anomaly.

Sedimentary sinks developing on either side of a rising granite cored mountain range could be attributed to the isostatic response of the lithosphere to loading, thus testifying to a considerably thickened crust. The predominance of a southern foredeep suggests that this crustal thickening process was effected by an uneven crustal underflow the northward directed one having been the more important. Thickening must of course have preceded intrusion and so it seems reasonable to attribute the inferred northward underflow to the late Lower Carboniferous backfolding phase extensively documented from the SAX-MOL border region (e.g. Schwan 1974; F2 of Stein 1988).

Compositionally the granites are slightly metaluminous to strongly peraluminous, the leucogranites being associated with Sn-W mineralizations (Tischendorf 1970, 1986). Such Hercynian type granites would fit a backthrusting or hypercollision setting of granite formation (Lameyre 1988). The backthrusting may be attributed to wedging on a primarily northwest-facing thrust system related to the deformation of the Variscan externides or to an independent northward A-subduction of the Bohemicum.

In either case the western boundary of a "Bohemicum microplate" should then be represented by a left-lateral transcurrent fault zone. In this context it should be noted that the western extremity of the gravity anomaly exhibits a left-lateral swing where the Erzgebirge granite belt founders and gives way to a host of smaller and shallower granites coeval yet compositionally different. Many of these granites are NNW-SSE aligned in a belt extending southward up to the Pfahl and Donau faults. Presumably, this granite belt is an indication of a transcurrent fault zone. Transcurrent faulting may create tensional domains and localized pressure release at depth allowing for the infusion of fluids and heat and thus inducing melting (Lameyre 1988). Some of these Upper Carboniferous granites are overprinted by low-grade mylonites of the Pfahl Fault Zone exhibiting clear indications for right-lateral transcurrent movement unrelated to the Variscan contraction at the northwestern edge of the Bohemicum. It is debatable whether this right-lateral shear zone overprints the postulated older left-lateral transcurrent zone or if both displacements are concurrent due to oblique collision. So the postulated left-lateral transcurrent fault zone seems veiled by younger processes and the following section thus attempts to synthesize indications for such a lineament.

4 A Transcurrent Zone Across the Saxothuringian - Moldanubian boundary

The presently favoured interpretations use a collisional suture model to explain the boundary between the Saxothuringicum (SAX) and the Moldanubicum (MOL). This cryptic suture is supposed to run beneath the Tertiary Eger Graben at the southern edge of the Erzgebirge and its associated gravity low. Furthermore, it is assumed to continue beneath the ZEV and farther W beneath the Permo-Mesozoic cover to join the Baden Baden - Lubine - Pays du Bray Fault. The Münchberg gneiss massif, ZEV and the zone of Tepla-Taus are regarded as nappe units of which at least the Münchberg massif and the ZEV were subjected to relatively large transports. Whereas the Münchberg massif can be well distinguished as a tectonic klippe on very low-grade Paleozoic, the relationship of the ZEV with its base is less obvious. Presumably, the ZEV has been overthrust as MP crystalline rock onto HT-LP crystalline rock of the Moldanubian Proterozoic in the late Lower Carboniferous.

In the north the structure of the ZEV consists of a NW-SE striking and in its southern part N-S striking metablastically strongly sheared monoclinal structure. Down to 4000 m the well is situated in the east limb of such a synformal element. From gravimetric profiles across the ZEV with well-known densities of paragneisses and amphibolite the possible structure at the drilling site can be inferred (Sect. 2.2 and Figs. 8, 9). A structure of that size seems inconsistent with the initial hypothesis of an overthrust to NW or SE even if postmetamorphic post-nappe transverse folding took place. Structurally, the ZEV fits into the zone between the Franconian Line and the western margin of the Bohemicum (zone of Tepla-Taus) which is 50 km wide, monoclinal or folded with a W vergency. Into this zone syn- and post-crystalline granite intruded with intrusion axes parallel to the tectonic trend. The transverse zone with an apparently very deep-reaching E-W contraction perpendicular to the SAX/MOL suture and large NNW-trending faults found in the environment of the drilling give cause to reevaluate these elements. Is this only a late Variscan overprinting or is this zone of considerable geodynamic significance and can it be traced back into the Proterozoic?

In Fig. 15 some characteristic properties of this transverse zone bordered by two marked fault lines are given. As eastern boundary the western margin of the Bohemicum distinctively emerges. At the W and N border of the Bohemicum metamorphism increases within a medium-pressure series and the tectonic ENE-WSW trend of the Tepla Barrandian changes to NW-SE. This tract is called the zone of Tepla-Taus. Both in the N and S of the zone of Tepla-Taus there is a large gabbro-amphibolite-serpentinite complex. The northern complex is the complex of Marienbad and the southern one that of the Hoher Bogen/Kydne. At the western margin of the zone of Tepla-Taus is a large fault system which contains the 40-km-long quartz vein of the Bohemian Pfahl. To the north the system continues in the Marienbad fault which forms the eastern margin of the NNW extending Eger basin (with Pliocene and Quarternary) and cross-cuts the Eger graben. From the Hoher Bogen northwards this tectonic line can be continuously traced over 100 km. A field of mineral springs with, among others, the spas Marienbad, Franzensbad, Bad Brambach and Bad Elster stretches over 50 km parallel to this direction through the Tertiary basin farther southwards into the Tepla Barrandicum and northwards into the Saxothuringicum. At the Marienbad fault the Erzgebirge gravity minimum decreases dramatically. The roots of the granite of Karlsbad which reach down to a depth of >15 km form a large vertical step westwards over which the granite continues at minor depths into the Fichtelgebirge and farther southwards to the Falkenberg granite and thus to the eastern margin of the ZEV. West of the zone of Tepla-Taus the complex granite body of Bor of about 5-7 km width intruded (diorites, granodiorites, granites). The flanks of the Bor granite are accompanied by faults and quartz veins. The granite intruded into the eastern flank of a NNW-SSE trending antiform with a 60-80° eastward dip. Pegmatite, aplite, lamprophyre and dioritic porphyry veins of up to 5 km in length run through and accompany the magmatic body in NNW direction as well. The age of the different intrusions is given with 500-250 m.y. Thus, intrusion of a much higher age than in the Erzgebirge (300-310 m.y.) and in the Karlsbad pluton (340-320 m.y.) must have participated. The alteration of gneiss connected with Sn-W greisenization is dated at 280 m.y. and quartz veins are dated at 265 m.y. with Mesozoic overprinting in two stages (185 ± 15 and 150 ± 20 m.y.). In the faults of the exocontacts and in the fault system of the granite body numerous uranium mineralizations and deposits occur (Zadny, Chodov, Vitkov, Dylen, Svata Anna), which have been investigated in detail (Arapov et al. 1984).

All these mineralizations are bound to mylonite and cataclastic zones of up to >2 m in thickness which are characterized by graphitization and polyphase alterations.

Except for the lack of uranium mineralization, formation and alteration of these graphitized cataclastic zones are largely equivalent to those found in the KTB well as well as the uranium structures near Mähringen (Dill 1982, 1983) in continuation of the western flank of the Bor granite into the Federal Republic of Germany.

West of the Marienbad fault the Erzgebirge gravity minimum turns to the SSE and follows the transverse element. This minimum is produced by the granites of Leuchtenberg, Flossenbürg, Marktredwitz in NNW-SSE structures, a large swarm of NNW-SSE trending granite veins between Tirschenreuth and Waidhaus as well as the granites of Steinwald and Falkenberg.

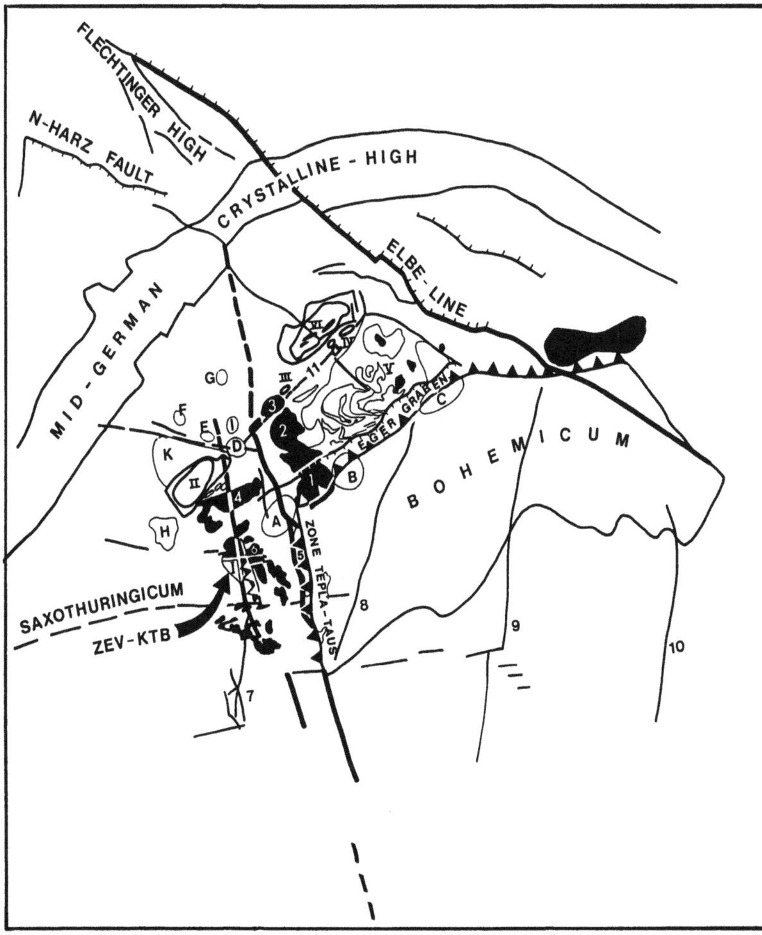

Fig. 15. The Oberpfalz fault block and related structural elements
I ZEV; II Münchberg klippe; III Wildenfels; IV Frankenberg-Hainichen; V E Erzgebirge; VI Granulitgebirge; A Seismic zone of Marienbad; B Duppau volcanic complex; C Central Bohemian volcanic complex; D Eichigt granite; E Pottiga granite; F Sparnberg granite; G Auma granite; H Bayreuth gravity low; I Vogtland arcs; K Hof gravity high. 1-6 outcropping granites (black); 7 Miocene Naab drainage system; 8 Plzn trend; 9 Blanic furrow; 10 Boskowic furrow; 11 Central Saxony Lineament

To the north the continuation of the transverse zone may be determined by the Eibenstock granite. Presumably the transverse zone continues between the gravity minimum of Eichigt and Milhoslov (Mühlessen) and farther northwards between the gravity minimum of Bergen and Kirchberg (Fig. 3). Farther to the north the Permian eruptive centre of the Erzgebirge basin and finally the Permian eruptive centre of Halle may suggest the continuation of a deep-seated discontinuity.

This dominance of NNW elements between Upper Carboniferous and post-Tertiary is persistent in time over 320 m.y. (or possibly even 500 m.y. in the Bor granite) and therefore cannot be a local event. The existence of narrow, long-lived NNW-SSE structural zones in the Central European region has long been discussed (Richter-Bernburg 1968, 1974; Pilger 1964; Lange and Steiner 1971; Mohr 1969; Werner 1988 and others). Between the E-W directed Erbendorf Line in the north and the E-W directed Luhe Line in the south, the area of the transverse zone can be called the Oberpfalz transcurrent fault block. It is a structural segment of at least 50 km in width and more than 100 km in length, the internal structure of which originates in the Carboniferous with NNW folding, shear, mylonitization and steep structures, platelike NNW trending granite intrusions and first cataclastic zones with alteration and beginning mineralization phases. The structure may control Permian NNW porphyry veins, a Jurassic reactivation with further mineralizations and alterations as well as Tertiary volcanics, the depression of the Tertiary Marienbad basin and numerous other smaller NNW Pliocene basins and depressions in the Quarternary and microseismicity even today.

This Oberpfalz transcurrent fault zone does not at all end at the Erbendorf fault but continues into the GDR where it can be mainly identified in the Paleozoic slate mountain and in the Permosilesian as well as in the fault pattern of the Mesozoic-Tertiary cover and as ascending paths for Carboniferous-Permian and Tertiary magmas. According to Lange and Steiner (1971) this is the structural zone Plauen-Halle which continues from the Vogtland slate mountains into the western end of the Gera offset to the western end of the Paschleben offset. Lange and Steiner present a large number of elements of the Saxonian tectonics, magmatic events, paleogeographic phenomena, mineralizations and morphologic structures which mark this zone. Most strikingly, the granitic intrusion axes as well as the ascending paths of the Permosilesian porphyries and porphyrites are bound in this direction. South of the complex Hoher Bogen - Kydne the continuation is not as clear, but can be recognized in the Runding mylonite zone and other small elements. But this direction can also be traced in the fault pattern of the molasse basin to the Alps.

The western border of the Oberpfalz transcurrent fault zone does not emerge that clearly from the map, but is nevertheless evidenced by numerous elements of which the Oberpfalz lineament is the most important. It was discovered as a structure of 30 km in length south of the Luhe Line in the course of uranium prospecting by interpretation of satellite photographs, geochemical and geological investigations as well as in prospecting work. A systematic large-scale fault analysis in the environment of the KTB evidenced well the continuation of the Oberpfalz lineament to the north at the western border of the NS-trending section of the ZEV. The ZEV and the immediate environment of the well is crossed by the Oberpfalz lineament which, being accompanied by Tertiary basalts and a small Miocene basin, then crosses the granite of Marktredwitz west of the Steinwald and which is also characterized by NNW-turning quartz veins, redwitzite bodies, pegmatites and rotated amphibolite bodies. To the north the element continues in the quartz porphyries of the Fichtelgebirge and can also be recognized at the western border of the Münchberg gneiss massif as well as in the Thuringian slate mountains. There is also evidence for 100 km of this western border of the Oberpfalz transcurrent zone.

West of the Oberpfalz transcurrent fault block in its northern part, the Münchberg gneiss massif is situated which is engulfed by the large gravity maximum of Hof. With its striking square outline the gravity maximum of Hof is a tectonically delimited block which has been referred to the "assyntic stage" by Plaumann. The gravity picture can be explained in terms of the Bohemian rift lithology characterized by basic volcanics, the ZEV lithology and that of the Münchberg gneiss massif. The Oberpfalz transcurrent zone may also be significant for the gravity block of Hof.

At the points of intersection of the Eger graben and the transcurrent fault block there is the most important microseismic focal area in the Eger graben (CSR) and the Vogtland (GDR). According to Trappe et al. (1988), this microseismicity is related to strong crustal reflectivity as a function of heat flow of the Eger graben. The mapping of the overall crustal reflectivity in the DEKORP NE-SW profiles shows that there is a sharp NNW border for a high reflectivity in the E and no border in the W. This border runs closely eastwards of the KTB drilling site and would be congruent with the Oberpfalz lineament. This line would then also have to be considered the W border of the heat-flow anomaly of the Eger graben. At the point of intersection of the central Saxonian lineament with the transcurrent fault block as well there are geological pecularities as the "Plauen arcs" in the Lower Paleozoic slate belt. It is thus assumed that there is a transverse zone in the Variscan basement which has been active periodically at least since Cadomian time until today. Possibly, this element first had the function of a transform in the Cadomian which served for the western plate boundary when the Bohemian terrane moved northwards. The eastern boundary is presumably situated at the eastern margin of the Bohemicum where it probably continues in the marked NW boundary of the Bober-Katzbach mountains which was later overprinted by the main Sudetic marginal fault.

The granite belt of the Erzgebirge - Iser - Riesengebirge presumably originated from crustal thickening, i.e. a subductive process must have taken place. Apitz et al. (1987) suggest a southward subduction at the Central German Crystalline High and the granites were produced by large-scale contraction. Due to the arclike configuration of the gravity minimum and the boundary of Saxothuringian and Bohemicum being equivalent with the southern margin of the granite belt, we assume a northward subduction of Bohemicum or crustal fragments below the Bohemicum. This does not exclude obduction of parts of the Bohemian onto the Saxothuringian. The subduction zone itself is today hidden by the Eger graben. Granulites and serpentinites in the Eger area are related to this event. In a gravity model by Apitz et al. (1987), showing the granite of Eichigt along the boundary between GDR and CSR across the southern border of the gravity minimum, evidence is provided for differences in the crustal composition between the above transcurrent zone and the Erzgebirge basement. As opposed to the transcurrent zone, under the Erzgebirge there is a thick body of d = 2,75. This relatively light, deep body, e.g. could be equivalent to the subducted Bohemian paragneisses. Also, the density model designed by Apitz et al. (1987) shows a much more differentiated crustal structure under the small granite body of Eichigt in our transcurrent zone than under the Erzgebirge.

The western boundary of the Bohemicum must also have been subducted or overthrust, to explain the MP metamorphism in the zone of Tepla-Taus. As the controversial model of the ZEV is related to this situation it must be discussed again. Did the ZEV as a supracrustal nappe move westwards after the obduction of the zone of Tepla-Taus, being part of this zone? Or did it originate in another deep shear zone in the north or south? Or was it obducted at the western margin of the transcurrent zone from the previously deeply subducted Bohemicum? Within the transcurrent zone the Moldanubian Proterozoic occurs with an HT-LP metamorphic assemblage which does not seem to occur in the ZEV and the zone of Tepla-Taus. This is considered an argument in favour of the supracrustal nappe model by most authors. The HT assemblage subjected to the deep-reaching Carboniferous E-W shortening of the transcurrent zone should then have been overthrust by the ZEV and refolded immediately before the intrusion of the granites.

Some authors, however, assume that the ZEV does not represent a folded klippe but rather a window into an MP Bohemian basement. In order to take into account the geochronological observations, we suggest another optional concept for the ZEV as an element of a subducted MP Bohemian basement that was obducted by transpressive displacements during an early stage of transcurrent faulting, the MP assemblages thus escaping the thermal front of the closely following HT metamorphism.

5 Regional Structure and Geodynamic Concept - a Summary

A first analysis of the regional gravity pattern presented above and a review of the granites, fault systems and a number of probably related geologic indications added to the structural situation found in the KTB pilot borehole led us to reconsider the current geologic concept.

Many lines of argument at least indicate a transcurrent fault block about 50 km wide accompanying the western border of the Erzgebirge gravity low and extending further north as well as south in a lineamentlike fashion (Fig. 16). This NNW trending zone should be at least as old as the MP-metamorphism of the Zone Tepla-Taus and probably as old as the mafic complexes of Hoher Bogen-Kydne and Marienbad. The transcurrent zone has however, seen reactivation in some form or other during the Variscan, the Mesozoic, the Tertiary and the present time. The original position of the transcurrent faults may have been determined by the transform faults to the Lower Paleozoic Rheic ocean.

The transcurrent fault block is cut by WNW to EW trending faults with possibly left-handed offset. Some larger elements of this group are the northern rim of the gravity high of Hof with an eastward extension into the northern contact of the Karlsbad Granite. This contact is later intruded by the Eibenstock Granite. Other elements are the mylonite zones in the

Falkenberg Granite and the Luhe Line. These elements have been active at least since the Upper Carboniferous as they influence granite intrusion and Upper Carboniferous sedimentation. The NW-SE trending Franconian Line - Donau Fault system interferes with the transcurrent zone so that it seems younger than the latter.

The inversions of the Erzgebirge Gravity Low strongly suggest that large masses of granitoids at depth cause the anomaly. Such a granite belt would of course have had a complex history of intrusion and differentiation. The 3-D inversion shows for the Karlsbad pluton several funnel-shaped roots which may be interpreted as feederlike structures. The intrusion mechanism could be visualized as a series of sandwichtype plutonic bodies forming consecutively and merging into each other to build a batholith. To produce such large amounts of peraluminous melts, in our opinion a crustal thickening process by a northward intracrustal subduction was necessary. The Permo-Carboniferous basins to the N, E and S of the Erzgebirge Granite Belt, attaining thicknesses in excess of 2 km, indicate isostatic compensation. Thus crustal thickening was more probably the cause of melting in the crust than mantle upwelling and lithosphere stretching (Lorenz and Nicholls 1976).

The interpretation of the gravity anomaly leads to an order of magnitude estimate of the volume of the granite belt of 10^5 km^3. Similar estimates are, for example, reported by Vigneresse (1988) for the late- to post-orogenic granites of Cornwall and Nova Scotia so that the volume of the Erzgebirge Granite Belt would not seem anomalous.

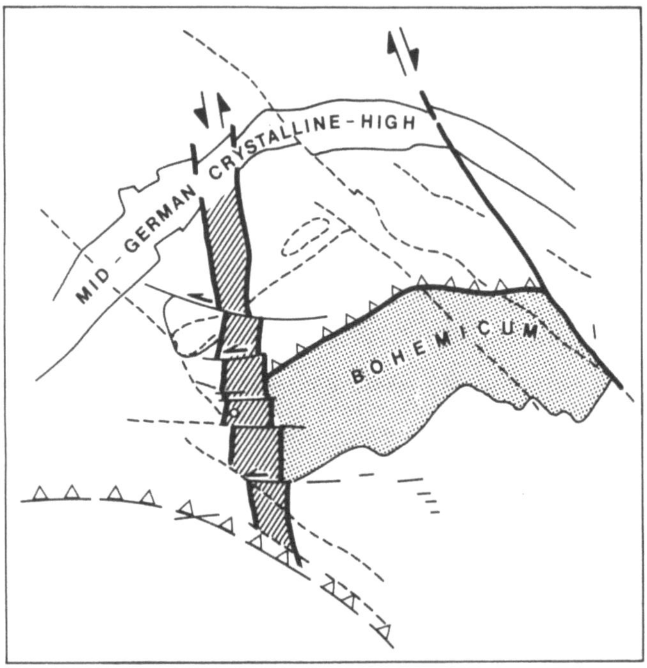

Fig. 16. Conceptual sketch map of Bohemian terrane and associated transcurrent zones. Bohemicum (dots), transcurrent fault block (striped), transcurrent faults along the eastern and western terrane border and direction of dip inferred, younger strike-slip faults (broken lines), KTB location (circle); postulated subduction (sawtooth on downgoing side).

If we correctly assume a northwestward subduction of the Bohemicum causing the Erzgebirge Granite Belt, then strike-slip zones must have existed to the east and west of the leading edge of this terrane. The transcurrent fault block of the Upper Palatinate may once have been such a zone which worked as the western transform boundary of the Bohemicum Terrane. If so then the SAX/MOL suture west of this zone must be offset towards the south. Thus terrane tectonics in the Paleozoic could imply a much more complex origin and geometry of the SAX/MOL suture than assumed so far. Also, it would mean that the KTB site is not simply situated in a collision zone but rather within a zone of transcurrent deformation. The observed NW-SE fold trends between the ZEV and the monocline of the granite of Bor could then be the result of transpressive shortening possibly due to rotation of the Bohemicum Terrane during collision. Formation of the Upper Palatinate granites follows closely or even overlaps with the 320 m.y. HT/LP metamorphism and it may be asked whether the latter could also be related to the transcurrent zone at the western edge of the Bohemian terrane. Also, the Variscan MP-metamorphism seems to be controlled by the northwestern and western border of the Bohemicum for reasons that are still poorly understood. If it is due to subduction, transform displacement along or within the Upper Palatinate transcurrent Fault Block could possibly lead to exhumation of MP-units like the ZEV without the need for supracrustal nappe transport. Such hypotheses certainly deserve closer examination in the context of a suspected Bohemicum Terrane.

Low-angle thrust structures, compressional or extensional, are currently emphasized in tectonic interpretations based on DEKORP's seismic reflexion profiling of the Central European Variscides (Behr & Heinrichs 1987; Weber & Vollbrecht, this volume; Eisbacher et al. 1989). However, it is quite possible that the steep transcurrent faults are present, yet not imaged on seismic reflexion data. Badham (1982) and lately Rajlich (1987) have presented arguments for the existence of large transcurrent shear zones and their interpretation by the concept of a "strike-slip orogen".

If the above hypothesis of a transcurrent fault zone in the Upper Palatinate can be confirmed, the KTB would provide direct access to an important structural element of Paleozoic tectonic evolution of Central Europe. This could represent a unique opportunity to study a transform fault zone within the continental basement that may contain a record of long-lasting tectonometamorphic processes during the better part of the Paleozoic.

ACKNOWLEDGEMENTS.

Continued funding of DEKORP by the Federal Ministry of Research and Technology (BMFT) is gratefully acknowledged. Plotting of the isoline maps has been assisted by the staff of the GWDG, in particular Mr. Wagenführ. Mrs. Chr. Engel-Suhrau helped with the English translation of some sections. Professor Soffel (Munich) is thanked for a constructive review.

REFERENCES

Andreas D, Enderlein F, Michael J (1974) Siles und Rotliegendes im Thüringer Wald und seinem südlichen Vorland, pp 356-445, in: Hoppe W, Seidel G (eds) Geologie von Thüringen, Haak, Leipzig

Apitz, E et 'al (1987) Komplexe geophysikalisch-geologische Untersuchungen zur Krustenstruktur der DDR auf der Grundlage refraktionsseismischer Messungen. Freiberger Forschungshefte, Leipzig C 425: 9-43

Arapov J A et al (1984) Ceskoslovenská loziska Uranu (Ceskoslovensky uranovy koncern, Pribram 1984) Ceskoslovensky uranovy prumipl, Praha

Arthaud F, Matte Ph (1977) Late-Paleozoic strike-slip faulting in southern Europe and northern Africa: result of a right-lateral shear zone between the Appalachians and the Urals. Geol Soc Amer. Bull 88: 1305-1320

Bachmann G H, Große (1989) Struktur und Entstehung des Norddeutschen Beckens. - Geologische und geophysikalische Interpretation einer verbesserten Bouguer-Schwerekarte. Veröff Niedersächs Akad Geowiss 2 (in press)

Badham J P N (1982) Strike-slip orogens - an explanation for the Hercynides. J geol Soc London, 139: 493-504

Bojkowski K, Porzycki, J (1983) The essentials of coal basin formation. In: Bojkowski K, Porzycki J (eds.) Geological Problems of coal basins in Poland. Geol Inst., Warsaw, pp 7-54

Bücker Chr (1986) Die Anomalien der Schwere im Bereich der kontinentalen Tiefbohrlokation (KTB-Oberpfalz) und ihre Interpretation. Diss, pp 1-130. LMU München

Bundesanstalt für Geowissenschaften und Rohstoffe (BGR) 1981 (eds) Geologische Übersichtskarte CC 6334 Bayreuth, Hannover

Conrad W (1980) Potentialfeldanomalien und geologischer Tiefenbau im Bereich des Südwestrandes der Osteuropäischen Tafel. Z geol Wiss Berlin 8: 439-448

Cordell L, Henderson R (1968) Iterative Three-Dimensional Solution of Gravity Anomaly Data Using a Digital Computer. Geophysics 33, 4: 596-601

Dill H (1982) Geologie und Mineralogie des Uranvorkommens am Höhensteinweg bei Poppenreuth (NE-Bayern) - Ein Lagerstättenmodell. Geol Jahrb Hannover D 50: 1-83

Dill H (1983) Plutonic mobilization, sodium metasomatism, propylitic wall-rock alteration and element partitioning from Höhensteinweg uranium occurrence (northeast Bavaria, F.R.G.). Uranium 1: 139-166

Dill H (1986) Sedimentpetrographie des Stockheimer Rotliegendbeckens, Nordostbayern. Geol Jahrb D88: 3-76

Eisbacher G H, Lüschen E, Wickert F (1989) Crustal-scale thrusting and extension in the Hercynian Schwarzwald and Vosges, Central Europe. Tectonics 8: 1-21

Elias M, Uhmann J (1968) Density of the Rocks in Czechoslovakia. Geol Survey of Czechoslovakia Prague

Forman J, Obr F (1977) Beitrag zur Kenntnis des Untergrunds des Sokolov- und Cheb-Beckens (Zusammenfassung des tschechischen Textes). Sbor geol ved Rada Prague G 29: 114-115

Große S (1959) Ergebnisse gravimetrischer Messungen im West- Erzgebirge. Z Geophys Berlin 24: 321-325

Große S, Oelsner Ch, Bremer H (1961a) Über die gravimetrische Vermessung des Vogtlandes und des Erzgebirges. Z Angew Geol Berlin 7: 357-362

Große S, Kopf M, Sonntag K (1961b) Ergebnisse der Gravimetermessungen im Westerzgebirge. Freiberger Forschungshefte Berlin C110: 57-102

Große S (1967) Methodische Untersuchung zur Anwendung von Potentialfeldtransformationen für die stockwerkgegliederte Interpretation von Schwerefeldern. Diss Bergakademie Freiberg

Große S, Bachmann G H, Dohr G (1986) Interpretation der verbesserten Bouguer-Schwerekarte Norddeutschlands nach Prof. Hermes. Forsch-Ber z Forschungsprogramm d dt Erdölindustrie: Geophysik z Untersuchung d tieferen Untergrundes in Norddeutschland. Hannover (unpubl)

Hermes H J (1986a) Calculation of pre-Zechstein Bouguer anomaly in northwest Germany. First Break, 4, No. 11

Hermes H J (1986b) Gravimetrie Norddeutschland. Verbesserte Bouguersche Schwerestörung. Geophysical Interpretation, Hamburg (unpubl)

Holub V, Skocek V, Tasler R (1975) Palaeogeography of the Late Paleozoic in the Bohemian Massif. Paleogeogr. Palaeoclim. Palaeoecol. 18: 313-332

Ibrmajer J (1981) Geological interpretation of gravity maps of Czechoslovakia. Geophysical Synthesis in Czechoslovakia. Bratislava pp 135-148

Katzung G (1982) Explanatory notes to the lithotectonic molasse profile of the Central European Depression (northeast German Depression), G.D.R. Veröff Zentralinst Phys Erde AdW DDR 66: 209-224

Kopf M, Große S, Sonntag K (1961) Dichtebestimmungen an den Gesteinen des Westerzgebirges. Freiberger Forschungshefte Berlin C 110: 7-53

Kulick J, Leifeld D, Meisl St, Pöschl W, Stellmacher R, Strecker G, Theuerjahr A, Wolf M (1984) Petrofazielle und chemische Erkundung des Kupferschiefers der Hessischen Senke und des Harz-Westrandes. Geol. Jahrb., D68, pp 223

Lameyre J (1988) Granite settings and tectonics. Rend Soc Ital Mineralogia Petrologia 43/2: 215-236

Lange H, Tischendorf G, Pälchen W, Klemm I, Ossenkopf W (1972) Fortschritte der Metallogenie im Erzgebirge. B. Zur Petrographie und Geochemie der Granite des Erzgebirges. Geologie 4/5: 457-493

Lange P, Steiner W (1971) Eggische Strukturlinien im geologischen Bauplan der DDR. Geologie, Berlin 20, 3: 212-235

Lemcke K. (1988) Geologie von Bayern I. Das bayerische Alpenvorland vor der Eiszeit. Stuttgart, Schweizerbart, 165 pp

Lorenz V, Nicholls I A (1976) The Permo-Carboniferous basin and range province of Europe: an application of plate tectonics. In: Falke H (ed) The continental Permian in Central, West and South Europe. NATO Adv Study Inst Ser C, Math Phys Sci, Utrecht, pp 313-342

Malkowski M (1987) The Mesozoic and Tertiary basins of the Bohemian Massif and their evolution. Tectonopysics 137: 31-42

Mayerova M, Nakladalova Z, Ibrmajer I, Herrmann H (1985) Areal distribution of the Moho discontinuity in Czechoslovakia based on results of DSS profile measurements and quarry blasts. Sbor konference cs geofyziku v Ceskych Budejovicich

Mohr K (1969) Zur paläozoischen Entwicklung und Lineamenttektonik des Harzes, speziell des Westharzes. Clausthaler Tekt H, Clausthal-Zellerfeld 9: 19-110

Müller K (1988) Anwendung moderner gravimetrischer Interpretationstechniken zur Erkundung der Kruste von Süddeutschland. Diss Free Univ Berlin

Nachtmann W, Wagner L (1987) Mesozoic and early Tertiary evolution of the Alpine foreland in Upper Austria and Salzburg, Austria. Tectonophysics 137: 61-76

Neuzilova M, Vejnar Z (1966) The geology and petrography of rocks of the Kladruby massif (summary of Czech text). Sbor Geol Ved Rada G Prague 11: 26-29

Pietsch K (1962) Geologie von Sachsen. Berlin, DVW, 870 pp

Plaumann S (1982) Die Schwereanomalie im Bereich der Münchberger Gneismasse und ihre Interpretation. Z dt geol Ges 133: 39-50

Plaumann S (1986) Die Schwerekarte der Oberpfalz und ihre Bezüge zu Strukturen der oberen Erdkruste. Geol Jahrb E33: 5-13

Plaumann S (1987) Karte der Bouguer-Anomalien in der Bundesrepublik Deutschland 1 : 1 500 000. Geol Jb Hannover 40: Reihe E

Plaumann S (1988) Kompilation einer Schwerekarte für die westliche CSSR, das Erzgebirge (DDR) und Nordostbayern. Z dt geol Ges, Hannover, 139: 177-190

Polansky J (1978) Geofyzikalni vyzkum slavkoskeho lesa. Sbornik geol ved 15: 7-27

Rajlich P (1987) Variszische duktile Tektonik im Böhmischen Massiv. Geol Rundschau 76/3: 755-786

Richter-Bernburg G (1968) Saxonische Tektonik als Indikator erdtiefer Bewegungen. Geol Jb Hannover 85: 997-1020

Richter-Bernburg G (1974) Das Rheinische Element im Bilde Westeuropas. Nachr Ges Wiss Göttingen, math-naturwiss Kl, Berlin IV, 3: 23-28

Schröder B (1988) Outline of the Permo-Carboniferous basins at the western margin of the Bohemian Massif. Z geol Wiss 16: 993-1001

Schubert H, Conrad W (1981) Beitrag zur Interpretation ausgewählter geophysikalischer Anomalien Mitteleuropas unter geodynamischen Aspekten. Z geol Wiss Berlin 9: 1179-1189

Schwan W (1974) Die Sächsischen Zwischengebirge und Vergleiche mit der Münchberger Gneismasse und anderen analogen Kristallinvorkommen im Saxothuringikum. Erlanger Geol Abh 99: 189 pp

Siemens G (1953) Die Schwerekarte der DDR. Freiberger Forschungshefte C7

Sokolowski St, Cieslinski St, Czerminski J (eds) (1970) Geology of Poland. Geol Inst, Warsaw

Stein E (1988) Die strukturgeologische Entwicklung im Übergangsbereich Saxothuringikum/Moldanubikum in NE Bayern. Geol Bavarica 92: 5-131

Tischendorf G, Wasternack J, Bolduan H, Bein E (1965) Zur Lage der Granitoberfläche im Erzgebirge und Vogtland. Z Angew Geol Berlin 11 (8) 410-423

Tischendorf G (1970) Zur geochemischen Spezialisierung des westerzgebirgischen Teilplutons. Geologie 19: 25-40

Tischendorf G (1986) Beziehungen zwischen variszisch-ensialischem Magmatismus und Metallogenese im Erzgebirge. Z geol Wiss 14: 17-25

Trappe H, Wever Th, Meissner R (1988) Crustal reflectivity pattern and its relation to geological provinces. Geophys Prospect 36: 256-281

Trusheim F (1964) Über den tieferen Untergrund Frankens - Ergebnisse von Tiefbohrungen in Franken und Nachbargebieten 1953-1960. Geol Bavarica 54: 92 pp

Vejnar Z, Neuzilova M, Syka J (1969) Geology and petrography of Bor massif.(Summary of Czech text). Vestnik Ustr ust geol Prague 44: 256

Vigneresse J-L (1988) Forme et volume des plutons granitiques. Bull Soc géol France 8, IV, 6: 897-906

Watznauer A (1954) Die erzgebirgischen Granitintrusionen. Geologie 3: 688-706

Weber K, Vollbrecht A (eds) (1986) Ergebnisse der Vorerkundungsarbeiten Lokation Oberpfalz. Report presented at 2nd KTB Colloq Seeheim/FRG, 186 pp

Werner W (1988) Synsedimentary faulting and sediment-hosted submarine hydrothermal mineralization - a case study in the Rhenish Massif, Germany. Göttinger Arb Geol Paläont Göttingen 36: 206 pp

The KTB Pilot Hole: Tectonic Setting, Technical Data and First Results

R. Emmermann

Institut für Geowissenschaften und Lithosphärenforschung der Justus-Liebig-Universität, Senckenbergstr. 3, 6300 Giessen, FRG

INTRODUCTION

The Continental Deep Drilling Program of the Federal Republic of Germany (Kontinentales Tiefbohrprogramm der Bundesrepublik Deutschland, KTB) and the closely associated German Continental Seismic Reflection Program (Deutsches Kontinentales Reflexions-seismisches Programm, DEKORP) are major German contributions to the understanding of the Central European crust. This crustal region is distinguished by its position between three Precambrian continental blocks of Pangea, namely Eurasia, North America-Greenland and Africa, and despite its relatively young age of about 500 Ma has a rather complex history. Formation of the Central European crust started in the early Paleozoic during an important rifting episode in Cambro-Ordovician time which resulted in the break-up of Gondwana, development of Variscan microplates and opening of at least some of the Variscan basins. Rifting dismembered an old continental crust previously generated in the course of several Precambrian orogenies. In the subsequent collision between Gondwana and Laurentia/Baltica the Variscan belt of Europe was formed by welding together the Variscan microplates and by incorporation of proto-oceanic basins, shelf areas, accretionary wedges, island arcs and back-arc troughs through compressive, subduction-related processes. This period of plate convergence and crustal shortening began in late Ordovician and continued until Upper Carboniferous. It was followed by a phase of renewed crustal thinning accompanied by intense rhyolitic volcanism in connection with extensional, wrench- and shear tectonics that generated intramontanous basins, graben structures and pull-apart basins. During the Alpine orogeny, the region was subjected to compressional and transpressional tectonics which were followed by another stage of rifting and graben formation in the Tertiary.

This repeatedly reactivated Central European crust is relatively thin (25-30 km) and is extremely heterogeneous both laterally and vertically. It is characterized by regionally variable, but generally high heatflow, pronounced seismic reflection patterns with lamination structures in the lower crust (especially in the young rift regions), the occurrence of high- and low-velocity layers and zones of high electrical conductivity at different depths, penetrative deformation, impregnation with numerous late-Variscan granites and an abundance of late- to post-Variscan hydrothermal mineralizations.

The joint effort of KTB and DEKORP, is to systematically investigate this crustal type which differs fundamentally from those sampled by the research boreholes of Kola (USSR) and Gravberg (Sweden). The main task of DEKORP is to acquire and interpret carefully selected seismic reflection profiles and to reconstruct major geological structures in three dimensions. The aim of the KTB is to sink a superdeep borehole in a geologically and geophysically distinct segment of the Variscan basement in order to:

- investigate the structure, composition, physical properties and thermodynamic state of a representative portion of this crustal type and the processes presently operating in it
- determine the nature of geophysical structures and heterogeneities
- probe the deep crustal stress field and the brittle-ductile transition
- interpret geochemical and geophysical gradients and discontinuities in terms of paleoconditions and paleoprocesses
- decipher the crustal architecture and reconstruct its geological-geodynamic evolution.

The scientific rationale for the KTB and the major questions to be tackled by a superdeep borehole were outlined by Emmermann (1986) and Behr and Emmermann (1987). After the major objectives of the KTB had been agreed upon, two basement regions, the Oberpfalz and the Schwarzwald, were selected as potential drilling targets. Operational considerations played a major role in the final choice of the Oberpfalz as the drill site (Emmermann and Behr, 1987). Of particular significance were the high temperature gradients and near-surface heatflow values of the Schwarzwald, which suggested that temperatures of around 300°C might be reached at a depth of only 7 - 8 km. Such high temperatures would make both drilling and logging of the hole very difficult and would have a considerable impact on the budget provided for the project.

LOCATION AND TECTONIC SETTING OF THE DRILL SITE

The Oberpfalz is situated at the western margin of the Bohemian Massif, the largest coherent surface exposure of basement rocks in Central Europe (Fig. 1). The actual drill site was chosen close to the intersection of KTB reflection profile 8502 and the DEKORP 4 line, a few kilometres south of the boundary between the Saxothuringian and Moldanubian, the two major basement units of the internal zone of the Variscan fold belt. In the Oberpfalz this boundary is called Erbendorf line and according to the results of the pre-site investigations is interpreted as a prominent thrust plane within a broad suture zone formed by the closure of an early Palezoic, probably oceanic rift basin (Franke this volume; Weber and Vollbrecht this volume).
In the Oberpfalz region three tectono-metamorphic units can be distinguished which are separated from Permo-Mesozoic sediments in the west by the Franconian line, a steeply northeastward-dipping wrench fault. The Moldanubian (MO) represents a typical basement complex and is made up chiefly of high-grade polymetamorphic gneisses that were reactivated during the Variscan orogeny. The Saxothuringian (ST) predominantly consists of lower Paleozoic metasediments and metavolcanics that were only affected by a monophase metamorphism with the metamorphic grade of the rocks increasing from lower greenschist-facies in the north to upper amphibolite-facies at the contact with the Moldanubian basement. The Zone of Erbendorf-Vohenstrauss, abbreviated as ZEV, where the

drill site is located, is a medium-pressure metamorphic basement unit that is made up chiefly of paragneisses with intercalations of metabasic rocks (see also Fig. 1). The drill site lies some 3 km east of the town of Windischeschenbach in a segment of the ZEV which is bordered to the southwest by the Franconian line and to the northeast by the Fichtelnaab fault, two elements of a prominent fault system that are responsible for considerable down-faulting of the basement at the western margin of the Bohemian massif.

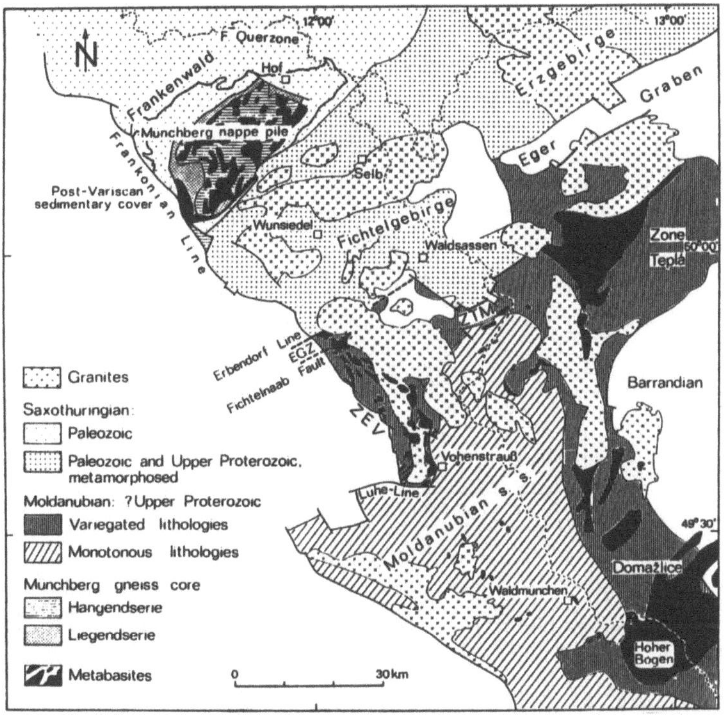

Fig.1. Schematic geological map of the ST/MO basement at the western margin of the Bohemian Massif (from Schüssler et al. 1989)

Particularly important for the reconstruction of the geodynamic evolution of the entire Oberpfalz region is the Münchberg Massif (MM) which has been convincingly shown to be an allochthonous nappe complex (Franke this volume). This interpretation is supported by an inversion in the metamorphic grade of the rocks of this unit: kyanite-bearing gneisses overlie amphibolite- and greenschist-facies rocks which rest upon fossiliferous Paleozoic.

In the course of the pre-site investigations it became increasingly apparent that rocks of the ZEV, which have similar lithologies, metamorphic grades and radiometric ages can be regarded as equivalents of the respective subunits of the Münchberg Massif and thus in all probability also constitute a medium-pressure nappe pile. Petrological and geochronological studies indicate the following P-T-time path for the three tectono-metamorphic units (Fig.2):

(1) The Moldanubian basement underwent, at least, three stages of metamorphism, starting with a high-pressure (HP) metamorphic event, followed by a medium-pressure (MP) overprint and ending with an intense late Variscan low-pressure metamorphism.
(2) Münchberg Massif and ZEV are also polymetamorphic basement units and appear to have a pre-Variscan history similar to the Moldanubian basement. However, both units obviously escaped the late-Variscan LP/HT-metamorphism. Their last metamorphic record is a MP event which occurred at around 380-390 Ma.
(3) The Saxothuringian rocks, on the other hand, display a monophase development and were only affected by the last, late Variscan LP/HT-metamorphism which welded together the Moldanubian and Saxothuringian basement.

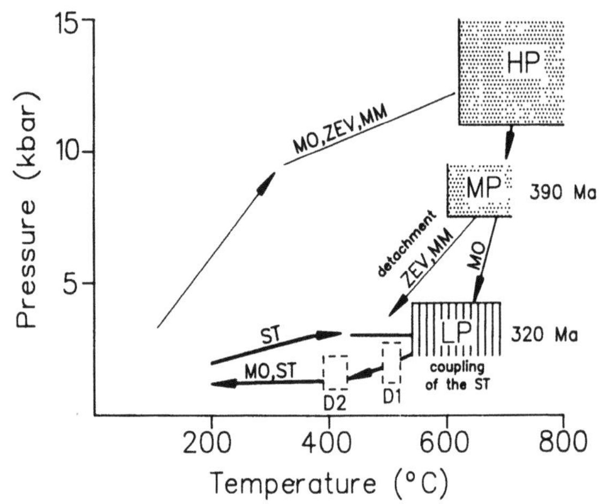

Fig. 2. P-T-time path of the three tectono-metamorphic units of the Oberpfalz region

In summary, the available data suggest the following geodynamic model: The Saxothuringian rocks were originally deposited in the previously mentioned early Paleozoic rift basin. During the Variscan orogeny these rocks were subducted towards the southeast and overridden by Moldanubian basement. Both units were welded together during late Hercynian LP-metamorphism. Münchberg Massif and ZEV represent supracrustal nappe complexes thrust over Saxothuringian and Moldanubian during or shortly after this metamorphism took place. The subsequent phase of orogenic uplift and decompression is indicated by emplacement of numerous late- to post-orogenic granites which were intruded into all three basement units (Fig. 3).

This geodynamic model which anticipates a tectonic stacking of markedly different basement units appears to correlate well with the seismic crustal image (Weber and Vollbrecht this volume). The most important geophysical results obtained are (see Fig. 4):
- There is an excellent correlation between tectonic structures derived from surface geology and the seismic reflection pattern of the upper crust. For example, the nappe complex of the Münchberg Massif can be clearly recognized as a bowl-shaped trough structure.

- In general, upper and middle crust are distinguished by a relatively high seismic reflectivity with a predominance of south dipping reflectors. In the Moldanubian basement these reflectors extend down towards the Moho which lies at a depth of 28 to 30 km.

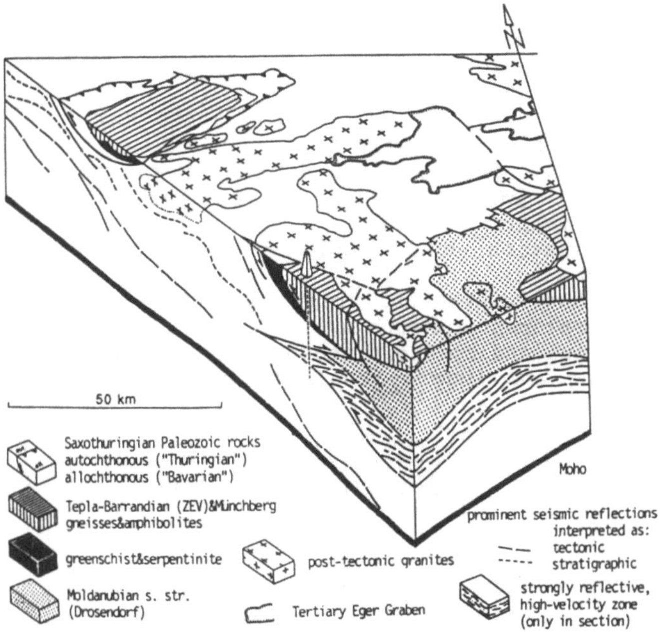

Fig. 3. Three-dimensional geotectonic model of the crustal structure in the Oberpfalz region (Franke 1989)

The drill site is located above a crustal segment of unusually high seismic reflectivity with a number of prominent seismic reflectors between a depth of 8 and 12 km (Figs. 3 and 4). The bottom of this highly reflective zone is marked by a high-velocity body ($v_p \geq 7$ km/sec) which appears to coincide with a zone of high electrical conductivity (Gebrande et al. this volume; Schmoll et al. this volume; Haak this volume). The opportunity to drill into this structure, to reveal the nature of seismic reflectors and to test the model of crustal architecture outlined above are the major attractions of the Oberpfalz location.

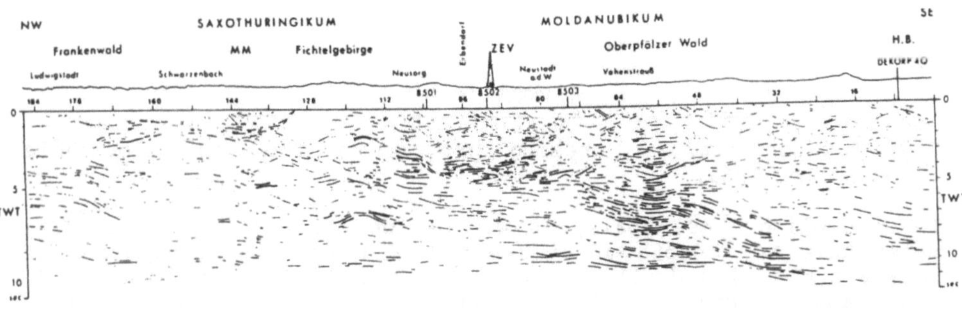

Fig. 4. Line drawing of the DEKORP 4 profile (from Behr 1987)

TECHNICAL DATA

The drilling strategy of the KTB calls for completion of a pilot hole before starting the main borehole which is planned to reach a final depth of at least 10 km. The pilot hole was spudded on September 22, 1987 and was terminated on April 4, 1989 at a depth of 4000.1 m. The main borehole, located some 200 m E of the pilot hole, will be started in summer 1990 and is scheduled to reach 10 km by the end of 1994.

The aims of the pilot hole were to:
- gather extensive geoscientific data by means of continuous coring, downhole logging, tests and analyses of core material, cuttings and drilling fluid
- collect information about problem sections and technically critical horizons
- investigate the temperature field and improve predictions of expected temperatures at depth
- minimize coring and borehole measurements in the main borehole
- test drilling and logging tools to be used in the main borehole.

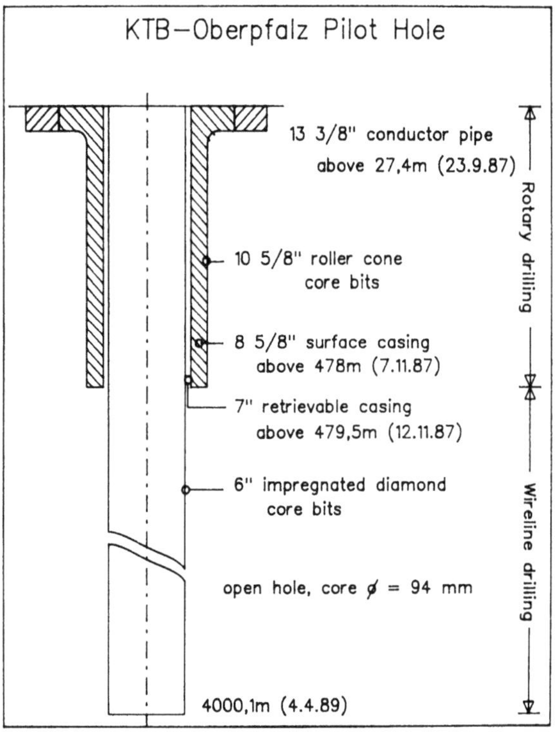

Fig. 5. Condition and casing program of the KTB pilot hole (Rischmüller 1989)

To accomplish these goals, it was decided to drill a 6-inch-diameter hole to a depth of at least 3000 m using a combination of rotary and mining drilling techniques (Rischmüller and Chur 1987; Chur and Sperber 1988). Mining drilling technique involves drilling at high rotational speeds with a double tube wireline core barrel system and thin-kerf diamond core bits. This technique allows wireline retrieval of the inner core barrel and core. Recovery during all phases of wireline coring was excellent and exceeded 97 %.

Figure 5 presents an overview of the condition of the borehole. The uppermost section (above 29.5 m) was drilled with a 17 1/2" roller cone bit, and a 13 3/8" conductor pipe was set with a casing shoe at 27.4 m. Rotary drilling with 10 5/8" roller cone core bits was used down to 478 m. Core recovery within this interval was only 42.8% corresponding to 192.7 m, with practically no recovery in the first 80 m. Following an extensive geophysical logging and testing program, an 8 5/8" surface casing was set above 478 m.

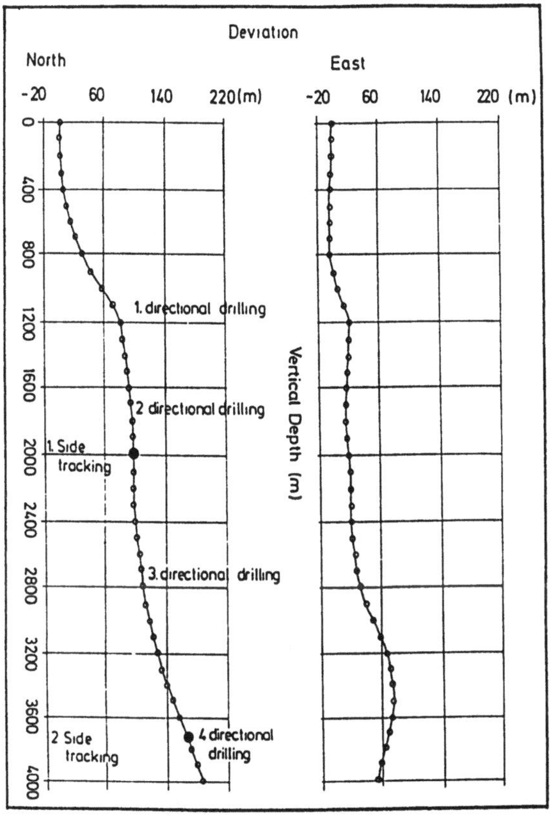

Fig. 6. Trajectory of the KTB pilot hole (from Rischmüller 1989)

Below 478 m the borehole is open. Wireline coring technique using a specially manufactured 5 1/2" drill string and 6" diamond core bits was applied below this depth except for four phases of directional drilling, which were necessary to reduce the angle of deviation (Fig. 6). About 3145 m of excellent quality core with a diameter of 94 mm were obtained by this technique. Average core length was 3.5 m and a total of 60 impregnated thin-kerf diamond core bits were used with the average life time being 48 m and the average rate of penetration being 1.68 m/h. The total rig time of

560 days includes 112 days of drilling/coring, 112 days of tripping drill string, 90 days of tripping core barrel, 73 days of logging, testing and sampling, 62 days of directional drilling, 40 days of fishing/milling, 28 days of miscellaneous, 28 days of rig equipment repair and 15 days of circulating/reaming (Rischmüller 1989).

On April 4, 1988, during a phase of directional drilling, the drill string got stuck at 1998.3 m and all attempts to free it failed. Because a "fish" had been lost in the hole, the hole had to be plugged and a sidetrack was started at 1709 m and successfully completed at 1802 m. The first part of the hole above 1998.3 m is designated KTB OBERPFALZ VB 1; the new hole below the kick-off point is called KTB OBERPFALZ VB 1a. Due to loss of bottom hole assembly at 3893 m, after an unsuccessful fishing job a second sidetrack was necessary at 3767 m. Figure 6 shows the trajectory of the pilot hole and indicates the intervals of directional drilling and side tracking (Rischmüller 1989).

SCIENTIFIC RESEARCH PROGRAM

Scientific data are collected in a field laboratory (Fig. 7) at the drill site and from accompanying special investigations at universities and other research institutions. The Deutsche Forschungsgemeinschaft (DFG) currently funds some 92 KTB-related research projects with a budget of about 7 million DM per year. This money is provided by the Federal Minister of Research and Technology (Bundesminister für Forschung und Technologie, BMFT) who also funds 15 research and development projects (F+E-Projekte) in the field of engineering and drilling technology. Altogether, about 300 scientists from 64 institutes are presently involved in the KTB.

Fig. 7. KTB drilling rig and field laboratory

Establishment of a field laboratory at the drill site had a high priority since the very first KTB discussions. During the preparatory phase of the program, a working group organized under the direction of this author defined the tasks to be accomplished in the lab and determined the equipment and personnel necessary to carry out the work. It was agreed that the main purpose of the field laboratory is to collect extensive geoscientific data on cores, cuttings, drilling muds, drilling fluids and gases recovered from both the pilot hole and the main borehole. In particular properties should be measured which:

- are necessary for quick operational decisions concerning drilling, sampling and testing
- have to be determined on a quasi-continuous scale as a function of depth
- are time-dependent and have to be recorded as soon as possible
- are necessary for correlation with data obtained by borehole measurements
- are needed for proper sample selection and serve as basic information for all individual research projects.

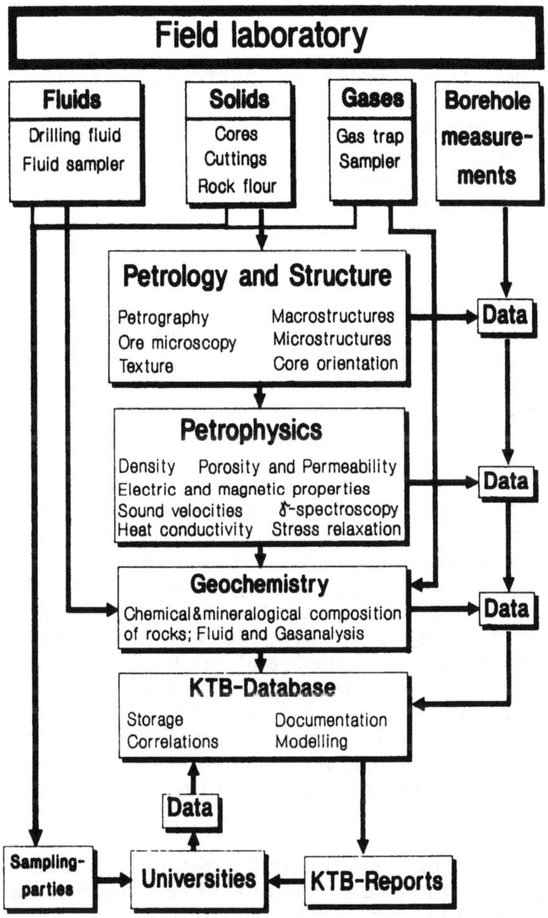

Fig. 8. Sampling and working scheme of the KTB field laboratory

To meet these requirements a comprehensive scientific program is carried out on-site which includes:

- structural, petrographical and mineralogical investigations on cores and cuttings
- preparation of a lithologic log and a first interpretation of geological structures
- determination of major and trace elements on core material, cuttings, drilling mud and drilling fluid
- on-line analysis of gases dissolved in the drilling fluid
- measurements of physical properties of cores, cuttings and drilling muds.

Figure 8 summarizes the tasks of the field laboratory and its organizational structure. In addition to the specified investigations the field laboratory is responsible for distribution, management and archiving of samples, and for regular publication of all scientific results. Progress reports are published every three months and cover borehole sections of 500 m length (Emmermann et al. 1988, 1989). Thus far, four "Sampling Parties" have been held and over 4000 samples have been prepared and distributed. The staff of the field laboratory currently includes 3 permanently employed assistants of the Project Management, 18 scientists and 16 technicians. These personnel are assigned to four working groups: Geology/Petrology, Geochemistry, Geophysics, and Data Processing.

The petrophysical investigations performed at the drilling site include determination of density, seismic velocity (vp), natural remanent magnetization, susceptibility, electrical resistivity, thermal conductivity, porosity and permeability. An important experiment carried out immediately after core retrieval is the measurement of anelastic strain recovery using a specially designed strain measurement apparatus.

A new X-ray diffraction technique is used to qualitatively and quantitatively determine the mineral composition of cores, cuttings and drilling mud separated from the drilling fluid by means of a centrifuge (Emmermann et al. 1989). X-ray fluorescence analysis using a fully automated and computerized XRF spectrometer (Siemens SRS 303) is used for determination of major oxides and selected trace elements on core samples, cuttings and drilling mud.

A new type of drilling fluid was successfully used for the first time in the pilot hole (Herold et al. 1988). This drilling fluid consists of water with about 2 wt.% Dehydril HT, a lithium, sodium, magnesium silicate with a clay mineral-like structure, which yields a solid-free, thixotropic fluid with a high carrying capacity. An extensive analytical program for a quasi-continuous fluid investigation has been developed and up to 12 cations are analyzed at intervals from 1 to 3 m using an ICP atomic emission spectrometer (ARL 3580). Additionally, Cl^- and SO_4^{2+} are determined with a chromatography system. The results obtained show that continuous fluid analysis allows immediate and reliable detection of inflow horizons and even provides qualitative information as to the composition of the waters (see below).

A specially designed gas mass-spectrometer (Vacuum Generators) is used for the qualitative and quantitative determination of gases released from the drilling fluid. Using an on-line technique, 10 gas components (N_2, O_2, Ar, CO_2, H_2, CH_4, C_2H_6, C_3H_8, He and H_2S) are determined routinely by this fully automatic system which operates round the clock. Despite a relatively complex interpretation of the gas data, the "gas logs" have been very sensitive indicators of problem horizons and permeable zones. The on-line gas data, therefore, proved to be very useful for quick operational decisions concerning positioning of drill-stem tests and fluid sampling.

FIRST RESULTS

As was expected from surface geology the two dominant rock types recovered by the pilot hole are paragneisses and amphibolites. Granites, which are widespread in the target area and which crop out north and east of the drill site, have not been encountered. This confirms the interpretations based on geochemical investigations and detailed gravimetric and magnetic surveys, that the granite bodies are flat, tabular intrusions with eastward dipping contacts (Fig. 9). The drill site itself is located near the centre of strong positive magnetic and gravimetric anomalies which are, as has been confirmed by the pilot hole, produced by underlying amphibolite bodies.

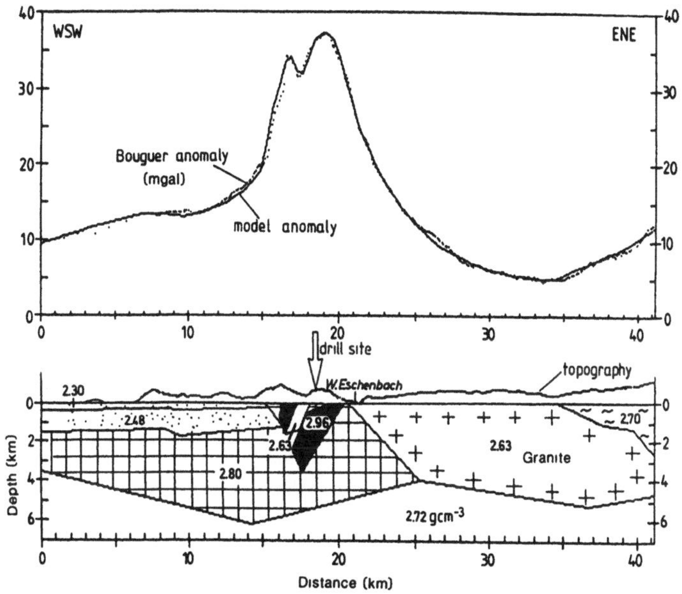

Fig. 9. Lithostratigraphic model of the drill site area resulting from gravity measurements (from Weber and Vollbrecht 1986)

The lithostratigraphic profile of the penetrated basement section is characterized by a succession of paragneisses and intercalated metabasic rocks with paragneisses making up about 70% of the entire drilled sequence (Fig. 10). Small lamprophyric and aplitic intrusions, which belong to the dyke sequence of the late-Variscan granites,cut the sequence at various depths.

Seven lithologic units can be distinguished in the drilled sequence (see also Fig. 10):

Unit 1. Above 460 m the borehole penetrated a variegated sequence of garnet-bearing biotite-plagioclase gneisses (sometimes containing K-feldspar and hornblende) and garnet amphibolites intercalated with calcsilicates and a marble horizon. This unit which is strongly affected by retrograde alteration under greenschist-facies conditions, represents a former volcano-sedimentary succession.

Unit 2. Between 460 and 1160 m a sequence of rather uniform garnet-biotite-plagioclase gneisses with varying amounts of kyanite and/or sillimanite was penetrated. Locally, these rocks contain small amounts of K-feldspar and muscovite. Accessories are zircon, monazite, apatite (sometimes accumulated in layers), tourmaline and graphite.

Unit 3. In the interval between 1160 and 1610 m, a sequence of metabasic rocks consisting of garnet amphibolites and metagabbros with subordinate intercalations of meta-ultramafic rocks was recovered. These meta-ultramafic rocks are invariably associated with the metagabbroic portions and occur as irregular fragments or layers of up to 6 m thick.

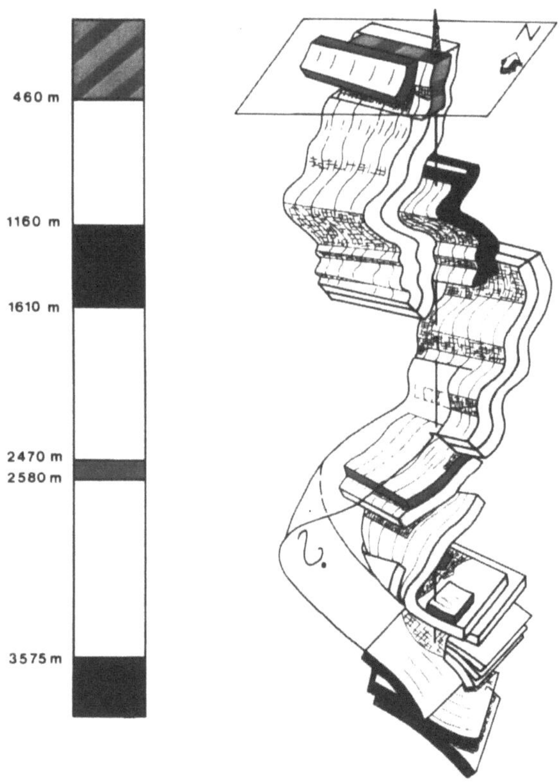

Fig. 10. Lithostratigraphy and structural interpretation of the penetrated basement section (0 - 4000 m)

Unit 4. Between 1610 and 2470 m, another sequence of kyanite- and/or sillimanite-bearing garnet-biotite-plagioclase gneisses was penetrated which are very similar to those of Unit 2. These rocks are cross-cut by lamprophyric and aplitic dykes at various intervals, particularly between 2040 and 2350 m.

Unit 5. From 2470 to 2580 m, a succession of garnet-bearing hornblende-biotite gneisses grading into garnet amphibolites was penetrated. Petrographically, these rocks resemble the amphibolites recovered from Unit 1. This is also supported by the occurrence of frequent intercalations of small layers of calcsilicates.

Unit 6. A third sequence of uniform kyanite- and/or sillimanite-bearing garnet-biotite-plagioclase gneisses identical to those of Units 2 and 4 was penetrated between 2580 and 3570 m.

Unit 7. Below 3575 m, a metabasic rock series similar to that of Unit 3 was recovered. This series includes garnet amphibolites, metagabbros and small amounts of meta-ultramafic rocks.

Aluminosilicate-bearing garnet-biotite-plagioclase gneisses constitute the dominant rock type drilled so far. Application of the garnet-biotite geothermometer to these rocks yielded temperatures of formation between 660°and 710°C (Reinhardt et al. 1989). From the almandine content of the garnets (Xalm = 0.75-0.70), the pressures of the garnet-biotite equilibration could be estimated to be in the range of 6-8 kbar (Fig. 11). Detailed petrological investigations of the garnets and the age relationships between kyanite and sillimanite suggest that the P-T path of the paragneisses is rather complex and that garnet-biotite equilibration could either mark the peak of prograde metamorphism or the thermal peak during decompression.

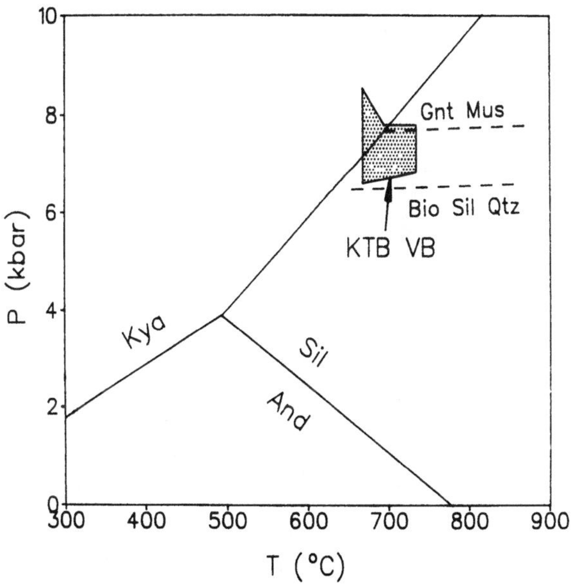

Fig. 11. Geothermobarometric data obtained on paragneisses from the KTB pilot hole

Up to three generations of garnet which formed as a result of a prograde reaction sequence can be distinguished (Reinhard et al. 1989). The earliest stage of metamorphism is indicated by large garnet porphyroblasts which probably formed at a low metamorphic grade through a reaction of chlorite + muscovite + quartz. A second generation of garnet which might have been produced by a staurolite breakdown reaction forms broad concentric overgrowths on the early-formed garnet. In most samples, a third garnet generation is observed as small anhedral grains which are the result of a further pressure and/or temperature increase.
Despite the abundance of kyanite in the matrix of the paragneisses the only Al_2SiO_5-polymorph found as inclusions in (the second generation) garnet is sillimanite. Hence, the prograde P-T path entered the kyanite stability field after crossing the sillimanite field, presumably due to an increase in pressure. A second generation of sillimanite occurs as sillimanite-biotite aggregates around dissolved garnets and was probably produced by breakdown of garnet + muscovite to form sillimanite + biotite + quartz, a reaction which marks the decompressive stage of late metamorphism. Thus, the first petrological results point to a rather complex P-T-time path of the paragneisses which presumably cannot be explained by a single "clockwise" P-T loop (Reinhardt et al. 1989).

Chemically, these paragneisses are relatively uniform and, on the basis of their SiO_2/Al_2O_3-and K_2O/Na_2O-ratios, were derived from (clayey) greywackes (Fig. 12). Their chondrite-normalized REE distribution patterns show only a rather limited variation and are distinguished by small negative Eu anomalies and relatively low HREE abundances (Fig. 13).

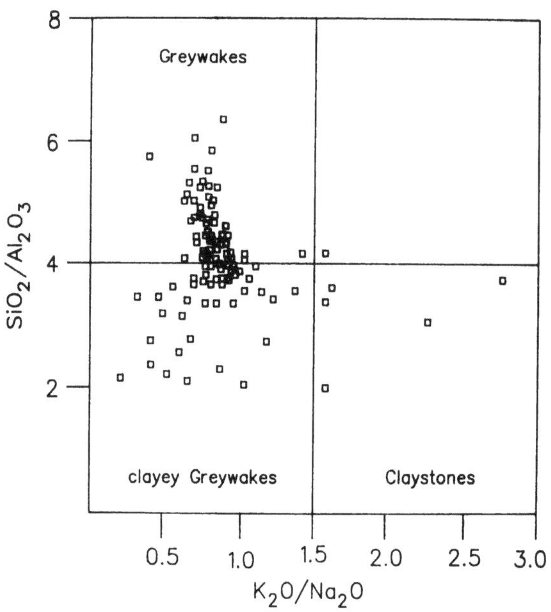

Fig. 12. Major oxide ratios of the paragneisses as indicators of the sedimentary starting material (Müller et al. 1989)

The metabasic rocks include amphibolites and metagabbros. The amphibolites are often rich in garnet and display variable textures from massive to strongly foliated. They frequently contain layers or lenses of calcsilicates and grade into paragneisses (Units 1 and 5). These rocks are composed of plagioclase, hornblende and unzoned garnet and occasionally clinopyroxene, biotite and quartz. Typical accessories are ilmenite, rutile, sphene, pyrite and chalcopyrite. All observations indicate that the amphibolites represent basaltic lavas or tuffs erupted contemporaneously with the rapid sedimentation of the greywackes. According to the phase relationships shown in Fig. 14, these rocks were metamorphosed under amphibolite-facies conditions.

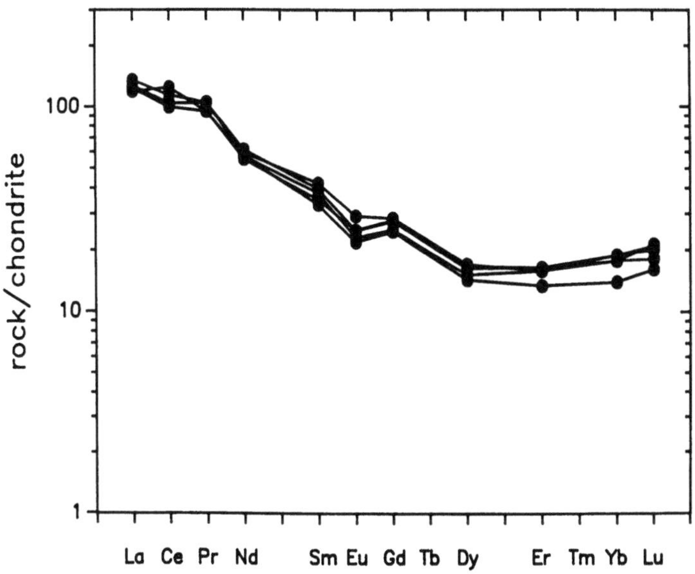

Fig. 13. Chondrite-normalized REE distribution patterns of paragneisses from the KTB pilot hole

The metagabbros are massive medium- to coarse-grained rocks which are distinguished from the amphibolites by relict magmatic textures. These textures and the intimate association of the metagabbros with meta-ultramafic rocks suggest mafic layered intrusions as protoliths. These rocks obviously experienced an early granulite-facies metamorphism under static conditions as is indicated by the development of garnet corona reaction textures around clinopyroxene and by symplectic intergrowths of clinopyroxene + plagioclase after probable omphacitic clinopyroxene. This high-pressure paragenesis was transformed by a later amphibolite-facies metamorphism which led to replacement of clinopyroxene by green hornblende and formation of plagioclase rims by decomposition of garnet in contact with hornblende. Chemically, the metagabbros differ from the amphibolites by having higher Cr and Ni and lower Zr, P and alkali contents. Representative analysis of the various rock types recovered are given in Table 1.

Table 1 Chemical and mineralogical composition of typical basement rocks

	a)	b)	c)	d)	e)	f)	g)
SiO_2	63.0	58.6	47.4	48.8	46.9	41.5	52.8
TiO_2	1.07	1.39	2.64	1.46	1.13	0.99	0.93
Al_2O_3	16.0	15.5	15.4	16.6	17.6	11.3	14.6
Fe_2O_3t	7.54	8.40	12.0	9.60	9.40	13.7	7.50
MnO	0.14	0.14	0.18	0.20	0.15	0.19	0.11
MgO	2.59	3.35	5.93	6.13	9.12	19.1	8.84
CaO	2.70	5.35	9.11	8.86	9.85	7.84	5.71
Na_2O	3.16	3.14	2.44	3.83	2.62	0.44	1.96
K_2O	2.77	2.06	1.42	0.49	0.55	0.11	2.39
P_2O_5	0.16	n.d.	0.49	0.49	0.13	0.19	n.d.
Rb	80	67	36	<10	10	<10	94
Sr	254	353	315	338	316	28	473
Y	35	36	n.d.	42	22	20	20
Zr	252	242	179	193	88	107	217
Ni	48	52	128	53	71	515	224
quartz	31	23	5	7	-	-	19
plagioclase	47	34	37	45	41	-	44
K-feldspar	-	5	4	-	-	-	-
white mica	-	4	-	-	-	-	-
biotite	19	14	3	1	-	-	11
chlorite	3	7	12	1	5	46	16
amphibole	-	12	39	39	34	54	10
garnet	-	-	-	7	13	-	-
pyroxene	-	-	-	-	7	-	-

a) sillimanite-bearing garnet-biotite-plagioclase gneiss (2479.5m)
b) chloritized hornblende-biotite gneiss (2526.2m)
c) amphibolite (380.6m)
d) garnet amphibolite (1294.8m)
e) metagabbro (1285.7m)
f) meta-ultramafic rock (1400.0m)
g) lamprophyre (2230.9m)

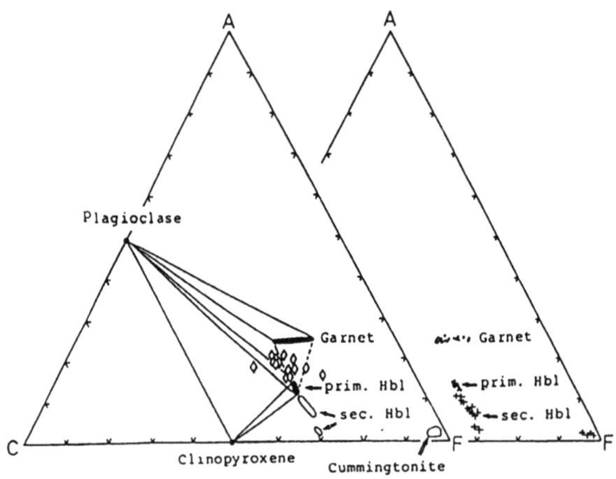

Fig. 14. ACF diagrams of amphibolites from Unit 1. On the right microprobe data of selected minerals are presented (Patzak et al. 1989)

The meta-ultramafic rocks (Table 1) occasionally recovered from Units 3 and 7 are mostly hornblendites with highly varying contents of talc, chlorite and phlogopite. These rocks, interpreted as former cumulates of layered gabbroic intrusions, were presumably derived from olivine pyroxenites as is suggested by relict diopsidic pyroxene and rare pseudomorphs after olivine.

Paragneisses and metabasic rocks differ systematically in their $\delta^{18}O$ (SMOW) whole rock data. The values obtained so far on paragneisses are typical for metapelites and vary between + 10.0‰ and 13.0‰. The metabasic rocks have $\delta^{18}O$ whole rock values of between + 6.7‰ and 9.7‰ with the metagabbros clustering around + 7.4‰. The data suggest that an external pervasive fluid phase which would have led to a homogenization of the $^{18}O/^{16}O$ ratios was not present during the peak of metamorphism (Hoernes et al. 1989).

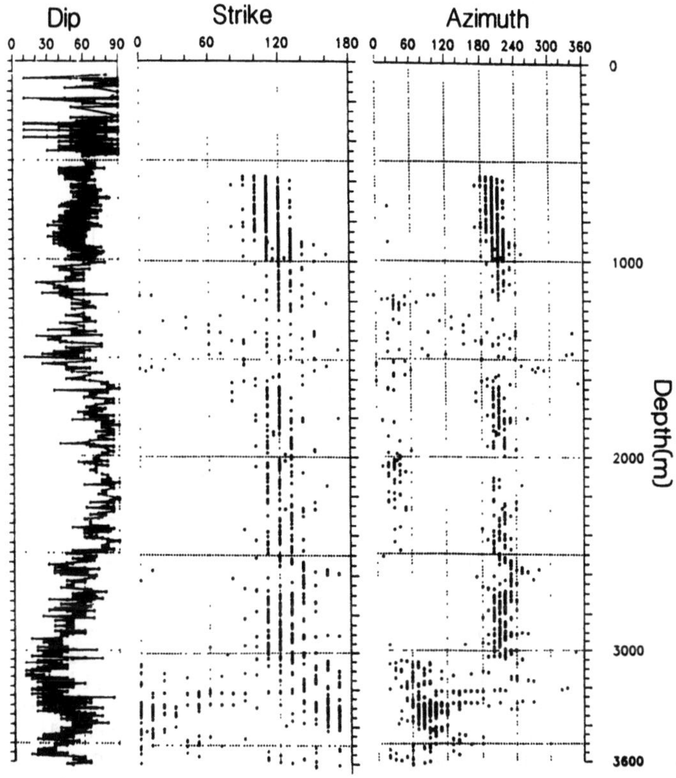

Fig. 15. Dip, strike and azimuth of foliation

Although it was known from surface geology that the ZEV rocks underwent a complex deformation history characterized by penetrative foliation and multi-stage post-metamorphic cataclasis both the steep inclination of the structures and the intensity of the cataclastic deformation of the basement penetrated by the pilot hole were a great surprise.

The foliation dips about 70-90° in the uppermost 500 m and between 1600 and 2300 m but is almost flat at a depth of around 3200 m. The average dip is 60° and there is no indication that there is a systematic decrease with depth as was suggested by the geological starting model. The foliation strikes WNW-ESE and preferentially dips to the SSW above 3200 m and NNE below that depth (Fig. 15). Evaluation of the structural data and reconstruction of the tectonic situation yielded a rather complicated picture and suggest that the pilot hole might have penetrated a huge fold structure (see Fig. 10).

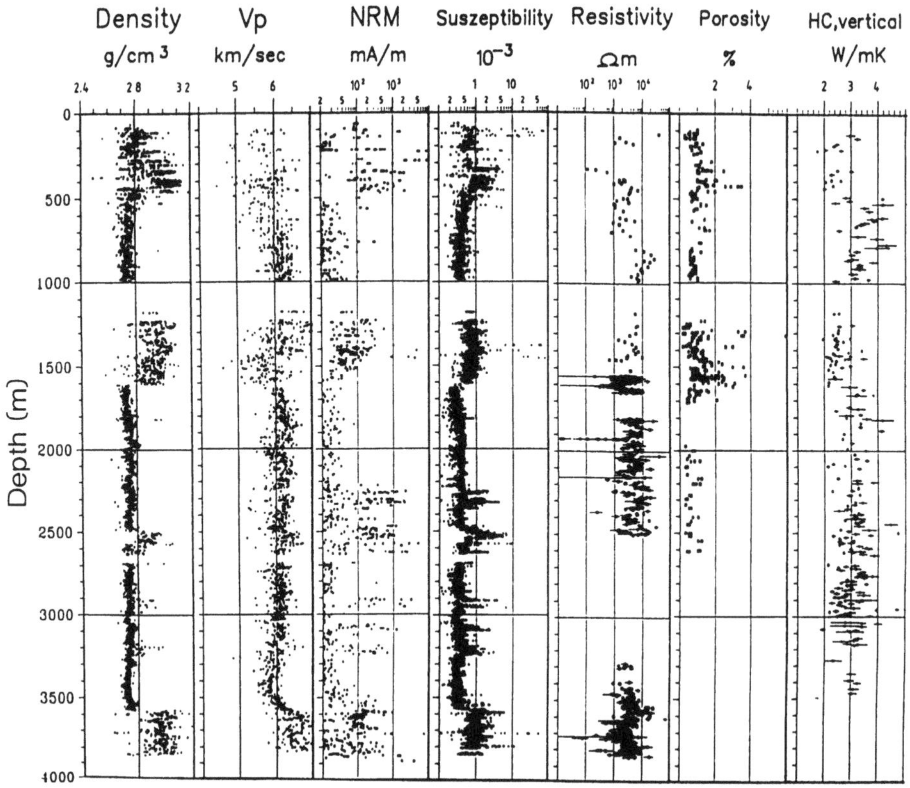

Fig. 16. Petrophysical data obtained on core material

The lithostratigraphy of the drilled basement section is reflected in the downhole variations of the petrophysical rock properties. Fig. 16 summarizes the data obtained on cores and cuttings by the geophysical working group of the field laboratory (Lippmann et al. 1989). As seen from these plots, there is an excellent correlation between lithology and density. The paragneisses are distinguished by very uniform density values of about 2.75 g/cm^3 whereas the metabasic rocks display average densities of 3.0 g/cm^3 with a relatively broad scatter of between 2.8 and 3.2 g/cm^3.

Except for the paragneiss/amphibolite transition at 3575 m, the p-wave velocities do not correlate as well with the lithology. On average, the values are slightly above 6 km/s and vary between 5 and 7 km/s. In many cases, there are only minor velocity differences between paragneisses and metabasic rocks. However, the p-wave velocities show a marked anisotropy which is especially pronounced in the strongly foliated paragneisses but is also detected in massive metagabbros. Preliminary investigations indicate that the in-situ velocity anisotropy of the paragneisses is around 10% (Lippmann et al. 1989).

The natural remanent magnetization is a very sensitive indicator of the distribution of magnetic ores (magnetite and pyrrhotite). The downhole plot, therefore, clearly distinguishes between paragneisses with lower values and metabasic rocks with higher values of magnetization. However, between 2400 and 3000 m, there are strong magnetic anomalies which are confined to paragneisses and supposedly due to pyrrhotite enrichments. The magnetic suceptibility which was recorded every centimetre is the physical parameter with the highest density of data. Characteristic susceptibility values of the paragneisses and metabasites are 0.4 x 10^{-3} SI and 0.8 x 10^{-3} SI, respectively. Unusually high magnetic susceptibilities of up to 0.15 SI were recorded in amphibolites at about 120 m (Lippmann et al. 1989).

The measured electrical resistivities vary between 12 Ωm (graphite containing shear zones) and > 40.000 Ωm (gneisses, marble). Average resistivity values for the paragneisses are ≈ 4000 Ωm and for the metabasic rocks ≈ 2200 Ωm. These differences are probably due, at least in part, to the lower porosity of the paragneisses. The porosity values obtained on plugs are about 1% in the gneisses and up to 4% in the metabasic rocks.

The thermal conductivity shows a pronounced anisotropy and is generally higher in the paragneisses (values usually between 3 and 4 W/mK) than in the metabasic rocks (values between 2 and 3 W/mK).

Major portions of the recovered basement rocks are affected by cataclastic deformation. The sequence is frequently cut by steeply dipping cataclastic shear zones up to 5 m in width, the formation of which was connected with an intense alteration of the host rocks under lower greenschist-facies conditions. Chloritization and serizitisation are widespread phenomena (see also Fig. 17). These shear zones often contain relatively high amounts of graphite, which locally is conspicuously enriched in veins up to several centimeters thick. On-line gas mass-spectrometry carried out in the field laboratory proved to be a very effective means of tracing graphite-containing shear zones. When approaching such a zone, the gas composition of the drilling fluid changed considerably and especially the contents of methane and helium become strongly enriched (Fig. 17).

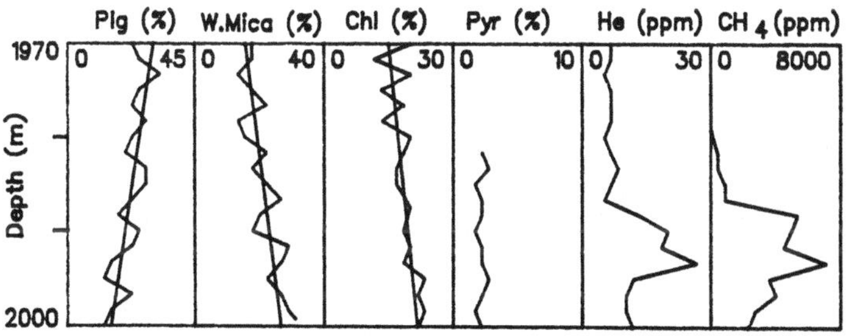

Fig. 17. Changes in mineralogical composition of drilling mud and composition of gases when approaching a shear zone

Despite the cataclastic deformation and a relatively intense veining of the basement rocks, practically no fluid inflow was observed in the borehole above 3400 m. It, therefore, was a surprise when just below this level the first open fractures were encountered which appear to become more abundant with depth. On-line analysis of the drilling fluid revealed that these fractures contain gas and saline waters (Fig. 18).

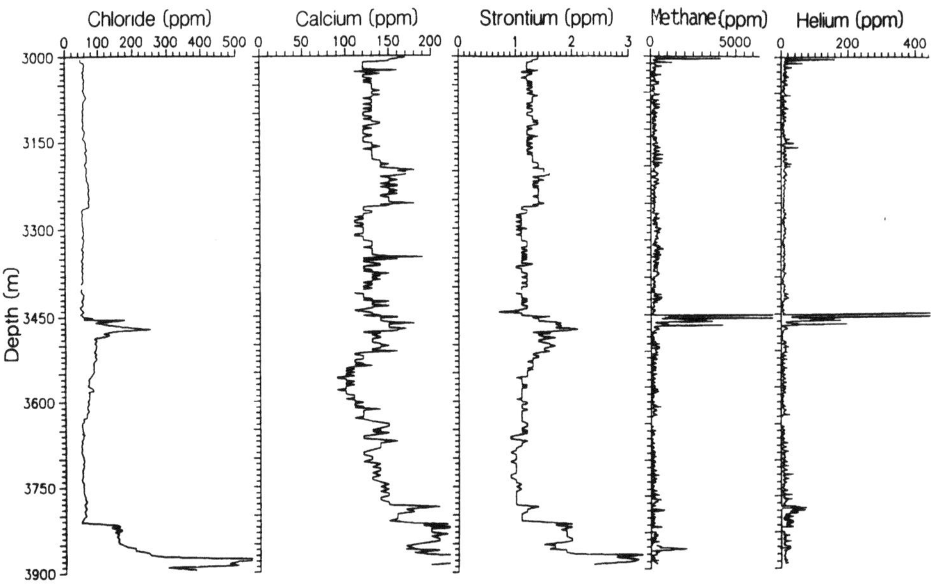

Fig. 18. Selected chemical data obtained on drilling fluid and gases

Evaluation of the chemical data and a drill-stem test yielded the following minimal concentrations for the deep crustal waters: Ca 18.000 mg/l, Na 16.500 mg/l and Cl 50.000 mg/l (Fritz et al. 1989). These data indicate that the fluids are Ca-Na-Cl brines whose composition has been determined largely by rock-water interaction. The $\delta^{18}O$-values of these fluids (-8 to -6 ‰) are lower than those of typical formation waters or hydrothermal solutions and suggest reaction temperatures below 150° C. The composition of the gas phase collected by fluid sampling at 3447 m is:

N_2	69.6 (%)	O_2	4.0 (%)
Ar	0.3 (%)	He	0.6 (%)
CO_2	0.5 (%)	H_2	4.3 (%)
H_2O	4.2 (%)	CH_4	16.0 (%)
C_2H_6	0.2 (%)	C_3H_8	0.07 (%)

Also unexpected was the early occurrence of core disking first observed at 3575 m which indicates strong horizontal stresses at these depths.

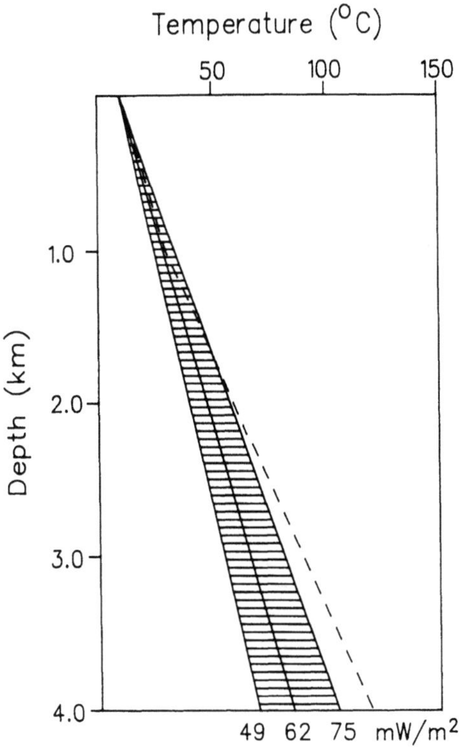

The greatest surprise encountered thus far, however, is the measured temperature profile. Figure 19 shows the actual temperatures obtained and the lower and upper limits predicted from pre-drilling geothermal investigations (Burkhardt et al. this volume). The expected temperature gradient of about 22°/km was only found in the uppermost 500 m of the hole. Below that depth the temperature gradient increases to reach values of between 28° and 30° C/km (see also Fig. 22).

Fig. 19. The temperatures encountered in the KTB pilot hole

MAJOR UNSOLVED PROBLEMS

The results obtained thus far already require modification of pre-existing models and pose a number of new scientific questions. Among these are:

(1) What is the reason for the unexpectedly steep inclination of structures in the entire drilled section and how deep do the shear zones reach? Are we dealing with a local phenomenon associated with the Fichtelnaab fault system or does the western margin of the Bohemian Massif represent an old and repeatedly reactivated deformation zone between two large crustal blocks? In the latter case, the Variscan thrust and nappe structures could have been overprinted by late- and post-Variscan transpressional and wrench-tectonics which might have caused steep monoclinal folding and intense shear deformation of the ZEV rocks and even their substratum. At present, there is no indication that the bottom of the postulated ZEV nappe has been reached and it appears that the starting tectonic model has to be modified or even replaced by an alternative model.

(2) Have seismic reflectors been penetrated in the pilot hole and, if so, what is their nature? Pre-stack migration of the DEKORP 4 data in the vicinity of the drill site and evaluation of the first VSP experiments (Kästner at al. 1989) confirms that the main borehole will encounter quite a number of prominent reflectors which are especially abundant in the depth range of between 8 and

12 km (Fig. 20). The data also suggest that the pilot hole already penetrated a reflective horizon at a depth of around 3400 m which roughly coincides with the first occurrence of fluid-filled fractures.

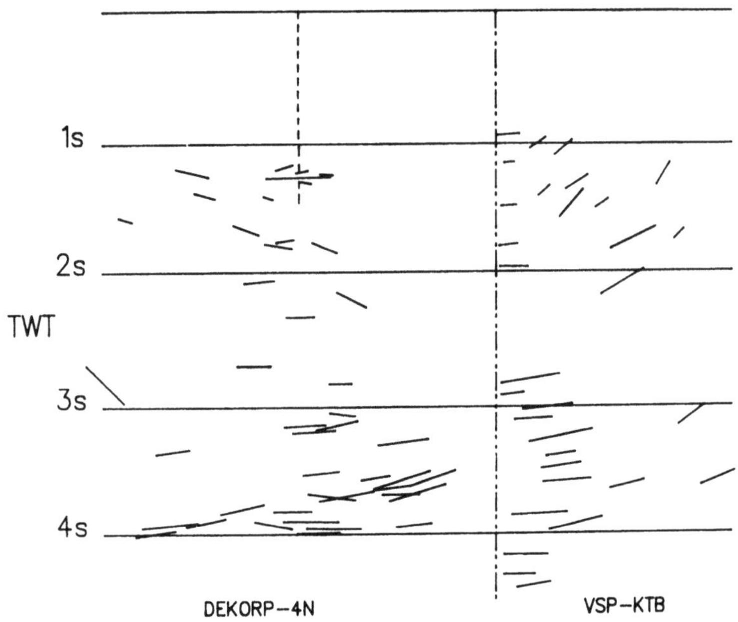

Fig. 20. Pre-stack migration of DEKORP 4 data in the vicinity of the drill site (on the left) and evaluation of a VSP experiment carried out 2200 m (on the right)

(3) What is the depth distribution of the principal in-situ stresses? Relaxation experiments on recovered core material yielded first data on the stress regimes at greater depth and allow a rough estimate of the principal in-situ stresses. To obtain information on the time-dependent strain retardation and to estimate magnitude and orientation of the principal in-situ stresses, core samples of different lithology (length of sample 100 mm, diameter 94 mm) were measured with a multi-component dilatometer. Depending on rock type and texture, the values of maximal and minimal radial principal retarded strains (e_1 and e_2) and the vertical strain (e_v) are quite different (Table 2).

Knowing the orientation of the samples, the absolute strain directions can be determined. As seen from Table 2 (Huenges et al. 1989), the orientation of maximum horizontal principal strain varies between 20 and 38 N°E which is perpendicular to the strike of foliation (110 - 130 N°E). In contrast to the gneiss samples which have a distinct foliation that apparently controls the orientation of maximum principal horizontal strain (θ), the fine-grained lamprophyric rocks are homogeneous and nearly isotropic. If the orientation of maximum principal horizontal strain in these rocks is assumed to correspond to the maximum in-situ stress direction it can be concluded that at a depth of about 2230 m the direction of maximum horizontal stress is NW-SE (mean value 155° ± 10°). From the strain retardation data obtained on the lamprophyre the following values for the principal in-situ stresses were derived: σ_H = 63 MPa; σ_h = 47 MPa; and σ_V = 61 MPa. This means that the vertical stress (σ_V) in this depth range is the intermediate stress and σ_H is about 1.4 σ_h (Wolter and Berkhemer 1989).

Table 2. Values of principal strain retardation and orientation of maximum principal horizontal strain (Θ)

Rock types	depth (m)	e_1 [μm/m]	e_2 [μm/m]	e_V [μm/m]	Θ (N°E)
paragneiss	1871	398	44	92	38 ± 10
paragneiss	2036	248	65	201	20 ± 10
lamprophyre	2051	312	103	147	172 ± 15
lamprophyre	2231	637	197	593	155 ± 10
paragneiss	2426	201	38	68	20 ± 10

(4) What is the source of the gases and saline waters? Among the gases analyzed so far methane and helium are of special interest since they are reliable indicators of inflow horizons and even graphite-containing shear zones. As is seen from Fig.21 the shear zone at 2000 m is distinguished from the inflow zones at 3400 and 3800 m by significantly different CH_4/He-ratios of the gases dissolved from the drilling fluid. The markedly higher methane contents of the shear zones are due to graphite enrichment. The open fractures at 3400 m are developed in paragneisses which have higher methane and helium contents than the metabasic rocks. Therefore, the different concentrations and ratios found for the fractures at 3800 m, which occur in metagabbros, are probably due to primary differences. The results obtained suggest that there is no hydraulic connection between the two fracture systems at 3400 and 3800 m (Erzinger et al. 1989). Preliminary determinations of the 3He/4He-ratio of the helium sampled at about 3455m point to a significant mantle component of about 20 % (Heusser et al. 1989).

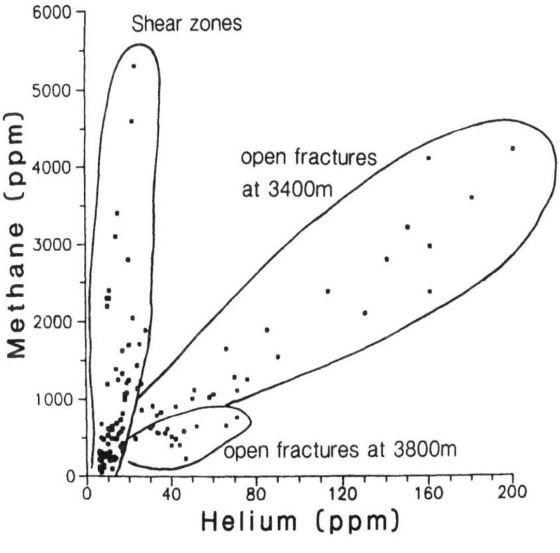

Fig. 21. Methane and helium contents in shear zones and fractures

(5) What is the source of graphite found in shear zones and how was it formed? The intense hydration of the paragneisses in connection with graphite enrichment and the occasional intimate intergrowth of graphite with chlorite and calcite indicate the presence of an H_2O- and CO_2- rich fluid and suggest a graphite deposition according to the equation: $CH_4 + CO_2 = 2C + 2H_2O$.
The carbon isotope data obtained so far are still ambiguous. The values measured ($\sigma^{13}C$ -19 to -22 %) are intermediate between those typical for carbon derived from mantle CO_2 and carbon derived from organic matter (Hoefs pers. comm.). Therefore, the ultimate origin of the carbon and its mechanism of transportation and precipitation are still unclear.

(6) What is the reason for the low electrical resistivity and the high self-potential anomaly observed around the borehole? Geoelectrical measurements carried out in the immediate vicinity of the drill site revealed the existence of a prominent shallow electrical conductor between 300 m and at least 1500 m depth and yielded the highly surprising result that the drill site is located on a very large and unusually high electrical self-potential anomaly about -600 mV (Haak et al. 1989). Graphite might decrease the electrical resistivity of metamorphic rocks by several orders of magnitude if it forms coherent films along grain-boundaries. Among the questions to be investigated, therefore, is the importance of graphite in producing these phenomena.

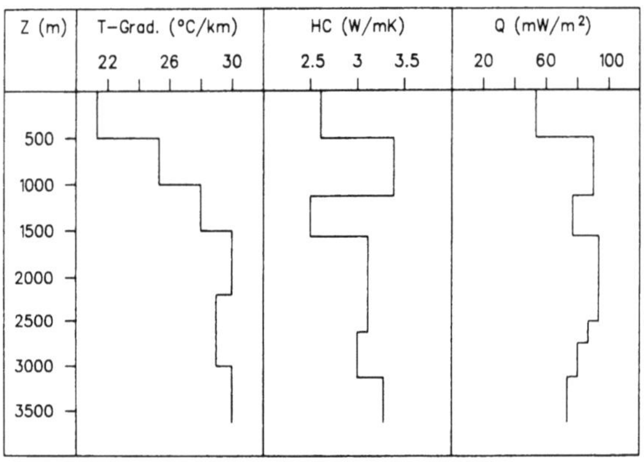

Fig. 22. Downhole distribution of temperature gradient, thermal conductivity and heatflow

(7) What are the reasons for the unexpected temperature profile? A first interpretation of the temperature gradient downhole, the thermal conductivity of the rocks and the resulting heatflow (Fig.22) suggests that the temperatures within the upper 500 m of the basement might have been systematically lowered by cooling of the crust through ground water convection (Burkhardt et al. 1989). The higher heatflow values observed at greater depths are presently interpreted to be the result of a higher than expected basal heatflow component. Among the questions to be investigated, therefore, is the role of the Tertiary Eger graben (see Fig. 1) and its associated alkalibasaltic magmatism as a source of the higher than predicted heatflow values.

REFERENCES

Behr HJ (1987) KTB und kontinentale Krustenforschung - warum ein wissenschaftliches Tiefbohrprogramm? KTB-REPORT 87-1, Projektleitung KTB im NLfB, Hannover, pp 1-15

Behr HJ, Emmermann R (1987) Scientific objectives and site- selection studies of the Continental Deep Drilling Program of the Federal Republic of Germany (KTB). In: Behr et al. (eds) Observation of the Continental Crust through Drilling II. Springer, Berlin Heidelberg New York London Paris Tokyo, pp 85-120

Burkhardt M, Erbas K, Giese P, Haack U, Hornamand H, Huenges E, Stiefel A, Wilhelm H, Zoth G, Buntebarth G, Schulz R (1989) Das vorhergesagte und das gemessene Temperaturprofil. In: Emmermann and Giese (eds) KTB REPORT 89-3, Projektleitung KTB im NLfB, Hannover, pp 216-242

Chur C, Sperber A (1988) The drilling concept of the Continental Deep Drilling Program of the Federal Republic of Germany (KTB). In: Boden and Eriksson (eds) Deep drilling in crystalline bedrock, (vol 2) Springer, Berlin Heidelberg New York London Paris Tokyo, pp 224-234

Emmermann R (1986) Das Deutsche Kontinentale Tiefbohrprogramm. Geowissenschaften in unserer Zeit 4: 19-33

Emmermann R, Behr HJ (1987) Location for super-deep borehole confirmed. Tectonophysics 139: 339-340

Emmermann R, Dietrich HG, Heinisch M, Wöhrl T (1988) Tiefbohrung KTB Oberpfalz VB. Ergebnisse der geowissenschaftlichen Bohrungsbearbeitung im KTB-Feldlabor (Teufenbereich von 0 - 480 m). KTB REPORT 88-1, Projektleitung KTB im NLfB, Hannover

Emmermann R, Dietrich HG, Heinisch M, Wöhrl T (1988) Tiefbohrung KTB Oberpfalz VB. Ergebnisse der geowissenschaftlichen Bohrungsbearbeitung im KTB-Feldlabor (Teufenbereich von 480 - 992 m). KTB REPORT 88-2, Projektleitung KTB im NLfB, Hannover

Emmermann R, Dietrich HG, Heinisch M, Wöhrl T (1988) Tiefbohrung KTB Oberpfalz VB. Ergebnisse der geowissenschaftlichen Bohrungsbearbeitung im KTB-Feldlabor (Teufenbereich von 992 - 1530 m). KTB REPORT 88-6, Projektleitung KTB im NLfB, Hannover

Emmermann R, Dietrich HG, Heinisch M, Wöhrl T (1988) Tiefbohrung KTB Oberpfalz VB. Ergebnisse der geowissenschaftlichen Bohrungsbearbeitung im KTB-Feldlabor (Teufenbereich von 1530 - 1998 m). KTB REPORT 88-9, Projektleitung KTB im NLfB, Hannover

Emmermann R, Dietrich HG, Heinisch M, Wöhrl T (1989) Tiefbohrung KTB Oberpfalz. Ergebnisse der geowissenschaftlichen Bohrungsbearbeitung im KTB-Feldlabor (Teufenbereich von 1709 - 2500 m). KTB REPORT 89-2, Projektleitung KTB im NLfB, Hannover

Emmermann R, Lauterjung J, Stroh A (1989) Das lithostratigraphische Profil der KTB-Vorbohrung bestimmt durch röntgenographische Phasenanalyse von Bohrklein. In: Emmermann und Giese (eds) KTB REPORT 89-3, Projektleitung KTB im NLfB, Hannover, pp 152-164

Erzinger J, Heinschild HJ, Figgemeier C, Samel M (1989) Ergebnisse der Gasanalytik in der KTB-Vorbohrung. In: Emmermann und Giese (eds) KTB-REPORT 89-3, Projektleitung KTB im NLfB, Hannover, pp 42-45

Franke W (1989) Tectonostratigraphic units in the Variscan belt. Geol. Soc. Amer. Spec. Paper 230: 67-90

Fritz P, Lodemann M, Rauert W, Pekdeger A, Pusch G (1989) Zusammensetzung von Tiefenwässern in den offenen Klüften der KTB-Vorbohrung. Poster. In: Emmermann and Giese (eds) KTB-REPORT 89-3, Projektleitung KTB im NLfB, Hannover

Haak V, Stoll J, Winter H (1989) Why is the electrical resistivity around the KTB-hole so low? In: Emmermann and Giese (eds) KTB REPORT 89-3, Projektleitung KTB im NLfB,Hannover, pp 54-60

Herold CP, Müller H, v Tapavicza S (1988) A new high temperature stable drilling mud additive for geological and deep drilling operations. In: Boden and Eriksson (eds) Deep drilling in crystalline bedrock (vol 2). Springer, Berlin Heidelberg New York London Paris Tokyo, pp 306-314

Huenges E, Bücker Ch, Wolter KE, Wienand J, Rauen A, Lippmann E (1989) Deep Drilling KTB Oberpfalz VB. Results of the Geoscientific Proceedings in the KTB -laboratory. In: Emmermann et al.(eds) KTB REPORT 89-2, Projektleitung KTB im NLfB, Hannover

Heusser E, Kirsten T, Richter H, Oehm J (1989) Edelgasisotopensystematik im Bohrkernabschnitt 500 m - 1500 m. Poster. In: Emmermann and Giese (eds) KTB REPORT 89-3, Projektleitung KTB im NLfB, Hannover

Hoernes S, Grimmeisen W, Hoefs J, Simon K (1989) Größenordnung der Fluid-Gesteins-Wechselwirkung abgeleitet aus Sauerstoff- und Wasserstoff-Isotopendaten der Gesteine und Minerale der KTB-Vorbohrung. In: Emmermann and Giese (eds) KTB REPORT 89-3, Projektleitung KTB im NLfB, Hannover, pp 81-82

Kästner U, Bram K, Hubral P, Macdonald C, Rühl T, Sandmeier KJ (1989) Seismic investigations at the KTB drilling site. In: Emmermann and Giese (eds) KTB-REPORT 89-3, Projektleitung KTB im NLfB, Hannover

Lippmann E, Bücker C, Huenges E, Rauen A, Wienand J, Wolter KE (1989) Gesteinsphysik im KTB-Feldlabor: Messungen und Ergebnisse. In: Emmermann and Giese (eds) KTB REPORT 89-3, Projektleitung KTB im NLfB, Hannover, pp 120-129

Müller H, Tapfer M, Emmermann R, Wimmenauer W (1989) Die Paragneise im Profil der KTB-Vorbohrung. In: Emmermann and Giese (eds) KTB REPORT 89-3, Projektleitung KTB im NLfB, Hannover, pp 61-66

Patzak M, Okrusch M, Röhr C (1989) Metabasite der KTB-Vorbohrung: Vorläufige Ergebnisse bis 1610 m. In: Emmermann and Giese (eds) KTB REPORT 89-3, Projektleitung KTB im NLfB, Hannover, pp 67-80

Reinhardt J, Kleemann U, Blümel O, Schreyer W (1989) Geothermobarometry of metapelites as a key to the pressure and temperature history of the ZEV (Zone von Erbendorf-Vohenstrauss), NE Bavaria. In: Emmermann and Giese (eds) KTB REPORT 89-3, Projektleitung KTB im NLfB, Hannover, pp 24-32

Rischmüller H, Chur C (1987) Technical concept and status of planning for the wells of the Continental Deep Drilling Program (KTB). In: Behr et al. (eds) Observation of the Continental Crust through Drilling. Springer, Berlin Heidelberg New York London Paris Tokyo, pp 136-147

Rischmüller H (1989) The German Continental Deep Drilling Project (KTB) - An Overview - . Paper presented at the 28th International Geological Congress, Washington, D.C. USA

Schüssler U, Richter P, Okrusch M (1989) Metabasites from the Oberpfalz target area, Bavaria - Geochemical characteristics and examples of mobile behaviour of "immobile" elements. Tectonophysics 157: 135-148

Weber K, Vollbrecht A (1986) Ergebnisse der Vorerkundungsarbeiten Lokation Oberpfalz. 2. KTB-Kolloquium Seeheim /Odenwald

Wolter KE, Berkhemer H (1989) Retardierte Entspannungsdeformation an KTB Bohrkernen. In: Emmermann and Giese (eds) KTB REPORT 89-3, Projektleitung KTB im NLfB, Hannover, pp 245-249

MIX
Papier aus verantwortungsvollen Quellen
Paper from responsible sources
FSC® C105338

If you have any concerns about our products,
you can contact us on
ProductSafety@springernature.com

In case Publisher is established outside the EU,
the EU authorized representative is:
Springer Nature Customer Service Center GmbH
Europaplatz 3, 69115 Heidelberg, Germany

Printed by Libri Plureos GmbH
in Hamburg, Germany